Handbook of Photomask Manufacturing Technology

Handbook of Photomask Manufacturing Technology

edited by

Syed Rizvi

Nanotechnology Education and Consulting Services
San Jose, California, U.S.A.

Taylor & Francis
Taylor & Francis Group

Boca Raton   London   New York   Singapore

A CRC title, part of the Taylor & Francis imprint, a member of the
Taylor & Francis Group, the academic division of T&F Informa plc.

Cover and spine pictures courtesy of Photronics, Inc.

Back cover picture courtesy of DuPont Photomask, Inc.

Published in 2005 by
CRC Press
Taylor & Francis Group
6000 Broken Sound Parkway NW, Suite 300
Boca Raton, FL 33487-2742

© 2005 by Taylor & Francis Group, LLC
CRC Press is an imprint of Taylor & Francis Group

No claim to original U.S. Government works
Printed in the United States of America on acid-free paper
10 9 8 7 6 5 4 3 2 1

International Standard Book Number-10: 0-8247-5374-7 (Hardcover)
International Standard Book Number-13: 978-0-8247-5374-0 (Hardcover)

This book contains information obtained from authentic and highly regarded sources. Reprinted material is quoted with permission, and sources are indicated. A wide variety of references are listed. Reasonable efforts have been made to publish reliable data and information, but the author and the publisher cannot assume responsibility for the validity of all materials or for the consequences of their use.

No part of this book may be reprinted, reproduced, transmitted, or utilized in any form by any electronic, mechanical, or other means, now known or hereafter invented, including photocopying, microfilming, and recording, or in any information storage or retrieval system, without written permission from the publishers.

For permission to photocopy or use material electronically from this work, please access www.copyright.com (http://www.copyright.com/) or contact the Copyright Clearance Center, Inc. (CCC) 222 Rosewood Drive, Danvers, MA 01923, 978-750-8400. CCC is a not-for-profit organization that provides licenses and registration for a variety of users. For organizations that have been granted a photocopy license by the CCC, a separate system of payment has been arranged.

Trademark Notice: Product or corporate names may be trademarks or registered trademarks, and are used only for identification and explanation without intent to infringe.

Library of Congress Cataloging-in-Publication Data

Catalog record is available from the Library of Congress

Taylor & Francis Group
is the Academic Division of T&F Informa plc.

Visit the Taylor & Francis Web site at
http://www.taylorandfrancis.com

and the CRC Press Web site at
http://www.crcpress.com

I dedicate this book to my parents who have instilled in me the value of strength of character with humility.

Foreword

In the year 2000, I along with Dr. Robert Doering from Texas Instruments had the honor of editing a book titled *Handbook of Semiconductor Manufacturing Technology*, which was published by Marcel Dekker. The purpose of that book was to serve as a reference for practitioners and developers of semiconductor manufacturing technology (SMT). I was pleased to see that the book served that purpose well. The book was a good mix of chapters on state-of-the-art technology and chapters that focused on the means and fundamentals of SMT.

Considering that the book had over 35 chapters covering a wide spectrum of SMT it is apparent that the chapters had to be concise and avoid in-depth discussion of any kind. It was a challenging task of balancing the amount of material covered and brevity of the contents, especially when some of the topics were the driving forces of the state-of-the-art technology. Now, in the year 2004, follow-up books on critical topics addressed in the earlier work have become necessary.

The ever-shrinking features continue to be the yardsticks of the progress in chip manufacturing technology. Here the lithographic exposures system and photomasks are key players. Today, the industry is preparing to meet the challenges of 100- and 65-nm nodes. While it may be true that with the current 4× or 5× exposure systems the features on the mask will be proportionally larger than their desired values on the wafer, the use of resist and the serif feature on today's mask may be required to be significantly smaller and in some cases could be even close, if not equal, to the smallest features on wafers. In order to meet the upcoming challenges it has become necessary to furnish the industry with comprehensive materials on mask making that could not have been covered in just one single chapter of the earlier book addressing a wide spectrum of SMT. *Handbook of Photomask Manufacturing Technology* is the result of such an endeavor.

During the last decade mask making has evolved into a very sophisticated technology. The availability of such a book will be of great value to mask makers. Moreover, this book will also be of value to mask users and those working in other areas of chip manufacturing so that they can appreciate the complexities of this technology and use their expertise in design and processing to optimize the mask constraint without sacrificing the functionality and specifications of the final chip.

A million-dollar mask set is not too far into the distant future, and the benefits from those masks could be maximized by technical breakthroughs in their development and innovations in their usage.

This book, in my opinion, will help the industry reach its objectives and goals.

YOSHIO NISHI
Stanford University
Stanford, CA

Preface

Photolithography, which was later referred to as microlithography, and now as nanolithography, has unquestionably been one of the major driving forces behind the progress made in the semiconductor technology. Since its early days, the semiconductor industry has strived to make features smaller and smaller so that the maximum number of devices could be packaged into the smallest possible area on a chip. It is in this miniaturization of features that photolithography has made its contribution, and where the advancement in mask making has played its role.

The term *mask* is an abbreviation of the term *photomask*, which is more descriptive of its functions; due to its brevity, however, mask is the term more often used within the semiconductor industry. Outside the sphere of the semiconductor industry, it is preferable to use the term photomask because the term mask can be attributed to masks used in areas having no relation to the semiconductor industry.

Many books have been written addressing all areas of semiconductor technology. Several books have focused on specific areas within semiconductor technology, including photolithography. However, photolithography itself consists of multiple areas of discipline such as optics, resists, pattern transfers, and of course photomasks (or simply masks). There are quite a few books that cover all of these various subjects, but few books published exclusively on photomasks.

Considering the degree of sophistication and complexities that today's photomasks present, it has become necessary to provide the industry with a book devoted to photomask technology. This book answers that need.

This book is written for professionals involved in the manufacturing of conventional photomasks as well as those involved in the R&D activities of leading-edge mask technology. Although a considerable portion of the book focuses on optical masks and their supporting technologies, several chapters have been devoted to masks being developed for *next-generation lithography* (NGL). There are four major NGL contenders, namely, extreme ultraviolet lithography (EUVL), electron projection lithography (EPL), ion projection lithography (IPL), and x-ray lithography (XRL), which will be addressed in this book. Some of the chapters and sections are very basic and should be of special interest to newcomers to mask technology. A sincere effort has been made to present the chapters in a sequence that follows the steps of the process flow in mask fabrications. However, due to the complexity and diversity of mask making, absolute adherence to this scheme was not always possible.

Chapter 1 gives the reader an overview of all aspects of the current mask technology, along with synopses on the various types of masks that are candidates for NGL. This chapter is completely self-contained and will be valuable to anyone wanting to know about masks without going into detail. Fabrication managers and professionals working in fields other than mask making can derive from this chapter a true appreciation of mask technology with which they may find themselves dealing, either directly or indirectly.

Chapters covering related topics have been grouped together in major sections identified by their subject matters.

Section II, Mask Writing, consists of Chapter 2 through Chapter 5. This section describes machines that do the writing, and software for preparing the data, which in turn will instruct the machine what and how to write. Currently, the two types of writers in the market are electron beam (e-beam) and laser. E-beam writers have evolved over three decades and spanned into machines with various writing schemes, the use of which is dictated by the complexity of patterns and throughputs. The laser writer represents a relatively new technology and is only now beginning to be widely employed, especially where throughput is an important factor. After Chapter 2 on data preparation, the following three chapters in this section focus on mask writers. Chapter 3 gives a general overview of various writers before discussing in depth the details of e-beam and laser writers.

Chapter 6 through Chapter 8 are grouped under a section titled Optical Masks. This section starts with an overview followed by chapters on conventional and leading-edge optical masks. Chapter 6 describes the different stages of masks that started from the early $2'' \times 2''$ emulsion masks and now has resulted into today's $6'' \times 6''$ and $9'' \times 9''$ PSM and OPC masks.

Section IV, NGL Masks, consists of five chapters (Chapter 9 through Chapter 13). Chapter 9 provides a summary of all NGL masks, which sets the stage for the following four chapters in which NGL masks are described in considerable detail.

Section V, Mask Processing, Materials, and Pellicles, consists of Chapter 14 through Chapter 19. These chapters cover topics related to mask materials, mask processing, and the pellicalization of masks, the final step in mask making. Section VI, Mask Metrology, Inspection, Evaluation, and Repairs, encompasses Chapter 20 through Chapter 30. Some of the topics in this section are covered in two or more chapters because topics such as CD metrology and mask inspection require more extensive background knowledge with regard to principle, theory, and definitions.

Last, but not least, is Chapter 31 on modeling and simulation (M&S). In the early days, M&S was not considered an area in which mask users or suppliers were particularly interested. Even a decade ago, M&S was viewed as a tool valuable only in device design and fabrication engineering. Today, M&S is considered an integral part of overall mask-making technology.

Each chapter in the book is independently written, hence a certain degree of overlaps in material is possible. This is especially true where an overview precedes a set of chapters with related topics. Such overlaps, however, will give the reader multidimensional and thus more valuable insights into the subjects that otherwise would not have been possible. The overview chapters are especially valuable to readers interested in a broader picture of the topic without going into details.

SYED A. RIZVI

About the Editor

Syed Rizvi, editor of this volume, is a veteran with over 40 years' experience in the semiconductor industry. He has worked for companies such as Texas Instruments, MOSTEK (now STM Electronics), and Photronics, and also served in consortia and think tanks such as SEMATECH and SRC (Semiconductor Research Corporation). Rizvi also ran education and training programs and web-casts from an internet company, Semicon-Bay, which he helped start up during the late 1990s. Currently, he is the owner and president of Nanotechnology Education Consulting Services in the Silicon Valley. The company's mission is to use education and training as tools for facilitating seamless transitions of high-tech industry from its current microelectronics phase to the nanotech phase that will integrate the state-of-the-art science and technology of moletronics and spintronics with biotech sciences for the next-generation intelligent chips and devices.

Syed Rizvi has coauthored several books, and published and presented numerous papers around the world in the areas of metrology and microlithography. He holds a B.Sc. from Patna University (India), an M.Sc. in nuclear physics from the University of Karachi (Pakistan), and an M.S. in solid state physics from the Northeastern University, Boston (Massachusetts).

In addition to his high-tech involvement, Rizvi also serves on the board of the Institute for Animals and Society (IAS), a think tank devoted to advancing the moral and legal status of animals on this planet. He is a vegetarian.

Acknowledgments

This book has 31 chapters contributed by 47 authors from around the world. I personally thank each and every one of them for taking time from their busy schedules to write their chapters. Without their timely submissions, this book would not have been possible.

This book is an outgrowth of a chapter on mask making that I wrote in an earlier book entitled *Handbook of Semiconductor Manufacturing Technology* published by Marcel Dekker, edited by Dr. Yoshio Nishi and Dr. Robert Doering. I deeply appreciated the invitation from the editors to contribute to that chapter, which has led to the development of this new book on the same topic. Of course, I cannot overlook the invitation from Marcel Dekker to write this book and I thank them for it; without their assurance I would not have been able to take the first step toward completing the task.

I extend my very special gratitude to Taylor & Francis for honoring the agreement made by Marcel Dekker to publish this book.

During the preparation of *Handbook of Photomask Manufacturing Technology*, I have had many useful discussions, critiques, and constructive inputs from a number of professionals and from my former coworkers and managers.

Soon after receiving the invitation from Marcel Dekker to write this book, I conferred with Dr. Robert Doering regarding his experience in editing a book of this magnitude. Dr. Doering's inputs were extremely helpful in my decision in accepting the challenge to write this book.

Jeff Dorsch, a former editor of *Electronic News*, helped me in preparing the initial outlines and proposal for the book. I am very thankful to Jeff for his assistance.

My thanks to John Duff and John Skinner who shared with me the manuals of the discourses they have given at photomask conferences, which I found very helpful in organizing the structure of my book.

Douglas Van Den Broeke, my former manager and also a coauthor of my chapter in the earlier book, gave me many ideas during the development of the materials for this book. I am very thankful to him for his advice and inputs.

I also convey my thanks to my nephew Saeed Raghib, a software engineer, who was always there to help me whenever I experienced a computer breakdown or software glitches during my work.

As I mentioned earlier, there are 47 authors who contributed to this book. It was quite a feat to identify the right authors from all over the world. Many of those who recommended and helped me in making contact with the prospective authors include: Bill Almond, Nagesh Avadhany, Heinrich Becker, Hans Buhre, Manoj Chacko, Giang Dao, Paul DePesa, Ben Eynon, Toshiro Itani, Vishnu Kamat, Shinatro Kawata, Hartmut Kirsch, Kenich Kosugi, Susan Lippincott, Tom Novak, Leif Odselius, Osamu Okabashi, Buno Patti, Jim Pouquette, Srinivas Raghvendra, Wolfgang Staud, Yoshio Tanaka, Hideo Yohihara, and many more. My thanks to all of them and also my apologies to those whose names have not been mentioned.

I have not been able to enumerate all my friends and colleagues who expressed special interest in this book and whose continued encouragement has been of great value in seeing it to its completion. My apologies to them. A partial list of those friends and colleagues is: Shahzad Akbar, Farid Askari, Cathy Baker, Mike Barr, Ron Bracken, Chris

Constantine, Noel Corcoran, Grace Dai, Roxann Engelstad, Bernard Fay, Manny Ferreira, Pat Gabella, Michael Guillorn, Cecil Hale, Maqsood Haque, Dan Herr, Asim Husain, Nishrin Kachwala, Ismail Kashkoush, Birol Kuyel, David Lee, Chris Mack, Dan Meisburger, Kent Moriya, Kent Nakagawa, Diane Nguyen, Mark Osborne, Tarun Parikh, Kristine Perham, John Petersen, Paul Petric, Jean Shahan, Bill Waller, Jim Wiley, and Stanley Wolf.

I am very much honored that Dr. Yoshio Nishi accepted my invitation to write the foreword, which is an invaluable introduction to the *Handbook*.

My special thanks to my seven cats for their understanding when I was not able to give the attention they deserved. My three laptops and one desktop were never enough to go around among them. There were some who seemed to prefer to lie down on the keyboard that I was working on at any particular time. Switching from computer to computer without disturbing them, while keeping my train of thought, turned out to be quite a matter of synchronicity and mindfulness.

SYED A. RIZVI

List of Contributors

Paul J.M. van Adrichem Synopsys, Mountain View, California, USA

Ebru Apak NANOmetrics, Inc., Milpitas, California, USA

Sergey Babin Abeam Technologies, Castro Valley, California, USA

Min Bai Department of Electrical Engineering, Stanford University, Stanford, California, USA

Joerg Butschke Institute for Microelectronics Stuttgart, Stuttgart, Germany

Dachen Chu Department of Electrical Engineering, Stanford University, Stanford, California, USA

Andreas Erdmann Fraunhofer-Institute of Integrated Systems and Device Technology, Erlangen, Germany

Shirley Hemar Mask Inspection Division, Applied Materials, Santa Clara, California, USA

Richard Heuser Micro Lithography, Inc., Sunnyvale, California, USA

Ray J. Hoobler NANOmetrics, Inc., Milpitas, California, USA

Christian K. Kalus SIGMA-C, München, Germany

Frank-Michael Kamm Infineon Technologies, Dresden, Germany

Kurt R. Kimmel IBM Microelectronics Division, Albany Nanotech Facility, Albany, New York, USA

Hal Kusunose Lasertec Corporation, Yokohama, Japan

Randall Lee FEI Company, Peabody, Massachusetts, USA

Michael Lercel IBM Microelectronics, Hopewell Junction, New York, USA

Florian Letzkus Institute for Microelectronics Stuttgart, Stuttgart, Germany

Vladimir Liberman Lincoln Laboratory, Massachusetts Institute of Technology, Lexington, Massachusetts, USA

Hans Loeschner IMS Nanofabrication GmbH, Vienna, Austria

Takayoshi Matsuyama Lasertec Corporation, Kohoku-ku, Yokohama, Japan

Wilhelm Maurer Mentor Graphics Corporation, San Jose, California, USA

David Medeiros IBM Corp. T.J. Watson Research Center, Yorktown Heights, New York, USA

A. Meyyappan AtometX, Santa Barbara, California, USA

Sylvain Muckenhirn AtometX, Montgeron, France

Masatoshi Oda NTT–AT Nanofabrication Corporation, Kanagawa Prefecture, Japan

Shane Palmer Lithography External Research, Texas Instruments Inc., Dallas, Texas, USA

Fabian Pease Department of Electrical Engineering, Stanford University, Stanford, California, USA

Michael T. Postek National Institute of Standards and Technology, Metrology Gaithersburg, Maryland, USA

James Potzick National Institute of Standards and Technology, Gaithersburg, Maryland, USA

Benjamen Rathsack Texas Instruments Inc., Dallas, Texas, USA

Syed A. Rizvi Nanotechnology Education and Consulting Services, San Jose, California, USA

Anja Rosenbusch Mask Inspection Division, Applied Materials, Santa Clara, California, USA

Christer Rydberg Micronic Laser Systems AB, Täby, Sweden

Norio Saitou Hitachi High-Technologies Corporation, Tokyo, Japan

Hisatak Sano Dai Nippon Printing Co. Ltd, Chiba-ken, Japan

Frank Schellenberg Mentor Graphics Corporation, San Jose, California, USA

Michael T. Takac Hitachi Global Storage Technologies, San Jose, California, USA

Ching-Bore Wang Micro Lithography, Inc., Sunnyvale, California, USA

C. Grant Willson University of Texas at Austin, Departments of Chemistry and Biochemistry and Chemical Engineering, Austin, Texas, USA

Masaki Yamabe Semiconductor Leading Edge Technologies, Inc., Ibaraki-ken, Japan

List of Contributors

Pei-yang Yan   Intel Corporation, Santa Clara, California, USA

Yung-Tsai Yen   Micro Lithography, Inc., Sunnyvale, California, USA

Makoto Yonezawa   Lasertec Corporation, Yokohama, Japan

Hideo Yoshihara   NTT–AT Nanofabrication Corporation, Kanagawa Prefecture, Japan

Nobuyuki Yoshioka   SELETE, Ibaraki-ken, Japan

Andrew G. Zanzal   Photronics, Brookfield, Connecticut, USA

Axel Zibold   Carl Zeiss Microelectronic Systems GmbH, Jena, Germany

About the Book

The challenges before the semiconductor industry are to make smaller and smaller features so that the maximum number of functions can be packed into the smallest space on a chip. Current state-of-the-art technology is pursuing the fabrication of sub-90 nm features, and microlithography, now referred as nanolithography, is playing a very significant role. One of the areas on which the achievement of microlithography is heavily dependent is the advancement in mask technology. The technology for making advanced masks is complex and challenging, and the trend continues. There are numerous reasons why the publication of a book on mask technology is timely and will serve the industry's needs:

1. The mask market has grown by 14% during the last year, amounting to $2.4 billion that is, 20% of the micro- and nanolithography world market.
2. The cost of a high-end mask set is projected to reach the $1 million mark in the coming years.
3. The industry will soon reach its crossroad of next-generation lithography (NGL) after the use of optical lithography ceases to be feasible technologically and economically.
4. There has never been a book published exclusively on the topic of mask technology.

Handbook of Photomask Manufacturing Technology is a comprehensive work that summarizes all areas of mask technology into one single volume. It has 31 chapters contributed by 47 authors of international repute representing academia, government, national labs, and consortia around the world. Chapters on related topics are grouped under major sections preceded by introductory chapters. Thus the book will be of value to those who want an overview of certain aspects of mask technology without going into details, and to those who want in-depth knowledge of specific areas of mask technology.

Contents

Section I Introduction

1. **Introduction to Mask Making** 3
   Andrew G. Zanzal

Section II Mask Writing

2. **Data Preparation** 19
   Paul J.M. van Adrichem and Christian K. Kalus

3. **Mask Writers: An Overview** 43
   Sergey Babin

4. **E-Beam Mask Writers** 59
   Norio Saitou

5. **Laser Mask Writers** 99
   Christer Rydberg

Section III Optical Masks

6. **Optical Masks: An Overview** 135
   Nobuyuki Yoshioka

7. **Conventional Optical Masks** 157
   Syed A. Rizvi

8. **Advanced Optical Masks** 163
   Wilhelm Maurer and Frank Schellenberg

Section IV NGL Masks

9. **NGL Masks: An Overview** 193
   Kurt R. Kimmel and Michael Lercel

10. **Masks for Electron Beam Projection Lithography** 199
    Hisatak Sano, Shane Palmer, and Masaki Yamabe

11. **Masks for Extreme Ultraviolet Lithography** 231
    Pei-yang Yan

12. **Masks for Ion Projection Lithography** 271
    Syed A. Rizvi, Frank-Michael Kamm, Joerg Butschke, Florian Letzkus, and Hans Loeschner

13. **Mask for Proximity X-Ray Lithography** 305
    Masatoshi Oda and Hideo Yoshihara

Section V Mask Processing, Materials, and Pellicles

14. **Mask Substrate** 321
    Syed A. Rizvi

15. **Resists for Mask Making** 325
    Benjamen Rathsack, David Medeiros, and C. Grant Willson

16. **Resist Charging and Heating** 341
    Min Bai, Dachen Chu, and Fabian Pease

17. **Mask Processing** 367
    Syed A. Rizvi

18. **Mask Materials: Optical Properties** 377
    Vladimir Liberman

19. **Pellicles** 395
    Yung-Tsai Yen, Ching-Bore Wang, and Richard Heuser

Section VI Mask Metrology, Inspection, Evaluation, and Repairs

20. **Photomask Feature Metrology** 413
    James Potzick

21. **Optical Critical Dimension Metrology** 433
    Ray J. Hoobler

22. **Photomask Critical Dimension Metrology in the Scanning Electron Microscope** 457
    Michael T. Postek

23. **Geometrical Characterization of Masks Using SPM** 499
    Sylvain Muckenhirn and A. Meyyappan

24. **Metrology of Image Placement** 531
    Michael T. Takac

25. **Optical Thin-Film Metrology for Photomask Applications** 551
    Ebru Apak

26. **Phase Measurement Tool for PSM** 577
    Hal Kusunose

27. **Mask Inspection: Theories and Principles** 589
    Anja Rosenbusch and Shirley Hemar

28. **Tool for Inspecting Masks: Lasertec MD 2500** 599
    Makoto Yonezawa and Takayoshi Matsuyama

29. **Tools for Mask Image Evaluation** 607
    Axel Zibold

30. **Mask Repair** 629
    Randall Lee

Section VII Modeling and Simulation

31. **Modeling and Simulation** 649
    Andreas Erdmann

Index 693

Section I

Introduction

1
Introduction to Mask Making

Andrew G. Zanzal

CONTENTS
1.1 Introduction 3
1.2 How Masks are Fabricated 5
    1.2.1 Data Preparation 5
        1.2.1.1 Transformation 5
        1.2.1.2 Augmentation 6
        1.2.1.3 Verification 6
    1.2.2 Front End of Line 6
        1.2.2.1 Mask Writing 7
        1.2.2.2 Process 7
    1.2.3 Back End of Line 7
        1.2.3.1 Defect Inspection 8
        1.2.3.2 Defect Repair 8
        1.2.3.3 Pellicle Application 8
1.3 The Technology History of Masks 8
    1.3.1 The Contact Print Era 8
    1.3.2 The 1× Projection Era 10
    1.3.3 The Wafer Stepper Era 13
    1.3.4 The Subwavelength Era 14
1.4 The Future of Masks 15

1.1 Introduction

A mask can be defined in simplest terms as a pattern transfer artifact, wherein exists a patterned surface on a substrate material. The patterned surface is transferred to form an image of the pattern on a receiving substrate, which is subsequently fabricated into electronic, electromechanical, or mechanical devices. In most cases, the transfer artifact substrate is highly transparent to the change agent of the pattern transfer (usually light), while the patterned media on the mask substrate is less transparent or opaque to the change agent. The transmitted image may pass through a lens, with either demagnification or 1:1, or it may be contact printed directly to the receiving substrate (Figure 1.1). Typical masks used by the semiconductor industry over the past 25—30 years have been comprised of a thin (80–100 nm) layer of chrome on a glass or fused silica substrate. The chrome is patterned using a coating of photosensitive material, known as photoresist, which is resistant to chrome etching. After selective exposure with energy, the photoresist

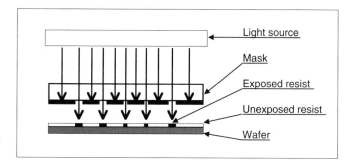

FIGURE 1.1
Schematic of wafer exposure using a mask.

is chemically developed, leaving voids where an etching chemical may be applied to remove unwanted chrome.

Masks, as used in the semiconductor and related industries that have adopted semiconductor fabrication methods, have their origin in traditional printing and lithography. For the purpose of this publication, *mask* refers to the pattern transfer artifact as described above when used in the fabrication of microelectronic or similar micro devices. In this genre, masks have also become commonly referred to as photomasks or reticles, almost interchangeably. Traditionally, a mask or photomask was a pattern transfer artifact that contained the complete pattern content of a single layer of a full semiconductor wafer, which could be printed in a single exposure without any optical demagnification. A reticle also contains a single layer of pattern data but only for a small part of the wafer. The reticle image could be projected, with or without demagnification, onto the small part of the wafer (Figure 1.2). After the formation of this single image by exposure, the wafer would move on a stage, and then a second image would be printed, until several subsequent areas of the wafer were printed with the pattern content from the reticle. Semiconductor and related industries have migrated over the years to using mostly reticles, but the interchangeable use of the terms reticle, photomask, and mask has blurred the traditional distinction.

As a class of pattern transfer artifacts masks can be used in multiple ways. They can be used with broadband or single wavelength light (radiation) sources. They can have transmitting or reflecting substrates. The pattern media can be fully opaque or partially transmitting for transmission masks, or absorptive for reflective masks. The substrate and media can be engineered and processed in such a way as to change the phase

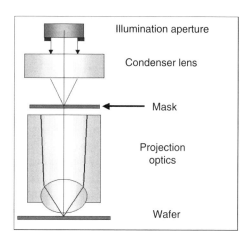

FIGURE 1.2
Mask in a typical projection/reduction lithography system.

Introduction to Mask Making 5

of the radiating wavelength. The substrates can be of rigid contiguous media or thin membranes with discontinuities. Masks can be used in direct hard contact or in projection with a particle-free "mini-environment" known as a pellicle. These and other options will be highlighted in the subsequent chapters.

The most common masks in use today are those with an image media of chromium that is sputtered on a fused silica substrate. The chromium is highly opaque to the radiating wavelength (usually 248 or 193 nm), while fused silica is highly transparent. The masks will most likely be used on a tool that prints multiple fields in an expose-and-repeat fashion. The mask will have a pellicle attached to the media coated surface. A pellicle is a transparent membrane that is stretched across a metal frame and subsequently attached to the mask using an adhesive. The pellicle keeps any particles from contaminating the patterned surface and provides a suitable distance to keep away such particles out of the printing focal plane. A mask as described in this paragraph is alternatively referred to as a binary intensity mask (BIM) or chrome-on-glass (COG) mask.

1.2 How Masks are Fabricated

Masks have been made by a variety of means since the 1960s, some of which will be described from a historical perspective later in this chapter. For the purposes of this section, however, the flow described is typical of processes used in producing BIMs today. The details of the process are often proprietary to the practitioner, and, as such, intimate details are omitted from this chapter. More complex flows that are required to produce more exotic masks either in use today or contemplated for the future, are described in subsequent chapters.

Today's mask patterns are defined using electronic design automation (EDA) software tools on computer automated design (CAD) systems. This operation is most often performed outside the mask arena, so the output from the design activity is the starting point of mask fabrication. The fabrication process can be subdivided into three broad categories of manufacturing activities: data preparation, front end of line (FEOL), and back end of line (BEOL). Figure 1.3 is a highly simplified diagram of the process flow and is meant to broadly convey the general steps employed. The reader should be aware that the actual processes can include loops or repetitions of many of the steps in the diagram

1.2.1 Data Preparation

The data preparation process is comprised of three broad process steps performed once the mask data files are received from the circuit designer. These process steps can be broadly categorized as transformation, augmentation, and verification.

1.2.1.1 Transformation

Incoming mask data from circuit designers can be either passed directly to the FEOL step or any number of data file manipulation steps can be performed. These include transforming the data from the input format to a format compatible with the mask writing tool used in FEOL processes. This transformation process is commonly referred to as fracturing and may or may not include the addition of process bias to compensate for linear differences in feature sizes between the mask process and the desired final wafer result. Advanced masks

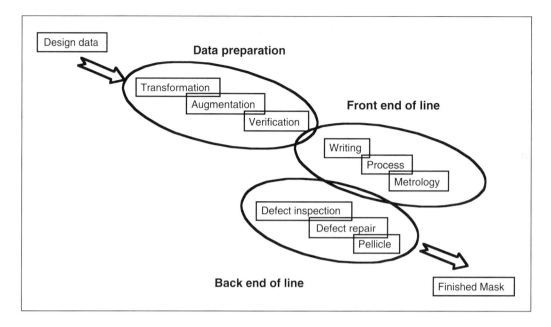

FIGURE 1.3
Simplified mask-making flow.

may also require differential biasing where compensations are made based on local patterning attributes. In addition to preparing data files for mask writing, data files used to inspect the finished mask are also included in the transformation process.

1.2.1.2 Augmentation

Other functions that may occur in the data preparation stage include the addition of optical proximity correction (OPC) features and the addition of standard mask patterns, such as machine-readable barcodes, human-readable labels, alignment marks, and metrology cells. Additionally, this step often includes composing a set of instructions, known as a job deck, that tell the mask writer where to put the content from multiple mask files onto the mask.

1.2.1.3 Verification

To varying degrees, based on project complexity and customer requirements, automated and manual verifications are performed. These steps are performed to minimize the number of masks that must be remanufactured and to cycle through more costly processes in FEOL and BEOL. Some of these processes include automatic comparison of the fractured data to the input file and may also include local or remote on-screen viewing by customers of the mask fabrication facility. A trend towards automating many of the verification processes has been apparent over the last several years

1.2.2 Front End of Line

The FEOL processes include the mask writing step, chemical processing, and metrology. Metrology, or measurement, may be inserted many times in the flow to ensure the final outcome and detect failures in the flow as early as possible to avoid adding unnecessary cost by performing operations on a mask that will subsequently be rejected.

1.2.2.1 Mask Writing

Today's masks are generally written on either of the two classes of write tools. Electron beam (e-beam) writers precisely direct a focused stream of electrons onto the mask substrate while controlling the position of those electrons through the use of an interferometer controlled stage. Laser writers essentially perform in the same fashion but use laser generated photon energy rather than negatively charged electrons. The optics and mechanics of how these electrons or photons are delivered to the substrate surface are distinctly different for virtually every writer employed in the industry. These differences will be discussed in later chapters.

1.2.2.2 Process

The energy delivered by the mask writer to the substrate surface is intended to react with a coating on the chromium. The coating, generically referred to as resist, is engineered to be sensitive to either e-beam or laser exposure. Resist is a chemical polymer that is usually cross-linked at the molecular level when exposed to the radiation wavelength for which it is designed.

1.2.2.2.1 Develop

The locally cross-linked molecules become either sensitive or insensitive to chemical developers used in post-write processing steps. When the develop step removes exposed resist, the process is referred to as positive-working, and when leaving behind exposed resist, negative-working.

1.2.2.2.2 Etch

After development the mask moves on to the etch step. In this part of the process, the surface of the mask that has been left uncovered by resist becomes exposed to the etching chemistry. The resists are engineered to withstand the etching process and at the very least stand up to the etch chemistry with a removal rate that is slower than the removal rate of the underlying chromium. Etching can be accomplished by using liquid (wet) or plasma (dry) etch chemistry. After complete removal of unwanted chromium by etching, the mask is stripped of all remaining resists.

1.2.2.2.3 Metrology

Masks are subjected to metrology throughout the manufacturing process, but most rigorously at the postdevelop and postetch stages. Postdevelop measurements of critical dimensions (CDs) are taken to ensure that the develop process has not over or under shot the final desired CD. Often an iterative develop process can ensure the likelihood of final CDs' meeting the specification. Iterative etching processes are also used but are less common. Once a sample of CDs is found to be within the desired final outcome after etching, customer specific sampling plans are generally employed to assure compliance with specifications. Another form of metrology that is often employed is registration or position metrology. This metrology is used to ascertain that elements of the mask are at intended locations relative to other customer-identified features.

1.2.3 Back End of Line

BEOL process is generally performed to ensure quality of the outgoing mask and to protect the mask from particles in transit to the user, and through its useful life. These steps include defect inspection, defect repair, and pellicle application. The impact of cleaning processes should not be ignored or understated as the mask may go through cleaning cycles before and after each inspection and repair, and a rigorous final cleaning process just prior to pellicle application.

1.2.3.1 Defect Inspection

Masks are inspected for defects on automated tools that scan the mask surface and constantly compare the physical mask to a reference image. Any anomaly detected that shows a difference within preset limits between the mask and the reference image is flagged as a potential defect, and its location is recorded for later review and classification by an inspection operator. The reference image can either be an intended identical pattern from another area of the mask, known as die-to-die inspection, or a digital representation on the intended mask image that was prepared in the transformation phase of data preparation.

1.2.3.2 Defect Repair

Once found, defects that are large enough to print need to be repaired. Printability assessment through lithography simulation has not become a universally accepted practice, resulting in an industry that tends to repair every defect it detects. Repair is accomplished on advanced masks using both focused ion beams and nano-machining with atomic force microscope (AFM) tipped repair tools. More mature masks are still repaired using laser tools.

1.2.3.3 Pellicle Application

After rigorous final cleaning and assessment to ensure that there are no particulate contaminants or chemical stains, a pellicle is attached to protect the mask surface from subsequent contamination in shipment and through the masks' useful life. The mask is inspected with both reflected and transmitted light after the pellicle is attached to ensure that the area under the pellicle has remained defect- and contaminant-free throughout the pellicle application process.

1.3 The Technology History of Masks

The use of masks to print semiconductor wafers goes back to the very early days of the formation of the industry. In the intervening decades, the mask industry has continually adapted its product offering to meet the needs of its semiconductor-producing customers. The changes in mask product offerings were usually driven by the development in wafer lithography. The industry has progressed from 1× contact print lithography through 1× projection to reduction lithography. In the most recent era beginning 1999, masks, and the lithography processes that support masks, have entered the subwavelength era where the line widths printed are often far smaller than the wavelength of light used to print them.

1.3.1 The Contact Print Era

In the early days of the semiconductor industry, masks were almost exclusively used in contact print applications. Masks used in this fashion were placed in intimate contact with the wafer, which was subsequently exposed with ultraviolet light. In the process of making intimate contact, difficulties arose in the area of creation of defects on the mask, which printed on subsequent wafers. Alignment with previous layers was also a problem, caused by thermal stability of mask materials and lack of other than visual or manual alignment.

The first masks in this era were actually not on the rigid glass substrates we know of today, but were actually made from either a material called rubylith or photographic film. Rubylith is a two-layer polymer-based material comprised of light-blocking red film laminated to a clear polymer base. In the contact print era, rubylith material was widely used in the graphic arts and adapted easily to the fledgling semiconductor industry. The softer red layer could be cut with a knife without damaging the tougher clear layer. After cutting, the unwanted red areas could be peeled away and a mask comprised of clear and opaque areas was created. The rubylith material could be printed directly on the wafer or copied onto a film, with or without reduction, and then printed. Obviously, this technique had little merit for any level of production volume, but in early R&D stages it was adequate for proof of concept demonstrations.

To enhance mask precision and manufacturing volume, the industry evolved to one where rubylith continued to be used but the circuits were cut into the material at high magnifications (often 200×). The resultant representation of the circuit was called "artwork." The artwork was photo reduced (often 20:1) onto emulsion glass substrates usually resulting in a 10× reticle. Metrology in those days consisted of checking the rubylith feature sizes with a steel scale and an eye loupe. Through this era, the precision and error rate of rubylith cutting were greatly enhanced by the development of a tool called a coordinatograph, which has an effective light table to which the rubylith was affixed (Figure 1.4). This tool had mechanical stops in the X and Y directions to which coordinates could be applied to precisely guide the length and direction of the cut. Some coordinatographs also had rotating light tables so that even angled features could be cut. Later versions of coordinatographs used primitive computer numerical control to make the precise cuts required to generate even more complex artwork.

The 10× reticle produced from the photo-reduced artwork was used in a tool called a photorepeater, which reduced the mask image to the final desired size while simultaneously printing it on a photographic emulsion coated glass substrate. This "camera" would expose, step to the next location, and then repeat the process until the mask was completed. The photorepeater also became known as a step and repeat camera and was a precursor to the wafer stepper that is used in most semiconductor fabrication methods today. The resultant mask from the photorepeater became a master mask from which many copies were made.

In this era, mask materials moved from the initial mylar-based films with rubylith or photographic emulsion to soda-lime glass substrates with photographic emulsion. Initially used in wafer contact printing, masks had a finite life, and many copies were made to support production lines making large volumes of devices. Emulsion photomasks first supported the industry; however, the emulsion coating was relatively soft and did not stand up long to contact printing. Each print added additional defects to the mask, in some cases rendering it useless after printing just a few wafers.

To mitigate these defect issues, the industry migrated over time to so-called hard surface materials that consisted of sputtered or evaporated metal films that had appropriate optical properties and could be cleaned repeatedly. Although not totally impervious to added defects over time, hard surface photomasks were far less susceptible to damage as compared to emulsion masks during the contact printing process, and they could be cleaned of particles. The two films the industry settled upon were iron oxide and chromium. Iron oxide provided the benefit of being transparent in the visible wavelength

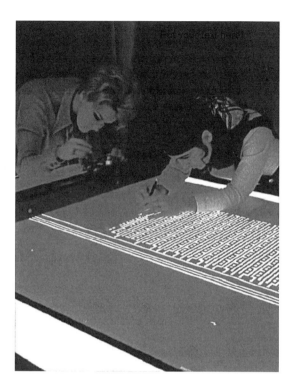

FIGURE 1.4
Operators cutting rubylith artwork. (Photo courtesy of Intel Corp.)

while being opaque at the exposure wavelength, allowing the user to see through the mask when aligning to a previous layer. The drawbacks of iron oxide were that the material processed poorly, was prone to excessive defects, and provided such non-uniform results, and that the semiconductor device specifications quickly outpaced its usefulness as a viable material. Chromium provided a much more uniform result and to this day remains the dominant material for mask making.

The key technology breakthroughs in this era from the 1960s to the early 1970s were the eventual development and migration to hard surface photomasks. More significant though, was the development of a tool called an optical pattern generator (Figure 1.5). This class of tools, developed by companies like David Mann Company and Electromask, effectively replaced rubylith as a means of creating artwork by allowing the production of 10× reticles directly on emulsion glass substrates. The tool had a computer-controlled stage that allowed for the movement of the substrate in the X or Y direction. An aperture in the optical path allowed for a variable shaped slit from a wide range of height and width values. The optics and aperture were mounted on a turret that could be rotated in 0.1° increments, allowing for angled features as well. These tools appeared on the scene in the late 1960s and reached their full popularity in the mid-1970s, at a time when the complexity of leading edge devices had gone beyond the ability of operators to cut them in rubylith. These tools allowed the industry to move beyond the physical mask content and subsequent verification limitations that had been imposed by using rubylith.

1.3.2 The 1× Projection Era

In the mid-1970s, the need for masks with longer lifetimes became apparent to the industry. Using large quantities of contact print masks was difficult to manage in a production line, but more significant were the yield losses associated with contact

Introduction to Mask Making 11

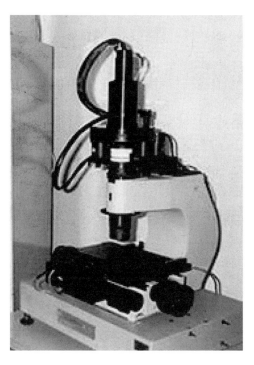

FIGURE 1.5
Mann3000 Pattern Generator. (Photo courtesy of the University of Notre Dame.)

printing due to defects on masks printing on wafers and contact with wafers damaging the masks. In this era, Perkin-Elmer Corporation developed the Micralign projection aligner and through its deployment precipitated several profound changes to the business of making masks that have continued forward to this day, many of which are discussed in later chapters.

The Micralign eliminated contact printing by using a system of mirrors and lenses to project the mask image onto the wafer. The mask never came in contact with the wafer, thereby eliminating the defect issues associated with contact printing. Another feature of the Micralign was the ability of the tool to automatically align to a previously printed layer through the use of specially designed and located alignment marks.

In this era, several changes in substrate materials were ushered into the industry. Since masks did not have to be cleaned as often, they were used for many more wafer exposures leading to increases in temperature at the mask plane. Over time, the thermal properties of soda-lime glass led to wafer alignment issues, which forced the industry to migrate to glass substrate materials that had a lower thermal coefficient of expansion. The industry settled on borosilicate glass materials, which had thermal properties similar to those of silicon wafers. Towards the end of the 1× projection era, wafer exposure wavelengths migrated to near 365 nm, creating the need for a substrate with higher transmission properties than borosilicate. The industry turned to synthetic quartz, or fused silica, which had very high transmission properties all the way through 193 nm, and remarkable thermal stability. Fused silica remains, to this day the substrate of choice for leading-edge mask techonology.

Some of the breakthrough technology advancements that occurred in this era included automated defect inspection, laser repair, and pellicle protection. The first automated defect inspection machine appeared on the scene in early 1978. Developed by KLA Instruments, the KLA 100 tool (Figure 1.6) performed defect inspection by scanning and comparing adjacent fields on the photomask intended to be identical. When an anomaly

FIGURE 1.6
KLA 101 Die-to-die automated defect inspection system. (Photo courtesy of KLA-Tencor.)

between the two inspected fields was detected, the location was recorded and stored for later review and classification by a machine operator. This automation replaced the tedious visual inspection method whereby an operator would spend his or her day scanning masks through a microscope and counting defects. Other benefits, in addition to the reduction of tedium, was the ability to replace sampling plans with 100% inspection and the ability to use the tool for process improvement through rapid collection of results in defect reduction experiments.

Of course, finding defects was only half of the story. The desire to achieve defect-free masks could only be accomplished with reasonable yield through the use of repair tools. The first laser repair tools from companies like Quantronix and Florod appeared in the late 1970s. These systems used laser pulses to remove unwanted chrome spots from the mask. Later versions included the ability to repair voids in the chrome through laser-assisted chemical vapor deposition. Although primitive by today's standards, these tools proliferated rapidly through the industry and continue to be used for trailing edge mask products even today.

Once defects were found and repaired, maintaining their low-to-zero defect condition became the next focus area. Again the industry responded by developing a hardware addition to the mask called a pellicle. A pellicle is essentially a thin transparent membrane that is stretched over a metal frame. The purpose was to effectively seal the surface of the mask to keep out particles. The height of the frame was designed specifically to keep any particles that landed on the membrane surface out of the focal plane of the exposure system, thereby keeping the particles from printing on the wafer. Companies like Advanced Semiconductor Products (ASP), Mitsui, and Micro Lithography Incorporated (MLI) introduced pellicles commercially in the early 1980s. Pellicles have become an integral part of photomasks and today's masks are rarely used without them.

Through this era, device complexity continued to increase to the point where the optical pattern generators that were used to create the 10× reticles, from which 1× projection masks were made, were unable to keep up with the increasing density. Pattern generation

Introduction to Mask Making 13

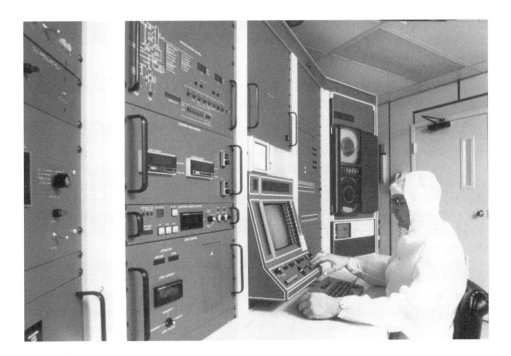

FIGURE 1.7
Operator at MEBES-III console. (Photo courtesy of Photronics, Inc.)

files soon eclipsed several hundred thousand exposures, and reticles could take as long as 48 h to expose. To mitigate this development, a faster method of writing these reticles was required. The answer was the development of the e-beam writer, which came out of Bell Laboratories to be commercialized in the late 1970s by ETEC Corporation (Figure 1.7). Dubbed the Manufacturing Electron Beam Exposure System (MEBES), this tool provided the ability to write reticles for use on photorepeaters at much faster throughput rates than possible with optical pattern generators. The added benefit of this tool was that it was significantly over-engineered for the task of making photorepeater reticles and could be used to make 1× masks directly from a data file. This led to the ability to create masks with die sizes that were not constrained by the photorepeater lens field, as well as having masks without repeating features, if desired. The resolution and flexibility of the early MEBES and its successors served the industry as the platform of choice well into the mid-1990s. Since these tools were capable of manufacturing 1× masks for leading edge devices, their inherent abilities precipitated an era that later came to be referred to by the industry insiders alternatively as "the mask maker's holiday" or the "5× holiday."

1.3.3 The Wafer Stepper Era

Through the later stages of the 1× projection era, the semiconductor industry migrated rapidly to wafer steppers as the lithography process of choice, from companies like GCA, Nikon, and Canon. Much of the reason for the adoption of wafer steppers was due to difficulties in printing in the 1 μm and below regime with 1× projection aligners. With reduction steppers these dimensions could be achieved with wide process windows and robust manufacturing yields. Another benefit was that defect-free masks could be easily achieved from within the existing mask infrastructure, which at the time was geared to produce 1× masks. The earliest wafer steppers were 10× tools, so the reticles needed to support lithography were effectively 10 times easier to make from a resolution and CD control standpoint. Later, even as wafer steppers migrated to 5×, the relaxation of specification criteria brought on by the magnification change from 1× was still profound. Given an infrastructure of tools capable of much finer precision than that needed by its customers, the mask industry focused on the service aspects of the business, including reducing cycle time, reducing defects, improving mask data handling and transfer, and implementing quality improvement systems.

Through this so-called holiday there was considerable consolidation in the merchant mask industry and divestiture of captive facilities. Cycle times, often 2 weeks in the 1× optical mask making days, were reduced to 3–4 days in routine cases and less than 24 h on a rush basis. Mask orders no longer required the magnetic tapes be shipped to the mask provider but, instead, the use of dedicated dial up lines, later to be supplanted by Internet transfer, flourished. Encouraged by their customers, mask makers embraced statistical process control and quality systems, such as ISO 9000.

The industry also implemented new innovations from its supplier base, many of which are covered in detail in later chapters. High throughput laser mask writers from companies like Ateq and later Mirconic Laser Systems supplanted many MEBES tools for all but the most technically challenging requirements. KLA introduced inspection systems that compared the mask to the data file from which it was produced allowing for defect inspection on masks without repeating patterns. Interferometer-based metrology tools from Nikon and Leica were developed to measure position accuracy and led to the ability of mask makers to match masks successfully from different mask writers. Focused ion beam repair tools, from companies like Seiko Instruments and Micrion, augmented laser repair systems. These tools did far less surface damage to the mask and allowed for repair of smaller defects with greater precision.

As mentioned in the previous paragraphs, the 5× holiday was not a holiday at all but an era that provided the opportunity for the industry to focus on time-to-market issues, BEOL inspection and metrology, implementation of quality systems, and consolidation. By the end of this era (*ca* 1994) the number of merchant mask suppliers in the world was reduced through consolidation to four major and about half dozen lesser players. The growth of these leading players towards achieving critical mass for attracting adequate capital investment proved to be of major consequence as the industry entered the subwavelength era in the late 1990s.

1.3.4 The Subwavelength Era

The mask industry is currently in the subwavelength era and will remain firmly therein through at least the rest of the first decade of the 21st century. This era is defined by the relative wavelength used by customers on their wafer exposure systems compared to the feature sizes being printed on the wafer. Over the years, since the first wafer steppers were introduced, stepper exposure wavelengths have continued to migrate downward from their initial state at 436 nm, through 365 and 248 nm, to today's state-of-the-art 193-nm exposure systems. Once feature sizes migrate below wavelength, lithographers need to apply techniques to enhance the imaging to achieve the wafer features. Some of these techniques can be applied at the wafer end of the stepper, such as special photoresists and antireflection coatings, while some might be applied to the illumination end of the stepper, such as off-axis or dipole illumination. At the reticle plane, enhancements can be made to the mask to improve the printed wafer results. These reticle treatments, collectively known as resolution enhancement techniques (RET), include a variety of techniques [phase shift masks (PSM) and OPC, discussed in greater detail in later chapters] to shift the phase of light and number of methods to correct feature shape loss during reduction.

There was little need to use RET at 436 nm as the transition to 365-nm steppers was relatively painless for the semiconductor industry. RET, as a means of advancing leading edge semiconductors, was implemented briefly in North America at 365 nm before giving way to the 248-nm wavelength. In Japan, however, RET at 365 nm was used in production for a longer period as a means of extending the useable life of wafer steppers. With the introduction of 248-nm exposure systems and the subsequent delay in availability of 193-nm systems, RET has quickly become a requirement for virtually all critical layers in a set of masks for advanced semiconductor devices.

With the near-term roadmap for the semiconductor industry indicating 193 nm, and possibly 157 nm, as the wavelengths of choice at least through the current decade, reticles have become and will remain a key enabling part of the lithography process. Feature sizes are expected to be nearly a one third of the wavelength by the year 2007, driving up the content and complexity of masks. Creating reliable and cost-effective methods for manufacturing the challenging masks of today and those contemplated in the future require significant investment in R&D and capital equipment costing hundreds of millions of dollars. Clearly, the industry will benefit from the visionary management at companies like Photronics and Dupont Photomask, and to a lesser extent at Dai Nippon and Toppan, that brought about the global consolidation of many small independent and captive mask-manufacturing organizations*. These larger organizations tend to have greater access to the capital resources required to extend mask technology.

Mask making in the subwavelength era requires the mask maker to possess a broader product offering to address the wide variety of RET, and one can infer that a broader set of core skills and new tools are required as well. Writers for advanced critical layer masks are now mostly high voltage e-beam systems with variable shaped beams and vector scanning positioning systems. Unlike earlier raster scan systems, which scanned the entire mask, these systems address and write only parts of the mask that need to be exposed. Complex software and algorithms control the size, shape, and energy of each exposure. Shape control is critical in providing the high fidelity OPC needed to provide adequate masks. To enhance the speed of these tools, both positive and negative chemically amplified resists are widely used. These new high voltage vector scan variable shaped e-beams are discussed in greater detail in later chapters. Mask makers also have moved into an era where new materials with precise optical properties for phase shifting require multilevel exposure, as well as different etch techniques, than those used in previous eras. In addition to new materials, tools and processes, knowledge and competence in lithography, simulation is also becoming more recognized in mask development organizations as critical to the rapid and robust implementation of ever more complex RET. In today's semiconductor economy, masks are no longer viewed as a commodity but as an integral part of a complete system.

1.4 The Future of Masks

The direction that mask making may take in the future beyond the year 2010 is unclear, as approaches to lithography after the 193-nm wavelength or perhaps 157-nm wavelength

---

* Editor's Note: Since the time the chapter was written, Dupout Photomask has announced plan to be acquired by Toppan.

are yet to be proven production worthy. Certainly, significant effort in developing lithography systems for the postoptical era has narrowed down the choices from among many candidates, but as often those that were thought to be dead are revived for another round of consideration. These postoptical or next generation lithography (NGL) approaches include extreme UV (EUV), electron projection lithography (EPL), ion projection lithography (IPL), proximity x-ray lithography (PXL), nano-imprint, and one might also argue direct write. With the exception of the latter, all of the choices require masks that are radically different in many ways from those that are used today and all except nano-imprint (a recently added option) are discussed in greater detail in later chapters. There are pockets of support around the globe for each of these technologies, though considerable effort continues to be invested in EUV and EPL. Although it is too soon to predict if any of these technologies will become as dominant as optical lithography has been, it is not difficult to envision that they will be used at least in a mix-and-match mode for applications for which they are best suited, such as PXL or EPL for contact holes.

Direct write maskless lithography looms on the horizon as a potential contender, but throughput and uniformity issues may relegate this technology to low volume applications, such as ASIC devices. The inherent benefit that has been embedded in masks since they were first used, through to the present day, is their ability to allow parallel exposure of massive amounts of content and to do so repeatedly wafer after wafer. I believe that the semiconductor industry will continue to leverage this attribute as long as there is no economically attractive alternative.

Section II

Mask Writing

2
Data Preparation

Paul J.M. van Adrichem and Christian K. Kalus

CONTENTS
2.1 Introduction 19
2.2 Mask Data Preparation Flow 20
    2.2.1 Layout Interchange Formats 21
    2.2.2 Boolean Mask Operations... 22
    2.2.3 Applying OPC and PSM 23
    2.2.4 Mask Data Creation... 25
    2.2.5 Passing of Mask Order Information 26
    2.2.6 Measurement Setup File Creation... 26
    2.2.7 Mask Data Preparation Ownership 26
2.3 The Jobdeck Concept... 27
2.4 Mask-Writing Principles... 28
    2.4.1 Raster Scanning Writing Principle.. 29
    2.4.2 Vector Scanning Writer Principle.. 30
    2.4.3 Variable Shape Beam Writer Principle 30
2.5 Trends in Mask Technology 30
2.6 Pattern Fidelity and Quality of Fractured Data...................................... 32
    2.6.1 Grid Snapping 33
    2.6.2 Line-Width Accuracy as Function of Data Slicing 34
    2.6.3 Scanfield Stitching 34
    2.6.4 Nonorthogonal Edge Approximation 35
    2.6.5 Laser Proximity Effects 36
    2.6.6 Electron Beam Proximity Effects... 37
    2.6.7 OPC Model Calibration 38
    2.6.8 Loading Effect Correction 39
2.7 Mask Data Processing Runtimes... 40
References 41

2.1 Introduction

After the successful completion of an integrated circuit (IC) design, the production phase of an IC can be started. To make the production of ICs possible, the photolithographic imaging process of the wafers needs masks. This chapter covers data preparation for the manufacturing of these masks. It first describes the general flow followed by some trends that can be observed in this area. There is another section on the quality impact of mask data on the final mask.

The mask-writing tool requires input data, and this input data format is usually not the same as the format that comes out of the design process. The design layout output format is usually an interchange format rather than a machine specific format. From a high level the objective of the mask data preparation (MDP) step is to convert this interchange format to a machine specific format. There are different mask-writing tools, and most of these tools have their own formats. Moreover, mask inspection tools and mask-measuring tools also require mask data dependent setup files.

There is more data that needs to be written on the mask than just the design data. The spaces between the actual ICs are used to saw the wafers in order to split them in individual dies. Wafer fabs normally use this saw-lane area for test pattern placements to monitor the wafer processing, and so these saw-lane patterns need to be created and written on the mask too. Moreover, alignment patterns are required for the alignment process on the wafer to align a pattern relative to the previous wafer-imaging step.

In general, apart from just the conversion of the design data, patterns need to be created for these saw lanes, alignment patterns, barcodes, and other patterns. This mask pattern synthesis is usually referred to as *frame-generation* and results in a so-called mask *jobdeck*.

During the conversion from the interchange format to the machine specific data or fracturing, the data can be compensated for a number of systematic errors that occur somewhere in the whole image transfer process from design data to image on the wafer. These errors may be caused by distortions that occur during the writing process, the mask development and etching process, and wafer imaging process. This compensation of data for systematic errors is getting a lot of attention recently. It improves the final mask quality at the cost of data manipulation time and quite often data volume.

2.2 Mask Data Preparation Flow

Although there is no formal definition of what MDP includes and different explanations do exist, one can regard it as the data link between IC design and mask making. After design completion, which is usually marked by a successful design rule check (DRC) run and a layout versus schematic (LVS) run, data needs to be created that enables the manufacturing of one or more masks, including all the required verification and measurements. Note that the total number of masks for a state-of-the-art process can easily exceed 30.

Figure 2.1 shows the complete flow of the mask data after the IC design. In some cases, these steps may not always be separable and are sometimes merged into one single step. In those cases the splitting up of the total MDP operation is rather theoretical than practical.

Most of the input data to the MDP is expressed in some layout interchange format. The output file is in a machine specific format, dedicated for the mask-writing tool that will be used for a particular layer. Because the machine capabilities differ from machine to machine, as well as the required specification for the different layers, a set of masks is almost never entirely written on one and the same tool. Hence, per mask layer, a different format needs to be generated.

Data Preparation 21

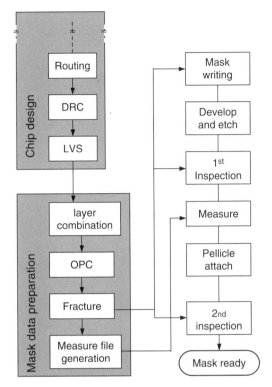

FIGURE 2.1
A simplified data preparation and mask manufacturing flow. Note that some steps in this data flow are sometimes combined, and some steps are interchangeable.

2.2.1 Layout Interchange Formats

A dedicated CAD environment is normally used to create the layout of an IC. The output of such a design phase is the layout of an IC and is normally represented in some layout interchange format. This format describes the layout in terms of geometrical elements, such as rectangles and polygons.

In the mid-1970s, Calma Company sold a CAD system Graphic Design System II (GDS2), which exports data in a published standard GDS2 stream format, also known under the name "Calma Stream" format [1]. Today it is the *de facto* industry standard for general IC layout description and interchange. The GDS2 format is a binary file format, which means that it cannot be loaded into a text editor directly. While transferring a GDS2 file (via FTP), a binary file transfer mode must be selected, otherwise the file becomes irreversibly corrupted.

A feature of the GDS2 format is the ability to use *hierarchy* in the layout description. Instead of describing every individual rectangle or polygon, it is possible in GDS2 to describe a cell and to refer to this cell in another cell using a single reference or even an array reference as shown in Figure 2.2. These cells can be single devices or a part thereof, or functional cells like NAND gates, memory cells, etc. For highly repetitive designs like memories the advantage is obvious.

In 2002, a new layout interchange format for ICs was proposed by SEMI [2] under the name "Open Artwork System Interchange Standard" (OASIS). The intention is to replace the GDS2 stream format. The main objectives of this format are to make the layout description file more compact and to overcome other shortcomings of the GDS2 format. There are lots of similarities between these two formats. Both formats can handle hierarchy, different layers, and data types. The main differences of OASIS with respect to GDS2 are:

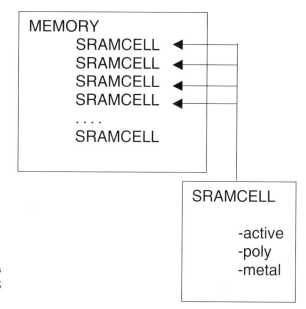

FIGURE 2.2
The GDS2 format can contain cell definitions and cell placements in other cells, thus using hierarchy to describe a layout.

- OASIS is much more compact and hence a smaller file size.
- In the OASIS format there is more room for future expansion of the format.

Apart from these two formats, a few other formats are sometimes used, especially in very dedicated applications. The vast majority of the design layouts are still interchanged from one package to another in GDS2 stream format, that is, the default input format to the total data preparation flow, and in the definition mentioned earlier.

2.2.2 Boolean Mask Operations

The first step that can be identified in the MDP flow is an operation in which the designed layers are combined and manipulated in order to obtain the physical mask data. For the more advanced processes, there is a tendency towards more complicated operations in this layer derivation step. Also, processes with built-in process options, like nonvolatile memory blocks, tend to have more derived layers than a more standard process.

These layer derivation manipulations normally consist of:

- Binary pattern operators, such a Boolean AND, OR, MINUS, and XOR between layers.
- Selection of design data based on certain criteria, such as feature area, size, shape, etc. These are sometimes referred to as unary pattern operators. Sizing of data can also be regarded as a unary pattern operator.

There can be various reasons why mask data is derived instead of directly designed. For one, the derived layers may be difficult to check, so a simpler check and a derivation of mask data are preferred. In some cases, mask data layers can be complementary, and two mask layers are derived from the same design layer. Quite often new processes evolve from older technologies, which makes these layer derivations something that is inherited and extended rather than set up from scratch. It is always preferable to maintain

Data Preparation

design methodologies and flows as much as possible. Previously designed cells and blocks are preferably reused instead of redesigned from scratch. One should realize that creating these designs is a complicated and time-consuming process. It represents a lot of value, and reusing such cells can leverage this investment. In other words, a layer derivation step can make the reuse of designs and design building blocks possible. Layer derivation can also be useful when an existing circuit design must be transferred to another (i.e., newer) process, thus avoiding the need to redesign it.

Given a certain layer operation and certain layout details, small notches, spikes, small lines, or spaces may be generated. For the purpose of the mask, these small features may not be relevant. However, if these features are not resolved during the mask-making process, the mask inspection step is usually quite problematic and may even cause higher mask manufacturing cost, and yield loss. In general, if these small features can be avoided, mask inspection issues can be avoided. There is a way to remove these small gaps or spikes by applying a small upsize and subsequent downsize of the data. A common term for these operations is *de-slivering* as shown in Figure 2.3. Such de-sliver operations are quite common in layer derivation steps. It is clear that the amount of upsize and downsize needs to be chosen carefully, or else data may become damaged.

Nonorthogonal edges are the ones that are usually overlooked and are thus the most problematic areas.

2.2.3 Applying OPC and PSM

After layer derivation, some of the layers are corrected for optical proximity effects (OPE) using one or more resolution enhancement techniques (RET). Out of the many RETs that are developed, the following are the most common:

- Optical proximity correction (OPC)
- Sub-resolution assist features
- Attenuated phase-shifting mask (PSM)
- Alternating PSM

Refer to Chapter 8 on RET for details on the technique itself. This section only considers the impact of OPC on MDP.

The result of the OPC step is that straight data elements with a few vertices are converted into data with a lot of vertices and line segments. It is clear that this addition of vertices and jogs yields a huge data volume increase (see Figure 2.4). Depending on the aggressiveness of the OPC factors, 10× file size increase is not uncommon. During an OPC run, usually, several layers are taken as an input. For example, when OPC is applied

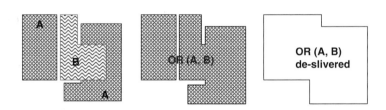

FIGURE 2.3
In this example, a logical OR between layer A and layer B generates small gaps that might be too small to make and inspect the mask. A small upsize and downsize can "clean" the data.

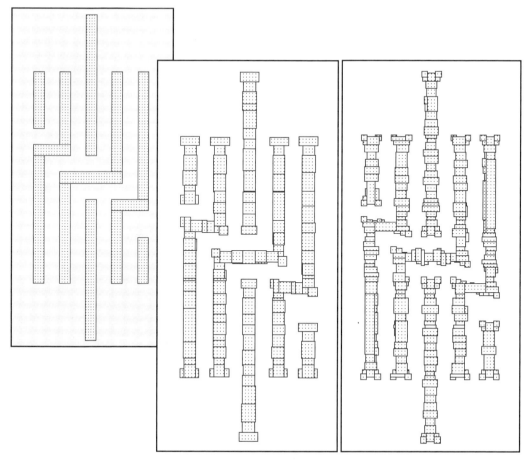

FIGURE 2.4
With the increase of OPC aggressiveness, a steep increase in the number of vertices is observed. It is clear that this further increases the file size of already big files. The image on the background is the reference image without any OPC applied.

on a poly-silicon mask data, the gate areas (the logical AND of the poly data and the active data) may receive a different OPC compared with the remaining poly data. This is because different criteria and specifications apply for both poly areas.

PSM adds another complexity because multiple layers are required to build a single mask. So not only does this require several input layers into this RET correction step but also it produces multiple output layers at the same time. A PSM is almost always companied with an OPC step, so these two output layers are OPC-corrected as well.

Adding the relatively complex and quite often iterative algorithm of OPC tools, it is clear that the OPC runtime can be significant. These runtimes sometimes exceed the actual writing time and inspection time. The following techniques address OPC runtimes on either the hardware or software side:

- Parallel processing (distributed processing and multithreading)
- Streamlining the OPC job
- Using existing hierarchy or creating hierarchy in the data

Data Preparation 25

Speeding up of OPC jobs requires knowledge of the tool and the data. Not all approaches work equally well on all OPC tools. Also whether hierarchy can be used in an OPC run depends on both the OPC tool and on data.

2.2.4 Mask Data Creation

After OPC is applied, data need to be created in a format that can be read by the mask writer and inspection tools. These formats are far from standardized and basically every writing tool has its own format. This is partly because these tools can be quite different in the way they operate. The way the data is organized in these formats usually reflects the way these machines work. Also, during the fracture step, data manipulations can also be carried out through:

- Scaling
- Sizing of data
- Rotation of data
- Pattern mirror
- Tone reversal

Scaling maybe used to compensate for the reduction factor in the lens of the exposure system. Sizing of data may be required for mainly two reasons: to compensate for mask process bias; to compensate for wafer process bias. Note that a sizing operation can also be carried out during Boolean operations.

The pattern mirror is required, because the writing of the mask is done with the chrome-side of the mask facing up, and the exposure of the mask is done with the chrome-side down. Figure 2.5 shows an example of the most common single layer operations.

Given the fact that OPC usually yields large files, it is clear that fracture tools will have to be able to handle these large files, and runtime is a major concern in many cases.

Along with the fracture step, data may be corrected for electron-beam (e-beam) proximity effects, or other systematic errors that may occur in the mask writing or processing steps. Refer to Section 2.6 for details on e-beam proximity effects. Since this correction is to compensate a systematic mask-writing error, the data for the inspection tool can be inherently different from the write data, apart from the obvious format difference.

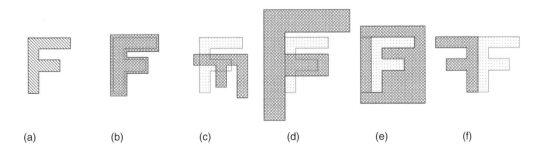

FIGURE 2.5
Operations that can be carried out during the generation of the physical mask data. Starting from the original data (a), sizing of data (b), rotation (c), scaling (d), tone-reversal (e), and mirroring (f).

2.2.5 Passing of Mask Order Information

When physical mask data is generated, it is usually handed over to another group that takes care of the mask manufacturing. Along with the mask data, additional order information and instructions are passed. This information may include shipping data; file names of patterns and quality specification. To avoid misinterpretation of this order information, a standard is being developed. This SEMI-P10 [3] electronic order format contains well-defined keywords, which must be listed in a specific order. Note that this SEMI-P10 order does not contain layout information other than placement of pattern files on the mask. Due to the fixed format of these SEMI-P10 order files, they are well suited for automated order handling.

2.2.6 Measurement Setup File Creation

The last step in the data preparation is the generation of measurement setup files. In the last few years, a steep increase in the number of measurements on the masks has been observed. Over 100 measurements are not unusual for high-end masks. With such a high number of measurements it has become impossible to locate all measurement sites just by looking through the microscope on the tool.

There are mainly two types of measurements performed on masks:

- Dimension control measurements commonly known as critical dimension (CD) control measurements
- Placement accuracy or registration measurements

With the generation of a measurement setup file, it is possible to locate the features that need to be measured either manually, automatically, or in a combination. The output of this operation is a tool specific setup file that can be loaded in the measurement tool along with the mask. Usually, such a measurement setup file is just a text file with coordinates and measurement instruction. Sometimes these setup files also contain images, which are used for position alignment just before an actual measurement.

2.2.7 Mask Data Preparation Ownership

In this section, the MDP flow is described in detail. It does not mention which groups normally carry out these steps. In fact large differences in approach are observed. Sometimes a group is only responsible for an OPC step, and in other cases a single group is responsible for almost the entire flow as described in this chapter. Apart from simple historical facts and the way companies are organized, several items are usually considered when assigning certain tasks to certain groups, such as:

- Data integrity
- Mask data quality
- Cycle time
- Mask cost and writing time
- Flexibility with respect to the writing tool

When masks are manufactured by an external company, the decision on where to place the "hand-over" point is even more important. It is important to consider the whole flow and look at all the relevant items from the list mentioned above and not just focus on a single one.

2.3 The Jobdeck Concept

Patterns on a mask are not always unique. In fact, quite often, mask patterns are placed on a mask several times, either as separate instances or in an array. When the number of these repetitions is sufficient and when the data volume of these patterns is relatively big, it pays off to create some sort of hierarchy, which is, in fact, the main idea of a jobdeck as depicted in Figure 2.6. A jobdeck file is nothing more than a table with one or more references to pattern files, and placement information, either a single coordinate or an array reference. Note that the jobdeck format for different writing tools is not the same. Here is an example of a MEBES jobdeck:

```
SLICE EDIT,17, $TED/DEMO.JB
* JOBDECK WRITTEN BY CATS: (NULL) AT WED NOV 07 15:32:05 2002
*!
*!GROUP COMMANDS
*!
SCALE 0.25
*!
*!ALPHA, RETICLE, REPEAT AND SIZING COMMANDS
*!
*!OPTION COMMANDS
OPTION AA=0.25
*!
*!TITLE AND ORIENT COMMANDS
*!
*!
*!
*!CHIP AND ROWS COMMANDS
*!
CHIP N1,
$ (A, BIRD$$$-$$-50)
ROWS 24200/62700,6,6000
ROWS 32200/44700,6,6000
ROWS 32200/86700,4,6000
ROWS 120200/44700,11,6000
ROWS 128200/62700,6,6000
*!
CHIP N2,
$ (A, BIRDREV-$$-50)
ROWS 32200/80700
ROWS 60200/77700
ROWS 80200/38700
ROWS 80200,2,8000/56700,5,6000
ROWS 80200/116700
ROWS 112200/74700
*!
END
```

Apart from pattern placements, a jobdeck can also contain various parameters and instructions for the mask-writing tool, such as mirroring and scaling of pattern data,

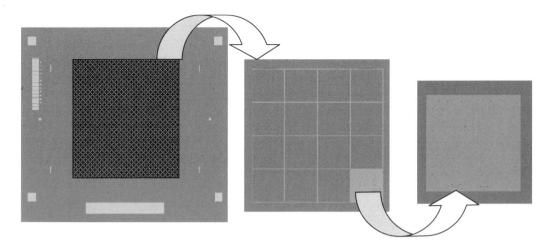

FIGURE 2.6
Quite often patterns appear more than once on most masks. This opens the possibility of using a hierarchy and having a more compact and efficient way of describing the whole mask in terms of data.

correction factors, beam setup parameters, subfield or scanfield placement information, etc. How the mask writer actually writes the pattern is not always the same as how the jobdeck is built up. Some mask-writing tools indeed follow the sequence of the jobdeck, and there can be a direct relation between the writing time of the pattern and the way the jobdeck is built up.

For writing tools that write patterns in the sequence of the jobdeck, it is assumed that the individual patterns can be written without influencing each other. When, for example, an e-beam is used to write the patterns, one has to be aware that while writing one pattern, scattered electrons of the writing process can reach the area of other patterns, thus causing pattern fidelity errors.

It is clear that some knowledge of the writing tool, which is used to write the mask, is required to set up the best possible jobdeck. Moreover, an unfortunate laid-out jobdeck not only can have an adverse impact on the writing time, but also on the pattern fidelity and thus on the mask quality.

There is another group of writing tools that merge the different patterns in the jobdeck again. This local and temporarily flattening process is usually done during the actual writing of the mask to continue taking advantage of the more efficient data representation of the jobdeck.

Since the resolution can differ per pattern file in a jobdeck, it is possible that some grid snapping occurs during the actual writing of a jobdeck. Some machines have a fixed spot size so that, even when writing a single pattern, snapping can occur. For masks with very tight specifications in terms of fidelity, this snapping can take up a significant part of the total error budget.

2.4 Mask-Writing Principles

In order to optimize the quality of a mask, prior knowledge of the writing tool is required. Apart from differences in the actual data format, there are big differences in the exposure system used in the various mask-writing systems. Data optimized for one writing tool can

Data Preparation 29

work out quite unfortunately for another tool, again apart from the actual data format. The overall mask quality is getting very critical in the total imaging process of the wafer, which is why mask quality is a subject of increasing importance. Knowledge of critical data details can favor a certain writing tool and/or writing strategy over another.

One of the first big differences among writing tools that can be observed is the use of either a laser or an e-beam for actual exposure. It is clear that an e-beam has the advantage of a much smaller wavelength and does not have the inherent wavelength limitation of a laser. E-beam systems, however, have to operate in vacuum systems, which makes it a more complicated machine altogether.

In this section, different writing techniques that are used in the various mask-writing systems are discussed.

2.4.1 Raster Scanning Writing Principle

A raster scanning mask writer scans the whole mask area and switches a beam "on" and "off" to image the pixelized pattern, similar to the TV principle. The beam can be either a laser or an e-beam.

Since the pattern is converted into pixels, the way the data is divided into blocks does not affect the final pixel configuration. Therefore, the way the patterns are sliced does not affect the quality of the mask. The mask quality in terms of line-width control or CD control is purely a function of machine properties, such as the speed and reproducibility of the beam switch.

In most raster scanning tools, the mask area is written using a combination of stage travel and beam deflection, whereby a small rectangular block of data is written using the raster scanning principle. These rectangular blocks are usually referred to as scanfields or main-fields. Figure 2.7 shows an example of data that is divided into scanfields. The way these scanfields are overlaid with the actual data differs from machine to machine. In some systems, the scanfield positions can be controlled, whereas in others these are hardware determined. It is possible that a pattern falls right on the interface of scanfields. In that case, this pattern is written as two individual parts separated into two different scanfields and thus separated in time. This implies that the pattern fidelity, such as line-width accuracy (CDcontrol), is a strong function of this scanfield-stitching accuracy. One may want to position critical features such that they do not cross scanfield boundaries.

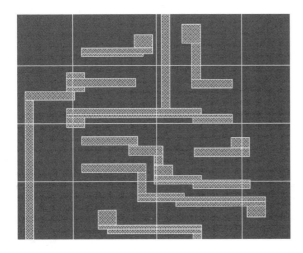

FIGURE 2.7
The data for raster scanning mask writer tools are divided in blocks. When these so-called scanfields divide a pattern, this pattern is written in different scanfields. This can lead to linewidth variation.

Although in mask writing, this is hardly a practical approach to improve mask quality, as it can be useful in certain cases.

Looking at the writing time of a pattern using the raster scanning approach, the pixel size is by far the quantity with the biggest impact. Note that the pixel size of the raster scan is not per definition the same as the beam size.

2.4.2 Vector Scanning Writer Principle

A vector scanning writing tool has a lot of similarities with a raster scanning tool. The total mask-writing area is still partitioned in rectangular scanfields. But now, instead of rasterscanning the whole scanfield area, only the actual data blocks (rectangles and trapezoids) are exposed.

During mask writing, pixelizing still takes place in the end. Similar to the raster scanning principle, the way the data is sliced in the scanfields does not have an impact on the final pixel assignment and hence on the line-width control. Also from a scanfield-stitching standpoint, the vector scanning and raster scanning techniques are equivalent. In both tool types this is a point of concern.

Compared with raster scanning tools, vector scanning tools do not usually show as high a writing time increase as a function of smaller pixel size. Where the writing time on a raster scanning tool is inversely proportional to the pixel size, the writing time on a vector machine is proportional to the area that has to be written and inversely proportional to the pixel size. Because of the scanning nature of the vector scanning tool, the way data is sliced has little or no effect on the final result after writing and processing of the mask.

2.4.3 Variable Shape Beam Writer Principle

A variable shape machine differs from a vector scanning machine in the way the final data elements are exposed on the mask. Vector scanning and raster scanning machines scan the element, whereas the variable shape machine is able to expose a complete data element in a single shot. These elementary blocks are usually rectangles and triangles or trapezoids. The whole data pattern is divided into these elementary blocks. For tools using this principle it does make a difference on how data is organized and sliced. For example, there is usually a direct relation between the number of elementary data blocks (rectangles and trapezoids) and the writing time.

The way data is sliced affects not only the writing time but also the dimension accuracy. The size of the elementary blocks is subject to a certain nonlinearity, which can have an adverse effect on the CD uniformity. Because of these nonlinearity issues, narrow shots should be avoided when such a mask writer is used.

2.5 Trends in Mask Technology

Just as the whole microelectronics industry is rapidly changing so is mask making. It is clear that the developments in this microelectronics industry largely drive the trends in the mask industry and thus in the MDP or "fracturing" arena [3]. The term fracturing comes from the operation where data is flattened and written out in single data units (mostly rectangles) that can be printed on the mask.

In the initial phase of the microelectronic development, the pattern transfer onto the wafer was done with masks that were in physical contact with the wafers (1× masks). Hence, the images on these 1× masks have the same dimension as on the wafer. With the introduction of the wafer steppers, a reduction lens was introduced for the pattern transfer, and the patterns on the mask were scaled accordingly.

Officially with the venue of wafer steppers we speak of reticles instead of masks. However, on the "work-floor" the term "masks" is commonly used for both reticles and masks. For the sake of simplicity this habit is followed in this chapter.

So the wafer stepper introduction resulted in an upscale of features on the mask because of the reduction lens that is used in these systems. Common scale factors are 4× and 5×, but also 2.5× and 10× systems are in use. One of the consequences of the introduction of a reduction system was the reduction of mask error contribution to the final printed wafer image because of this scaling property. This effect turned the mask into a commodity good, where only prices and delivery times were the remaining issues. However, since then, the continuing reduction of feature sizes in high-end semiconductor devices narrowed down the wafer error budgets at the same pace. Through the shrinking of wafer error budgets, the error budgets for the masks are shrinking as well. For the more advanced processes, the mask errors already take up a significant part of the total error budget and are challenging the capabilities of the mask-making technologies. One of the symptoms of these challenges is the steep price increase of masks with increasing quality grades.

Apart from the reduction of mask error budgets, OPC constitutes another major challenge for both mask making and MDP. Refer to Chapter 8 on resolution enhancement for details on OPC. The need for the OPC arises from the fact that feature sizes printed on the wafer are smaller than the wavelength, as shown in Figure 2.8. This causes corner rounding and other image deformation during the exposure process. This deformation is generally referred to as OPE. Since these effects are well predictable through models, it is possible to correct the data for OPE, which in essence is what we call OPC. OPC is a data manipulation step between design and fracture. Figure 2.9 shows an example of an OPC operation. The number of corners is increased, which makes mask making more complicated, and mask inspection in particular. From a data preparation standpoint, it is clear

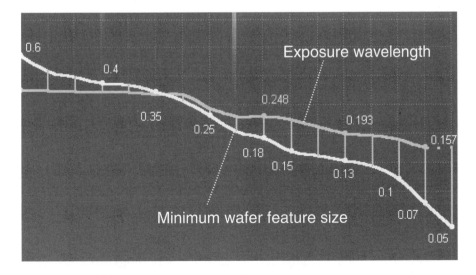

FIGURE 2.8
Printing features with dimensions smaller than the wavelength leads to image deformations (OPE) and needs to be corrected for.

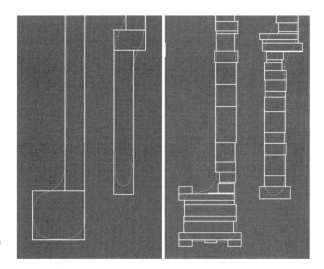

FIGURE 2.9
OPC adds many jogs, which challenges both mask making and data processing.

that these OPC operations cause a steep increase in data volume. The amount of correction, and hence the number of corners and jogs, is clearly a function of the severity of the OPE, which in turn is a function of the ratio between the exposure wavelength and the printed feature size. For MDP, OPC has the following consequences:

- Increase in pattern complicity
- Higher data volume
- Mask manufacturability is difficult to judge

There are other RETs like phase shifting in different flavors, which complicate the data preparation flow. These techniques sometime require multiple data files to produce a single structure on the mask.

In the past, the MEBES format was a *de facto* standard for mask data. These machines raster scanned the pattern data with either a laser or an e-beam. Errors in the mask were caused by placement errors and limited switch speed of the machine. When fracturing a certain polygon or a set of polygons, there exists a single and unique configuration that the MEBES employs for its outputs. The placement of the MEBES rectangles and trapezoids is either right or wrong. In the venue of variable shape beam (VSB) mask-writing tools, the after-fracture result may look different. Certain configuration of rectangles and trapezoids can yield a more accurate result than another. This quality of fracture is not only a function of the data but also of the writing tool. Since the mask quality plays an important role in the wafer imaging process, the quality of fracturing has a direct impact on the final wafer image quality.

2.6 Pattern Fidelity and Quality of Fractured Data

As stated before, the general quality of the mask has a big impact on the final wafer image fidelity. For high-end processes, the mask error contribution on the final wafer image errors is an issue of increasing concern. Out of all quantities that affect the pattern fidelity, the most critical ones are:

Data Preparation

- Line-width control, commonly referred to as CD control
- Pattern placement accuracy

The way mask data is prepared and configured can have impact on these quantities and hence on the quality of the mask. This section discusses MDP items that can have an impact on mask quality in general.

2.6.1 Grid Snapping

One of the most important MDP (or fracturing) parameters that have a big impact on mask quality is grid and grid rounding (grid snapping). Grid snapping can create a deviation from the original designed line width, which causes a CD error even before the mask is physically written. Grid snapping can happen in various steps in the data preparation and mask writing.

The input data itself can have off-grid points to begin with. The GDS data, which is in many cases the starting point of the MDP procedure, ultimately consists of polygon and path definitions. When the polygon coordinates are all on a certain grid, grid snapping can easily be avoided by using this grid as the final fracturing grid. A *path* definition consists of a set of centerline coordinates plus a certain path width. Avoiding grid snapping in such a structure not only requires the centerline coordinates to be on-grid but also the width of the path needs to be a multiple of the double-grid (or the half-width of a multiple of the grid). Under these conditions the perimeter of such a structure can be on this particular grid. However, if there is a non-orthogonal section in a path, this perimeter is almost never exactly on this grid in the direct vicinity of this oblique section (Figure 2.10). In other words, in case of a nonorthogonal path definition in GDS, grid snapping is inevitable near the nonorthogonal edges.

A less obvious cause of grid snapping can occur during mask writing in case the mask-writing tool has a fixed address unit and the data is not fractured to this grid or an integral multiple of it.

During manipulations of data, little sizing of data area is sometime applied. Special care has to be taken with the nonorthogonal edges in case of such sizing manipulations, as this can lead to off-grid situations. In any case the grid-snap or rounding behavior needs to be controlled. Grid snapping, in general, either on itself or in combination with sizing constitutes one of the biggest sources of error during MDP.

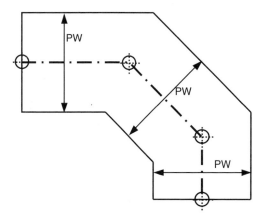

FIGURE 2.10
The basic PATH definition consists of a set of centerline coordinates plus a path width (PW).

It can happen that a sloped line is crossing several scanfield boundaries. In such a situation most formats require that at the scanfield boundaries these sloped lines are on the fracture grid. This grid snapping can cause a straight line to end up as separate lines joint at the scanfield boundaries with an almost 180° angle. If data containing such lines is re-input and re-fractured several times, the placement of these scanfield boundaries with respect to the data will require special attention.

2.6.2 Line-Width Accuracy as Function of Data Slicing

In some data formats the way input data is sliced in elementary blocks is not unique. In other words in such formats there are several ways the input data can be sliced. Depending on the writing methodology of the tool, difference in slicing can lead to differences in actual line-width accuracy. If a rectangle is divided into two elements, the total line-width error can be different from the case where the data is not split up (see Figure 2.11). What the actual difference is in terms of line-width error on the mask is a function of many things. If the data is raster scanned in the end, there may be no difference at all. It is clear that a good understanding of the mask-writing tool and data is required to achieve the best possible data configuration. It can also make a difference on how the slicing is done when the original design intention is observed. If it is known that slicing a certain feature has an adverse impact on the line-width accuracy, one could try to place the cut lines outside the most sensitive area. Sometimes the data is such that exposing the non-data and using a negative-tone resist can yield a better line-width control. If, for example, a line is formed by printing the surrounding of this line, the final line-width control is different compared with printing the line itself. Note that the use of negative-tone resist will require all data to be reversed in tone, including alignment keys, barcode, text, etc.

2.6.3 Scanfield Stitching

For almost all writing tools the total exposure area is divided into rectangles, usually referred to as scanfields. The size of the scanfields can vary from a few microns to several thousands of microns. Most formats use these scanfields to be able to cover the whole exposure area with a combination of stage travel and beam deflection. The stage moves

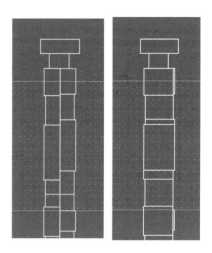

FIGURE 2.11
Different slicing of input data can yield different line-width errors on the written mask.

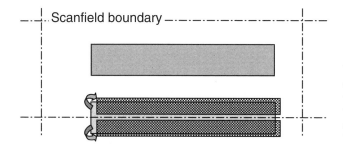

FIGURE 2.12
When a scanfield runs through a data structure, this structure may get broken up and written as two individual patterns. This splitting of data may have an adverse impact on the dimension control of such a structure.

the mask and places the right scanfield in the exposure area, and the beam deflection "writes" the content of the scanfield.

If a data structure is placed exactly on a scanfield boundary, this structure could be written in two separate steps as shown in Figure 2.12. Scanfield stitching causes problems that many mask-writing tools have to deal with. Whether measures have to be taken to deal with scanfield stitching really depends on the specification versus the introduced error, which is usually a function of the machine placement accuracy. There are several measures one can take to reduce the impact of scanfield stitching on the final pattern fidelity, including:

- One could make the scanfields overlapping one another and avoid slicing of features of a size smaller than or equal to the overlap amount. A feature that is partly in the overlapping zone is completely written in the adjacent scanfield.
- In case of overlapping scanfields, one could write the features in the overlapping zone with only 50% of the exposure dose. When writing the adjacent scanfield, this feature is again written with 50% of the dose. This approach improves scanfield stitching but does not improve the relative placement accuracy of the scanfields.
- By positioning a scanfield outside critical areas, the impact of scanfield stitching is avoided altogether. This approach is not always practical for mask writing, but when writing small patterns this is a simple and useful technique to overcome scanfield-stitching problems.

In some writing tools the scanfield size is fixed and cannot be altered at all, while with other formats both the size and the position can be manipulated.

Some scanning laser mask-writing tools scan the whole mask from one end to the other. With such an exposure technique, scanfield stitching occurs only on the boundaries of the individual stripes.

2.6.4 Nonorthogonal Edge Approximation

As mentioned, nonorthogonal path segments could give rise to grid snapping near the sloped edges. Also during the conversion to the mask data formats, other deformations of the sloped edges can occur. This deformation is very tool specific, and hence data format specific. Some tools simply approximate nonorthogonal edges with pixels (Figure 2.13). If the pixel size is large, this step may be significant enough to make it visible during the print-down of the mask. Using a smaller pixel size will reduce this problem, although there is usually a writing time trade-off when using a smaller pixel size.

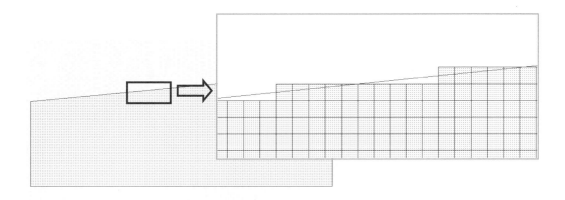

FIGURE 2.13
Some mask-writing tools approximate sloped lines. The size of the step is equal to the pixel size of the system.

Some variable-shape writing tools can only print orthogonal and 45° edges. In case of non-45° features, edges can be approximated with a combination of 45° segments and orthogonal segments to follow the sloped edge as close as possible.

2.6.5 Laser Proximity Effects

For the actual exposure of the mask either a laser or an e-beam is used. Both these exposure methods have their limitations and hence an impact on the final pattern fidelity. One of the main limitations of a laser is the limited optical resolution (Figure 2.14). The wavelengths of lasers that are being used for mask exposures are in the range of 400–200 nm. Features smaller than or in the same order of magnitude of the wavelength are deformed to some extent during mask writing. One of the most obvious symptoms of this image deformation is corner rounding. This corner rounding of mask data is similar to the corner rounding experienced on the wafer. In case of a mask-scaling factor, all dimensions on the mask are bigger by this amount.

One of the measures that can be taken to compensate this optical proximity on the mask is to use similar measures as the OPC on the wafer. For example, one could correct the mask data prior to writing by using serifs to improve the corner fidelity. Clearly, for the mask inspection normal uncorrected data are used.

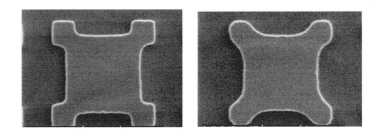

FIGURE 2.14
Limited resolution of a laser typically causes a more pronounced corner rounding compared to e-beam masks.

2.6.6 Electron Beam Proximity Effects

When an e-beam is used for mask writing, the writing process itself no longer limits the resolution. However, the high acceleration voltage, which is used to create the e-beam, gives the electrons a high kinetic energy and causes them to scatter with the atoms of the resist and underlying material. This phenomenon is commonly known as the e-beam proximity effect [4]. The distance the electrons can reach with respect to the entry point is a strong function of the initial kinetic energy and hence the acceleration voltage. But also the composition of the total film stack, in which the electrons are injected, influences the scatter range. The proximity range is in the order of several tens of microns for acceleration voltages of 50–100 kV, as shown in Figure 2.15.

The e-beam proximity can be modeled as a weighted sum of two Gaussian functions:

$$f(r) \frac{1}{\pi(1+\eta)} \left\{ \frac{1}{\alpha^2} \exp\left(-\frac{r^2}{\alpha^2}\right) + \frac{\eta}{\beta^2} \exp\left(-\frac{r^2}{\beta^2}\right) \right\}$$

In this proximity function, one Gaussian curve is relatively narrow and high and models the so-called forward scattering, whereas the second is shallower but with a much longer range and models the back-scattered electrons. The weighting factor η can run from 0 to 1 and defines the ratio between the two Gaussian curves.

Alternatively, empirically determined proximity functions are used. These functions can be determined either by plain measurements or by using a Monte Carlo electron trajectory simulator.

The actual dose that gets injected in the resist layer can now be calculated by convoluting the proximity function with the data.

$$d(x, y) = f(r) \otimes p(x, y)$$

By turning this function around, it is possible to correct the data for e-beam proximity effects. During the process of e-beam proximity correction (EBPC), the convolution is normally solved as a multiplication in the frequency domain using Fourier transforms [5]. The most common way of correcting the data for e-beam proximity is to modulate the

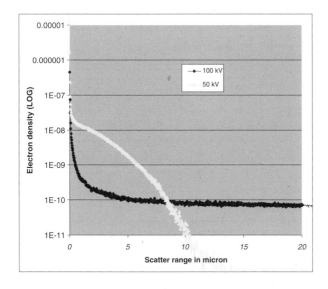

FIGURE 2.15
The e-beam scattering behavior is a strong function of the initial kinetic energy.

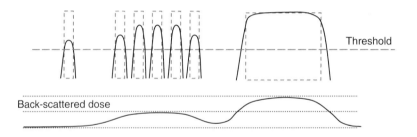

FIGURE 2.16
The backscattering yields a modulation of the effective dose as a function of pattern density. The correct dose for an isolated small line will overexpose an array of such small lines, as well as a wide line.

actual writing dose to compensate to backscatter dose variation as shown in figure 2.16. Areas with a high pattern density will get a relatively low dose, and more isolated and smaller features will get a relatively high dose. Alternatively, edges of features could be modified to compensate for e-beam proximity.

EBPC is a CPU-intensive operation. Moreover, the increase in output file size can be substantial. Because of the relatively long range of e-beam proximity effects, relatively large amounts of data have to be considered. It can happen that data patterns, although placed within the e-beam proximity range, physically reside in different files, for example, when a jobdeck is used to combine patterns. When applying EBPC on one file, the presence of the other pattern data has to be taken into account. The simplest way to deal with interacting files is to simply merge them into one, although this can lead to large file sizes. Another downside of dose modulation is that the writing process may slow down if a large range of dose values is used, although this effect is very machine specific. An approach to reduce CPU time is to exploit the hierarchy in the layout [6,7]. Methods have been developed to use the hierarchy despite the fact that electrons are scattered across cell borders.

A less computational way of applying proximity correction is to compensate the backscattered electrons with a background dose, which is written in a second pass. This technique is known as GHOST [8]. It uses a second writing pass with the inverted write data, which is written with a defocused beam. This wide beam is configured in such a way that together with the inverted image it yields a dose, which is the opposite of the backscattered dose of the first writing pass. So the GHOST technique compensates the effect of backscattered electrons by adding a background dose proportional to the inverse of the pattern density. The downside of this technique is the need for a second writing pass.

2.6.7 OPC Model Calibration

When OPC is applied on mask data, a model is normally used to apply the corrections and to verify the corrections afterwards. These OPC models are usually semiempirical, which means that the basis of the model is a mathematical model plus a tuning mechanism, namely, model calibration [9]. For this model calibration process, test patterns are being printed and measured. The objective of the model calibration is to capture the behavior of the whole wafer patterning process as shown in Figure 2.17. Because a mask is used to image the test patterns on the wafer, the systematic errors on the mask, such as corner rounding and proximity effects, are captured in the calibration process. It is clear that any change in the mask process that affects the mask pattern fidelity will have consequences for the accuracy of the model. Such changes in the mask process can thus hamper the OPC accuracy, even if the change itself only improves the mask pattern fidelity.

Data Preparation

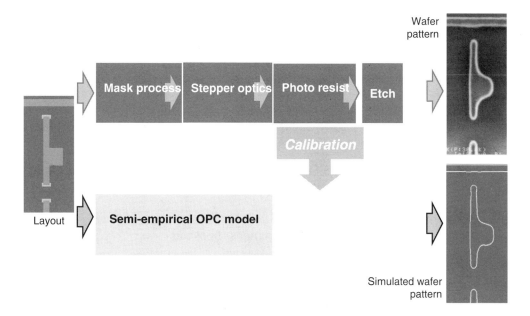

FIGURE 2.17
OPC uses a semi-empirical process model. The mask process is part of this calibration process and should not be changed once calibrated.

Because of the reduction lens that is normally used in a wafer stepper, the mask corner rounding contribution to the total corner rounding on the wafer is usually small. Still any changes in the mask process should be made with great caution for all mask types, and for OPC-corrected layers in particular.

2.6.8 Loading Effect Correction

In order to meet the tight CD specification, all systematic errors are under continuous investigation. Contributions to CD errors that were neglected in the past because of the relative magnitude are now investigated and modeled to find ways for correction.

Dry-etching has been introduced in mask making for improved pattern fidelity and process control. One of the phenomena observed with dry-etching is the etch rate dependency upon the amount of etchable surface exposed to the etching plasma. The more material that needs to be etched, the more the etching plasma will be depleted, and hence reduce the etch rate, that is, the etch-loading effect [10–13].

The observed result of this dry-etch loading effect is a variation of line-width as a function of pattern density. The effects are comparable with e-beam proximity, but then with a range that is usually much larger. The kernel function that is commonly used to describe this etch loading effect is a two-dimensional Gaussian:

$$k_\beta = \frac{1}{\pi \beta^2}\, e^{-r^2/\beta^2}$$

The convolution between the data and the kernel function yields a bias parameter:

$$b = b_0 - c\{k_\beta(r) \otimes p(x, y)\}$$

This bias parameter can be used to precorrect the data for etch loading effects.

The amount of correction is typically less than 20 nm. This correction is performed after OPC and before EBPC, or laser proximity correction.

2.7 Mask Data Processing Runtimes

In the previous process generations, there has been a steep increase in MDP complexity. There are several reasons for the increase in complexity, including:

- Introduction of RET
- More and more complex layer combinations (Booleans)
- Increase of number of masks per process
- Different mask-writing tools
- Data volume increase because of feature size reduction
- EBPC
- Other corrections for systematic process variations

All these individual items give rise to a data volume increase and a huge increase in MDP run-time, as shown in Figure 2.18. The total conversion time from DRC to read-to-write data experienced today is sometimes exceeding the total times needed to process the mask. On the other hand, the requirements for the total run-times for the final mask, including mask making and data processing, are getting tighter, which leave less time for the MDP. In this section, different techniques and approaches are discussed to speed up the total MDP time. How beneficial these techniques are, heavily depends on the software package and the data itself.

There are different ways to reduce the MDP runtimes, apart from the most obvious way, which is the use of a faster computer. Runtime improvement can be achieved by optimizing the flow of data manipulation. Some operations are inherently faster than others. When large intermediate files are generated and re-input to combine with other large files with some Boolean operation, such a Boolean operation can be quite costly in terms of CPU times.

When the data manipulation flow can be set up in such a way that slower operations are avoided, a speed-up of the total process can be achieved.

Another way of speeding up data processing is possible through process parallelization. There are several ways through which this can be accomplished. One way is to have multiple CPUs in a single machine. Another approach is to simply split up the total task over several machines in a cluster. Even a combination of these two approaches is possible [14,15].

The way it is configured in the different data processing tools can differ to a large extent, and also some data manipulation steps are better suited for parallelization than others. It is clear that the performance improvement through process parallelization is highly tool specific, machine specific, and even data specific.

If there is a lot of repetition in data, such as cells that are placed many times, it may be advantageous to manipulate data in a hierarchical way, so that the scanning is on a cell-by-cell basis rather than through the entire layout which disregards the hierarchy of the data. Whether a hierarchical data manipulation approach is advantageous depends on

Data Preparation

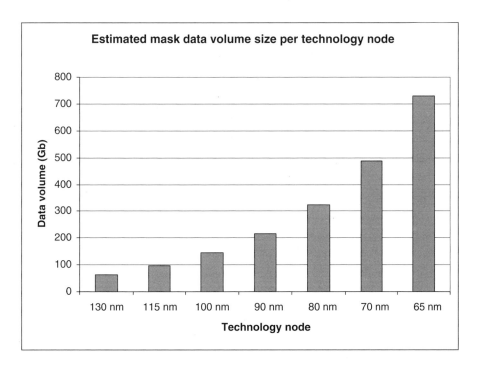

FIGURE 2.18
The estimated mask data volume increases exponentially as a function of technology node. *Source*: ITRS roadmap [3].

many things. Apart from the tools themselves, the amount of hierarchy in the data file clearly has a big influence. Moreover, there may be a lot of hierarchy in a data file, but this hierarchy may not always be usable. For example, if cells, although identical, overlap with other data, they cannot be handled as equivalent cells.

Proximity effect correction, either optical or e-beam, does a correction as a function of the surroundings. Hence, if a cell is placed in two different surroundings, a hierarchical correction is not always possible or at least the hierarchy needs to be modified. Proximity correction has the tendency to reduce the amount of hierarchy in the data. Because of the much larger range of e-beam proximity effects, this reduction of hierarchy gets even worse with hierarchical EBPC, which makes hierarchical approach for EBPC not very beneficial.

References

1. http://www.vsi.org/resources/techdocs/contech/gdsii.pdf.
2. http://www.semi.org.
3. 2001 International Technology Roadmap for Semiconductors, http://public.itrs.net.
4. T.H.P. Chang, Proximity effect in electron beam lithography, *J. Vac. Sci. Technol. B.*, 12, 1271–1275 (1975).
5. H. Eisenmann, T. Waas, and H. Hartmann, PROXECCO – proximity effect correction by convolution, *J. Vac. Sci. Technol. B.*, 11 (6), 2741–2745 (1993).

6. A. Rosenbusch, C.K. Kalus, H. Endo, Y. Kimura, and A. Endo, On the way to the 1 gigabit: demonstration of e-beam proximity effect correction for mask making, *Proc. SPIE*, 3236 (1998).
7. C.K. Kalus, W. Rößl, U. Schnitker, and M. Simecek, Generic hierarchical engine for mask data preparation, *Proc. SPIE*, 4754 (2002).
8. M. Gesley and M.A. McCord, 100 kV GHOST electron beam proximity correction on tungsten x-ray masks, *J. Vac. Sci. Technol. B.*, 12 (6), 3478–3482 (1994).
9. S.P. Wu et al., Process calibration with 2D semi-empirical resist modeling: capturing both linewidth variations and line-end shortenings, vol. 3678–34.
10. H.J. Kwon, D.S. Min, P.J. Jang, B.S. Chang, B.Y. Choi, K.H. Park, and S.H. Jeong, Dry etching of Cr layer and its loading effect, in: H. Kawahira (ed.), *Photomask and Next-Generation Lithography Mask Technology VIII, Proc. SPIE*, 4409, 382–389 (2001).
11. J.Y. Lee, S.Y. Cho, C.H. Kim, S.W. Lee, S.W. Choi, W.S. Han, and J.M. Sohn, Analysis of dry etch loading effect in mask fabrication, in: G.T. Dao and B.J. Grenon (eds.), *21st Annual BACUS Symposium on Photomask Technology, Proc. SPIE*, 4562, 609–615 (2002).
12. Y. Granik, Correction for etch proximity: new models and applications, in: C.J. Progler (ed.), *Optical Microlithography XIV, Proc. SPIE*, 4346, 98–112 (2001).
13. H.J. Kwon, D.S. Min, P.J. Jang, B.S. Chang, B.Y. Choi, and S.H. Jeong, Loading effect parameters at dry etcher system and their analysis at mask-to-mask loading and within-mask loading, in: G.T. Dao and B. Grenon (eds.), *21st Annual BACUS Symposium on Photomask Technology, Proc. SPIE*, 4562, 79–87 (2002).
14. G.M. Amdahl, Validity of single-processor approach to achieving large-scale computing capability, in *Proceedings of AFIPS Conference*, Reston, VA, 1967, pp. 483–485.
15. Gustafson, J.L., Reevaluating Amdahl's law, *Commun. ACM*, 31 (5), 532–533 (1988).

3

Mask Writers: An Overview

Sergey Babin

CONTENTS

3.1 Introduction 43
3.2 Raster Scan Systems 45
3.3 Vector Scan Systems 46
3.4 Variable Shaped Beam Systems 46
3.5 Raster-Shaped System 47
3.6 Cell-Projection Systems 48
3.7 Novel Concepts of E-Beam Systems 48
    3.7.1 Microcolumn 49
    3.7.2 Arbitrarily Shaped Beam System 49
    3.7.3 Raster Multibeam System 49
    3.7.4 Multicolumn Cell Lithography System 50
    3.7.5 DiVa 50
    3.7.6 Multicolumn Multibeam System 50
    3.7.7 MAPPER 51
    3.7.8 DEAL 51
3.8 Comparative Evaluation of E-Beam Mask Writing Systems 51
    3.8.1 Accuracy 51
        3.8.1.1 Beam Edge 51
        3.8.1.2 Butting Error 52
        3.8.1.3 Proximity Effect Correction 52
        3.8.1.4 Resist Heating 53
        3.8.1.5 Fogging, Charging, Substrate Heating 53
    3.8.2 Throughput 53
        3.8.2.1 Settling Times 53
        3.8.2.2 Diagonal Lines 54
        3.8.2.3 Beam Current Limitations 54
3.9 Laser Pattern Generators 55
    3.9.1 Raster Scan LPGs 55
    3.9.2 Matrix Exposure LPG 56
References 56

3.1 Introduction

This chapter provides an overview of today's mask writers. The chapter also lists the state-of-the-art mask writers that are in the early stages of their development. Detailed descriptions and characterizations of the commonly used systems will appear in later chapters 3 & 4 of this book.

Modern maskmaking systems employ focused electron beams or laser beams to fabricate patterns on photomasks. Mask writers can differ from each other in a number of ways: in the number of beams; the shape, energy, or wavelength of the beam; in their writing strategies. We will outline the basic principles of these different systems and compare them in terms of their accuracy and throughput.

In the early days of semiconductor manufacturing, photomasks were fabricated by manually drawing the magnified patterns; the image of the pattern was then optically reduced to the desired size and transferred onto emulsion glass plates, which were used as photomasks. Later on, with the advent of computer technologies, manual drawings were replaced by computer-controlled processes. A computer-controlled knife blade was used to cut patterns on polymer opaque films. The undesired segments of the film were then peeled off with a pair of tweezers. These large patterns were then reduced by a photo-camera, resulting in emulsion plates bearing patterns of the right size.

Emulsion patterns were later replicated to a chrome layer because chrome-layered masks were less susceptible to damage during contact printing with wafers than were the emulsion masks. Contact wafer printing was used in manufacturing before steppers were developed. Compared to an emulsion mask, a chrome mask could make contact prints on a significantly larger number of wafers before damage would occur to the mask. Over time, technology developed to pattern chrome directly without the intermediate step of an emulsion pattern.

Optical and electron beam (e-beam) mask generators came into use and replaced knife-based mask fabrication processes. When writing with light beams, the size and shape of the light from a mercury lamp was defined by computer-controlled mechanically adjustable aperture blades. The blade edges defined the edges of pattern features. A mask blank was put on a stage where its movement was synchronized with aperture blades to ensure that features were exposed at the right locations on the plate. In later years, laser beam systems were developed that provided higher energy density and allowed for more accurate pattern generation compared to mechanically controlled optical systems.

Electron beam lithography (EBL) came to mask manufacturing in the late 1970s, when MEBES® systems were built. These systems were originally developed by AT&T. Focused e-beams irradiated electron-sensitive polymer in desired areas, which created the required patterns. For decades, the unique features of EBL systems — easily programmable computer control, high accuracy, and relatively high throughput — have positioned these systems as the main tools to fabricate critical masks.

Mask fabrication was driven by general needs of the microelectronic industry: make features smaller, place more features on the mask, and fabricate them with higher accuracy. According to Moore's law, the density of transistors in microcircuits will double every two years. New principles of maskmaking have been invented and refined gradually to comply with this trend, and so have the new technologies of maskmaking.

The only exception was in the late 1980s, when the demand for development in the maskmaking technology industry received a break for a few years. The reason was the adoption of commercial wafer steppers with magnification of 1:10. Masks at this magnification were much easier to fabricate compared to masks at 1:1 and 1:5. Larger patterns allowed more tolerance in feature size and defects on the masks. The quality of the then existing pattern generators greatly exceeded requirements for 1:10 masks. As demand for maskmaking equipment went down, companies producing the equipment were on the brink of extinction.

Essential components of mask pattern generators are data path, control electronics, precision stage, and beam delivery system (e-beam column or a laser beam system). Here

Mask Writers: An Overview

we shall consider writing strategies and their impact on system performance; then we will make a comparative evaluation of various systems. Two major parameters of mask pattern generators will be considered in detail: accuracy of a written pattern and system throughput.

3.2 Raster Scan Systems

Raster scan systems like MEBES and Exara (both are from Etec Systems, an applied materials company) utilize single Gaussian beam to write a pattern. The E-beam moves over a mask blank in a raster fashion like a beam on a TV screen. The system is equipped with a beam blanker that can be programmed to selectively expose or not-expose portions of resist to generate the required pattern, see Figure 3.1a.

In a raster e-beam system, the beam repeatedly scans a single line, while the stage holding a mask blank moves continuously under the beam. Thus, a stripe of features is composed by the raster scanned lines. The stripes form the mask pattern.

Because of the simplicity of this principle, the system has many advantages. It can be easily calibrated and tested; these tasks are one-dimensional. Errors, such as scan non-linearity, can be calibrated out with a high precision. The scan can be relatively long (~1 mm) and still have high precision. Data preparation for the system is relatively easy: a flat bit-map format describes edges of pattern features in each scan. This format called MEBES data format became one of the industry standards.

The throughput of raster system depends on blanking frequency and on beam address size. Targeted write time for all generations was 6 h/mask of the highest quality. The address size, which is a distance between two beam spots, is a matter of a design grid of a pattern. The higher required accuracy of a pattern would require smaller address size. For many generations of raster systems, the address size has been decreasing along with increases in blanking frequency and beam current. In modern commercial systems, the blanking speed is typically 320 MHz.

Later on at some point, increasing blanking speed became a technical challenge. To comply with the throughput requirements at address grid of less than 10 nm, raster strategy was modified to a gray scale multipass writing scheme. Smaller grid size of a pattern can be achieved without reducing beam address size when using gray scale

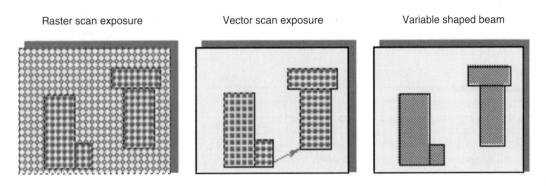

FIGURE 3.1
Raster scan, vector scan, and VSB exposure strategies. It takes 500 flashes in raster writing, 132 in vector scan, and 4 in VSB flashes.

writing [1]. A fraction of the dose can be delivered by the beam near the edge of a feature. Because of this, the boundary of the feature is shifted by a fraction of beam address size. In addition, while applying multipass writing, each pass can be shifted by half the address size in a four-pass writing thus further decreasing the minimum grid size without compromising the throughput.

To improve edge blur and to further decrease manufacturable grid size, the technique of per-pixel deflection of the beam was developed. During exposure time of a single pixel, the beam can be deflected toward a desired direction, such as toward the corner of a feature. Per-pixel deflection was initially used to decrease beam blur in a direction of scan that arises due to the movement of the beam during the exposure of a single pixel. Even though the exposure time is about 3 ns short, the beam travels a noticeable distance and contributes a few nanometers to the edge blur. The retrograde per-pixel deflection the fixes position of the beam during its exposure and allows for a smaller grid size.

The throughput of raster system does not depend on resist sensitivity. As a rule, Gaussian beam system is able to deliver enough current for any reasonable resist sensitivity. All the current goes into one spot. This differs from variable shaped beam (VSB) or cell-projection systems, where the beam current is distributed over a flash of maximum size and a only fraction of it is used at any single flash.

The throughput is also independent of the pattern. The beam has to fly over the entire mask independent of pattern coverage.

3.3 Vector Scan Systems

Systems using vector scan strategy were designed to avoid unnecessary scanning over areas between pattern features. A beam starts writing from an edge of a feature; scanning [2] only goes within the boundaries of the feature. When the exposure is complete, the beam is blanked and deflected directly to an edge of the next feature, as shown in Figure 3.1b, and the process is repeated.

Writing a single feature does not require any beam blanking; and therefore, the speed of scanning can be very high: 500 MHz writing speed was used in a commercial EBES system developed by AT&T and commercialized by Lepton, Inc. Such a strategy offers throughput advantages in the case of sparse patterns. Such an advantage, however, is greatly diminished where pattern coverage is relatively high. Detailed comparison of vector scan system with other systems in terms of throughput and accuracy will be given in the later sections.

While widely used in R&D, vector scan systems are not common in maskmaking. Examples of vector scan systems are VB-6 of Leica Microsystems and JBX-9300 of Jeol.

3.4 Variable Shaped Beam Systems

The systems described earlier use a small spot of a focused beam, which is then scanned to fill in the pattern features that need to be exposed. A VSB system forms the beam into a shaped beam that is larger in size than the Gaussian spot and exposes the entire feature of a mask or a significant part of the feature in a single shot.

The concept of the shaped beam was first developed at IBM [3]. In this system, the beam originating from the source is first made to pass through a squared aperture that gives the beam a square shape. The beam then goes through a second squared aperture. Between the two apertures there is a deflector that is used to move the beam along the x and y axes so that part of the beam is obstructed by the second aperture. This gives the beam the shape and the size that is needed for the task at hand, see Figure 3.1c.

After the beam size and shape are formed, two levels of beam control are used in VSB systems to route the beam to a desired location on the mask blank:

- The beam is positioned in a subfield by a separate deflector; the size of the subfield is typically 24–80 μm.
- The subfield itself is positioned inside a major field using a subfield deflector, which is normally a magnetic type; the size of a major field is about 1–2 mm.

So, three levels of deflection are required for system calibration, each in two dimensions.

In the VSB systems, the stage can move either in a continuous or step-and-write mode according to major fields. A continuously moving stage offers higher throughput because of its lower overhead.

While writing Manhattan-type patterns is natural for VSB systems, writing diagonal figures requires decomposition of a shape into rectangles. The minimum size of the rectangle is comparable to the specified edge roughness of a pattern. This problem may result in a considerable slowdown of the system throughput: a pattern with 3% of the features as diagonal lines can take twice the writing time as compared to the ones that do not have such lines. Moreover, a mask with an optical proximity correction (OPC) requires writing a significant amount of small rectangles, which also slows down the speed of the system. It is not uncommon that high-end designs require 20–30 h of writing a single mask when using VSB system.

Dwell times in VSB can be controlled on flash-by-flash basis. This provides an opportunity for proximity effects correction by varying the dose of each flash. The dose proximity correction generally does not increase data volume as in the case of correction by geometry modification. Compared to GHOST correction used in raster systems, both dose and shape correction methods provide higher contrast to the image.

One of the problems associated with vector systems is resist heating. The temperature increase in a local area of exposure changes resist sensitivity and leads to distortion of critical dimensions. The area of a subfield in VSB is comparable to an area of heat transfer at 20–100 kV energies, and therefore high temperature can be reached in a local area. Resist heating is one of the main reasons why throughput of VSB system is restricted.

Examples of VSB systems are EL-4 developed at IBM, EBM-4000 of NuFl are (former division of Toshiba), JBX-9000 of Jeol, AEBLE of Etec Systems, HL-7000 of Hitachi, and SB-350 of Leica Microsystems.

3.5 Raster-Shaped System

A system that combines the principles from both VSB and raster systems was developed at Etec Systems [4]. Here, like in the case of VSB, the beam is formed using two apertures and a deflector between them. The beam is then deflected in a direction perpendicular to the stage motion. Blanking and beam shaping are performed according to a pattern to be written. In this way, the beam still has to fly over the entire pattern independent of its coverage, but unlike the raster system with a Gaussian beam, in this case, more pixels can be exposed in one single shot. Thus, the throughput of raster shaped system is considerably higher compared to the traditional raster system. The speed increase is measured by N/F, where N is the number of pixels in a single VSB of maximum size, and F is a frequency factor, which measures how much lower the beam-blanking speed of raster shaped system is compared to a raster system. The maximum beam size should be comparable to the minimum feature size on the mask; this differs from a VSB system, where this limitation does not exist.

The calibration of the raster system is relatively simple compared to a VSB system — it excludes deflections over one coordinate and the deflection of subfields. Every additional deflection step leads to butting error, which will be discussed later.

Raster shaped system offers another huge advantage over VSB system in terms of resist heating. Due to a long scan, heat has time to diffuse before an e-beam passes nearby again, and therefore local temperature remains low compared to that in VSB.

3.6 Cell-Projection Systems

Cell-projection systems are similar to those of VSB systems except that they utilize exposure of a few pattern features in a single shot [5]. In a pattern like a DRAM, the pattern elements are repeated millions of times. Multiple repetitive elements can be combined into a group. An aperture is fabricated as a stencil mask with a pattern of this group. When a wide e-beam is passing through the cell-pattern, the resulting e-beam is shaped accordingly. Multiple elements of a pattern are exposed in one single shot.

Multiple various groups can be fabricated on one and the same aperture, allowing for a choice between them while writing. In this way, exposure of certain elements of a pattern goes fast. In reality, mask patterns involve large areas of nonrepetitive features. They are written by the system exactly in the way used in VSB system. However, because beam current is distributed over an area of larger magnitude, the writing speed in VSB mode is considerably slower. An overall throughput advantage, if any, is significant only for specific patterns. These systems are not widely used; a few of them are used to write prototypes of memory devices directly on the wafer.

Resist heating is potentially a serious issue here. Proximity effects correction is only possible on a coarse level because an area of approximately $5 \times 5\,\mu m$ is exposed as a single block without modification of dose and shape for features within the block.

3.7 Novel Concepts of E-Beam Systems

In addition to the systems described so far, there are many more newer systems in early stages of their development promising to meeting future challenges in maskmaking and direct write. A few concepts, described later, were developed recently. Of these the multibeam systems are of interest. Total beam current and the corresponding throughput of conventional e-beam system are limited by the stochastic Coulomb interaction in the beam path, which leads to beam blur and therefore to loss of resolution and CD control. The only way to overcome this problem is to form and control multiple e-beams that do

Mask Writers: An Overview

not have a common crossover point. Multibeam systems offer an advantage of writing a pattern by individually addressed parallel beams; the throughput in such writing can be increased considerably compared to a single beam system.

3.7.1 Microcolumn

The concept of microcolumn system was invented at IBM and later developed at Etec Systems [6]. The system consists of an array of separate closely spaced columns, each with a Shottky field emitter. The pattern is written in parallel. Each column works in a raster mode deflecting the beam over about 100-μm-wide stripes. The distance fabricated between the columns was 2 cm. This concept is scalable to place microcolumns over the entire wafer area.

The electrostatic lenses, apertures, and deflectors are built using MEMS technology. A microcolumn is designed to work at low voltages of 1–2 kV. The resist is a double layer resist with the top imaging layer and buffer bottom layer.

3.7.2 Arbitrarily Shaped Beam System

VSB systems lose their throughput advantages when writing patterns that involve diagonal lines and patterns with multiple OPC features. The concept of an arbitrarily shaped beam (ASB) uses the formation of beams of arbitrary shapes not restricted to rectangles and triangles. In writing a complex pattern, flashes of arbitrary shapes can reduce writing time by filling in a feature using fewer shots.

The concept is displayed in Figure 3.2. The originally designed pattern (Figure 3.2a) should be modified using OPC to print correctly on the wafer. The ideal shape of a corrected pattern (Figure 3.2b) is often replaced by simplified Manhattan-type polygons (Figure 3.2c) because mask writers cannot print "ideal" OPC features. Fracturing these simplified figures into flashes results in a large number of shots; therefore, printing takes a long time. Examples of fracturing are shown for a VSB system (Figure 3.2d), raster shaped system (Figure 3.2e), and an ASB system (Figure 3.2f). An ASB system requires considerably fewer flashes to print a pattern and also improves accuracy of an OPC pattern, making it closer to "ideal."

The ASB system uses four apertures in the beam-shaping module, each of which cuts a wide beam at a certain angle and position [7]. Electrical deflectors in the beam-shaping module are similar to VSB; apertures are round or polygons. The system can use a magnetic field in its beam-shaping deflector or aperture assembly to ensure low aberration over a wide deflection area.

3.7.3 Raster Multibeam System

In the Etec Systems' approach to build a raster multibeam system, multiple laser beams are used to generate a pattern on a photocathode plate [8]. They scan in a raster mode similar to that in a laser pattern generator. Electrons generated by photocathode are collimated, accelerated, and demagnified to form an array of beamlets at resist plane.

The system offers advantages of fast pattern generation using multiple laser beams and at the same time, a high accuracy and resolution achievable by e-beams. The accuracy is not limited by diffraction of light in this system.

The laser pattern generation aspect of this system and the data path can be adopted from commercial ALTA laser pattern generator.

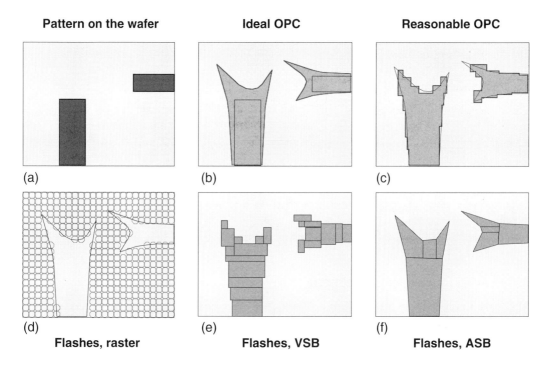

FIGURE 3.2
Pattern on the wafer (a) and corresponding features on the mask after OPC — ideal (b) and reasonable (c). To print these mask features, a system with an ASB requires significantly fewer flashes compared to other systems (d–f): it takes 500 raster flashes, 21 VSB flashes, and 8 ASB flashes. Also, accuracy of a pattern written by ASB is much higher, including ability to fabricate ideal OPC features.

3.7.4 Multicolumn Cell Lithography System

In a multicolumn cell system, under development at Advantest, VSBs are employed to the writing in parallel [9]. Each beam is formed by six electromagnetic lenses. Each lens represents an array of openings corresponding to the number of beams. The distance between the beams is 25 mm, and the number of beams is 16 in the current version. Shields are used between electrostatic electrodes to prevent interference between columns.

3.7.5 DiVa

Distributed variable shaped beams (DiVa) are used in a concept created at IBM [10]. The system utilizes either a planar cathode or a photoemitter to form multiple beams. The beams are individually formed to desired sizes using shaping apertures and deflectors. A common, uniform, axial magnetic field transfers an image of all beams onto a substrate with a magnification of 1:1. Deflection of all formed beams is done in parallel, while blanking is individual.

3.7.6 Multicolumn Multibeam System

A system developed by Ion Diagnostics is a combination of multicolumn approach with a multibeam, so-called M×M [11]. This system utilizes a number of columns extending over

Mask Writers: An Overview

the entire area of the wafer or a mask. Multiple beams write simultaneously in each column. In this approach, the throughput is independent of the wafer size. Two hundred and one columns with 32 beams each are proposed in the design to write a 300-mm wafer in 120 s. An array of micromachined cold field emitters and focus lenses is used to deliver beams.

3.7.7 MAPPER

The multiaperture pixel-by-pixel enhancement of resolution (MAPPER) concept was developed by MAPPER Lithography, BV. It targets high speed direct write for up to 20 wafers/h [12]. This massively parallel EBL utilizes 13,000 beams. All beams are scanned over the wafer using one common deflector. Each of the beams is blanked independently according to its separate datastream. E-beams are Gaussian and each writes a pattern in a narrow stripe, composing the entire chip pattern.

Beam focusing and blanking subsystems are made using micromechanical (MEMS) technology. Modern high speed data transmission systems are used to feed rasterized pattern data to each channel.

3.7.8 DEAL

Digital electrostatically focused e-beam array lithography (DEAL) is under development at Oak Ridge National Laboratory [13]. This approach uses addressable field emission arrays integrated into a logic and control circuitry implemented on a wafer. Carbon nanofibers are used as electron emitters. Electrostatic focusing lenses, blanking, and accelerating electrodes are integrated on the same wafer with emitters as a stock of prefabricated layers. The design goal is to implement 3 million beams over an area of $1\,cm^2$, this will allow for maskless writing at a speed of a few tens of wafers per hour.

3.8 Comparative Evaluation of E-Beam Mask Writing Systems

In the evolution of maskmaking, the raster scan systems were developed first, followed by vector scan systems; then later on, came VSB and cell-projection systems. Nevertheless, for the past two decades till the end of the 1990s, the raster scan MEBES systems remained the major maskmaking tool for critical masks. For many experts in the field, it remains unexplainable as to why the "obviously advantageous" variable shaped and cell projection systems did not compete seriously with raster scan systems for such a long time.

In the following sections, major concepts of the systems will be discussed in the light of accuracy and throughputs.

3.8.1 Accuracy

3.8.1.1 Beam Edge

The major advantage of high voltage VSB systems over raster systems is their ability to provide a sharp edge to the electron beam. The edge blur directly translates into CD-variation.

Raster systems in principle could use a fine beam with a comparable beam edge blur. However, the address size would have to be decreased proportionally, which would lead to impractically long writing times. For example, vector scan systems with a narrow beam, like VB-6 (Leica Microsystems), are able to write patterns of exceptionally high quality; however, they are not used in maskmaking because of low throughput.

High accuracy of raster systems is provided by the averaging out of all kinds of errors due to multipass writing. The technique of multipass can also be employed in vector systems (both raster shaped and variable shaped); however, this contributes to noticeable increase in writing time because of overheads like settling times involved in vector writing.

3.8.1.2 Butting Error

Butting errors occur in areas where pattern features are to be split up at the boundaries of subfields, stripes, etc.

Errors can happen in raster scan systems at frequencies corresponding to the butting of stripes; the errors are one-directional. In vector scan systems, butting errors can happen at a number of specific levels:

- Major fields, about 1–2 mm
- Minor fields, about 24–80 μm
- Subminor fields, in some systems, about 2–5 μm
- VSB flashes, about 0.5 μm, they add to line edge roughness

All these errors are two-dimensional in vector systems.

A diagram of butting errors is shown in Figure 3.3 as a function of special frequencies. Raster system suffers much less from butting errors. Also, raster scan systems are easier to calibrate and correct for any scan-related errors than vector systems. In a multipass writing, butting errors can be decreased using overlapping stripes and subfields.

3.8.1.3 Proximity Effect Correction

Proximity effects due to electron scattering lead to CD variation that depends on local pattern density. In a raster system, proximity correction can be addressed without time-consuming data preparation. When using GHOST technique, an additional writing pass is made using a defocused e-beam. The data for this pass is the same pattern but is written in a reversed tone. This flattens out a backscattered dose all over the pattern. If the writing

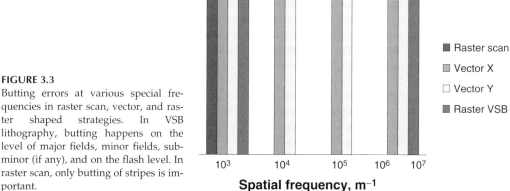

FIGURE 3.3
Butting errors at various special frequencies in raster scan, vector, and raster shaped strategies. In VSB lithography, butting happens on the level of major fields, minor fields, subminor (if any), and on the flash level. In raster scan, only butting of stripes is important.

is done in four passes, proximity correction adds one more pass, which is 20% of total exposure time.

Vector scan systems utilize dose correction for every written shape. The dose correction is normally more accurate than GHOST correction and does not decrease aerial image contrast; however, depending on the implementation, the number of shapes to be written can increase by an order of magnitude thus increasing the writing time.

3.8.1.4 Resist Heating

Resist heating is one of the major issues in error budget. This effect arises from local temperature rise in resist, which leads to change of resist sensitivity and corresponding CD variation. High and medium voltage VSB systems (>10 keV) especially suffer from this problem. The typical area of heat accumulation is in the order of 30–100 µm at 50 kV, which is about the area of a subfield in a VSB system. AEBLE system was capable of writing at 10–100 A/cm^2 beam current density; however, it was used at the lowest value because of resist heating. The maximum flash size was historically decreased from 10 µm in earlier systems to 5 µm, to 2.5 µm, and then to 1 µm because of heating. Multipass writing also helps to decrease heating. All these methods slow down system throughput. Positive chemically amplified resists are advantageous when encountering resist heating because of their high sensitivity and relatively low response to heating. No solution for heating correction has yet been developed.

In raster system, this problem is smaller by an order of magnitude. This is because of the long scans; over scan duration, the heat has time to dissipate.

3.8.1.5 Fogging, Charging, Substrate Heating

These effects are not attributed to certain writing strategy but rather to specific implementation of subsystems. Fogging is an effect of additional irradiation of resist by electrons arising from the bottom of objective lens. These are the third-generation electrons: primary electrons produce backscattered electrons from the mask; they reach the bottom of objective lens and generate new electrons that come back to the resist and produce additional undesired exposure over an area of a few millimeters. Resist charging is mainly a function of resist properties and electron energy. Placement errors due to substrate heating depend greatly on reticle mounting to the stage.

3.8.2 Throughput

Raster scan and raster shaped systems are independent of the pattern to be written. The write time of pattern is solely a function of address size. The historic trend has been to keep the writing time of raster systems within 6 h for any newest generation of masks.

The throughput of systems using vector strategies like vector scan, variable shape, and cell projection depends greatly on pattern and on write overheads.

3.8.2.1 Settling Times

The write overheads include settling time between consequent flashes, between subfields, and between major fields. While a flash is moved by a variable distance to every new pattern feature, settling time is needed in order to achieve accurate position of the beam. In raster scan and raster shaped systems, settling times are short because of their repeatability and short distances. This is different in other systems. In vector scan systems, the settling time is much longer compared to the dwell time. If the number of flashes is large enough, raster system can outperform vector scan system.

Overheads also involve calibration time, loading of the mask blank, and data transfer time if the system is limited by data speed. These overhead times can be roughly counted as a constant for a given system. The write time in vector scan systems is a function of a number of e-beam flashes that constitute the pattern. The numbers of flashes vary greatly from pattern to pattern: a contact layer can be written in 1 h whereas the interconnect layer can take as long as 30 h on a VSB system.

3.8.2.2 Diagonal Lines

An important parameter is the decomposition of a pattern into flashes. Rectangular flashes are normally used to approximate diagonal lines of a pattern. In a VSB system, they give rise to a large number of flashes. If 3% of a pattern consists of diagonal lines, the total amount of flashes may be doubled.

Small features like OPC force the system to use smaller size of flashes resulting in a significant increase in the number of flashes and write time. An example of data fracturing in VSB, raster shaped, and arbitrary shaped beam systems is shown in Figure 3.2d–f.

Diagonal lines are not an issue with raster scan and shaped scan systems; neither do they cause any major problem with arbitrary shaped beam systems.

3.8.2.3 Beam Current Limitations

The write time also depends on the resist sensitivity. Total beam current is distributed evenly over the maximum flash area allowed in a system. In VSB, each shot of a variable size is only a fraction of the maximum allowable shot size. Thus, for such systems, utilization of current can be poor. It is even worse in a cell projection system, especially when writing nonregular features. In a cell projection system, the maximum flash area is much larger than in VSB and, therefore, the throughput of a cell projection system is lower than VSB, with an exception where the cell projection system is writing regular features like memory layers.

In case of raster scan and vector scan systems, since they deliver a beam into a single Gaussian spot, current limitations are less of a problem.

Figure 3.4 shows the throughput of VSB as a function of beam current density for different cases of pattern coverage. A certain write speed and number of diagonal lines

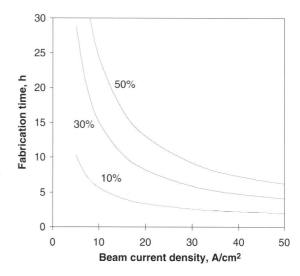

FIGURE 3.4
Writing times of VSB systems as a function of beam current density and pattern coverage. Area to be written, percentage of diagonal lines, and resist sensitivity were fixed. In raster scan and raster shaped systems, the writing time does not depend on these parameters and is about 6 h for high-end masks.

are assumed. At high beam currents, the write time approaches an overhead time; at low beam currents, most of the time is consumed by the actual exposure time of the resist.

The throughput of laser systems, to be addressed later, is often higher than that of e-beam systems. However, quality of masks from e-beam system is normally better than the ones from a laser system. Today a typical practice is to use e-beam systems for low volume high-end masks and use the laser system for the high volume lower-end masks.

3.9 Laser Pattern Generators

E-beam writer has been extensively used for maskmaking for decades, but its throughput has always been an issue. In the mid-1980s, laser pattern generators (LPGs) were introduced; now they are used at maskmaking facilities around the globe.

LPGs use single or multiple laser beams and employ raster mechanism to write pattern on masks. In this respect, they are similar to e-beam raster scan systems, but unlike e-beam systems LPGs do not require a vacuum for their operations. Moreover, there is no need for proximity correction associated with scattering of electrons in EBL and for grounding the substrate to dissipate the charge. The LPG mask fabrication process benefits from advances in high resolution, high-contrast photoresists originally developed for wafer production.

In general, laser systems provide higher throughput, better reliability and lower cost of ownership compared to e-beam systems, while e-beam systems offer better resolution. Also, laser pattern generators can be used to fabricate larger sized photomasks for display industry and for custom applications.

3.9.1 Raster Scan LPGs

LPGs use a continuous-wave laser beam where the stage moves at a constant speed and a beam scans in a direction perpendicular to the stage motion. Scanning is provided by a rotating polygonal mirror or by acousto-optical deflector (AOD). In this way, the scanned angle is converted to a spatial displacement at the mask blank, and a stripe of a pattern is written. The modulation of the beam is controlled by varying the radio-frequency power to acousto-optical modulator (AOM). Similar to EBL systems, the stripes are lined up so that the entire mask pattern is generated.

Examples of laser single beam systems are Micronic LPG, and Heidelberg Instruments DWL systems. Multibeam LPGs are ALTA and custom optical reticle engraver (CORE) from Etec Systems and Omega from Micronic Laser Systems. Five parallel individually blanked laser beams are used in Omega. Combining digital and analog modulation at AOM, writing on a fine address grid is achieved. Omega uses AOD for all beams. Data path flexibility, including opportunity to write directly from GDSII file, and real-time corrections are attractive features of Omega systems.

Eight beams write in parallel in Etec Systems' CORE LPG. The system was later redesigned to ALTA systems that use larger number of beams, enhanced platform, and improved data path.

In ALTA systems [14], 32 parallel laser beams are created by a beamsplitter. Beams pass through an array of AOMs and are deflected by a rotating 24-facet polygonal mirror that creates scanning of the beams as by a brush. Four-pass and eight-pass writings are used. In each pass, features are printed with a different polygon facet and different portion of lens field to average out systematic errors. Gray levels of exposure in each pixel, as well as an offset of writing grids in multiple passes, ensure a fine pattern address grid and edge placement. Humidity-compensated focus system keeps constant the level of humidity in the photoresist during writing to ensure better CD control. Typical time to write a photomask is 2 h. ALTA systems are the most widely used laser systems in maskmaking industry.

Laser source, in newer generations of LPG laser source, is gradually being replaced by those with a shorter wavelength to address the need for higher resolution. Typical pulsed (1–10 kHz) excimer lasers developed for optical steppers are incompatible with raster pattern generators because of the use of continuous-wave lasers in raster LPG architectures. Lasers with shorter wavelengths are being developed specifically for LPGs.

Significant improvements in resolution have been achieved by working with shorter wavelengths. For example, an argon-ion ultraviolet (UV) laser with a wavelength of 363.8 nm is used in ALTA 3500, which results in 270 nm FWHM diameter of Gaussian spot on the mask blank. In ALTA 4000, an argon-ion laser with wavelength doubling delivers a deep UV 257-nm beam; the final spot size is about 50% smaller. This is small enough to write subquarter micron assist features and serifs on the mask.

3.9.2 Matrix Exposure LPG

A new approach for exposure has been developed by Micronic Laser Systems [15]. In its Sigma systems, a spatial light modulator (SLM) is based on an array of micromirrors. Each mirror is tilted individually modulating the beam according to pattern data; the scattered light is blocked by an aperture. A single block of mirrors is capable of modulating simultaneously about 1 million beams. This SLM works actually as a computer-controlled reticle in a microstepper. The light from a DUV laser is reflected from an SLM and is focused on a mask blank through a high numerical aperture objective lens.

This system uses pulsed laser illumination. The SLM is reloaded with a new pattern data during delay between laser pulses. The exposures or arrays are synchronized with continuously moving stage.

The system provides potentially high throughput. It can also achieve high resolution by incorporating some of the techniques developed for optical steppers because, in principle, matrix exposure LPG is similar to optical steppers. Resolution enhancement techniques like OPC and phase shift, high-power pulsed short wavelength laser sources, and resist technologies developed for wafer steppers can be directly applied to this type of mask writing system.

References

1. A. Murray, F. Abboud, F. Raymond, and C. Berglund, *J. Vac. Sci. Technol. B.*, 11 (6) 2391 (1993).
2. M.G.R. Thomson, R. Liu, R.J. Collier, H.T. Carroll, E.T. Doherty, and R.G. Murray, *J. Vac. Sci. Technol. B.*, 5 (1) 53 (1987).
3. H.C. Pfeiffer, *J. Vac. Sci. Technol.*, 15 (3) 887 (1978); also H.C. Pfeiffer, *IEEE Trans. Electron Devices*, ED-26 (4) 663 (1979).
4. Y. Nakayama, S. Okazaki, N. Saitou, and H. Wakabayashi, *J. Vac. Sci. Technol. B.*, 8 (6) 1836 (1990); also *J. Vac. Sci. Technol. B.*, 10, 2759 (1992).
5. L.H. Veneklasen, H.M. Kao, S.A. Rishton, S. Winter, V. Boegli, T. Newman, G. Bertuccelli, G. Howard, P. Le, Z. Tan, and R. Lozes, *J. Vac. Sci. Technol. B.*, 19 (6) 2455 (2001).

6. L.P. Muray, J.P. Spallas, C. Stebler, K. Lee, M. Mankos, Y. Hsu, M. Gmur, and T.H.P. Chang, *J. Vac. Sci. Technol. B.*, 18 (6) 3099 (2000).
7. S. Babin, *J. Vac. Sci. Technol. B.* 22(6) 2004 (to be published).
8. S.T. Coyle, D. Holmgren, X. Chen, T. Thomas, A. Sagle, J. Maldonado, B. Shamoun, P. Allen, and M. Gesley, *J. Vac. Sci. Technol. B.*, 20 (6) 2657. (2002)
9. A. Yamada and T. Yabe, *J. Vac. Sci. Technol. B.*, 21 (6) 2680 (2003).
10. T.R. Groves and R.A. Kendall, *J. Vac. Sci. Technol. B.*, 16 (6), 3168 (1998); also D.S. Pickard, C. Campbell, T. Crane, L.J. Cruz-Rivera, A. Davenport, W.D. Meisburger, R.F.W. Pease, T.R. Groves, *J. Vac. Sci. Technol. B.*, 20 (6) 2662 (2002).
11. E. Yin, A. Brodie, F.C. Tsai, G.X. Guo, and N.W. Parker, *J. Vac. Sci. Technol. B.*, 18 (6) 3126 (2000).
12. P. Kruit; High throughput electron lithography with the MAPPER concept, *J. Vac. Sci. Technol. B.*, 16 (6) 3177 (1998).
13. L.R. Baylor, D.H. Lowndes, M.L. Simpson, C.E. Thomas, M.A. Guillorn, V.I. Merkulov, J.H. Whealton, E.D. Ellis, D.K. Hensley, and A.V. Melechko, *J. Vac. Sci. Technol. B.*, 20 (6) 2646 (2002).
14. M. Bohan, C. Hamaker, and W. Montgomery, *Proc. SPIE*, 4562, 16 (2001).
15. T. Sandstrom and N. Eriksson, *Proc. SPIE*, 4889, 157 (2002).

4

E-Beam Mask Writers

Norio Saitou

CONTENTS

4.1 Introduction 60
    4.1.1 Role of E-Beam Writer in Lithography System 60
    4.1.2 Mask Technology Roadmap 61
4.2 Development History of E-Beam Writer 62
    4.2.1 The History of Evolution 62
    4.2.2 An Example of the System 63
4.3 System Structures and Features 65
    4.3.1 Stage Movement and Scanning Mode 65
        4.3.1.1 Stage Movement 65
        4.3.1.2 Scanning Mode 65
    4.3.2 Acceleration Voltage 66
        4.3.2.1 Coulomb Effect 66
        4.3.2.2 Resist Sensitivity and Current Density 67
        4.3.2.3 Proximity Effect 67
    4.3.3 Probe Forming System and Beam Shapes 68
        4.3.3.1 Point Beam System 68
        4.3.3.2 Variable Shaped Beam and Cell Projection 69
    4.3.4 Relationship Among Resolution, Accuracy, and Throughput 70
4.4 Technology for Throughput and Resolution 71
    4.4.1 Electron Gun 71
    4.4.2 Beam-Shaping Lens System 72
    4.4.3 Objective Lens System 75
    4.4.4 Multistage Deflection Architecture 76
    4.4.5 Digital to Analog Converter 78
4.5 Technology for Image Placement Accuracy 79
    4.5.1 Stage and Chamber Mechanics 80
        4.5.1.1 Stage and Writing-Chamber Materials 80
        4.5.1.2 Chamber Structure 80
        4.5.1.3 Stage Structure 81
    4.5.2 Temperature Balance 81
    4.5.3 Mask Flatness 82
    4.5.4 Superpose Writing 82
4.6 Technology for Critical Dimension Accuracy 83
    4.6.1 Beam Shaping Function 84
        4.6.1.1 Triangle Beam 84
        4.6.1.2 Dual Shaping Deflector 85
    4.6.2 Proximity Effect Correction 85

	4.6.2.1 Dose Correction Method 86
	4.6.2.2 Shape Modification Method... 89
	4.6.2.3 GHOST Method 89
	4.6.2.4 Multilayer Resist Method 89
4.6.3	Multiple-Scattering Effects in Optical Column... 90
	4.6.3.1 Antireflection Plate... 90
	4.6.3.2 Fogging Effect Correction... 91
4.7 Data Preparation.. 92
4.8 Commercially Available Systems... 94
4.9 Summary... 94
References 95

4.1 Introduction

4.1.1 Role of E-Beam Writer in Lithography System

Electron beam (e-beam) can easily be focused into nanometer diameter by electromagnetic or electrostatic lenses and can be steered by electromagnetic or electrostatic deflectors. So the e- beam can define patterns of a large-scale integrated circuit (LSI) with very high resolution in a resist. Tools that provide a focused e-beam have played an important role in the semiconductor industry for many years.

The role of e-beam lithography (EBL) in semiconductor device manufacturing is illustrated in Figure 4.1. The initial pattern data of LSI is generated using a computer-aided design (CAD) system. The CAD data is converted to EBL data, and the EBL system

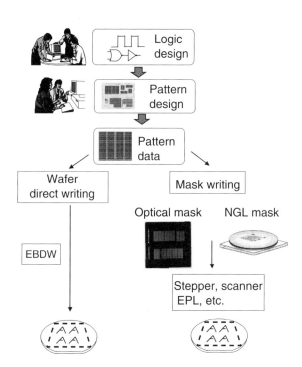

FIGURE 4.1
Role of e-beam writer in lithography system.

produces the final pattern. EBL is used in two ways: (1) mask writing, a key component of conventional optical lithography system; and (2) direct writing on semiconductor substrate. Mask is sometimes called reticle. Here, we use the word "mask". In the present integrated circuit manufacturing, optical lithography is in the main stream. In this case, the circuit pattern is contained in a chromium layer on a glass plate called a photomask. The mask is illuminated with light of a single wavelength and its image is projected onto the wafer. The image of the mask is typically reduced by a factor of 4, making it much easier to pattern the mask accurately. However, optical lithography is approaching its fundamental limits. To prolong the use of optical lithography, the wavelength needs to be shortened progressively. Optical lithography using excimer laser ArF ($\lambda = 193$ nm) and F_2 ($\lambda = 157$ nm) have now been developed. The masks used here require much smaller pattern. And in some next generation lithography (NGL), a thin membrane mask is used instead of a glass plate. The candidates for the NGLs are extreme ultra violet (EUV), electron projection lithography (EPL), proximity x-ray lithography (PXL), proximity electron lithography (PEL), and so on. These NGLs require highly accurate masks and the e- beam is thought to be the only one way to make such masks.

4.1.2 Mask Technology Roadmap

The minimum feature size of semiconductor devices has decreased dramatically in recent years. Miniaturization has been supported by advancement in lithography technology. Mask making is one of the most challenging technologies in the area of lithography. The International Technology Roadmap for Semiconductors (ITRS) presented the new version of semiconductor road map [1] in "2002 Update Conference" on December 4, 2002, at the Tokyo International Forum, Japan. The requirements of advanced optical masks presented there are shown in Table 4.1; the pattern size for optical proximity correction (OPC) is approaching to almost the same as the main pattern size in the node of 90 nm and beyond — even in the case of 4× masks. For example, mask requirement of the minimum pattern for 90-nm node in 2004 has been 106 nm. The requirements of accuracy for image placement (IP) and critical dimension (CD) are almost similar to the requirements at the wafer level. For example, mask requirements of IP and CD for 90-nm node are 19 and 4.2 nm, respectively. IP is 1/5, CD is 1/20 of the node.

To gain higher accuracy of a mask, sustained improvements are necessary in mask writers, mask blanks, resists, and development and etching technologies. In this chapter, e-beam mask writers of the past, present, and future are described in detail.

TABLE 4.1

Requirements for Advanced Optical Mask (From [1] International Technology Roadmap for Semiconductors, 2002 Update Conference, Dec. 4, 2002, Tokyo International Forum, Japan)

Year of production	2002 115 nm	2003 100 nm	2004 90 nm	2005 80 nm	2006 70 nm	2007 65 nm
Mask minimum image size (nm)	300	260	212	180	160	140
Mask OPC feature size (nm), clear	230	200	180	160	140	130
Mask OPC feature size (nm), opaque	150	130	106	90	80	70
Image placement (nm, multi-point)	24	21	19	17	15	14
CD uniformity (nm)						
Isolated lines (MPU gates) binary	6.1	5.1	4.2	3.7	3.4	2.5
Isolated lines (MPU gates) ALT	8.5	7.2	5.9	5.1	4.8	4
Contact/vias	6.9	6.1	5.3	4.8	4.3	3.2
Linearity	17.5	15.2	13.7	12.2	10.6	9.9

4.2 Development History of E-Beam Writer

4.2.1 The History of Evolution

EBL systems have various system architectures depending on scanning method, stage moving strategy, and beam shape. To achieve higher throughput and higher accuracy, various systems were constructed according to the combination of the beam shape, the scanning methods, and the stage moving strategies. The throughput and accuracy of e-beam systems have improved drastically over the last 35 years. The system has become more and more complex to maintain the writing speed in spite of scaled device dimensions and the associated reduction in the pattern address unit. In particular, beam shape and stage movement have evolved as shown in Figure 4.2, where the historical development of e-beam systems is illustrated schematically.

In 1965, a scanning electron microscope was used to write patterns for integrated circuits [2,3]. Since then, many research organizations have developed EBL systems. The first generation of the e-beam system [4,5] appeared in the 1970s and was based on the scanning electron microscope. The beam shape was pointed and the stage was moved one after another over the deflection field and hence referred to as the step-and-repeat mode. As the current density of the point beam is distributed in Gaussian, it is sometimes called Gaussian beam. This era was the dawn of LSI where the minimum feature size of LSIs was a few micrometers.

In the 1980s, the minimum size of LSIs was less than 1 μm. To achieve higher throughput, the stage movement was changed from step-and-repeat mode to a continuous movement mode; and also the point-shaped beam in the 1970s was replaced by the variable shaped beam (VSB) [6–9].

In the 1990s, the minimum size had reached to 0.35 μm; and the more complex projection methods, e.g., the cell projection [10–12], block exposure [13], and character projection [14], appeared to meet the challenges. The concept of projection beams is shown in Figure 4.3, where the beams are formed by imaging shaped-apertures. Projection beams were considered to be advanced-shaped beams. Projection systems were

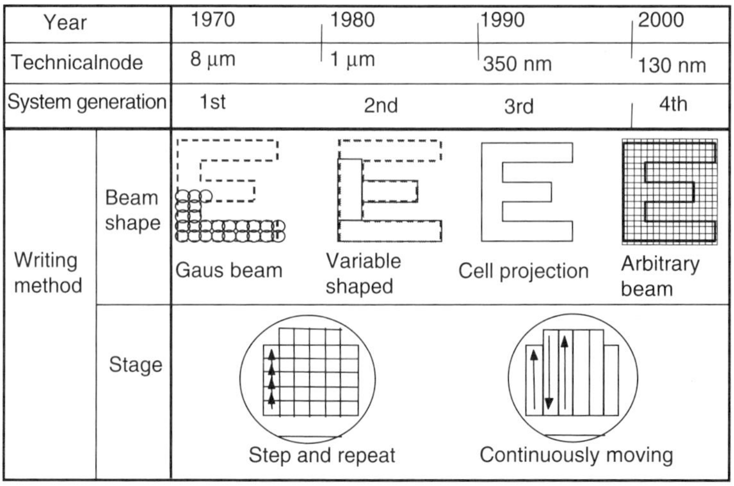

FIGURE 4.2
Equipment evolution history.

E-Beam Mask Writers

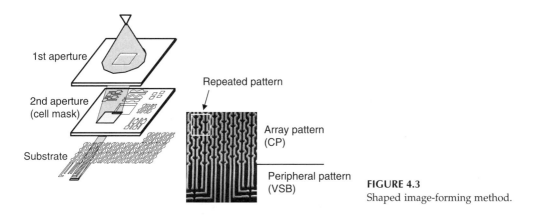

FIGURE 4.3
Shaped image-forming method.

originally developed for higher throughput, and they were also expected to generate oblique patterns with accurate CD.

Since 2000, several types of multibeam systems have been studied [15–17]. These new technologies will, however, need more time to mature.

4.2.2 An Example of the System

An architectural example of an e-beam lithography system is shown in Figure 4.4. Figure 4.5 shows a picture of an electron optical column installed in a clean room. The system

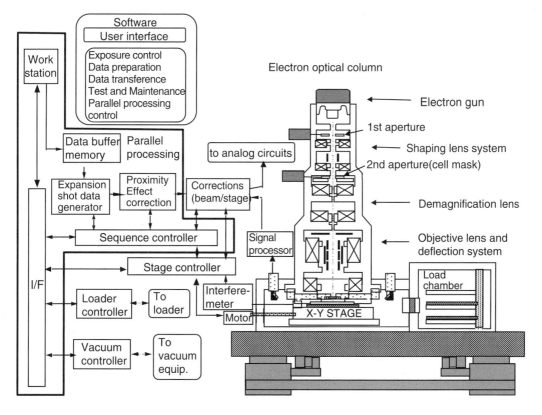

FIGURE 4.4
An architecture example of an e-beam lithography system.

FIGURE 4.5
An example of mask writing e-beam system.

consists of four subsystems: (1) electron optical column, (2) mechanical system, (3) control electronics, and (4) software.

1. The electron optical column produces the high-resolution focused e-beam and deflects it on the substrate. The column is comprised of an electron gun, illumination optics, a beam shaping optics, and a set of focusing and objective lenses that include deflectors. The column should be kept at a high vacuum of 10^{-6} to 10^{-8} Pa.
2. The mechanical system consists of mask stage and loading system. The stage moves in the vacuum under the control of a laser interferometer. The high precision X–Y stage has a low distortion table structure. The temperature of the mask loader unit is stabilized to reduce the mask temperature change and to provide better positioning accuracy.
3. A workstation controls the overall system that includes the mechanical systems. Control electronics drives the deflectors inside the electron optical column to position the shot beam correctly with reference to the position of the mask from the stage position during the continuous stage movement. For higher throughput, the parallel processing function is equipped to support simultaneously the huge amount of data for exposure and the data for proximity effect correction (PEC).
4. Software handles the data preparation, exposure control, and the overall diagnosis and maintenance. The data preparation software controls data conversion.

Exposure control is the main control program for the EBL system. This program delineates the patterns on the substrate using e-beam data. The e-beam data consists of a job management program, an e-beam system control program, and data library management program. The hierarchical data from the workstation is fractured into shot data. Then the PEC system adjusts the shot time depending on the exposed pattern area density. Very fine address size like 1 nm, and reconstruction of subfield pattern data are also utilized. The third software structure is the system evaluation and maintenance program.

The EBL system is so large and complicated that the selection of the system strategy and optimization of each subsystem become very important in their design. In Section 4.3, the EBL system structures and features are discussed.

4.3 System Structures and Features

This section addresses system structures and a variety of features that involve stage, scanning mechanism, acceleration voltage, probing mechanism, and also presents the relationship and possible trade-offs among the three competing entities, namely, throughput, resolution, and accuracy.

4.3.1 Stage Movement and Scanning Mode

Stage movement and scanning mode are interrelated and are presented in the following two subsections.

4.3.1.1 Stage Movement

In the EBL system, highly accurate beam deflection is required. The scanning field size of deflector is generally smaller than a few square millimeters due to optical aberration and distortion limit. Without stage movement, it is impossible to delineate the pattern on the whole substrate surface, such as a mask.

Two stage moving methods are illustrated in Figure 4.2. The first is the step-and-repeat (S&R) method. The stage moves to the next field and writing begins again when the stage stops. The S&R method results in loss of time because the pattern is not delineated during the moving time of the stage. The second is the continuous stage moving method, in which writing is carried out while the stage is moving in one direction. In this method, the deflection aberration problem tends to be less serious than S&R method because the field size may be smaller. The continuous moving method can contribute to high throughput because the mechanical overhead time is short compared to the S&R method. However, the field-stitching error problem is serious because the quantity of field stitching increases with decreasing field size. A continuous moving stage is complicated in terms of system control and stage mechanics, but the electron optical column design is simpler than that of S&R because the deflection field is smaller.

4.3.1.2 Scanning Mode

An e-beam is scanned in two different ways: (a) raster scanning mode and (b) vector scanning mode. Both methods are shown in Figure 4.6.

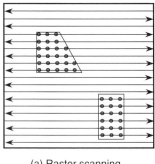 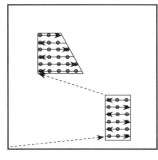

FIGURE 4.6
Beam scanning method. (a) Raster scanning. (b) Vector scanning.

In raster scanning mode, the deflector scans an overall field. The e-beam is "blanked off" in the nonpattern region and is "unblanked" in the pattern region only. The raster scanning method has the advantage of simplifying the system structure. The pattern writing time of raster scanning does not depend on pattern complexity because whole region of the substrate surface is uniformly scanned. As in the raster scanning method the beam shot time is constant, it is difficult to use a wide range of resist sensitivity. This raster scanning method is usually combined with the continuous moving stage. The design of electron optics is relatively easy because of the small field size. Since there is a large number of stripe stitching, the stage movement should be controlled accurately.

In the vector scanning mode, the deflector scans only the pattern region. It is possible to use a wide range of resist sensitivity because the beam shot time is an arbitrary variable. For the same reason, this method has the advantage of PEC. The vector scanning method also has the advantage of shortening the writing time because it is not necessary to scan the nonpattern region.However, the field size should be larger to get higher throughput. Keeping the deflection distortion small and correcting it accurately are the major issues in vector scanning electron optics design.

Both types of scanning modes can be used in case of point beam. However, in the case of shaped beam, vector scanning method is generally used.

4.3.2 Acceleration Voltage

The acceleration voltage V should be selected carefully. The acceleration voltage of the system relates to the throughput because resist sensitivity and current density depend on the acceleration voltage. The acceleration voltages also related to resolution, CD accuracy, and IP accuracy of the e-beam system due to the Coulomb effect and proximity effect.

4.3.2.1 Coulomb Effect

The space-charge effect is a phenomenon that deteriorates the beam edge sharpness. The beam blur due to the Coulomb force between electrons is proportional to $IL/V^{3/2}$, where I is the beam current, and L is the optical path length [18]. A larger beam current is better for increasing throughput but the beam resolution decreases with current. In a VSB system, the current changes shot to shot during pattern writing, and therefore the beam resolution can be different for each shot. Strictly speaking, the focus position must be changed shot to shot. As the fast refocusing shot to shot makes the e-beam system complicated, most VSB systems adopt high acceleration voltage and limit the maximum current depending on the accuracy. However, the cell projection system is able to have the refocusing function for each cell pattern, because the time for refocusing can be negligible as a cell is usually shot repeatedly.

4.3.2.2 Resist Sensitivity and Current Density

Lower acceleration voltage seems to be better for higher throughput because the resist sensitivity is roughly inversely proportional to V. However, higher current density can be obtained with the acceleration voltage because the brightness is proportional to V. Though the beam current increases proportional to V, the beam blur caused by the Coulomb effect finally decreases with $V^{1/2}$ for the same exposure time. Very highly sensitive, chemically amplified resist (CAR) have recently been used. So, higher acceleration voltage can be accepted from the viewpoint of the throughput.

4.3.2.3 Proximity Effect

Incident e-beams are broadened in materials as a result of electron scattering. The Monte Carlo method turned out to be very effective to simulate the scattering effects in e-beam lithography [19]. An example of the simulation results of electron scattering in Si substrate is shown in Figure 4.7. In this simulation, 100 electrons were impinged perpendicularly to one point of the surface, and the trajectories of these electrons were projected in x–y plane. From the figure, it turns out that beam scattering in the target limits the resolution in e-beam lithography. The penetration depths of electrons are 0.7, 2, 10, and 20 μm for 5, 10, 30, and 50 kV, respectively. Its increase is proportional to $V^{3/2}$.

It is clear that the lateral spread of the electrons at the surface is almost the same as with the penetration depth in case of 10 kV or lower voltage. In higher than 30 kV, the lateral spread is negligibly small. The situation is almost the same with resist coated mask where the resist thickness is 0.3 to 0.4 μm. The scattering phenomena of the e-beam are generally divided into two effects. One is forward-scattering effect and the other is backward-scattering effect as shown in Figure 4.8.

Forward scattering is less during high voltage. It gives rise to an intraproximity effect. The pattern fidelity to the beam shape is generally better using high acceleration voltage for submicron isolated pattern as shown in Figure 4.9.

The backward scattering gives rise to interproximity effect between patterns, shown in the same Figure 4.9. The backward-scattering effect is larger for higher acceleration voltage.

Acceleration voltage lower than 10 kV causes charging-up in resists, which deteriorates the position accuracy. It needs thin resist process but thin resist has defect problems. At

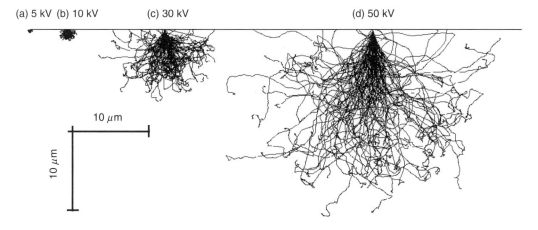

FIGURE 4.7
Computer simulated e-beam trajectories in Si substrate.

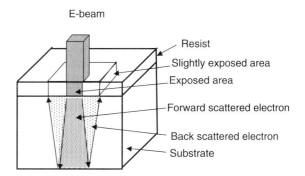

FIGURE 4.8
Mechanism of proximity effect.

the 100-nm node, the most suitable acceleration voltage might be 50 kV as adopted mostly in commercially available e-beam mask writers.

The scattering phenomenon causes the proximity effect in e-beam pattern writing. This phenomenon can gravely affect the patterns when they are close together in a mask. The correction method for the proximity effect will be discussed in detail in Section 4.6.2 of this chapter.

4.3.3 Probe Forming System and Beam Shapes

The following two subsections will focus on point beams and shaped beams.

4.3.3.1 Point Beam System

The earliest EBL system was based on scanning e-beam microscope technology, in which a source image of the electron gun is focused on a sharp round spot on the specimen. This is usually called the point beam, and here the intensity distribution is Gaussian. A typical optical column is shown in Figure 4.10(a). The combination of the second and the third

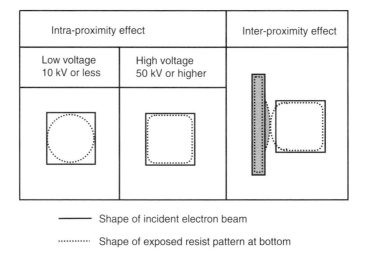

FIGURE 4.9
Proximity effect occurs in usual mask writing. Intra-proximity effect occurs in submicron hole pattern writing at low voltage. Interproximity effect is serious at high acceleration voltage.

E-Beam Mask Writers

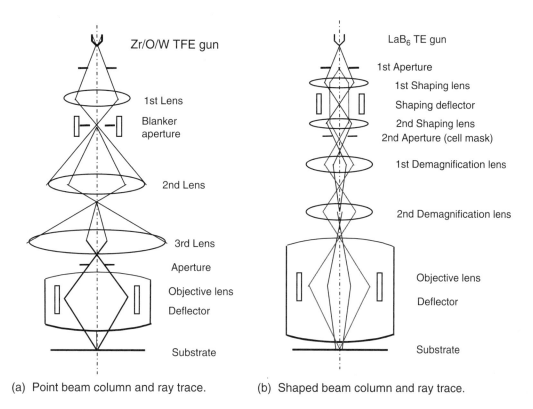

(a) Point beam column and ray trace.   (b) Shaped beam column and ray trace.

FIGURE 4.10
Schematic diagram of electron optics and its ray trace.

lens makes the zoom lens system, which controls the beam diameter without focal point change.

Many point beam systems have been developed so far [20–22]. However, the throughput of the point beam system is still low even if a high brightness electron gun is used as stated in Section 4.4.1. To obtain a point beam system with higher throughput than shaped beam system, it is necessary to develop a multibeam system and this will take some time. At present, the point beam system is mainly used for high-resolution application.

4.3.3.2 Variable Shaped Beam and Cell Projection

During the second EBL era, a VSB method was developed. Figure 4.2 shows how the character "E" can be delineated by four VSB shots, while a large number of shots would be necessary for a spot beam. Its probe forming optical column is illustrated in Figure 4.10(b). In this method, a beam shaping lens system is necessary. A rectangular beam with a variable size is made by passing through two square apertures. The e-beam emitted from an electron gun illuminates the first square aperture. The current density of the VSB is determined by illumination conditions changed by a condenser lens system. The image of the first aperture is focused on the second aperture by two projection lenses. The shaping deflectors change the overlapping of two apertures. The overlapped image is de-magnified by 1/25 to 1/100 through two demagnification lenses and an objective lens and focused on the substrate. Then, the beam spot with the arbitrary length and width is finally projected on to a substrate on a shot-to-shot basis depending on the pattern data. The source image of the electron gun is not focused on the substrate.

In the third generation, several unit cell patterns are defined in a second stencil mask. The patterns suitable for forming various cell beams are arrayed around the central square aperture. The pre-shaped square beam of the first aperture is deflected onto one of the unit cells, and the cell is completely covered by the square beam. The final e-beam has the complex shape of a unit cell. Both the VSB and cell patterns are demagnified to 1/25 to 1/100 size through two demagnifying lenses and an objective lens system. The entire second aperture, including the cells, is made of a thin silicon stencil structure using conventional wafer fabrication process [23]. The accuracy required for the cell-mask is much less than for 4× optical masks or 1× proximity x-ray masks.

4.3.4 Relationships among Resolution, Accuracy, and Throughput

There are three parameters, namely, resolution, throughput, and accuracy that are important for the characterization of an e-beam system. Here accuracy is defined as the delineated positional accuracy per unit length on the substrate. Figure 4.11 is a three-dimensional representation of three separate e-beam systems, where each system shows its strength in one of the three parameters. The first system (a) has a high resolution of the order of 10 nm;, the second system (b) is strong in accuracy;, while the third system (c) shows a throughput that supersedes the other two. The volume of the solid triangle defined by these parameters can be regarded as the performance capacity of these systems. It is noticeable that the volumes of these three solid triangles are almost equal to each other. It can therefore be surmised that the performance capacity of a system is a measure of the technology level of the era or of the system manufacturing company. This performance capacity has grown almost 1000× during the past 30 years and is still growing.

These systems have their own niche for their applications. The first type of system with nanometer resolution may be good for R&D activities, the second type with high IP accuracy can be valuable in mask making, while the third type of system can add significant contribution to direct wafer writing where throughput continues to be the main issue. In the first system, the thermal field emission (TFE) gun and the point beam

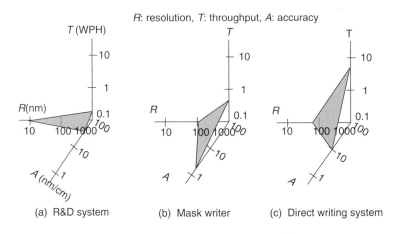

FIGURE 4.11
Three types of e-beam system.

method will be mainly used. In the second and the third systems, the thermionic LaB$_6$ gun [thermionic emission (TE)] and the VSB methods are mainly utilized. The mask writing system (b) now requires almost the same resolution and accuracy as the direct writing system (c). Thus, the throughput of the system (c) becomes almost the same as the throughput of the system (b).

4.4 Technology for Throughput and Resolution

Since the EBL system was first built about three decades ago, throughout its development there has been a history of continued focus on how to get higher resolution and higher throughput. Compared with the beam current of the order of 10^{-10} A in the scanning electron microscope, the EBL system requires the beam current of the order of 10^{-6} A to obtain a reasonable throughput. The resolution or the beam's blur of the EBL system may be about one third of the minimum pattern size. It is usually not so high compared to the scanning electron microscope. For example, the EBL system for the 90-nm node requires a beam blur of 30–40 nm. The design problem of the system is how to achieve larger current while maintaining this beam blur. The main issues for higher throughput are: (1) development of electron gun, (2) beam-shaping technologies, (3) objective lens system, and (4) beam deflection technologies. These are discussed here in that order.

4.4.1 Electron Gun

There are two types of electron guns used in the EBL system. One uses the TE cathode and the other uses the TFE cathode. These two cathodes are compared in Table 4.2.

The main differences between them are in their operation, vacuum condition, and temperature. The currently used TFE cathode is composed of a tungsten needle covered with Zr/O, which reduces the Schottky barrier height and it is operated at around 1800 K. At this temperature the Zr/O will migrate and activate the emitter surface. The thermal field emitter Zr/O/W [24] is operated in a vacuum of better than 10^{-7} Pa. On the contrary, the thermionic LaB$_6$ cathode [25,26] is operated in a vacuum of better than 10^{-5} Pa

TABLE 4.2

Comparison of Electron Sources

Characteristics	Thermionic Emitter LaB$_6$	Thermal Field Emitter Zr/O/W-TFE
Operation vacuum	$\leq 10^{-5}$ Pa	$\leq 10^{-7}$ Pa
Operation temperature	1900 K	1800 K
Energy broadening	2.5 eV	1.5 eV
Brightness B (at 50 keV)	5×10^5 A/cm^2 sr	5×10^7 A/cm^2 sr
Virtual source size d	10 μm	0.05 μm
Emission semi-angle θ	3 mrad	10 mrad
Maximum beam current $I_{max} = \pi/2 \cdot B(d\theta)^2$	7 μA	~200 nA
Maximum current density for 5 μm square beam	30 A/cm^2	1 A/cm^2
Application	Shaped beam system	Point beam system

The brightness B, the virtual source size d, and the emission semi-angle θ are also different for the two cathodes. The brightness of LaB_6 is 5×10^5 A/cm^2 sr at 50 kV, that is, 1% of the Zr/O/W. The virtual source size of LaB_6, however, is about 10 µm, that is, 100× larger than that of Zr/O/W. The emission semi-angle θ of LaB_6 is 3 mrad and that of the Zr/O/W is 10 mrad. The emittance ε, defined by the product of d and θ, is 60× larger in the LaB_6 than in Zr/O/W. The maximum beam current obtained from the cathode is proportional to $B(d\theta)^2$ for larger beam size. Thus, the LaB_6 cathode is effective for shaped beams.

In a single beam system, the problem is how to obtain a small spot size and a large beam current. Let us calculate the relationship between beam diameter and its current. The size of the beam diameter is affected by the image of source size, spherical aberration of the lens, chromatic aberration, and diffraction aberration. In the case of TE, the image size of crossover and spherical aberration are the main determinants for the diameter. The beam current of TE is the product of $B\alpha^2$ and the area of crossover image, where B is the brightness and α is the beam semiangle at the substrate. In the case of TFE, the source size is usually very small. The main consideration for the diameter is spherical aberration and diffraction. The beam current is the product of $\pi\alpha_E^2$ and $dI/d\Omega$, where α_E is the emission semi-angle from the gun and $dI/d\Omega$ is the angular emission current. The calculated results are shown in Figure 4.12. In this figure, the minimum spot diameter of about 5 nm comes from the diffraction aberration $1.2\lambda/\alpha$. From this figure, it turns out that the current is proportional to $d_b^{8/3}$ in TE for large beam size region and is proportional to $d_b^{2/3}$ in TFE. For smaller beam region, TFE is able to give a larger beam current. The crossover point is around at the diameter of a few 100 nm. This means that TE is more effective for area beam, such as VSB, and TFE is more effective for small spot beam.

4.4.2 Beam-Shaping Lens System

The beam-shaping lens system is necessary in many cases. This system is illustrated in Figure 4.10(b). In this section, we will introduce the beam-shaping lens system for VSB and CP beams. The function of this lens system is to transfer the first aperture image onto the second aperture using deflectors. The system example is shown in Figure 4.13(a) [27]. It has 21 cell patterns.

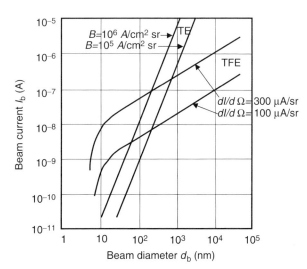

FIGURE 4.12
Relationship between beam diameter vs. beam current in TE and TFE.

E-Beam Mask Writers

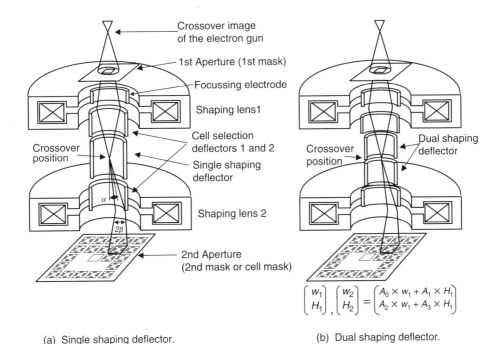

(a) Single shaping deflector.

(b) Dual shaping deflector.

FIGURE 4.13
Beam-shaping lens system.

This lens system consists of the first mask aperture, the first shaping lens, a focusing electrode, a pair of cell-selection deflectors, a single shaping deflector, the second shaping lens, and the second mask aperture. The deflection center of the pair of selection deflectors coincides with the deflection center of the single shaping deflector. The e-beam from the crossover point above the first shaping lens illuminates the first mask aperture. The crossover is focussed at the deflection center. The first mask aperture image is focused onto the second mask aperture through the first and the second shaping lenses with the magnification of 1×. The overlapping part between the first image and the second aperture can be changed by the beam deflection. The current density is kept constant when the beam size changes because the crossover image is located at the center position of the shaping deflector.

The overlapping with the square aperture at the center of the second mask makes the VSB, using a single shaping deflector, which is an electrostatic quadrupole. On the contrary, if a pair of selection deflectors moves the first image to a pattern around the center of the second aperture, a cell beam can be made. This pair of cell-selection deflectors is electrostatic octupole. The overlapped image is demagnified by around 1/25 through the reduction lenses and an objective lens. If the maximum size of the shaped beam is $5\,\mu m^2$ on the specimen, the aperture size of the second mask will be $125\,\mu m^2$.

The cell-selection and shaping deflectors are not for common use. The cell-selection deflector requires a large deflection distance, and the shaping deflector requires a high-speed deflection. This is the reason why two kinds of independent and separate deflectors are necessary and that these should have a common deflection center. The shaping deflector has a settling time of 100 ns, and the deflection distance is $125\,\mu m$. The cell-selection deflector has a settling time of a few microseconds and the deflection distance is $300\,\mu m$.

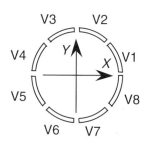

FIGURE 4.14
Applied voltage to the eight electrodes.

V1 = VX+0.4VY
V2 = 0.4VX+VY
V3 = −0.4VX+VY
V4 = −VX+0.4VY
V5 = −VX−0.4VY
V6 = −0.4VX−VY
V7 = 0.4VX−VY
V8 = VX−0.4VY

To keep the current density uniform in a cell beam, the optical aberration due to the cell-selection deflector should be small. Electrostatic octupole deflectors are adopted because of smaller aberration than in electrostatic quadrupole. To make uniform electric field around the axis, the applied voltage to the eight electrodes is shown in Figure 4.14. The aberration δ of the crossover due to cell-selection deflection is expressed in Equation (4.1):

$$\delta = C_s\{\alpha^3 + \alpha^2\beta(2e^{i\theta} + e^{-i\theta}) + \alpha\beta^2(2 + e^{2i\theta}) + \beta^3\} \quad (4.1)$$

where, C_s is the spherical aberration coefficient of the second shaping lens, α the deflection angle (corresponds to the deflection distance on the second mask), β the beam semi-angle (corresponds to the cell size), and θ is the angle around the axis. The first is the shift term of the crossover position and can be corrected using a pair of deflectors. The second is the field curvature and the astigmatism. The third and the fourth are coma and spherical aberration on the axis. It is difficult to correct these two aberrations.

The second is of a serious concern because it is proportional to the square of the deflection distance. The field curvature can be corrected dynamically by a focusing electrode. Astigmatism can also be corrected by superimposing the correction voltage to a pair of octupole deflectors. The simulation result is shown in Figure 4.15. Using this correction function, the aberration is decreased to 0.1 μm, which is, negligibly small compared to the crossover diameter of about 10 μm. This value is adequate for 65 nm mask writing.

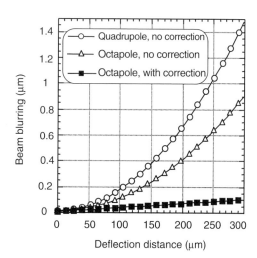

FIGURE 4.15
The aberrations occurring in the crossover. The crossover diameter in the center of shaped deflector is about 10 μm. The uniformity of the current density depends on this aberration.

4.4.3 Objective Lens System

The objective lens system that consists of a lens and a deflector is one of the most important subsystems that relates to the throughput, the accuracy, and the resolution. Earlier, the deflection technique, used in scanning microscopy, was applied to the EBL system. The deflector was arranged after, or before, the final lens as shown in Figure 4.16(a) and (b). The postdeflection system (a) is low in resolution due to the long focal length. The predeflection system (b) has a small field due to the large off-axis aberration. In EBL, the objective lens system is required to obtain high resolution and a larger field in order to have high accuracy and high throughput. It turned out that an in-lens deflection system developed by Pfeiffer [28] met the contradictory requirements. By superposing the lens and deflection fields, the lens and deflector can be designed so that some of the critical deflection aberrations of the deflector and the corresponding off-axis aberrations of the lens compensate each other as shown in Figure 4.16(c). The chromatic deflection error, for instance, is compensated by the opposed lens dispersion as shown in Figure 4.17. The low-energy electrons deflected more than high-energy electrons by the deflector [Figure 4.17(a)] are brought back larger by the lens field, and hence the chromatic aberration can be compensated. In the example of Figure 4.17(b), the objective lens system consists of one lens and one deflector. For the purpose of minimizing deflection aberration, nowadays, more complicated objective lens systems are designed for the EBL systems. In those systems, plural lenses and deflectors are used to compensate many of the following aberrations.

The aberration of the objective lens system is expressed as the polynomials [29] of the beam convergence semi-angle α and the deflection distance r. The odd terms only appear from the symmetry. The deflection distance corresponds to the deflection angle β as shown in Figure 4.16(a). The terms higher than the third order are known as negligibly small in the aberration theory of optics. The aberration δ is expressed by Equation (4.2).

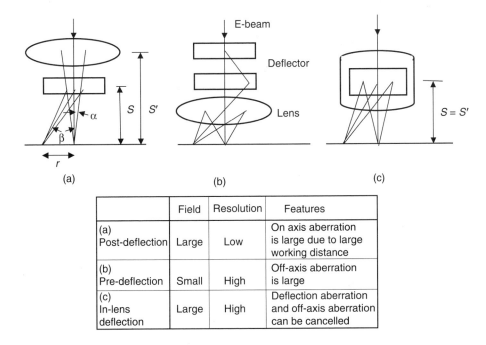

FIGURE 4.16
Three types of the objective lens with deflection.

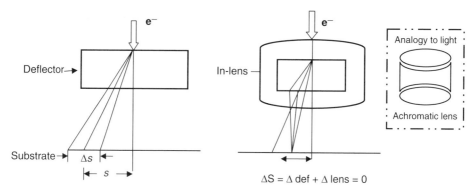

FIGURE 4.17
The chromatic aberrations and the deflection system.

$$\delta = C_{ac}\alpha\Delta V/V + C_{tc}r\Delta V/V + C_s\alpha^3 r + C_k\alpha^2 r + C_a\alpha r^2 + C_f\alpha r^2 + C_d r^3 \quad (4.2)$$

where, ΔV is the energy broadening of e-beam, V is the acceleration energy, and the coefficients mean as follows: C_{ac}, axial chromatic aberration coefficient; C_{tc}, chromatic aberration coefficient due to deflection; C_s, spherical aberration coefficient; C_k, coma due to deflection; C_a, astigmatism due to deflection; C_f, field curvature; C_d, deflection distortion.

These coefficients can be calculated as the integral of the superposed field of the in-lens system [30]. In the lens design, the appropriate structure of lens and deflector are assumed at first. Then, the electric and magnetic fields are calculated by computer simulations, and the aberration coefficients are calculated using these fields. The parameters, such as the lens geometry and the strength, are optimized to compensate for the aberrations and the landing angle. Various systems have been designed and used in various EBL systems. It is theoretically possible to produce the objective lens system with low aberration, vertical landing, and large convergence angle if many optical elements are used. An example with four-stage deflection yokes is reported to have a high resolution at high current density [31]. In the practical viewpoint, however, the number of optical elements should be as small as possible because error in the manufacturing causes another aberration. The deflector is the most critical in manufacturing. A system example with two lenses and two deflectors [32] is shown in Figure 4.18. Where ϕ is the diameter and L is the length of the deflector. In this system, the simulated total aberration is compared with that of the one-deflector system [33]. The deflection aberration at the corner of the 2-mm field is 35 nm, almost the same as that at the center and is reduced by 35%, compared to the one-deflector system.

4.4.4 Multistage Deflection Architecture

Most recent systems utilize multistage deflection architecture to obtain a high precision and a high-speed deflection at the same time. The reason is that it is difficult to scan a whole field with high speed by one deflector from the viewpoint of the circuit. A scanning field is divided into subfields in a two-stage deflection system as shown in Figure 4.19. A high-precision deflector scans the beam between subfields and a high-speed deflector

E-Beam Mask Writers

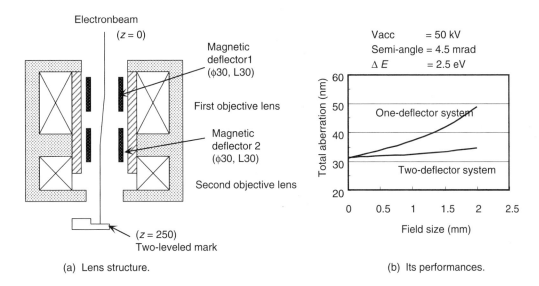

FIGURE 4.18
Schematic diagram of two-lens, two-deflector objective lens and their performances.

scans the beam within a subfield. Some systems adopt a three-stage deflection architecture [12]; here the subfield is divided into several micro-fields or sub-subfields.

Both an electromagnetic force and an electrostatic force can be used to deflect an e-beam. The electromagnetic force is generated from a deflection coil, and the electrostatic force is generated from a deflection plate. The advantages of an electromagnetic deflection are that (1) the coil can be set outside the vacuum, (2) a large deflection distance is possible by a large number of coil-turns, and (3) the chromatic deflection aberration is half of the electrostatic. The disadvantage of electromagnetic deflection is that it tends to be affected by the hysteresis and the eddy current. On the contrary, although electrostatic deflection has high scanning speed, the deflection plate has to be installed within a vacuum. E-beam tends to be affected by the contamination on the surface of the plate and by the noise deriving from its own circuits. Earlier, an electromagnetic deflection system was widely used in almost all e-beam systems, but since the development of

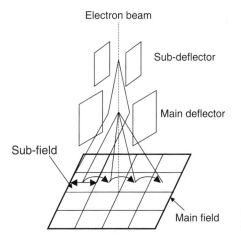

FIGURE 4.19
Multistage deflection architecture.

vacuum technology and analog circuit technology in the recent sub-100-nm mask writer, an electrostatic technology is used.

4.4.5 Digital to Analog Converter

The deflection circuits consist of the digital to analog converters (DACs) and the output amplifiers. It is now necessary to reduce the beam positioning address unit to 1 nm, which is the least significant bit (LSB), 1 nm. To obtain an effective throughput, the deflection field size should be of the order of 1 mm. These factors produce a tremendous increase in the number of resolution bits to as high as 20 bits as shown in Figure 4.20. The figure shows the relation between the field size and the bits number of DACs at 1-nm LSB. Additionally, the requirement for the linearity is at the level of ± 0.5 LSB. It is essential to minimize the amount of time spent within circuitry. Extremely high speed has inherently a negative impact on high resolution; and therefore multideflection architecture is presently adopted in most EBL systems. Several kinds of deflection circuits are required in an EBL system, such as the main deflector, the subdeflector, the shaping deflector, and the cell-selection deflector. There may be high-speed DACs in the field of consumer electronics. However, those DACs cannot be applied to EBL systems because the required specifications are quite different. In consumer electronics, the bit numbers are 8–10, and the output voltage is 0–1 V. The characteristics of the DACs used in EBL are: (1) the large number of bits, 12–14, even in the subdeflector, (2) bipolar output voltage larger than ± 10 V, and (3) their short settling time. There are no commercially available DACs to meet such demands. The available DACs now are summarized in Table 4.3 which shows the present technical level [34].

One of the bottlenecks to the higher writing speed is settling time in case of VSB system. At the present stage, the settling time between beam shots is about 100 ns. The typical shot numbers of a 130-nm mask are said to be 10 G shots. The total settling time will become 4000 s in a four-pass writing. This time is shorter than resist exposure time of 10,000 s in case of the resist sensitivity of 10 μC/cm^2 and the current density of 10 A/cm^2. However, in case of a 70-nm mask, the shot numbers will become 100–200 G shots due to the OPC. The total settling time will exceed 40,000–80,000 s, 11–22 h. It will be 4–8 times longer than resist exposure time. To develop the settling time of 10 ns or so, the transistor device itself,

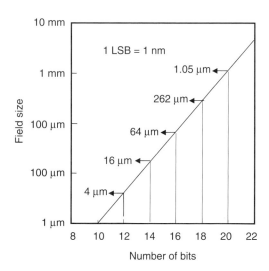

FIGURE 4.20
Deflection distance vs. bits number of DACs.

E-Beam Mask Writers

TABLE 4.3

The Specifications Available for DACs

Number of Bits	Output (V)	Linearity	Settling Time	Applications/Load (PF)
14 bit	± 30	< ± 0.5 LSB[a]	50 ns	Shaping deflector/∼ 10
16 bit	± 30	< ± 0.5 LSB	50 ns	Sub-deflector/∼10
18 bit	± 125	< ± 0.5 LSB	500 ns	
20 bit	± 320	< ± 0.5 LSB	10 μs	Main deflector/∼100

[a] LSB: least significant bit.

used in DAC circuit and its assembling or packaging technology, must be developed. The demands for the EBL systems are increasing in the industry, and it may be recognized that extremely high-speed DAC circuits should be developed by collaboration.

4.5 Technology for Image Placement Accuracy

Mask requirement of IP for 90-nm node are 19 nm. Keeping the IP within 19 nm in a mask plate with such a large area as 6 in.^{2} (150 mm^2) seems to be very difficult for the EBL system. In addition to the EBL system, there are many IP error factors; the blank itself, the measurement tool, and the supporting method. They are summarized in Table 4.4. In such a small IP error region, the error analysis needs special care. For thin mask blanks like NGL stencil mask, IP accuracy is influenced by the special factors, such as membrane stress control, high aspect trench etch, stencil pattern distortion, deep or large area silicon etch. The mask support will produce a distortion due to the mechanical clamping and gravity force. The difference between mask writer and IP measurement tool supporting methods creates IP errors.

Most EBL systems recently adopted the minimum address unit of 1 nm or less in their design. From this point of view, there are no problems in IP errors. The IP errors in EBL will basically occur due to the relative position change between e-beam and the substrate. The position change will occur by (1) residual correction error of deflection distortion in

TABLE 4.4

Image Placement Error Factors and its Reduction Method

Process	Factor	Method
EB writer	Writing accuracy	Objective lens system improvement
		Stage structure improvement
		Temperature balance
		Mask flatness
Blanks	Blanks flatness	Blanks flatness improvement
	Cr tensional force	Low stress blanks
		Film coating conditioning
Measurement method	Differences of standard point and measurement process	Adoption of stepper method
Measurement tool	Measurement accuracy	Distortion correction
	Cassette cramping	Tool management
		Temperature environment control
		Mask cramping method

electron optics, (2) stage and chamber mechanics, (3) temperature imbalance, (4) mask distortion by outer force, and (5) fluctuation of magnetic field inside the electron optical column. Here, the technologies from (1) to (4) will be discussed. Technology (5) has been discussed in earlier literature [35].

4.5.1 Stage and Chamber Mechanics

The mechanical system is composed of the specimen stage and the chamber mechanics in the vacuum. The mechanical system is a very important subsystem in performance and in cost of the EBL system. The stage positions x and y are measured by laser interferometer at the unit of less than 1 nm. However, there are many sources which can deteriorate the measured position.

4.5.1.1 Stage and Writing-Chamber Materials

For constructing the mechanical systems, only a limited number of materials are available. The stage materials must be nonmagnetic because the e-beam would be deflected by the magnetic field change from moving magnetic materials. The stage also is to be made of non-out gassing materials because it is in a high vacuum. In continuous writing mode, the stage moves in the stray magnetic field from the objective lens; and hence to avoid the eddy current [36] generated in the stage, the top stage materials are not allowed to be electrically conductive [37]. On the contrary, the stage surface must be made of electrically conductive materials to avoid charging from the incident e-beam. In some systems, metal-coated ceramics are used in the top stage [38].

Even if the temperature of the system is kept constant, within 0.01°C, the mechanical system will expand or shrink about 100 nm if it is made of iron. This shows that a small change in temperature can degrade the IP accuracy. The writing chamber of the recent EBL system is made of low-thermal expansion materials, such as an alloy of iron and nickel [39]. The thermal-expansion coefficient of this alloy is smaller than 1/10 of stainless steel. This material plays a role of shielding effect of the stray magnetic field.

4.5.1.2 Chamber Structure

The main structure of an EBL system consists of an electron optical column and a vacuum chamber. The mechanical vibration comes from the rotation of the vacuum pump, the impact force of opening and closing valves, floor vibration, and stage movement. Mechanical vibration is transmitted from the vibration sources to the column structure. The vibration sources excite the natural modes. The modes generate a relative displacement between the beam and the substrate. The beam positioning error is caused on the substrate. It is difficult to reduce the vibration effect because the behavior is often transient. To clarify the natural frequency and modes of such a structure, the modal analysis of the system is done by assuming the geometrical structure and materials of the column and the chamber.

The typical natural modes are shown in Figure 4.21. The first mode is column swing in the X or Y direction. The second mode is vertical vibration of the column in the Z direction. And the third mode is pitching the rotation of the column around the X or Y axis. These modes except for the second mode, can affect the optical path of the beam. When the mechanical vibration is transmitted to the structure, the structure is often excited at its own natural frequencies. At this moment, the beam positioning error is caused on the substrate. Although the vibration amplitude and damping time of the structure differ at each frequency, the actual beam vibration is a mixture of each beam vibration.

The error effect has been decreased by stiffening the EBL structure. The best structure is decided by the modal analysis for a combination of several structures [40,41]. The first

E-Beam Mask Writers

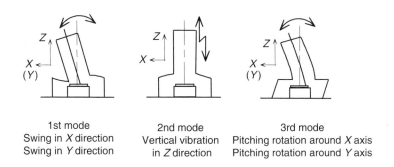

FIGURE 4.21
Typical natural modes on electron optical column motion.

is the optimization of the top plate thickness of the stage chamber. The second is the rib structure applied to the bottom of the column, such as square rib, radial ribs, and a cylinder with a larger diameter.

The stiffening the structure makes the system heavy and enormous. So, the real time correction method for residual mechanical vibration is also required [42].

4.5.1.3 Stage Structure

The structure of the conventional X–Y stage has a three-stage structure. It consists of the base, X table, and Y table. The driving forces are directly added to X table and Y table and cause stage deformation. There is a problem that the distortion of the Y stage influences bar mirror for the stage position measurement in the vacuum. The effect of the top table is shown in Figure 4.22. Therefore, the stage position is measured with an error of 20–30 nm. To avoid this distortion, a top table is added on the Y table so that the distortion might not influence a bar mirror [43]. The amount of deformation could be reduced to less than 10 nm.

The mechanical system will play an even more important role for highly accurate EBL systems in the future. A stage with no mechanical contact, using air bearing in a vacuum may be developed.

4.5.2 Temperature Balance

Temperature control of the EBL system is very important [44]. The thermal-expansion coefficient of quartz glass is $0.35 \times 10^{-6}/°C$, which is a very small number. However, a 6-in. glass plate will expand 5.3 nm if the temperature rises 0.1°C as shown in Figure 4.23.

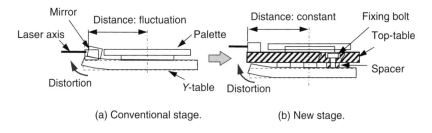

FIGURE 4.22
X–Y stage structure. The conventional X–Y stage was a three-stage structure of the base, the X table, and the Y table. A top table is added on the Y table in the new stage so that distortion might not influence a bar mirror. (From Ref. [43].)

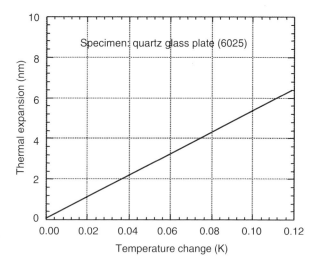

FIGURE 4.23
Temperature change versus thermal expansion.

To avoid room temperature changes, the EBL system itself is usually installed in a subchamber with a constant temperature. The temperature control system by a liquid flow, such as water-cooling, is adopted in objective lens, stage, writing chamber, and loader unit. So, a temperature change in the mask is mainly caused by (1) the initial temperature difference between the plate and the stage, (2) the heating effect induced by the e-beam exposure, (3) the friction energy due to stage motion, and (4) the adiabatic expansion during fast evacuation. If the mask temperature before writing is severely different from the stage temperature, the pattern writing is usually waited out until the temperature regains the equilibrium. To compensate for the temperature rise or drop due to these factors, some EBL systems are equipped with a heater in the load chamber. Thus, the temperature of the specimens is controlled within $\pm 0.01°C$.

4.5.3 Mask Flatness

The supports of a mask will produce distortion due to the mechanical clamping and gravity force [45,46]. Generally, the mask plate is supported at three or four points. The 6025 quartz mask, that is, 150 mm^2 with 6.4 mm thickness, has the weight of 150 g. If the mask is supported at three points, the mask surface will bend down about 0.2 μm due to the gravity force. If the pattern is written in this concave mask, the IP error after flattening will be about 26 nm. If the height map is known in advance, this error can be corrected [47]. Many of the next generation masks have a membrane or stencil structures. So the surface distortion is created easily by the clamping force. For the NGL masks, the cassette management between the tools becomes very important.

4.5.4 Superpose Writing

Stitching is inevitable in EBL because the field size is usually smaller than the chip size. Although the stitching error is essentially an IP error, it sometimes causes CD error when the patterns exist at field and subfield boundary. The IP accuracy is related to many factors, such as the deflection distortion error, the optical column vibration originating from the stage movement and floor vibration, the beam drift by charging up, and so on. A system designed to minimize the above errors is very essential to obtain high IP

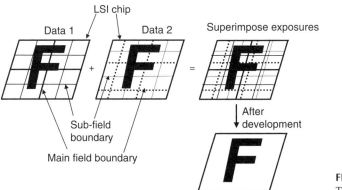

FIGURE 4.24
The principle of multiple exposures.

accuracy. Here, another method is discussed to obtain high accuracy of the field and subfield stitching.

To reduce random error occurring during e-beam writing, multiple exposures turned out to be a very effective method. Originally, the method was tried to reduce the effect of the flicker noise of the beam current. In this experiment, the superposition of 20 times was tried, and the improvement of CD accuracy was proved [48]. Then, the superpose-exposure of four fields combined with the half-field stage movement turned out to reduce the field-stitching error and to improve the IP accuracy [49]. In 1993, Ohki et al. [50] found that the multiple exposures statistically blur the field stitching. In the experiment, the field and subfield sizes are varied so as to expand the seamless connection of fields to any level of deflection and obtain the best results. The principle is shown in Figure 4.24, where the main- and subfield boundaries are varied. In this example, the main- and sub-field boundaries consist of two different e-beam pattern data transcribed from the same CAD data. In the experiment, four sets of e-beam data were prepared for the same chip with the main- and subfield size combinations of (500, 25), (700, 28), (900, 18), and (1000, 25) μm, respectively. For the sake of simplicity, the field size is usually assumed to be an integer multiple of the subfield size. In this experiment, the stitching accuracy improved to about twice that obtained with the conventional exposure method. In practical application, completely seamless connections are not attained. The field and subfield sizes are mostly common. The subfield positions are shifted in the field. Even in such cases, multiple writing improves the stitching accuracy to a ratio of 60–70% to that obtained by the conventional single exposure.

In this method, the dose of each exposure is determined simply by dividing the optimal dose by the number of multiple exposures. One of the drawbacks of this method is the increase in writing time. Another is that several sets of e-beam data should be prepared so that the field boundaries do not coincide with each other. However, most commercially available systems adopt the multiple exposure methods [51–53].

4.6 Technology for Critical Dimension Accuracy

Mask requirements of CD accuracy are 4.2 nm for isolated line and 13.7 nm for linearity in 90-nm node in 2004. Keeping the CD of such small values in a mask plate with such a large area as 6 in.^{2} (150 mm^2) seems to be very difficult for EBL system.

Most e-beam mask writers adopt high acceleration voltage of 50 kV. The sensitivity of resist goes down with the acceleration voltage. CAR usually has high sensitivity for e-beam. The CARs need postexposure baking, where the temperature uniformity is very important, and the process control within a mask becomes more and more difficult. These are discussed in chapter 15 of this handbook. In this chapter, the CD error factors caused by EBL system will be discussed.

The CD error factors due to EBL system are: (1) beam size correction error, (2) electron-scattering effect in substrate material, (3) electron multiple-scattering in optical column, and (4) stitching error. The electron scattering in substrate material causes the proximity effect. The proximity effect deteriorates CD uniformity and linearity. The electron multiple-scattering in optical column affects the uniformity within mask. Here, the technologies from (1) to (3) are discussed. The stitching error is discussed in Section 4.5.4.

4.6.1 Beam-Shaping Function

The CD error using a VSB occurs from beam shot stitching error. That is caused by beam rotation errors and beam size errors. The shaped beam is observed using backscattered electrons from a fine heavy-metal particle on a Be or Si plate [54], or transmitted electrons through a small hole [55]. Using this signal, the rotation and the size can be measured [56]. When a shaped beam is rotated, the line profile obtained by beam scanning will incline at the side edge. The aperture rotation is adjusted so that the line profile of the side edge becomes uniform. The rotations of the first and the second shaping apertures can be controlled to within ± 1 nm in the beam size for a 1-μm^2 beam.

Two problems in rectangular beam method become noticeable in pattern writing at less than 100 nm. One is in oblique pattern writing. The oblique patterns have to be divided into many small rectangular patterns. The CD accuracy of the oblique patterns becomes worse because the pattern is "stair shaped." Moreover, this decreases the throughput. Another is the uniformity of the current density in beam size change. It is difficult for the two deflection centers to mechanically coincide with each other. The change of the current density and nonuniformity in the VSB occur from noncoincidence with two shaping deflection centers of the width and the height.

4.6.1.1 Triangle Beam

The triangular-shaped beam, which is one example of an oblique-shaped beam, has been applied to delineate oblique patterns [57,58]. Hattori et al. [57] successfully demonstrated the improvement of edge roughness. This method uses a home base shaped aperture as the second mask. In this method, the pattern flexibility is limited because it is only applied to 45° patterns as shown in Figure 4.25(b).

Another triangular beam method is a modification of the cell-projection method [59]. The square and triangular apertures are present in the second mask, and the e-beam is shaped by these apertures. The electron optical column used here is shown in Figure 4.10(b). The image of the second mask is demagnified to 1/25 and projected onto the substrate. Figure 4.25(a) shows an example of the second mask. This mask consists of four units. Each unit has one square aperture and four rotated triangular apertures. Triangles of 30°, 45°, and 60° are present and can be projected onto the substrate. These angles of oblique patterns can be delineated. Other angles of triangles are delineated by making the appropriate angle of apertures. The length of the square is 62.5 μm on the mask. Thus, the maximum beam size is 2.5 μm on the substrate. The method of beam formation is different from the CP method where the cell aperture is completely covered by the image of the first aperture. In shaping triangular beams, the e-beam is deflected by the

E-Beam Mask Writers

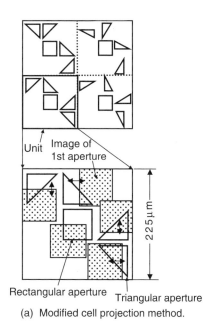

FIGURE 4.25
The second mask for the triangular shaped beam.

shaping deflector and partially irradiates the triangular aperture in the second mask. The beam blurring and the shaped beam distortion are negligibly small for 2.5-µm beam on the substrate because the deflection distance is small compared with the cell selection.

4.6.1.2 Dual Shaping Deflector

The ability relevant to shaping beam size has a great impact on its line width or CD accuracy. Shaping deflectors have been employed in the VSB systems in order to generate a desired size of e-beam. In order to keep the current density constant when the beam size changes, the first projection lens forms the crossover image at the center position of the shaping deflector as mentioned in Section 4.4.2. To reduce an aberration caused within the shaping lens system, the dual quadrupole electrostatic shaping deflector has been utilized [60]. The conventional shaping deflector is just a single quadrupole electrostatic one. When the electron optical crossover positioned in the vertical center of the deflector is a little off from the point where it should be, the e-beam path under the second mask-aperture gets further deviated from the axis. Then, the aberration is caused by the off-axis magnetic field in the second projection lens. In contrast, the dual shaping deflector consists of a pair of deflectors vertically placed as seen in Figure 4.13(b). It is capable of adjusting the crossover position by changing the applied voltage between the first and second deflectors and it can fix the e-beam path. The applied voltage (w_2, H_2) to the second deflector is slightly different from the applied voltage (w_1, H_1) to the first deflector. The relationship is shown in Figure 4.13(b).

4.6.2 Proximity Effect Correction

The proximity effect occurs from the scattered electrons in the resist and substrate as mentioned in Section 4.3.2.3. The scattered electrons expose some resist outside of the intended exposure area as shown in Figures 4.7 and 4.8. These electrons degrade the CD uniformity and the CD linearity for the patterns less than 2 μm on the substrate. The deposited energy distribution is given in Equation (4.3) as the well-known double Gaussian expression. The first term is forward scattering in the resist, and the second term is the backward scattering from the substrate:

$$f(r) = \frac{1}{1+\eta}\left\{\frac{1}{\beta_f^2}\exp\left(-\frac{r^2}{\beta_f^2}\right) + \frac{\eta}{\beta_b^2}\exp\left(-\frac{r^2}{\beta_b^2}\right)\right\} \quad (4.3)$$

where r is the distance from the incident point, β_f the forward-scattering range, β_b the backward-scattering range, and η is the ratio of the backward-scattering energy to the forward-scattering energy. The forward- and the backward-scattered electrons both contribute to the proximity effect. The proximity effect is corrected by using dose correction, shape correction, equalization of background dose, and multilayer resist techniques. These methods are summarized in Table 4.5.

4.6.2.1 Dose Correction Method

The deposition energy due to pattern exposure is obtained by integrating Equation (4.3) over the pattern. The result can be expressed as a combination of the error functions. Correcting the effect usually requires time-consuming calculations. For efficient computation, several algorithms have been developed. One of them is the strategy for pattern partitioning by Parikh [61] that is summarized as follows:

1. PECs are attempted on a given pattern.
2. The pattern quality is assessed at numerous sample points throughout the patterns.
3. If the pattern quality fails to satisfy certain "goodness" criteria at particular sampling points, such points and the associated regions are subdivided from the rest of the shape.
4. Proximity correction is attempted again on this partitioned pattern.
5. This procedure can be repeated until the quality criteria are satisfied.

Every surrounding pattern has interaction with each other. In this calculation method, the number of the combinations becomes numerous in actual LSI pattern, and the calculation time becomes too long.

TABLE 4.5

PEC Methods

Method	Principle	Characteristics
Dose correction	Exposure doses are corrected for all patterns so that the accumulated energy becomes constant	• Long calculation time • Recent development of the dedicated fast hardware enable calculation time shorter
Shape modification	Shape sizes are corrected for all patterns so that the accumulated energy becomes constant	Long calculation time
GHOST exposure	Equalize the dose in non-pattern area by reverse pattern exposure by blurred beam	• Throughput drop due to the normal and the reverse pattern exposure • Narrowing the process latitude
Multi-layer resist	Thick bottom-layer of light materials absorbs backscattered electrons from the substrate	Process cost up

E-Beam Mask Writers

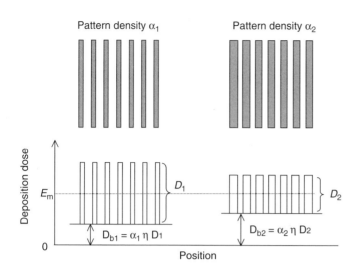

FIGURE 4.26
Determination of the optimum doses.

A high-energy e-beam simplifies the correction process. The forward- and backward-scattering ranges β_f and β_b of 50 kV electrons are about 0.05 and 10 µm in silicon substrate, respectively. This indicates that the effect of forward scattering is negligibly small, and the effect of backward scattering within a small region of about 5 µm is roughly constant. If the pattern size is sufficiently smaller than β_b, the details of the pattern become irrelevant but the global information is relevant. The simple approaches are the pattern-area-density technique [62] and the representative-figure technique [63].

The principle of the pattern-area-density technique one is shown in Figure 4.26. Let us suppose two-pattern-area density regions, α_1 and α_2. The incident doses are D_1 in region α_1 and D_2 in region α_2. The deposited doses that come from backscattered electrons from nearby patterns are $D_{b1} = \alpha_1 \eta D_1$ and $D_{b2} = \alpha_2 \eta D_2$, respectively. At an exposed point, the total deposited dose is the sum of the incident dose and the backscattered dose. If the average dose E_m becomes constant at any pattern area density as shown in Figure 4.26, the next relationship will be satisfied:

$$E_m = D_{b1} + D_1/2 = D_{b2} + D_2/2 \tag{4.4}$$

From this equation, the exposure dose D at the pattern density α must satisfy the following relation:

$$D = D_0/(1 + 2\alpha\eta) \tag{4.5}$$

where D_0 is pattern density-independent constant that is the dose in the zero pattern density, namely, isolated line pattern.

This technique is implemented in hardware consisting of area-density map memory, fast multipliers, and arithmetic and logic units. The hardware and the data flow are shown in Figure 4.27. At first, the virtual exposure is executed. The LSI patterns are partitioned with a mesh of fixed size, where the backscattered energy is constant. The pattern area density is then calculated for each small region. After smoothing the area densities in two dimensions, these data are stored in the map memory. For actual

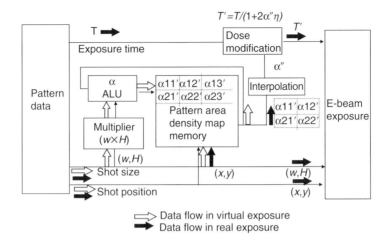

FIGURE 4.27
Configuration of correction hardware.

exposure on substrate, each shot time T' at the pattern density α is modified from T_0 according to Equation (4.6). T_0 is the shot time at the zero pattern density that is the isolated pattern. Since this correction is processed during the exposure, the throughput is not reduced.

$$T' = T_0/(1 + 2\alpha\eta) \qquad (4.6)$$

In this simple method, errors occur due to drastic change of pattern density as shown in Figure 4.28. For the line pattern with designed CD of 150 nm, simulated CD shows 18 nm in range by the above first-order correction. The error is considered due to sudden change of

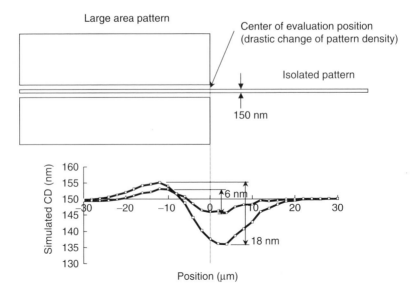

FIGURE 4.28
Improvement of the proximity effect.

pattern density at the end of the large area pattern. In order to reduce such an error, higher-order correction scheme has been developed [64]. In the above case, the corrected shot time $T'(x)$ is calculated with Equation (4.11) if $Q_1(x) = \alpha(x)$, where the shot time is adjusted depending on the pattern density α. In case of new PEC, Q_1 is given by Equations (4.7)–(4.11) to consider the second-order correction using approximate equation:

$$Q_0(x) = \int f(x - x')\alpha(x')\,dx' \tag{4.7}$$

$$Q_{11}(x) = \alpha(x)/(1 + 2\eta Q_0(x)) \tag{4.8}$$

$$Q_{12}(x) = \int f(x - x')Q_{11}(x')dx' \tag{4.9}$$

$$Q_1(x) = (1 + 2\eta Q_0(x))Q_{12}(x) \tag{4.10}$$

$$T'(x) = T_0/(1 + 2\eta Q_1(x)) \tag{4.11}$$

In order to obtain the new dose map $Q_1(x)$, $Q_{11}(x)$ as intermediate map 1 and $Q_{12}(x)$ as intermediate map 2 are successively calculated from the above dose map $Q_0(x)$. These calculations are processed with hardware circuit controlled by digital signal processor (DSP). The most advantageous use of this new PEC scheme is that the similar basic hardware platform for conventional method by Equation (4.6) can be used. Owing to the parallel processing function for the stripe exposure and the PEC map preparation in the subfield reconstruction process, no overhead time for this new PEC dose map preparation is needed once the stripe exposure starts, and the high throughput is maintained. In Figure 4.28, the error due to sudden change of pattern density is reduced from 18 to 6 nm by this method.

4.6.2.2 Shape Modification Method

The pattern shape modification method [65] attempts to compute dimensions of exposed shapes such that the shapes developed in the resist will have the designed dimension. Almost the same, long, computation time with dose correction is necessary to calculate the required shape modification. The magnitudes of the required shape changes are small fractions of the minimum feature dimension within the pattern. As a result, fine address spacing is essential and even minimum features must be composed of several scans of a small beam. Though this method has become less popular, it takes notice of the mask of EPL because dose correction method is not applicable there [66].

4.6.2.3 GHOST Method

This method is proposed by Owen and Rissman [67] based on the equalization of background dose. In this method, the reverse patterns are exposed with a defocused beam in order to equalize the dose of nonpattern area. In a raster scanning system, this method may be used because there is no need for complex calculation. However, in vector scanning system, it is not easy to generate the reverse pattern. One of the drawbacks of this method is the necessity for a second exposure that results in low throughput. The narrowing of process latitude is another drawback of this method.

4.6.2.4 Multilayer Resist Method

This method is a kind of processing technique [68]. Under the thin top layer, there is a relatively thick layer composed of low atomic number materials. The e-beam exposes the top resist. The second layer significantly reduces the proximity effect on the top layer resist. Also, no lengthy and costly computation is required. The only issues are considerable increase of the processing cost and time. Thus, this method is used in special applications such as nanometer device development.

4.6.3 Multiple Scattering Effects in Optical Column

It is pointed out that the backscattered electrons cause severe dose variations over the x-ray membrane mask and degrade the line-width accuracy [69]. It was also pointed out that the multiple scattered electrons have the same effect [70] in normal mask. The first experimental set-up is shown in Figure 4.29. The incident electrons are focused onto a Si wafer coated with resist. Three electron-scattering materials under the objective lens were investigated. A copper, aluminum or carbon plate was set at the lower plane of the objective lens. The electrons were injected directly at one point on the wafer coated with 0.6 μm PMMA. The dosage corresponds to the exposure dose for a 5-cm square. After developing the PMMA resist, the interference patterns were observed. In this experiment, the exposed area was the region with a radius of about 15 mm. The interference patterns in Figure 4.30(a), (b), and (c) correspond to a copper, aluminum, and a carbon plate, respectively. The resist thickness measurement showed that the maximum quantity of resist is removed in the case of a copper plate. For the carbon plate the quantity was the smallest. Figure 4.30(d) shows a carbon cage experiment. Here, the same amount of electrons was irradiated into a small carbon cage in the center of the wafer, and the copper plate was used as a scattering material. The backscattered electrons were completely trapped, and there was no observed resist thickness reduction.

Different from the proximity effect, these electrons produce a long-range background exposure in the resist and they are now called fogging electrons. This effect is not severe in the case of low-density pattern like a hole pattern. However, global CD variation due to this effect becomes a few tenths of a nanometer when the pattern density becomes large. It has seriously affects the minimum feature size of 100 nm or so. There are two methods to reduce the line-width accuracy by fogging effect. One is the adoption of scatteringless structure, and the other is the dose correction depending on the pattern density. Most recent EBL systems combine these two methods.

4.6.3.1 Antireflection Plate

Selecting the material of an antireflection plate made of low atomic number can reduce the multiscattered electrons. Carbon or beryllium has been proposed because the backscattering coefficient is low, and adopting a scatteringless structure of its surface also has

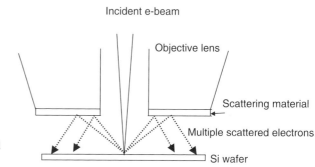

FIGURE 4.29
Experimental setup for multiple scattered electrons.

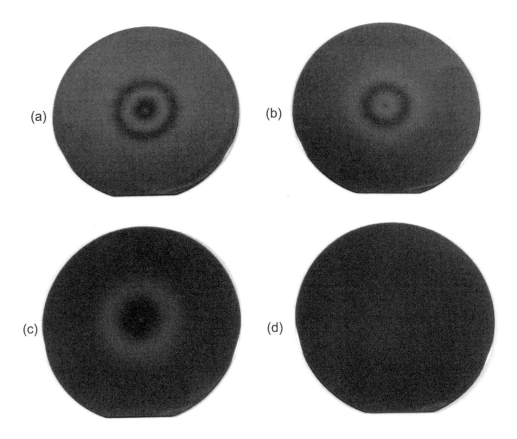

FIGURE 4.30
The interference patterns from the exposed residual resist by multiple scattered electrons: (a) copper, (b) aluminum, (c) carbon, and (d) cage. The wafer diameter is 75 mm. (From Ref. [70].)

been proposed like a rugged surface. Recently, a more effective structure was proposed that is a parallel-hole array structure with a honeycomb pattern as shown in Figure 4.31 [71]. The reflectance remains about 42% of that of a plane structure. Foggy electron absorber reduces global CD uniformity down to 10 nm.

4.6.3.2 Fogging Effect Correction

The correction system [72,73] of the fogging electron exposure effect is equipped with almost all EBL systems at present. Residual dose error by foggy electron is derived using exposure area calculation, and this dose correction is mixed with PEC. A fogging effect correction (FEC) table is made by actual measurement of fogging caused by a standard beam. Then, an exposure area map is calculated from a given mask pattern. Then, a fogging effect distribution map is deduced. At the time of writing, pattern exposure time is modified to make uniform pattern dosage. This scheme is similar to that of PEC. As the fogging distribution is broad and simple, the correction cell size is as large as 1 mm or so. In case of PEC, the cell size needs to be as small as 2 μm in case of 50 kV e-beam.

In addition to the EBL system, the CD errors also come from the resist process. The patterns are different from each mask in size and area. The resist developing and etching process introduces loading effect, which depends on the patterns. The loading effect causes global CD uniformity error. The loading effect is also corrected using this FEC

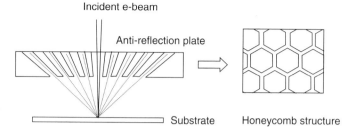

FIGURE 4.31
Structure of the antireflecting plate. (From Ref. [71].)

system. It is estimated that the global CD accuracy of 6 nm in 6025 mask is obtained using this correction method.

4.7 Data Preparation

The data preparation is discussed in chapter 2 of this handbook. Here, only the main function of the data preparation will be stated and the issues are also mentioned.

The data flow in an e-beam system is shown in Figure 4.32. The pattern data is the output from the CAD system in one of the various types of format, such as GDSII, STREAM, CALMA, and so on. They are converted to the e-beam data and the inspection data. The data format is different for each system. The main functions of the data conversion software are decomposition of the basic patterns, overlap elimination, resizing, PEC, mirror-pattern preparation, tone reversal, and so on. Since the conversion process is so complex and the data volumes so large, a very high-speed processing is required. To handle a large number of LSI patterns quickly, a system of hierarchical processes [74] is usually constructed.

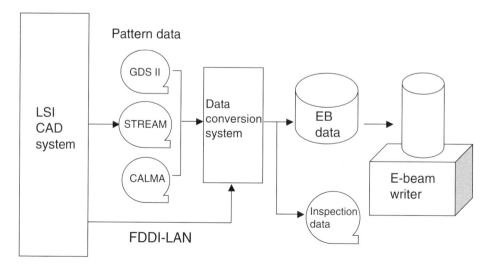

FIGURE 4.32
Data flow in e-beam lithography system.

E-Beam Mask Writers

After the generation of 130-nm node, various resolution enhancement technologies are developed to prolong the life of optical lithography. In addition to shortening the optical wavelength, phase-shifting technology and optical PEC technology have been developed. In the phase-shifting technology, the volume of the pattern data is generally growing, and the number of masks will be doubled. The optical proximity effects are the phenomena when exposed patterns are deformed from the original mask patterns. The effect is corrected by adding small bias patterns to the original patterns. The data volume is growing exponentially due to the OPC. For example, the data volume for the correction grid size of 2 nm increased by about 20 times larger compared with non-OPC pattern data.

The storing and transferring such large data volumes will be the serious issues in mask writer systems. To avoid this problem, applying the capabilities, both adapting CATS intermediate file and the storage area network (SAN) data processing architecture [60], has known to shorten the total time consumption for an exposure by about 40% compared to the predecessor. The SAN architecture will enable a fast transferring within the system. With the conventional mask writers, users had to wait for the completion of transactions both transferring and duplicating the large machine format data before starting data processing and exposure. Once the vast volume of data in the machine format went through the predecessor system, the entire transactions of transferring and processing got stuck, and the user had to wait for the termination of the drawing transaction. In contrast, within the system shown in Figure 4.33, the PC cluster converts data in parallel and transfers the processed data to the system within the speed of 1 Gbps. The system can even transfer and process data simultaneously while implementing an exposure. This means the system is virtually able to draw an unlimited volume of pattern data. SAN architecture utilizes a fiber channel technology and it allows the system to transfer data with the speed of 1 Gbps.

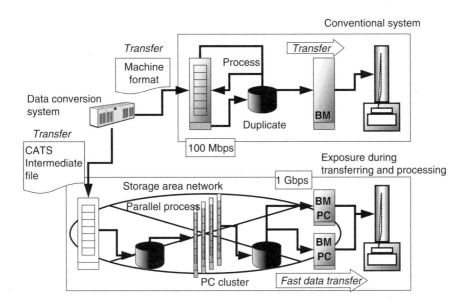

FIGURE 4.33
Schematic diagram of transactions within SAN data processing architecture and the conventional system.

4.8 Commercially Available Systems

Table 4.6 shows the specifications of commercially available latest model from each company in the year 2003. The three systems are vector scanning and one is a raster-scanning system. The accelerating voltage becomes 50 kV in all systems. The address unit is down to 1 nm in all systems, from the previously required 10–20 nm. All systems adopt shaped-beam strategies. Most of the electron guns except MEBES adopt the thermionic LaB_6 cathode, and the maximum beam sizes are of a few micrometer square. The MEBES system, however, uses thermal field emitter Zr/W/TFE, and the maximum beam size is as small as a quarter micrometer square. The announced CD accuracy and IP accuracy are better than 5 and 12 nm, respectively. These performance values may be thought of as just the references because they are not taken under the same conditions and the resist processes are different.

4.9 Summary

EBL will play more and more crucial role in mask making. The commercially available EBL systems at present are announced to meet the requirements for 90-nm node production and 65-nm node developments. The next target system should meet the requirements for 45-nm node, where the CD and IP accuracy and the throughput are unbelievable

TABLE 4.6

Commercially Available E-Beam Mask Writing Systems for 90-nm Node

	Hitachi Hitechnologies	New Flare Technology	JEOL	ETEC
Latest system	HL-7000M	EBM-4000	JBX-3030MV	MEBES RSB
Writing method				
Electron gun	LaB_6	LaB_6	LaB_6	Zr/W/TFE
Beam shape	VSB+CP	VSB (triangular)	VSB (triangular)	VSB
Scanning method	Vector	Vector	Vector	Raster
Stage motion	Continuous	Continuous	Step and repeat	Continuous
Multiple exposure	2–4 passes	4 passes	2 passes	4 passes
Acceleration (kV)	50	50	50	50
Address unit (nm)	1	1.25/1	1	1
Max mask size (in^2)	178	230	178	152
Performances				
Local CD (nm)	3.7 (3 σ)	3.7 (3σ)	8 (3σ)	4 (3σ)
CD linearity (nm)	6.6 (p-p)	—	15 (p-p)	14 (range)
Global IP (nm)	11 (3σ)	8.6 (3σ)	12 (max)	12 (3σ)
Source	PMJ 2003[a]	EIPBN2003[b]	PMJ 2003	To be announced

[a] PMJ 2003: Presented at Photomask Japan 2003 at Yokohama.
[b] EIPBN2003: Presented at EIPBN 2003 at Tampa.

values. ITRS technology road map shows that a red brick wall stands in the way of development, which means that the manufacturing solutions are not known. In order to achieve the specifications, there are many technical challenges in the field of electron optics, mechanical systems, electrical-control systems, and data handling. The components in electron optics are the electron gun, shaping lens system, and objective lens system. It will increase more and more of the EBL system's cost. The current systems are single-beam systems. In the future, multibeam systems may be developed. In the EB writers mechanical system it will become very important to create ideal operating environments with no temperature fluctuation, vibration, or electrical noise. New loading and chucking systems will be necessary to handle the fragile membrane or stencil masks for NGL. In the electrical control system, the high speed DACs with low electrical noise also will be required. As for the data, the biggest issue is the handling of the exponentially growing number of data volumes.

It may be that these challenges will be approaching their limits. The solution will need to be explored in conjunction with processes.

References

1. International Technology Roadmap for Semiconductors, 2002 Update Conference, Dec. 4, 2002, Tokyo International Forum, Japan.
2. R.K. Matta, D. Green, and M.W. Larkns, The use of the scanning electron microscope in the fabrication of an integrated circuit, in: *ECS Spring Meeting, Extended Abstracts of Electrothermics and Metallurgy Div.*, vol. 3, No. 1, 1965, pp. 32–33.
3. Y. Tarui, S. Denda, H. Baba, S. Miyauchi, and K. Tanaka, An electron beam exposure system for integrate circuits, *J. Electro Commun. Soc.*, 51-C (2), 74–81 (1968) (in Japanese).
4. D.R. Heriotte, R.J. Collier, D.S. Alles, and J.W. Stafford, EBES: a practical electron lithographic system, *IEEE Trans. Electron Device*, ED-22, 385–392 (1975).
5. B. Liebman, Quality assurance procedures for MEBES, *J. Vac. Sci. Technol.*, 15 (3), 913–916 (1978).
6. E. Goto, T. Soma, and M. Idesawa, Design of a variable aperture projection and scanning system for electron beam, *J. Vac. Sci. Technol.*, 15 (3; May/June), 883–886 (1978).
7. H.C. Pfeiffer, Variable spot shaping for electron-beam lithography, *J. Vac. Sci. Technol.*, 15 (3; May/June), 887–890 (1978).
8. M.G.R. Thomson, R.J. Collier, and D.R. Herriott, Double-aperture method for producing variably shaped writing spots for electron lithography, *J. Vac. Sci. Technol.*, 15 (3; May/June), 891–895 (1978).
9. M. Fujinami, T. Matsuda, T. Takamoto, H. Yoda, T. Ishiga, N. Saitou, and T. Komoda, Variably shaped electron beam lithography system EB55: 1. System design, *J. Vac. Sci. Technol.*, 19 (4; Nov./Dec.), 941–945 (1981).
10. Y. Nakayama, S. Okazaki, N. Saitou, and H. Wakabayashi, Electron beam cell projection lithography: a new high–throughput electron beam direct-writing technology using a special tailored Si aperture, *J. Vac. Sci. Technol. B.*, 8 (6; Nov./Dec.) 1836–1840 (1990).
11. N. Saitou, S. Okazaki, T. Matsuzaka, Y. Nakayama, and M. Okumura, EB cell projection lithography, in: *JJAP Series 4, Proceedings of the International MicroProcess Conference 1990*, 1990, pp. 44–47.
12. Y. Sakitani, H. Yoda, T. Todokoro, Y. Shibata, T. Yamazaki, K. Ohbitsu, N. Saitou, S. Moriyama, S. Okazaki, G. Matsuoka, F. Murai, and M. Okumura, Electron-beam cell projection lithography system, *J. Vac. Sci. Technol. B.*, 10 (6; Nov./Dec.), 2759–2763 (1992).
13. N. Yasutake, Y. Takahashi, Y. Oae, A. Yamada, J. Kai, H. Yasuda, and K. Kawashima, 'NOWEL-2' Variable-shaped electron beam lithography system for 0.1 μm patterns with refocusing and eddy current compensation, *Jpn. J. Appl. Phys.*, 31, 4241–4247 (1992).

14. K. Hattori, R. Yoshikawa, H. Wada, H. Kusakabe, T. Yamaguchi, S. Magoshi, A. Miyagaki, S. Yamasaki, T. Takigawa, M. Kanoh, S. Nishimura, H. Housai, and S. Hashimoto, Electron beam direct writing system EX-8D employing character projection exposure method, *J. Vac. Sci. Technol. B.*, 11 (6), 2346–2351 (1993).
15. T.H.P. Chang, M.G.R. Thomson, E. Kratschmer, H.S. Kim, M.L. Yu, K.Y. Lee, S. Rishton, and S. Zolgharnain, Electron-beam microcolumns for lithography and related applications, *J. Vac. Sci. Technol. B.*, 14 (6), 3774–3781 (1996).
16. H. Yasuda, S. Arai, J. Kai, Y. Ooae, T. Abe, S. Maruyama, and T. Kiuchi, Multielectron beam blanking aperture array system SYNAPSE-2000, *J. Vac. Sci. Technol. B.*, 14 (6), 3813–3820 (1996).
17. M. Muraki and S Gotoh, New concept for high-throughput multi electron beam direct write system, *J. Vac. Sci. Technol. B.*, 18 (6), 3061–3066 (2000).
18. A.V. Crewe, Some space charge effects in electron probe devices, *Optik*, 52, 337–346 (1978).
19. N. Saitou, Monte Carlo simulation for the energy dissipation profiles of 5–20 keV electrons in layered structures, *Jpn. J. Appl. Phys.*, 12 (6), 941–942 (1973).
20. N. Saiotu, S. Hosoki, M. Okumura, T. Matsuzaka, G. Matsuoka, and M. Ohyama, Electron optical column for high speed nanometric lithography, *Microelectronic Eng.* 5, 123–131 (1986).
21. F. Abboud, D. Alexander, T. Coreman, A. Cook, L. Gasiorek, R. Naber, F. Raymond, and C. Sauer, Evaluation of the MEBES 4500 reticle writer to commercial requirements of 250 nm design rule IC devices, *Proc. SPIE*, 2793, 438–451 (1996).
22. H. Takemura, H. Ohki, and M. Isobe, 100 kV high resolution e-beam lithography system, JBX-9300FS, *Proc. SPIE*, 4754, 690–696 (2002).
23. H. Satoh, Y. Nakayama, N. Saitou, and T. Kagami, Silicon shaping mask for electron-beam cell projection lithography, *Proc. SPIE*, 2254, 122–132 (1994).
24. L.W. Swanson and N.A. Martin, *J. Appl. Phys.*, 46, 2029 (1975).
25. A.N. Broers, *J. Vac. Sci. Technol.*, 16 (6), 1692 (1979).
26. N. Saitou, S. Ozasa, and T. Komoda, Variably shaped electron beam lithography system, EB55:II Electron optics, *J. Vac. Sci. Technol.*, 19 (4), 1087–1093 (1981).
27. ASET Annual report, Research Report in the Fiscal Year 1996, vol. 1, 1998, pp. 22–55 (in Japanese).
28. H.C. Pfeiffer, New imaging and deflection concept for probe-forming micro fabrication systems, *J. Vac. Sci. Technol.*, 12 (6; May/June), 1170–1173 (1975).
29. E. Munro, Calculation of the optical properties of combined magnetic lenses and deflection system with superimposed fields, *Optik*, 39, 450–466 (1974).
30. H.C. Chu and E. Munro, Numerical analysis of electron beam lithography system. Part III: Calculation of the optical properties of electron focusing systems and dual-channel deflection systems with combined magnetic and electrostatic fields, *Optik*, 61, 121–145 (1982).
31. Y. Takahashi, A. Yamada, Y. oae, H. Yasuda, and K. Kawashima, Electron beam lithography system with new correction techniques, *J. Vac. Sci. Technol. B.*, 10 (6), 2794–2798 (1992).
32. H. Ohta, Y. Sohda, and N. Saitou, Design and evaluation of an electron objective lens system with two lenses and two deflectors, *Jpn. J. Appl. Phys.*, 41, 4127–4131 (2002).
33. Y. Sohda, N. Saitou, H. Itoh, and H. Todokoro, An objective lens system for e-beam cell projection lithography, *Microelectronic Eng.*, 23, 73–76 (1994).
34. H. Furukawa, H. Ikehata, and T. Kikuchi, High performance analogue circuits of e-beam systems, in: *Abstract of 4th International Workshop on High Throughput Charged Particle Lithography*, p-5, 2000.
35. H. Ohta, Y. Someda, Y. Sohda, N. Saitou, S. Katoh, and H. Itoh, Double shielded objective lens for electron beam lithography system, *Proc. SPIE*, 3997, 667–675 (2000).
36. R. Spehr, Eddy currents in a wafer moving in the magnetic field on an electron lens, *Microelectronic Eng.*, 9, 263–266 (1989).
37. H. Tsuyuzaki, N. Shimazu, and M. Fujinami, High speed flat guide ceramic stage for electron beam lithography system, *J. Vac. Sci. Technol. B.*, 4 (1), 280–284 (1986).
38. N. Saitou, Electron beam lithography – present and future, *Int. J. Jpn. Soc. Prec. Eng.*, 30, 107 (1996).
39. Y. Hattori et al., Solution for 100 nm -EBM-4000-, *Proc. SPIE*, 4754, 697–704 (2002).

40. H. Matsukura, T. Tsutaoka, and K. Nakajima, Reduction in beam positioning error by modification of dynamic responses in electron beam direct writing system, *J. Vac. Sci. Technol. B.*, 8 (6; Nov./Dec.), 1863–1866 (1990).
41. H. Ohta, T. Matsuzaka, N. Saitou, K. Kawasaki, T. Kohno, and M. Hoga, Stitching error analysis in an electron beam lithography system: column vibration effect, *Jpn. J. Appl. Phys.*, 32, 6044–6048 (1993).
42. H. Tsuji, H. Ohta, H. Satoh, K. Nagata, and N. Saitou, Correcting method for mechanical vibration in electron beam lithography system, *Proc. SPIE*, 3096, 104–115 (1997).
43. T. Nakahara, K. Mizuno, S. Asai, Y. Kadowaki, K. Kawasaki, and H. Satoh, Advanced e-beam reticle writing system for next generation reticle fabrication, *Proc. SPIE*, 4066, 594–604 (2000).
44. H. Ohta, T. Matsuzaka, N. Saitou, K. Kawasaki, K. Nakamura, T. Kohno, and M. Hoga, Error analysis in EBL system – thermal effects on positioning accuracy, *Jpn. J. Appl. Phys.*, 31, 4253–4256 (1992).
45. R. Hirano, K. Matsui, S. Yoshitake, Y. Takahashi, S. Tamamushi, Y. Ogawa, and T. Tojo, Reticle flexure influence on pattern positioning accuracy for reticle writing, *Proc. SPIE*, 2512, 235–241 (1995).
46. S. Yoshitake, K. Matsuki, S. Yamasaki, R. Hirano, S. Tamamushi, Y. Ogawa, and T. Tojo, Analysis of pattern shift error for mask clamping measured by Nikon XY-3I, *Proc. SPIE*, 2512, 242–252 (1995).
47. T. Komagata, H. Takemura, N. Gotoh, and K. Tanaka, Development of EB lithography system for next generation photomasks, *Proc. SPIE*, 2512, 190–196 (1995).
48. K. Iwadate and T. Matsuda, Mask pattern fabrication by e-beam multiple exposure, in: *Extended Abstracts of 35th Spring Meeting of Japan Society of Applied Physics and Related Societies*, 30p-H-8, 1988 (in Japanese).
49. M. Asaumi and T. Yamao, Fine pattern fabrication by e-beam multiple exposures, in: *Extended Abstracts of 35th Spring Meeting of Japan Society of Applied Physics and Related Societies*, 30p-K-7, 1989 (in Japanese).
50. S. Ohki, T. Matsuda, H. Yoshihara, X-ray mask pattern accuracy improvement by superimposing multiple exposure using different field sizes, *Jpn. J. Apply. Phys.*, 5933–5940 (1993).
51. T. Tojo et al., Advanced electron beam writing system EX-11 for next-generation mask fabrication, *Proc. SPIE*, 3748, 416–425 (1999).
52. T. Komagata, Y. Nakagawa, N. Gotoh, and K. Tanaka, Performance of improved e-beam lithography system JBX-9000MV, *Proc. SPIE*, 4409, 248–257 (2001).
53. A. Fujii et al., Advanced e-beam reticle writing system for next generation reticle fabrication, *Proc. SPIE*, 4409, 258–269 (2001).
54. M. Nakasuji and H. Wada, *Abstract of Microcircuit Engineering Conference*, Cambridge, 1978.
55. Y. Someda, Y. Sohda, H. Satoh, N. Saitou, A new detection method for the 2-dimensional beam shape, *Proc. SPIE*, 3997, 676–684 (2000).
56. K. Suzuki, S. Matsui, and Y. Ochiai (eds.), *Sub-Half-Micron Lithography for ULSIs*, Cambridge University Press, Cambridge, 2000, pp. 136–137.
57. K. Hattori, O. Ikenaga, H. Wada, S. Tamamushi, E. Nishimura, N. Ikeda, Y. Katoh, H. Kusakabe, R. Yoshikawa, and T. Takigawa, Triangular shaped beam technique in EB exposure system EX-7 for ULSI pattern formation, *Jpn. J. Appl. Phys.*, 28, 2065–2069 (1989).
58. H.C. Pfeiffer, D.E. Davis, W.A. Enichen, M.S. Gordon, T.R. Groves, J.G. Hartley, R. J. Quickle, J.D. Rockrohr, W. Stickel, and E.V. Wever, EL-4, A new generation electron-beam lithography system, *J. Vac. Sci. Technol. B.*, 11 (6), 2332–2341 (1993).
59. Y. Someda, Y. Sohda, and N. Saitou, Triangular-variable-shaped beams using the cell projection method, *J. Vac. Sci. Technol. B.*, 14 (6; Nov./Dec.), 3742–3746 (1996).
60. M. Tanaka et al., Technological capability and future enhanced performance of HL-7000M, in: *The Proceedings of Photomask and Next Generation Lithography Mask Technology*, Proc. SPIE, 5130, 287–296 (2003).
61. M. Parikh and D.E. Schreiber, Recent development in proximity effect correction techniques, in: *Proceedings of the Symposium on Electron and Ion Beam Science and Technology, Ninth International Conference 1980*, 1980, pp. 304–313.

62. F. Murai, H. Yoda, S. Okazaki, N. Saitou, Y. Sakitani, Fast proximity effect correction method using a pattern area density map, *J. Vac. Sci. Technol. B.*, 10 (6; Nov./Dec.), 3072–3076 (1992).
63. T. Abe, S. Yamasaki, R. Yoshikawa, and T. Takigawa, II Representative figure method for proximity effect correction, *Jpn. J. Appl. Phys.*, 30 (3B), L528–L531 (1991).
64. A. Fujii et al., Advanced e-beam reticle writing system for next generation reticle fabrication, *Proc. SPIE*, 4409, 258–269 (2001).
65. N.D. Wittels and C. Youngman, Proximity effect correction in electron-beam lithography, in: *Proceedings of the Symposium on Electron and Ion Beam Science and Technology, 8th Conference*, 361, 1978.
66. M. Osawa, K. Takahashi, M. Sato, H. Arimoto, K. Ogino, H. Hosino, and Y. Machida, Proximity effect correction using pattern shape modification and area density map for electron-beam projection lithography, *J. Vac. Sci. Technol. B.*, 19 (6; Nov./Dec.), 2483–2487 (2001).
67. G. Owen and P. Rissman, Proximity effect correction for electron beam lithography by equalization of background dose, *J. Appl. Phys.*, 54 (6), 3573–3581 (1983).
68. J.B. Kruger, P. Rissman, M.S. Chang, Silicon transfer layer for multilayer resist systems, *J. Vac. Sci. Technol.*, 19 (4), 1320–1324 (1981).
69. K.K. Christenson, R.G. Viswantan, and F.J. Hohn, x-ray mask fogging by electrons backscattered beneath the membrane, *J. Vac. Sci. Technol. B.*, 8 (6), 1618–1623 (1990).
70. N. Saitou, T. Iwasaki, and F. Murai, Multiple scattered e-beam effect in electron beam lithography, *Proc. SPIE*, 1465, 185–191 (1991).
71. M. Ogasawara et al., Reduction of long range fogging effect in a high acceleration voltage electron beam mask writing system, *J. Vac. Sci. Technol. B.*, 17 (6), 1618–1623 (1999).
72. T. Komagata, Y. Nakagawa, N. Gotoh, and K. Tanaka, Performance of improved e-beam lithography system JBX-9000MVII, *Proc. SPIE*, 4409, 248–257 (2001).
73. Y. Hattori, et al., Solution for 100 nm -EBM-4000-, *Proc. SPIE*, 4754, 697–703 (2002).
74. K. Koyama, O. Ikenaga, T. Abe, R. Yoshikawa, and T. Takigawa, Integrated data conversion for the electron beam exposure system EX-8, *J. Vac. Sci. Technol. B.*, 6 (6), 2061–2065 (1988).

5

Laser Mask Writers

Christer Rydberg

CONTENTS

5.1 Laser Pattern Generators 100
    5.1.1 The Making of a Photomask 100
    5.1.2 History 100
    5.1.3 Current Trends 102
5.2 Components of a Laser Pattern Generator 104
    5.2.1 Data Path 105
        5.2.1.1 Address Grid 105
        5.2.1.2 Multiple Exposure Techniques 105
    5.2.2 Focus Control 105
    5.2.3 Climate Chamber 106
5.3 Raster-Scan Laser Pattern Generators 106
    5.3.1 Architecture 106
    5.3.2 Formation of Aerial Image 107
    5.3.3 Scan Separation 107
    5.3.4 Stripe Boundaries 108
    5.3.5 Gray Scaling 109
    5.3.6 Acousto-Optics 110
    5.3.7 Beam Modulation 111
    5.3.8 Beam Scanning 111
    5.3.9 Focusing the Beam onto the Photomask 113
    5.3.10 Multibeam Strategies 113
    5.3.11 Synchronization of Image Formation and Stage 114
5.4 SLM Image Formation 115
    5.4.1 The Lithographic Exposure Tool, the Stepper 115
        5.4.1.1 Imaging 115
        5.4.1.2 Numerical Aperture 116
        5.4.1.3 Illumination 118
        5.4.1.4 Diffraction Effects on the Image 118
    5.4.2 A Programmable Mask — the SLM 122
        5.4.2.1 Piston Micro-Mirrors 123
        5.4.2.2 Pivot Micro-Mirrors 124
    5.4.3 Architecture 125
    5.4.4 Formation of Aerial Image 126
    5.4.5 The SLM Chip 127
    5.4.6 Calibration 128

	5.4.7 The Difference between a Phase-Modulating SLM and a Transmission Mask 128
5.5	Current Products and Outlook 129
	5.5.1 Outlook 129
References 130

5.1 Laser Pattern Generators

This chapter will discuss the technology and principles behind photomask pattern generators based on exposure by light, or more specifically, laser. There are two main groups of laser pattern generators — raster scan tools and spatial light modulator (SLM) tools. After a general introduction, the subsystems of laser pattern generator are discussed. Later the image forming properties of the two groups of pattern generators are presented separately, and the chapter ends with an outlook.

5.1.1 The Making of a Photomask

The making of a photomask is the transfer of a two-dimensional geometrical pattern defined in a data base (CAD file) file to an etched image. The etched image is in most cases a thin chrome film or other opaque material, but the image could also be etched in a transparent material as with a phase shifting mask.

The maker of a mask uses a laser pattern generator to expose a layer of photosensitive material (photoresist), according to a two-dimensional binary geometric pattern. The photons introduce chemical changes in the photoresist making up a latent image. The latent image is then developed to create a three-dimensional relief image in the resist. Etching the chrome, partly covered by photoresist, transfers the two-dimensional image on to the chrome.

To characterize a pattern generator, it is in most cases sufficient to analyze the irradiance distribution — the aerial image. A further study takes into account the effects related to the latent image, such as focus change along the depth of the photoresist film, standing waves in the photoresist, and also transmittance change of photoresist with exposure.

The steepness of the photoresist wall depends on the gradient in the aerial image. Due to the resist threshold even sloping photoresist walls produce a sharp edge on the final photomask. Despite this, steep photoresist walls in the aerial image are important because they are less sensitive to process variations and decrease the statistical deviation of the chrome edges in the final image. Further, due to the resist threshold fine irradiance structures in the aerial image not located on the edges will not be present in the image in the chrome film (Figure 5.1).

To place an edge at the correct position several factors must be controlled. The irradiance in the aerial image is not allowed to fluctuate randomly from position to position; the system must be dose stable. A system that is out of focus will produce an aerial image with low gradients making the system sensitive to process variations. Furthermore, distortion in the aerial image will displace the edges as well.

5.1.2 History

The earliest laser pattern generators exposed the photomask with a single focused laser beam that moved over the surface of the photomask. Bell Labs and what later became

Laser Mask Writers

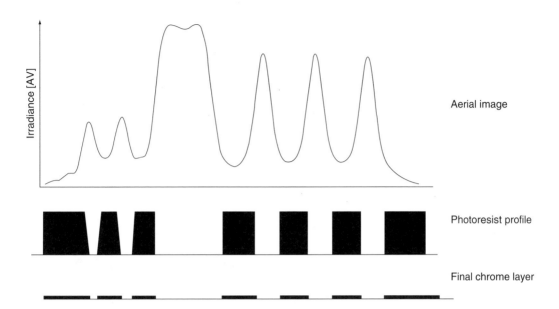

FIGURE 5.1
Top figure: the aerial image—irradiance distribution. *Middle figure*: the photoresist profile after illumination and processing. *Bottom figure*: the resulting opaque layer (usually chrome) on the final photomask.

Micronic Laser Systems AB made early developments in the 1970s [1,2]. The trend of ever-smaller features was followed by a smaller size of the focused spot. The decreased spot size meant increased exposure time to print the same area. To counter the loss of throughput in the systems, the use of multiple beams writing different parts of the pattern in parallel was introduced. A pattern generator, very advanced for its time, was built by TRE Semiconductor Equipment Corporation writing with 16 beams in parallel using an acousto-optic deflector (AOD) to create a crosswise motion of the beams [3]. The TRE system was the first pattern generator of the raster scan type. Today the majority of laser pattern generators are of this type. Later developments of this technology have been made by Texas Instruments, Heidelberg, and ATEQ, later Etec Systems, Inc.— an Applied Materials company [4]. The first commercial industry system bringing the raster scan technology to the commercial mask making market was the CORE™ system from Etec Systems, Inc.

Some modern laser-based photomask-writing platforms are based on the lithographic exposure tool, a stepper or scanner. The lithographic exposure tool is used to project the image of the photomask onto the wafer, which is also reduced in the projection by a factor of 4. The image on the wafer is made up of several exposures of an identical photomask. In a pattern generator for making photomasks, the same imaging process is used but instead of a fixed photomask imaged onto a wafer, a dynamic mask, an SLM is imaged onto a photomask. This architecture enables a printing strategy where the full image is made up of several individual "stamps." To achieve a stamp that can change dynamically, different technologies can be used. One simple dynamic arrangement is the grating light valve (GLV™) by Silicon Light Machines, shown in Figure 5.2 [5]. The GLV™ is a one-dimensional SLM arrangement of reflecting ribbons that can be used in combination with a scanning movement. The device is already being used by Agfa and Dai Nippon Screen (DNS) for computer-to-plate offset lithography in the graphics arts industry. Moreover, the high modulation of the GLV™ allows the simpler one-dimensional array configuration. Other dynamic masks are the Digital Light Processors

FIGURE 5.2
GLV® from Silicon Light Machines. By flexing the ribbons a phase change is introduced in the diffracted field, i.e., a dynamic grating. (Courtesy of Silicon Light Machines.)

(DLP), shown in Figure 5.3, used by Texas Instruments in projectors and by NimbleGen in gene-chip direct writing [6,7]. The DLP is a micro-mirror device, where light is specularly deflected by pivoting individual mirrors. The first production system for writing photomasks with dynamic SLM imagining is the Sigma product family from Micronic Laser Systems. The Sigma systems use an SLM with pivot micro-mirrors operating in diffraction mode developed by the Fraunhofer IMS and IPMS Institutes, shown in Figure 5.4 [8,9].

5.1.3 Current Trends

So far the semiconductor industry has followed the pace outlined by Moore's law stating that the number of transistors on a single chip increases exponentially with time [10]. A significant part of this growth has been due to improvements in performance and increased resolution of the lithographic exposure tools (stepper or scanner), making smaller and smaller image features possible. The current trend is that the image feature shrinkage outpaces the increase of resolution capabilities in the lithographic exposure tool. This has the consequence that the lithographic exposure systems must operate closer to the resolution limit, where the performance of the tool is limited by diffraction. A diffraction limited optical system does not image perfectly due to information loss. This information loss is deterministic and can to some extent be compensated for in the

Laser Mask Writers 103

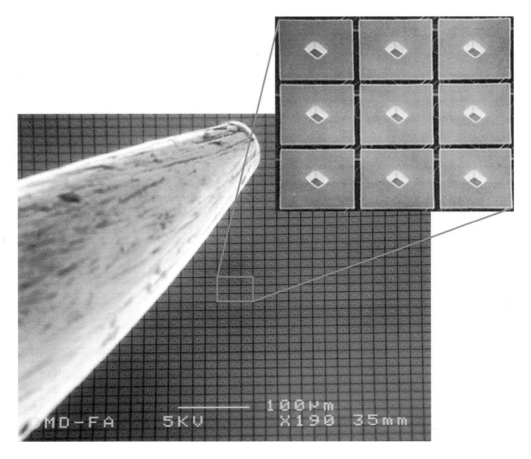

FIGURE 5.3
DLP made by Texas Instruments. Each micro-mirror is individually addressed and specularly deflects the illuminating light. Used in projectors and as direct writing of gene-chips by NimbleGen. To give a hint of the small dimensions the tip of a needle has been inseted in the picture.

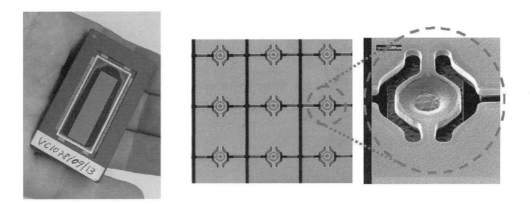

FIGURE 5.4
SLM chip developed by the Fraunhofer IMS and IPMS institutes. Each micro-mirror is individually addressed and pivoted around the center axis. The mirrors diffractively control the partly coherent field. The device is used by Micronic Laser Systems in the Sigma photomask pattern generators.

photomask. This puts higher demands on the photomasks when resolution enhancement techniques (RET), such as sub-resolution assist features (SRAF), phase-shift masks (PSM), and optical proximity correction (OPC), are introduced. Further, adding SRAF and OPC to pattern data can make the data sizes grow significantly, and the data path must be dimensioned accordingly, increasing the cost of the pattern generator.

Due to the scaling factor of 4 between photomask and wafer, one could expect that the resolution demands on a photomask pattern generator are four times lesser than in a stepper. However, this is not true for advanced photomasks, where RET complicates the simple scaling relationship between photomask and wafer.

The mask error enhancement factor (MEEF) or in some literature, mask error factor (MEF) is a metric on how errors in the photomask scale when imaged in an exposure tool, such as a stepper [11,12]. The MEEF is defined as the line width error at wafer level divided by the line width error at the photomask level multiplied by the exposure tool magnification. The MEEF can in some cases be as high as 5–10 putting tight requirements on the statistical properties of a photomask.

5.2 Components of a Laser Pattern Generator

All laser pattern generators are composed of the same basic components, a laser source, an image formation system, a focus control system, and an interferometer controlled X–Y stage (Figure 5.5). All contamination- and temperature-sensitive components are enclosed

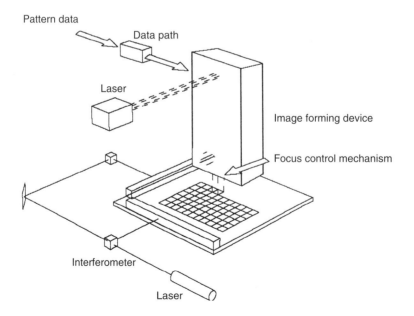

FIGURE 5.5
The components of a laser pattern generator. Data path, laser, image formation mechanism, focus control mechanism, and interferometer controlled X–Y stage. Components sensitive to ambient conditions are enclosed in a climate chamber.

in a climate chamber. The functionality of the data path, the focus control mechanism, the X–Y stage, and the climate chamber are similar for all image formation architectures and will be discussed in this chapter.

5.2.1 Data Path

The data path accepts treated CAD data, usually in a vector format with hierarchic structure. This input is converted to a data stream that is tailored to control the image formation subsystem. The vector pattern data is rasterized into a gray-scale bitmap. The bitmap is often rasterized with area coverage algorithms. A pixel that is partly covered by a feature will receive a gray-scale value that is proportional to the area covered.

5.2.1.1 *Address Grid*

In a high throughput laser pattern generator, there is a need to place edges with accuracy better than the pixel grid. This is possible with the use of gray-scaling and multiple exposures technique.

In a pattern generator three grids coexist; the pattern data is expressed on a virtual data grid or design grid; the pixels are located on the pixel grid; the address grid is the pixel grid extended through techniques, such as multiple exposure and gray-scaling. The address grid limits the resolution of the edge placement, and edges in the data will round to the address grid — grid snapping. For example, a pattern generator with a 10-nm address grid would not be able to print a 505-nm wide line, since it would automatically be truncated to either a 500-nm or a 510-nm wide line.

5.2.1.2 *Multiple Exposure Techniques*

Multiple exposures, or multipass writing, decrease the sensitivity to both stochastic and systematic deviations [13]. Statistically, several consecutive exposures of the same pattern will average out noise in the imaging process.

It is possible to apply an offset to the placement of the pixel grid before dividing the data into pixels. If multiple exposures are done with an offset of a part of a pixel between each exposure, the address grid will be extended. Further, the impact of systematic stitching error can be decreased if the locations of the stripe borders are displaced between each exposure.

5.2.2 Focus Control

The focus control mechanism maintains focus by keeping a fixed distance between the photomask and the objective lens. The topology of the photomask deviates from an ideal flat surface, and the focus control mechanism must be able to compensate and keep a fixed distance to the photomask surface.

The principle used to measure the distance between lens and photomask is airflow measurement. A pod is placed close proximity to the photomask and the airflow through a small nozzle in the pod is measured. The flow is a function of the distance between the pod and the photomask. To remove the correlation with ambient air pressure, a reference is used. The reference pod is placed at a fixed distance to a reference surface. Either electronically controlled piezo-electric materials or magnetic coils can be used to make the mechanical adjustment of the distance between the lens and the substrate.

5.2.3 Climate Chamber

The pattern generator is self-contained in a climate chamber, where environmental variables are controlled and monitored. The air in the climate chamber is flowing as closely to a laminar flow as possible at a pressure slightly higher than the ambient pressure to prevent contaminants from entering the pattern generator. Before the air is fed into the climate chamber it will pass through an environmental module where it is filtered and adjusted to the right temperature.

5.3 Raster-Scan Laser Pattern Generators

Raster-scan pattern generators are the workhorses of the photomask shops. The technology is mature and known to be both stable and fast. This chapter will describe the image formation technology and also to some extent describe technologies where the two major product families, the ALTA® series from Etec Systems, Inc., and the Omega® series from Micronic Laser Systems differ from each other [14,15].

5.3.1 Architecture

The raster-scan pattern generator writes the pattern with one or several laser beams of an approximate Gaussian shape. The beams scan over the mask surface while being amplitude-modulated according to pattern data. The writing time for a pattern is essentially proportional to the size of the pattern area and not to the complexity or number of features on the pattern. Above a limit, set by the data handling capabilities, the relation between pattern areas and writing time breaks down. The data handling capabilities of the pattern generator is designed to ensure that the system operates below this limit.

As illustrated in Figure 5.6 and Figure 5.7, the two pattern generators consist of the following major components: a laser, a beam splitter, a modulator, either a rotating polygon or an acousto-optic deflector (AOD) creating a crosswise scanning motion of the beams, a reduction lens and an X–Y stage.

The beam splitter divides the light from the laser into several beams for increased capacity by writing different parts of the pattern in parallel. Each beam is individually amplitude modulated in the acousto-optic modulator (AOM). The modulation is controlled by input from the data path. The area to be exposed is divided into stripes of equal width. The width of a stripe, or a scan stripe, is typically a few hundred micrometers. During exposure the X–Y stage moves the photomask at a constant speed along the stripe. At the same time the focused spots scan in a direction perpendicular to the movement of the X–Y stage. A rotating polygonal mirror or an AOD creates the crosswise scanning motion of the spot (Figure 5.8).

The spot is moved continuously in the scan direction, and discrete scans are added in the stripe direction. After reaching the end of a stripe and before the start of the next stripe, the X–Y stage moves one increment in the direction of the scan and returns to the start position in the stripe direction.

Due to the small speed vector component from the movement of the photomask, a scan, placed perpendicular to the X–Y stage motion, will not end up perpendicular on the photomask. This small angular deviation from an orthogonal behavior can be compensated for by a slightly tilted scan line, denoted as the azimuth angle.

Laser Mask Writers 107

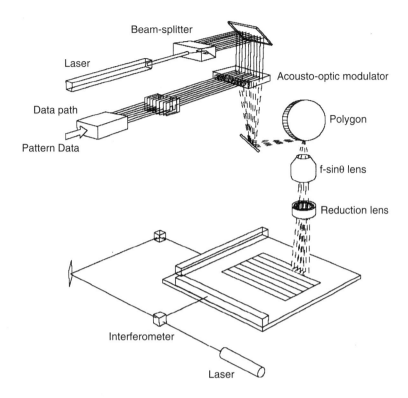

FIGURE 5.6
The ALTA raster scan tool from Etec Systems, Inc., an applied materials company.

5.3.2 Formation of Aerial Image

The different beams, in the case of a multibeam system, are separated and the respective fields do not interfere: the imaging is incoherent. Hence, the aerial image, $I(x,y)$, shown in Figure 5.11, can as a first approximation be calculated as a convolution between two functions:

$$I(x,y) = g(x,y) \otimes I_{spot}(x,y)$$

where "$\otimes$" is the convolution.

The first function, $g(x,y)$, shown in Figure 5.10, yields a Dirac response for coordinates illuminated by the center of the spots, and in each point the gray-scale intensity is factored in, all according to the output of the data path. The second function $I_{spot}(x,y)$, shown in Figure 5.9, is the irradiance distribution of the spot, which is close to a Gaussian distribution.

In Figure 5.10, $g(x,y)$ represents a cross. The convolution of $g(x,y)$ and the spot irradiance distribution $I_{spot}(x,y)$ in Figure 5.9 gives the aerial image in Figure 5.11.

5.3.3 Scan Separation

The ratio between the spot size and the separation of two adjacent scans is an important factor in deciding the image quality. A large separation of the scans in relation to a given spot size can exhibit effects, such as statistical sensitiveness, bad behavior of gray levels,

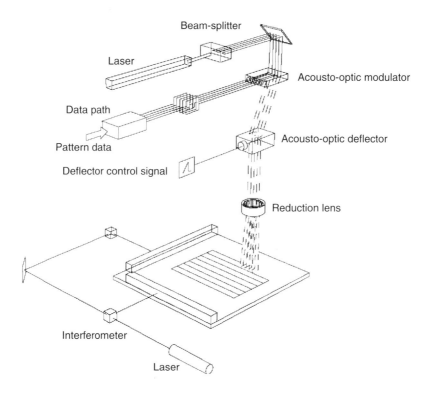

FIGURE 5.7
The Omega raster scan tool from Micronic Laser Systems AB.

and in extreme cases edge roughness on or broken features along the stripe direction. On the other hand smaller separation of the scans results in a lower throughput.

The distance between two adjacent scans in a specific tool is usually fixed. This is due to the fixed spacing between the multiple beams, and the fact that a tool with variable pitch must be able to control the azimuth angle flexibly. However, there are some multibeam pattern generators on the market where the user may choose between two scan separations.

5.3.4 Stripe Boundaries

Distortions of edges and image features crossing a stripe boundary are referred to as a stitching or a butting error. To minimize the impact of the stitching errors various techniques are used, in practice often in combination with multiple exposure techniques.

By smoothly blending two adjacent stripes, a few percent of the scan length overlap into the adjacent stripe [16]. In the overlap area, the end of the scans blend with the beginning

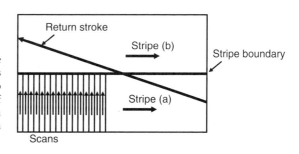

FIGURE 5.8
Raster scan writing principle. The X–Y stage moves the photomask so that crosswise scans created in a deflector device are placed next to each other creating a stripe. After exposures of the last scan in stripe (a) the X–Y stage performs a return stroke and start exposure of the first scan in stripe (b).

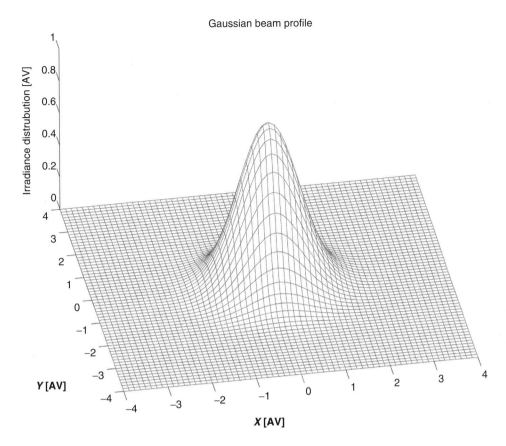

FIGURE 5.9
The irradiance distribution of the focused spot, $I_{spot}(x,y)$. The distribution is a first approximation Gaussian in shape.

of the scans in the next stripe, the rise and the fall of all scans are modulated as smooth functions while keeping the total irradiance constant.

Another technique to reduce stitching error is to alternate stripe border position for consecutive scans — dovetailing, shown in Figure 5.12 [17].

5.3.5 Gray Scaling

The edge of an image feature can be shifted a part of a scan spacing by using a scan with an intensity level that is intermediate between on and off, often referred to as a gray level intensity [18].

Illustrated in Figure 5.13 is the aerial image of a feature cut along the stripe direction. The aerial image for each individual scan and the total sum is shown. In this particular feature, scan numbers 2–7 are modulated as "on." Scan number 8 in the sum is modulated to a value in-between on and off, and the edge of the feature can be displaced by an amount smaller than the scan separation. This effect is used to increase the address grid resolution. The effect is nonlinear, and a lookup-table transforms the desired edge displacement to gray-scale value. Furthermore, the edge displacement is a function of irradiance and process, and thus a lookup-table needs to be created for each dose and process [19].

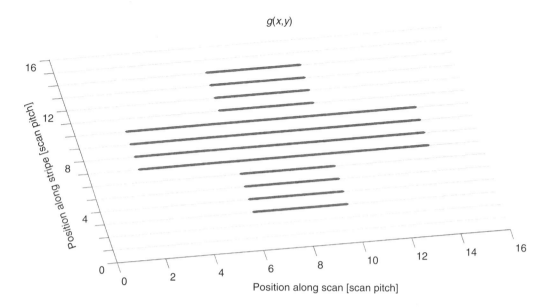

FIGURE 5.10
The function $g(x,y)$ yields a Dirac response multiplied with gray-scale value for coordinates illuminated by the center of a spot. In this example of a cross, the dashed lines represent a gray-scale value of zero (no irradiance) and the solid lines represent a gray-scale value of unity (full irradiance).

One undesired effect of using intensity modulation is that the energy gradient at the edge in the aerial image is lower when using an intermediary gray-scale value than when the scan making up the edge of a pattern feature is of full intensity. A less steep slope is the same as loss of process latitude. To assure that all feature edges end up with equal process latitude multiple pass writing can be introduced, and a feature edge will receive contributions from different gray levels from each pass.

Other novel approaches exist for dealing with address grid issues, for example, extra deflectors, but are not used in high-end pattern generators.

5.3.6 Acousto-Optics

Acousto-optical elements are central in controlling the beams in the raster scan image formation. The acousto-optic devices used in raster scan pattern generators are made up of a crystal with a piezoelectric acoustic transducer. The transducer is fed by an electric signal creating an ultrasonic sound wave through the crystal. The sound wave will displace the molecules from their equilibrium positions in the crystal, introducing mechanical strain. The strain will in turn induce periodic local changes in the index of refraction, i.e., a moving volume grating.

The so-called Bragg condition relates the beam of light into the crystal, the ultrasonic sound wave, and the first-order diffracted beam. At the Bragg condition light refracted from the sound nodes in the crystal will act constructively creating a high efficiency in one specific order, see Figure 5.14.

The Bragg condition is given under the condition:

$$\sin(\theta) = \frac{\lambda_{\text{light}}}{2 \cdot \lambda_{\text{sound}}}$$

Laser Mask Writers

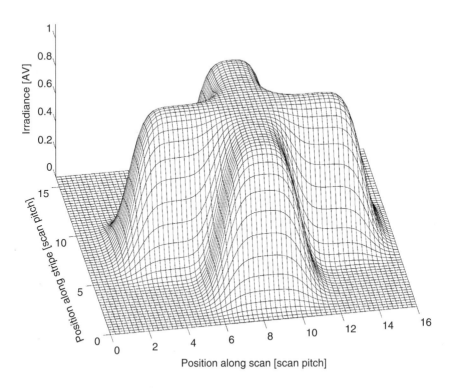

FIGURE 5.11
Simulated aerial image of a cross, exposed with a raster scan pattern generator. The aerial image is calculated as convolution of $g(x,y)$ (shown in Figure 5.10) and $I_{spot}(x,y)$ (shown in Figure 5.9).

The Bragg angle is small as seen in the expression above. A systematic treatment can be found in Refs. [20,21].

5.3.7 Beam Modulation

The amplitude modulation of the beams in a raster scanner pattern generator is done in an AOM. The AOM is an acousto-optic device operating at the Bragg angle and at a suitable sound amplitude level. The light in the first order will be proportional to the intensity of the sound wave. If the amplitude of the sound is too high, saturation will occur and the AOM acts as a switch. Either the first order or the zero order can be used to expose the substrate.

5.3.8 Beam Scanning

There are two major kinds of different physical principles used to achieve the scanning motion of the beams; rotating mirror polygon and AOD.

The dominating technique is the use of high-speed rotating polygonal mirrors. Typically, a 24-facet polygonal mirror is used to create the scanning motion. This technique has a very long scan length and also a stable dose over the scan.

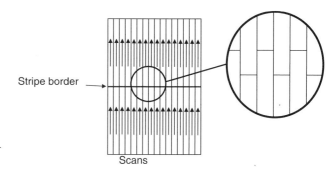

FIGURE 5.12
Alternating stripe border position, dovetailing of stripes.

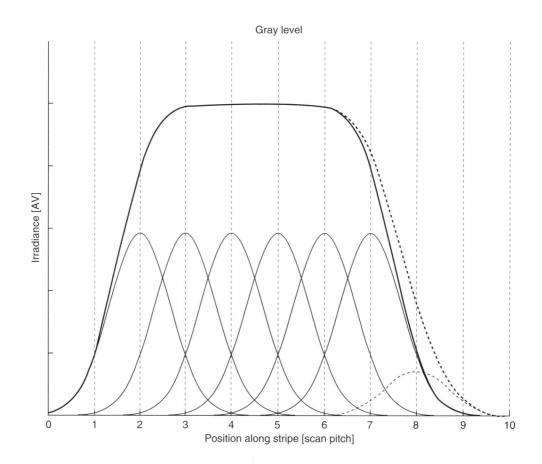

FIGURE 5.13
Gray-scaling illustrated on an aerial image cut along the stripe direction. By changing the amplitude of the scan at the right edge of the feature (number eight) it is possible to move the edge of the feature by increments smaller than the separation between scans. The effect is nonlinear and needs to be calibrated for different process conditions.

The AOD is an acousto-optic device working close to the Bragg condition. Here the ultrasonic sound wave is frequency modulated with the intention to change the angle of the diffracted light in the first order. In an acousto-optical deflector, nonlinearity can be compensated for electronically, creating an optical system that behaves in a linear fashion. Angular displacement is not a linear function of frequency, but feeding the piezoelectric

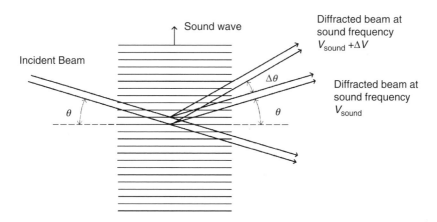

FIGURE 5.14
Principle of an acousto-optic device. Light is diffracted off the periodic structure in the crystal induced by the sound wave. By modulating the amplitude of the sound wave the relative irradiance between diffracted and non-diffracted beam can be controlled. By changing the frequency of the sound wave the angle of the diffracted light can be changed.

acoustic transducer with a nonlinear signal can compensate and make the behavior of the system linear. Further, modulation of the sound intensity can compensate for intensity variations in the deflected light due to frequency-dependent diffraction efficiency. The deflection angle created by an AOD is relatively small.

5.3.9 Focusing the Beam onto the Photomask

The sharpness of the beam is of importance when imaging features, such as corners, thin lines, and small contact holes. It is a fundamental physical property that there is a constraint to which extent it is possible to focus the beam on the photomask [22,23]. Diffraction smears out the image and puts a limit to the resolution. Limiting factors are both the design of the lens system of the imaging system and the wavelength used.

5.3.10 Multibeam Strategies

Historically, multibeam pattern generators were introduced to increase throughput. Today pattern generators with multiple beams are considered to be standard. In multibeam writing, each beam is individually modulated, and all beams share a common scanning mechanism. The beams scan in parallel and are separated in the stripe direction. To avoid interference patterns between two spots on the photomask, a distance separating the beams is maintained.

In the ALTA 4000™, the number of beams is 32, divided into two subgroups with 16 beams each [24]. Within each group, the spacing of the beams is six times the separation between two adjacent scans. The two groups are separated nine times the scan grid separation. The X–Y stage moves a distance corresponding to 32 times the scan grid in between the start of each collective scan for all the 32 beams. The whole system acts as a brush that fills out the full pattern.

Using multiple beams introduces additional error sources, and the need for calibration increases. Possible error sources that need attention are, for example, intensity variations between the beams, separation errors, and variations in beam shape.

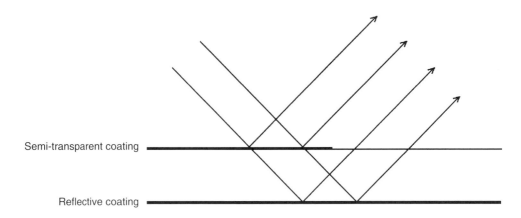

FIGURE 5.15
Part of a beam splitter apparatus. The beam splitter consists of several mirrors aligned to double the numbers of beams for each reflection.

Two kinds of beam splitter devices are used in high-end pattern generators, partly reflecting surfaces and diffractive optical elements (DOE).

In Figure 5.15, a scheme with partly reflecting surfaces used to generate beams of equal intensity is illustrated [25]. Each reflective device consists of a coating that reflects half of the intensity of the light, while the other half is transmitted and encounters a totally reflective layer on the backside of the device. In this way the number of beams is doubled for each reflection.

A DOE is a device with a surface engineered to modify a beam through diffraction [26]. The shape of the surface is etched so that the incoming light is diffracted in multiple discrete directions. The surface topology controls both the number of beams and the angular separation of each generated beam.

5.3.11 Synchronization of Image Formation and Stage

During the exposure of a stripe, the X–Y stage will strive to move the photomask according to ideal exposure placement; due to several factors there will still be stochastic deviations that can be measured with the interferometers. All pattern generators have systems to handle these deviations.

A deviation in the X–Y stage speed will give a deviation of the placement in the stripe direction. In pattern generators equipped with a rotating polygon mirror, the angular momentum is large and does not allow for anything other than a constant revolving speed. The servos controlling the X–Y stage are synchronized to the rotation of the polygon mirrors. A mirror mounted on piezoelectric electrodes that change the position of the beams in the stripe direction compensates for deviations of the X–Y stage motion relative to the mirror rotation. In the AOD there are no moving mechanical objects, and a scan is free to start as soon as the X–Y stage is in the correct position.

The X–Y stage will also have a stochastic deviation in the direction perpendicular to the movement. A raster scan tool with mirror polygon control the timing of the signal to the modulators, a delay of the data stream moves the pattern along the scan. A raster scan pattern generator with an AOD can control the frequency function fed to the deflector. By offsetting the frequency function, placement along the length of the scan can be controlled.

5.4 SLM Image Formation

This chapter gives an introduction to laser pattern generators of the SLM type. The SLM pattern generators are based on the same imaging architecture as the lithographic exposure tool, the stepper or scanner. For all practical purposes, the SLM pattern generators can be seen as a lithographic exposure tool with a programmable mask. For this reason, an introduction to the lithographic exposure tool is included.

5.4.1 The Lithographic Exposure Tool, the Stepper

A stepper exposes a photomask image onto a substrate, such as a silicon wafer. The image of the photomask is usually demagnified by a ratio of 4. The image of the photomask is exposed onto the wafer several times creating identical copies side-by-side on the substrate. This method is suitable for mass production of identical components (Figure 5.16).

Is there any way possible to use the same imaging process for making photomasks? To write a photomask, a mask where an arbitrary image can be generated is desired. Programmable masks do exist and are called SLM and will be discussed later in this chapter.

5.4.1.1 Imaging

In a lithographic exposure tool, the illuminating light is diffracted by the photomask. In projection lithography, the diffracted light propagates to the Fraunhofer or far-field region (Figure 5.17). The far field can also be realized in the focal point of a positive lens [27]. For a spatially coherent field, the irradiation distribution is proportional to the square of the absolute value of the field.

The field in the far field can be calculated via a simple Fourier transform of the field instantaneously after the photomask. With this knowledge it is a simple task to analyze the behavior of the diffracted light in the far-field region (Figure 5.18).

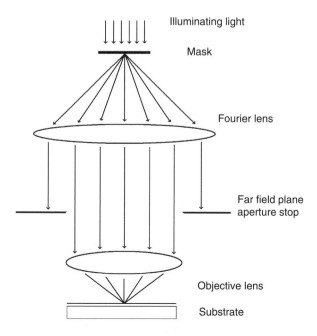

FIGURE 5.16
A lithographic exposure tool, such as a stepper or scanner. The illuminating light can in most cases be set to a partial coherence suitable for the specific pattern to be exposed. The illuminating light is diffracted off the mask and collected by a Fourier lens. An objective lens then creates an image of the mask on the substrate.

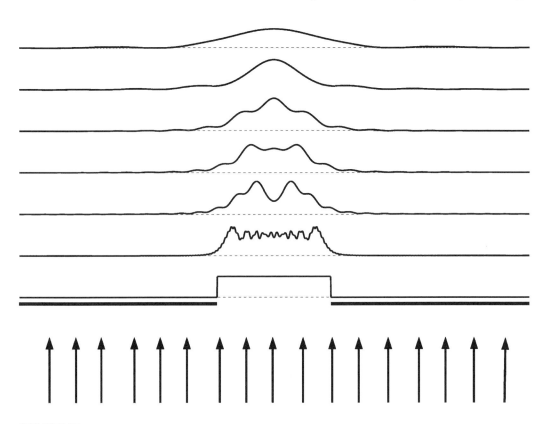

FIGURE 5.17
Diffraction pattern from a slit at different distances. The intensity is shown both at proximity (Fresnel) region and far-field (Fraunhofer) region.

5.4.1.2 Numerical Aperture

Due to the finite extent of the objective lens, some of the information in the Fourier spectrum will be lost. This loss is more significant for small feature patterns because a pattern with smaller features will create a diffraction pattern with a larger angular spread. To be able to quantify the information loss due to diffraction, the numerical aperture (NA) is introduced. The NA is dependent of the optical system and can be used as a metric on how much information in the original mask is transferred to the image by the optical system. The NA is defined as sinc of the maximum half angle in the image plane times the refractive index of the medium where the imaging takes place. The imaging is usually done in air, but in immersion lithography a liquid with a higher index of refraction is introduced in the space between lens and the substrate giving a higher NA:

$$\mathrm{NA} = n_{\mathrm{ambient}} \sin(\alpha_{\mathrm{out}})$$

Due to geometric constraints, $\sin(\alpha_{\mathrm{out}})$ cannot become larger than unity (Figure 5.19).

The NA of the system relates to the diffraction of the pattern in the mask plane through the magnification of the system. The objective lens reduces the image on the photomask with a demagnification. The objective lens can only accept angles diffracted by the photomask corresponding to the NA of the lens divided by the demagnification factor:

$$n_{\mathrm{air}} \sin(\alpha_{\mathrm{acceptance\ angle}}) = \frac{\mathrm{NA}}{\mathrm{demagnification}}$$

Laser Mask Writers

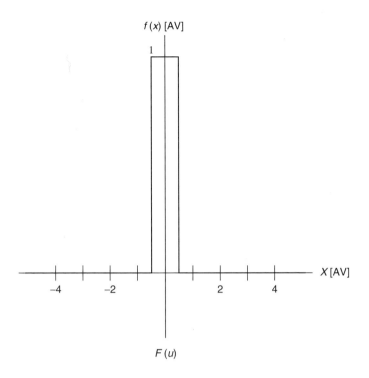

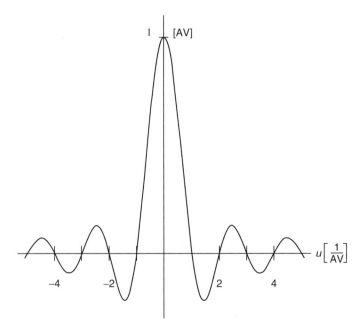

FIGURE 5.18
Fourier transform pair. A square pulse has the sinc function as its Fourier transform.

For large angles of θ, the scalar treatment of light is not valid, and vector treatments taking polarization effects into account must be considered [28]. The scalar treatment breaks down because the polarization parallel to the plane of incidence at high angles requires vector addition; see Wierner's [29] experiment.

A periodic pattern will produce a discrete far-field diffraction pattern (see Figure 5.20). To make up an image on the substrate, the limiting aperture must accept at least two of

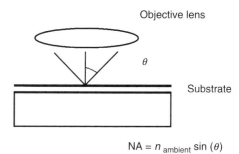

FIGURE 5.19
Definition of NA; sine for the maximum half angle times the index of refraction of the ambient medium.

$$NA = n_{ambient} \sin(\theta)$$

the discrete orders. This sets a fundamental resolution limit of the system for periodic structures.

5.4.1.3 Illumination

Different illumination conditions exist. The mask can be illuminated with a wave of normal incidence, where the field of the light is perfectly correlated at all points on the mask surface. This is called spatially coherent illumination or just coherent illumination for short. Incoherent illumination means the opposite: the electromagnetic fields at two separate points on the mask surface are not correlated. At partially coherent illumination the correlation of the field at two points is dependent on the distance between the points. One method of creating partially coherent illumination is by utilizing an angular spread of incoherent sources, i.e., the size and shape of the illumination aperture [30]. This can mathematiclly be viewed as applying the van Cittert-Zernike theorem [27, 30].

Different illumination conditions give different behavior of the final image. An incoherent illumination can print a periodic pattern with a smaller period than coherent illumination. However, by looking at the contrast in the aerial image for a few illumination conditions it can be seen that different illumination conditions might be favorable for different kinds of features (Figure 5.21).

The Rayleigh formula for resolution is defined as

$$D = k_1 \frac{\lambda}{NA}$$

where D is the width of a feature, and k_1 is a factor dependent on many factors, such as illumination and photoresist [31]. Applying the Rayleigh formula on a periodic pattern to analyze an illumination condition shows that the minimum achievable k_1 for fully incoherent illumination is 0.25. For coherent illumination the resolution is half of this with a k_1 of 0.5. The theoretical resolution limit for a periodic pattern is $k_1 = 0.25$. It can be approached by off-axis illumination and or by phase-shift photomasks.

5.4.1.4 Diffraction Effects on the Image

An ideal diffraction limited optical system, such as a lens system free from aberrations, produces a scaled image of the object. The effect of diffraction on the image is to convolve the ideal image with the Fraunhofer diffraction pattern of the limiting aperture of the system [32].

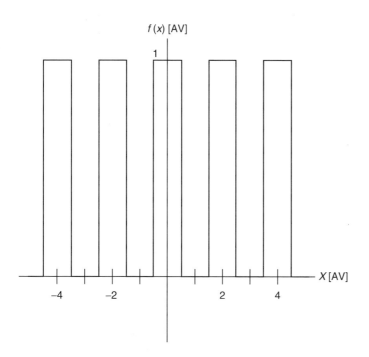

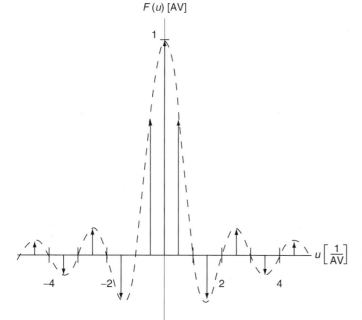

FIGURE 5.20
Fourier transform pair. An infinitively long square function produces discrete orders. Mathematically it is described as Dirac pulses with a sinc envelope.

The function describing the wave front, both at the far field and the image on the photomask, can be obtained by a set of transforms of the wave front at the mask. In the case of low NA it is possible to use a scalar light model and the set of transforms is simplified. The principles are the same so for the remainder of this chapter a scalar approximation valid for low NA will be used.

For the case of cohehernt illumination let the complex function $G_0(x,y)$ denote the wave front right after it has been diffracted at the mask. The wave front at the far field is $B(u,v) = F\{G_0(x,y)\}$, where the operator F is the Fourier transform

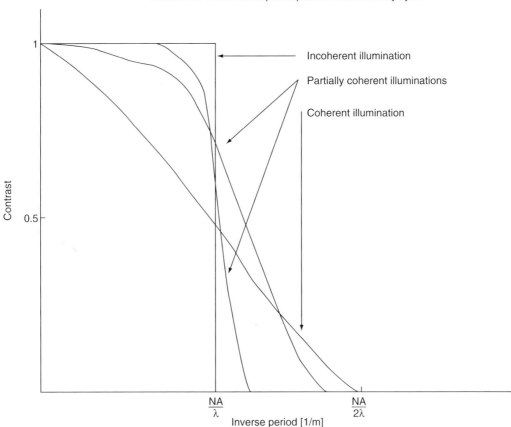

FIGURE 5.21
Contrast shown as function of inverse period for different illumination conditions.

$$F\{G_0(x,y)\} = \iint\limits_{-\infty}^{\infty} G_0(x,y) \cdot e^{-i \cdot 2 \cdot \pi \cdot (u \cdot x + u \cdot y)} dx\, dy$$

and u and v are the spatial frequency coordinates in the far field. The objective lens will low pass the spatial information of the mask, and the filter can be expressed as an aperture function:

$$H(u,v) = \begin{cases} 1, \sqrt{u^2 + v^2} \leq \frac{NA}{\lambda} \\ 0, \sqrt{u^2 + v^2} > \frac{NA}{\lambda} \end{cases}$$

The radius of the aperture function $H(u,v)$ is equal to the diffracting limit of the total imaging system. The limit is set by the objective lens.

The expression for the wave front on the substrate is

$$G_1(x,y) = F^{-1}\{H(u,v)F\{G_0(x,y)\}\}$$

where the operator F^{-1} represents the inverse Fourier transform

$$F^{-1}\{K(u,v)\} = \iint_{-\infty}^{\infty} K(u,v) \cdot e^{i \cdot 2 \cdot \pi \cdot (u \cdot x + v \cdot y)} du\, dv$$

By applying the convolution theorem, the field on the photomask can be expressed as a convolution with the point spread function (PSF) of the system:

$$G_1(x,y) = F^{-1}\{H(u,v)F\{G_0(x,y)\}\} = F^{-1}\{H(u,v)\} \otimes G_0(x,y) = \text{PSF}(x,y) \otimes G_0(x,y) \quad (5.1)$$

Calculating $F^{-1}\{H(u,v)\}$ for a circular aperture gives

$$\text{PSF} = F^{-1}\{H(u,v)\} = \frac{\text{NA}}{\lambda} J_0\left(2\pi \frac{\text{NA}}{\lambda} r\right)/r$$

where J_1 is the Bessel function of the first kind and zeroth order, further

$$r = \sqrt{x^2 + y^2}$$

as illustrated in Figure 5.22.

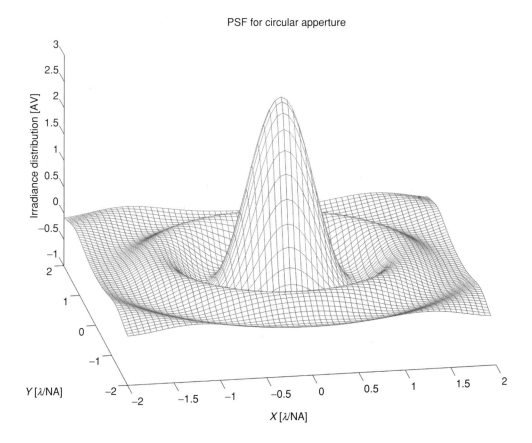

FIGURE 5.22
The PSF is defined as the field in the image plane in response to a point. For an optical system with a circular aperture the PSF is a Bessel function of the first order.

The aerial image or irradiance distribution is then proportional to the absolute value of the field squared:

$$I(x,y) = |G_1(x,y)|^2 = |F^{-1}\{H(u,v)\} \otimes G_0(x,y)|^2$$

For partially incoherent illumination conditions the situation is a little more complex. As stated earlier partially coherent illumination can be achieved by illuminating the mask with an angular spread of incoherent illumination. Looking at one infinitesimal area element at a time each area element can be treated as a coherent illumination with an angular offset (u^T, v^T) from the normal axis:

$$G_1(x,y) = F^{-1}\left\{H(u,v) F\left\{G_0(x,y) \cdot e^{-i \cdot 2 \cdot \pi \cdot (u^T \cdot x + v^T \cdot y)}\right\}\right\} = \text{PSF}(x,y) \otimes G_0(x,y) \cdot e^{-i \cdot 2 \cdot \pi \cdot (u^T \cdot x + v^T \cdot y)}$$

The aerial image is proportional to

$$I(x,y) = |G_1(x,y)|^2 = \int\int S(u,v) \cdot |\text{PSF}(x,y) \otimes G_0(x,y) \cdot e^{-i \cdot 2 \cdot \pi \cdot (u^T \cdot x + v^T \cdot y)}|^2 \, du^T \, dv^T \quad (5.2)$$

where $S(u,v)$ defines the illumination. For partially coherent illumination created by means of circular disc:

$$S(u,v) = \begin{cases} \frac{1}{\pi(\sigma \cdot \text{NA}/\lambda)^2} & \text{if } \sqrt{u^2 + v^2} \leq \sigma \cdot \text{NA}/\lambda \\ 0, \text{ otherwise} \end{cases}$$

where σ is the filling factor of the system.

For coherent illumination $S(u,v) = \delta(u,v)$ and equations (5.1) and (5.2) become equivalent.

5.4.2 A Programmable Mask — the SLM

The idea of using a SLM is to be able to replace the static photomask in the stepper with a programmable mask. If other properties, such as high refresh rates, low cross talk between pixels, high contrast, gray-scaling, and stability over time, are fulfilled, it is possible to use the dynamic SLM in a stepper-like system for writing photomasks. There is also potential to use the same principles for a system to expose the wafer directly without using any photomask at all — a direct-write system.

A SLM is a device that can control a light field through modulation of the amplitude, phase, or polarization. An example is an LCD, where an electric field controls the polarization of light passing through a liquid. In practice an LCD is not a realistic alternative for an SLM in a pattern generator due to slow refresh rates giving low throughput.

Micro-mechanical SLMs are more suitable. An example of an amplitude-modulating SLM is the DMD micro-mirror construction from Texas Instruments (see figure). In this device, the tilting mirrors deflect light away from the aperture of the projection optics. Gray-scaling, necessary for photographic quality, is achieved by time modulation. NimbleGen uses this device for direct-writing of gene-chips.

An interesting group of SLMs are the phase-shifting SLMs where interference effects are used. The main idea is that the phase of the illuminating field is modulated in cells covering an area. Either interference between the cells or interference within the cell itself cancels the light in a certain direction. Due to the use of interference phenomena some constraints have to be put on the illuminating field. It is not possible to use incoherent

Laser Mask Writers

illumination, and with partially coherent illumination the field must have some correlation over the area that is intended to interfere. The design and alignment of the illumination and projection optics are more delicate than for the DMDs, but the reward is an image that is equivalent to the image from a physical reticle in the lithographic exposure tool.

An early example of a phase modulating SLM was a reflecting film suspended on the surface of a viscoelastic layer [33]. By electrostatic force the reflective film was made to flex down, creating a distortion in the smooth reflective surface on controlled locations. Due to the displacement of the surface, the path length difference changes and destructive interference is created. A direct-write tool was built on this device [34].

5.4.2.1 Piston Micro-Mirrors

Another SLM of the phase-modulating type is the piston arrangement, shown in Figure 5.23 [35]. In the piston arrangement, each mirror in a two-dimensional matrix can, by some mechanism, be set at an individual height. With this arrangement the average phase change over each individual mirror can be controlled. The average wave front changes over one mirror as the mirror moves downward can be illustrated in the complex plane (Figure 5.24).

To emulate a transmission mask only positive real values are used. This can be achieved with a scheme where two mirrors work in pair and by displacing every second mirror by one-half of the illuminating wavelength minus the displacement of

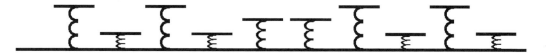

FIGURE 5.23
SLM of piston mirror type. A phase change of 2π corresponds to a mirror displacement of $\lambda/2$.

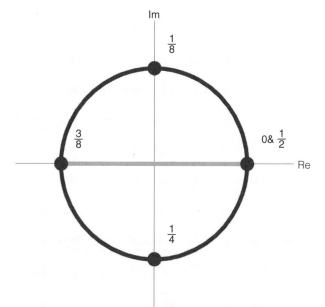

FIGURE 5.24
Average change of the wave front over the extent of one piston-type mirror. The numbers indicate the piston displacement expressed as parts of the wavelength of the illuminating light. Two mirrors working in complex conjugates will have an average normalized field change on the real axis.

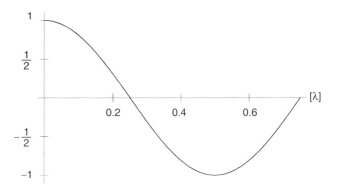

FIGURE 5.25
Average change of the wave front over two piston mirrors working as complex conjugates. See Figure 5.25 for a representation in the complex plane of one mirror.

the first; two adjacent mirrors then make up a complex conjugate pair and the sum is real (Figure 5.25).

Areas in the image that should be exposed correspond to mirrors that are unperturbed, and areas in the image that should not be exposed correspond to alternating mirrors of, for example, the complex amplitudes $0 - i$ and $0 + i$. The alternating mirrors will not send any energy in the specular direction, and the energy will be distributed in diffraction orders. The placement of the diffraction orders in the spatial spectrum is inversely proportional to the pitch of the mirrors. If the mirror pitch is small enough, the diffraction orders will not be collected by the objective lens.

With the piston mirror arrangement and a data path with two mirrors working in pair it is possible to emulate any real intermediate value of on and off. It is even possible to have a field that is 180° out of phase, such as 3/8 in Figure 5.25. The implication of this is that a strong PSM could be emulated. However, the strong phase effects inherent in piston mirrors make them susceptible to through-focus artifacts, unless the individual mirrors are made very small.

5.4.2.2 Pivot Micro-Mirrors

The arrangement with mirrors that pivot around a center axis is shown in Figure 5.26. An advantage is that the phase remains constant for all pivot angles, which gives stable image properties. It can also be noted that a part of the negative side of the real axis can be reached, where the field is phase-shifted by 180° (Figure 5.27 and Figure 5.28). The intensity of the phase-shifted light reaches a maximum of 4.7% compared with the non-phase shifted light [36]. This can be used to improve edge acuity by the same principles as in an attenuated PSM.

Pivoting mirrors can be modified to create more negative amplitude, making the pivot mirror principle suitable for both weak and strong phase shifting [42].

FIGURE 5.26
SLM of pivot mirror type. Displacing the tip of a micro-mirror by $\lambda/4$ gives a linear phase difference of 0 to 2π over the extent of the mirror, i.e., extinction of the field.

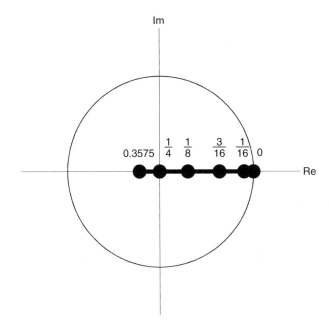

FIGURE 5.27
Average change of the complex amplitude of the wave front over the extent of one pivot mirror. The numbers indicate displacement of the tip of the mirror expressed as parts of the wavelength of the illuminating light. See Figure 5.29 for the real axis.

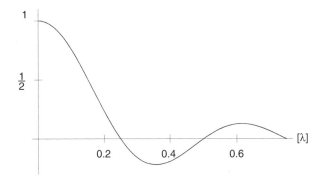

FIGURE 5.28
Average change of the wave front over one pivot mirror as function of displacement of the tip of a mirror expressed as part of a wavelength. The change is real for all tilt displacements. See Figure 5.28 for a representation in the complex plane.

5.4.3 Architecture

The only commercially available pattern generator with SLM architecture for photomask production is the Sigma product family from Micronic Laser Systems, shown in Figure 5.29 [37].

The Sigma family uses SLM architecture of the pivot mirror type. In one configuration, the Sigma7300™, the SLM has 2048 × 512 micro-mirrors that are 16 × 16 µm each and have a projected image on the photomask of 80 × 80 nm. The device works in diffraction mode, not specular reflectance, and needs to deflect the mirrors by only a quarter of the wavelength (62 nm at 248 nm) to go from the fully on state to the fully off state. To create a fine address grid the mirrors are driven to on, off, and 63 intermediate values. A pulsed excimer laser is used and the SLM is illuminated with partially coherent illumination.

The pattern is stitched together from millions of images from the SLM chip. Flashing and stitching proceed at a rate of 2000 stamps per second. While writing the X–Y stage moves continuously in the stripe direction (Figure 5.30).

The programmable mask loads the pattern data to be exposed. When the stage passes over the correct position, the laser is pulsed and projects the content on the programmable

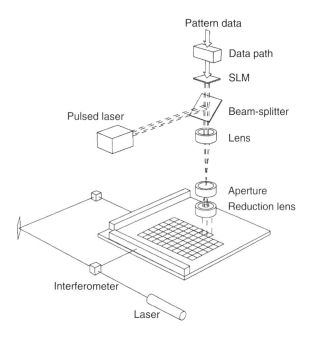

FIGURE 5.29
The Sigma SLM tool from Micronic Laser Systems AB.

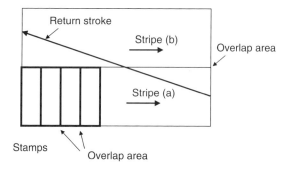

FIGURE 5.30
Sigma SLM writing principle. The X–Y stage moves the photomask and at the correct position the excimer laser fires placing stamps next to each other creating a stripe. After exposure of the last stamp in stripe (a) the X–Y stage performs a return stroke and starts exposure of the first stamp in stripe (b).

mask on to the photomask. The laser pulse is short enough, in the range of 20 ns, to freeze the image. The pattern is divided up into several stripes of equal width. Stamps along the stripe direction expose each stripe while the X–Y stage moves continuously. Both adjacent stamps and adjacent stripes are printed with a small overlap to ensure pattern quality at the boundaries. Because the imaging process within a stamp is partially coherent, but the stamps are incoherent to each other, even an image printed with perfect placement needs to use blended overlaps to "dilute" the edge discontinuity.

5.4.4 Formation of Aerial Image

The imaging system in the Sigma pattern generator is very similar to the imaging system of a stepper. The optics low-pass filters the SLM image and the individual mirror structure of the SLM is not resolved, due to that the pitch of the mirrors is smaller than the resolution of the optical imaging system.

The aerial image of the projection of the SLM onto the substrate can for a partial coherent illumination be calculated using Equation (5.2). One example is the SLM set to expose a cross in Figure 5.32. The PSF of the optical system is shown in Figure 5.32. The

Laser Mask Writers

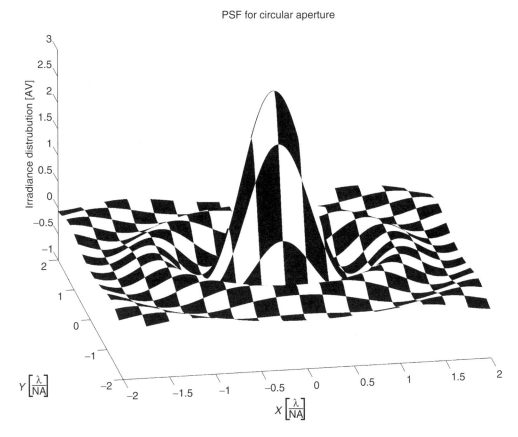

FIGURE 5.31
PSF of the Sigma optics with the size of the individual mirrors indicated. Note that the projection of individual mirrors is smaller in comparison of the resolution of the system; hence internal structure of the mirrors will not be resolved.

PSF is defined as the field at the substrate created by an infinitely small point source in the SLM plane. The projected mirrors are shown as alternating black and white to indicate the size relative to the PSF. By convolving the PSF with the wave front at the SLM plane according to Equation (5.2) and using a partial coherent illumination, the aerial image in Figure 5.33 is obtained.

5.4.5 The SLM Chip

The micro-mirror SLM module is made using MEMS and CMOS processes. First an electronic chip is created in a standard CMOS process. On top of the electronic a matrix of individual micromirrors with flexing hinges are created by use of MEMS processes. The micromirrors, hinge structure, and support posts are formed in an aluminum alloy.

Underneath the mirror matrix is an addressing system for each individual mirror. Each mirror has a transistor and is addressable in a matrix structure, similar to a TFT display. An electrostatic force is created that causes the mirror to be slightly deflected. The pivot angle applied to each micro-mirror is in the range of a few milliradians.

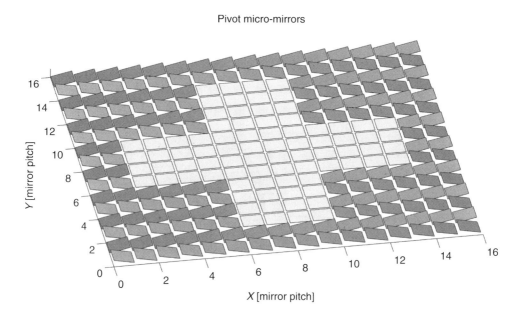

FIGURE 5.32
An SLM of the pivot type with the micro-mirrors arranged in the alternating row layout. The SLM is set to expose a cross.

5.4.6 Calibration

The mirrors need to be individually calibrated. A CCD camera, sensitive to the exposure light, is placed in the optical path in a position equivalent to the image on the mask blank. The SLM mirrors are driven through a sequence of known voltages and the camera measures the response. A calibration function is determined for each mirror, to be used for real-time correction of the gray-scale data on a pixel-by-pixel basis during writing.

5.4.7 The Difference between a Phase-Modulating SLM and a Transmission Mask

The difference between projection of a fixed transmission chrome mask used in a stepper and an SLM of the pivoting micro-mirror type is small. There are two effects that make the SLM different from a photomask of transmission type.

One effect relates to the fact that the SLM use distinct cells that modulate the light like pixels. The rasterization process converts the vector pattern data to a gray-scaled bitmap. Like in any rasterized system the effect is a filtering process that slightly degrades the image quality. By using displaced multipass writing this effect can be reduced and smoothed for a uniform image. It is further possible to implement prefiltering of the bitmap in the data path to increase pattern fidelity.

Another effect is due to the finite size of the SLM cells. In the case of an SLM of the pivot type SLM, a mirror with an applied pivot angle represents areas that are not intended to receive any exposure. Still the convolution of a pivoted mirror with the PSF will leave a small residue. For moderate mirror sizes the residue is several magnitudes smaller than the field corresponding to untilted mirrors. The size of the residue is also dependent on the mirror layout. An advantageous layout is to have every second row of mirrors pivoting in different direction. The alternating row mirror layout is shown in Figure 5.32.

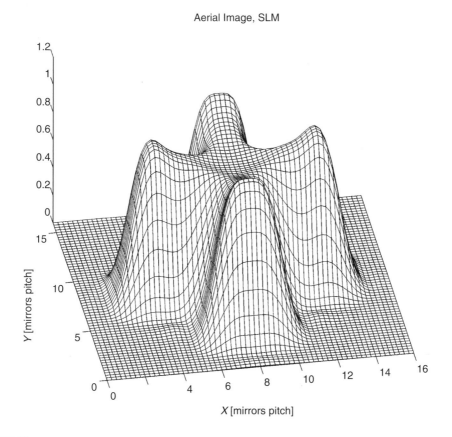

FIGURE 5.33
Simulated aerial image of a cross, exposed with pattern generator of the SLM pivot mirror type and with a partially coherent illumination. See Figure 5.11 for a raster scan pattern generator comparison.

5.5 Current Products and Outlook

For the semiconductor market there are two major suppliers of laser mask writers. Both suppliers have raster scan products with wavelengths geared towards i-line processing (Table 5.1). At the next lower wavelength the choice of architecture differs. Etec Systems, Inc., has continued to build on the raster scan concept while Micronic Laser Systems has introduced an SLM architecture with micro-mirrors.

5.5.1 Outlook

How could the optical pattern generators improve to keep up with Moore's law and the increasing demand put on the photomasks due to low k_1 imaging? It is reasonable to expect that the basic concepts of raster scan and SLM imaging could be incrementally improved with lower wavelengths and higher NA. There is also room for further improvements through several now exotic technologies. Some examples that might be a part of mainstream photomask manufacturing in the future are: immersion lithography, off-axis illumination, longitudinal polarization, apodization or pupil filtering and most likely

TABLE 5.1

Current Products

	ALTA 3900	**Omega6600**	**ALTA 4000**	**Sigma7300**
Manufacturer	Etec, an Applied Material Company	Micronic Laser Systems AB	Etec, an Applied Material Company	Micronic Laser Systems AB
Imaging System	Raster scan with rotating polygon	Raster scan with AOD	Raster scan with rotating polygon	SLM imaging with pivoting micro-mirrors
Laser	Continuous-wave Ar-ION laser @ 363.8 nm	Continuous-wave Kr-ION laser @ 413 nm	Frequency-doubled continuous-wave Ar-ION laser @ 257 nm	Pulsed KrF laser @ 248 nm
Parallelism	32 beams	5 beams	32 beams	2048 * 512 micromirrors
Exposures	4	1 or 2	4	4
Photoresist	Standard I-line	Standard I-line	Chemically amplified resist	Chemically amplified resist

several other advanced wave front engineering technologies will be in the hands of future mask makers [38–41].

References

1. F.L. Howland and K.M. Poole, Overview of new mask-making system, *Bell System Technical Journal*, 49 (9), 1997–2009 (1970).
2. G. Westerberg, Device for Generating Masks for Microcircuits, United States Patent, 3 903 536, Sept. 2, 1975.
3. D.B. MacDonald, M. Nagler, C. Vanpeski, and T.R. Whitney, 160 MPX/SEC laser pattern generator for mask and reticle production, *Proceedings of the Society of Photo-Optical Instrumentation Engineers*, 470, 212–220 (1984).
4. P.A. Warkentin and J.A. Schoeffel, Scanning laser technology applied to high speed reticle writing, *Proc. SPIE*, 633, 286–291 (1986).
5. O. Solgaard, F.S.A. Sandejas, and D.M. Bloom, Deformable grating optical modulator, *Optics Letters USA*, 17 (9; 1 May), 688–690 (1992).
6. R.J. Gove, V. Markandey, S.W. Marshall, D.B. Doherty, G. Sextro, and M. DuVal, High definition display system based on digital micromirror device, in: *Signal Processing of HDTV, VI, Proceedings of the International Workshop on HDTV '94*, 1995, pp. 89–97.
7. S. Singh-Gasson, R.D. Green, Y.J. Yue, C. Nelson, F. Blattner, M.R. Sussman, and F. Cerrina, Maskless fabrication of light-directed oligonucleotide microarrays using a digital micromirror array, *Nature Biotechnology*, 17, 974–978 (1999).
8. H. Lakner, P. Durr, U. Dauderstaedt, W. Doleschal, and J. Amelung, Design and fabrication of micromirror arrays for UV lithography, *Proc. SPIE*, 4561, 255–264 (2001).
9. P. Durr, A. Gehner, U. Dauderstadt, Micromirror spatial light modulators, in: *Proceedings of the 3rd International Conference on Micro Opto Electro Mechanical Systems (Optical MEMS), MOEMS 99*, 1999, pp. 60–65.
10. G.E. Moore, Cramming more components onto integrated circuits, *Electronics*, 38 (8; Apr.), 114–117 (1965)
11. J. van Schoot, J. Finders, K. van Ingen Schenau, M. Klaassen, and C. Buijk, Mask error factor: causes and implications for process latitude, *Proc. SPIE*, 3679 (pt. 1/2), 250–260 (1999).

12. A.K. Wong, R.A. Ferguson, and S.M. Mansfield, The mask error factor in optical lithography, *IEEE Transactions on Semiconductor Manufacturing* 13 (2; May), 235–242 (2000).
13. H.C. Hamaker, G.A. Burns, and P.D. Buck, Optimizing the use of multipass printing to minimize printing errors in advanced laser reticle writing systems, in: *15th Annual Symposium on Photomask Technology and Management, Proc. SPIE*, 2621, 319–328 (1995).
14. M.J. Bohan, H.C. Hamaker, and W. Montgomery, Implementation and characterization of a DUV raster-scanned mask pattern generation system, *Proc. SPIE*, 4562, 16–37 (2002).
15. P. Liden, T. Vikholm, L. Kjellberg, M. Bjuggren, K. Edgren, J. Larson, S. Haddleton, and P. Askebjer, CD performance of a new high-resolution laser pattern generator, *Proc. SPIE*, 3873 (pt. 1/2), 28–35 (1999).
16. A. Thuren and T. Sandström, Method and Apparatus for the Production of a Structure by Focused Laser Radiation on a Photosensitively Coated Surface, United States Patent, 5,6635,976, Jun. 3, 1997.
17. P.C. Allen, M.J. Jolley, R.L. Teitzel, M. Rieger, M. Bohan, and T. Thomas, Laser Pattern Generation Apparatus, United States Patent 5,386,221, Jan. 31, 1995.
18. A. Murray, F. Abboud, F. Raymond, and C.N. Berglund, Feasibility study of new graybeam writing strategies for raster scan mask generation, *Journal of Vacuum Science and Technology B. (Microelectronics Processing and Phenomena)*, 11 (6; Nov./Dec.), 2390–2396 (1993).
19. C.A. Mack, Impact of graybeam method of virtual address reduction on image quality, *Proc. SPIE*, 4562, 537–544 (2002).
20. A. Yariv, *Quantum Electronics*, 3rd ed., John Wiley & Sons, Inc., New York, 1988, pp. 327–338.
21. A. Korpel, *Acousto-Optics*, 2nd ed., Marcel Dekker, Inc., New York, 1997.
22. P. Kuttner, Image quality of optical systems for truncated Gaussian laser beams, *Optical Engineering* 25 (1; Jan.), 180–183 (1986).
23. H. Haskal, Laser recording with truncated Gaussian beams, *Applied Optics*, 18 (13; July), 2143 (1979).
24. M.J. Bohan, H.C. Hamaker, and W. Montgomery, Implementation and characterization of a DUV raster-scanned mask pattern generation system, *Proc. SPIE*, 4562. pp. 16–37 (3/2002).
25. P.C. Allen, P.A. Warkentin, Beam Splitting Apparatus, United States Patent, 4,797,696 Jan. 10, 1989.
26. J. Turunen and F. Wyrowski, *Diffractive Optics for Industrial and Commercial Applications*, Akademie Verlag GmbH, Berlin, 1997.
27. J.W. Goodman, *Introduction to Fourier Optics*, 2nd ed., McGraw-Hill, New York, 1996 (Chapter 5.2).
28. M. Born, E. Wolf, *Principles of Optics*, 7th ed., Cambridge University Press, Cambridge, 1999 (Chapter 8).
29. O. Wierner, *Ann. d. Physik.*, 40, 203 (1890).
30. M. Born and E. Wolf, *Principles of Optics*, 7th ed., Cambridge University press, Cambridge, 1999 (Chapter 10.4.2).
31. Lord Rayleigh, *Philosophical Magazine* 5 (8), 261 (1879).
32. J.W. Goodman, *Introduction to Fourier Optics*, 2nd ed., McGraw-Hill, New York, 1996 (Chapter 5.3).
33. H. Kiick, W. Doleschal, A. Gehner, W. Grundke, R. Melcher, J. Paufler, R. Seltmann, and G. Zimmer, Deformable micromirror devices as phase modulating high resolution light valves, in: *Sensors and Actuators, Proceedings of Transducers '95 and Eurosensors IX*, 1995.
34. R. Seltmann, W. Doleschal, A. Gehner, H. Kück, R. Melcher, J. Paufler, G. Zimmer, New system for fast submicron optical direct writing, *Microelectronic Engineering*, 30 (1–4; Jan.), 123–127 (1996).
35. Y. Shroff, Y.J. Chen, and W.G. Oldham, Optical analysis of mirror based pattern generation, *Proc. SPIE*, 5037, 70 (2003).
36. T. Sandstrom and N. Eriksson, Resolution extensions in the Sigma7000 imaging pattern generator, *Proc. SPIE*, 4889 (2002).
37. U. Ljungblad, T. Sandstrom, H. Buhre, P. Duerr, and H. Lakner, New architecture for laser pattern generators for 130 nm and beyond, *Proc. SPIE*, 4186, 16–21 (2001).

38. M. Switkes, M. Rothschild, R.R. Kunz, S. -Y. Baek, D. Cole, and M. Yeung, Immersion lithography: beyond the 65 nm node with optics, *Microlithography World*, 12 (2; May) (2003).
39. B.J. Lin, Off-axis illumination-working principles and comparison with alternating phase-shifting masks, *Proc. SPIE*, 1927 (pt. 1), 89–100 (1993).
40. Eric S. Wu, Balu Santhanam, S.R.J. Brueck, Grating analysis of frequency parsing strategies for imaging interferometric lithography, *Proc. SPIE*, 5040, 1276–1289 (2003).
41. R. von Bunau, G. Owen, and R.F.W. Pease, Depth of focus enhancement in optical lithography, *Journal of Vacuum Science and Technology B.*, 10 (6; Nov./Dec.), 3047–3054 (1992).
42. U. Ljungblad, H. Martinsson, T. Sandström. Phase Shifted Addressing using a Spatial Light Modulator. Micro and Nano Engineering International Conference. 2004

Section III

Optical Masks

6

Optical Masks: An Overview

Nobuyuki Yoshioka

CONTENTS
6.1 Optical Lithography 136
   6.1.1 Contact/Proximity Lithography 137
   6.1.2 Projection Lithography 137
6.2 Optical Mask 139
   6.2.1 Function of Optical Mask in Optical Lithography 139
   6.2.2 Classification of Optical Masks 140
   6.2.3 Resolution Enhance Techniques for Optical Masks 141
   6.2.4 Mask Material 141
      6.2.4.1 Glass Substrate 141
      6.2.4.2 Pattern Film Material 141
   6.2.5 Mask Fabrication Process 143
      6.2.5.1 Mask Process Flow 143
      6.2.5.2 Process Technology 145
   6.2.6 Pellicle 147
6.3 Quality of Optical Mask 149
   6.3.1 Critical Dimension 149
   6.3.2 Image Placement 152
   6.3.3 Defect 152
   6.3.4 Phase and Transmittance Measurements 153
References 155

The objective of this chapter is to give the reader a general background on optical masks that can be put under two distinct classifications:

1. Optical masks that represent matured technologies
2. Optical masks that represent the advanced technologies

Here matured technologies relates to the ones that are well established on the manufacturing lines and are used for making low-end high-volume masks. The advanced technologies on the other hand are employed to produce high-end state-of-the-art masks that require innovations and are significantly more expensive to make than their low-end counterpart. Today the main challenge before the industry is to make the high-end masks and still remain cost-effective. In a latter chapter of this book the various types of advanced optical masks will be revisited and be discussed in greater details. However,

I believe that the various types of advanced optical mask are described in ch. 8. in order to appreciate the evolutionary steps leading to advanced mask, a chapter on matured technology precedes the chapter on advanced mask in this book.

The nature and construction of an optical mask is dictated by the type of optical lithography in which a mask is used; and hence it seems pertinent to give the reader a brief description of the different types of lithography techniques where those types of mask are used.

6.1 Optical Lithography

Optical lithography is an important processing step in the fabrication of semiconductor chips. The process utilizes ultraviolet (UV) lights to form circuit patterns on semiconductor wafer as is shown in Figure 6.1. The first step in the process of optical lithography is to coat an already preexisting film (e.g. oxide, nitride, etc.) on a semiconductor wafer

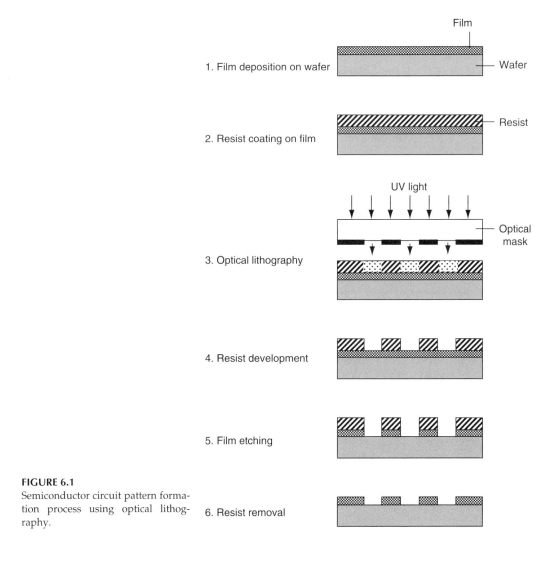

FIGURE 6.1
Semiconductor circuit pattern formation process using optical lithography.

Optical Masks: An Overview

with a light sensitive material that is called photoresist. The next step is to expose the resist using a patterned mask. The wavelength of the light employed for exposing the resist can vary from UV (436 nm) down to deep UV (157 nm). The resist is then developed which then reproduces the mask pattern on the resist film. This pattern is then transferred on the preexisting film on the wafer by etch removal of the part of the film not covered by the resist. Finally, all resists are removed leaving behind an etched pattern on the wafer that is a replica of the pattern on the original mask.

Optical lithography can be classified in terms of the types of exposure techniques used for imaging a mask pattern on wafers; which are contact/proximity type and projection type as described in the following subsections.

6.1.1 Contact/Proximity Lithography

Figure 6.2 shows contact and proximity exposure techniques. Contact exposure method has been used from the very early days of semiconductor fabrication [1]. In the contact mode, the mask makes an actual contact with the wafer coated with resist — and hence the term "contact mode." This approach, however, causes mask damage resulting in transfer of defects on wafer. In order to resolve this problem, a method was developed where a very small gap (of the order of a few microns) could be maintained between the wafer and mask. This technique was named as "off-contact" mode or "proximity mode." The latter term now is more commonly used in the industry. This proximity method, however, gives relatively poor resolution because of the diffraction of light around the edges of features.

In both the types of exposure (contact and proximity), the size on mask pattern is designed to be the same as the required size on the wafer. This type of imaging is referred as 1× (or 1:1) lithography.

6.1.2 Projection Lithography

In order to obtain both, high resolution and longer mask life, the techniques of projection exposure were later developed. In the projection exposure, the mask patterns are imaged on wafer through a set of projection optics consisting of mirrors or lenses. Figure 6.3 shows a 1:1 projection optics using 1× mask where the features of the mask are of the same size as the features of the wafer [2]. Because of the 1× mask, the specs on the feature tolerance on mask remain the same as they would be on wafer. The lithographic tolerance

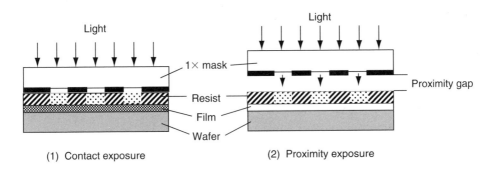

FIGURE 6.2
Optical lithography by contact exposure and proximity exposure with 1× magnification mask.

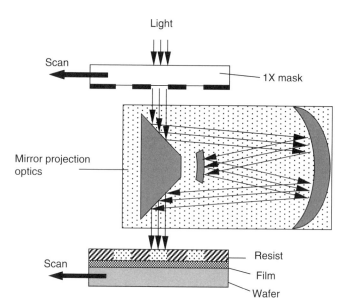

FIGURE 6.3
Optical lithography by projection method with 1× magnification mask.

on mask feature depends strongly on the process capability of mask making and puts significant constraints on the limitation of 1:1 projection optics.

The problem was then addressed by introducing the reduction–projection optics as shown in Figure 6.4 [3]. In this method, masks with larger features (e.g., 2×, 5×, or even

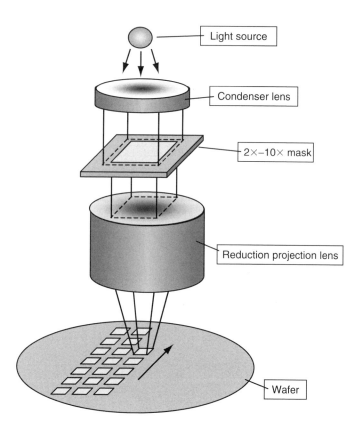

FIGURE 6.4
Reduction projection lithography.

Optical Masks: An Overview

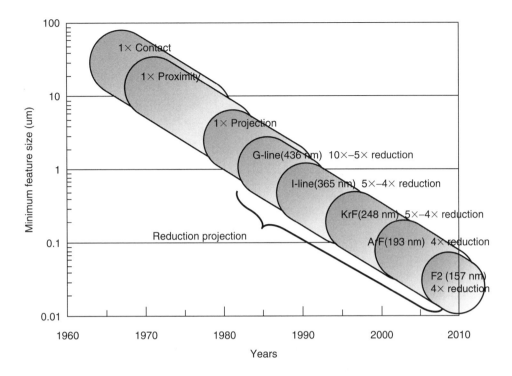

FIGURE 6.5
Trends of lithography technologies and exposure wavelength.

10× magnifications) were employed for printing on wafers. Specs on larger features are more forgiving since small errors on mask when printed on wafer can be reduced to the point of disappearance. The reduction–projection methods are widely applied for LSI fabrication under 1-um node.

In order to meet the challenge of resolving smaller and ever shrinking features, the industry has also been employing shorter and shorter exposure wavelengths. Figure 6.5 shows the trends of lithography and exposure wavelength.

6.2 Optical Mask

6.2.1 Function of Optical Mask in Optical Lithography

A schematic structure of a 6-in. optical mask used in reduction optics is shown in Figure 6.6. The optical mask is made of glass (quartz) substrate with chrome (Cr) pattern on it. In order to protect the pattern from contamination and foreign particles, the mask is mounted with a pellicle — also shown in the figure. The making of the original patterns on the mask is extremely critical and requires great accuracy and zero tolerance for defects.

Mask technology can be seen as a refinement of the early days of printing technology for publishing books and papers. The final semiconductor product strongly depends on the quality of the original mask, and therefore mask making is regarded as one of the key technologies in semiconductor manufacturing.

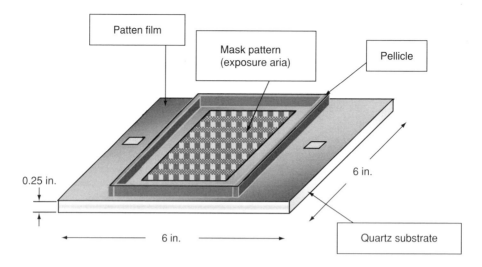

FIGURE 6.6
Schematic structure of optical mask (6-in. reduction mask: reticle).

6.2.2 Classification of Optical Masks

Optical masks can be classified according to the types of exposure in which it is used, as is shown in Table 6.1. For contact and proximity exposure, 1× masks are used. The sizes of glass plate correspond to the wafer sizes. 1× masks are also used in 1× projection exposure systems. In all cases of projection printing, nowadays, pellicles are used for protecting the patterns for contamination.

The size and the thickness of glass substrate for 1× masks are usually 2–7 and 0.06–0.15 in., respectively. For reduction–projection exposure systems, the glass substrates are of 5 and 6 in.^{2} with thickness of 0.09 and 0.25 in., respectively. The masks with magnification of 2×, 2.5×, 4×, and 5× are chosen depending upon the reduction optics of the exposure system.

The optical masks used on reduction exposure systems are also called "reticles."

TABLE 6.1

Categories of Optical Mask for Exposure Types.

Exposure Type	Contact Printing	Proximity Printing	Projection 1X	Reduction	
Mask magnifications	1X	1X	1X	2X~10X	
Substrate size	2″~4″	2″~4″	5″~7″	5″~6″	
Substrate material	HTE	HTE	LTE	Qz	
	MTE	MTE	Qz		
Pellicle	none	none	use	use	
Pattern Type	Binary	Binary	Binary	Binary	PSM
Pattern Material	Emulsion Cr	Emulsion Cr	Cr	Cr	Cr/Qz
					MoSiON
					etc.

HTE: High Thermal Expansion
MTE: Medium Thermal Expansion
LTE: Low Thermal Expansion
Qz: Synthetic Quartz

6.2.3 Resolution Enhance Techniques for Optical Masks

Facing the challenges to produce ever shrinking features the industry has been making significant progress in various areas of lithography that involve shorter wavelengths, improved optics, as well as innovation in mask designs.

Using shorter wavelength to print smaller feature has always been pursued for achieving smaller features. Figure 6.7 shows a relationship between minimum feature size and the exposure wavelength [4]. The figure also points out that at some point the minimum feature size becomes smaller than exposure wavelength. This area is called sub-wavelength lithography and requires improved optics, and innovation in mask designs that are commonly referred as resolution enhancement techniques or simply RET.

The two areas of improvement in mask designs are phase shift masks (PSM) and mask with optical proximity corrections (OPC).

Figure 6.8 shows typical PSM [5,6]. Patterns of PSM are formed by using shifter material, such as SiO_2. The shifter patterns are introduced to shift phase of exposure light by 180°. The resolution is enhanced by interference with light, which passes through the shifter, and light, which does not pass through the shifter. The alternating PSM (Alt.PSM) uses this principle. Attenuated PSM (Att.PSM) is formed by using translucent shifter material, such as MoSiON [7] and CrON [8]. The translucent shifter allows only very small amount of light to pass through, which being 180° out of phase with the unobstructed light, creates sharp edges to the feature resulting in enhanced resolution.

The masks with OPC utilizes a technology that corrects the pattern shape change, which occurred by optical proximity effect [9]. The correction is done by adding correction patterns. Figure 6.9 shows an example of OPC patterns and the printability performance. The detail of PSM and OPC technology are discussed in later chapters of the book.

6.2.4 Mask Material

6.2.4.1 Glass Substrate

The types of glass substrates used in optical masks are summarized in Table 6.2. High thermal expansion (HTE) glass, such as soda lime, has been used for contact exposure during the early days of semiconductor fabrication. Low thermal expansion (LTE) glass was developed to improve the overlay accuracy in proximity exposure and 1× projection exposure. However, as shown in Figure 6.10, the LTE glass cannot be used when the wavelength is shorter than 400 nm because of its reduced transmittance at shorter wavelengths. Therefore, quartz glass, which has high transparency for shorter wavelength lithography, was used for mask substrate. The synthesized quartz glass can allow the transmission of light down to 200 nm. For 193- and 157-nm lithography, new technology to shift to the short wavelength has been applied to the quartz glass fabrication [10,11].

6.2.4.2 Pattern Film Material

Category of mask pattern materials is summarized in Table 6.3 [7,8,12–16]. Emulsion masks were the first masks to be used for exposing silicon wafers in the early 1960s. Emulsion masks were like the black-and-white photographic negatives on glass substrate. These contained patterns representing circuits to be printed on wafers. The emulsion masks, as well as emulsion reticles, were later replaced by Cr on glass (CoG) materials, since Cr has higher chemical mechanical strength than the emulsion. In today's technol-

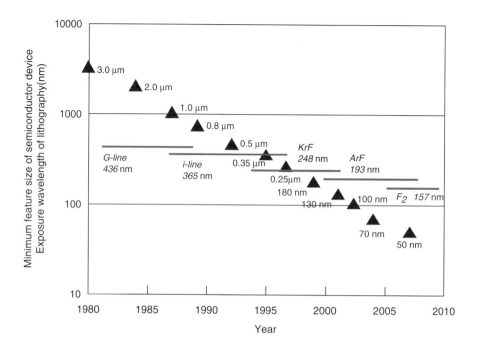

FIGURE 6.7
Relationship between minimum feature size of semiconductor device and exposure wavelength of lithography. [Adapted from Numerical Technologies (now Synopsys Inc.).]

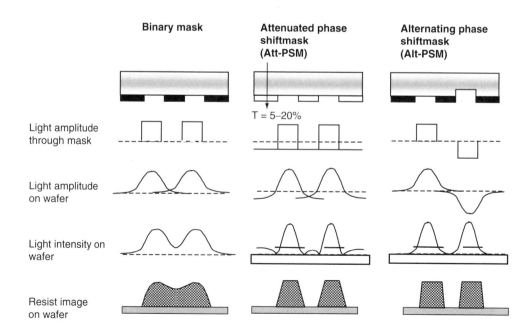

FIGURE 6.8
Typical PSM: (1) Att.PSM; (2) Alt PSM.

Optical Masks: An Overview

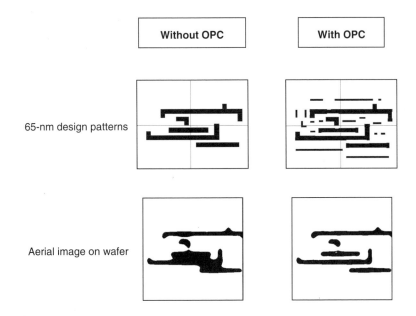

FIGURE 6.9
An example of OPC performances obtained by lithography simulation (simulation condition: 157 nm lithography; NA = 0.85).

ogy, Cr continued to be used for binary mask and the alternating PSM. For attenuated shifter, single layer films, such as MoSiON and CrON, have been developed. The phase and the transmittance of these shifters depend on n and k (real and imaginary part of refractive index), which can be controlled by condition of the film spattering process.

6.2.5 Mask Fabrication Process

6.2.5.1 Mask Process Flow

Figure 6.11 shows the fabrication process flow for 1× mask, reduction mask, and PSM.

In the case of the 1× mask for contact exposure and proximity exposure, the masks were made by using a master reticle and photorepeaters as mentioned earlier. These patterns on masks were made by transferring larger patterns from another mask called a "reticle." This transfer required systems with 10× reduction optics that could expose an array of smaller images over the desired area of a glass substrate that then becomes the final mask (also called photomask). This task is accomplished by a sequence of "expose and step" operation repeated over the entire array area. These optical systems were called photorepeaters. The fabrication of reticle itself required a different type of technique. In early days, the patterns on the reticle were made by exposing the photosensitive surface of the reticles by shining light through series of apertures of various shapes. These shapes consisted of rectangles and triangles of various sizes and configurations, which then constituted the desired pattern representing the circuit or device layout. These systems were called pattern generators. Later on, with the advancement in e-beam technology, the pattern generators were replaced by e-beam writers. Nowadays, reticles or "masks for reduction optics" are made by direct-write with e-beams or laser writers. The PSM, such as alternative types, require two exposure processes in order to form the shifter patterns and the Cr patterns.

TABLE 6.2

Properties of Glass Substrates that are used in Optical Mask. (The Table was Modified from Original Table 1.1, p. 15, of: I. Tanabe, Y. Takahana, M. Hoga, Introduction of Photomask Technologies, 1996, Printed in Japan, ISBN4-7693-1149-4 C3055. Data courtesy of Asahi Glass Company and HOYA Corporation.)

			HTE	HTE	MTE	MTE	LTE	LTE	ULTE
Glass Materials			Soda Lime AS	White Crown AW	Bobsilicate AX	Non-akali NA-40	Alumino Silicate	Alumino Silicate	Synthetic Quartz QZ
							AL	LE-30	
Composition	SiO_2		72.5	70	72	55	62.5	59.5	100
	Al_2O_3		2	0.5	5	14	14.5	15	—
	B_2O_3		—	—	9	—	4.5	6	—
	RO		12	13.5	7	29.5	17	17.5	—
	R_2O		13.3	15.7	6.8	—	1.5	2	—
Thermal Properties	Thermal expansion coefficient (50~200C) ($\times 10^{-7}$/C)		81	93	49	41	37	37	6
	Softening point	(C)	740	708	790	904	900	909	1600
	Glass transition point	(C)	562	521	558	—	673	688	—
	Annealing point	(C)	554	533	571	721	685	699	1120
Refractive Index (Nd at 436 nm)			1.52	1.52	1.50	1.56	1.53	1.53	1.46
Mechanical Properties	Specific gravity		2.49	2.56	2.41	2.87	2.54	2.57	2.20
	Young modulus	(kg/mm^2)	7300	7200	7050	9130	8450	8510	7340
	Shear modulus	(kg/mm^2)	3020	2900	3000	3700	3470	3470	3160
	Poison Ratio		0.21	0.23	0.18	0.23	0.22	0.20	0.16
	Knoop Hardness	(kg/mm^2)	540	530	650	665	640	630	650
Weight loss (D I water 95C 40hr)		(mg/cm^2)	0.016	—	0.003	0.064	0.046	0.054	0.001
Weight loss (0.01N HNO_3 95C 20hr)		(mg/cm^2)	0.002	0.003	0.001	0.040	0.030	0.059	0.000
Weight loss (5%NaOH 80C 1hr)		(mg/cm^2)	0.042	0.057	0.009	0.085	0.100	0.126	0.032
Electric Properties	Bulk resistivity	(Ω/cm)	7.7	9.6	8.0	16.0	10.6	10.6	12.5
	Dielectric constant (1 M Hz, R.T.)		7.5	7.1	5.9	—	6.1	—	4.0

HTE (High Thermal Expansion)
MTE (Medium Thermal Expansion)
LTE (Low Thermal Expansion)
ULTE (Ultra Low Thermal Expansion)
AS, AW, AX, AL: Asahi Glass Company (AW, AX, AL: not available on commercial at 2003)
LE-30, NA-40: HOYA Corporation

Optical Masks: An Overview

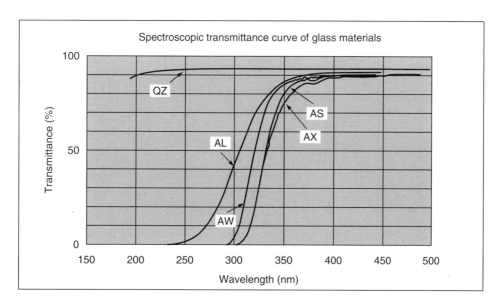

FIGURE 6.10
Spectral transmittance of optical mask glass substrate. AS, AW, AX, AL: Asahi Glass Company (AW, AX, AL: not available on commercial at 2003); QZ: synthetic quartz. (Modified from original Figure 1.5, p. 16, of: I. Tanabe, Y. Takahana, M. Hoga, Introduction of Photomask Technologies, 1996, Printed in Japan. Data by courtesy of Asahi Glass Company.)

6.2.5.2 Process Technology

The process technologies required for making the optical mask are summarized in Figure 6.12. The first step here is to expose the resist film coated on mask blanks using e-beam. This exposure is made in accordance with the design intended to be written on the mask. Following this step, the resist film is developed resulting in the creation of the desired pattern on resist. Next, the Cr film exposed by the opening of resist windows is etched away by wet or dry (plasma) processes. Finally, the undeveloped resist is also removed leaving behind Cr pattern on the glass substrate resulting in the final mask (or reticle).

The mask is then checked for quality, which involves defect inspection and metrology. Defects can be hard defects or soft defects. Hard defects involve appearance of Cr in the areas that needed to be clear, or the absence of Cr from areas where Cr needed to be present. In either case, the defects can be repaired if the defects are small in size. The repairs are made by tools, such as focused ion beam (FIB) or laser repair tools. In case of FIB repair, the opaque defect is repaired by local etching with FIB, and in case of clear defects, carbon films are patched on them by the FIB assisted deposition [17]. Soft defects relate to contamination, such as foreign particles on the plate, or stains and residues left from the previous processing. These can be removed by subsequent cleaning of the plate.

In the area of metrology the masks are to be checked for size of the critical features known as critical dimension (CD) and also for ensuring that features are at the correct location on the layout. This area is called the metrology of image placement or simply IP metrology.

More on quality of masks will be explained in Section 6.3.

TABLE 6.3
Properties of Mask Pattern Materials. (The Table was made by using the Source by Courtesy of ULVAC Coating Corporation and Konica Corporation.)

Mask Pattern Type		Binary			Att-PSM				Alt-PSM	
		Emulsion	Hard mask Regular Thickness	Hard mask Thin Thickness	I-line (365 nm)	KrF (248 nm)	ArF (193 nm)	F2 (157 nm)	Added Shifter	Etched Shifter
Material		Silver halide salt	Cr	Cr	MoSiON CrON	MoSiON CrF	MoSiON CrF	Shade SiON TaSiON ZrSiON	Shifter SOG Cr	Quartz Cr
			MoSi				ZrSiON			
Thickness		4–6 μm (After development)	ex. 100~100 nm	ex. 550 nm, 700 nm						
Optical characteristics	Optical Density	Over 3.5 (tangsten light source)	3.0 around 450 nm	3.0 at 248 nm	depend on material and wave length					
	Reflectance	7–10% (365 nm)	10~15% at 436 nm	10~15% at 248 nm	—	—	—	—		
	Transmittance	—	—	—	4%, 6%, 8%	3%, 6%	6%	6%		
	Phase value	—	—	—	180 degree, 175 degree inapplicable				180 degree	
Etching characteristics	Wet etch rate *1	—	1.7~2.7 nm/sec depend on wet etching condition	—	around 1 nm/sec strongly depend on dry etching condition					
	Dry etch rate *2	—	0.4~0.9 nm/sec strongly depend on dry etching condition	—						

*1 Etchant: Ceric Ammonium Nitrate $(NH_4)_2Ce(NO_3)_6$ 165 gr
(at 20C) Perchbric Acid $HClO_4$ 42 ml
DIwater H_2O
Total 1,000 ml

*2 Etchant Gas: ex. $Cl_2/O_2 = 4/1$ for Cr material

Optical Masks: An Overview

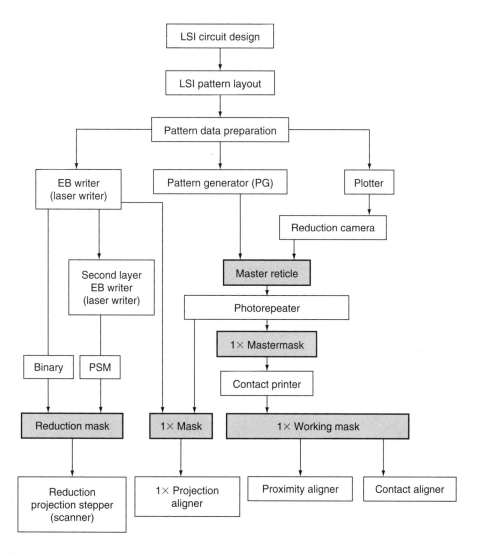

FIGURE 6.11
Fabrication process flows for 1× mask, reduction mask, and PSM. (Modified from original Figure 1.8, p. 19, of: I. Tanabe, Y. Takahana, M. Hoga, Introduction of Photomask Technologies, 1996, Printed in Japan.)

6.2.6 Pellicle

Even when a mask is checked out for cleanliness it is still susceptible to contamination and for this reason mask surface has to be protected from foreign particles by the use of pellicles [18,19]. Figure 6.13 shows the structure of mask mounted with pellicle. The pellicle film, which is transparent to the exposure light, is stretched on frame 6.3 mm in height as shown in Figure 6.13. The particle on film is not printed on wafer, since the film surface remains out of focus in the path of the imaging optics. It is necessary that the pellicle films meet the criteria of high transparency for exposure light, mechanical strength, and cleanliness, etc. The film materials for pellicle are summarized in Table 6.4 [20,21]. Nitro-cellulose based films were used as pellicle film until 365-nm lithograph. Recently, the amorphous fluoropolymer is used for lithography of 248 nm and beyond.

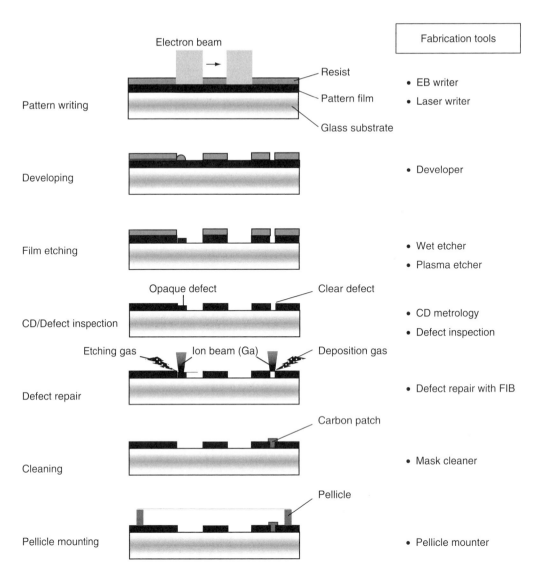

FIGURE 6.12
Process technologies of optical mask.

FIGURE 6.13
Structure of mask mounted pellicle (6-in. reticle).

Optical Masks: An Overview

TABLE 6.4

Materials for Pellicle. (Courtesy of Mitsui Chemicals, Inc and Asahi Glass Company.)

Pellicle Type	Nitrocellulose		Fluorinated Polymer
Film structure	Single layer	Both side AR	Single layer
Film material	Nitrocellulose(NC) 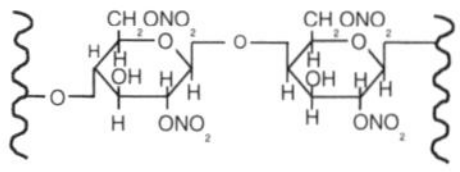	AR/MC/AR MC: Modified Cellulose AR: Fluorinated polymer	Amorphous fluorinated polymer 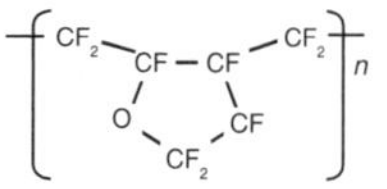
Exposure wavelength	G-line(436nm)	G-line(436nm) I-line(365nm)	KrF(248nm) ArF(193nm)
Film thickness	0.87μm@436nm	1.2μm@436nm 1.2μm@365nm	0.82μm@248nm 0.83μm@193nm
Transparency characteristic	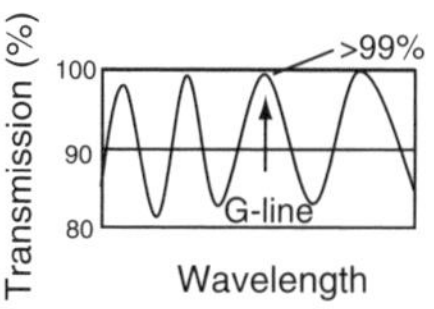 >99% @ G-line	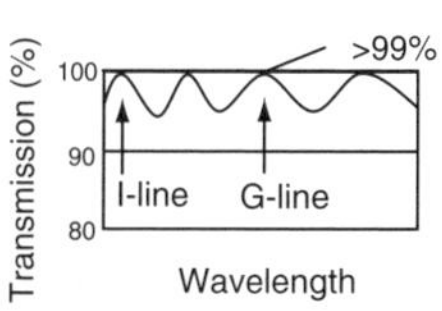 >99% @ I-line, G-line	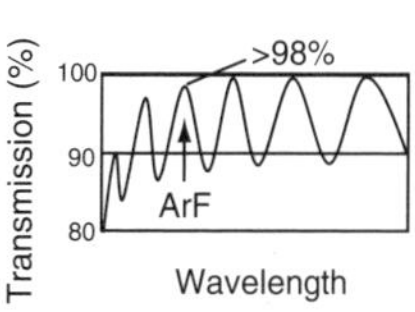 >98% @ ArF

6.3 Quality of Optical Mask

Figure 6.14 shows the quality factors in the case of Att-PSM; they are CDs, image placement, and defects. In addition, the figure also refers to phase error and transmittance that are not considered in binary masks. The quality factors are to be ensured by employing the following methods.

Table 6.5 shows trends in the quality factors. In order to meet the specifications, significant improvement in process and tools has been made to assure the availability of high-quality masks.

Sections 6.3.1–6.3.4 address the quality factors in more detail.

6.3.1 Critical Dimension

The CDs on mask are an important factor and influence the CDs on wafer. Since CDs depend on patterning processes, such as writing, development, and etching, they are

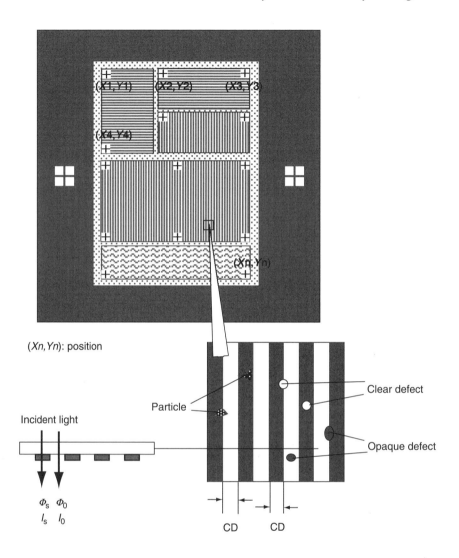

FIGURE 6.14
Mask quality items.

controlled by management of these process and tool conditions in mask production. Furthermore, in order to improve CDs, high-performance process technologies and tools are essential. Moreover, CD measurement is also important to ensure that their values meet the specs. Category of CD measurement tools is summarized in Table 6.6. For CD measurements, there are optical methods, electron beam methods, and mechanical methods. Optical methods can be transmittance, as well as reflective type, and are usually employed for CD measurements. Figure 6.15 shows schematic optics of transmittance type measurement system (Leica LWM 240) [23]. The CDs are collected by measuring the images using 240-nm wavelength light through the mask. Furthermore, to obtain measurement using the printability of image on wafer, an aerial image monitor system has been developed [24]. In recent days, the measurement of very small patterns, such as

TABLE 6.5
Trends of the Quality Items for Reticle in Reduction Projection Lithography. (The Table was made by using the Data from Original Table 3.1, p 80, of: I. Tanabe, Y. Takahana, M. Hoga, Introduction of Photomask Technologies, 1996, Printed in Japan, ISBN4-7693-1149-4 C3055. Data Courtesy of Asahi Glass Company and the data from ITRS [22].)

Design Node (μm)	2	1.3	0.8	0.5	0.3	0.2	0.15	0.13 (130 nm)	0.09 (90 nm)	0.065 (65 nm)	0.045 (45 nm)
Magnification	5X	←	←	←	←	4X	←	←	←	←	←
Patten CD (μm)	± 0.4	± 0.25	± 0.15	± 0.10	± 0.05	± 0.025	± 0.016	± 0.010	± 0.007	± 0.005	± 0.003
Pattern Placement (μm)	± 0.5	± 0.3	± 0.2	± 0.12	± 0.07	± 0.05	± 0.03	± 0.025	± 0.019	± 0.014	± 0.011
Pattern Defect (μm)	3.0	2.0	1.2	0.8	0.5	0.2	0.12	0.10	0.07	0.05	0.035

TABLE 6.6

Category of CD Measurement Tools.

Category	Optical		Electron Beam	Mechanical
	Transmission	Reflection	SEM	AFM
Measurement method	Incident light ↓↓↓↓ → Transmittance Image → Image Sensor	Incident light → Reflection Image → Image Sensor	Scanning electron beam, Detector, Secondary electron	Stylus scan
Signal shape	⎍	⎍	∧∧	⎍

OPCs, are made by low energy CD-SEMs [25]. As a mechanical method, CD measurement system using atomic force microscopy (AFM) was developed that can measure a 3-D shape of mask pattern [26].

6.3.2 Image Placement

Image Placement error is defined as a shift in the position of a feature from its design value. The image placement error can be caused by factors, such as error in writing and warp of substrate, etc. Therefore, the placement error is minimized by controlling the writer and substrate material. As shown in Figure 6.14, the placement measurement is done by measuring the positions of each cross marks placed in pattern chips as a measurement method. The mark positions are measured by using placement measurement system as shown in Figure 6.16 [27]. The system detects placement error using high accuracy stage equipped with laser interferometer.

6.3.3 Defect

Since defects on masks result in defects on wafer, it is essential that masks be free from all defects.

In the early days of semiconductor manufacturing, defects inspection was used to be done visually. In today's technology, defect quality is checked by defect inspection tools. Figure 6.17 shows methods of mask-pattern defect inspection tools. The earlier inspection tools used to employ a method of comparison between two chips by using images formed by the transmitted lights [28]. Later inspection systems compared the

Optical Masks: An Overview 153

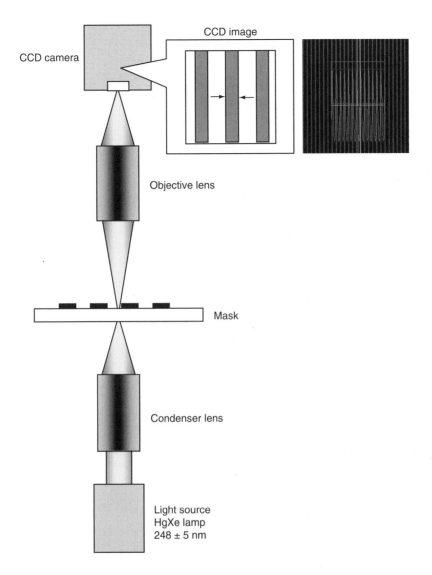

FIGURE 6.15
Pattern CD measurement system. (Courtesy of Leica Semiconductor Systems K.K.)

chip pattern with the design layouts [29]. Moreover, inspection systems using reflected light were also developed in order to detect particles on Cr features, since they could not be detected by transmitting lights [30]. Ideally, a good inspection system should be equipped with reflective, as well as transmissive, optics to be able to detect defects on the opaque, as well as transparent, region of the mask. This would ensure a 100% defect-free mask.

6.3.4 Phase and Transmittance Measurements

For PSM the detection of phase error is a key factor to ensure the mask quality. The phase error is defined as the deviation in phase from the prescribed phase angle of 180°. In the case of Att-PSM, the transmittance of the shifter is also an important factor. In order to

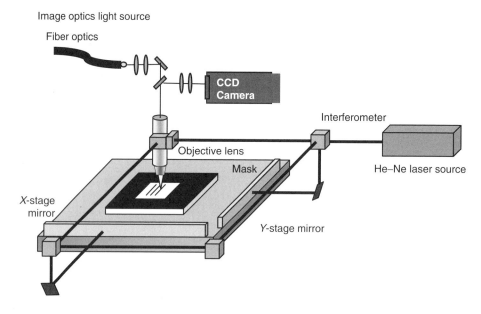

FIGURE 6.16
Pattern placement measurement system (simplified diagram). (Courtesy of Leica Semiconductor Systems K.K.)

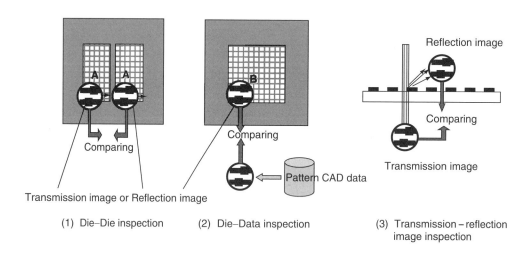

FIGURE 6.17
Methods of defect inspection tools.

measure the phase error and transmittance of shifter, phase and transmittance measuring systems have been developed [31,32]. Figure 6.18 shows an optical schematic of the Lasertec phase and transmittance measurement system. The Lasertec's systems have made a major impact on the implementation of PSM and are widely used for all generations of lithography such as 365, 248 and 193 nm.

Optical Masks: An Overview 155

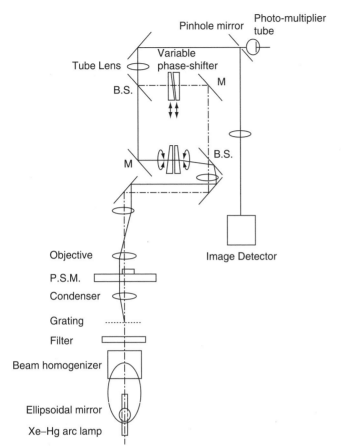

FIGURE 6.18
Phase measurements system for 248 nm wavelength. (Courtesy of Lesertec Corporation.)

References

1. L.F. Thompson, C.G. Willson, and M.J. Bowden, *Introduction to Microlithography*, 2nd ed., ACS Professional Reference Book, 1994, pp. 21–30.
2. D.A. Markle, A new projection printer, *Solid State Technol.*, 17 (6), 50–53 (1974).
3. J. Roussel, Step-and-repeat wafer imaging, *Solid State Technol.*, 21 (5), 67–71 (1978).
4. Figure (Subwavelength Gap) opened to the Public from Numerical Technologies (at present, Synopsys Inc.) in Seminar report: Advanced Lithography. Semiconductor European, March 25, 1999.
5. M.D. Levenson, N.S. Viswanathan, and R.A. Simpson, Improving resolution in photolithography with a phase-shifting mask, *IEEE Trans. Electron Devices*, ED-29, 1828–1836 (1982).
6. T. Terasawa, N. Hasegawa, H. Fukuda, and S. Katagiri, Imaging characteristics of multi-phase-shifting and halftone phase-shifting masks, *Jpn. J. Appl. Phys.*, 30, 2991–2997 (1991).
7. N. Yoshioka, J. Miyazaki, H. Kusunose, K. Hosono, M. Nakajima, H. Morimoto, Y. Watakabe, and K. Tsukamoto, Practical attenuated phase-shifting mask with a single-layer absorptive of MoSiO and MoSiON for ULSI fabrication, in: *International Electron Devices Meeting 93*, 1993, pp. 653–656.
8. M. Nakajima, N. Yoshioka, J. Miyazaki, H. Kusunose, K. Hosono, H. Morimoto, W. Wakamiya, K. Murayama, Y. Watakabe, and K. Tsukamoto, Attenuated phase-shifting mask with a single-layer absorptive of CrO, CrON, MoSiO and MoSiON, *Proc. SPIE*, 2197, 111–121 (1994).
9. F.M. Schellenberg, H. Zhang, and J. Morrow, SEMATECH J111 project: OPC validation, *Proc. SPIE*, 3334, 892–911 (1998).

10. M. Takeuchi, Y. Shibano, and S. Kusama, Properties of our developing next generation photomask substrate, *Proc. SPIE*, 3748, 41–51 (1999).
11. Y. Ikuta, S. Kikugawa, T. Kawahara, H. Mishiro, N. Shimodaira, H. Arishima, and S. Yoshizawa, New silica glass for 157 nm lithography, *Proc. SPIE*, 3873, 386–391 (1999).
12. Y. Watakabe, S. Matsuda, A. Shigetomi, M. Hirosue, T. Kato, and H. Nakata, High performance very large scale integrated photomask with a silicide film, *J. Vac. Sci. Technol. B.*, 4 (4; Jul./Aug.), 841–844 (1986).
13. H. Mohri, M. Takahashi, K. Mikami, H. Miyashita, N. Hayashi, and H. Sano, Chromium-based attenuated phase shifter for DUV exposure, *Proc. SPIE*, 2322, 288–298 (1994).
14. T. Matsuo, K. Ohkubo, T. Haraguchi, and K. Ueyama, Zr-based film for attenuated phase shift mask, *Proc. SPIE*, 3096, 354–361 (1997).
15. T. Motonaga, M. Ohtsuki, Y. Kinase, H. Nakagawa, T. Yokoyama, H. Mohri, J. Fujikawa, and N. Hayashi, The development of bilayer TaSiOx-HTPSM, *Proc. SPIE*, 4409, 155–163 (2001).
16. O. Nozawa, Y. Shioya, H. Mitsui, T. Suzuki, Y. Ohkubo, M. Ushida, S. Yusa, T. Nishimura, K. Noguchi, S. Sasaki, H. Mohri, and N. Hayashi, Development of attenuating PSM shifter for F2 and high transmission ArF lithography, *Proc. SPIE*, 5130, 39–50 (2003).
17. K. Hiruta, S. Kubo, H. Morimoto, A. Yasaka, R. Hagiwara, T. Adachi, Y. Morikawa, K. Iwase, and N. Hayashi, Advanced FIB mask repair technology for ArF lithography, *Proc. SPIE*, 4066, 523–530 (2000).
18. V. Shea, and W.J. Wojicik, Pellicle Cover for Projection Printing System, U.S. Patent 4131363, Dec. 26, 1978.
19. T.A. Brunner, C.P. Ausschnitt, and D.L. Duly, Pellicle mask projection for 1:1 projection lithography, *Solid State Technol.*, May, 135–143 (1983).
20. S. Shigematsu, H. Matsuzaki, N. Nakayama, and H. Mase, Development of pellicle for use with ArF excimer laser, *Proc. SPIE*, 3115, 93–103 (1997).
21. M. Nakamura, M. Unoki, K. Aosaki, and S. Yokotsuka, New low K dielectric fluoropolymer suitable for interlayer insulation materials, in: *International Conference on Multichip Modules (ICMCM)'92 Proceedings*, 1992, pp. 264–269.
22. International Technology Roadmap for Semiconductors, 2001 edition, Dec. 2001, pp. 241–257.
23. G. Schlüter, G. Scheuring, J. Helbing, S. Lehnig, and H.-J. Brück, First results from a new 248 nm CD measurement system for future mask and reticle generation, *Proc. SPIE*, 4349, 73–77 (2001).
24. R.A. Budd, J. Staples, and D. Dove, A new mask evaluation tool, the microlithography simulation microscope Aerial Image Measurement System, *Proc. SPIE*, 2197, 530 (1994).
25. I. Santo, M. Ataka, K. Takahashi, and N. Anazawa, Calibration and long-term stability evaluation of photo mask CD-SEM utilizing JQA standard, *Proc. SPIE*, 4889, 328–336 (2002).
26. Y. Tanaka, Y. Itou, N. Yoshioka, K. Matsuyama, and D. Dawson, Application of atomic force microscope to 65 nm node photomask, in: *Digest of Papers, Photomask Japan 2004*, 2004, 139–140.
27. C. Bläsing, Pattern Placement Metrology for Mask Making, Technical Programs on SEMICON Europe in March 1998.
28. P. Sandland, Automatic inspection of mask defects, *Proc. SPIE*, 100, 26–35 (1977).
29. I.A. Cruttwell, A fully automated pattern inspection system for reticles and masks, *Proc. SPIE*, 394, 223–227 (1983).
30. F. Kalk, D. Mentzer, and A. Vacca, Photomask production integration of KLA STARlight 300 system, *Proc. SPIE*, 2621, 112–121 (1995).
31. H. Kusunose, A. Nakae, J. Miyazaki, N. Yoshioka, H. Morimoto, K. Murayama, and K. Tsukamoto, Phase measurement system with transmitted UV light for phase-shifting mask inspection, *Proc. SPIE*, 2254, 294–301 (1994).
32. H. Takizawa, H. Kusunose, N. Awamura, T. Ode, and D. Awamura, Transmittance measurement with interferometer system, *Proc. SPIE*, 2793, 489–496 (1996).

7

Conventional Optical Masks

Syed A. Rizvi

CONTENTS
7.1 Introduction 157
7.2 Classification of Optical Masks 158
    7.2.1 Conventional Optical Mask 159
    7.2.2 Advanced Optical Masks 159
7.3 Making Conventional Masks 159
    7.3.1 Starting Material 159
    7.3.2 Mask Writing 160
    7.3.3 Mask Processing 160
    7.3.4 Mask Qualification 160
        7.3.4.1 Mask Inspection and Repair 160
        7.3.4.2 Metrology 160
    7.3.5 Pellicles 160
7.4 General Comments 161

7.1 Introduction

This chapter is a prelude to Chapter 8 that addresses *Advanced optical masks*. In order to fully appreciate the innovation and technologies that led to the development of the advanced optical masks, it is necessary to know something about the conventional optical masks behind the backdrop of which these technologies and innovation took place.

Unlike Chapter 6, which gives an overview of optical masks in general, this chapter focuses exclusively on conventional optical masks, whereas Chapter 8 provides a detailed description of advanced optical masks technologies.

Chapter 1 in this book has given an overview on "all types" of masks. The connotation behind this "all types" of mask requires knowledge of the evolution of masks in the semiconductor industry.

In the early 1960s, when the semiconductor industry was in its infancy, the masks consisted of glass plates with emulsion patterns, which were not much different from the black and white negatives used in photography in its early days. Later on, the emulsion patterns on glass plates were replaced by chrome pattern on glass (CoG) plates because the chrome material was less prone to damage when mask and wafer had to be brought into intimate contact during exposure.

But as the requirement on mask quality became more stringent even the CoG masks could not meet the specifications, since the contact printing on wafers could still cause damage to the masks, although to a lesser degree.

The answer to mask damage was sought in "off-contact," also known as "proximity" printing, where the mask is brought in close proximity of the wafer without making a real contact between the two. This approach was a form of shadow casting of the mask features on the wafer that required a tradeoff between mask's life and resolutions. With continuous demands for higher resolutions the practice could not be continued.

The next step was then to project the image of the entire masks on the wafer through some forms of 1:1 optics, where the desired resolution could be obtained without the mask coming into direct contact with wafers. Early systems had utilized refractive optics consisting of set of lenses, but to project the image of the entire mask and still attain high resolution required large-diameter and almost perfect lenses that were not available in those days. In fact very few such machines were built and their use was very limited. The 1:1 projection printing really took off when Perkin-Elmer introduced its first "all reflective" scanning-mirror optics, which is capable of imaging the entire mask on the wafer and also attaining desirable resolutions. Use of mirror also did away with chromatic aberration encountered in lens optics. The scanning mirror technology lasted almost a decade until wafer steppers were introduced into the fabs. Wafer stepper technology was essentially an extension of the fully matured technology of mask steppers that were routinely applied for making masks in those days.

The masks consisted of an array of dies that were made by stepping de-magnified images of reticle across the entire mask area. Typically, the ratio of the reticle image to the image of die on wafer used to be 10:1. In case of wafer stepper also, the practice of 10:1 reduction continued for sometime. In today's technology, most machines utilize stepping, as well as scanning, mechanism for image transfer with 5:1 and 4:1 reduction ratios.

The evolution of optical lithography also caused changes in the design and structure of optical mask. Although the transformation from emulsion mask to chrome mask could be viewed as significant change in mask structure, the most radical change in mask making came about with the introduction of PSM and OPC that are referred as advanced optical masks in this book. The masks that do not employ these techniques are referred as conventional masks, which form the subject of this chapter.

Beyond the optical masks there are a number of nonoptical masks over the horizon belonging to various nonlithographic techniques referred as next generation lithography.

Although advanced optical masks remain as the focus for many chip manufacturers, the masks that do not incorporate OPC or phase shifting in their structure continue to be used by those manufacturers for their noncritical levels and claim a major share of the mask market. These masks are to be regarded as the basis on which the advanced mask technology is shaped and is labeled by the author as "conventional masks."

7.2 Classification of Optical Masks

The masks can be classified in a number of ways, and the most common set of the criteria for classification is shown in Figure 7.1. The classification of optical masks is discussed in the following.

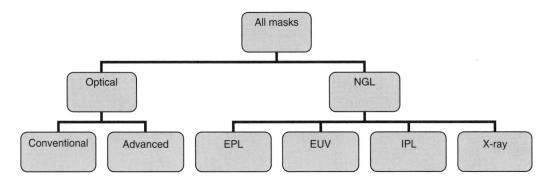

FIGURE 7.1
Classification of masks.

7.2.1 Conventional Optical Mask

This is the basic mask structure and represents a generic mask, since it consists of CoG mask. The CoG mask can be of two types. It can be a light field mask with CoG background, or it can be dark field mask which means features are etched into the chrome film covering the entire mask.

7.2.2 Advanced Optical Masks

The masks with OPC are essentially CoG masks as mentioned earlier, except that these have some "very small features" (called serif, scattering bars, assist, etc.) added to the already existing features to offset the diffraction effects that occur as the industry moves towards the smaller features. Hence, in principle, the processing of OPC mask is not different from the processing of conventional mask except that OPC masks require more refinement and precision in their making.

The structure of phase shift masks may be just CoG masks, or some other phase shifting material (e.g., MoSi) on glass. There are also phase shift masks where the features are simply etched into glass to induce phase shifting during the printing. These masks have no chrome or any phase shift material on them. However, during the processing of these masks, chrome film can be employed as will be mentioned later in chapter 8.

7.3 Making Conventional Masks

Conventional masks form the basis for all optical masks, since fabrication of all other types of optical masks incorporates the basic steps employed in the making of conventional masks.

7.3.1 Starting Material

The starting material in mask making continues to be glass plates coated with a film of chrome that is covered with another film of some kind of photosensitive or electron beam-sensitive material known as photoresist or simply resist. The final material is known as mask blanks.

7.3.2 Mask Writing

The mask blank is then mounted on the stage of an electron beam writer or laser writer depending upon the type of the resist it is coated with. The beams are scanned and manipulated to selectively expose the resist to form desired patterns on the resist film after the plates are developed.

7.3.3 Mask Processing

Develop: The exposed plates are developed using the chemicals that can dissolve out the exposed resist as in the case of positive resist, or by using the chemicals that can dissolve out the unexposed resist in the case of negative resist. In either case, a resist pattern can become visible after the develop cycle. This resist pattern is formed on the chrome surface exposing parts of the chrome to be etched in the next step.

Etch: The plate is then subjected to the etch process, where the exposed chrome is etched away leaving the glass surface behind. The unexposed chrome remains protected under the resist. The etching can be done by wet chemicals or dry plasma process.

Resist removal and final cleanup: In the next and final step, the resist is removed and the plate is sent through a cleanup operation. The resist can be removed using wet chemicals, as well as oxygen plasma that is regarded as the dry process.

7.3.4 Mask Qualification

7.3.4.1 *Mask Inspection and Repair*

Once the mask is cleaned it is ready for inspection and repair. There are a number of machines in the market that can do inspection. The machines for repairing opaque defects simply zap the defect with the high power laser, whereas clear defects are repaired by spot deposition of gallium or any material with the properties similar to that of gallium.

7.3.4.2 *Metrology*

Two important parameters to be measured before the mask can be qualified are size of any prescribed critical feature known as critical dimension (or simply CD) and the position of features referred as image placement (IP) measurement.

For CD metrology, there are many systems in the market, but only few of them can measure IP. Leica IPRO is one of them.

Thin film measurement is also important, and that is typically employed for measuring thicknesses of chrome and resist at the beginning of the process, at the time of fabrication of mask blanks.

7.3.5 Pellicles

The final step in the mask fabrication flow is the mounting of pellicle on mask to protect its surface from particles in the ambient that may fall on the mask surface and contaminate it.

7.4 General Comments

The discussion here describes the fabrication of conventional optical masks, and the process is generic in nature. Deviation form the process can occur especially when making the state-of-the-art optical masks, which has been explained in detail in Chapter 8. The content of this chapter due to its basic nature sets the stage for Chapter 8.

8
Advanced Optical Masks

Wilhelm Maurer and Frank Schellenberg

CONTENTS

8.1 Resolution Enhancement Technologies 164
8.2 Alternating Phase Shift Masks 165
    8.2.1 History and Basic Functionality 165
    8.2.2 Manufacturing of alt.PSM 168
    8.2.3 Layout Considerations for alt.PSM: phase conflicts 171
    8.2.4 Special Cases of alt.PSM 173
8.3 HT.PSM 175
    8.3.1 Basic Functionality 175
    8.3.2 Manufacturing of HT.PSM 175
    8.3.3 High Transmission HT.PSM and 3tone.PSM 177
8.4 Subresolution Assist Features 179
    8.4.1 Basic Principle and Application 179
    8.4.2 Manufacturing and Metrology Issues of SRAF 181
8.5 Optical Proximity Effect Correction 182
    8.5.1 Basics and OPC Strategies 182
    8.5.2 Manufacturing and Metrology Issues 184
8.6 Outlook for RET 186
References 186

As lithographers constantly strive towards smaller feature sizes in accordance with Moore's law, all components of the lithographic pattern transfer process are being pushed to their extreme limits. Exposure tools are using wavelengths as short as possible, and lenses are closing in on the NA = 1.0 limit. Resists are constantly being improved in contrast and other essential properties. In addition, resolution enhancement techniques (RET) have been developed that, by manipulating the patterns on the mask, engineer the wavefront of light passing through the mask. This allows high volume processing closer and closer to the limits of resolution of a lithography system. For this reason, mask makers will need to learn all they can about these new techniques and their impact on mask making.

8.1 Resolution Enhancement Technologies

More than 130 years ago, Ernst Abbe established that the resolution limit of a "reasonably well built" imaging system (in his case, a microscope) depends only on the numerical aperture of the objective lens and the wavelength of the illuminating light, if this illumination is perpendicular to the image/object plane [1]. In lithographic imaging systems, the ratio of the minimum resolved feature over the optical resolution factors identified by Abbe has gained wide use over the last ten years [2]. This ratio is commonly known as k1 factor and is defined as

$$k1 = \text{min.feature}/\text{opt.resolution} = \text{min.feature}/(\lambda/\text{NA})$$

with the min.feature being half of the smallest pitch to be printed, and the optical resolution being ratio of wavelength λ over NA, the numerical aperture of the imaging lens.

So the ever-increasing demand to print smaller features than enabled by the optical resolution of the exposure tools provided can be described as a constant decrease of k1. Table 8.1 shows this decrease since 1990 in semiconductor manufacturing [3].

For the basic optical concept used in lithography exposure tools (commonly called "steppers," although most of them nowadays perform step and scan operation), a k1 of 0.25 is generally regarded as the theoretical limit [4]. Beyond that, the light emerging from the stepper lens does not show any variation in intensity (aerial image) over the image field, provided that the whole image contains only minimum features.

Approaching this limit, the aerial image decreases rapidly in contrast, and in its fidelity relative to the mask image. To counteract the first effect, numerous strategies of RET have been developed and are the subject of extensive research and development efforts [5]. To compensate for the latter effect, which is called the optical proximity effect in analogy to the previously known proximity effect of imaging using electrons, optical proximity correction (OPC) was established [6].

One special aspect of the lack of fidelity of a low k1 pattern transfer process is that the linearity is also compromised. A small change in a mask dimension can be amplified into a serious change at the wafer. This leads to another factor that has tremendous impact on mask making. Since it seems to enhance the mask errors, it has been named mask error enhancement factor, or MEEF [7]:

$$\text{MEEF} = \delta(\text{wafer linewidth})/\delta(\text{mask linewidth}) \qquad (8.1)$$

TABLE 8.1

k1 for Different Semiconductor Device Generations

Device Generation (nm half pitch)	Exposure Wavelength (nm)	Numerical Aperture of Stepper Lens	k1
500	365	0.5	0.68
250	248	0.5	0.5
200	248	0.6	0.48
180	248	0.63	0.46
140	248	0.75	0.42
110	193	0.75	0.43
90	193	0.85	0.40
70	157	0.75	0.33
50	157	0.85	0.27

Advanced Optical Masks

An ideal linear process has an MEEF of 1.0. Obviously, adoption of processes that increase MEEF puts serious requirements on mask specifications in order to control the overall process. And, a process in which MEEF is reduced below 1 is a process highly tolerant of mask errors. Certain phase shifting masks (PSM) can reduce MEEF below 1 and in some cases allow a $k1 = 0.25$. This can double the effective resolution of a lithography system.

Certain RET strategies do not have any impact on mask making. One prominent and widely used example is off-axis illumination [8]. Here, the angle of the light illuminating the reticle is controlled, allowing for certain diffracted orders to be emphasized in the imaging system. This is generally viewed as an RET implemented through the lithography system and is beyond the scope of a mask making text.

The RET and OPC strategies that do have an impact on mask making can be grouped into three main categories:

- RET that require etching into the mask substrate
- RET that use different mask substrates than the established chrome on quartz
- RET that demand higher resolution at mask making

The first two groups are the PSM, with alternating PSM (alt.PSM) as the most prominent member of the first group, and halftone PSM (HT.PSM) as the most prominent of the second group.

Although many of the PSM approaches have been predicted to have superior lithographic imaging quality, only HT.PSM made it into the high volume production of integrated circuits. The reason for this is that this PSM option provides its benefit with minimal overhead to the user: all the phase technology is embedded in the fabrication of the mask layers themselves. This indicates how important it is not only to concentrate on the technical benefit of an option but also to take potential issues into account. So we will try in the following to indicate also the major issues of each particular PSM option. However, an exhaustive treatment of the cost of ownership of PSM options is not the subject of this task.

Sub-resolution assist features (SRAF) [9,10] and OPC are two very common techniques in the third group. Printing at low k1 — with or without any other RET strategy — demands OPC. The lower k1 a given RET strategy allows, the more aggressive the OPC strategies have to be.

Combinations of different RET options, such as PSM and off-axis illumination (always followed by OPC), are routine ways of applying these technologies. The specifications for these masks are also often tighter, forcing mask making equipment to be used at the limits of its capability.

8.2 Alternating Phase Shift Masks

8.2.1 History and Basic Functionality

Phase shift masks (PSM, also called phase shifted masks or phase shifting masks) not only change the intensity distribution of the transmitted light as do a regular (chromium-on-glass, COG) mask but also (intentionally) vary the phase of the transmitted light at some parts of the mask. The basic principle of imaging a phase transition is not new. Lord Rayleigh's treatment of light from neighboring phase-shifted apertures was documented at the end of the 19th century [11], and even Ernst Abbe's lectures, as transcribed by Lummer in 1910 [12], describe the imaging expected for PSM in detail. The superior image quality achievable by transformation of phase differences of the object into amplitude differences of the final image (the human eye is incapable of detecting phase differences) has been widely used in phase contrast microscopy for nearly a century, and phase holograms have long been known as the most efficient kind of holograms.

A holographic storage project by Hänsel in the mid-1970s created combined amplitude and phase masks that are essentially equivalent to modern PSM [13]. However, the application of phase shifting for improved resolution in lithography, initially proposed by Hank Smith of MIT [14] and also by M. Shibuya at Nikon [15], was first realized in the early 1980s by M.D. Levenson at IBM [16,17]. Levenson fabricated the first lithographic phase masks by overcoating a normal mask with PMMA and patterning that layer with electron beam exposure. His papers also described the theory of imaging with phase masks and have become one of the most cited papers in lithography today.

Since then an almost incomprehensible number of PSM options have been proposed. These are usually evaluated first conceptually through simulation with process modeling tools, such as ProLith [18], Solid [19], or EM Suite [20]. Most of those tools now also provide additional, more accurate modeling of the mask topography itself, which was first introduced by TEMPEST [21]. Although the modeling of imaging properties is the most common method for comparing and evaluating the expected performance of various OPC and PSM approaches, the topic of simulation is treated elsewhere in this volume.

Alternating phase shift masks (alt.PSM, also called alternating aperture PSM) improve the process window of a narrow dark feature by providing a 180° phase transition between the two bright features defining this dark feature. Figure 8.1 shows the basic principle of alt.PSM.

The light transmitted by the two openings has a 180° phase difference, so the electric field changes sign at the transition. Light intensity is the square of the electrical field.

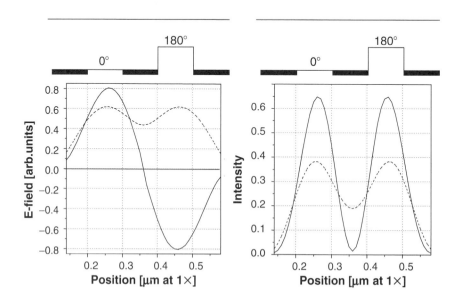

FIGURE 8.1
Aerial image simulation for a photomask with two line openings of 100 nm width, separated by a 100-nm chromium line. If the light transmitted by one opening has a phase difference of 180° to the light transmitted by the other opening (full line), the electrical field (left figure) changes sign, and the contrast (right figure) is much higher than without a phase difference (dashed line). Simulation for 193 nm, 0.75 NA, and a coherence factor (σ) of 0.3.

Advanced Optical Masks

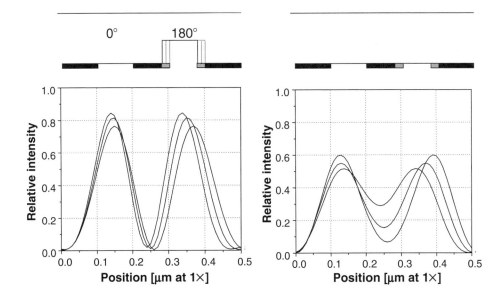

FIGURE 8.2
Intensity distribution comparison between alt.PSM (left) and regular chromium mask (right) printing a dark feature for three feature widths (120, 100, and 80 nm) between two 120-nm openings. Aerial image simulation for 193 nm, 0.75 NA, and a coherence factor (σ) of 0.3.

Since the square of zero remains zero, the intensity at the transition is also zero — regardless of how narrow the lateral width of the transition is. In contrast to alt.PSM, the minimum intensity of two openings of a COG mask increases as the width of the dark feature between the openings decreases. Figure 8.2 compares this behavior for alt.PSM and COG mask.

Figure 8.3 illustrates a particularly attractive feature of PSM. As the width of the dark feature between two openings with opposite phase on an alt.PSM decreases, the width of the printed feature converges to a constant value dictated by the imaging system, no matter how small the feature width is on the mask. This means that the MEEF is driven

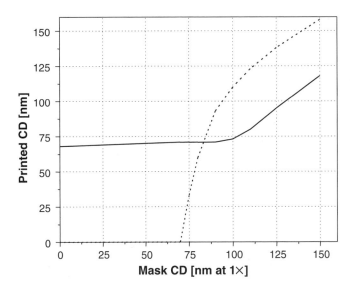

FIGURE 8.3
Printed line width of a dark mask feature between two 150-nm openings as function of feature width. As the mask feature width decreases, the printed width decreases rapidly without phase contrast (dashed line) but approaches a constant width with phase contrast (solid line). Simulation for 193 nm, 0.75 NA, and a coherence factor (σ) of 0.3.

essentially to 0. For COG masks at these dimensions, however, small variations in the feature width on the mask are magnified into significant width variations in the printed image, until at some minimum resolvable dimension, catastrophic imaging failure occurs. These high MEEF processes — current 90-nm processes have to live with MEEF numbers as large as 5 — demand tighter mask specifications (= higher cost, lower throughput).

Because of its small MEEF, alt.PSM found its first broad application in printing the gate level of microprocessors [22]. Variation in the width of the active gate structures (= the effective transistor length) is one of the major limiting factors for the maximum clock speed of the processor.

8.2.2 Manufacturing of alt.PSM

The first alt.PSM were manufactured by coating a COG mask with an e-beam resist, which is transparent, and which provides the 180° phase shift by its appropriate thickness. E-beam exposure was used to leave material only on these openings that require phase shift. This was the technique first used to fabricate phase masks [16]. However, the alt.PSM generated by this process could not sustain the rigorous cleaning processes necessary to deliver masks with zero defects. Various other added materials (e.g., spin-on glass) were evaluated. Today, almost all alt.PSM are made by selectively etching into the quartz substrate of regular COG masks [23]. Figure 8.4 shows an example of the manufacturing of a contemporary alt.PSM.

An alt.PSM requires at least two lithographic patterning steps. Usually — as indicated in Figure 8.4 — the first step defines all mask openings, and the second step defines only the structures created as phase features by etching into quartz. Using this strategy, and using an etch process optimized for high selectivity between chromium and quartz, the requirements to the second step with regard to alignment and resolution can be substantially relaxed, since the chromium patterned with the first high-resolution exposure serves as the actual mask for the etching step. In this case, the much cheaper laser writer can be used to delineate the mask openings with 180° phase difference.

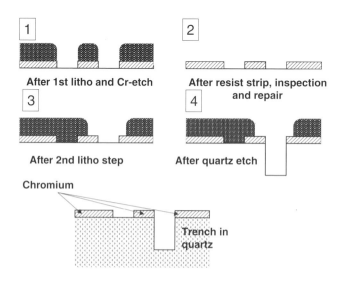

FIGURE 8.4

Manufacturing of an alt.PSM with phase shifter etched into chromium. For a nominal 180° phase shift, the etch depth for quartz is $\lambda/(2(n-1))$, with n as the refraction index of quartz.

Advanced Optical Masks

The image printed by an alt.PSM is not only a function of the width of the features on the mask but is also critically determined by other mask parameters, such as phase or the detailed structure of the etched trenches. So the most straightforward strategy to measure and qualify an alt.PSM would be to expose a wafer using the mask on the tool where it will be used, and to measure the printed image. Obviously, it is impossible to have in a mask house all tools and processes of its customers, and such a strategy also requires too much effort to be done in the wafer fabrication. In addition, the assessment of the printed image just shows the final effect and can only suggest which parameter of the mask may be out of specification. So the mask house only can measure and control all these mask parameters that determine the quality of the printed image, in order to finally provide a usable mask.

The most prominent of these parameters obviously is the phase shift between adjacent openings. If the phase shift is generated by etching of the mask blank material (quartz), which usually is very homogeneous, this difficult phase measurement can be substituted by a much easier measurement of etch depth. Atomic force microscopy (AFM) can provide the required accuracy for this measurement. As it is the case for CD (critical feature width), phase measurements have to be done on many locations on the mask, and mean and uniformity of the measured results are the two parameters to be controlled. Simulation and experiment determine the required specification. Results of $\pm 1°$ for both mean to nominal (usually 180°) and 3sigma uniformity have been reported, which are clearly sufficient for today's 90-nm technologies [24].

In addition to the phase measurement, the measurement of feature width also needs particular consideration regarding its calibration. The response of a CD metrology tool typically used in mask making (be it an optical tool or an SEM) depends on the actual topology of the measured object. Most important, the width of features printed on a wafer using an alt.PSM also critically depends on the actual three-dimensional structure mask. Since no calibrated standard is available, high-resolution SEM images of cross-sections of masks are used to correlate the results delivered by the CD metrology tool to actual feature sizes.

By the same token, the topography of the mask — the actual morphology of the etched edges in particular — needs sufficient control and specification. With alt.PSM finally ready to become a standard production item of mask houses, these facts become of higher concern to mask makers and mask users, and draw the attention of standardization efforts.

Although these individual measurements provide an indication that the phase mask has been correctly fabricated, they cannot be practically applied with great resolution throughout the mask area. For that, mask inspection is required, and any defects that are found must be repaired before a mask is shipped. For phase masks, this becomes harder than inspection and repair for COG masks, because of the variety of defects that may occur with the addition of phase shifting structures. In addition to the defects common in COG masks (discussed elsewhere in this book), alt.PSM has at least two more defect types: quartz bumps and quartz divots [25]. In most cases, these two are further differentiated into bumps and divots of arbitrary height/depth, and into those, whose height corresponds to a phase shift of 180°.

As was mentioned earlier about feature width, a quartz bump and a quartz divot, which appear to have the same size in the review microscope of a defect inspection tool, will have different sizes when assessed by an SEM and, in particular, very different sizes in the image printed using that mask. So the assessment of the effective size of a defect on an alt.PSM has been the main application of the aerial image measurement system (AIMS), a microscope which simulates the imaging process of a lithographic exposure tool [26]. Recently, a defect inspection tool has been constructed, which uses the optical setup of an AIMS to detect defects [27]. This tool works only in die-to-die inspection mode, so it requires two identical chips on one mask. It will be interesting to see the introduction of this tool into production and also its extension to die-to-database inspection; the latter will demand additional effort in fast image simulation.

Defects consisting of excess (not sufficiently etched) quartz can — at least in principle — be repaired either by ion-beam assisted etching or by a mechanical repair tool, which is based on the AFM basics. Until now, the repair of missing quartz defects has been tried, but no successful repair has been demonstrated. So adjusting the mask manufacturing strategy towards avoiding defects of missing quartz is the only way to avoid yield loss and intolerable low throughput.

The most challenging demand to the manufacturing process of alt.PSM is made by the issue of phase balancing [28]. As indicated earlier, two mask openings with a nominal 180° phase difference and of the same physical size (as assessed by an SEM, a CD measurement tool, or an AFM) will in most cases not print with the same size on a wafer. The electromagnetic field transmitted by a trench is quite different than the field transmitted by just an opening between two chromium features — as Figure 8.5 clearly shows.

Figure 8.6 illustrates the consequences on the wafer. Even when the size of each opening is adjusted by biasing, so that they print at the center of the process window of the printing process with the same size at best focus, the sizes will show a diverging behavior for different defocus settings of the exposure tool. The left part of Figure 8.6 shows these relationships.

In simulation, equivalent behavior of the printed images of the two openings can be described by assuming a phase difference between the two openings different from 180°. It appears, as if the phase difference generated by the mask topology is changed into an "effective phase." Further investigation shows that this effective phase is a function of the mask geometry like feature width and feature pitch.

Combining an optimized slope and geometry of the phase edges with under-etching the quartz beyond the chromium, and with an additional etch step, which also etches into the quartz at all mask openings, an almost perfect phase balance over defocus could be achieved by concerted efforts of simulation and process development for a given mask geometry, as demonstrated in the right part of Figure 8.6 [29]. Although the same

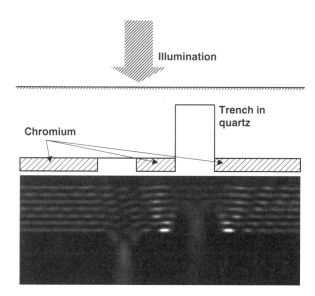

FIGURE 8.5
Two-dimensional intensity ($|E|^2$) plots of light transmitted through an alt.PSM with two openings, each 80 nm wide, separated by 80 nm aerial image simulation of EM fields for $\lambda = 193$ nm and normal incidence.

Advanced Optical Masks

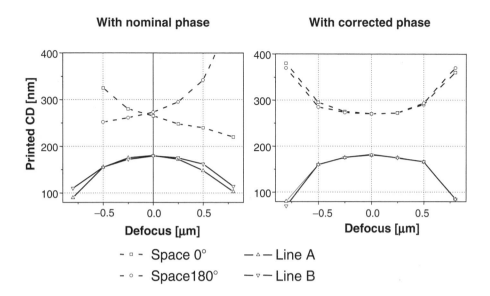

FIGURE 8.6
Illustration of intensity imbalance. *Left*: The width of spaces (dashed lines) printed for an alt.PSM is significantly different for the 0° and the 180° openings through defocus, even when the etch depth is the nominal 180°, and when the openings have been biased to correct for the intensity imbalance. *Right*: Optimizing the etch depth and sidewall structure corrects for this. The width of the dark features (solid lines) between the openings is unaffected and does not need optimization. [Adapted from U. Griesinger, R. Pforr, J. Knobloch, C. Friedrich, *Proc. SPIE*, 3873, 359–369 (1999).].

parameters can be used to minimize the phase in-balance for arbitrary geometries, this effect will need attention in the error budget of the lithography process [30].

An alternative manufacturing method of alt.PSM has been to first etch the phase topography into the quartz, and then deposit chromium over all mask structures with constant thickness. When the chromium is then patterned to define the dark and the transparent features on the mask, it is done such that all phase edges remain covered by chromium. It has been shown by simulation and experiment that this so-called "phase phirst" strategy has much less phase imbalance [31]. It remains to be shown if this is the case for all arbitrary geometries.

8.2.3 Layout Considerations for alt.PSM: Phase Conflicts

Data management provides additional challenges for the mask maker. An alt.PSM generally must be patterned twice, once for patterning apertures and a second time to define which apertures are to be phase shifting. Some phase shifting schemes require at least four writing steps, making the writing process much more expensive [32].

Layout adaptation is required for PSM because certain conventional circuit structures cannot be easily adapted to the requirements of phase assignment. In fact, it is impossible to provide a 180° phase difference between all small features for arbitrary layouts. A simple example of a T-structure is shown in Figure 8.7.

If all lines of the T are small enough to require support by phase in printing, there is no simple phase assignment scheme that allows this to occur. This is a good example of a so-called "phase conflict" [33]. The simplest solution of this problem is to restrict the layouts to be printed by alt.PSM to contain no such structures provoking phase conflicts. This can,

FIGURE 8.7
Layout with a T consisting of lines with a width requiring phase assistance to print with sufficient process window.

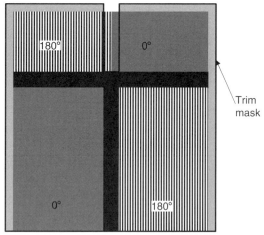

FIGURE 8.8
Solution to the phase conflict at the center of the T of Figure 8.7. In order to support printing of all lines with phase assistance, a dark line is created by the 0°/180° phase transition above the T. This line has to be cleared by an exposure of the second, the so-called trim mask.

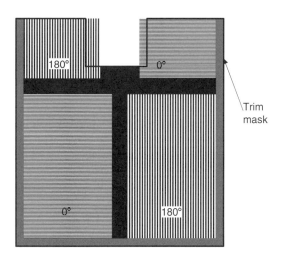

FIGURE 8.9
Solution to the phase conflict at the center of the T of Figure 8.7 by increasing the line width at the center portion of the T to a width not needing phase assistance. This considerably relaxes the requirements for line width and placement accuracy of the trim mask.

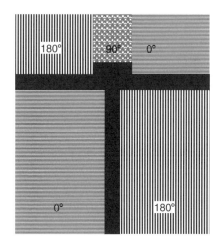

FIGURE 8.10
Solution to the phase conflict at the center of the T of Figure 8.7 by using a partial shifter. A region with intermediate phase can eliminate the need for a trim mask.

for example, done by requiring a design rule check with appropriate design rules for each layout before it is released for mask making.

Another strategy is to arrange the phase structures in a way that provides the phase difference for almost all the features except for the center of the T, and to use a second exposure with another mask — usually called "trim mask" — to expose away the

Advanced Optical Masks

unintended dark feature caused by the additional phase transition [34]. This strategy is shown in Figure 8.8. The data processing to parse the layout into data for two masks represents an additional task that must be managed, and the algorithms for doing this correctly are not trivial. Automated software packages for phase assignment incorporating double exposure are the most common solution to this problem. But even then, any variation in mask dimensions or in the overlay precision at the exposure of both masks leads to a printing failure at the center of the T.

To minimize that risk, a third strategy was developed, which changes the original layout into a layout that can be printed without phase conflicts [35]. For the case of the T, the center of the T is widened for a certain length in order to allow space for a phase transition, which is again exposed away by the trim mask exposure, as shown in Figure 8.9. In this strategy — and in many similar options —, the widened part of the T is defined by the trim mask.

A fourth strategy, which was a few years ago quite popular, avoids the second exposure of the trim mask. To do so, multiple phase regions are generated in order to split unintended 180° phase transitions into partial phase transitions of smaller height, which do not print as readily as 180° [32,36]. This is illustrated in Figure 8.10 and was the initial solution proposed to deal with phase conflicts. This strategy can only be applied to layouts that need phase shifting at rather few spots, and which have a lot of open space in between [37].

Although the second and the third strategies deliver very similar results, their impact to the overall strategy called "design for manufacturability" is considerably different. However, these tradeoffs are more often those that require the cost of ownership of double exposure on a stepper versus mask cost to be evaluated, which is a topic beyond the scope of this chapter. The shrinking pitches of contemporary ICs, along with the costs of multiple writing steps for the partial shifter approach, have rendered this technique impractical.

8.2.4 Special Cases of alt.PSM

From all the different possibilities to creatively use phase to improve the contrast of the printed image, we would like to pick a few, which are either currently widely discussed, or which are even used for very special applications.

In Section 8.1.1, we indicated that a 180° phase transition without chromium prints at a finite width, which depends only on the lithographic setup and on the geometry (primarily the width) of the two phase regions. This so-called *chromeless PSM* (also sometimes known as "phase edge PSM") — together with over-etch strategies — have been used to generate transistors with the smallest gates. Currently, the record is the creation of 9-nm gates using a 248-nm stepper with a 0.75 NA lens [38]. To print all minimum features at the same width, the size of the phase regions has to be identical. This requirement together with the requirement to remove the second phase transition (phase areas have to be limited by closed loops) by the trim exposure have limited the application of chromeless PSM to very special cases. The performance advantage for those special cases can be very attractive; however, the first commercial product fabricated with PSM used this technique to fabricate GaAs transistors in 1991 [39].

A completely transparent area with multiple, dense phase edges can efficiently diffract light at a wide angle. If the spacing between edges is smaller than wavelength/2NA for the stepper, all the transmitted light will be diffracted away from the lens pupil, and none will reach the wafer. This area, even though transparent, therefore prints as opaque. This phase approach to create dark regions with dense phase edges can also be called as "Chromeless PSM" [40,41] — although "shifter shutter" is also used. The additional complexity of converting opaque regions to all phase layouts, and the susceptibility of defects causing a light leak in these areas, has made this a relatively unattractive approach to phase shifting.

Using this approach to form a single line, by having only two phase transitions in close proximity (separated by < wavelength/2NA of the stepper lens), has found some application. Although a single dark line formed by two closely spaced phase edges does not have a useful process window using conventional illumination, or even with highly coherent light — as required to provide the maximum benefit of alt.PSM, when the same mask geometry is illuminated by off-axis illumination, a dark line with high contrast is printed. If the distance between the two phase edges increases, the width of the printed line increases slowly providing a lithographic pattern transfer with low MEEF (see Section 8.1). This behavior of the so-called chromeless phase lithography (CPL) [42] mask is shown in Figure 8.11.

As a certain distance between the two phase edges is exceeded, the width of the printed line decreases again, and the aerial image looses contrast drastically [43]. In order to print lines with larger width, either SRAF (see the following) or chromium has to be added to the mask, resulting in either a rim PSM (when the chromium is added between the two phase edges) or the so-called CL PSM (when the chromium is added outside of the two phase edges). Since the common features of these PSM is a phase edge, chromium, and off-axis illumination, the term PCO.PSM has been coined. Figure 8.12 shows the different options for PCO.PSM.

Since PCO.PSM provide many degrees of freedom (each feature width can be varied individually), they have shown — at least in simulation — to be able to print arbitrary layouts with a process window almost as high and an MEEF almost as low as alt.PSM. This lack of rigid layout constraints and the lack of a second exposure have made PCO currently a rather promising subject of evaluation.

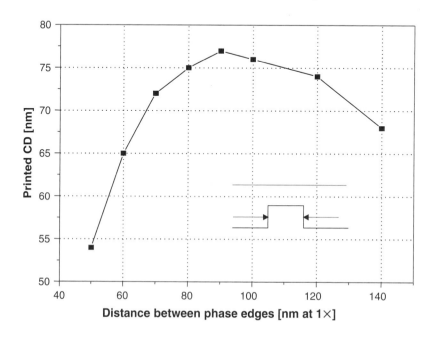

FIGURE 8.11

Printed CD as a function of distance between two Cr-less phase edges (= mask CD). Aerial image simulation for $\lambda = 193\,nm$, $NA = 0.75$, annular illumination ($\sigma = 0.8$–0.9).

Advanced Optical Masks

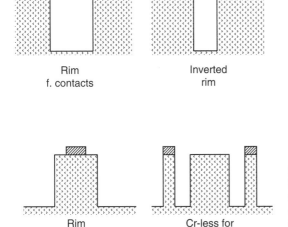

FIGURE 8.12
Options of PCO.PSM. All drawings are cross-sections of the structures etched into quartz and chromium. The typical etch depth for quartz is the nominal amount for 180° phase shift $\lambda/(2(n-1))$, with n as the refraction index of quartz.

8.3 HT.PSM

8.3.1 Basic Functionality

All of the phase mask techniques described so far require that the layout data be parsed into two sets, one defining apertures on the mask and the other defining phase shifts. This complicates layout and data handling for alt.PSM and increases costs.

Some of the advantages of phase shifting can be realized, however, without complicated data parsing. HT.PSM, also known as attenuated PSM, uses an attenuating layer that is not completely opaque but allows the transmission of some light. This transmitted light is not bright enough to actually expose resist but is also phase shifted by 180° relative to the light passing through the clear apertures, causing interference at the edges of the features. The interference again forces intensity to be 0 at the crossing point, increasing contrast. This was originally proposed as a technique for x-ray lithography but was quickly applied to optical lithography as well [44]. The principal mechanism is illustrated in Figure 8.13.

This phase shifting effect comes entirely from the optical properties of the partially transparent material used on the mask, so the problems of data parsing and quartz etching encountered for alt.PSM simply do not occur. This has made this the most widely applied option of phase shifting.

8.3.2 Manufacturing of HT.PSM

HT.PSM has been made using various materials, mostly ternary compounds (e.g., metal silicides/nitrides) that allow an independent adjustment of phase and transmission. The most common material is a co-sputtered layer of molybdenum and silicon (MoSi), although MoSiON films are also available [45]. Mixtures of various chrome oxides have also been used for I-line HT.PSM materials [46]. Commercial mask blanks are available with transmissions of 4% and 8% for use with I-line steppers, and 6% for DUV (248 and 193 nm) applications. Tuning the material composition and ratios can allow adjustment of the optical properties, and experimental blanks with transmission values anywhere

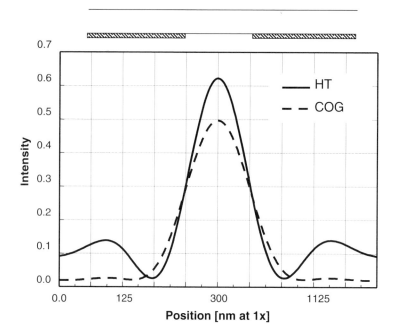

FIGURE 8.13
Imaging of a 125-nm contact hole in a chromium (dashed line) and in a HT.PSM (solid line), simulated for $\lambda = 193\,nm$, $NA = 0.75$. The HT.PSM produces higher contrast but also higher sidelobes (secondary intensity maxima).

between 4% and 20% have been produced. The mask house now can buy mask blanks with the appropriate material guaranteed to fulfill demanded specifications. This commercial availability has simplified the manufacturing of HT.PSM considerably.

HT.PSM is processed very much like normal COG masks: exposure and development of a resist, followed by transfer of the resist pattern into the absorbing layer, either with wet or dry etching. For HT.PSM, dry etching is far more common. In practice, dedicated etchers may be used to avoid any contamination between COG masks and the HT.PSM materials, and different plasma settings and etch recipes may be used to tune the process for optimum results.

The adoption of HT.PSM does not come without an additional effort, however. Since the scribe lines on wafers (the lines between the chips, where they will be cut to yield single dies) are generated by overlapping adjacent exposures of the same mask, the features in the scribelines have to be delineated in 100% opaque material. So the blanks for HT.PSM manufacturing must have not only the HT material but an additional top layer of opaque chromium. Two writing steps are then used to make the mask. The first is a step to pattern the chrome, with the pattern transferred into the MoSi film through a subsequent dry-etch step. Then, a second writing step must be used to remove the chromium in the active area of the chip while leaving it in the scribelines. Typically, this can be done using a process of relatively low resolution and with rather relaxed specifications for the overlay to the pattern definition process in the HT material. This is therefore much faster and less expensive than the multiple writing steps at high specification that are needed for alt.PSM.

Unfortunately, the transmission of the HT film depends drastically on wavelength. This is shown in Table 8.2 for a MoSi HT.PSM material tuned for use with 193-nm steppers [47].

TABLE 8.2

Typical Transmission Values for a 193-nm HT.PSM Blank [47]

Wavelength (nm)	Transmission (%)
193	6.0
248	26.1
365	53.2
436	56.8
633	63.3

It is not an issue for mask metrology if the measurement of feature width is performed by an SEM, or by an optical tool using the same wavelength as the lithography tool the HT.PSM is made of. Since in the latter case the metrology tool has to be calibrated against an SEM anyway, also metrology tools using wavelengths not too different from the exposure wavelength can be used.

The wavelength dependence of the transmission of the HT film creates more significant problems for mask inspection. Although a HT material may be tuned for 6% transmission at 193 nm, the transmission can be >50% at the wavelengths used at mask defect inspection (365 or 248 nm). Contrast and reliability of the inspection process is therefore a concern, especially if the mask contains other regions, as mentioned earlier, that are completely opaque. However, mask inspection tools have been able to compensate this loss in the contrast of the image at inspection by sophisticated optical setups (e.g., combining transmitted and reflected light) and/or by highly sophisticated defect detection algorithms.

A defect on HT.PSM consisting of excess HT material can be removed by the same methods developed for the removal of chromium. Defects consisting of missing HT material can be repaired by the ion-assisted deposition of materials (mostly carbon), which have the same transmission as the HT material. However, in most cases the deposited material does not produce perfect 180° phase shift. This is tolerable for relatively small defects in large dark areas. But for large defects or defects in areas with dense pattern repair is not possible. So the mask making processes are optimized in a way to avoid defects of missing HT material; the specification for the mask blank regarding holes in the HT material is tightly controlled.

The final mask cleaning process has to be qualified as not harmful neither to the HT material nor to the material deposited in mask repair. Despite these slight differences and the second, low-resolution exposure, HT.PSM is fabricated by a process with a great deal in common with regular COG mask processing. This is one of the major reasons why HT.PSM gained such a broad acceptance.

8.3.3 High Transmission HT.PSM and 3tone.PSM

HT.PSM with higher transmission (up to 25%) provides an even larger lithographic process window [48]. This makes them quite appealing for certain lithographic applications, and this is an ongoing topic of research. However, this benefit is gained at the cost of an increase in the intensity of the so-called sidelobes. These are interference patterns on the wafer, which can become intense enough to cross the exposure threshold of the resist — in particular, when bright features, such as contact holes, are at a pitch where adjacent sidelobes overlap. Sidelobes that print can even appear using HT.PSM with only

6% transmission under certain mask bias and illumination conditions. Figure 8.14 shows an example of that situation.

One method to suppress the printing of sidelobes is to leave chromium patches over the HT material at the spots sidelobes are expected to print. These HT.PSM are usually called 3tone.PSM (or tri-tone PSM), since their optical transmission has (in the active areas) three levels. An example of the aerial image intensity simulation for such an arrangement is shown in Figure 8.15.

Since most mask blanks are shipped with both Cr and HT layers for the scribelines and outer alignment marks, as mentioned earlier, the blanks and processing for this are already in use. But, for processing the 3tone.PSM, the second writing step now becomes a high-resolution step, and overlay with the conventional masks becomes more important.

In addition, the creation of 3tone.PSM requires a software tool to identify the locations of potential sidelobes in arbitrary layout and to generate the layout for the chromium patches. It also puts considerable demands to the resolution and alignment of the second lithography process delineating the chromium pattern. Furthermore, the three levels of transmission provide a challenge for mask inspection tools, which — up to now — only had to deal with 100% and 0% intensities in the pattern area. Although all these issues have been demonstrated to be manageable, these additional drawbacks have kept 3tone.PSM as a topic of process development only.

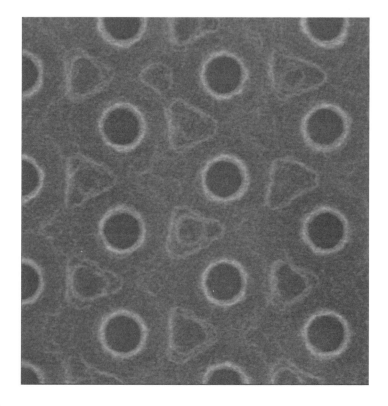

FIGURE 8.14
Wafer SEM image showing the result of printing a contact layer with a HT.PSM at nonoptimized process settings. The resist clearly shows the effect of sidelobes. (Courtesy of R. Pforr, Infineon Technologies, Dresden.)

Advanced Optical Masks

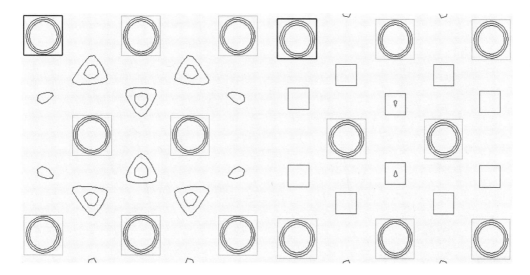

FIGURE 8.15
Aerial image simulation of an array of 150-nm contacts exposed with a HT.PSM (left) and with a 3tone.PSM (right). The chromium spots in the right layout suppress almost all sidelobes, without affecting the process window of the contacts (as can be seen from the distance between the different intensity levels). Simulation for 193 nm, 0.75 NA, and a coherence factor (σ) of 0.3.

8.4 Subresolution Assist Features

8.4.1 Basic Principle and Application

A widely used rule-based RET involves the addition of SRAF. Sometimes also called scattering bars (although diffraction and not scattering is the phenomenon they utilize) these techniques add features to the mask layout that are smaller than the imaging resolution of the lithography system. Masks made with these features will have particular diffractive properties that can improve image fidelity and process windows. The challenge for the mask maker is that, being by definition sub-resolution, they are significantly smaller than the other pattern features in the mask (at least 30%) and will stress the capabilities of the manufacturing and inspection equipment.

Most lithographic processes work best for a (moderately) dense pattern, and isolated features often suffer from an insufficient depth of focus. One widely used RET option — off-axis illumination — in particular, enhances the printing of features with specific pitches [49]. In order to extend these benefits also for isolated features, SRAF have been developed [9]. These features are large enough to provide isolated layout features with a dense neighborhood in the aerial image but are small enough that they do not themselves print. By strategically placing these around isolated features, they diffract light like dense features, matching dense and isolated imaging behavior [10]. In an optical system designed to enhance dense process windows, this allows isolated features to be enhanced as well.

Figure 8.16 shows a typical layout before and after the addition of SRAF. Figure 8.17 shows the imaging properties of an aerial image of an isolated line with and without SRAF, using off-axis illumination. Note the increase in slope at the resist threshold due to SRAF.

180   Handbook of Photomask Manufacturing Technology

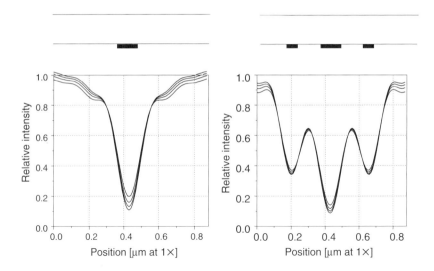

FIGURE 8.16
Aerial image simulation of the intensity distribution of an isolated 120-nm line over defocus (up to 0.2 μm) without SRAF (left) and with SRAF (right). Note that the variation around the printing threshold (0.25–0.32) is significantly less for the line with the SARF, indicating a superior process window.

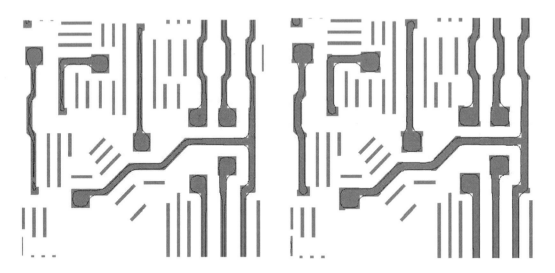

FIGURE 8.17
Layout with SRAF before OPC (left), and after OPC (right). The simulated printed image (outlined contour) clearly shows that SRAF alone do not provide a complete correction solution.

SRAF are both applied on bright-field masks (like the gate level) and dark-field masks (like contact levels), and in both COG masks and cases for PSM. In fact, an early application of phase shifting used phase shifted SRAF to increase pattern fidelity and depth of focus [50,51]. SRAF are routinely applied with HT.PSM. HT.PSM with high transmission and PCO.PSM in particular need SRAF to print arbitrary layouts.

SRAF can be applied by simple rules, but in fact the insertion of SRAF can be fairly complex. There are various placement approaches for different two-dimensional layout situations, and the use of two or even more SRAF have been shown to be useful for some situations. Although SRAF provide an optically dense neighborhood to isolated structures, it is not a very effective method to compensate the influence of different neighborhoods to the width of features (the so-called proximity effects). In a typical mask data preparation flow, SRAF are first assigned to all features small and isolated enough to need them, and then proximity effects are compensated (more about that latter step can be found in the following section). Most data preparation procedures for SRAF also involve a "clean-up" step to resolve conflicts that might have arisen from the insertion of SRAF using conflicting rules.

Strategies how to optimize the size and placement of SRAF simultaneously with the selection of the optimum off-axis illumination conditions for a given set of layouts within the constraints of a given lithographic tool-set and process are definitely beyond the scope of this treatise. It should be only noted that all these parameters are all interrelated, and that — although difficult to measure and control — mask parameters are a critical component of the whole system. And therefore any change in the mask process bears the risk that this delicate balance between parameters may tip towards undesired lithographic performance. Except for very specialized cases, mask specifications for lithography with off-axis illumination and SRAF have to be tighter than for regular masks.

8.4.2 Manufacturing and Metrology Issues of SRAF

The most obvious challenge in making a mask with SRAF is the significant increase in resolution demanded from the mask making process. As a rule of thumb, for a process with k1 < 0.4, the dimensions of the SRAF are about 40% smaller than the minimum feature of the mask. This forces all fabrication procedures to essentially jump a generation for the production of SRAF masks. Since the printability drops more rapidly for smaller k1, the trend towards smaller k1 has somewhat eased the resolution demands for SRAF.

As features on a mask get smaller, they usually show a systematic deviation in width from nominal. This nonlinearity in feature width can be a major issue for the usability of a mask with SRAF, especially if no special measures are taken to keep it constant over the mask and from mask to mask. It is highly advisable to measure and specify the width of SRAF in mask making, in particular, when SRAF of different sizes are used on the same mask. For this measurement, a potential non-linearity of the linewidth measurement tool has to be taken into account.

The increase in mask process resolution demanded by SRAF is of course also an issue for mask inspection and repair. SRAF make it impossible to detect mask defects by a series of width measurements, a method that has been widely used before. But also conventional defect detection algorithms have received major upgrades until a mask with SRAF could be inspected without false defects at high sensitivity settings.

Ion beam mask repair tools implant a considerable dose of Ga ions into mask structures for imaging and for repair. The dose can be high enough to change the electronegativity of the chromium so that it can no longer sustain certain aggressive mask cleaning steps. Although not a problem for larger Cr features, where repairs may be a small portion of a feature, SRAF may be so small that the repair sites detach on cleaning. Because the size and placement of the SRAF can be important, repair by deposition of an SRAF can be extremely difficult, and the mask more often than not must be rewritten.

8.5 Optical Proximity Effect Correction

8.5.1 Basics and OPC Strategies

As already indicated in the previous sections, one of the major issues of lithography at low k1 is the nonlinearity of the pattern transfer process. A linear pattern transfer implies that all features in the printed image are congruent to the features on the mask but just demagnified by a constant factor — the demagnification factor of the stepper lens — and biased by a constant amount — the process bias.

It has been known for quite a long time that features printed by electron-beam lithography undergo a distortion due to the electron scattering in resist and substrate. The smaller a feature, the more relative dose it loses. But two features close together do not lose that much dose because some of the electrons scattered by one feature contribute to the dose for the other feature and vice versa. For this reason, this effect has been called the e-beam proximity effect [52]. Solutions to alter pattern data to precompensate for these effects have been in use ever since these were first characterized and are now part of any standard e-beam exposure procedure. These e-beam effects have been well documented elsewhere and are described in chapter 20 of this book.

Effects that appear similar to these proximity effects have also been observed for optical lithography. Since light diffraction is the major contribution to the nonlinearity of a pattern transfer process by optical lithography, the commonly used name for this effect is optical proximity effect [53,54]. Consequently, the correction of layout data for the nonlinearities of a pattern transfer by optical lithography has been called OPC.

Initially, optical phenomena dominated the effects that required correction. However, the methodology of OPC soon was generalized to include resist processing and even etch effects. Mask making artifacts can also be included in the collection of effects as well. For this reason, the OPC acronym has been generalized to stand for optical and process correction [55]. Depending on the type of model used, OPC is either called rule-based OPC, or model-based OPC (when a behavioral model is used).

In either case, the first step of OPC is to assess the effect on the wafer by printing a mask with a test pattern, and measuring the wafer results. These are then used to calibrate a description of the process behavior and to describe the nonlinearities of the process. This description then becomes the process "model." This model can be as simple as a set of a few rules (such as "nested lines print 20 nm smaller than isolated lines"), or it can be a rather sophisticated simulation (behavioral) model, which describes the imaging process in some detail. This can be a simple simulation, such as an aerial image using only wavelength, NA, and illumination conditions, or something more complex that includes a vectorial description of the light propagation in resist, parameters to describe lens aberrations, and various diffusion models for the resist process.

There are also two approaches to the correction itself. Correction can either consist of inverting the model — both for the rule model (i.e., make all nested lines 20 nm larger than the isolated lines), and for the simulation model, or of an iterative process, where a trial correction is created by moving the mask features systematically, and simulation checks after each move to see how far the printed image is from the nominal image. Both the trial suggestion and the simulation engine itself must be calibrated against the test pattern measurements. The results of such an approach are illustrated in Figure 8.18.

Another aspect of OPC — particularly important for mask making — is the fragmentation of the mask pattern after OPC. The simplest strategy is to only shift the already existing feature edges of the pattern. However, this cannot correct for an abrupt change in

Advanced Optical Masks

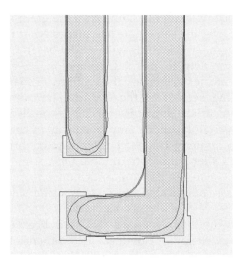

FIGURE 8.18
Principle of OPC: the edges of nominal dark feature (shaded area) are broken up into segments (straight lines), which are moved in order to adjust layout. Ideally, the printed image of the adjusted layout (smooth lines) will correspond to the nominal feature outline within predefined tolerances.

proximity of a long line. So OPC tools have developed over time rather sophisticated strategies how to fragment the mask pattern. Figure 8.19 shows some options [56].

It should be noted that the degree of fragmentation is independent of the complexity of the process description (simple rule or complex simulation). However, the simultaneous

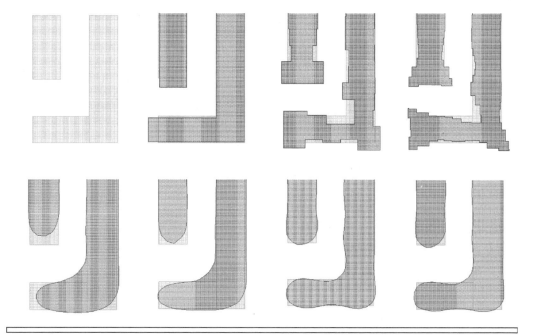

FIGURE 8.19
Different OPC fragmentation options leading to different correction quality and different mask complexity requirements. [Adapted from C. Dolainsky, W. Maurer, *Proc. SPIE*, 3501, 774–480 (1997).]

increase in fragmentation complexity over time with the gradual switch from rule-based OPC to model-based OPC over the same time has suggested — not always correctly — an association of complex fragmented layouts with model-based OPC.

Process correction is not new to printing technology. Serifs have been added to the corners of letters for centuries to increase the visibility of sharp, high spatial frequency edges through the imaging system of the eye and remain in most fonts used for printing today. One of the earliest strategies for OPC was also the addition of serifs at corners [57–59]. This can certainly lead to improvements in corner fidelity. However, in dense pattern areas, this strategy can lead to unwanted bridging and shorts, so it cannot be recommended. Other simple, traditional OPC strategies are specific applications of biasing, in which features that are anticipated to print smaller are enlarged appropriately [60,61]. More complicated approaches using solely rules have also been demonstrated [62].

The theoretical work behind model-based OPC has been established for over two decades and is a natural extension of nonlinear two-dimensional image theory [57,63]. However, the ability to solve layout problems with the complexity found in modern ICs was not possible before the development of very fast, very large computers, and model-based software [64,65] has been generally available only since 1994. However, the adoption of model-based OPC in production has typically occurred much later. The major driver for this change has been the impossibility to develop and maintain the highly complex rule sets necessary for rule-based OPC on complicated layouts.

The data manipulations for OPC can be quite complex and sophisticated, requiring significant data preparation resources to be dedicated to making the corrected layout. In addition to their strategies in modeling, in correction, and in fragmentation, OPC software tools also differ in how they deal with the hierarchy of the layout. This influences, in particular, the size of files delivered to the mask house, and consequently the time and effort for data preparation and for mask writing. File size reductions > 10× have been observed by changing the OPC tool and the software tools used for other data manipulations to tools with a superior hierarchy treatment.

Double exposure techniques — as required, e.g., for some kinds of alt.PSM (see Section 8.2.3) — present a further challenge for OPC, since features on two masks need to be corrected simultaneously to generate the optimum printed image. A strategy called "Matrix OPC," which is based on the matrix of influence of all mask features to the printed features, has been developed to address this complex situation [66].

Although it is a straightforward fact, it should be pointed out explicitly again that OPC can only correct for the predictable systematic errors in the pattern transfer process. Nonsystematic errors, random process fluctuations, and errors that change over time — in particular, in metrology, but also incremental process drifts — cannot be corrected by OPC.

8.5.2 Manufacturing and Metrology Issues

Process stability is the most critical demand by OPC to mask making. The major requirement OPC demand from mask making is process stability. Especially for a process with a high MEEF, any difference between the mask used to build the OPC model and the actual product mask bears a high risk of substantially reducing the benefits of OPC.

If the selected OPC option contains a high level of feature fragmentation, then the mask process must deliver a resolution high enough to truly reproduce this fragmentation. Otherwise, all the increase in data volume and the associated increase in writing time for that mask are wasted. The critical question is now, how to define "truly reproduce."

Advanced Optical Masks

It is straightforward that specifications for CD (CD mean to nominal, CD uniformity, CD linearity) need to reflect the expected precision of OPC. In addition to these one-dimensional mask specifications (they are measured on long lines = one-dimensional structures), two-dimensional parameters need to be specified and controlled by measurement. Until now, the primary strategy to do so has been to measure and assess (sometimes even just by visual inspection of SEM pictures) test structures, which the mask maker and the mask user have agreed on to use for that purpose.

One of the two-dimensional parameters intended to specify the usability of a mask for OPC has been corner rounding. The intuitive definition of this parameter is easy: fit a circle to a high-resolution image of the corner, and determine the circle's radius. However, close evaluation shows that corners on mask structures are rounded by more than one process, and therefore more than one corner radius can be fitted, as demonstrated in Figure 8.20 [67]. Since the printed image in lithography is averaged over an area of (wavelength divided by lens NA) squared, area is the important parameter. So it seems reasonable to characterize corner rounding by measuring the area loss relative to the area of a straight corner.

As it was the case for SRAF, the small mask features delivered by an OPC strategy with high level of feature fragmentation is an issue for mask inspection. Contemporary algorithms used by the mask inspection tools are capable of handling this issue, as long as the differences between the nominal features and the features on the mask remain sufficiently constant over one given mask and are significantly smaller than the minimum defect size to be detected.

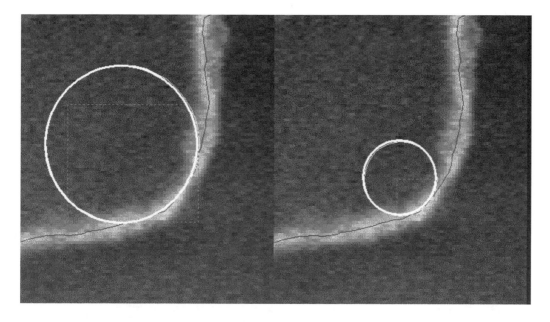

FIGURE 8.20
Corner-rounding measurements by fitting a circle to different parts of the same corner of a printed resist feature. Different "corner radii" can be extracted, depending on how much of the corner outline is taken for the fit.

8.6 Outlook for RET

Alt.PSM, HT.PSM, SRAF, and OPC, as they were described here, have been the major RET strategies implemented on the mask, and have found fairly broad acceptance in optical lithography for high-volume manufacturing of integrated circuits, especially in mask making for processes smaller than 250 nm. For each successive IC generation with smaller dimensions, more mask layers must be fabricated using various RET (for the 130-nm generation, as many as eight masks may use some form of RET). It has been mentioned that many more variations of these three strategies exist as were described. Some of them have the potential to change mask making completely, such as pixelization schemes, where the mask features are just — more or less — random pattern of pixels with two or more intensity levels of transmitted light.

Masks for near-field holography — a lithographic pattern transfer option once heavily discussed — have a rather similar appearance.

Other strategies involve the simultaneous exposure of two mask patterns, which are conveniently placed on both sides of a mask blank. If adopted for production, this RET strategy will certainly demand a radical change in the tooling and handling processes of mask making.

Next generation lithography (NGL) has been seen by many as the logical next step after optical lithography becomes too complex and expensive — among other reasons also because of RET. However, more detailed evaluations of EUV lithography — currently the most likely NGL option — reveal that this pattern transfer also introduces systematic nonlinearities, which have to be corrected by strategies similar to those used by OPC. And even PSM have already been proposed for EUV to provide enough process window for printing 30 nm and smaller features, where EUV will finally be applied, according to current wisdom.

Other NGL options, such as ion projection lithography or x-ray techniques, employ masks on thin films. These masks exhibit systematic distortions that depend on the feature density, and which can be simulated and corrected using methods developed for OPC.

It may be therefore no exaggeration to state that RET is here to stay. Mask making has demonstrated a remarkable capacity to adapt to the requirements of RET in the past. It seems very likely that this flexibility will be needed as long as lithographic techniques evolve.

References

1. E. Abbe, Beiträge zur Theorie des Mikroskops und der mikroskopischen Wahrnehmung, *Archiv für Mikroskopische Anatomie*, 9, 413–468 (1873).
2. B.J. Lin, Where is the lost resolution? *Proc. SPIE*, 633, 44–50 (1986).
3. The International Technology Roadmap for Semiconductors, Lithography Module, http://public.itrs.net/.
4. C. Mack, The natural resolution, *Microlithography World*, 8 (1), 10–11 (1998).
5. A.K.K. Wong, *Resolution Enhancement Techniques in Optical Lithography*, SPIE Press, Bellingham, WA, 2001.
6. A.K.K. Wong, *Resolution Enhancement Techniques in Optical Lithography*, SPIE Press, Bellingham, WA, 2001, pp. 91–115.

7. W. Maurer, K. Satoh, D. Samuels, and T. Fischer, Pattern transfer at k1 = 0.5: get 0.25 um lithography ready for manufacturing, *Proc. SPIE*, 2726,113–124 (1996).
8. A.K.K. Wong, *Resolution Enhancement Techniques in Optical Lithography*, SPIE Press, Bellingham, WA, 2001, pp. 71–90.
9. J.F. Chen and J.A. Matthews, Mask for Photolithography, U.S. Patent #5,242,770, issued September 7, 1993.
10. J. Garofalo, C. Biddick, R.L. Kostelak, and S. Vaidya, Mask assisted off-axis illumination technique for random logic, *J. Vac. Sci. Technol. B.*, 11, 2651–2658 (1993).
11. Lord Rayleigh, On the theory of optical instruments, with special reference to the microscope, *Philos. Mag.*, 42, 167–195 (1896).
12. O. Lummer and F. Reich, Die Lehre von der Bildentstehung im Mikroskop von Ernst Abbe, Friedrich Vieweg und Sohn, Braunschweig, Germany, 1910.
13. H. Hänsel and W. Polack, Verfahren zur Herstellung einer Phasenmaske Amplitudenstruktur, DDR Patent #26 50 817, issued November 17, 1977.
14. D.C. Flanders, A.M. Hawryluk, and H.I. Smith, Spatial period division — a new technique for exposing submicrometer-linewidth periodic and quasiperiodic patterns, *J. Vac. Sci. Technol.*, 16, 1949–1952 (1979).
15. M. Shibuya, Projection Master for Transmitted Illumination, Japanese Patent Publication No. Showa 57-62052, published April 14, 1982, and Japanese Patent Showa 62-50811, issued October 27, 1987.
16. M.D. Levenson, N.S. Viswanathan, and R.A. Simpson, Improving resolution in photolithography with a phase-shifting mask, *IEEE Trans. Electron Devices*, ED-29, 1828–1836 (1982).
17. M.D. Levenson, D.S. Goodman, S. Lindsey, P.W. Bayer, and H.A.E. Santini, The phase-shifting mask II: imaging simulations and submicrometer resist exposures, *IEEE Trans. Electron Devices*, ED-31, 753–763 (1984).
18. Prolith is offered by KLA-Tencor, http://www.kla-tencor.com.
19. Solid is offered by Sigma-C, http://www.sigma-c.de/.
20. EM Suite is offered by Panoramic Technology, http://panoramictech.com/.
21. TEMPEST was originally developed at UC Berkeley, http://www.eecs.berkeley.edu/~neureuth/.
22. C. Spence, M. Plat, E. Sahouria, N. Cobb, and F.M. Schellenberg, Integration of optical proximity correction strategies in strong phase shifter design for poly-gate layer, *Proc. SPIE*, 3873, 277–287 (1999).
23. C. Pierrat, A. Wong, and S. Vaidya, Phase-shifting mask topography effects on lithographic image quality, in: *International Electron Devices Meeting (IEDM)*, Technical Digest, San Francisco, 1992, pp. 53–56.
24. C. Brooks, M. Buie, N. Waheed, P. Martin, P. Walsh, and G. Evans, Process monitoring of etched fused silica phase shift reticles, *Proc. SPIE*, 4889, 25–31 (2002).
25. L. Liebmann, S. Mansfield, A. Wong, J. Smolinski, S. Peng, K. Kimmel, M. Rudzinski, J. Wiley, and L. Zurbrick, High resolution ultraviolet defect inspection of darkfield alternate phase reticles, *Proc. SPIE*, 3873, 148–161 (1999).
26. The AIMS tool is offered by Carl Zeiss, http://www.zeiss.de.
27. S. Hemar and A. Rosenbush, Inspecting alternating phase shift masks by matching stepper conditions, in: *Proceedings of the 19th European Mask Conference on Mask Technology for Integrated Circuits and Micro-Components*, Sonthofen, 2003, pp. 173–178.
28. G. Wojcik, J. Mould Jr., R. Ferguson, R. Martino, and K.K. Low, Some image modeling issues for I-line, 5× phase shifting masks, *Proc. SPIE*, 2197, 455–465 (1994).
29. U. Griesinger, R. Pforr, J. Knobloch, and C. Friedrich, Transmission & phase balancing of alternating phase shifting masks (5×) — theoretical & experimental results, *Proc. SPIE*, 3873, 359–369 (1999).
30. W. Maurer, C. Friedrich, L. Mader, and J. Thiele, Proximity effects of alternating phase shift masks, *Proc. SPIE*, 3873, 344–349 (1999).
31. M.D. Levenson, J.S. Petersen, D.G. Gerold, and C.A. Mack, Phase phirst! An improved strong-PSM paradigm, *Proc. SPIE*, 4186, 395–404 (2000).

32. J. Nistler, G. Hughes, A. Muray, and J. Wiley, Issues associated with the commercialization of phase shift masks, *Proc. SPIE*, 1604, 236–264 (1991).
33. L. Liebmann, I. Graur, W. Liepold, J. Oberschmidt, D. O'Grady, and D. Rigaill, Alternating phase shifted mask for logic gate levels, design and mask manufacturing, *Proc. SPIE*, 3679, 27–37 (1999).
34. H. Jinbo and Y. Yamashita, Improvement of phase-shifter edge line mask method, *Jpn. J. Appl. Phys.*, 30, 2998–3003 (1991).
35. L. Liebmann, J. Lund, F.L. Heng, and I. Graur, Enabling alternating phase shifted mask designs for a full logic gate level, *J. Microlith. Microfab. Microsyst.*, 1, 31–42 (2002).
36. T. Terasawa, N. Hasegawa, H. Fukuda, and S. Katagiri, Imaging characteristics of multi-phase shifting and halftone phase shifting masks, *Jpn. J. Appl. Phys.*, 30, 2991–2997 (1991).
37. G. Galan, F. Lalanne, P. Schiavone, and J.M. Temerson, Application of alternating type phase shift mask to polysilicon level for random logic circuits, *Jpn. J. Appl. Phys.*, 33, 6779–6784 (1994).
38. M. Fritze, B. Tyrrell, D.K. Astolfi, D. Yost, P. Davis, B. Wheeler, R. Mallen, J. Jarmolowicz, S.G. Cann, H.Y. Liu, M. Ma, D.Y. Chan, P.D. Rhyins, C. Carney, J.E. Ferri, and B.A. Blachowicz, 100-nm node lithography with KrF? *Proc. SPIE*, 4346, 191–204 (2001).
39. K. Inokuchi, T. Saito, H. Jinbo, Y. Yamashita, and Y. Sano, Sub-quarter micron gate fabrication process using phase-shifting mask for microwave GaAs devices, *Jpn. J. Appl. Phys.*, 30, 3818–3821 (1991).
40. K. Toh, G. Dao, R. Singh, and H. Gaw, Chromeless phase shifting masks: a new approach to phase shifting masks, *Proc. SPIE*, 1496, 27–53 (1990).
41. H. Watanabe, Y. Todokoro, Y. Hirai, and M. Inoue, Transparent phase shifting mask with multistage phase shifter and comb shaped shifter, *Proc. SPIE*, 1463, 101–110 (1991).
42. D. Van Den Broeke, J.F. Chen, T. Laidig, S. Hsu, K. Wampler, R. Socha, and J. Petersen, Complex 2D pattern lithography at lambda/4 resolution using chromeless phase lithography (CPL), *Proc. SPIE*, 4691, 196–214 (2002).
43. A Torres and W. Maurer, Alternatives to alternating phase shift masks for 65 nm, *Proc. SPIE*, 4889, 540–550 (2002).
44. Y.C. Ku, E. Anderson, M. Schattenburg, and H.I. Smith, Use of a pi-phase shifting x-ray mask to increase the intensity slope at feature edges, *J. Vac. Sci. Technol. B.*, 6, 150–153 (1988).
45. HT.PSM blanks are provided by Hoya, www.hoya.com.
46. F.D. Kalk, R.H. French, H.U. Alpay, and G. Hughes, Attenuated phase shifting photomasks fabricated from Cr-based embedded shifter blanks, *Proc. SPIE*, 2254, 64–70 (1994).
47. M. Ushida, H. Kobayashi, and K. Ueno, Photomask blank quality and functionality improvement challenges for the 130 nm node and below, *Yield Management Solutions*, 3, 47–50 (2000).
48. R. Socha, W. Conley, X. Shi, M. Dusa, J. Petersen, F. Chen, K. Wampler, T. Laidig, and R. Caldwell, Resolution enhancement with high-transmission attenuating phase-shift masks, *Proc. SPIE*, 3748, 290–314 (1999).
49. R. Socha, M.V. Dusa, L. Capodieci, J. Finders, J.F. Chen, D.G. Flagello, and K.D. Cummings, Forbidden Pitches for 130 nm lithography and below, *Proc. SPIE*, 4000, 1140–1155 (2000).
50. M. Prouty and A. Neureuther, Optical imaging with phase shifting masks, *Proc. SPIE*, 470, 228–232 (1984).
51. T. Terasawa, N. Hasegawa, T. Kurosaki, and T. Tanaka, 0.3 micron optical lithography using a phase shifting mask, *Proc. SPIE*, 1088, 25–33 (1989).
52. W. Moreau, *Semiconductor Lithography: Principles, Practices, and Materials*, Plenum Press, New York, 1988, pp. 437–446.
53. A.E. Rosenbluth, D. Goodman, and B.J. Lin, A critical examination of submicron optical lithography using simulated projection images, *J. Vac. Sci. Technol. B.*, 1, 1190–1195 (1983).
54. P. Chien and M. Chen, Proximity effects in submicron lithography, *Proc. SPIE*, 772, 35–40 (1987).
55. C. Dolainsky and W. Maurer, Application of a simple resist model to fast optical proximity correction, *Proc. SPIE*, 3501, 774–480 (1997).
56. W. Maurer, C. Dolainsky, T. Waas, and H. Hartmann, *Proximity Correction in Optical Lithography by OPTISSIMO*, GMM Fachbericht 21, VDE Verlag, Berlin, Offenbach, 1997, pp. 161–167.
57. B.E.A. Saleh and S. Sayegh, Reduction of errors of microphotographic reproductions by optimal corrections of original masks, *Opt. Eng.*, 20, 781–784 (1981).

58. T. Ito, M. Tanuma, Y. Morooka, and K. Kadota, Photo-projection image distortion correction for a 1 μm pattern process, *Denshi Tsushin Gakkai Ronbunshi*, J68-C, 325–332 (1985) [translated in Electronics and Communications in Japan Part II: Electronics, 69, 30–38 (1986)].
59. A. Starikov, Use of a single size square serif for variable print bias compensation in microphotography: method, design, and practice, *Proc. SPIE*, 1088, 34–46 (1989).
60. Y. Nissan-Cohen, P. Frank, E.W. Balch, B. Thompson, K. Polasko, and D.M. Brown, Variable proximity corrections for submicron optical lithographic masks, in: *1987 Symposium on VLSI Technology: Digest of Technical Papers*, Karuizawa, Japan, 1987, pp. 13–14.
61. N. Shamma, F. Sporon-Fieder, and E. Lin, A method for the correction of proximity effects in optical projection lithography, in: *Interface'91, Proceedings of the 1991 KTI Microelectronics Seminar, San Jose, CA*, 1991, pp. 145–156.
62. O. Otto, J. Garofalo, K.K. Low, C.M. Yuan, R. Henderson, C. Pierrat, R. Kostelak, S. Vaidya, and P.K. Vasudev, Automated optical proximity correction: a rules-based approach, *Proc. SPIE*, 2197, 278–293 (1994).
63. B.E.A. Saleh and K. Nashold, Image construction: optimum amplitude and phase masks in lithography, *Appl. Opt.*, 24, 1432–1437 (1985).
64. M. Rieger and J. Stirniman, Using behavior modeling for proximity correction, *Proc. SPIE*, 2197, 371–376 (1994).
65. N. Cobb, A. Zakhor, and E. Miloslavsky, Mathematical and CAD framework for proximity correction, *Proc. SPIE*, 2726, 208–222 (1996).
66. N. Cobb and Y. Granik, Model-based OPC using the MEEF matrix, *Proc. SPIE*, 4889, 1281–1292 (2002).
67. W. Maurer, V. Wiaux, R. Jonckheere, V. Philipsen, T. Hoffmann, S. Verhaegen, K. Ronse, J. England, and W. Howard, OPC aware mask and wafer metrology, *Proc. SPIE*, 4764, 175–181 (2002).

Section IV

NGL Masks

9

NGL Masks: An Overview

Kurt R. Kimmel and Michael Lercel

CONTENTS
9.1 Introduction 193
9.2 Electron Projection Lithography 195
9.3 Extreme Ultraviolet Lithography 196
9.4 Ion Beam Projection Lithography 196
9.5 Proximity X-Ray Lithography 196
9.6 Low-Energy Electron Beam Proximity Lithography 197
9.7 Nanoimprint Lithography 197
9.8 Summary 197

9.1 Introduction

"NGL" or "next generation lithography" entered the semiconductor industry vocabulary decades ago, as a category to encompass the various emerging and contemplated lithography techniques being developed as replacement for the long-standing optical projection technology that defines microlithography today. The deployment of any of the NGL technologies has been far later than most expert prognosticators in the field imagined. The continued ability of optical lithography technologies to provide imaging needs at justifiable cost has surprised many and crushed the hopes of several technology ventures. A long-standing joke in the industry applies the parody marketing tag line, "The Technology of the Future" to any NGL technology a speaker wishes to bash. It reflects the industry's fascination, for decades with the perception of a perpetually forecasted end of the optical lithography era and pursuit of a still unreached dawning of the NGL era.

Eventually, the universal laws of science do prevail, and, clearly, optical lithography will have a finite limit of application when looking purely at the technology capability. The pertinent industry question will be whether another business case emerges such that advances in nonlithography-related areas (design, materials, processing methods, etc.) will sufficiently promote further evolution of microelectronics and obviate the need to deploy one or more of the NGL technologies, thus further extending the usefulness of optical lithography. Or, perhaps the nonlithography-related microelectronics manufacturing factors will reach their own fundamental limits before lithography does, and lithography will no longer be the limiting technology factor. Of course all these assessments of where and when optical lithography will find its demise are made in the context of a balanced technology (capability) versus business (financial) decision.

Regardless of the timing or business case, the generic challenges facing NGL technologies are staggering. Nevertheless, the next four chapters assume certain failure of optical lithography and present the main NGL technologies in development today and recently. "NGLs" addressed in this book are electron beam projection lithography (EPL), extreme ultraviolet lithography (EUV ~ 13.5 nm wavelength), ion projection lithography (IPL), and proximity x-ray lithography (PXL ~ 1 nm wavelength). Low-energy electron beam projection lithography (LEEPL) and imprint lithography (a.k.a. step-and-flash), are introduced in this chapter but are not addressed later in this book.

While talking of NGLs it would also be appropriate to mention another important lithography technology in development: direct patterning of the device substrate using electrons or photons. This is collectively know as "mask-less lithography" (or ML2) and is also not addressed in the book for obvious reasons.

The differences in the overall imaging methods for NGL versus optical technologies lie in the radically different mask materials and architectures. The ever-shrinking imaging requirements (at any demagnification) and the lethality of some gases or monolayers as new classes of defects pose problems that are driving sophisticated — and expensive — solutions. Also, a crucial mask accessory that is not feasible for any of the NGL technologies is a pellicle to protect the mask from defects postmanufacture. The full impact of this seemingly small missing element has not been fully assessed as resources continue to be focused on the more fundamental functional issues such as resolution, CD control and image placement error. The NGL technologies also create new pattern transfer challenges for resists due to new contrast, sensitivity, and proximity situations associated with each of the mask structure/illumination pairs. Table 9.1 summarizes some of the primary differences in NGL technology and mask physical parameters. The pertinence and details of each of these mask types and parameters are discussed in subsequent chapters.

Although the NGL technologies employ significantly different types of sources for exposing wafers, they share some basic fundamental traits. The exposure energies are much higher than those used in optical lithography (approximately few eV) and greater than the bonding energy of materials, so the radiation exposure energy cannot pass through a solid material without some attenuation or change in beam profile. Therefore, all the mask formats require a fundamentally different approach to transmission. This may include a thin membrane, a reflective surface, or the use of a stencil mask. This introduces new challenges to the mask for flatness, film stress, mask distortion, and uniformity. The higher energy exposure, however, puts an opposite demand on the absorber material. The patterning layer must have sufficient absorption or scattering (determined by its thickness and the exposure energy) but remain thin enough to avoid high aspect ratios that are difficult to pattern into the mask. The absorption or scattering of the radiation can also lead to a heat buildup in the mask that must be dissipated.

In addition, the higher exposure energies and lower beam numerical apertures preclude the use of an organic defect protective layer (the pellicle) out of the focal plane. As such, reticle defect protection schemes are a challenge for all the NGL technologies. New approaches of keeping defects off of the reticle and ensuring that they do not return are required.

Nanoimprint lithography is the one exception to the higher exposure energy concerns. However, the same concerns about reticle protection apply, and the near-perfect image transfer fidelity leads to near-perfect defect printing.

The following sections give brief introductions of six NGL technologies, the first four of which are described in detail in the subsequent chapters of this book.

TABLE 9.1

Summary of Primary NGL Mask Attributes

	EPL	EUV	IPL	PXL	LEEPL	Imprint
Wavelength	100 keV Electrons	13.5 nm	75 keV He+ Ions	~1 nm	2 keV Electrons	Not applicable
Type	Transmission	Reflective	Transmission	Transmission	Transmission	Mold
Substrate	Silicon	Quartz	Silicon	Silicon	Silicon	Quartz
Blank	Membrane	Reflector	Membrane	Membrane	Membrane	Quartz
Structure	With struts	Buffer absorber	Single field	Single field	With struts	Substrate only
Absorber Material	Si membrane	TaN	Si membrane	TaSi alloy	Si absorber	Polymer filler
Transmission Material	Hole or ultra-thin SiN + C membrane	MoSi reflector stack	Hole in membrane	SiC or C (diamond) membrane	Hole in membrane	Quartz substrate
Magnification	4×	4×	4×	1×	1×	1×
Challenge 1	Distortion	Blank defects	Distortion	Resolution for 1×	Resolution for 1×	Resolution for 1×
Challenge 2	Field stitching	Flatness	Ion damage	Distortion	Distortion	Mask defects
Challenge 3	Defect protection	Defect protection	Defect protection	Defect protection	Defect protection	Defect protection

9.2 Electron Projection Lithography

Electron projection lithography (EPL) has a fundamental attraction of resolution derived from the extremely short effective wavelength afforded by using electrons as an illumination source. Using charged species also provides the opportunity to do global scale pattern distortion correction by deflecting the beam appropriately. This is a crucial advantage for EPL considering that the delicate, tensile membrane substrate and perforated mask structure create a mask inherently prone to distortion. Recent work in Japan and Europe has created a continuous membrane mask structure, where the transmissive areas of the mask are extremely thin (~20 nm) membranes, which help reduce distortions imposed by a perforated structure and resolve the so-called "donut problem." The donut problem is present for all stencil technologies and refers to a pattern where an island of opaque material exists. A completely disconnected island structure is physically impossible to build in a stencil mask structure. The solution is to split the troublesome pattern into two complementary patterns that additively create the desired island. The severe penalty for this is that two masks are needed for every one layer that contains an island or other similar feature.

EPL development is centered in Japan with Nikon leading the commercialization effort. EPL mask technology benefits from the extensive membrane mask development work in Japan and elsewhere to support PXL.

Primary issues for this technology are: in-plane pattern distortion control, mask inspection, field-to-field stitching, and defect protection postmask manufacture.

9.3 Extreme Ultraviolet Lithography

Extreme ultraviolet lithography (EUVL) was introduced originally as a 4× demagnification alternative to 1× projection x-ray lithography and was referred to as "soft-x-ray reflective lithography." Later, it changed names to the less accurate but more easily marketed "extreme ultraviolet" to distinguish itself from PXL, which had begun its fall from favor due to the challenges imposed by the 1× mask architecture. EUVL has ascended in just the last few years to be the primary NGL choice for most technologists and International SEMATECH even established an annual EUVL lithography symposium in 2001 to further collect and focus on the reports from the sizable resources working on this technology.

Primary issues for this technology are: fabrication of the complex, multilayer reflective mask blanks at reasonable cost, mask defect repair, defect protection postmask manufacture, generation of sufficient photon power for reasonable scanner throughput, and contamination of the optical path including the mask.

9.4 Ion Beam Projection Lithography

Ion beam projection lithography (IPL) was proposed as an alternative charged-particle technology affording pattern distortion control capability but imposing additional mask challenges. The ions, having appreciable mass and energy, physically erode the mask over time. This was addressed by applying a carbon buffer layer on top of the mask to absorb the ions, act as a discharge layer, and provide additional thermal dissipation capacity. Although the mask is 4× magnification, it is a perforated stencil type which is, therefore, prone to in-plane distortion and to the donut problem described in the EPL section earlier. Overall, this technology did not accumulate the resources and support necessary to bring it to a commercial state, and the technology now continues to develop as a backup technology or a possible alternative for meeting special needs.

Primary issues for this technology are: in-plane pattern distortion control for the membrane structure, erosion from the impinging ions, and defect protection postmask manufacture.

9.5 Proximity X-Ray Lithography

PXL has one of the longer histories in NGL and reached a peak of development resource and interest in the mid-1990s being aggressively pursued in both the U.S. and Japan. Although multiple chip fabrication demonstrations were achieved, the simultaneous challenges imposed by the mask being 1× and the competitions from ever-advancing optical extension techniques made x-ray become unjustifiable for the mainstream microelectronics industry. Originally, membranes were silicon and the absorber was gold electroplated into a resist mold patterned on the membrane. This structure evolved over time to become silicon carbide then diamond membrane with a tantalum–silicide absorber, patterned with a more traditional subtractive etch process. Inspection is done by electron beam and repair by focused ion beam. The technology is currently available commercially from JMAR-SAL using a collimated plasma point source of x-rays and a vertically oriented mask exposure stage. Masks are not commercially produced but are available in limited quantities from multiple sources worldwide.

Primary issues for this technology are: resolution requirements for the 1× mask fabrication, in-plane pattern distortion balance and control for the absorber + membrane, and defect protection post mask manufacture.

9.6 Low-Energy Electron Beam Proximity Lithography

Low-energy electron beam proximity lithography (LEEPL) is a variant of EPL, which seeks to take advantage of the much simpler electron optic column design required for a low energy, 1× magnification projection system. This makes the exposure system much less costly but imposes some new challenges on the mask. Foremost, the mask at 1× becomes much more difficult to fabricate than the mask at 4× utilized in EPL. Because the electrons have so little energy, the mask must be a perforated structure to provide sufficient contrast and, consequently, the continuous membrane structure being developed for EPL is not feasible. However, the lower energy electrons are easier to deflect so that pattern distortion correction is more easily achieved. LEEPL development is centered in Japan and is a natural adjunct to the EPL technology work there.

Primary issues for this technology are: resolution requirements for the 1× mask fabrication, in-plane pattern distortion control for the perforated mask structure, field-to-field stitching, and defect protection postmask manufacture.

9.7 Nanoimprint Lithography

Nanoimprint lithography involves a direct physical transfer of a pattern from a template to a transfer material on the wafer surface. For this reason, nanoimprint is entirely different from the radiative exposure methods. The resolution and pattern fidelity are determined almost entirely by the mask (or template). The few-nanometer scale resolution is a tremendous advantage but also a nuisance in that any defects or imperfections on the mask are faithfully transferred to the wafer surface. The mask therefore is a 1× version of the pattern and requires strict control of image size, pattern distortion, and defects.

Primary issues for this technology are: patterning of the 1× mask, defect control and repair on the mask, and level-to-level overlay.

9.8 Summary

Clearly, no single NGL technology has overwhelming appeal and each has its own distinct advantages and challenges. At a fundamental level, the technical requirements must be balanced against the business constraints of cost to develop, implement, and maintain a technology. Since lithography has become such a significant component of the overall microelectronics manufacturing cost structure, fabricators will likely choose imaging solutions that are optimized for their business cost sensitivities.

For example, a fundamental difference in business cases is mask utilization, that is, the number of wafers or chips exposed per mask over the product lifetime. High utilization fabrications, such as microprocessor and memory, will have stronger justification for technologies that are more mask cost intensive but offer high resolution and technology extendability. EUVL may fit the needs best in that case for critical layers while for foundries, having relatively smaller quantity, shorter-lived products may find that EPL or even ML2 on appropriate levels to offer the best value.

In any case, users will optimize the cost of their imaging needs commensurate with their technical needs, and this may mean that more than one imaging solution will find commercial implementation. The risks are high and very difficult to assess for all cases. The evidence of this is reflected in the wide array of NGL technologies being pursued and the substantial investments behind each. Within any particular imaging technology, generally the mask cost is the primary driver of the cost of ownership. This explains why the majority of lithography infrastructure development resources are actually directed to the mask sub-component of the technology.

While the cost of developing a single NGL technology is staggering, developing two or more to completion may be impossible. A primary goal of microelectronics fabrication managers and their lithography strategists will be to avoid having to make this dicey choice of an NGL for as long as optical lithography can viably serve the industry needs.

10
Masks for Electron Beam Projection Lithography

Hisatak Sano, Shane Palmer, and Masaki Yamabe

CONTENTS

10.1 Introduction	200
10.2 Masks for Character Projection Lithography	200
10.3 Masks for Electron Beam Projection Lithography	201
10.3.1 Mask Structures, Materials, and Processes	203
10.3.1.1 Stencil Membrane	203
10.3.1.2 Continuous Membrane	206
10.4 Masks for Proximity Electron Lithography	208
10.5 Mask Making	211
10.5.1 Fabrication of Blanks	211
10.5.2 Patterning	211
10.5.3 Trench Etching	212
10.6 Cleaning	214
10.6.1 Wet Cleaning	214
10.6.2 Dry Cleaning	215
10.7 Metrology	215
10.7.1 Image Placement	215
10.7.2 Critical Dimension	216
10.8 Inspection and Repair	217
10.8.1 Printability of Defects	217
10.8.2 Defect Inspection	218
10.8.2.1 Detection of Defects by Transmission Image	218
10.8.2.2 Optical Detection of Defects	218
10.8.2.3 Electron Beam Detection of Defects	219
10.8.2.4 Electron Beam Inspection	219
10.8.3 Particle Inspection	220
10.8.4 Repair	221
10.8.4.1 FIB Repair	221
10.8.4.2 E-Beam Repair	223
10.9 Clean Container	223
10.10 Data Preparation	224
10.10.1 Data Preparation for EPL Masks	224
10.10.2 Complementary Split	226
10.10.3 Data Preparation for PEL Masks	227
10.11 Summary	227
References	228

10.1 Introduction

In electron beam projection lithography (EPL), a flood of electrons illuminates a mask that is then imaged by a lens system to produce a pattern at a surface. The mask is the essential element to provide contrast of patterns through the use of low-scatter region replicating the transparent or clear features region on a mask and high-scatter region replicating the opaque or absorbing feature on the same mask.

Analogous to optical lithography, the imaging system in electron projection lithography may contain a demagnification or reduction of the mask pattern. For proximity electron lithography (PEL) systems, including low-energy electron-beam proximity-projection lithography (LEEPL™) systems, the mask features to be imaged are patterned at 1:1 magnification. Other tools, such as the projection reduction exposure with variable axis immersion lenses (PREVAIL™) and scattering with angular limitation projection electron lithography (SCALPEL™) systems operate using a 4:1 reduction of mask to image pattern. Character or cell projection (CP) tools, which can also be considered as another category of EPL, typically utilize larger demagnifications, e.g., ≥10:1, and operate with a much smaller exposure area at the image plane. Masks used for CP tools invariably involve stencil type masks, i.e., the features defining the pattern are completely absent or removed. One consequence of using a stencil mask is a restriction on the types of geometries that can be defined on the mask. For example, a donut-shaped feature could not be made without the use of a complementary mask. To overcome this restriction, another type of mask used in EPL is the membrane mask, which consists of a thin low-scattering material for the transparent region (to the flood of electrons) and a high-scatter (absorbing) region on the membrane to define the opaque features on the mask. In the following section, we begin the discussion with the character projection mask.

10.2 Masks for Character Projection Lithography

All of the character (or cell) projection (CP) lithography tools use an aperture mask, which has stencil structures defined on a thin silicon (Si) membrane. Character projection tools are a natural extension of direct write variable-shaped e-beam systems. In CPL, the unique shapes and patterns are formed from a uniform e-beam flood exposure of an aperture that contains the highly repetitive patterns (such as the SRAM) and then printed as a single "character" on the write surface (mask or wafer). Figure 10.1 illustrates the difference in the projection between the variable-shaped beam method and the CP method [1]. The CP method uses multiple shapes (defined as the character) printed by one shot to increase the throughput. Table 10.1 summarizes the requirements for the three different CP tools used today or under development and associated mask technology [2–4]. Each of these tools has maskless pattern generation capability, as well as the CP capability, and may be used for small volume production and prototyping. The Toshiba CP tool is under development. The substrates for CP aperture masks are typically silicon-on-insulator (SOI) wafers. A buried oxide layer (BOX) is used in the manufacturing of the mask as an etch stop. The systems that are made by Hitachi High-Technologies and ADVANTEST involve large magnification, 25:1 and 60:1, respectively, which greatly simplifies the patterning (the write and etch) of the stencil mask. Figure 10.2 shows photographs of CP aperture masks for (a) Hitachi HT tool and (b) ADVANTEST tool.

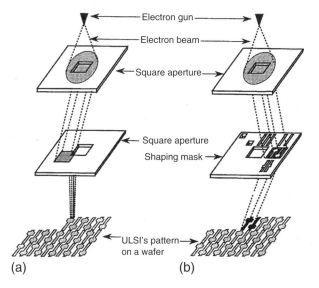

FIGURE 10.1
Comparison of principles between (a) variable-shaped beam method and (b) character projection method. [From H. Satoh, Y. Nakayama, N. Saitou, and T. Kagami, *Proc. SPIE*, 2254, 122–132 (1994).]

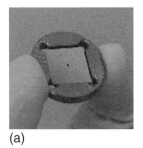
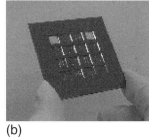

FIGURE 10.2
Photographs of CP aperture masks for (a) Hitachi HT tool and (b) ADVANTEST tool. The mask for Hitachi HT tool is mounted on a holder. (Courtesy of Toppan Printing Co., Ltd., Tokyo, Japan)

TABLE 10.1

Requirements for CP Lithography (Hitachi HT: Hitachi High-Technologies) [*Source*: Refs. [2–4]]

	Hitachi HT [2]	ADVANTEST [3]	Toshiba [4]
Machine type	H900D	F5112	—
Acceleration voltage (kV)	50	50	5
Magnification	25	60	10
Exposed area on wafer ($\mu m \times \mu m$)	5 × 5	5 × 5	5 × 5
CP aperture mask			
Area of character ($\mu m \times \mu m$)	125 × 125	300 × 300	50 × 50
Number of characters per block	21	100	400
Number of blocks per mask	25	16	10–15
Mask size (mm × mm)	11 × 11–13 × 13	50 × 50	5 × 10
Stencil membrane material	Si	Si	Si
Stencil membrane thickness (μm)	15–20	20	0.5–2.0

10.3 Masks for Electron Beam Projection Lithography

The method of exposure for an e-beam stepper is schematically illustrated in Figure 10.3(a) [5]. In this system, the mask moves parallel to the *y*-axis, while the beam scans

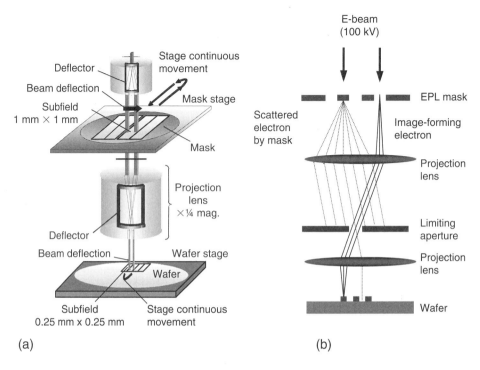

FIGURE 10.3
Principles of (a) exposure of e-beam stepper with PREVAIL column [from S. Kawata, N. Katakura, S. Takahashi, and K. Uchikawa, *J. Vac. Sci. Technol. B.*, 17, 2864–2867 (1999)] and (b) contrast formation.

across the x-axis (in steps) to expose the 20 subfields (SFs). The subfields are reduced (by $4\times$ demagnification) and stitched together by accurate deflections to create the full exposure field. Figure 10.3(b) illustrates the principle of contrast formation by a stencil membrane in the EPL tool, where a central limiting aperture rejects electrons that are scattered away by the membrane, and the nonscattered electrons are allowed to pass through to form image. The left drawing in Figure 10.4(a) shows an EPL mask made from 200-mm wafer with two arrays of 42×102 membranes. Each membrane area is 1.13×1.13 mm. In the current EPL system proposed by Nikon Corporation, the outermost membranes in the arrays are reserved for alignment marks leaving two arrays of 40×100 membrane areas for the primary pattern. This system also uses a demagnification exposure of 1/4, which results in a 20×25-mm field size per noncomplementary mask. If the complementary mask is required for the pattern, the maximum of chip size is 10×25 mm or 20×12.5 mm. The drawing in Figure 10.4(b) describes how the patterns on adjacent membranes are put together to form the desired pattern on the wafer. Currently, several mask vendors can offer 200-mm format masks with device patterns at 70-nm design rules. Figure 10.5 shows a photograph of the "Anaheim" [6–8] pattern, which represents the 70-nm rule system-on-chip (SoC) device designed by Semiconductor Leading Edge Technologies, Inc. (Selete) [6]. Figure 10.5 shows (a) the front side and (b) the backside of 10×12.5 mm field gate level of the complementary mask set needed to pattern the CPU, logic, DRAM, and SRAM devices of the Anaheim pattern [8]. Also, Table 10.2 contains the latest list of EPL requirements set by "International Technology Roadmap for Semiconductor" [9]. The most challenging task for EPL masks is the image placement (IP) accuracy requirement, which is <10 nm for the 65-nm node.

Masks for Electron Beam Projection Lithography

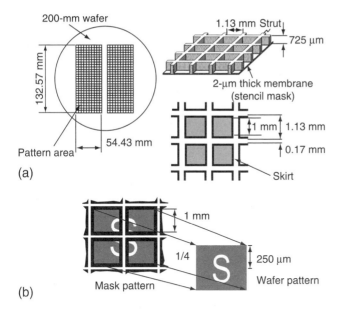

FIGURE 10.4
Illustration of (a) 200-mm wafer format of a stencil mask and (b) subfield stitching of mask pattern in exposure.

FIGURE 10.5
Photographs of a complete mask, showing (a) front and (b) back. The mask pattern "Anaheim" was designed by Selete Inc., Tsukuba, Japan. [From H. Fujita, T. Takigawa, M. Ishikawa, Y. Aritsuka, S. Yusa, M. Hoga, and H. Sano, *Proc. SPIE*, 5256, 826–833 (2003).]

10.3.1 Mask Structures, Materials, and Processes

In this section, we describe the two types of mask structures and the processes that have been used to build the stencil and the continuous membrane EPL masks.

10.3.1.1 Stencil Membrane

A preferred starting material for the stencil (membrane) type mask is a low-stress 1.5 to −2.0-μm-thick silicon membrane. It is strong enough to support itself, extending over the 1.13 × 1.13 mm square area, and it can efficiently scatter the electrons away from the limiting aperture. For the mask minimum feature of 98 nm for 65-nm node (as shown in Table 10.2), the aspect ratio of the opening is about 15 to 20. Such an opening is still manufacturable.

There are two ways of making Si membranes. One is from SOI wafers, and the other involves deposition by sputtering [6]. There are two processes of mask making, namely, wafer-flow process and membrane-flow process as shown in Figure 10.6. Both processes start with an SOI wafer or Si/etch-stop layer/Si wafer. Two patterning steps are required to build the EPL mask: e.g., in the membrane-flow process [Figure 10.6(b)],

TABLE 10.2

EPL Mask Requirements

Year of Production	2006	2007	2010
Node (nm)	70	65	45
Magnification	4	4	4
Mask image size for DRAM 1/2 pitch (nm)	280	260	180
Mask minimum image size (nm)	112	98	70
Image placement (nm, multi-point)			
Non-linear error in sub-field	11	10	7
CD uniformity (nm, 3sigma)			
Isolated lines (MPU gate, nm)	4.5	4	2.5
Dense lines (DRAM 1/2 pitch, nm)	11.5	10.5	7.5
Contact/vias	13	11.5	8
Linearity (nm)	11	10	7
CD mean to target (nm)	6	5.5	4
Pattern corner rounding (nm)	45	40	28
Defect size (nm)	55	50	35
Pattern sidewall angle (degrees)	90	90	90
Pattern sidewall angle tolerance (+ degrees)	0.2	0.2	0.2
Scatterer/stencil LER (3σ, nm)	5	4	3
Maximum mask resistivity Ω cm		20	
Substrate form factor	10	200 mm diameter, 0.725 mm thick	
Mask substrate flatness (μm, p-v)	10	5	
Mask flatness within a subfield (μm, p-v)	11	1	

Source: Semiconductor Industry Association, International Technology Roadmap for Semiconductors: 2002 Update, Austin TX, International Sematech, 2002, http://public.itrs.net

patterning to define the mask struts and the second patterning to define printed layout in a membrane state. The latter step is generally referred to as patterning the "mask blank." Ultimately, the EPL mask process will be similar to the optical mask process in that the mask vendor will begin with a prequalified mask blank to obtain a much shorter process time.

When an SOI wafer is employed as a starting substrate, the dry etch for strut formation places a requirement for the BOX thickness. It must be >0.8 μm to adequately protect the membrane during the dry etch for the current process. Since the BOX layer has a strong compressive internal stress (250–300 MPa), the Si membrane released after this process step tends to have a weak compressive stress of several mega-Pascal, depending on the way of manufacturing the SOI wafer and the BOX thickness. In order to keep the membrane flat, the membrane has to have an adequate amount of tensile stress. However, a large tensile stress tends to induce large IP error. Therefore, the maximum tensile stress is determined to satisfy IP requirement in Table 10.2. For the 65-nm node, the maximum tensile stress of the film should be <5 MPa [10]. It is a common practice to dope boron (B) or phosphorus (P), the atom of which has a smaller diameter than of atomic Si, into the film to adjust the tensile stress. The relation between the internal stress and the boron doping concentration is reported to depend on the bonding procedure of the materials examined [11–13], and the stress of 5 MPa corresponds to a concentration of $4–5 \times 10^{18}$ atoms/cm^3 in one case [11]. The effect of doping species and doping methods on the internal stress was examined [12]. It was found that phosphorus has wider latitude of concentration than boron as shown in Figure 10.7, and that the thermal diffusion method is superior over the ion implantation method in its depth uniformity.

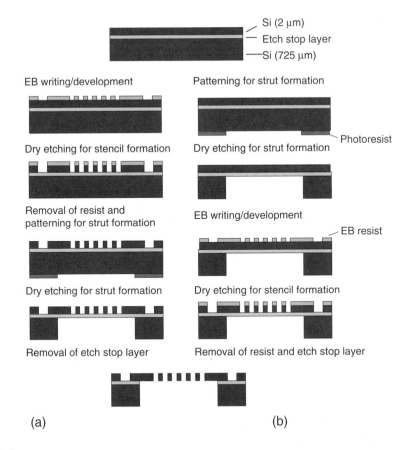

FIGURE 10.6
Representative steps in (a) wafer-flow process and (b) membrane-flow process. Both processes start with an SOI wafer or Si/etch stop layer/Si wafer.

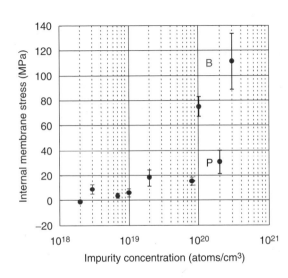

FIGURE 10.7
Relationship between internal stress and impurity boron or phosphorus concentration of Si membrane of SOI wafer due to the thermal diffusion method. [From N. Katakura, S. Takahashi, M. Okada, S. Shimizu, and S. Kawata, *Proc. SPIE*, 4562, 893–901 (2002).]

The other method that has been used for an etch stop is to utilize a dual layer of sputtered Si accompanied with the sputtered chromium nitride (CrN_x) layer. This process has been reported to exhibit excellent property in obtaining low internal stress films (0–10 MPa) [6,14]. The process flow is basically the same as shown in Figure 10.6 but replaces the BOX etch stop with CrN_x. The property of CrN_x will be discussed in Section 10.5.1.

10.3.1.2 Continuous Membrane

To minimize electron blur in e-beam steppers and scanners, stencil membranes are the preferred mask structure; however, many patterns are not possible with the single stencil mask and to avoid the complexity of complementary masks, continuous membranes have been employed in systems, e.g., SCALPEL. The SCALPEL mask consists of a 100 to 150-nm-thick low-stress silicon nitride (SiN_x) membrane and a bi-metal layer of 25 to 50-nm-thick tungsten (W) and 5-nm-thick chromium (Cr) as scatterer. By using a continuous membrane mask there is no need for complementary split. However, this structure and material tend to have disadvantages of low-contrast, low-beam transmission, and the addition of a blur caused by the generation of secondary chromatic peak by plasmon coupling in the membrane. After screening suitable materials for the membrane, Yamashita et al. [15] found that diamond-like-carbon (DLC) films are excellent candidates as supporting membranes even at thicknesses <60 nm and also a scatterer film when they are thicker than 600 nm. They recommended a tri-layer structure, DLD/etch stop layer/DLD, dubbed Light On The Ultimate (EPL) System (LOTUS). A membrane structure consisting of a 600-nm-thick DLC scatterer, a 15-nm-thick CrN_x etch stop layer, and a 30-nm-thick supporting membrane has been developed [16]. Figure 10.8 shows an SEM photograph of (a) donut-shaped scatterer patterns and (b) a cross-sectional drawing of this structure. This stack of layers were sputter deposited on a 100-nm-thick CrN_x etch stop layer and a 725-μm-thick Si substrate, where the CrN_x layer works an etch stop for Si backetching. This type of mask may also be called an ultrathin membrane (UTM) mask. The zero-loss electron aperture transmittance was measured to be 41%, 62%, and 70%,

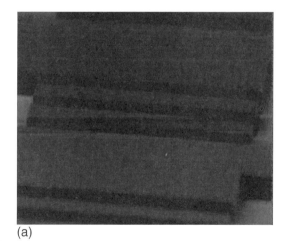

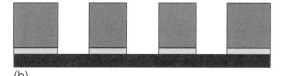

FIGURE 10.8
SEM photograph and a cross-sectional drawing of a continuous DLC membrane mask: (a) 600-nm-thick donut-shaped scatterer patterns supported by a 44-nm-thick DLC membrane, and (b) a stack of DLC scatterer, CrN_x etch stop, and DLC membrane (top). [From I. Amemiya, H. Yamashita, S. Nakatsuka, M. Tsukahara, O. Nagarekawa, *J. Vac. Sci. Technol. B.*, 21, 3032–3036 (2003).]

TABLE 10.3

Comparison of Mask Structures for EPL Systems

Parameters	LOTUS	SCALPEL	Stencil
Mask type	DLC membrane	SiN membrane	Si stencil
Structure	DLC/CrN$_x$/DLC	W/Cr/SiN	Si/opening
Membrane thickness	15–35 nm	100–150 nm	—
Beam transmission	50–70%	30–40%	100%
Energy spread by mask	Approx. 20–22 eV	Approx. 22 eV	0 eV
Scatterer thickness	300–600 nm	30–50 nm	2 μm
Mask split	Unnecessary	Unnecessary	Necessary

Source: The authors of H. Yamashita, I. Amemiya, E. Nomura, K. Nakajima, and H. Nozue, *J. Vac. Sci. Technol. B.*, 18, 3237–3241 (2002), modified by the authors.

respectively, for DLC membrane samples of the thicknesses of 44, 23, and 17 nm. A comparison of the structures and performances of EPL masks is summarized in Table 10.3 by Yamashita et al. [15].

Other types of UTM masks have also been proposed [10], including a tri-layer UTM mask and a bi-layer UTM mask. These UTM masks are frequently referred to as a SiN-C-UTM mask and a C-UTM mask. Figure 10.9 shows the typical mask fabrication process

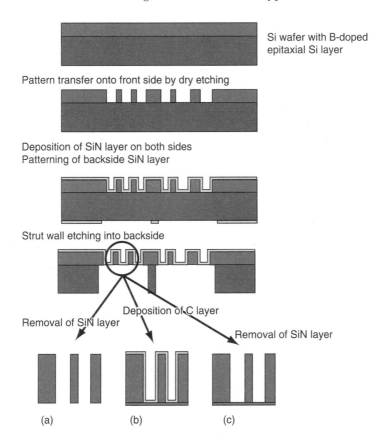

FIGURE 10.9
EPL mask fabrication process for three possible types of masks: (a) Si stencil mask, (b) SiN-C-UTM (ultrathin membrane) mask, and (c) C-UTM mask. [Courtesy of Team Nanotec GmbH, Villingen-Schwenningen, Germany. Modified from P. Reu, C.-F. Chen, R. Engelstad, E. Lovell, T. Bayer, J. Greshner, S. Kalt, H. Weiss, O. Wood II, and R. Mackay, *J. Vac. Sci. Technol. B.*, 20, 3053–3057 (2002).]

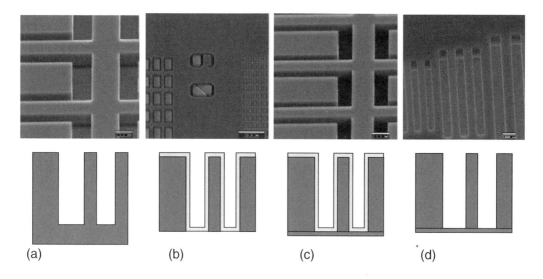

FIGURE 10.10
SEM photographs and cross-sectional drawings of substrate after process steps described in Figure 10.9: (a) pattern transfer into the front by dry etching, (b) strut wall etching into back, (c) SiN+C-UTM (ultrathin membrane) mask, and (d) C-UTM mask. (Courtesy of Team Nanotec GmbH, Villingen-Schwenningen, Germany.)

for three possible types of masks. The mask strut formation for both UTM masks is similar to the wafer-flow process depicted in Figure 10.6(a). The SiN-C-UTM mask is typically composed of a 1500-nm-thick, boron doped, epitaxial Si layer (with an internal stress of 5 MPa), 45 nm-thick Si_3N_4 layer (100 MPa), and 5-nm-thick carbon layer (-240 MPa, i.e., compressive). In order to form the C-UTM mask, the ultrathin carbon membrane is introduced to support isolated Si scatterers when the Si_3N_4 layer is removed. In this case, the carbon membrane has to be 20 to 40-nm thick. Several groups have reported on the successful fabrication of SiN-C-UTM and C-UTM masks. Figure 10.10 shows SEM photographs and cross-sectional drawings of the masks shown in Figure 10.9.

10.4 Masks for Proximity Electron Lithography

Figure 10.11 shows the basic principle of the proximity exposure tool, LEEPL, where a 2-kV e-beam of 2–20 μA scans over an exposure field of 40 mm × 40 mm with the mask placed 30–50 μm above the wafer. By using the distortion correction deflector, the distortion of the mask is partly corrected in the image on the wafer. In Figure 10.12, the SEM photographs show (a) stencil features of 50 nm L & S and (b) 70 nm logic patterns formed on a silicon carbide (SiC) membrane. Figure 10.13 shows three formats of LEEPL masks: (a) NIST-like format [17], (b) and (c) 200-mm wafer format [18], and (d) 6025 format [19]. In the NIST-like format, a 100-mm diameter wafer is bonded to a 125-mm diameter metal frame [17]. In a 200-mm wafer format, a 200-mm wafer is used without any frame bonding. In the 6025 format, a 200-mm wafer is cut and bonded to a ceramic or Si frame to make it photomask-e-beam-writer compatible. Each format can accept various kinds of exposure fields (or windows), such as those shown in Figure 10.13. They are (a) a single large (30 × 30 mm) window, (b) nine large (24 × 24 mm) windows, (c) and

Masks for Electron Beam Projection Lithography 209

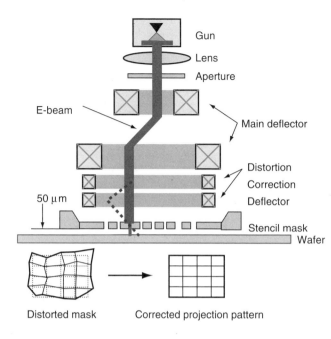

FIGURE 10.11
Principle of proximity electron-beam lithography (PEL or LEEPL), showing the ability to correct distorted mask patterns by deflection of the beam.

(d) multiple small (1.55 × 1.55 mm) windows called COmplementary Stencil Mask On Strut-supports (COSMOS) Type 1 (which will be explained in the next paragraph).

Since the feature on the LEEPL mask is 1× of the pattern to be formed on the wafer, i.e., 50–70-nm wide, the thickness of the stencil membrane is limited to be <1000 nm for fabrication reasons. This thin membrane in turn limits the maximum membrane size because the gravitational sag gets large. Such a consideration leads one to a structure with dozens of smaller membranes (or windows). Figure 10.14 shows a structure named COSMOS [20]. There are two types of COSMOS, as shown in Figure 10.14, where (a) Type 1 and (b) Type 2 are designed for dry etching and wet etching, respectively, in the backside Si etching. White and gray parts correspond to membranes and struts, respectively, in both structures. The struts for Type 2 are wide for wet etching. As struts are placed in complementary position, multiple exposures of the four quadrants surrounded

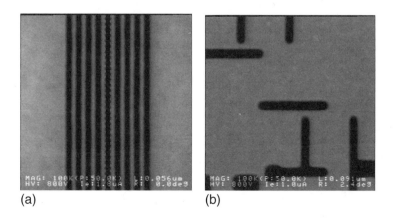

FIGURE 10.12
Patterns on SiC membrane of LEEPL mask: (a) 50-nm L/S pattern, and (b) 70-nm logic pattern. (Courtesy of LEEPL Corp., Mitaka, Japan)

210 Handbook of Photomask Manufacturing Technology

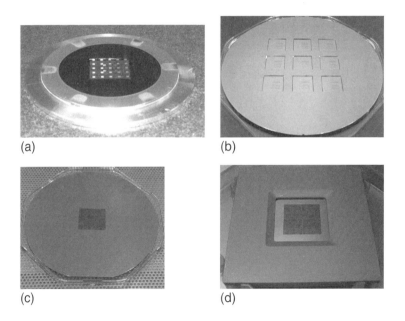

FIGURE 10.13
LEEPL mask formats: (a) NIST-like format, (b) 200-mm wafer format, (c) 200-mm wafer format, and (d) 6025-format (back view). [Courtesy of (a) NTT Advanced Technology Corp., Atsugi, Japan (b) HOYA Corp., Tokyo, Japan (c) Toppan Printing Co., Ltd., Tokyo, Japan and (d) Dai Nippon Printing Co., Ltd., Tokyo, Japan

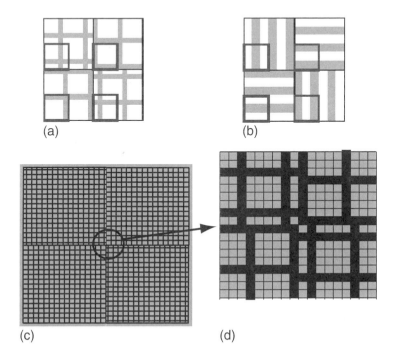

FIGURE 10.14
COSMOS types of LEEPL mask: (a) COSMOS Type 1, (b) COSMOS Type 2, (c) COSMOS Type 1 with 32 × 32 windows, (d) blow-up of the center of (c), showing struts (bold lines) and unit cells (defined by narrow solid lines). [From K. Koike, S. Omori, K. Iwase, I. Ashida, and S. Moriya, *Proc. SPIE*, 4754, 837–846 (2002).]

Masks for Electron Beam Projection Lithography 211

by the bold lines in (a) and (b) yield one complete pattern. Figure 10.14(c) and (d) shows the whole image of COSMOS Type 1 and its center area, respectively.

Stencil membrane materials for LEEPL masks are selected among Si [18], SiC, undoped diamond [17], and boron doped diamond. The purpose of boron doping in diamond is to give it suitable electrical conductivity. Finite element model simulations were used to predict that <5 MPa of internal stress of the membrane is required to keep in-plane-distortion negligible [20].

10.5 Mask Making

The following discussion is based mainly on the membrane-flow process for EPL masks and dry etching of the backside of Si.

10.5.1 Fabrication of Blanks

One of the challenging tasks for EPL masks is to achieve good yield and throughput in defining the struts shown in Figure 10.4. Because a large area must be etched away on the wafer at great depths (725 μm), a special etching process is required, such as the time-multiplex plasma etching process also known as the Bosch process [21,22]. In this process, etching with SF_6/O_2 plasma and deposition with C_4F_8 plasma are switched.

For the etch stop layer a 0.8 to 1.0-μm-thick BOX (SiO_2) layer in the SOI wafer or 0.35-μm-thick CrN_x layer in $Si/CrN_x/Si$ is used. A 25 to 30-μm-thick resist layer works well as an etch mask, whereas SiO_2 or metal layer masks are also applicable. To obtain clear membranes for all subfields, lower etch nonuniformity [(max − min)/(2 × mean)] with higher selectivity or etch rate ratio (Si/etch stop layer) has to be realized. The target values for the etch nonuniformity is smaller than 8% and a selectivity >300 to the 1-μm-thick BOX layer. Several successful etch processes have been reported [7,8]. Figure 10.15 shows SEM views of the struts formed and a photograph of a completed blank. Ashing the resist, removal of the SiO_2 layer by wet etching in a buffered hydrofluoric acid, cleaning, and inspection yield a blank. The high compressive internal stress (−250 MPa) of SOI wafers creates a couple of problems.

Another process that has been used successfully to fabricate EPL mask struts is to employ a sputtered low-stress ($\pm$ 10 MPa) CrN_x film as the etch stop layer [6]. The CrN_x film provides excellent etch selectivity (Si/CrN_x) of 1000 to 2000, which reduces the requirement for the nonuniformity of dry etching. With its high selectivity to fluorine plasmas, the CrN_x layer can remain as an etch stop layer for Si trench etching.

10.5.2 Patterning

Thin membrane materials offer a distinct advantage over thick chrome-on-quartz masks during the e-beam patterning process. The proximity effects caused by the primary and secondary electrons are reduced due to the lower number of scattering atoms below the resist, i.e., a thin membrane. This provides an improved pattern in the resist with decreased image blur and a net increase to resolution [7]. Both positive and negative tone chemical amplified resists of 200 to 300-nm thickness have been used in e-beam writing of stencil masks. An example is shown in Figure 10.16(a). Issues relating to

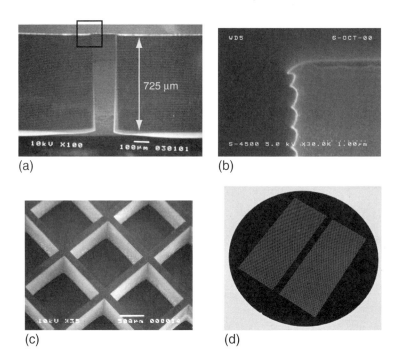

FIGURE 10.15
Views of mask blank: (a) cross-section of a strut, (b) blow-up of the top of the strut, showing "ripples," (c) struts formed, and (d) finished mask blank. [From H. Fujita, T. Takigawa, M. Ishikawa, Y. Aritsuka, S. Yusa, M. Hoga, and H. Sano, *Proc. SPIE*, 5256, 826–833 (2003).]

membrane mask blanks are reported [23]. Current photomask writer JEOL JBX-9000MVII and wafer writer ADVANTEST F5112M can accommodate 200-mm wafer formats. A new generation of e-beam writers are needed that can accommodate 200 mm wafer formats for 65-nm node EPL masks and 90-nm node applications for LEEPL masks.

Since the patterned area is small for LEEPL masks, the 6025 format will be one solution because it is accepted by e-beam writers for photomask making [19].

10.5.3 Trench Etching

Both the trench-etch process that has been used in silicon wafer processing and the Bosch process have been reported as acceptable for forming openings to membrane scatterer (or absorber) material. The Bosch process has been previously reported in the fabrication of ion beam lithography [24] and other early research and development of masks [25]. On the other hand, the trench-etch process accepted in IC wafers is based on the simultaneous process of etching and passivation. One example of the gases commonly used is HBr/SF_4. Figure 10.16(b)–(d) show stencil Si structures formed by such a trench-etch process [8]. In another example, the trench is formed by using reactive ion etching (RIE) with high-density plasma with SF_6 as the main etching gas and CHF_3 as an assistance gas to control the pattern sidewall angle [6]. Fabrication of complete EPL masks at 70-nm design rules has been reported using such processes [6–8]. Figure 10.5 and Figure 10.16(c) and (d) show the whole and part of a test mask with the Anaheim pattern layout [8].

For continuous DLC membrane mask blanks, a stack of a 600-nm-thick DLC scatterer, a 15-nm-thick CrN_x etch stop layer, and a 30-nm-thick DLC supporting membrane

Masks for Electron Beam Projection Lithography 213

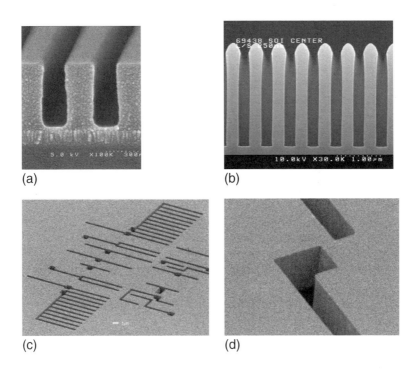

FIGURE 10.16
SEM images of stencil mask structures: (a) cross-section of 300-nm-thick resist, (b) cross-section of 200-nm lines and spaces after dry etching, showing passivation layer on top and sides of Si, (c) stencil Si structure, and (d) blow-up of Si stencil structure. The mask pattern was designed by Selete Inc., Tsukuba, Japan [From H. Fujita, T. Takigawa, M. Ishikawa, Y. Aritsuka, S. Yusa, M. Hoga, and H. Sano, *Proc. SPIE*, 5256, 826–833 (2003).]

represent the film stack to be processed. The DLC scatterer can easily be etched with oxygen RIE and has a high etch selectivity to the CrN_x etch stop layer [16]. Note that since resist is not durable against oxygen RIE, an insertion of an intermediate layer between the resist layer and the DLC layer is required. As with the thick stencil mask, a 30-nm-thick continuous DLC membrane mask has been successfully fabricated using noncomplementary 70-nm-rule SoC device pattern from the Anaheim layout as shown in Figure 10.17 [16].

FIGURE 10.17
Photograph of a 200-mm, continuous DLC membrane EPL mask with noncomplementary 70-nm-rule SoC device patterns. Two chips are placed on the right haft of the mask. [From I. Amemiya, H. Yamashita, S. Nakatsuka, M. Tsukahara, and O. Nagarekawa, *J. Vac. Sci. Technol. B.*, 21, 3032–3036 (2003).]

10.6 Cleaning

Cleaning methods similar to those for photomasks or wafers can be applied to EPL masks. However, EPL masks have several specific points to be concerned with while cleaning: (1) the stencil membrane is fragile, especially when features are formed, (2) the stencil patterns have high-aspect-ratio features, and (3) the strut structure prevents smooth flow of liquid or gas. Therefore, physical treatment should be mild; quick perpendicular movement of EPL masks in liquid or scrubbing with a brush is not allowed.

10.6.1 Wet Cleaning

Applying megasonic waves in liquid is allowed when their power is low. The durability of the EPL masks to conventional cleaning chemicals, such as sulfuric acid/hydrogen peroxide mixtures (SPM) and ammonium hydroxide/water/hydrogen peroxide (SC1), has been reported [26]. SPM tend to oxidize silicon surfaces while removing organic films, whereas SC1 simultaneously etches and oxidizes silicon surfaces, effectively removing contamination by undercutting the particle. Megasonics were applied in the SC1 bath at a low level of 25 W. Multiple (up to ten times) cleaning cycles to B-dosed Si and B-Ge-dosed Si membranes did not affect IP and image size. The effect of cleaning by megasonic vibrations was also reported as shown in Figure 10.18 [12]. In Figure 10.18(a) and (b), Al_2O_3 particles were removed by the megasonic cleaning under the conditions: power 300 W; cleaning liquid alkaline surfactant; pH 12; time 5 min. A Marangoni drying, where the mask is gradually extracted from the de-ionized water into the vapor of isopropyl alcohol and nitrogen, has been demonstrated to be useful to reduce the destructive effect

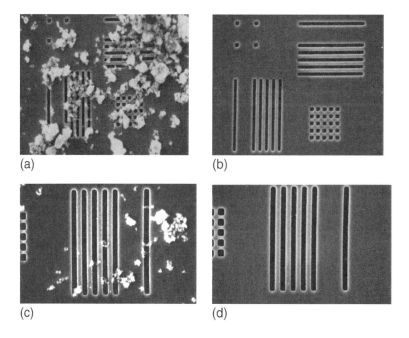

FIGURE 10.18
Effect of cleaning by megasonic and Ar aerosol methods. SEM images of Al_2O_3 contaminated samples are shown (a) before and (b) after megasonic cleaning and (c) before and (d) after Ar aerosol cleaning. [From N. Katakura, S. Takahashi, M. Okada, S. Shimizu, and S. Kawata, *Proc. SPIE*, 4562, 893–901 (2002).]

of the surface tension of the water on three-dimensional structures with high aspect ratio on the mask [27].

10.6.2 Dry Cleaning

Since no protective cover like a pellicle exists for e-beam masks, particles tend to attach to the masks during transportation or storage. It is necessary to remove particles before exposure. One candidate for such preexposure cleanings is an Ar aerosol cleaning technique [12] because it is a dry one. Ar aerosol is blown to the surface of the masks with N_2 carrier gas in a vacuum chamber. Figure 10.18(c) and (d) show the results of an Ar aerosol cleaning, where 0.1-μm Al_2O_3 particles on the surface and inside the through-hole pattern of the EPL mask are effectively removed. Selete has reported similar results with the use of N_2 aerosols. Another dry cleaning technique is the plasma mechanical activation and excitation of particle contamination (PLASMAX) process [28], where a mechanical vibration of the membrane induced in charged plasma does the work.

10.7 Metrology

Two main quality assurance items of EPL masks are IP and critical dimension (CD). Unlike conventional binary optical masks, the CD requirements for stencil masks are affected by the tolerances to the etched CD, such as slope and sidewall roughness.

10.7.1 Image Placement

For masks, IP error or registration error is defined as a deviation of the real position from the designed position of the whole pattern. In practice, scores of special marks called IP marks are inserted to form a grid and the x- and y-coordinates of the marks are optically measured by an IP metrology tool, such as Leica LMS IPRO. If the whole pattern is in one exposure field as in one case of PEL masks then this method can be applied. On the other hand, in the case of EPL masks, their patterns are placed on about 8000 SFs. Therefore, the IP is defined for each subfield. For example, the IP error in one subfield has been measured [7]. However, it is practically impossible to measure the IP error in all subfields. Considering the fact that the subfield size is small (1.0 mm × 1.0 mm), the IP is assumed to be well controlled within the specification once IP errors measured by using specially prepared qualification masks are within the specification. Since the mask may have a global deformation, displacement measurement marks are so placed on the struts that adjacent two of them work as alignment masks for the subfield between them. These marks are to be measured as fundamental data. The exposure tool uses the coordinate information to adjust the beam position.

It is well admitted that repeatable and identical chucking throughout the entire mask fabrication (including e-beam writing, IP measurement) and in the EPL exposure tool is important to minimize IP distortion [29]. A chucking for an e-beam writer and an IP metrology tool similar to the ones used in exposure tool are proposed.

There is one problem of IP measurement for EPL masks as compared with PEL masks. Figure 10.19 shows the configurations of the mask and chuck in the PEL and EPL exposure tools and an IP metrology tool. One noticeable restriction of the present IP

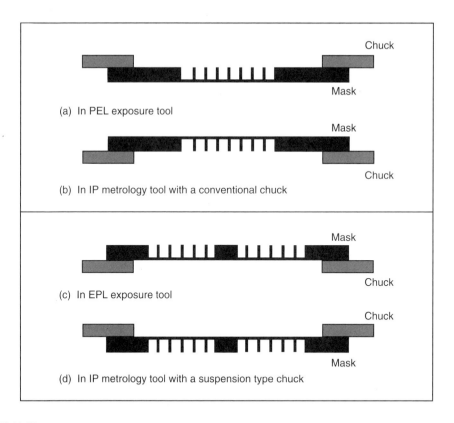

FIGURE 10.19
Schematic diagram of the configurations of the mask and the chuck in PEL and EPL exposure tools and IP metrology tool: (a) in PEL exposure tool, (b) in IP metrology tool with a conventional chuck, (c) in EPL exposure tool, and (d) in IP metrology tool with a suspension type chuck. [Modified from H. Yamamoto, T. Aoyama, N. Hirayanagi, and K. Suzuki, *Proc. SPIE*, 5037, 991–998 (2003).]

metrology tools is that the IP measurement is available only of the upper side of the sample. In the case of PEL, the mask is chucked on its backside both in the exposure tool [in Figure 10.19(a)] and in the IP metrology tool with a conventional chuck [in Figure (b)] although it is placed upside down in the exposure tool. Therefore, a proper gravity correction has to be done [30]. In the case of EPL, however, the mask is chucked upside down on its membrane side in the exposure tool [in Figure 10.19(c)]. If the IP metrology tool with a conventional chuck is employed, the mask is chucked right side up on its backside. The difference of the IP measurement and the IP distortion of the mask in the exposure tool can be correlated to the difference in the chucking methods, which results in a change to the surfaces, i.e., degree of mask bending [31]. However, a suspension type chuck, which holds the mask by its membrane surface right side up as shown in Figure 10.19(d), is effective in giving good correlation after gravity correction [31].

10.7.2 Critical Dimension

Scanning-electron microscopes designed for CD measurement (CD-SEMs for short), which usually operate in a reflection mode, are commonly used for CD measurement of photomasks with target CD below 1000 nm. Since no optical method is applicable for e-beam masks for 65-nm node, they are also used for CD measurement of EPL masks [7] and PEL masks [27]. However, it is desirable to measure CD of an opening of an e-beam stencil mask with image captured by transmission electrons. Recently, a dual-mode CD-SEM, which can operate in a transmission mode as well as in a reflection mode, has been developed and evaluated [32]. It has a stage with an opening and a detector behind the stage for transmission electrons in addition to a normal SEM unit. Their performance has been evaluated. Typical measurement conditions are an acceleration voltage of 5.5 kV with no bias voltage, and a current of 10 pA. At this acceleration voltage, even a thin (0.4 µm) PEL mask absorber of any material works as a perfect absorber. The short-term repeatability and long-term repeatability in the transmission mode are 1.4 and 1.2 nm (3sigma), respectively. Therefore, it is concluded that the measurement of CDs in the transmission mode is suitable for stencil masks. An application of such a dual-mode CD-SEM to an EPL mask was successful [8].

10.8 Inspection and Repair

10.8.1 Printability of Defects

To know the printability of various mask defects under suitable exposure conditions is very important because it determines the criteria of inspection and repair. Exposure tests were done on Nikon's experimental EPL column with masks with defects of various types and sizes [33]. Images of a negative-tone resist were compared between normal patterns of 400-nm lines and spaces on mask and patterns with programmed defects. Figure 10.20

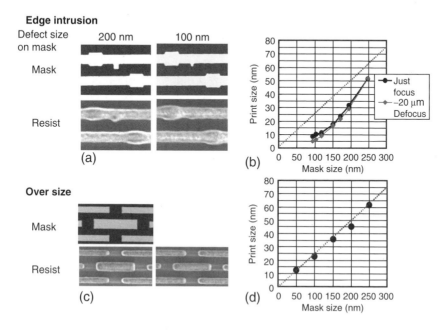

FIGURE 10.20
Printability results: (a) mask images (dark portion: scatterer) and resist images (bright portion) of edge intrusions on a line, (b) relation between print size and mask size of edge intrusions, (c) mask and resist images of oversized holes, and (d) relation between print size and mask size of oversized holes. [Modified from Y. Tomo, Y. Kojima, S. Shimizu, M. Watanabe, H. Jakenaka, H. Yamashita, T. Iwasaki, K. Jakahashi, and M. Yamabe, Proc. SPIE, 4688, 786–797 (2002).]

TABLE 10.4

Defect Printability Summary

Defect Type	Allowable Defect Size[a] (nm) Line	Hole
Pin dot	—	—
Pin hole	140	100
Edge extension	110	40
Edge intrusion	70	120
Corner extension	100	20
Corner intrusion	100	80
Over size	28	12
Under size	28	12
Elongation	28	28
Truncation	28	28
Misplacement (partial)	28	—
Misplacement (whole)	28	28
Edge extension (diagonal)	120	—
Edge intrusion (diagonal)	110	—

[a] Defect size: minimum size that should be detected by inspection system; determined by $\pm 10\%$ CD change criteria (7 nm for 70-nm node).
Source: J. Yamamoto, Y. Tomo, S. Shimizu, T. Iwasaki, and M. Yamabe, *Proc. SPIE*, 5037, 972–982 (2003), modified.

shows two examples of defect printability results, namely, for edge intrusion of a scatterer line and oversized holes. Note that the terminology in this text is different from that in ref. [33]. The edge intrusion has the same width and height, ranging from 100 to 250 nm on mask [in Figure 10.20(a)]. The two solid curves in Figure 10.20(b) show the printability results along with the expected curve (dotted line). If 10% margin is allowed for main resist image size of 100 nm, then 10 nm is allowed for the printed size. Figure 10.20(b) indicates the maximum allowable defect size to be 100–110 nm. For oversized holes [in Figure 10.20(c) and (d)], the printed size falls on the dotted line. This means that the fidelity of the exposure is excellent for oversized holes. Therefore, the allowable oversize on mask is 40%, i.e., 40 nm in this case. Table 10.4 lists allowable defect sizes for lines and holes for 70-nm node, summarizing the defect printability results.

Similar evaluation for PEL masks by using the LEEPL β-tool revealed that the critical defect sizes were 14.5 and 22.8 nm for the intrusions on the edges of 100-nm wide spaces and the 150-nm contact holes, respectively [27,34].

10.8.2 Defect Inspection

10.8.2.1 Detection of Defects by Transmission Image

For photomasks, the transmission image with light is used to detect defects of patterns. For e-beam stencil masks it is also essential to obtain the transmission image of defects to detect them.

10.8.2.2 Optical Detection of Defects

The concept of using an optical microscope for the inspection system has been examined [35]. Clear transmission image of 400 nm lines-and-spaces stencil pattern was obtained with deep ultraviolet (248 nm) light, showing particulate contaminations inside the openings. Simulations, however, predicted that with 257, 193, or 157 nm light, there would not be sufficient light to allow transmission inspection for defects of 80-nm contact holes [36].

Masks for Electron Beam Projection Lithography

Therefore, the feasibility of an inspection system based on optical microscopy depends on future development.

10.8.2.3 Electron Beam Detection of Defects

Detection of detects by transmitted electrons was done with a dual-mode SEM [37]. Figure 10.21 shows the images captured by (a) secondary electrons (in the reflection mode), (b) transmitted electrons (in the transmission mode), and (c) the composite image of both images for lines-and-spaces and contact holes. It should be mentioned that the contamination on the line in Figure 10.21(a) happens to be so thin that the edges of the line underneath it are clearly imaged. E-beam imaging has sufficient resolution for the detection purpose.

10.8.2.4 Electron Beam Inspection

The ability of detecting defects is a necessary condition but not a sufficient condition for an inspection system because an acceptable throughput and 100% detection ability for specified defects are required for it. A system based on a dual-mode SEM cannot give an acceptable throughput because its image acquisition rate is too low to cover the whole mask area in a practical duration of time. A new inspection system, named e-beam scanner, based on a combination of a transmission electron image and optical signal acquisition is being developed to detect defects on e-beam masks with design rules of 65 nm and below [38]. Table 10.5 lists the target specification of the system and Figure 10.22 shows the concept of the system and the structure of the electron-imaging camera. An e-beam with an acceleration voltage of 5 kV is focused on an e-beam mask by the illumination electron optics. After passing through the mask, the beam is magnified and focused on the multiline time delayed integration (TDI)-CCD camera by projection electron optics (EO). Actually the transmitted electron intensity profile is converted to optical signal by the TDI-CCD camera, which is composed of a scintillator plate, a fiber optic plate, and a TDI-CCD sensor as shown in Figure 10.22(b). Since the pixel size of the CCD sensor is 12 µm, the EO magnification of 240 and 400 is necessary to get a resolution

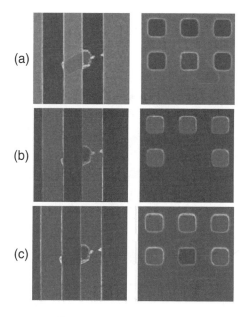

FIGURE 10.21
Images captured by (a) secondary electrons, (b) transmitted electrons, and (c) combined image on a dual-mode SEM. (Courtesy of HOLON Co. Ltd., Tokyo, Japan)

TABLE 10.5

Target Specifications of Mask Inspection System (Courtesy of Selete Inc., Tsukuba, Japan)

Mask	4 × stencil, 1 × stencil, other stencils	
Scan	Stage scan and e-beam scan	
Alignment	Optical and e-beam	
Mask loading	Palette	
Inspection mode	Die to database	
Pixel size	50 nm	30 nm
Acceleration voltage	5 kV	5 kV
EO magnification	240×	400×
Line length on mask	166 µm	100 µm
Throughput	4.6 h (at 80 cm^2)	2.5 h (at 21 cm^2)

of 50 and 30 nm, respectively. Figure 10.23(a) shows the scanning movement of an e-beam strip with continuous and stepping movements of the stage. The charge transfer rate of 100 kHz determines the scan speed, thus the throughput. The system is capable of inspecting both stencil EPL masks and PEL masks by changing the magnification. Figure 10.23(b) shows an image captured by the system, where a programmed defect on the center feature is clearly seen. The pixel size is 30 nm and the features are 160 nm wide. This system cannot be used for the inspection of continuous membrane masks because the continuous membrane stops almost all the incident electrons. Therefore, a new inspection system has to be developed for continuous membrane masks.

10.8.3 Particle Inspection

Another type of inspection system is a particle detector, which detects foreign materials (often called particles) on a blank or mask by detecting the light scattered by particles. It has not yet been established how to inspect particles attached on the backside of the membranes or strut walls.

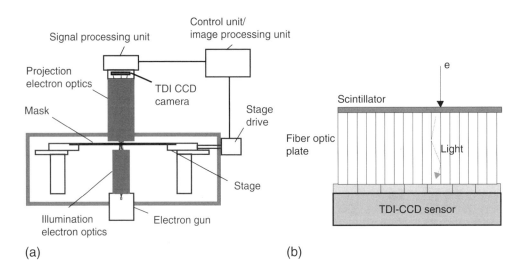

FIGURE 10.22
Concept of e-beam defect inspection system: (a) schematic drawing of the system, and (b) structure of TDI–CCD camera. (Courtesy of Selete Inc., Tsukuba, Japan)

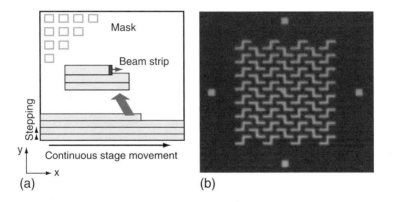

FIGURE 10.23
E-beam defect inspection system performance: (a) way of scanning the beam, and (b) image captured in 30-nm pixel mode. A defect is located on the one on the center of 160-nm-wide features. (Courtesy of Selete Inc., Tsukuba, Japan)

10.8.4 Repair

When an irregular feature is found by an inspection system or a particle detector and classified as a defect to be repaired, three types of repairing paths are accepted. If it is a particle or some residue on the mask surface, cleaning is the best path. If it is a feature that is in excess (i.e., an opaque defect), it has to be removed. If it is a feature that is missing (i.e., a clear defect), it has to be filled. A focused ion beam (FIB) technique has been used to repair opaque and clear defects on photomasks. This technique is also found to be useful to repair defects on e-beam stencil masks [39–41]. A method based on focused e-beams has recently been developed for repairing photomasks and PEL masks [41,42]. However, the three-dimensional structure of the e-beam stencil mask brings technical difficulties in repairing defects as compared with those on photomasks. Figure 10.24 shows the principle of repair by FIB or e-beam for opaque and clear defects on e-beam stencil masks.

10.8.4.1 FIB Repair

An experimental FIB system was set up to study the feasibility of FIB repair [43]. A fine Gaussian shaped beam of 30 keV Ga$^+$ ions at 5-nm image resolution was used. For an

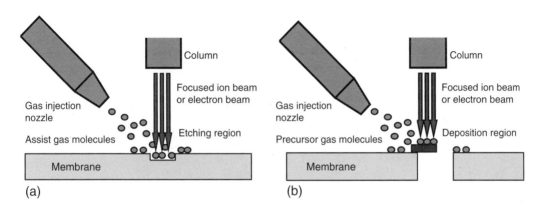

FIGURE 10.24
Principle of defect repair by FIB or e-beam: (a) opaque defect repair (with or without a help of assist gas in case of FIB etching), and (b) clear defect repair by deposition.

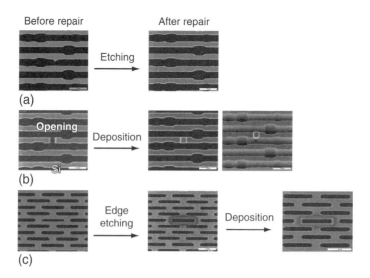

FIGURE 10.25
Repair of defects on EPL mask by FIB: (a) opaque defect repair by etching, (b) clear defect repair by deposition, and (c) opaque defect (undersized feature) repair by gas-assisted etching followed by deposition. The pitch of the patterns is 800 nm on mask. (Courtesy of SII Nano Technology Inc., Chiba, Japan)

opaque defect in Figure 10.25(a), only FIB was used because FIB itself has an ability to etch the membrane material, such as Si by sputtering. However, when assist gas molecules are supplied to the etching region, they help accelerate etching. They also help to form near vertical features, i.e., a sidewall angle of more than 89° (it was 87° without them). For a clear defect, diamond-like carbon is deposited to fill to the defect, where precursor gas molecules of a special hydrocarbon are supplied to the deposition region and decomposed by the energy of the ion beam into volatile hydrocarbon molecules and deposited carbon molecules. Figure 10.25 shows scanning ion microscope (SIM) photographs of EPL mask patterns (bright portions) before and after repair. In (a), a protrusion is removed by etching, in (b) an opening is filled by deposition, and in (c) an undersized slot is first enlarged by etching, and then shaped by deposition because the deposition gives better edge definition than the etching does. Note that the deposition in (b) makes a bridge over the programmed space, and it is not as thick as the membrane because the deposition hardly grows directly on the side of the space. To have a thicker bridge, an innovative

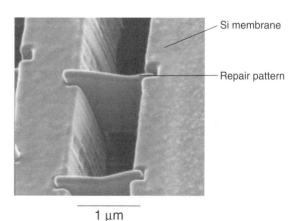

FIGURE 10.26
SIM image of the deposition on preetched slot patterns. [From S. Shimizu, S. Kawata, and T. Kaito, *Proc. SPIE*, 3997, 560–567 (2000).]

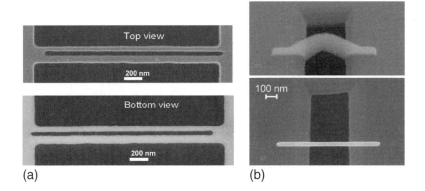

FIGURE 10.27
Etching and deposition by e-beam for showing repair capability: (a) 40 nm wide, 2-μm-long slit etched through 300-nm-thick SiC stencil PEL mask provided by LEEPL Corp., Mitaka, Japan and NTT Advanced Technology Corp., Atsugi, Japan and (b) deposition of a 50-nm wide Pt/C bridge across Si membrane space. (Courtesy of NaWoTec GmbH, Rossdorf, Germany)

technique is employed [39]. Figure 10.26 shows an SIM image of the deposition on the preetched slot pattern, which is an example of forming a narrow but tall bridge crossing a 1.0-μm-wide space. First, two slots are formed by FIB etching. Then, FIB deposition forms a bridge by starting from the bottom of each slot. The FIB repair method is also successfully applied to repair opaque and clear defects on a 70-nm node PEL mask.

10.8.4.2 E-Beam Repair

An experimental e-beam mask repair system has been built on the basis of a high-resolution scanning electron microscope with a GEMINI column and a variable pressure system [42]. It has a Schottky TFE gun. The suitable acceleration voltage for processing is around 1.0 kV with a current of 15–50 pA. The minimum spot size is smaller than 3 nm. A Pt containing organometallic species is selected as a precursor to obtain a deposition of Pt/C. Figure 10.27 shows two examples of e-beam repair. In (a), about 40-nm-wide, 2-μm-long slit was etched through a 200-nm-wide bar in a 300-nm-thick silicon carbide PEL stencil mask. In (b), a Pt/C bridge was deposited across the Si membrane space. The 45-nm-thick Pt/C deposition gives sufficient scattering contrast for 100 kV electrons of EPL and 100% attenuation for 2 kV electrons of PEL. Processing at low pressure is found to be effective for charge compensation. The potential of e-beam repair, however, has not been fully demonstrated yet.

10.9 Clean Container

For e-beam masks, there are no pellicles that protect the surface of the mask from particles. Therefore, it is an important issue how to protect the surface of an e-beam mask from contamination by particles during transport and storage. One solution is to keep one piece of e-beam mask in a clean container and open the container in a clean place like the transfer or loadlock chamber of an exposure tool. A container called standard mechanical interface (SMIF) pod is already employed in wafer and photomask handling. Therefore, it is natural to employ an SMIF pod for e-beam masks. Nikon Corp. proposed a mask case for its e-beam stepper that is slightly modified from the standard 200 mm SMIF

FIGURE 10.28
Photographs of SMIF pods: (a) top view of an atmospheric SMIF pod for Nikon's e-beam steppers, (b) bottom view, (c) top view of a vacuum SMIF (CAVS) pod for LEEPL's exposure tools, and (d) bottom view. [(a) and (b): K. Suzuki, *Proc. SPIE*, 4754, 775–789 (2002); (c) and (d): courtesy of LEEPL Corp., Mitaka, Japan].

pod to hold an EPL mask with its mask-side down [44]. Figure 10.28(a) and (b) shows photographs of the atmospheric SMIF pod with an EPL mask blank in it, taken from the front and back, respectively. The mask is kept at atmospheric pressure.

On the other hand, PEL takes another approach by employing a vacuum SMIF pod dubbed clean and vacuum system (CAVS) pod [45], which had originally been developed by TDK Corp. for its magnetic head manufacturing applications. The inside of a CAVS pod is kept in a low vacuum (100–1000 Pa). A special opener that opens the pod without breaking the vacuum has to be used. Figure 10.28(c) and (d) shows the photographs of the pod taken from the front and back, respectively.

10.10 Data Preparation

EPL masks and PEL masks need a special data preparation.

10.10.1 Data Preparation for EPL Masks

Several data conversion systems have been developed by three electrical design automation vendors [46–48]. Figure 10.29 shows a processing flow of a data conversion system

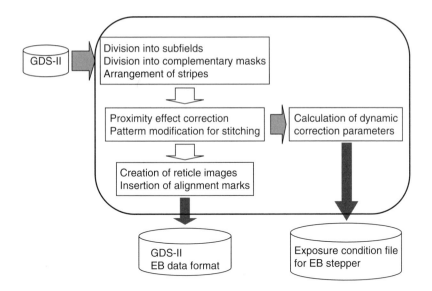

FIGURE 10.29
Data conversion flow chart for EPL masks.

for stencil e-beam masks. It includes: (1) division into subfields, (2) division into complementary masks, (3) e-beam proximity effect correction (EBPC or PEC), (4) pattern modulation for stitching, (5) generation of parameters for e-beam exposure tools, (6) creation of mask images with alignment marks, and (7) output of necessary data files. The conversion system takes a hierarchical GDS-II formatted file as input data. First the file structure is flattened, and the whole pattern is divided into subfields (subfield division), where the subfield has a basic area of 1000 × 1000 μm and a surrounding buffer area called a fuzzy boundary area, a typical width of which is 10 μm. Subfields are handled as independent cells in the layout library file. Then the patterns are to be split and dispatched into two complementary masks (complementary division or split). Since the stencil mask cannot have a donut-shaped pattern (opening), such a pattern should be split. Other particular patterns, such as long lines, spaces, and large openings, are structurally so weak that they tend to break in the fabrication process. Therefore, they have to be split based on a predetermined criterion. Then a PEC and a pattern modification for stitching have to be done, where a PEC is used to correct the effect of backscattered electrons to resist exposure because the effective radius of the backscattered electron is about 50 μm for 100 keV electrons. The mask biasing method is commonly employed [49]. When patterns are dense, the dose by the backscattered electrons is high. Therefore, the space width in the dense patterns should be corrected to be smaller, i.e., it should have a negative bias. Shape modulation for stitching at the stitching boundaries is useful to reduce the effect of mismatch on the resist image. Several methods have been proposed for it [25,47,48,50]. Then the dynamic correction parameters for each subfield are calculated for the exposure tool taking the space charge effect into account [51] and the data file is created. Creation of mask images and insertion of alignment marks follow. Finally, the processed data file is outputted in a GDS-II format [46] or in an e-beam format [47]. The data file for an inspection tool is also available.

10.10.2 Complementary Split

There are several rules for complementary split. The first rule is that the patterns must be structurally stable. The bending moment criterion is a useful judgment of structural stability [52]. The second is that the two masks should have almost the same pattern density and global homogeneity [29]. The third is that the connection points between the two mask patterns should be minimized. However, this rule is not so strong. The fourth is that connections at pattern-shape sensitive sites should be avoided, e.g., at gate lines. Let us try to understand the complementary split with an example. If there is a pattern composed of an array of four O features, there are several ways of complementary split for this pattern, and four of them are shown in Figure 10.30. The first rule is satisfied by all the ways of split assuming that the cantilever-shaped features in Split 4 are short enough to satisfy the bending moment criterion. The second is satisfied by Splits 2–4 but not by Split 1 even though it is the simplest. If the third is employed, Splits 3 and 4 are preferable to Split 2. The fourth is not employed in this case. Although each of the software systems from the three vendors work well for these exposures, further experiments with the exposure tool and masks are necessary to optimize the subfield division and complementary splitting methodology [41]. Figure 10.31 shows SEM photographs of complementary

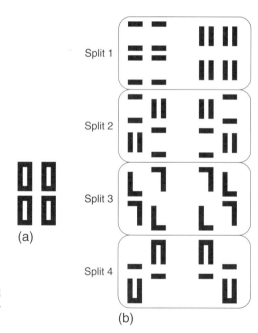

FIGURE 10.30
Example of complementary pattern splits: an original pattern (a) can be split into different pairs of complementary patterns (b).

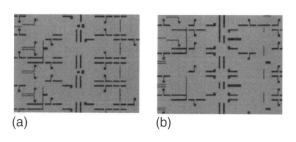

FIGURE 10.31
SEM images of a pair of complementary patterns: (a) complementary pattern A, and (b) complementary pattern B. [From H. Yamashita, K. Takahashi, I. Amemiya, K. Takeuchi, H. Masaoka, H. Takenaka, and M. Yamabe, *J. Vac. Sci. Technol. B.*, 20, 3015–3020 (2002).].

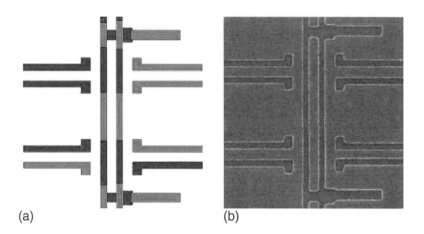

FIGURE 10.32
Complementary split and exposure result: (a) pattern split into A (black portion) and B (gray portion) masks, and (b) resist image obtained by exposure of masks A and B. [From H. Yamashita, I. Amemiya, K. Takeuchi, H. Masaoka, K. Takahashi, A. Ikeda, Y. Kurioki, and M. Yamabe, *J. Vac. Sci. Technol. B.*, 21, 2645–2649 (2003).].

masks for a gate layer [53]. Efforts are made to reduce the conversion time and output file volume. Parallel processing using a PC cluster is an established method of reducing the conversion time [46–48]. Reconstruction of hierarchical data structures after the subfield division is another effective method [46,53]. Figure 10.32 shows an example of the complementary split and an exposure result by a Nikon e-beam stepper NSR-EB1 [48]. The original pattern was split into two [dark and light patterns in Figure 10.32(a)]. Then the complementary masks with these patterns were exposed on the wafer by a Nikon e-beam stepper. The SEM photograph of the resist image in Figure 10.32(b) indicates a successful exposure.

10.10.3 Data Preparation for PEL Masks

PEL masks have one or several exposure fields as for COSMOS Type 1. When COSMOS Type 1 is employed, the original pattern is divided into an array of "processing" fields. Then the pattern in each field is split into 2–4 SFs on the mask, as shown in Figure 10.14, following the rules mentioned in the earlier subsection. This can be treated properly by specific data conversion systems [54,55]. The systems offer IP correction, where the IP error induced by the internal stress of the stencil membrane is calculated based on the pattern density and corrected in advanced. This correction can also be applied to EPL masks.

10.11 Summary

E-beam projection lithography and proximity electron lithography are expected to be adopted for production of semiconductor devices in a couple of years. Masks for them are under development and the content of this handbook is to be revised to cope with the progress of mask-making technology.

References

1. H. Satoh, Y. Nakayama, N. Saitou, and T. Kagami, Silicon shaping mask for electron-beam cell projection lithography, *Proc. SPIE*, 2254, 122–132 (1994).
2. Y. Shoda, H. Ohta, F. Murai, J. Yamamoto, H. Kawano, H. Satoh, and H. Itoh, Recent progress in cell-projection electron-beam lithography, *Microelectronic Eng.*, 67/68, 78–86 (2003).
3. A. Yamada and T. Yabe, Correction deviations in the shape of projected images in electron beam block exposure column, *J. Vac. Sci. Technol. B.*, 21, 2680–2685 (2003).
4. T. Nakasugi, A. Ando, R. Inanami, N. Sasaki, T. Ota, O. Nagano, Y. Yamazaki, K. Sugihara, I. Mori, M. Miyoshi, K. Okumura, and A. Miura, Maskless lithography: a low-energy electron-beam direct writing system with a common CP-aperture and the recent progress, *Proc. SPIE*, 5037, 1051–1058 (2003).
5. S. Kawata, N. Katakura, S. Takahashi, and K. Uchikawa, Stencil reticle development for electron beam projection system, *J. Vac. Sci. Technol. B.*, 17, 2864–2867 (1999).
6. I. Amemiya, H. Yamashita, S. Nakatsuka, I. Kimura, M. Tsukahara, S. Yasumatsu, and O. Nagarekawa, Fabrication of complete 8 in. stencil mask for electron projection lithography, *J. Vac. Sci. Technol. B.*, 20, 3010–3014 (2002).
7. H. Sugimura, H. Eguchi, T. Yoshii, and A. Tamura, 200-mm EPL stencil mask fabrication by using SOI substrate, *Proc. SPIE*, 5130, 925–933 (2003).
8. H. Fujita, T. Takigawa, M. Ishikawa, Y. Aritsuka, S. Yusa, M. Hoga, and H. Sano, 200-mm EPL stencil mask fabrication and metrology, *Proc. SPIE*, 5256, 826–833 (2003).
9. Semiconductor Industry Association, International Technology Roadmap for Semiconductors: 2002 Update, Austin TX, International Sematech, 2002, http://public.itrs.net.
10. P. Reu, C.-F. Chen, R. Engelstad, E. Lovell, T. Bayer, J. Greshner, S. Kalt, H. Weiss, O. Wood II, and R. Mackay, Electron projection lithography mask format layer stress measurement and simulation of pattern transfer distortion, *J. Vac. Sci. Technol. B.*, 20, 3053–3057 (2002).
11. F.-M. Kamm, A. Ehrmann, H. Schafer, W. Pamler, R. Kasmaire, J. Butschke, R. Springender, E. Haugeneder, and H. Loschner, Influence of silicon on insulator wafer stress properties on placement accuracy of stencil masks, *Jpn. J. Appl. Phys. Part 1*, 41, 4146–4149 (2002).
12. N. Katakura, S. Takahashi, M. Okada, S. Shimizu, and S. Kawata, EPL reticle technology, *Proc. SPIE*, 4562, 893–901 (2002).
13. H. Eguchi, T. Kurosu, T. Yoshii, H. Sugimura, K. Itoh, and A. Tamura, Low-stress stencil masks using a doping method, *Proc. SPIE*, 5256, 871–879 (2003).
14. I. Amemiya, H. Yamashita, S. Nakatsuka, T. Sakurai, I. Kimura, M. Tsukahara, and O. Nagarekawa, Stencil mask technology for electron-beam projection lithography, *Jpn. J. Appl. Phys. Part 1*, 42, 3811–3815 (2003).
15. H. Yamashita, I. Amemiya, E. Nomura, K. Nakajima, and H. Nozue, High-performance membrane mask for electron projection lithography, *J. Vac. Sci. Technol. B.*, 18, 3237–3241 (2000).
16. I. Amemiya, H. Yamashita, S. Nakatsuka, M. Tsukahara, and O. Nagarekawa, Fabrication of continuous DLC membrane mask for electron projection lithography, *J. Vac. Sci. Technol. B.*, 21, 3032–3036 (2003).
17. K. Kurihara, H. Iriguchi, A. Motoyoshi, T. Tabata, S. Takahashi, K. Iwamoto, I. Okada, H. Yoshihara, and H. Noguchi, Stencil masks for electron-beam projection lithography, *Proc. SPIE*, 4409, 726–733 (2001).
18. K. Yotsui, G. Suzuki, and A. Tamura, LEEPL mask fabrication using SOI substrates, *Proc. SPIE*, 5130, 942–950 (2003).
19. Y. Aritsuka, Y. Iimura, M. Hoga, and H. Sano, Development of LEEPL 6025 format mask blanks, *Proc. SPIE*, 5130, 951–957 (2003).
20. K. Koike, S. Omori, K. Iwase, I. Ashida, and S. Moriya, New mask format for low energy electron beam proximity projection lithography, *Proc. SPIE*, 4754, 837–846 (2002).
21. I. Johnson, H. Ashraf, J. Bhardwaj, J. Hopkins, A. Hynes, G. Nicholls, S. McAuley, and S. Hall, Etching 200 mm diameter SCALPEL masks with the ASE process, *Proc. SPIE*, 3997, 184–193 (2000).

22. W. Dauksher, S. Clemens, D. Resnick, K. Smith, P. Mangat, S. Rauf, P. Ventzek, H. Ashraf, L. Lee, S. Hall, I. Johnston, J. Hopkins, A. Chambers, and J. Bhardwaj, Modeling and development of a deep silicon etch process for 200 mm electron projection lithography mask fabrication, *J. Vac. Sci. Technol. B.*, 19, 2921–2925 (2001).
23. C. Magg, M. Lercel, M. Lawliss, R. Kwong, W. Huang, and M. Angelopoulos, Evaluation of an advanced chemically amplified resist for next generation lithography mask fabrication, *Proc. SPIE*, 4186, 707–716 (2001).
24. A. Ehrmann, S. Huber, R. Kaesmaier, A. Oelmann, T. Struck, R. Springer, J. Butschke, F. Letzkus, K. Kragler, H. Loeschner, and I. Rangelow, Stencil mask technology for ion beam lithography, *Proc. SPIE*, 3546, 194–205 (1998).
25. H. Sano, K. Morimoto, Y. Aritsuka, and H. Fujita, Impact of deformation of the edges of two complementary patterns on electron beam projection lithography mask making, *Proc. SPIE*, 4757, 799–804 (2002).
26. C. Thiel, L. Kindt, and M. Lawliss, Image placement distortions in EPL masks, *Proc. SPIE*, 4889, 1133–1142 (2002).
27. S. Omori, K. Iwase, K. Amai, Y. Watanabe, S. Nohama, S. Nohdo, S. Moriya, T. Kitagawa, K. Yotsui, G. Suzuki, and A. Tamura, Litho-and-mask concurrent approach to the critical issues for proximity electron lithography, *Proc. SPIE*, 5256, 132–142 (2003).
28. J. Festa, A. November, D. Bennett, R. Kasica, B. Bailey, and M. Blakey, Cleaning of SCALPEL next-generation lithography masks using PLASMAX, a revolutionary dry cleaning technology, *Proc. SPIE*, 3873, 916–926 (1999).
29. O. Wood II, P. Reu, R. Engelstad, E. Lovell, M. Lercel, C. Thiel, M. Lawliss, and R. Mackay, Reduction of image placement errors in EPL masks, *Proc. SPIE*, 5037, 521–530 (2003).
30. S. Omori, K. Iwase, Y. Watanabe, K. Amai, T. Sasaki, S. Nohama, I. Ashida, S. Moriya, and T. Kitagawa, On-site use of 1× stencil mask: control over image placement and dimension, *Proc. SPIE*, 5130, 958–969 (2003).
31. H. Yamamoto, T. Aoyama, N. Hirayanagi, and K. Suzuki, Distortion management strategy for EPL reticle, *Proc. SPIE*, 5037, 991–998 (2003).
32. M. Ishikawa, H. Fujita, M. Hoga, and H. Sano, Evaluation of a transmission CD-SEM for EB stencil masks, *Proc. SPIE*, 5130, 898–906 (2003).
33. Y. Tomo, Y. Kojima, S. Shimizu, M. Watanabe, H. Jakenaka, H. Yamashita, T. Iwasaki, K. Jakahashi, and M. Yamabe, Defect printability analysis on electron projection lithography with diamond stencil reticle, Proc. SPIE, 4688, 786–797 (2002).
34. S. Nohama, S. Omori, K. Iwase, Y. Watanabe, K. Amai, T. Sasaki, S. Moriya, and T. Kitagawa, State-of-the-art performance of stencil mask for LEEPL, *Proc. SPIE*, 5130, 970–978 (2003).
35. M. Okada, N. Katakura, and S. Kawata, Stencil reticle inspection using a deep ultraviolet microscope, *J. Vac. Sci. Technol. B.*, 20, 3025–3028 (2002).
36. J. Welsh, M. McCallum, and M. Okada, Optical inspection of EPL stencil masks, *Proc. SPIE*, 5037, 999–1008 (2003).
37. T. Okagawa, K. Matsuoka, Y. Kojima, A. Yoshida, S. Matsui, I. Santo, N. Anazwa, and T. Kaito, Inspection of stencil mask using transmission electron for character projection electron beam lithography, *Microelectronic Eng.*, 46, 279–282 (1999).
38. J. Yamamoto, T. Iwasaki, M. Yamabe, N. Anazawa, S. Maruyama, and K. Tsuta, EPL stencil mask defect inspection system using a transmission electron beam, *Proc. SPIE*, 5037, 531–537 (2003).
39. S. Shimizu, S. Kawata, and T. Kaito, Repair method on silicon stencil reticles for EB projection lithography, *Proc. SPIE*, 3997, 560–567 (2000).
40. M. Okada, S. Shimizu, S. Kawata, and T. Kaito, Stencil reticle repair for electron beam projection lithography, *J. Vac. Sci. Technol. B.*, 18, 3254–3258 (2000).
41. O. Wood II, W. Trybula, M. Lercel, C. Thiel, M. Lawliss, K. Edinger, A. Stanishevsky, S. Shimizu, and S. Kawata, Benchmarking stencil reticles for electron projection lithography, *J. Vac. Sci. Technol. B.*, 21, 3072–3077 (2003).
42. V. Boegli, H. Koops, M. Budach, K. Edinger, O. Hoinkis, B. Weyrauch, R. Becker, R. Schmidt, A. Kaya, A. Reinhardt, S. Braeuer, H. Honold, J. Bihr, J. Greiser, and M. Eisenmann, Electron-beam induced processes and their applicability to mask repair, *Proc. SPIE*, 4889, 283–292 (2002).

43. Y. Yamamoto, M. Hasuda, H. Suzuki, M. Sato, O. Takaoka, H. Matsw N. Matsumoto, K. Iwasaki, R. Hagiwara, K. Suzuki, Y. Ikku, K. Aita, T. Kaito, Y. Adachi, and Y. Yasaka, FIB mask repair technology for electron projection lithography, *Proc. SPIE*, 5446, 348–356 (2004).
44. K. Suzuki, EPL technology development, *Proc. SPIE*, 4754, 775–789 (2002).
45. A. Yoshida, H. Kasahara, A. Higuchi, H. Nozue, A. Ando, and N. Shimazu, Performance of the beta-tool for low energy electron-bream proximity-projection lithography (LEEPL), *Proc. SPIE*, 5037, 599–610 (2003).
46. K. Kato, K. Nishizawa, T. Haruki, and T. Inoue, EPL data conversion system EPLON, *Proc. SPIE*, 5130, 916–924 (2003).
47. M. Shoji and N. Horiuchi, EPL data conversion, *Proc. SPIE*, 5130, 907–915 (2003).
48. H. Yamashita, I. Amemiya, K. Takeuchi, H. Masaoka, K. Takahashi, A. Ikeda, Y. Kurioki, and M. Yamabe, Complementary exposure of 70 nm SoC devices in electron projection lithography, *J. Vac. Sci. Technol. B.*, 21, 2645–2649 (2003).
49. K. Ogino, H. Hoshino, Y. Maeda, M. Osawa, H. Arimoto, K. Takahashi, and H. Yamashita, High-speed proximity effect correction system for electron-beam projection lithography by cluster processing, *Jpn. J. Appl. Phys. Part 1*, 42, 3827–3832 (2002).
50. T. Fujiwara, T. Irita, S. Shimizu, H. Yamamoto, and K. Suzuki, High accurate CD control at stitching region for electron beam projection lithography, *Proc. SPIE*, 4343, 727–735 (2001).
51. K. Okamoto, K. Kamijo, S. Kojima, H. Minami, and T. Okino, New data post-processing for e-beam projection lithography, *Proc. SPIE*, 4343, 88–94 (2001).
52. H. Yamashita, K. Takeuchi, and H. Masaoka, Mask split algorithm for stencil mask in electron projection lithography, *J. Vac. Sci. Technol. B.*, 19, 2478–2482 (2001).
53. H. Yamashita, K. Takahashi, I. Amemiya, K. Takeuchi, H. Masaoka, H. Takenaka, and M. Yamabe, Complementary mask pattern split for 8 in. stencil masks in electron projection lithography, *J. Vac. Sci. Technol. B.*, 20, 3015–3020 (2002).
54. M. Shoji and N. Horiuchi, LEEPL data conversion system, *Proc. SPIE*, 5130, 934–941 (2003).
55. I. Ashida, S. Omori, and H. Ohnuma, Data processing for LEEPL mask: splitting and placement correction, *Proc. SPIE*, 4754, 847–856 (2002).

11
Masks for Extreme Ultraviolet Lithography

Pei-yang Yan

CONTENTS
11.1 Overview 232
11.2 EUVL Mask Blank Fabrication 233
    11.2.1 EUVL Mask Substrate Fabrication 234
        11.2.1.1 Material Requirement 234
        11.2.1.2 Surface Flatness Requirement 235
        11.2.1.3 Surface Roughness Requirement 236
        11.2.1.4 Surface Defect Requirement 237
    11.2.2 EUVL ML Blank Fabrication 238
        11.2.2.1 ML Deposition Processes and Tools 240
        11.2.2.2 ML Interface Engineering 241
        11.2.2.3 The Capping Layer 241
        11.2.2.4 ML Smoothing 242
        11.2.2.5 ML Thermal Stability 244
        11.2.2.6 ML Stress Control 245
        11.2.2.7 ML Defect Repair and Mitigation 246
    11.2.3 The Buffer Layer, Absorber Layer Stack, and Backside Conductive Layer 250
        11.2.3.1 The Buffer Layer 250
        11.2.3.2 The Absorber Layer Stack 250
        11.2.3.3 The Backside Conductive Coating 252
11.3 EUVL Mask Patterning 252
    11.3.1 E-Beam Resist Patterning 253
    11.3.2 Absorber Stack Etch 253
    11.3.3 Absorber Defect Inspection 253
    11.3.4 Absorber Defect Repair 255
    11.3.5 Buffer Layer Etch 257
    11.3.6 Buffer Layer Defect Inspection and Repair 257
    11.3.7 Mask Cleaning 257
11.4 EUVL Mask Protection 258
11.5 New EUVL-Specific Mask Processing Tools 259
    11.5.1 ML Deposition Tool 259
    11.5.2 Blank Inspection Tool 260
    11.5.3 EUVL Mask Reflectometer 260
    11.5.4 E-Beam Repair Tool 260
    11.5.5 EUV Aerial Image Monitor System 260

11.6 EUVL Reflective Mask Performance 261
    11.6.1 Mask Shadowing Effect 261
    11.6.2 Bossung Process Window Tilt and Focus Shift 263
11.7 EUVL Phase-Shift Masks.. 264
    11.7.1 Attenuated Phase-Shift Mask.. 264
    11.7.2 Alternating Phase-Shift Mask 266
11.8 EUVL Mask SEMI Standards.. 267
References 267

11.1 Overview

The extreme ultraviolet lithography (EUVL) extends optical lithography down to 32 nm and below regime by using short exposure wavelength in the range of 11 to 14 nm. One of the differences between the conventional optical lithography and EUVL is the strong material absorption at EUV wavelength. It leads to the requirement of vacuum environment in the EUV light path, all-reflective optics for the lithography tool, and reflective blank for the masks. These requirements represent disruptive technologies for EUVL and mask manufacturers.

EUVL mask fabrication faces several big challenges as for any other types of mask in next generation lithography (NGL) technology. The top challenges existed almost at every step of the mask blank fabrication. It includes low thermal expansion material (LTEM) development for mask substrate, defect-free substrate polishing with zero defect greater than 50 nm, substrate surface roughness less than 0.15 nm root-mean-square (RMS) value, substrate peak-to-valley flatness less than 30 nm, defect-free reflective multilayer (ML) blank deposition with defect density less than $0.003/cm^2$ at 30-nm defect size, ML defect inspection, ML damage-free defect repair, etc. After the mask is fabricated, mask handling poses another challenge for technology. As all solid materials absorb EUV light, no pellicle will be available for the EUVL masks. Mask contamination protection during transport and exposure, therefore, becomes very critical. The mask handling scheme may involve many aspects, such as new material selection and new design for mask handling robotics and carriers, possible active particle protection scheme during mask storage and transportation, local or full mask cleaning at the wafer fabrication, *in situ* mask inspection in the scanner tool, etc.

Due to the reflective nature of the EUVL mask as compared to the current transmission mask technology, it requires several additional steps and stringent specifications in mask substrate and blank fabrication processes. The current optical mask blank infrastructure does not support EUVL mask blank fabrication due to either the new steps or the stringent specifications. This technology needs new and improved fabrication tools. Establishing EUVL blank fabrication infrastructure, therefore, becomes crucial to the success of EUVL technology.

EUVL mask specifications, such as mask format, substrate materials and substrate finishing, ML, buffer layer, absorber layer, etc., are the first group standards among all the other established mask standards. SEMI P37-1102 standard, the specification for EUVL mask substrates, and SEMI P38-1103 standard, the specification for absorbing film stacks and ML on EUVL mask blanks, were established in 2001 and 2002, respectively. One important aspect of the mask substrate standard is the mask format. The SEMI P37-1102 standard specifies EUVL mask format to be a 6×6 in.^{2} with thickness of 1/4 in. This is consistent with today's optical mask blank format. The same format factor allows the current mask patterning technology to be extended to EUVL mask patterning with minimum impact on the current mask fabrication infrastructure.

In the past few years, much progress has been made in various areas of EUVL mask development. The mask blank substrate LTEM composition nonuniformity in ULE™ has drastically reduced [1]. The mask substrate surface finishing roughness less than 0.15 nm has been demonstrated [2]. The standard polishing technique has improved the mask blank flatness from 500 to 200 nm level. With additional surface finishing techniques, such as ion-beam figuring (IBF) and Megnetro-Rheological finishing (MRF), sub-100-nm substrate flatness on quartz and LTEM has been achieved [3,4]. ML deposition with no added ML defect at 90-nm particle size has been demonstrated [5]. However, to routinely achieve a defect level of below 0.05 defect/cm^2 at 100-nm polystyrene latex (PSL) sphere equivalent size, i.e., where the detected defect appears to be the same size as a PSL examined under the same inspection conditions, has been shown to be far more difficult. It may require a breakthrough in either deposition technology or deposition tool design and several years of continuous industrial effort to achieve the ultimate ML blank defect density of 0.003/cm^2 at 30-nm defect size.

In the following sections, EUVL mask fabrication steps, their challenges, various specifications, and the reasons behind EUVL mask fabrication will be addressed and discussed.

11.2 EUVL Mask Blank Fabrication

The EUVL mask fabrication process consists of mask blank fabrication and mask patterning. A schematic EUVL reflective mask fabrication process flow is given in Figure 11.1. According to Figure 11.1, the EUVL mask blank fabrication starts with a rigid LTEM substrate [as apposed to the membrane used for x-ray and electron beam (e-beam) lithography masks], followed by substrate polishing, front side high reflectance ML interference coating, buffer layer coating (may or may not be needed), absorbing layer coating, an antireflection over-layer coating, and backside conductive layer coating.* The EUVL mask patterning after blank fabrication follows the very similar steps as today's optical mask patterning process. In the case of an existing buffer layer, an additional buffer layer etch, pattern inspection, and buffer layer defect repair are needed after the absorber removal.

The format of an EUVL mask, according to the SEMI P37-1102 standard, is the same as that of today's conventional optical mask, e.g., 6 in.^{2} format with 1/4 in. thickness. This format allows the mask manufacturers to fabricate EUVL masks by utilizing their current optical mask processing capability with limited modification imposed by the EUVL mask-specific requirements and with a few additions in EUV-specific tools. The mask fabrication tools that need to be modified to meet EUVL mask specifics include e-beam patterning tool, registration metrology tool, and flatness metrology tool. These tools are required to have an electrostatic chuck (e-chuck) or vacuum chuck (see Section 11.2.1.2). The new EUVL-specific mask fabrication tools include ML deposition tool, substrate and ML blank defect inspection tool, e-beam repair tool, EUVL aerial image monitor system (AIMS), EUV reflectometer, etc.

In Sections 11.2.1 and 11.2.2, detailed steps and requirements of mask substrate fabrication and mask patterning will be discussed.

---

* The order of deposition for the backside conductive layer is not finalized.

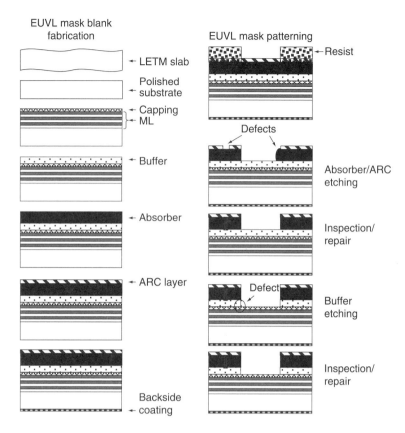

FIGURE 11.1
Schematics of EUVL mask fabrication process flow.

11.2.1 EUVL Mask Substrate Fabrication

According to SEMI P37-1102 standard for EUVL mask substrate, the EUVL mask substrates are composed of an LTEM, such as ULE or Zerodur®, which must have both low average and low spatial variation in thermal expansion. The surface finishing of the substrate is evaluated primarily in three aspects: surface roughness, surface flatness, and number of surface defects. The need for special substrate material other than the conventional quartz material and the additional stringent surface finishing requirements come from several unique EUVL-specific issues that will be discussed in the following sections.

11.2.1.1 Material Requirement

The LTEM requirement for EUVL mask substrate deals with the issue of mask thermal heating during EUV light exposure. The EUV light absorption by an EUVL mask consists of two portions, i.e., the light absorption in the mask absorber region and in the exposed ML region. The light incident on the mask absorber region will be absorbed nearly 100% by the mask. About 30% of EUV incident light on the reflector region or the exposed ML region will also be absorbed due to ML material absorption. The absorbed light energy in the mask will convert to heat energy and lead to mask temperature rise. The change in mask temperature can result in mask deformation, which, in turn, induces pattern placement error in wafer printing. The actual mask deformation for a given temperature change depends on mask substrate material properties, more specifically, the substrate material coefficient of thermal expansion (CTE). Simulation study has shown that when the mask surrounding boundary condition is the same, the induced pattern placement error on the printed wafer is about 36 times larger for Si wafer mask substrate than that of ULE substrate when using a pin chuck, and about 15 times larger for using a flat chuck. The placement error difference in Si and ULE mask substrates is a direct result of the two orders of magnitude larger than CTE that Si exhibits (2.5 ppm/°C as compared to 0.02 ppm/°C of ULE) [6]. SEMI P37-1102 standard requires EUVL mask substrate materials to have CTE in a range from 0 ± 5 ppb/°C with 6 ppb/°C total spatial variation to 0 ± 30 ppb/°C with 10 ppb/°C spatial variation for four different classes of materials.

Two potential commercially available LTEMs for EUVL mask substrate are ULE and Zerodur. ULE material is an amorphous silica (SiO_2) glass containing about 7.5 mol% titania (TiO_2). The Ti atoms substitute the Si atoms and form a solid solution rather than forming a separate phase if the titania content is kept below ~10 mol%. Current available premium grade ULE material has an average CTE in the range ± 10 ppb/°C. Zerodur material is a two-phase material comprising approximately 75% crystalline phase and 25% glass phase. The thermal expansion is negative for the crystal phase and positive for the glass phase. The overall thermal expansion can be adjusted to zero or near zero by controlling thermal treatment during manufacture. Current available Zerodur material has an average CTE in the range ± 15 ppb/°C and spatial variation of about 12 ppb/°C.

11.2.1.2 Surface Flatness Requirement

The main concern on EUVL mask substrate nonflatness-induced error in wafer printing is the pattern in-plane-distortion (IPD). The wafer IPD requirement imposed on the EUVL mask substrate flatness overweighs the defocus requirement, which drives the current optical mask substrate flatness specifications.

The IPD in wafer printing induced by the mask nonflatness is unique in EUVL due to the fact of reflective mask combined with the nontelecentric exposure tool design. This effect is illustrated in Figure 11.2. When the EUV light incidents on the mask with an oblique angle θ from the mask normal, the amount of the position shift Δx, or IPD, at the wafer due to mask nonflatness Δz along the mask thickness direction, or mask out-plane-distortion (OPD), can be simply estimated via geometrical optics as given in Equation (11.1):

$$\Delta x = \frac{\Delta z \times \tan(\theta)}{M} \tag{11.1}$$

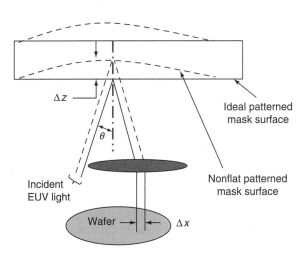

FIGURE 11.2
Illustration of wafer IPD in EUVL induced by mask nonflatness or OPD.

where M is the EUVL scanner reduction ratio. For a 4× reduction scanner with 5° light incident angle on the mask, the IPD induced at the wafer due to the OPD in the mask as described in Equation (11.1) is reduced to:

$$\Delta x \cong 0.023 \Delta z \qquad (11.2)$$

For a mask nonflatness of 50 nm, the corresponding IPD at the wafer will be 1.15 nm. The current SEMI P37-1102 standard specifies maximum EVUL mask substrate peak-to-valley (P–V) flatness to be 30 nm for the highest quality mask substrates in the mask quality area of 142 mm².

The attributes that contribute to the final mask OPD are originated from substrate nonflatness, mask substrate thickness variation, mask deformation due to ML, buffer layer, and absorber layer stress. To reduce the wafer IPD due to mask nonflatness, all the above-mentioned contribution factors need to be addressed.

The current global mask blank polishing technology, either batch polishing or single sample polishing, can extend the polishing finish from the current 500-nm flatness level down to 200-nm, or even 100-nm flatness level. It, however, is not sufficient to reach the 30-nm P–V flatness goal. Local flatness correction, such as surface-figuring techniques, may become necessary to pave the path to the 30-nm flatness level.

There are two potential surface-figuring techniques that have been demonstrated to achieve local flatness correction in the application of EUVL mask substrate finishing [3]. One is the MRF technology and the other is the IBF technology. Both technologies have many applications in optical lens fabrications. These surface non-flatness correction processes first require a completed blank surface interferogram, or mapping of the blank surface nonflatness via high precision interferometer. The measured interferogram will then be fed back to the correction systems, either the MRF or the IBF tools, to determine the location and the amount of nonflatness that needs to be corrected. It may take a few iterations of measurement and correction to achieve the required flatness. Both methodologies currently show a very slow throughput, in the order of an hour to a few hours per blank. Process improvements, such as incorporating *in situ* metrology and batch sample correction, can lead to a higher throughput. Another drawback of these figuring techniques is the increased local surface roughness on an order of a few angstroms. This roughness number exceeds the allowed 0.15 nm RMS spec value in SEMI P37-1102 standard. An additional polishing maybe needed to reduce this local roughness. The effect of substrate surface roughness on reflectivity and its specifications will be discussed in the next section.

11.2.1.3 Surface Roughness Requirement

EUVL mask substrate surface roughness is typically categorized according to its spatial frequency. The substrate high spatial frequency roughness (HSFR) causes EUV light to be scattered at large angles so that it is lost from the optical system. This, in effect, reduces the reflectivity of the mask. Roughness scattering light within the projection lens is often referred to as mid-spatial frequency roughness (MSFR). The roughness in the mid-spatial frequency range can also be characterized as a surface slope error. The effect of MSFR is twofold. First, the longer period mask slope errors coupled with defocus will produce image placement error at the wafer plane. Second, the slope errors on a scale near the resolution limit will contribute to the line edge roughness (LER) in wafer printing. The LER from mask slope errors worsens when the image is out of focus [7]. It was proposed by Gullikson [7] that for a 0.25-NA camera, the frequency range for MSFR and HSFR should be $10^{-6}/\text{nm} < f < 0.004/\text{nm}$ and $0.004/\text{nm} < f < 0.02/\text{nm}$, respectively.

ML smoothing, a deposition technique to smooth the rough substrate surface, is effective in reducing the mask substrate HSFR. However, the current processes for smoothing are not very effective for the MSFR or slope errors (also see Section 11.2.2.4). Therefore, reducing MSFR needs to be tackled at the substrate polishing step.

The SEMI P37-1102 standard specifies the EUVL substrate front surface slope error to be less than 1.0 mrad and substrate front surface HSFR to be less than 0.15 nm RMS.

11.2.1.4 Surface Defect Requirement

The mask substrate defects that ultimately affect the mask printing are those appearing on the substrate surface. Substrate material nonhomogeneity or defects inside the substrate will have no impact on the mask performance as the EUV light does not pass through the mask substrate. The material CTE inhomogeneity, however, is found to be related to the substrate polishing quality. It has been shown recently that CTE nonhomogeneity and defect in the LTEM material can lead to higher surface roughness and defect, such as striation (striae), which are the series of parallel scratches on the surface [8,9].

During ML deposition, the substrate surface defects or particles are covered by the ML stack. Depending on the deposition techniques and processes (e.g., ion-beam incident angle, secondary ion-beam-assisted deposition, etc.), the substrate defect can either be magnified or demagnified (smoothen) in both longitudinal and lateral directions by the ML. Examples of magnification and demagnification of a substrate defect via different deposition conditions are shown in Figure 11.3(a) and (b), where cross-section transmission electron microscopic (TEM) images of a defect nucleated by a 60-nm Au sphere grown at off-normal incidence and at near-normal incidence ML deposition are given, respectively [10].

The EUVL mask substrate front surface defect requirement in the mask quality area of 142 mm^2 is defined by SEMI P37-1102 standard. It allows no scratch and sleek with depth greater than 1 nm and no localized light scatters greater than 50 nm PSL equivalent size for top quality substrate. The localized scatter can be any isolated feature, such as particle

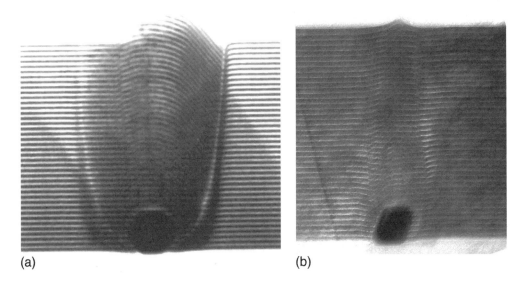

FIGURE 11.3
Cross-sectional TEM images for ion-beam sputtered Mo/Si multilayer on ~60-nm diameter Au spheres for (a) off-normal incidence deposition and (b) near-normal incidence deposition. Much greater smoothing is observed for the near-normal case. Ref 10.

or pit. The effect of ML smoothing on small substrate defects has been taken into account when the substrate defect size specification is defined in SEMI P37-1102 standard. The SEMI P37-1102 standard did not define the total allowed localized light scatters that are less than 50 nm. It is a requirement to be agreed upon between the user and the supplier.

11.2.2 EUVL ML Blank Fabrication

The ML interference coatings on the top of a mask substrate are necessary to make the mask blank reflective at EUVL wavelength. This ML is formed by depositing alternating layers of two materials that have different index of refraction. Typically, the two materials are of alternating high and low atomic number in order to maximize the difference in electron density. In addition, the two materials will have low EUV light absorption. Two of the ML coating materials that have been widely adopted for EUVL mask application are the silicon (Si) and molybdenum (Mo). The Si and Mo pair is chosen because of their relatively large index of refraction contrast. The Si layer has little absorption and acts as "spacer." The Mo layer has higher absorption and provides scattering. The ML deposition process can produce very stable ML structures with relatively smooth and compositionally abrupt interfaces. The thickness of the Mo and Si layers is determined by minimizing the absorption and maximizing the scattering. The period of the ML pairs, which leads to the successive scattering peaks, satisfies the modified Bragg's law [11]:

$$m\lambda = 2d \, \cos\theta \sqrt{1 - \frac{2\overline{\delta}}{1 - \cos^2\theta}} \qquad (11.3)$$

where m is the integer, d is the period of the ML pairs or d-spacing, λ is the EUV wavelength, θ is the light incident angle to the mask normal, and $\overline{\delta}$ is the bi-layer weighted δn. δn is defined as $1 - n$ with n the real part of the index of refraction.

A Mo/Si ML stack for EUV mask blanks at 13.4 nm typically consists of 81 thin film layers; 40 layer pairs of Mo and Si bi-layers plus one capping layer. The bi-layer is 6.9 nm with Mo thickness about 2.8 nm and Si thickness about 4.1 nm. The reflectivity of such an ML design reaches a maximum at about 40 to 50 pairs as shown in Figure 11.4, where the calculated ML reflectivity versus number of ML pairs is given. According to Figure 11.4, ML pairs lower than 40 will result in lower reflectivity. ML pairs higher than 40 can

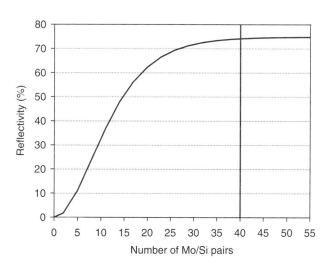

FIGURE 11.4
Simulation result of EUVL ML mask blank reflectivity at 13.4 nm versus number of Mo/Si pairs.

Masks for Extreme Ultraviolet Lithography

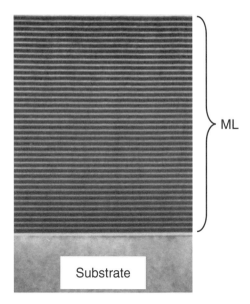

FIGURE 11.5
Cross-sectional TEM image of a 40-pair Mo/Si ML mask blank.

continuously lead to a maximum gain of about 1%. However, depositing more MLs will impact the deposition throughput and potentially increase the ML defect density due to longer deposition time. In Figure 11.5, a cross-section TEM of a 40-pair Mo/Si ML with 4.2-nm Si capping is given. The optimum ML optical properties are obtained when the individual layers are smooth, the transition between the different materials is abrupt, and the layer-to-layer thickness variation is maintained within 0.01 nm. The theoretical calculation predicts the maximum reflectivity of ~75% for an Mo/Si 40-pair ML. In practice, the ML reflectivity will be a few percent lower than the theoretical maximum value. The currently achieved maximum reflectivity is about 70 to 71% [12]. The primary factors that hinge the maximum reflectivity in fabrication are the interdiffusion at the Mo and Si interface and the substrate surface roughness.

In Figure 11.6, both the simulated and experimentally obtained ML blank EUV spectra are given. The ML in both simulation and experiment are based on 40 pairs of Mo/Si ML

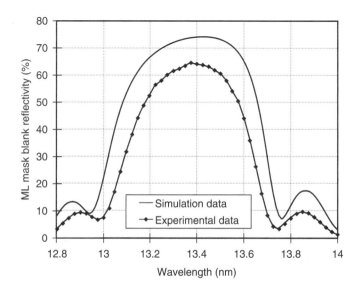

FIGURE 11.6
The simulated and experimentally obtained EUVL ML mask spectra of 40-pair Mo/Si ML.

with Mo and Si thickness of ~2.8 and ~4.2 nm, respectively. In the simulation, the Mo and Si interface is assumed to have atomically abrupt and smooth interfaces, i.e., no molybdenum silicide formation at the Mo–Si interface. The light incident angle to the ML for both simulation and experiment is 5°. The reduction in spectral full-wave-half-maximum (FWHM) bandwidth and peak reflectivity in the experimentally obtained ML spectrum as compared to that of an ideal one is the indication of silicide formation in the interfacing layers and possible surface roughness of the substrate.

The biggest challenge for EUVL mask blank fabrication is the defect reduction. Any defect on the substrate surface prior to ML deposition or any particle fall onto the blank during ML deposition can become a printable defect in wafer printing depending on the original defect size, its location in the ML stack, and the deposition method. To obtain defect-free ML yield of 60%, ML defect density of $0.003/cm^2$ for any printable defect size is required.

The ML defect reduction effort consists of producing ultraclean mask substrate for ML deposition, developing ultraclean ion-beam sputtering deposition system and processes, and developing inspecting metrology for finding the small ML defects. The repair of certain ML defects has been proposed [13,14] (also see Section 11.2.2.7). The performance of the repaired ML depends on the repaired defect size and its position inside the ML. To realize EUVL, the "defect-free" blanks (very low defect density such that the probability of defects landed on the critical area of the circuit is very small) essentially are needed.

11.2.2.1 ML Deposition Processes and Tools

The ML deposition process for EUVL optics and 6 in.^{2} EUVL mask blanks has been demonstrated by using techniques, such as magnetron sputtering [15–17] and ion-beam sputtering [18]. Magnetron sputtering has been widely used in EUVL optics deposition. It yields higher throughput than that of ion-beam sputtering deposition system. It also gives a good film thickness control, as well as high EUV reflectance when the substrate surface roughness is less than 0.15 nm RMS. The defect level, however, is relatively high. In the past, little effort has been given to defect reduction in the magnetron sputtering tools. It is primarily because that the EUV optics is less sensitive to the defects as the defects will be imaged out of focus. The defects on the mask, however, are right in focus during the lithographic imaging process. Defect-free deposition process for EUVL mask blank becomes essential. In the past few years, effort and progress has been made in developing a clean ion-beam deposition system with fully automated standard mechanical interface (SMIF). ML substrate HSFR and isolated bump defect smoothing have been demonstrated with ion-beam sputtering deposition system with and without secondary assisted ion-beam etching [10,19,20]. When the optics or the mask substrate has high HSFR, the peak reflectivity of the ion-beam sputtering produced ML can surpass that of the magnetron sputtering-system due to its ability to smooth the surface roughness.

To achieve an abrupt ML interface, a deposition temperature less than 150°C is also necessary (see also Section 11.2.2.5). Higher temperature promotes ML interface diffusion and interface silicide formation. Any interface mixing will result in ML reflectivity reduction. This temperature constraint excludes deposition methods, such as chemical vapor deposition (CVD), when high temperature is involved.

During ML deposition, ML coating uniformity also needs to be controlled tightly. ML coating nonuniformity will consume the exposure dose control budget. The ML reflectivity uniformity includes peak reflectivity uniformity and centroid wavelength uniformity. The ML peak reflectivity uniformity is affected by the local variations of ML surface contamination, the ratio of Mo to Si layer thickness, substrate roughness, Mo/Si interface roughness, Mo/Si interfacing layer width, Mo and Si film density, and ML period.

Among these factors, the ratio of Mo to Si layer thickness and ML period variations also impact the centroid wavelength uniformity.

In the SEMI P38-1103 standard, the maximum range of peak reflectivity for top quality blanks is specified as 0.5% and the maximum range of centroid wavelength for the top quality blanks is 0.06 nm.

11.2.2.2 ML Interface Engineering

In the ML deposition process, the interface of the Mo and Si layers, in general, is not a sharp interface. Metal tends to diffuse into Si to form metal silicide. As a result, a very thin layer of molybdenum silicide exists at the interface. The ML interface silicide formation can continue when the blanks experience high temperature. Molybdenum silicide formation at the Mo/Si interface is the primary reason that prevents us from achieving the ultimate maximum ML reflectivity. It also posts constraint for any high temperature mask patterning processing.

The molybdenum silicide interfacing layer thickness can be reduced when using low-pressure magnetron sputtering system with Xe gas [12]. But this does not solve ML poor thermal stability issue. Other method to reduce the interfacing layer thickness is to use an appropriate diffusion barrier between the Si and Mo. The first successful application of diffusion barriers in Mo/Si ML was implemented by a thin carbon layer on the Si-on-Mo and Mo-on-Si interfaces [21]. These MLs showed increased thermal stability but no reflectance increase over standard Mo–Si ML. Recently, interface-engineered Mo–Si ML with 70% reflectance at 13.5-nm wavelength were achieved with 50 bi-layers. These new MLs consist of alternating Mo and Si layers separated by a thin boron carbide layer (B_4C) [17]. Depositing B_4C on the Mo–Si interfaces leads to reduction in molybdenum silicide formation at the interfaces. The best results were obtained with 0.4-nm-thick B_4C layers for the Si-on-Mo interfaces and 0.25-nm-thick B_4C layers for the Si-on-Mo interfaces. As a result, the Mo–Si interfaces are sharper in the interface-engineered ML than those in the standard Mo–Si ML.

Adding B_4C as the interdiffusion barrier between the Mo and Si not only helps to prevent metal interdiffusion at room temperature but also helps to prevent metal interdiffusion at higher temperature, i.e., improves ML thermal stability [22].

It is still uncertain at this time whether ML interfacing engineering is needed in EUVL mask blank fabrication. The benefit of adding an interfacing barrier layer needs to be weighted by the complexity added in the deposition process.

11.2.2.3 The Capping Layer

The very top layer, or the finishing layer, on the ML stack is called the capping layer. The ideal capping layer needs first to be compatible with the ML stack design, i.e., after capping layer deposition, no severe degradation in ML reflectivity can occur. In general, it requires capping layer materials with a relatively low EUV light absorption. Second, the capping layer needs to be compatible with mask fabrication, for example, it should serve as the protection layer for ML during mask patterning, repair, cleaning, and exposure. It is unfortunate that the first requirement usually calls for a thinner capping layer as all solid materials absorb EUVL light. The second requirement calls for a thicker capping layer to ensure no damage to the ML stack during absorber and buffer layer etch, repair, and cleanings. Two potential capping layer materials are Si and Ru. Si is the most transparent solid material at 13.4-nm EUV wavelengths. Using Si as a capping layer not only simplifies the coating process (as Si is one of the elements used in the ML composition) but also allows thicker capping layer thickness due to its low EUV absorption.

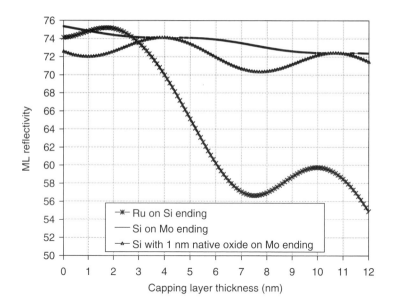

FIGURE 11.7
EUVL ML mask blank peak reflectivity at 13.4 nm versus capping layer thickness for Ru capping, Si capping, and Si capping with 1-nm native silicon oxide, respectively. Ref 42.

However, Si oxidizes easily, especially during EUV light exposure when small amount H_2O vapor exist. Additional oxide on the top of Si will reduce the blank reflectivity, as oxide is darker in EUV light than that of Si. Ru does not oxidize as easily as Si does. Using Ru as a capping layer will need an additional sputtering target, since Ru is not part of the ML composition. In addition, Ru is relatively dark compared to Si at the EUV wavelength. To minimize the impact of Ru absorption to the ML reflectivity, a thin Ru capping of less than 2.5 nm is required. Ru capping on the standard Mo/Si ML also may needs a barrier to suppress the interdiffusion between the Si and Ru layers [17,23]. In Figure 11.7, the simulation results of ML reflectivity of the Si capping layer, the Si capping layer with 1-nm native SiO_2, and the Ru capping layer for the standard 40-pair Mo/Si ML as a function of capping layer thickness are given, respectively. In all cases, the interdiffusion effect between Si and metal is not included in the modeling. In the case of Ru capping, when the Ru capping layer thickness is greater than 2.5 nm, the ML blank reflectivity starts to drop drastically as the capping thickness increases. The capping layer thickness for Ru needs to be chosen so that process requirement and performance (reflectivity) are not compromised.

11.2.2.4 ML Smoothing

ML smoothing refers to the ML deposition process that can mitigate the effect of small substrate surface defects and substrate surface roughness via smoothing of substrate surface imperfections during each layer deposition. As a result, the top 10, 20, or maybe 30 ML layers, depending on both the deposition process and the initial surface defect size or surface roughness, will not be affected by the substrate imperfection. The overall ML reflectivity or yield impact due to substrate defect and substrate surface roughness will be reduced.

The propagation of the substrate imperfection strongly depends on the substrate imperfection type (e.g., bumps or pits), the deposition tool, and the process conditions.

For example, bump defects and pits defects are usually not smoothed simultaneously with the same deposition condition. To smooth the surface roughness (HSFR), higher deposition energy can enhance the smoothing by increasing the motion of atoms on the surface, so that the atoms can move around to smooth the roughness of the substrate. However, higher ion energy also promotes Mo/Si interlayer mixing during ML deposition, resulting in ML reflectivity reduction. The ion-beam incident angle also plays a big role in ML quality, such as reflectivity uniformity and substrate defect smoothing. Ion-beam deposition with an oblique angle usually provides a better deposition uniformity. However, oblique incidence ion-beam deposition is less effective in substrate bump defect smoothing due to the shadowing effect at the defect (see Figure 11.3a).

Recent ML substrate bump defect smoothing study showed that using secondary ion-beam etching or polishing could smooth both substrate surface defect and roughness [19,24]. Secondary ion-beam polishing is more effective when only Si layers are polished during deposition. In Figure 11.8, a cross-sectional surface profile as measured by atomic force microscopy for an uncoated Au sphere of diameter 50 nm and for 50-nm diameter Au spheres coated with Mo/Si with and without ion-assistance is given. It showed that with ion assisted process, a substrate particle at 50-nm diameter could be smoothed to about 1 nm height, rendering it harmless [24]. With the assistance of the secondary ion polishing during ML deposition, the substrate defect smoothing effect becomes less sensitive to the incident angle for primary ion-beam deposition. A more uniform reflectivity film can be obtained at an oblique incident angle without jeopardizing the defect smoothing effect. It is found, however, that the peak reflectivity is reduced by about 1% with secondary ion-beam-assisted deposition. The amount of EUV light absorption due to additional Ar atoms implanted into the ML in the case of secondary Ar ion-beam-assisted deposition plus possible interfacial-layer-mixing due to secondary ion-beam

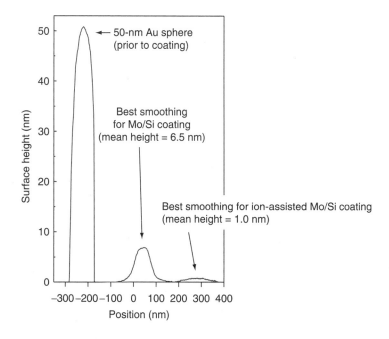

FIGURE 11.8

Cross-sectional surface profile as measured by atomic force microscopy for an uncoated Au sphere of diameter 50 nm and for 50-nm-diameter Au spheres coated with Mo/Si with and without ion assistance. (Courtesy of Paul Mirkarimi, Lawrence Livermore National Laboratory.)

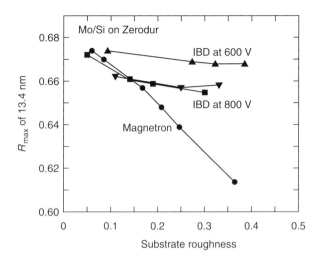

FIGURE 11.9
Peak reflectivity of Mo/Si ML coatings on the substrates of different roughness produced by ion-beam deposition with ion energy of 600 and 800 V without additional polishing. Ref 19.

bombardment accounts for the peak reflectivity reduction. Heavier ions, such as Kr or Xe, may have less incorporation into the ML and, therefore, could have less impact on ML reflectivity.

In figure 11.9, a plot of maximum ML blank reflectivity versus substrate surface roughness was given for different deposition technologies and parameters [19]. For ion-beam deposition at a relatively low voltage of 600 V, the dependence of peak reflectivity on substrate surface roughness is with the 0.5% range. For 800 V ion-beam deposition, the dependence of peak reflectivity on substrate surface roughness is slightly greater than 1%. However, in the case of magnetron sputtering deposition, the peak reflectivity linearly decreases as a function of the substrate surface roughness. It is shown in the 600 V ion-beam deposition case that when the deposition technique and parameters are optimized, substrate surface roughness up to 0.4 nm RMS will not affect ML blanks peak reflectivity more than approximately 0.5%.

11.2.2.5 ML Thermal Stability

ML thermal stability refers to its reflectivity stability when it experiences temperature rise. The ML reflectivity change at high temperature includes two aspects: (1) the ML peak reflectivity reduction and (2) the centroid wavelength shift towards to the shorter wavelength side. The primary mechanism that causes these ML changes at high temperature is the formation of molybdenum silicide interfacing layers. These molybdenum silicide interfacing layers tend to destroy the original optimized Mo/Si bi-layer design that minimizes the absorption and maximize the scattering. The reflectivity of the ML, therefore, decreases. The molybdenum silicide also presents less volume than that of Mo and Si layers. As a result, the ML period (d-spacing) decreases, resulting in ML centroid wavelength shift towards the shorter wavelength side. As the peak reflectivity shifts to the shorter wavelengths side, the drop in reflectivity at the desired wavelength could become drastic as the spectral bandwidth cutoff shifts towards the desired wavelength.

In the case of Mo/Si ML, silicide formation can occur at temperature as low as 150°C. The amount of silicide formed at the ML interface at lower temperature depends on the heating time. At higher temperature, e.g., >400°C, a complete silicide of the ML can be formed by seconds to minutes of heating. In Figure 11.10, experimentally obtained reflectivity spectra of a standard 40-pair Mo/Si ML after rapid thermal process (RTP) at different temperature and time are given. The results are also compared to those of the

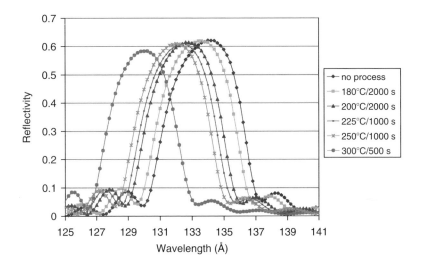

FIGURE 11.10
EUVL ML mask blank reflectivity spectrum variation as a function of temperature (rapid thermal anneal) and heating time. The centroid wavelength always shifts to the shorter wavelength side due to silicide-induced volume contraction or ML d-spacing reduction.

ML without heating. It is shown in Figure 11.10 that for a standard Mo/Si ML, the ML reflective spectrum shifted to the shorter wavelength side by 1 Å for RTP temperature of 180°C and time of 2000 s. The peak reflectivity drop is minimum in this case. For a standard Mo/Si ML with no interdiffusion barrier, a mask process temperature less than 150°C is recommended.

11.2.2.6 ML Stress Control

In Section 11.2.1.2, we have discussed the need and requirement for a very flat EUVL mask substrate. This mask flatness requirement applies equally to the final finished mask, for example, in addition to the mask substrate's nonflatness, the induced mask OPD during mask backside conductive metal layer, ML, buffer layer, absorber layer deposition, and mask patterning process will equally affect the registration of the printed mask pattern. Given a flat mask substrate, the sources of the final mask nonflatness are from the stress of mask backside coating, ML, buffer, and absorber layers. For example, the as-deposited stress of a typical high-reflectance EUVL Mo/Si ML mirror is in the order of −400 Mpa. The minus sign indicates that the stress is compressive. The ML stress is the combination of tensile stress from all Mo layers on the order of 300 Mpa and compressive stress from all Si layers on the order of −1300 Mpa [25]. There are several studies on ML stress reduction, such as varying the ML composition by adjusting Mo thickness, varying deposition conditions, annealing during deposition, annealing after deposition, spacer layer compensation, etc. Most of these methods, however, reduce the ML stress at the sacrifice of the ML reflectivity. For example, to reduce the Mo/Si ML stress to zero by varying Mo composition, the ML reflectivity will be dropped by 33%. Annealing, either after or during the deposition, increases the silicide formation at the interface and causes reflectivity reduction. The spacer layer compensation, i.e., using a spacer layer underneath the ML with a stress of sufficient magnitude and opposite sign to counteract the distortion of the ML, does not reduce the ML reflectivity provided that the spacer layer is as smooth as the substrate. However, the additional process step by adding the spacer layer not only complicates the mask blank fabrication process but also potentially adds more defects in the blank. In addition to the ML stress, the buffer layer and the absorber layer stress can either add up to the ML stress or partially cancel out the ML stress depending on whether the combined buffer and absorber stress is a compressive or tensor stress. After the mask patterning, some stress will be released in the etched region as compared to that of the nonetched region. This nonuniform stress release will result in local mask distortion. Even with a stress-matched buffer, absorber, and ML, the partially etching of the buffer and absorber layers, while keeping the ML untouched, will destroy the stress balance between the three film stacks.

The best way to control mask distortion, therefore, is to control the distortion contributions from each single element; that is, to achieve the substrate flatness first, then the near-zero stress of ML, buffer layer, absorber layer, and backside coating, respectively.

There are several options that exist to deal with the mask residual stress from ML, buffer, and absorber layers. The primary approach is to use the e-chuck in the exposure tool to flatten the mask. In order to implement e-chuck in the exposure tool, it is preferably to have the similar chucking force applied to the mask during mask e-beam write and during mask flatness metrology measurement. Other options include reducing the buffer and absorber thickness whenever possible for a given film stress. The buffer layer thickness can be reduced when the patterning performance, such as absorber etch selectivity and repair etch selectivity, of the buffer layer improves. The absorber film thickness is primarily governed by the absorber material absorption. High absorption material allows a thinner absorber layer. A thicker mask substrate would also help to reduce the mask distortion for a given film stress. It however, involves mask format change.

11.2.2.7 ML Defect Repair and Mitigation

An ML phase defect can be generated from a substrate defect or from a particle during ML deposition. The ML defect usually contains the phase and the amplitude error components. When the phase error has the dominant effect in mask printing, the defects are referred to as phase defects. When the amplitude error has the dominant effect, the defects are referred to as amplitude defects. The phase defects usually originated from the substrate defects or from the particles introduced at the early deposition stage. These defects involve the full or majority ML layers due to defect propagation from the bottom of the ML. The amplitude defects only involve the top ML layers. The more precise classification of the phase and amplitude defects can be done using a high-resolution actinic mask inspection microscope, such as AIMS. For a phase defect, the AIMS image will display a contrast reversal through focus provided the numerical aperture (NA) of the objective is sufficiently high, so that the amount of light scattered outside of the objective pupil is negligible. For an amplitude defect, the AIMS image will show a blur symmetrically through the focus. The aerial image simulations illustrating this through-focus effect for phase and amplitude ML defects are shown in Figure 11.11 [13]. ML defect repairs for both phase and amplitude defects were explored with limited cases [13,14]. For a defect embedded in the middle of the ML, both phase and amplitude play a role in mask printing. There will be no easy way to repair such a defect. ML defect mitigation by aligning the patterned absorber to cover such a defect can be applied. However, in the cases of multiple ML defects, simultaneous coverage of all defects may not be possible. Another ML defect mitigation scheme is the optical proximity correction (OPC) compensation scheme, which modifies the nearby absorber pattern to compensate the effect of ML defects [26]. In all cases of ML repair, completely restoring the ML to its defect-free level is unlikely. The ML repairs simply reduce the defect printability to an acceptable level. Details of each ML repair technique and mitigation method will be discussed in the next four sections.

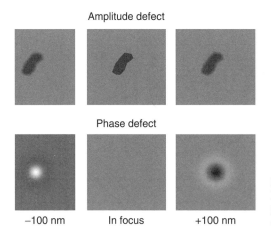

FIGURE 11.11
Illustration of the different through-focus behavior of phase and amplitude defects that can be exploited for defect characterization. Ref 13.

11.2.2.7.1 ML Phase Defect Repair

At the mask substrate, any pit or bump overacting with the ML will cause ML mismatch with ML of defect-free region. The phase difference of the ML, on top of the defect and that of away from the defect, is related to the original defect height. When the ML coverage is conformal, the relationship between the phase and the defect height can be estimated as:

$$\phi = \frac{4\pi}{\lambda} d \cos\theta \quad (11.4)$$

where λ is the exposure wavelength, d is the substrate defect height, and θ is the light incident angle to the mask normal. Such a defect is a phase defect. In the actual ML deposition process, the defect coverage by ML usually forms a Gaussian shape. The effect of the ML defect is less dependent on the original substrate defect height and shape but is more dependent on the physical size of the propagated defect that appeared on the top of the ML. This is understandable since the scattering cross-section of the defect is proportional to the defect size both in the vertical and lateral dimensions.

Since the ML phase defects originate at or toward the bottom of the ML stack, the standard defect removal repair technique will not work. When the ML mismatch in the defect region and in the defect-free region is in an order of a few nanometers, it is feasible to reduce the ML mismatch, or the impact of the ML defect, by reducing the ML period in the defect region via local heating [14]. The local heating promotes the silicide formation at Mo/Si interface and leads to less volume than that of nonheated Mo/Si ML region. The small reduction in each period will result in approximately 40 times the period reduction (for 40-pair ML stack) in the whole ML. As the height of the ML defect reduces, the mismatch between the defect and the defect-free regions will also reduce, as illustrated in Figure 11.12. It has been demonstrated that nanometer scale depth change can be controllably produced with no significant degradation to the reflectance of the coating using high current e-beam with a radius of 400 μm [14]. The limitation for this type of defect repair is that the initial ML defect height has to be within several nanometers. Higher defect requires more d-spacing reduction, which means more silicide formation. Silicide formation at ML interface, as discussed in Section 11.2.2.5, will reduce the ML peak reflectivity and will shift the centroid wavelength towards the shorter wavelength side. The ML reflectivity at the original designed wavelength can get reduced to an

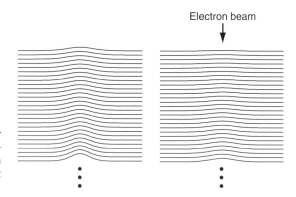

FIGURE 11.12
Schematic illustration of the repair method for EUVL mask blank phase defects: an e-beam enhances silicide formation in the Mo/Si ML, which results in a contraction that reduces the defect height appearing on top of the ML. Ref 14.

unacceptable level combining the effect of peak reflectivity reduction and the centroid wavelength shift.

11.2.2.7.2 ML Amplitude Defect Repair

The proposed method of repairing an ML amplitude defect is to physically remove the opaque spot, which could be an embedded particle near the ML top layers or any damage in the first 10 to 20 topmost ML, by using focused ion beam (FIB) or e-beam. The etched region then will be covered with a thin capping layer by local deposition. The process leaves behind a shallow crater in the ML as shown in Figure 11.13 [13]. The basic principle behind this repair is that after several topmost ML pairs are removed along with the opaque spot, the remaining ML pairs in the locally repaired region can still remain high. It has been shown previously that when an ML composed of 50 to 60 bi-layers, removing 20 pairs of topmost ML will only alter the reflectivity by about 1% (see Figure 11.4). When the number of original ML pairs and the allowable reflectivity variation are given, there will be a limitation on the depth of the amplitude defect that can be removed using this technique. In addition to the reflectivity reduction consideration when the ML pairs are being removed alone, there is also an impact on the reflectivity due to phase-shift introduced by milling out a small region of the reflective surface. To minimize the undesired impact due to removing some of the MLs during the repair, the crater profile must be carefully controlled to have very small slope as shown in Figure 11.13. The

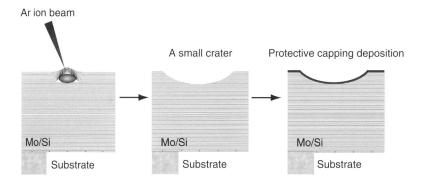

FIGURE 11.13
Illustration of amplitude defect removal using an FIB. After defect removal, a small crater remains in the multilayer. Ref 13.

calculation has shown that a crater with gently sloping sides will not print as easily as craters with steep sides, since there will be very little light scattered outside of the project lens pupil [13]. The dominant effect of the crater phase structure then becomes a defocus effect, which is negligible for shallow craters. In practice, the exposed Mo and Si boundaries will oxidize and cause reflectivity drop in the repaired region. To prevent oxidation after repair, *in situ* local ion-beam sputtering deposition of a passivation capping layer has been proposed and investigated [13]. The right capping layer material and material deposition thickness are essential to control the reflectivity variations across the repaired region.

11.2.2.7.3 ML Defect Compensation

EUV ML defect mitigation via OPC on the absorber features has been studied and demonstrated [26]. The concept of the method is similar to that of OPC used in the optical mask. When the desired pattern fidelity is interrupted by a defect, the modification of the mask features will be conducted to correct the distortion. A simple case of defect compensation would be a dark ML defect near a line structure as illustrated in Figure 11.14. The effect of this ML defect is to cause local line-width increase near the defect location in wafe printing. For a large defect, line bridging may occur. To correct the effect of this ML defect, mask line-width near the defect needs to be trimmed. The trimmed region will be printed narrower, while the ML defect nearby will make the same region to be printed wider. The net effect, when the absorber trimming is done properly, will yield an unchanged line width in the defect region. This defect compensation method is more applicable for the amplitude-dominated ML defects. The pure phase ML defect cannot be well compensated for with this method through defocus range.

In order to realize this defect compensation scheme, actinic inspection or AIMS view of the mask defect is necessary to first locate the defect location. The effect of the defect also needs to be determined via AIMS imaging in order to define the appropriate absorber trimming size and trimming location.

11.2.2.7.4 ML Defect Mitigation via Absorber Pattern Coverage

Using absorber pattern to cover limited ML defect is feasible. It is especially attractive in the case of dark field mask, such as a contact layer mask, where most mask areas are

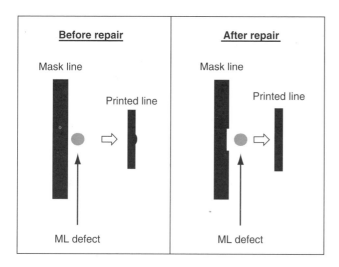

FIGURE 11.14
Illustration of optical proximity ML defect repair. In the repair process, a region of the absorber will be removed to compensate the patterning effect of an ML defect.

covered by the absorber patterns. The implementation of this method requires: (1) fiducial alignment mark on the ML; (2) defect size and location identification from ML defect inspection tool; (3) mask data matching to the ML defects such that via a correct amount mask pattern translation, maximum ML defects are covered by the absorber; (4) good e-beam alignment during e-beam patterning.

The effectiveness of this method depends on the locations of all the ML defects and the type of mask, e.g., clear field or dark field mask. In general, only very limited defects can be mitigated or covered simultaneously.

11.2.3 The Buffer Layer, Absorber Layer Stack, and Backside Conductive Layer

After ML fabrication, a buffer layer, an absorber layer, an antireflection coating (ARC) layer, and a backside conductive layer will be consecutively deposited onto the top of the ML. A buffer layer is used to protect the ML during the absorber layer etch and repair. It is not necessary to use a buffer layer if both absorber etch and repair have a sufficient high etch selectivity to the ML capping. The need for the ARC layer comes from mask defect optical inspection requirement. The absorber layer combined with its ARC is also referred to as the absorber stack. The ML blank backside metal coating, according to SEMI P38-1103 standard, is required to enhance the e-chucking force during all the processes that involve e-chuck.

11.2.3.1 The Buffer Layer

In order to effectively protect ML during absorber etch and repair, a buffer layer should have the following characteristics: (1) no pinhole defect; (2) high etch selectivity to the absorber etch and to the repair etch; (3) high etch selectivity to the ML capping layer during buffer layer removal; (4) relatively low EUV absorption such that a large buffer defect can be tolerated without repair; (5) good chemical cleaning durability such that no undercut will result during multiple mask cleanings.

Buffer layer needs to be deposited with low stress to minimize stress-induced mask. SEMI P38-1103 standard specifies the buffer layer film stress to be less than 200 Mpa. Any stress in the film can cause mask nonflatness, which will translate to the pattern IPD in wafer printing. Even though the use of e-chuck may reduce the global mask flatness greatly, it is desirable to minimize the film stress in the first place.

Several buffer layer candidates have been studied. They are silicon oxide (SiO_2), silicon oxynitride (SiON), carbon (C), chromium (Cr), and ruthenium (Ru) [27–33]. There are different performance pros and cons against the buffer layer requirements for each buffer material.

The specific requirements on the buffer layers also depend on the ML capping layer material. If an ML capping has high resistance or selectivity to the absorber layer etch during the mask patterning and repair, a buffer layer can be ultimately eliminated (also see Section 11.3.2).

11.2.3.2 The Absorber Layer Stack

Typical EUVL mask absorber layer stack consists of a metal, which absorbs the EUV light during mask exposure, and an ARC layer. The antireflection function of the ARC layer is to reduce the light reflection from the top of the absorber stack at the mask defect inspection wavelength, typically the deep ultraviolet (DUV) wavelength. In the case of a reflective EUVL mask, the inspection image contrast depends on the reflected light between the reflector region (ML) and the absorber region. In general, a metal absorber has relatively high reflectivity in the DUV wavelength range, e.g., from 150 to 250 nm. High DUV reflection in the absorber region will result in a low contrast for DUV optical inspection tool. To incorporate the mask optical inspection need, a low DUV reflectivity ARC is needed. The EUV light (13.4 nm) reflection at the absorber is typically less than 0.2% and is not a concern in mask exposure. Whether an ARC layer is needed depends on the mask inspection method. DUV optical inspection system usually will require an ARC layer, while an e-beam based inspection system or an actinic (at-wavelength) inspection system will not require ARC layer (see also Section 11.3.3).

The material requirement for the ARC layer not only needs to have low reflectivity at the inspection wavelength but also needs to satisfy certain mask fabrication constraints, which include: (1) Easy to etch with near-zero etch bias and a good critical dimension (CD) control. (2) High etch selectivity or resistance during buffer etch, since the ARC and the absorber pattern will be used as an etch hard mask at this etch step. No material removal during the etch, since any thinning of the ARC will result in a shift in minimum reflectivity valley, and therefore an increase in reflectivity at the inspection wavelength. (3) Low defect level. (4) Preferably that the ARC layer can be etched with the same etch process as that of the absorber layer to simplify the mask patterning process. (5) Good chemical cleaning durability such that no ARC material loss occurs during multiple mask cleanings.

Several EUVL mask absorber material candidates that include AlCu, Ti, TiN, Ta, TaN, and Cr have been explored [27,28,30,33–37]. The results indicated that TaN and Cr are the better candidates among the others evaluated. In Figure 11.15, a plot of EUV absorption at 13.4 nm as a function of absorber thickness for Cr and TaN films is given. It shows that TaN is slightly darker at EUV wavelength than that of Cr. Cr absorber has been used for many optical lithography generations. Further extending Cr mask absorber to EUV lithography presents minimum impact on the mask technology infrastructure. However, the current Cr absorber in optical masks are known to have larger etch bias and etch loading effect. These shortcomings may limit the extension of Cr absorber for the application of EUVL. TaN is a new film that has not been used in the current mask technology. The Ta-based metal compound, however, has been studied previously in x-ray mask technology. Its performance in EUVL mask fabrication and printing has been found compatible and comparable in many process steps and performance aspects to that of the Cr absorber.

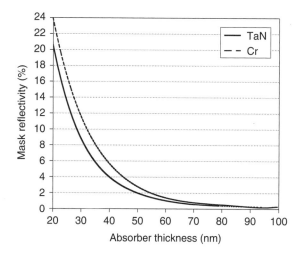

FIGURE 11.15
Normalized EUVL ML mask blank reflectivity at 13.4 nm versus absorber thickness for TaN and Cr absorbers, respectively. TaN and Cr absorbers with a thickness of greater than 70 nm will yield less than 1% residual reflectivity in the absorber region.

11.2.3.3 The Backside Conductive Coating

SEMI P38-1103 standard requires a backside conductive coating for the EUVL mask blanks. The backside blank conductivity is needed for the purpose of e-chucking that occurred in various process steps. The EUVL mask used in the exposure tool will inevitably have both the IDP and the OPD. These are the two major error sources for wafer IDP distortion. Even if the mask substrate meets the flatness of 30-nm requirement, buffer and absorber films meet the stress of less than 200-Mpa requirement; both IPD and OPD of the mask can still be large due to ML blank stress, buffer and absorber residual stress, and inconsistency in mask mounting during e-beam write and wafer exposure. When the mounting of the mask blanks in certain processes, such as e-beam write, flatness metrology, and wafer exposure, is consistent, the primary source of mask pattern IPD can be eliminated. The source of mask OPD may still exist unless the mask mounting in the exposure tool will also flatten the mask at the same time. This requirement will rule out the traditional optical mask mounting method. Since EUVL mask is a reflective mask, chucking the mask from its backside to flatten the mask becomes feasible. The traditional vacuum chucking will not work for the EUVL mask in the exposure tool, since the mask exposure is carried out in the vacuum environment to avoid light absorption in the air. One of the options is to use an e-chuck. In order to effectively flatten the mask with an e-chuck, a conductive mask backside is required. The properties of the conductive materials for EUVL mask backside are not specified in the SEMI P38-1103 standard; it is to be agreed upon between the user and blank manufacture.

Up to this point, an EUVL mask blank fabrication is completed. It constitutes the following steps: substrate material preparation, substrate polishing, ML deposition, possible ML repair, buffer layer (may or may not be needed), absorber layer stack, and backside conductive layer deposition.

11.3 EUVL Mask Patterning

EUVL mask patterning consists of many similar steps as that in the conventional optical mask fabrication and some additional EUVL-specific steps, for example, additional etch step for the buffer layer. Several EUVL mask patterning processes and tools that are similar to that of the conventional optical mask fabrication, however, need to be modified to accommodate EUVL mask patterning requirements. In Figure 11.16, a schematic EUVL mask patterning process flow is given. In the flow, the CD and registration metrology steps are omitted. The shaded boxes in the process flow indicate the processes or tools

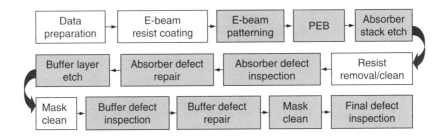

FIGURE 11.16
The detailed EUVL mask patterning flow chart. The shaded blocks indicate that the modules are either EUVL-specific or requiring modification to comply with EUVL-specific specs.

that need to be modified compared to those of the conventional optical mask patterning process. In the next few sections, the EUVL mask-specific process steps and requirements will be discussed in more detail. The nonEUVL-specific steps or requirements that are similar to those of the conventional optical mask will not be addressed.

11.3.1 E-Beam Resist Patterning

EUVL mask e-beam resist patterning is similar to that of the conventional optical mask. However, in order to comply with the mask distortion or flatness requirement, using an e-chuck to chuck the mask in the e-beam patterning tool and in the wafer exposure tool is highly recommended. If the mask is written while electrically chucked, it will experience a minimum stress-related pattern distortion during the mask exposure when the mask is chucked again with a similar e-chuck in the scanner exposure system.

To obtain the high-resolution resist patterning that is required for 32-nm and below generation nodes, 50 keV and higher voltage e-beam patterning tools and a chemically amplified resist are desirable. Another EUVL specific constraint in resist patterning is the resist postexposure bake temperature, it needs to be controlled below 150°C due to ML thermal instability at higher temperature.

11.3.2 Absorber Stack Etch

To obtain a tight mask CD control, EUVL mask absorber etch requires dry etch. The etch chemistry and the etch process depend on the absorber stack materials. If a Cr absorber and a $C_rO_xN_y$ ARC are used, the etch process will be similar to that of today's state-of-the-art optical mask absorber etch process, especially when the buffer layer is SiO_2. The Cr dry-etch chemistry used in today's optical mask patterning are combined Cl_2 and O_2 gases. The Cr etch selectivity to the quartz substrate is typically greater than 20. In the case of Cr absorber EUVL mask with SiO_2 buffer layer, etch selectivity greater than 20 has been demonstrated [36]. When the buffer layer material changes from SiO_2 to others, such as C or Ru, the etch selectivity and the etch profile may be impacted if the same process is used. In these cases, process reoptimization is required.

For the TaN absorber, the typical dry-etch chemistry is Cl_2. TaN etch using Cl_2 chemistry has been demonstrated with greater that 20 etch selectivity to the buffer SiO_2 layer [37].

Several ARC layer materials for TaN absorber have been investigated [33,38–41]. Some can provide very good antireflection effect which improves the DUV inspection contrast from 30% up to 90%. The most challenge aspect of the ARC layer is whether it can be etched with the same etch chemistry as that of the absorber and whether it can sustain the buffer layer etch when the ARC and the absorber are used as a hard etch mask. Any thinning of the ARC during buffer layer etch can impact the DUV inspection contrast at the given wavelength. If the EUVL mask patterning process can ultimately eliminate the buffer layer, the ARC layer will not be needed for the etch hard mask.

A recent study has shown that etching TaN/ARC absorber stack stopping on a Ru capped ML yielded an etch selectivity of 100:1 [42]. As a result, the buffer layer can be potentially eliminated in the process.

11.3.3 Absorber Defect Inspection

The mask inspection typically not only finds the absorber defects (hard defects) but also finds all the contaminants (soft defects), e.g., cleaning solution residues, resist residues, etc. In the case of EUV light illumination, many materials that are transparent to the visible or DUV light, such as organic residues, become partially or completely opaque. The question is that whether an actinic inspection for EUVL mask is needed. To answer this question, we need to either test all types of possible soft defects under both the optical and the actinic inspection systems or to have the mask tested in the real production line. Either case is not feasible at this time. Limited work has been done to understand the inspectability of certain soft defects under the EUV light inspection and DUV light inspection. In one study, the detection signal for the thin oxide defects is found to have very high contrast by either the optical inspection system or the e-beam inspection system [38]. There has not been any concrete data to show that certain soft defects that can be detected by the actinic method cannot be detected by the optical or e-beam inspection methods. Since the e-beam inspection tool, in general, has very low throughput, the optical inspection tool with improved resolution by reducing the inspection wavelength becomes primary mask inspection choice for EUVL mask inspection.

In an optical mask inspection system, defect detection sensitivity depends primarily on both the inspection wavelength and the signal contract between the patterned and the nonpatterned mask regions. In the case of a transmission optical mask, the inspection contrast is very high (>90%) as the inspection light either passes through the clear region (the quartz) or is blocked by the dark region (the chrome) as shown in Figure 11.17a. In the case of an EUVL reflective mask, the reflectivity in the clear region (ML region) is about 60% at the inspection wavelength range of 190 to 250 nm. The light signal contrast is primarily determined by the darkness of the absorber ARC as illustrated in Figure 11.17b. When the inspection light reflectivity from the absorber stack is high, the inspected images not only have the low contrast but also accompanied with undesired light signals as a result of interference between the reflected beam I_r in the reflector region (ML region) and I_a in the absorber region as shown in Figure 11.17b. These signals typically cause the so-called edge-ringing effect, e.g., an additional bright spike signal at the edge of a dark line region [40]. The edge-ringing effect will drastically degrade the mask defect inspectability. When an EUVL mask inspection contrast reaches 70%, the edge-ringing effect becomes negligible. The inspection algorithm used in the transmission mask inspection can be continuously extended to the EUVL reflective mask inspection [41].

When the absorber ARC is more reflective, the thickness of the buffer layer, e.g., in the case of SiO_2 or C, can also play a role in the inspection contrast. This is because the

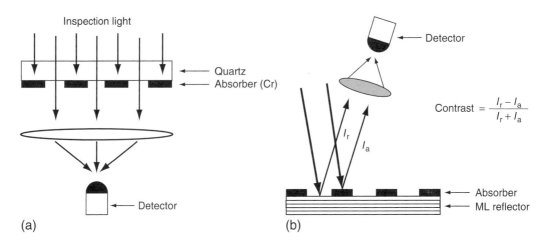

FIGURE 11.17
Illustration of optical mask inspection principle for (a) a transmission mask and (b) for a reflective mask.

inspection light absorption in the buffer layer is incomplete due to a thin buffer layer. The reflected light intensity in the buffer region will be modulated by the buffer layer thickness (thin film interference effect). As a result, the detection sensitivity can be greatly impacted when the buffer layer thickness is not optimized for the inspection wavelength.

In addition to all the EUVL mask-specific issues, the printable defect size becomes much smaller for the EUVL technology generation, e.g., about 25 nm for the 32-nm node generation. The mask defect inspection tool detection sensitivity needs to be greatly enhanced to meet the resolution requirement. It requires development for both the shorter wavelength inspection system and the new data analysis algorithm.

11.3.4 Absorber Defect Repair

During absorber patterning, two types of absorber defects can exist. They are the absorber intrusion defects (extra undesired absorber in the clear region) and the protrusion defects (absorber missing from its desired position). Both defects, depending on their size, need to be repaired before the next process step.

The key challenge in EUVL mask defect repair is to avoid the damage and the contamination to the ML stack underneath the buffer layer or underneath the absorber layer when the buffer layer is not used. Although missing of several layers of ML will not post a big impact to the ML reflectivity, the consequences from the exposed Mo layer can continue during the rest of fabrication process. For example, the molybdenum oxide formation in the exposed molybdenum can result in more EUV light absorption as compared to the Mo layer. The exposed Mo is also more vulnerable to the multiple chemical cleanings during the mask patterning process. The contamination in the ML blank induced by the repair has more severe effect in the reflective mask than in the conventional transmission optical mask, as the light will pass the contaminated region twice in a reflective mask. The need for a buffer layer between the ML and absorber layer in mask patterning is partially due to the EUVL mask absorber repair requirement.

The current optical mask repair uses laser repair and focused ion beam (FIB) repair. Both repair tools can either etch the absorber intrusion defects or patch the absorber protrusion defects. However, for EUVL mask repair, the laser repair is limited by its spatial resolution. Even in the case of advanced femtosecond pulsed laser repair tool, while its resolution is limited by the optical diffraction, the DUV optics can only provide resolution down to about 100 nm. In the case of FIB for EUVL mask repair, the limiting factor is the gallium (Ga) ion implantation into the ML. Ga ion implantation can cause ML reflectivity loss in two ways, i.e., to promote ML interface mixing and to absorb EUV light directly. In Figure 11.18, a plot of ML reflectivity as a function of Ga ion-beam voltage for three different buffer.

SiO_2 thicknesses are given when the ML is exposed for 60-s Ga ion-beam only (i.e., no etching). It clearly showed the impact of Ga ion implantation on the ML reflectivity [43]. In order to completely protect the ML, the buffer layer must be thicker than the ion-projected range plus the thickness loss during the repair for a given ion-beam voltage. For example, to prevent 30-kV Ga ion reaching the ML, 70-nm-thick SiO_2 is needed [44].

Using electron beam instead of Ga ion-beam for EUVL mask absorber defect repair has been investigated [45]. E-beam can provide high image resolution that is necessary for 32-nm technology node. In the defect etching process, it induces chemical reaction at the surface in the presence of proper gas chemistries. As a result, it has no or little impact to the underneath substrate.

Direct e-beam-induced deposition has long been explored to produce nanostructures, such as for maskless direct-write lithography. The mechanism for generating such

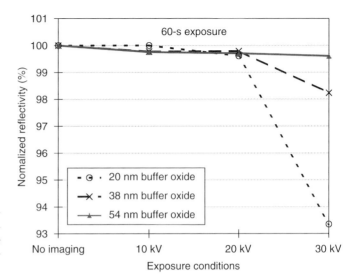

FIGURE 11.18
EUVL ML mask blank reflectivity change due to 60-s gallium beam exposure versus FIB operation voltage for three different buffer oxide thicknesses. Ref 43.

structures was to deposit the carbon atoms from e-beam-induced dissociation of hydrocarbons absorbed on the surface. Based on a similar mechanism, material removal is also possible if the process can generate species, such as free radicals or ions, that are reactive to the substrate atoms and produce volatile products. The electron-assisted material removal is a relatively pure chemical process with no physical sputtering of the material. The process can potentially lead to a very high etch selectivity to the substrate. E-beam-induced etching of Si, Si_xN_y, SiO_2, PMMA, and GaAs have been reported [46–49]. The e-beam repair on EUVL mask protrusion absorber defects has shown a very clean deposition boundary using e-beam-assisted platinum (Pt) deposition [45]. That is, no material spread during the deposition as usually seen in the case of FIB deposition. In Figure 11.19, an SEM comparison of e-beam deposited Pt and FIB deposited tungsten (W) is given. The material spread at the deposition boundary for FIB deposition is clearly seen (Fig. 11.19a). In the case of TaN absorber removal, e-beam-induced etching with XeF_2 gas has been demonstrated [45]. In Figure 11.20, an AFM image of e-beam etched 100-nm TaN EUV absorber on SiO_2 buffer layer is given. The underneath oxide loss in this case is about 2 nm. The average etch selectivity to the buffer SiO_2 layer is about 20:1. This e-beam etching process enables TaN absorber repair with a very thin buffer SiO_2 layer. It is believed that the e-beam repair technology will be able to provide the resolution, the edge placement precision, and damage-free repair solution to the EUVL mask needs.

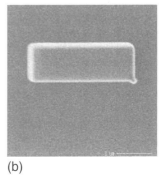

FIGURE 11.19
(a) SEM micrographs of FIB deposited W and (b) e-beam deposited Pt, indicating the clear difference that there is no spray in e-beam deposition. Ref 45.

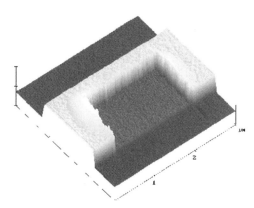

FIGURE 11.20
Three-dimensional AFM image showing the profile of a TaN absorber pattern after being etched with 1 keV electrons in XeF_2. Te surface where the etching stopped is smooth, with about 2-nm surface roughness as measured by AFM. Ref 45.

11.3.5 Buffer Layer Etch

Buffer layer etch is an additional etch step for EUVL mask fabrication as compared to that of the conventional optical transmission mask. In the buffer etch process, the patterned absorber stack will be used as the etch hard mask instead of resist. When the etch selectivity between the buffer etch and the hard mask is very high, the mask CD is defined primarily by the absorber stack etch. The buffer etch process optimization needs to focus on the high etch selectivity to the ML capping. In the case of thin buffer layer, wet-etching of buffer layer is also feasible.

The selection of buffer layer dry-etch chemistry depends on the buffer materials used. In the case of SiO_2 buffer material, the F_2 chemistry is usually used. For a Cr buffer, the etch chemistry will be the same as the Cr absorber etch. In this case, the absorber material has to be some other material, e.g., TaN. For a C buffer, O_2 plasma can be used. Ru buffer layer can be etched with Cl_2 combined with O_2 chemistry.

11.3.6 Buffer Layer Defect Inspection and Repair

Buffer layer defect inspection has different requirements and challenges as those of the absorber defect inspection. In general, the allowed buffer layer defect size can be much larger than that of the absorber defect, provided that the buffer layer is thin and the buffer layer material is less absorbent at EUV wavelength. The inspection algorithm may not be the same as that of the absorber defect, since the inspectability of an absorber defect is primarily determined by the contrast difference between the absorber defect region and the clear ML region. In the case of buffer layer defect, the contrast between the thin buffer layer defect and the clear ML region will be low. The defect inspection needs to be based on either the defect edge detection or phase detection.

Similar to that of the absorber repair, the buffer layer repair can be done also with e-beam repair tool. The key to the process is to develop the chemistry that yields high etch selectivity to the ML capping. Other option for the buffer layer defect repair is to use AFM-based repair by physically removing the defect using the AFM tip.

In Figure 11.21, the picture of a completed EUVL mask that was the first one used in the EUV engineering test stand (ETS) is given. The mask used ULE substrate, Cr absorber, and SiO_2 buffer layer.

11.3.7 Mask Cleaning

Mask cleaning after absorber etch and after absorber defect repair has less stringent requirements than that of after buffer etch when the ML is exposed. After the ML is

FIGURE 11.21
The first square EUVL mask with LETM substrate for the ETS, which was built by Virtual National Laboratory.

exposed, ML damage, such as pinhole generation and reflectivity degradation, as a result of cleaning becomes an important aspect that needs to be considered in all the potential EUVL mask cleaning techniques. The requirements for the final EUVL mask cleaning for 32-nm technology node can be summarized as follows:

1. Capable of removing all particles larger than 30 nm
2. Capable of removing organic contamination
3. No change of the absorber ARC reflectivity by more than 1%
4. No change to the mask CD or mask LER
5. No change to the ML reflectivity
6. Meeting environmental safety standards

Due to no pellicle protection for the EUVL mask, frequent mask cleans may be required. In this case, the above-mentioned requirements should also comply with the multiple mask cleanings.

The current optical mask cleaning technology is not compatible with the EUVL mask cleaning needs. Even with the current Cr absorber optical mask, multiple cleans will generate pinholes and degrade Cr ARC. The removable particle size is still much larger than what is required by the EUVL technology. It is obvious that the new cleaning technology needs to be developed for the EUVL mask.

11.4 EUVL Mask Protection

Due to lack of transparent material for EUVL mask pellicles, the EUVL masks will be handled without a pellicle. It is very crucial to keep the mask clean for mask handling after the final mask clean. The mask handlings include mask griping, storage in the carrier, shipping, removing from the carrier, loading into the vacuum loadlock, loading and mounting in the scanner mask stage, exposing and removing from the scanner, and returning to the storage.

The consideration for mask gripping includes the mask gripping contact area, the end-effector (gripper) design, the end-effector material, the mask contact area material selection, and the robotic handling design. The end-effector contact with the mask should not generate any particles. If the contact point is in the backside of the mask, the mask backside coating material also needs to be compatible with the mask handling requirement. The Mask carrier and storage design involve similar aspects as those of the end-effector, i.e., material consideration, mask contact with the carrier, and handling interface, etc. It is not clear at this time whether any active protection scheme is required for the mask carrier or storage. The requirement for the mask carrier and storage is not to generate any particle on the mask during mask storage and transportation.

The success of EUVL mask handling that meets the requirement directly impacts the wafer yield. EUVL mask handling is one of the high-risk areas in EUVL.

11.5 New EUVL-Specific Mask Processing Tools

Even though EUVL mask has the same mask format as that of the conventional optical mask, several conventional mask processing tool configurations still need to be modified to comply with EUVL mask processing requirement. Most of the tool configurations that need to be modified are due to EUVL mask flatness or mask registration requirement. The processing tools that require to be modified are: (1) The absorber film deposition tool. This tool needs to incorporate new deposition targets for new absorber material, such as TaN and buffer layer material. (2) The e-beam patterning tool, which needs to incorporate an e-chuck. The purpose of using e-chuck is to flatten the mask blank during e-beam write such that any pattern distortion due to mask nonflatness will be eliminated when it is chucked the same way in the scanner system. (3) The registration metrology tool. This tool needs to incorporate an e-chuck or a vacuum chuck. The chucking force for the mask registration metrology tool needs to be matched to that of the e-beam patterning tool and the scanner mask stage. (4) The absorber and buffer defect inspection tool. The current inspection tool is capable of detecting the signals at the reflective mode. However, new algorithms need to be developed for EUVL reflective mask defect detection, including edge contrast detection for thin buffer layer defects. (5) The flatness metrology tool, which needs to incorporate an e-chuck or a vacuum chuck for the mask blank and the finished mask flatness measurement. The chucking force needs to be matched to that of the scanner mask stage. For the substrate flatness measurement, it requires simultaneously measurement of the frontside flatness, the backside flatness, and the substrate thickness variation. E-chucking or vacuum chucking is not required for these measurements. Other improvements for the tool, such as flatness measurement precision down to subnanometer, are also required for EUVL mask application.

In addition to the current mask processing tool sets, some EUVL-specific tools also need to be added in. They will be discussed in Sections 11.5.1–11.5.5.

11.5.1 ML Deposition Tool

The low defect density ML EUVL mask blank deposition is the key to the success of the EUVL mask technology. To achieve low ML defect density, the cleanliness of many tool components is essential. These components include the substrate robotic handling system, ion source, Mo and Si sputtering targets, moving stage, deposition chamber, etc. The tool should also be able to deposit ML films with high film density, precise interface control, low film stress, good film uniformity, and repeatability. In the case of adopting the ML interfacing barrier or using a different capping layer other than Si, more than two sputtering target are needed. The current ion-beam ML deposition tool also has an assistion source for potential applications, such as low power etch for ML smoothing.

11.5.2 Blank Inspection Tool

EUVL mask blank fabrication requires substrate defect inspection prior to the ML deposition. The typical mask substrate defects include pits, bumps, or any scratches. Very small substrate defect that is in an order of 50 nm can become a printable defect in EUVL as discussed in Section 11.2.2.7. In the case of ML defects, an embedded ML defect with a bump height of a few nanometers appearing on the top of ML becomes printable in EUVL. The inspection tool sensitivity for EUVL mask substrate defects needs to be below 50-nm PSL equivalent size and for ML defects needs to be below 30-nm PSL equivalent size with bump height in an order of a few nanometers for the 32-nm technology generation [50]. Because of such a stringent inspection sensitivity requirement for EUVL mask application, the current optical inspection tools used for optical mask blank inspection will not be able to meet this specification. Recent evaluation on multibeam confocal inspection system with 488-nm wavelength demonstrated inspection sensitivity of 60-nm PSL equivalent on ML blanks [50]. Continuous development of the tool is needed to meet 30-nm and below defect inspection sensitivity.

11.5.3 EUVL Mask Reflectometer

EUVL mask reflectometer measures the mask reflectivity spectrum centered at the exposure tool wavelength. The measurements will provide the EUVL blank peak reflectivity, the FWHM of the mask reflectivity spectrum, and the mask spectrum centroid wavelength. In the mask fabrication process, EUVL reflectometer will be used for ML blank qualification at the blank fabrication step and for mask qualification after the mask patterning. The mask blank reflectivity uniformity and the centroid wavelength uniformity can be affected by the substrate surface roughness, ML deposition uniformity, and ML thickness control. The finished mask reflectivity uniformity and the centroid wavelength value can be affected by the buffer etch or the absorbed etch when no buffer is used and by any high temperature experienced in the mask pattering process. Any remaining buffer material or any etch induced ML capping layer damage will result in mask reflectivity nonuniformity. Capping layer surface oxidation or mask multiple cleaning may also cause nonuniform capping film change or loss, results in mask reflectivity nonuniformity.

11.5.4 E-Beam Repair Tool

E-beam mask repair tool uses well-focused e-beam to induce chemical reactions at the surface of the mask (see also Section 11.3.4). Its advantages include: (1) Low or no damage to the mask blank. (2) High repair selectivity to the ML blank, since the etch is entirely chemical (i.e., no physical sputtering). (3) High spatial resolution since surface charging is relatively low compared to FIB. It allows long imaging time. (4) Infrastructure exists today to build an e-beam system (column is similar to CD-SEM). The e-beam repair tool for EUVL mask is expected to be applicable for both material deposition and etching.

11.5.5 EUV Aerial Image Monitor System (AIMS)

AIMS is an actinic microscope used for mask defect review. In this application, AIMS images the defect under the conditions that closely mimic the actual lithographic projection system. In this manner, the aerial image projected by the AIMS will be essentially similar to the aerial image within the lithographic system. By analyzing the contrast, slope, and through-focus behavior of the aerial image in the presence of the mask defect, the printability of the mask defect in a lithographic system can be estimated (with an appropriate resist model). For an EUVL mask, AIMS is not only very useful to review any hard defects or contaminants during the final mask inspection but is also the key inspection step to review and determine the printability of the ML defects after ML blank inspection. In many defect mitigation schemes, such as direct ML defect repair and using absorber OPC to compensate ML defect, AIMS verification is essential. Manufacturing of EUVL AIMS tool represents a new technology from that of the visible or DUV AIMS as it requires all reflective ML optics designs similar to that of EUVL scanner. Other AIMS tool designs, such as zone plate design, require a solution to the chromatic aberration issue.

11.6 EUVL Reflective Mask Performance

EUVL mask is a reflective mask with oblique light incident. Its performance in lithographic printing will be impacted due to these characteristics as compared to the transmission mask. Two effects have been noted. The first one is the shadowing effect that will lead to mask print bias and mask pattern position global shift. The second one is the Bossung process window (a set of CD versus defocus curve at different exposure dose) tilt and the focus shift.

11.6.1 Mask Shadowing Effect

Mask shadowing effect in EUVL is due to the combination of oblique illumination, the reflective mask, and the mask topography. The mask shadowing effect will cause printed CD bias (will refer to as CD shadowing effect) and global pattern position shift. A schematic drawing of EUVL mask CD shadowing effect is given in Figure 11.22. A geometrical calculation of the printed lines and spaces and feature global position shift, shown in Figure 11.22, are given in Equations (11.5)–(11.7):

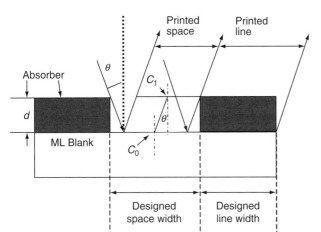

FIGURE 11.22
Geometrical optics illustration of EUVL mask shadowing effect: the printed space CD will be larger than that of designed CD and the printed line CD will be smaller than that of designed CD. The center position of the feature will also be shifted from C_0 to C_1.

$$\text{space CD(printed)} = \text{CD(designed)} - (2d \times \tan\theta) \times M \quad (11.5)$$

$$\text{line CD(printed)} = \text{CD(designed)} + (2d \times \tan\theta) \times M \quad (11.6)$$

$$\Delta\text{center} = d \times M \times \tan\theta \quad (11.7)$$

where CD and the pattern center position shift are measured at the wafer plane, M is the EUVL scanner reduction ratio, d is the buffer and absorber stack height, and θ is the light incident angle to the mask normal.

According to Equations (11.5) and (11.6), the mask space features will print smaller and the line features will print larger as compared to the designed CD. The difference is determined by the light incident angle to the mask normal and the mask absorber thickness (including the thickness of the buffer layer when exits).

In Figure 11.23, the calculated EUVL CD shadowing effect of three different absorber materials (with no buffer layer), TaN, Cr, and Ge, at 30-nm dense lines are given. The results are also compared to the geometrical calculation. The mask CD shadowing effect given in Figure 11.23 was measured as line CD versus absorber height. The line CD was determined by using 30% threshold of the normalized aerial image intensity level. It is shown in Figure 11.23, for a given absorber height, the printed CD depends on the mask absorber material. More specifically, it depends on the material index of refraction n and k, while geometrical optics predicts no dependence of CD shadowing effect on the absorber material's optical properties. TaN and Cr have a larger Δn (Δn is the difference of the real part of material index of refraction and that of the vacuum) as compared to that of Ge. The k values of TaN, Cr, and Ge are similar. By comparing the three curves in Figure 11.23, it is found that the CD shadowing effect of TaN and Cr is larger than that of Ge. As the absorber thickness increases, the difference in CD for different material becomes smaller.

When a buffer layer is used, the CD shadowing effect is further modulated by the buffer layer material properties [51]. The CD difference between no buffer and with buffer for the same absorber in the thinner absorber region is directly related to both n and k values of the buffer and the absorber materials.

The printing bias in EUVL mask is expected to be correctable through the mask design. The pattern position shift is not an issue since the effect is global.

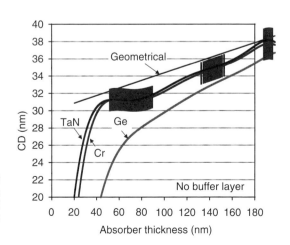

FIGURE 11.23
Printed CD as a function of the absorber height based on the geometrical optics calculation and based on the two-dimensional mask EM model simulation for TaN, Cr, and Ge absorbers with no buffer layer. Ref. 51

11.6.2 Bossung Process Window Tilt and Focus Shift

The Bossung curve asymmetry and focus shift effect in EUVL have been discussed previously [51–53]. When an EUVL mask with a given topography is considered, the Bossung curve of a line at a large pitch is tilted, and the best focus is shifted. This focus shift was found to be pitch dependent. From the periodic lines to the isolated lines, the focus shift increases as the pitch increases. The maximum focus difference observed by simulation between the periodic and isolated lines can be as much as 40 nm for the TaN absorber. In Figure 11.24 and Figure 11.25, plots of Bossung curves for dense and isolated 30-nm lines for TaN absorber are given, respectively. No buffer layer is used in the calculation. The absorber thickness in both cases is 100 nm. It is shown in Figure 11.24 that a small focus shift exists even for the periodic lines. The maximum focus shift difference between the isolated and dense lines is about 40 nm. This pitch-dependent focus shift will not be able to compensate via scanner focus adjustment for different pitch. It, therefore, posts a big impact on the total lithographic process window.

It is found that the amount of focus shift between the isolated lines and dense lines depends on the buffer and absorber material's n and k values [51]. The difference in the

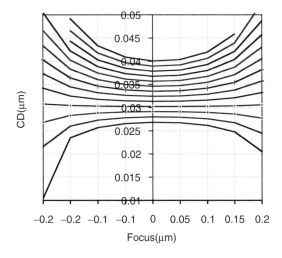

FIGURE 11.24
Simulated Bossung curves of 30-nm (1×) dense lines. The mask has 100 nm TaN absorber with no buffer layer. Other simulation parameters are: $\lambda = 13.4$ nm, NA = 0.25, partial coherence = 0.7. Ref 51.

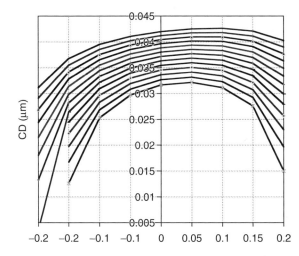

FIGURE 11.25
Simulated Bossung curves of 30-nm lines with pitch of 200 nm (1×). The mask has 100 nm TaN absorber with no buffer layer. Other simulation parameters are identical to that used in Figure 11.23. Ref 51.

real part of the refraction index between the absorber material and the vacuum (Δn) plays a major role as compared to that of the imaginary part of the refraction index (k). A material with small Δn, e.g., Ge, will have less focus shift and less Bossung curve tilt as compared to the cases of materials with larger Δn, e.g., TaN or Cr.

When a buffer layer is applied, the focus shift and the Bossung curve tilt effect will be modulated again by the buffer layer material's optical properties. The light diffraction will also occur at the edge or the boundary of both the buffer layer/ML and buffer layer/absorber layer. When a buffer layer with a relatively large Δn is used, the focus shift and the Bossung curve tilt effect will increase regardless the optical properties of the absorber material [51].

The primary cause of this effect is due to the existence of nonplanar geometrical features like the edges, corners, and vertices of the absorber structures. Simulation using rigorous two-dimensional frequency domain EM EUVL mask solver when the near-field diffraction is considered showed that neither the amplitude nor the phase of the electrical field at the top of the mask is a square wave. The electrical field that is diffracted at the mask edge interferes with the light bounced or reflected from the ML mirror. The electrical field distribution at the mask edge, therefore, is of the combined light from the light diffracted at the mask edge and the light reflected from the reflective substrate. As a result, the amplitude and phase of the electrical field at the mask edge will not have a sharp cut-off. It, instead, has a transition region that extends to the absorber dark region. These amplitude and phase errors at mask edge cause Bossung curve asymmetry for the lines at large pitch [53].

It has been found via simulation that there are several performance advantages of a dark field mask over a clear field mask in the case of EUVL [54]. No Bossung curve tilt and focus shift effect between the dense and isolated space aerial images is one of the advantages. The performance advantages of a dark field mask lead to a better process margin or CD control. However, to implement a dark field poly mask, a high contrast and high resolution negative resist needs to be developed.

11.7 EUVL Phase-Shift Masks

11.7.1 Attenuated Phase-Shift Mask

The current optical attenuated phase-shift mask is fabricated by replacing the Cr absorber layer with another material, which is synthesized such that at a given thickness, both phase and attenuation requirements are met. The attenuated phase-shift masks obtained in such a way are also referred to as the embedded phase-shift mask (EPSM). The patterning of such an EPSM mask is very similar to that of the conventional optical mask except that the absorber material is different and a different etch process is required. At EUV (13.4 nm) radiation, choosing a single material, which matches both the phase and the desired absorption is difficult, especially when a range of attenuation may be required for different applications. This difficulty is primarily due to both the high absorption of the absorber materials at 13.4-nm wavelength and very small difference in the real part of refractive index between the absorber materials and the vacuum. This index refraction difference determines the phase difference between the attenuated region and the reflective region.

It is, however, more feasible to use two materials to obtain the desired phase and absorption for EUVL attenuated PSM as there are four optical parameters and two thicknesses can be selected and optimized. The schematics of the two-material attenuated

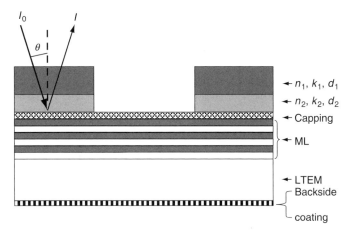

FIGURE 11.26
Schematics of two-material EUVL EPSM structure. The index of refraction and the thickness of the two films are n_1, k_1, d_1, and n_2, k_2, d_2, respectively.

PSM mask are given in Figure 11.26. The EUVL mask attenuation for a given thickness is governed by the absorption coefficient k (imaginary part of the material index of refraction) and film thickness d. The attenuation of the EUV beam in the absorber region after a round-trip reflection from the ML is, in the first order approximation, given by Equation (11.8):

$$I \cong I_0 \left| \left[-\left(\frac{2\pi}{\lambda}\right)\left(\frac{2k_1 d_1}{\cos\theta} + \frac{2k_2 d_2}{\cos\theta}\right) \right] \right|^2 \quad (11.8)$$

where I_0 is the incident EUV light intensity, I is the EUV light intensity reflected from the absorber region (round trip reflection), λ is the EUV light wavelength, k_1 and k_2 are the imaginary parts of the index of refraction of film 1 and 2, d_1 and d_2 are the thicknesses of film 1 and 2, and θ is the light incident angle to the mask normal.

Both the film thickness and the real part of refractive index difference between the absorber material and the vacuum govern the phase shift induced by the absorber. An attenuated PSM with a π phase shift will satisfy Equation (11.9):

$$\left| \frac{2\pi}{\lambda}\left(\frac{2\Delta n_1 d_1}{\cos\theta} + \frac{2\Delta n\, 2d2}{\cos\theta}\right) \right| = \pi \quad (11.9)$$

where $\Delta n_1 = 1 - n_1$ with n_1 the real part of index of refraction of film 1. $\Delta n_2 = 1 - n_2$, with n_2 the real part of index of refraction of film 2. In both Equations (11.8) and (11.9), the reflected light from thin film to thin film interface, and thin film to vacuum interface are ignored due to closely matched real part of refractive indices.

Solving Equations (11.8) and (11.9) for d_1 and d_2 yields:

$$d_1 = \cos\theta \left(\frac{-\lambda \Delta n_1 \ln\frac{I}{I_0}}{8\pi} + \frac{\lambda \alpha_1}{4} \right) \div (\Delta n_1 \alpha_2 - \Delta n_2 \alpha_1) \quad (11.10)$$

$$d_2 = \cos\theta \left(\frac{-\lambda \Delta n_2 \ln\frac{I}{I_0}}{8\pi} + \frac{\lambda \alpha_2}{4} \right) \div (\Delta n_2 \alpha_1 - \Delta n_1 \alpha_2) \quad (11.11)$$

For a given attenuation I/I_0, any combination of two films with index of refraction satisfies Equations (11.10) and (11.11) with positive d value can be used provided d has a reasonable thickness. This two-film approach is also consistent with absorber/buffer EUVL mask process flow. In the case of EUVL attenuated PSM, the top layer can be considered as the absorber layer and the bottom one the buffer layer. To meet the attenuation, phase, and process requirement, the film selection also needs to consider the process compatibility, etch selectivity of the two films, their proper film thickness ratio such that minimum repair buffer layer thickness is satisfied, etc.

11.7.2 Alternating Phase-Shift Mask

The EUVL alternating phase-shift mask (APSM) fabrication is different from that of the optical APSM where quartz is etched in the mask patterning step to create a phase step. In the case of EUVL APSM, the 180° phase step is generated in the mask substrate followed by Mo/Si ML deposition. The process starts in the very early stages of the mask blank fabrication. This phase step can be created either by directly etching into the substrate or by depositing a thin film layer and then patterning the thin film layer. The advantage of depositing a thin film layer is that the step control can be obtained via thin film deposition thickness control and via the thin film etch selectivity to the substrate. A schematic of EUVL APSM structure is given in Figure 11.27. The first few orders of phase steps, which are determined by $[\lambda/(4 \cos \theta)] (2m + 1)$, with θ the light incident angle to the mask normal and $m = 0, 1, 2, \ldots$, are viable in EUVL APSM fabrication. The smallest phase step needed to induce a 180° phase shift is $\lambda/(4 \cos \theta)$, or about 3.36 nm, for $\lambda = 13.4$ nm and $\theta = 5°$. In most of the cases, the thin film quality or smoothness increases as the film thickness increases. In other words, using the minimum step may not be ideal from process point of view. Simulation study for EUVL APSM with different phase step heights, which yield an effective 180° phase difference in the adjacent regions (i.e., 180°, or 180° plus multiples of 360°), was conducted [55]. It wass found that the image imbalance effect, which induces position and CD changes in the two adjacent lines, is relatively smaller and is negligible for the smallest step in the case of EUVL APSM. As the step heights increase, the image imbalance effect also increases. In the case of step height of 23.54 nm ($m = 3$), the image imbalance-induced space difference in the zero and 180° phase regions is about 1.5 nm for the 25-nm periodic lines. As the pattern pitch increases, the imaging imbalance effect diminishes.

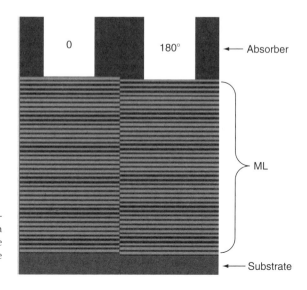

FIGURE 11.27
Schematics of an EUVL APSM structure configuration. In this APSM structure, the minimum step ($m = 0$) is used. The light reflected in the two reflector regions that is separated by the step is out of phase by 180°. Ref 55.

11.8 EUVL Mask SEMI Standards

Four EUVL mask standards have been proposed. They are: (1) Substrate standard — specification for extreme ultraviolet lithography (EUVL) mask substrates. This standard has been approved and published (SEMI P37-1102). (2) ML and absorbing stack standards — specification for absorbing film stacks and MLs on EUVL mask blanks. This standard has been approved and published (SEMI P38-1103). (3) Chucking standards — specification for mounting requirements and alignment reference locations for EUVL masks. This standard has been approved and published (SEMI P40-1103). (4) Reticle carrier standards. This standard is still in the defining stage (SEMI-3553).

Acknowledgments

I would like to thank Paul Mirkarimi and Eric Gullickson for their many useful discussions and for proofreading some of the sections in this chapter. Many thanks also go to Syed Rizvi, the editor of this book, for providing guidance and many useful inputs to this chapter.

References

1. K. Heckle, K. Hrdina, B. Ackerman, and D. Navan, Development of mask materials for EUVL, *Proc. SPIE*, 4889, 1113–1120 (2002).
2. K.L. Blaedel, S.D. Hector, J.S. Taylor, P.Y. Yan, A. Ramamoorthy, and P.D. Brooker, Vendor capability for uncoated low-CTE mask substrates for EUVL, *Proc. SPIE*, 4688, 767–778 (2002).
3. F. Ruggeberg, T. Leutbecher, S. Kirchner, H. Sauerbrei, K. Walter, F. Lenzen, and L. Aschke, Flatness correction of polished quartz glass substrates, in: *The Second International Symposium on Extreme Ultraviolet Lithography*, Antwerp, Germany, Oct. 2003.
4. C. Walton, M. Johnson, J. Taylor, D.W. Pettibone, and P.K. Seidel, Progress in EUV mask manufacture to meet P37/P38 specifications, in: *The Second International Symposium on Extreme Ultraviolet Lithography*, Antwerp, Germany, Oct. 2003.
5. J.A. Folta, J.C. Davidson, C.C. Larson, C.C. Walton, and P.A. Kearney, Advances in low-defect multilayers for EUVL mask blanks, *Proc. SPIE*, 4688, 173–181 (2002).
6. S.E. Gianoulakis, A.K. Ray-Chaudhuri, Thermal management of EUV lithography masks using low expansion glass substrates, *Proc. SPIE*, 3676, 598 (1999).
7. E. Gullikson, in: *The first International Symposium on Extreme Ultraviolet Lithography*, Dallas, Texas, Oct. 2002.
8. K. Hrdina, B.Z. Hanson, P.M. Fenn, and R. Sabia, Characterization and characteristics of a ULE Glass Tailored for the EUVL Needs, *Proc. SPIE*, 4688, 454–461 (2002).
9. I. Mitra, J. Alkemper, U. Nolte, A. Englel, R. Muller, S. Ritter, H. Hack, K. Megges, H. Kohlmann, W. Pannhorst, M. Davis, L. Aschke, and K. Knapp, Improved materials meeting the demands for EUVL substrates, *Proc. SPIE*, 5037, 219–226 (2003).
10. P.B. Mirkarimi and D.G. Sterns, Investigating the growth of localized defects in thin film using gold nanospheres, *Appl. Phys. Lett.*, 77 (14), 2243–2245 (2000).
11. D. Attwood, *Soft X-rays and Extreme Ultraviolet Radiation*, Cambridge University Press, Cambridge, 2000, p. 101.

12. M. Shriaishi, N. Kandaka, and K. Murakami, Mo/Si multilayers deposited by low-pressure rotary magnet cathode sputtering for extreme ultraviolet lithography, *Proc. SPIE*, 5037, 249–256 (2003).
13. A. Batty, P.B. Mirkarimi, D.G. Stearns, D. Sweeney, and H. Chapman, EUVL mask blank repair, *Proc. SPIE*, 4688, 385–394 (2002).
14. P.B. Mirkarimi, D.G. Stearns, and S.L. Baker, Method for repairing Mo/Si multilayer thin film phase defects in reticles for extreme ultraviolet lithography, *J. Appl. Phys.*, 91, 81–89 (2002).
15. T.W. Barbee Jr., S. Mrowka, and M.C. Hettrick, Molybdenum-silicon multilayer mirrors for the extreme ultraviolet, *Appl. Opt.*, 24, 883–886 (1985).
16. D.G. Stearns, R.S. Rosen, and S.P. Vernon, Fabrication of high-reflectance Mo–Si multilayer mirrors by planar-magnetron sputtering, *J. Vac. Sci. Technol.*, A9 (5), 2662–2669 (1991).
17. S. Bajt, J. Alameda, T. Barbee Jr., W.M. Clift, J.A. Folta, B. Kaufmann, and E. Spiller, Improved reflectance and stability of Mo/Si multilayers, *Opt. Eng.*, 41, 1797–1804 (2002).
18. P.A. Kearney, C.E. Moore, S.I. Tan, S.P. Vernon, and R.A. Levesque, Mask blanks for extreme ultraviolet lithography: ion beam sputter deposition of low defect density Mo/Si multilayers, *J. Vac. Sci. Technol. B.*, 15, 2452–2454 (1997).
19. E. Spiller, S. Baker, P. Mirkarimi, V. Sperry, E. Gullikson, and D. Sterns, High performance Mo/Si multilayer coatings for EUV lithography using ion beam deposition, *App. Opt.*, 42 (19), 4049–4068 (2003).
20. P.B. Mirkarimi, E.A. Spiller, D.G. Sterns, V. Sperry, and S.L. Baker, An ion-assisted Mo–Si Deposition process for planarization reticle substrates for extreme ultraviolet lithography, *IEEE J. Quantum Electron.*, 37, 1514–1516 (2001).
21. H. Takenaka and T. Kawamura, Thermal stability of Mo/C/Si/C multilayer soft x-ray mirrors, *J. Electron Spectrosc. Relat. Phenom.*, 80, 381–384 (1996).
22. S. Bajt, Private conversation.
23. S. Bajt, H. Chapman, N. Nguyen, J. Alameda, J. Robinson, M. Malinowski, E. Gullikson, A. Aquila, C. Tarrio, and S. Grantham, Design and performance of capping layers for EUV multilayer mirrors, *Proc. SPIE*, 5037, 236–248 (2003).
24. P.B. Mirkarimi, E.A. Spiller, S.L. Baker, and V. Sperry, Developing a viable multilayer coating process for extreme ultraviolet lithography reticles, *J. Microlithography, Microfabrication, Microsystems*, Vol 3, 139–145 (2004).
25. P. Mirkarimi, Stress, reflectance, and temporal stability of sputtering deposited Mo/Si and Mo/Be multilayer films for extreme ultraviolet lithography, *Opt. Eng.*, 38, 1246–1259 (1999).
26. A.K. Ray-Chaudhuri, G. Cardinale, A. Fisher, P.Y. Yan, and D. Sweeney, Method for compensation of extreme-ultraviolet multilayer defects, *J. Vac. Sci. Technol. B.*, 17, 3024–3028 (1999).
27. H.J. Voorma, E. Louis, N.B. Koster, F. Bijkerk, Fabrication and analysis of extreme ultraviolet reflection masks with patterned W/C absorber bilayers, *J. Vac. Sci. Technol. B.*, 15 (2), 293–298 (1997).
28. P.Y. Yan, G. Zhang, P. Kofron, J. Chow, A. Stivers, E. Tejnil, G. Cardinale, and P. Kearney, EUV mask patterning approaches, *Proc. SPIE*, 3676, 309–313 (1999).
29. E. Hoshino, T. Ogawa, M. Takahashi, H. Hoko, H. Yamanashi, N. Hirano, A. Chiba, M. Ito, and S. Okazaki, Process scheme for removing buffer layer on multilayer of EUVL mask, *Proc. SPIE*, 4066, 124–130 (2000).
30. P. Mangat, S. Hector, S. Rose, G. Cardinale, E. Tejnil, and A. Stivers, EUV mask fabrication with Cr absorber, *Proc. SPIE*, 3997, 76–82 (2000).
31. J. Wassib, K. Smith, P.J.S. Mangat, and S. Hector, An infinitely selective repair buffer for EUVL reticles, *Proc. SPIE*, 4343, 402–408 (2001).
32. B.T. Lee, E. Hoshino, M. Takahashi, H. Yamanashi, H. Hoko, N. Hirano, A. Chiba, M. Ito, T. Ogawa, and S. Okazake, EUV mask patterning using Ru buffer layer, *Proc. SPIE*, 4343, 746–753 (2001).
33. T. Shoki, M. Hosoya, T. Kinoshita, H. Kobayashi, Y. Usui, R. Ohkubo, S. Ishibashi, and O. Nagarekawa, Process development of 6-inch EUV mask with TaBN absorber, *Proc. SPIE*, 4754, 857–864 (2002).
34. P.Y. Yan, G. Zhang, P. Kofron, J. Powers, M. Tran, T. Liang, A. Stivers, and F.C. Lo, EUV mask absorber characterization and selection, *Proc. SPIE*, 4066, 116–123 (2000).

35. M. Takahashi, T. Ogawa, H. Hoko, H. Yamanashi, N. Hirano, A. Chiba, M. Ito, and S. Okazaki, Smooth, low-stress, sputtered tantalum and tantalum alloy films for the absorber material for reflective-type EUVL, *Proc. SPIE*, 3997, 484–495 (2000).
36. G. Zhang, P.Y. Yan, and T. Liang, Cr Absrober Mask for Extreme Ultraviolet Lithography, *Proc. SPIE*, 4186, 774–780 (2000).
37. P.Y. Yan, G. Zhang, A. Ma, and T. Liang, TaN EUV mask fabrication and characterization, *Proc. SPIE*, 4343, 409–414 (2001).
38. T. Liang, A.R. Stivers, P.Y. Yan, E. Tenjil, and G. Zhang, Enhanced optical inspectability of patterned EUVL mask, *Proc. SPIE*, 4562, 288–296 (2001).
39. J.R. Wasson, S.-I. Han, N.V. Edwards, E. Weisbrod, W.J. Dauksher, P.J.S. Mangat, and D. Pettibone, Integration of anti-reflection coatings on EUVL absorber stacks, *Proc. SPIE*, 4889, 382–388 (2002).
40. N. Bareket, S. Biellak, D. Pettibone, and S. Stokowski, Next generation lithography mask inspection, *Proc. SPIE*, 4066, 514–522 (2000).
41. D. Pettibone, A. Veldman, T. Liang, A. Stivers, P. Mangat, B. Lu, S. Hector, J. Wasson, K. Blaedel, E. Fishch, and D. Walker, Inspection of EUV reticles, *Proc. SPIE*, 4688, 363–374 (2002).
42. P.Y. Yan, G. Zhang, S. Chegwidden, P. Mirkarimi, and E. Spiller, EUVL mask with Ru ML capping, *Proc. SPIE*, 5256, 1281–1286, (2003).
43. P.Y. Yan, S.P. Yan, G. Zhang, P. Keaney, J. Richards, P. Kofron, and J. Chow, EUV mask absorber repair with focused ion beam, *Proc. SPIE*, 3546, 206 (1998).
44. T. Liang, A. Stivers, R. Livengood, P.Y. Yan, G. Zhang, and F.C. Lo, Progress in EUV mask repair using focused ion beam, *J. Vac. Sci. Technol. B.*, 18 (6), 3216 (2000).
45. T. Liang and A. Stivers, Damage-free mask repair using electron beam induced chemical reactions, *Proc. SPIE*, 4688, 375–384 (2002).
46. J.W. Coburn and H.F. Whinters, Ion- and electron-assisted gas-surface chemistry — an important effect in plasma etching, *J. Appl. Phys.*, 50 (5), 3189–3196 (1979).
47. S. Matsui, T. Ichihashi, and M. Mito, Electron beam induced selective etching and deposition technology, *J. Vac. Sci. Technol. B.*, 7 (5), 1182–1190 (1989).
48. K. Nakamae, H. Tanimoto, T. Takase, H. Fujioka, and K. Ura, *J. Phys. D.: Appl. Phys.*, 25, 1681 (1992).
49. D. Winkler, H. Zimmermann, M. Mangerich, and B. Traunner, E-beam probe station with integrated tool for electron beam induced etching, *Microelectronic Eng.*, 31, 141–147 (1996).
50. A. Stivers, T. Liang, M. Penn, B. Lieberman, G. Shelden, J. Folta, C. Larson, P. Mirkarimi, C. Walton, E. Gullikson, and M. Yi, Evaluation of the capability of a multi-beam confocal inspection system for inspection of EUVL mask blanks, *Proc. SPIE*, 4889, 408–417 (2002).
51. P.Y. Yan, The impact of EUVL mask buffer and absorber material properties on mask quality and performance, *Proc. SPIE*, 4688, 150–160 (2002).
52. C. Krautschik, M. Ito, I. Nishiyama, and K. Otaki, The impact of the EUV mask phase response on the asymmetry of Bossung curves as predicted by rigorous EUV mask simulations, *Proc. SPIE*, 4343, 392–401 (2001).
53. P.Y. Yan, Understanding Bossung curve asymmetry and focus shift effect in EUV lithography, *Proc. SPIE*, 4562, 279–287 (2001).
54. P.Y. Yan, Study of dark field EUVL mask for 45 nm technology node poly layer printing, *Proc. SPIE*, 4889, 1106 (2002).
55. P.Y. Yan, EUVL alternating phase shift mask imaging evaluation, *Proc. SPIE*, 4889, 1099–1105 (2002).

12

Masks for Ion Projection Lithography

Syed A. Rizvi, Frank-Michael Kamm, Joerg Butschke, Florian Letzkus, and Hans Loeschner

CONTENTS
12.1 Introduction 271
12.2 Stencil Mask Fabrication 274
12.3 Pattern Placement and Process-Induced Distortions 279
      12.3.1 Process-Induced Distortions 280
      12.3.2 Pattern-Induced Distortions 280
      12.3.3 Distortion Control 281
12.4 Metrology of Stencil Masks 283
      12.4.1 Tools for Measurement 284
      12.4.2 Test Mask 284
      12.4.3 Method of Measurement 285
      12.4.4 Image Placement Metrology 286
            12.4.4.1 Repeatability of IP Measurement 287
            12.4.4.2 Accuracy of IP Measurements 288
      12.4.5 CD Metrology 289
            12.4.5.1 CD Repeatability 289
            12.4.5.2 CD Uniformity 289
12.5 Defects Inspection Repairs 291
      12.5.1 Optical Inspection Tools 291
      12.5.2 E-Beam Inspection Tools 292
      12.5.3 Defect Repairs 293
12.6 Cleaning of IPL Masks 295
12.7 Mask Stability and Stitching Errors 297
      12.7.1 Mask Stability 297
      12.7.2 Effect of Pattern Stitching on CDs 299
12.8 Summary and Closing Remarks 303
References 303

12.1 Introduction

Ion projection lithography (IPL) has been regarded as one of the major contenders of the postoptical lithography, referred to as "next generation lithography" (NGL). Due to the larger mass of ions compared to electrons, the de Broglie wavelength is significantly smaller than in the case of electrons, giving the opportunity to design lithographic exposure systems with an extremely small numerical aperture (NA). As a consequence, much larger exposure fields can be achieved with IPL compared to electron-based projection technologies [1]. As an example, helium ions with an acceleration voltage of 100 keV have a de Broglie wavelength of $\lambda_{100\,keV\,He} = 5 \times 10^{-5}$ nm. With an NA of 10^{-5}, the diffraction limit of ion-optical resolution is $R = (1/2)\lambda/NA = 2.5$ nm at a DOF of $\pm \lambda/NA^2 = \pm 500\,\mu m$. The actual resolution and "effective" DOF of an IPL tool is, however, limited by electrostatic lens errors (geometric and chromatic blur) and by the stochastic Coulomb interaction between ions, mainly at the beam crossover in the ion-optical system (stochastic blur). In practice, a resolution of 50 nm has been achieved on 10× demagnifying small-field laboratory tools, while a full-field resolution of 75 nm lines and spaces has been achieved over an image field of $12.5 \times 12.5\,mm^2$ in a prototype 4× demagnification tool, with <50 nm resolution in parts of the field.

An IPL system is operationally similar to conventional steppers using UV and DUV sources. The schematic of one such 4× reduction tool is shown in Figure 12.1. It consists of an ion source, a multielectrode electrostatic ion-optics, a mask unit with automated handling system, a pattern-lock system, an off-axis optical wafer alignment system, and a vertical x–y wafer stage. The stepper is structured in a horizontal fashion, where mask and wafer are held vertically. The vertically mounted stencil mask is less distorted due to gravitational sag than with horizontal mounting. Additionally, particle contamination of mask and wafer is reduced. The system is equipped with cooled electrodes that keep the temperature of the mask stable during the ion beam exposure [2].

Another major advantage of ions compared to electrons is the strong interaction with matter. This interaction allows extremely fast resist systems with a smaller exposure dose (i.e., a shorter exposure time at a given beam current) and thus potentially a high

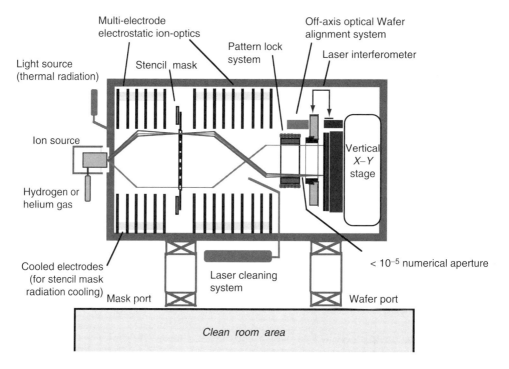

FIGURE 12.1
Principle of IPL. (IPL White Paper 1999.)

throughput. The latter is of high importance for industrial systems, since one of the main drivers of the lithography roadmap is cost reduction. A limitation for the resist speed and thus for throughput is given, however, by statistical fluctuations of the few ions exposing one resist pixel. These fluctuations, called shot-noise, are equivalent to local dose variations and lead to an increased line-edge roughness (LER). They pose a fundamental lower limit on the resist sensitivity that can practically be used [3].

The large interaction of ions with matter has a significant influence on mask architecture. Unlike in the case of optical masks, where materials can be found that are transparent down to wavelengths of 157 nm, ions are stopped or strongly scattered even by a thin layer of material. As a consequence, IPL requires the use of stencil masks, i.e., masks consisting of thin free-standing membranes with open sections (stencils), defining the pattern to be exposed. The membrane has to be rather thin, since the pattern definition by dry etching is typically limited by a maximum achievable aspect ratio. As an example, for 50-nm-wide lines on a wafer (equivalent to 200-nm lines on a mask at a reduction of 4×), a membrane thickness of 2 to 3 μm is required. Large membranes, consisting of crystalline silicon with diameter of 126 mm and as thin as 2 μm, have been fabricated successfully and have a sufficient mechanical stability for industrial handling.

On the scale of pattern dimensions, i.e., a few ten nanometers, the mechanical properties of large and thin membranes, however, are a severe challenge for placement and overlay accuracy. Unlike for electrons, no membrane masks with a continuous and mechanically stabilizing membrane layer can be used. As a consequence, the open stencils lead to local distortions of mask patterns due to mechanical stress-relaxation effects. For this reason, great care has to be taken during the mask production process to accurately control local stress and pattern distortions. Additionally, open stencils cannot support isolated island-patterns (called "donut-pattern"), which seem to require the use of two complementary masks. In this method, the pattern is split into two complementary sections with no donut-structure and is distributed over two masks. Alternatively the complementary pattern fields are placed onto one stencil mask, see Figure 12.2. Wafer exposure in this case was planned in an "IPL stitcher" mode, exposing a 50-mm × 50-mm

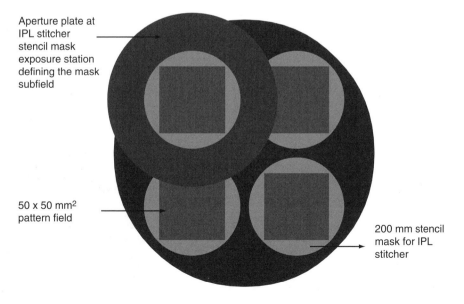

FIGURE 12.2
Using mask in IPL stitcher configuration. (IPL White Paper 1999.)

field of a 200-mm stencil mask with four fields [2]. The original pattern is restored on the wafer by subsequent exposure of both complementary pattern fractions. As a result, the throughput of the system is reduced due to double exposures. It has been demonstrated, however, that complex patterns, including donuts and oblique lines, can be formed by using a single stencil mask in combination with fourfold electrostatic step exposure (ESE) techniques. Such stencil masks are denoted as "ESE masks" [1]. With this method, the use of complementary masks can be avoided. However, due to the fourfold exposure, the pattern dimensions have to be reduced by a factor of 2, effectively requiring a 2× mask technology instead of a 4× mask.

While 200-mm stencil masks have been fabricated (Figure 12.3), most of the work was done with 150-mm stencil mask format.

IPL is the only sub-100-nm projection lithography technique, where the mask is not mechanically moved during exposure. It is, therefore, possible to envision *in situ* distortion compensation of mask patterns by applying localized heating, for example, with a properly scanned laser beam [4].

This chapter shall give a brief review of IPL-stencil mask fabrication and use. Section 12.2 describes the fabrication process, while Section 12.3 discusses the effects governing placement accuracy. Section 12.4 describes the tools and methods used for stencil mask metrology, while Section 12.5 addresses defect inspection and repair. Section 12.6 discusses cleaning techniques, and Section 12.7 covers the issue of stencil mask stability.

12.2 Stencil Mask Fabrication

Like for all other charged-particle lithography technologies, the fabrication and productive use of membrane-based stencil masks is the strongest deviation from conventional optical technologies. The introduction of a wafer-based mask technology with a form factor and materials other than that of standard optical photomasks requires a new tool-and-process infrastructure for the maskshops with a corresponding financial investment. For this reason, IPL-stencil mask technology tries to utilize as much of an existing wafer-processing infrastructure as possible by using silicon-on-insulator (SOI) wafers as mask blanks [5]. Although there still is the need to transfer tools and processes from wafer fabrication to the maskshop, the materials, as well as processes, are not unknown to the semiconductor industry. Nevertheless, the handling of thin and fragile membranes and the pattern placement errors resulting from the relatively poor mechanical stability still

FIGURE 12.3
A 200-mm stencil mask. [29]

pose a severe challenge for the application of this kind of mask-type in a productive environment.

A typical stencil mask for charged-particle applications consists of a three-layer stack, with one bulk layer that increases the mechanical stability and a membrane layer that carries the pattern information. Both layers are separated by an etch-stop layer, which enables the separate processing of the membrane definition and the membrane patterning. While several layer stacks have been proposed for charged-particle lithography, the material system with the highest compatibility to existing process infrastructure is the SOI wafer, consisting of a bulk silicon layer with a thickness of about 675 μm, a SiO_2 etch-stop layer with 0.3 μm thickness, and a Si-membrane layer with 2 to 3 μm thickness. SOI wafers are commercially available for advanced device applications and can either be fabricated by a wafer-bonding method or by oxygen ion implantation. Although the material is readily available in large quantities, the requirements for IPL-mask applications, especially with respect to stress homogeneity, are tighter than for device applications (see Section 12.3). For this reason, a modification of the existing SOI-production process will become necessary to fulfill the needs of stencil mask applications.

The general productive flow of an IPL mask is described by the block diagram in Figure 12.4. Starting with the data processing step, the original design pattern is processed by a pattern-split software. This software carries out the pattern-split procedure, in which the design is checked according to certain pattern-split rules. All donut-shaped features, as well as long lines, which could deteriorate the local mechanical stability, are split into two complementary layers. In the subsequent mask patterning process, these layers are transferred to either two separate complementary masks or separate subfields of one stencil mask. With an appropriate double exposure, the two complementary parts of the pattern are overlaid to reconstruct the original pattern. To decrease the impact of local critical dimension (CD) errors due to complementary pattern overlay errors, line-end structures are applied to the split patterns as well. These line-end features, also called "butting structures," have dimensions of typically about one half of the pattern CD and are of rectangular or triangular shape. They can be regarded as the charged-particle analog of optical pattern correction (OPC) structures of high-end optical photomasks. However, compared to advanced OPC patterns, which are already close to the wafer-scale CD of the main feature size, these "OPC-like" pattern modifications have a much

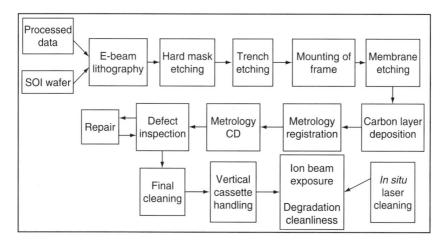

FIGURE 12.4
SOI stencil masks production.

more relaxed absolute CD (about 2× wafer-scale CD), as well as CD tolerance than in the optical case. In case of the ESE principle, corresponding data preparation for the symmetry reduction has to be worked out, reducing the mask-CD to about 2× wafer-scale CD.

After the data preparation step, the processed design pattern is transferred to the mask blank using a conventional mask electron beam (e-beam) writer with an appropriate wafer chuck. Generally, two mask fabrication flows can be used: a wafer-flow process, starting with a bulk SOI wafer for pattern transfer with subsequent membrane definition; or a membrane-flow process, in which the e-beam writing and pattern transfer is done on a previously fabricated membrane. The latter has the advantage of smaller process-induced distortions, since the stress relaxation occurring during membrane etch appears before the pattern is defined by e-beam exposure. However, the membrane-flow process requires the handling of the fragile membranes in all front-end-tools and therefore has a higher risk of damage to the membrane, resulting in a smaller process yield. For this reason, the wafer-flow process has been preferred for IPL-stencil mask fabrication, making it necessary to correct process-induced placement errors (see next section) but with the benefit of reducing the number of membrane-handling steps.

A more detailed process flow is shown in Figure 12.5. The sequence begins with an ion-implantation step, which adjusts the mechanical stress of the Si-membrane layer. Generally, a small tensile stress is required for sufficient mechanical stability. This stress should be large enough to flatten the membrane but small enough to minimize process-induced distortions. A typical membrane stress value is 4 to 5 MPa. On top of the resist-coated SOI wafer a 50-mm × 50-mm mask pattern is defined with an e-beam writer. Nowadays, mostly chemically amplified resists are used due to their high sensitivity, which results in minimized e-beam writing times. The resist mask pattern is transferred into the Si layer by a reactive ion dry etch process, using the buried SiO_2 layer as an etch-stop layer. For this Si trench etch process, a gas chopping etch technique has been applied. This technique enables highly anisotropic Si stencil features with an aspect ratio >20:1 and a sufficient selectivity to the resist mask below [6]. The basic concept of the gas chopping is a clear chronological separation of Si passivation and etching steps. The etching and passivation precursors are SF_6 and C_4F_8. At first the passivation precursor gas is

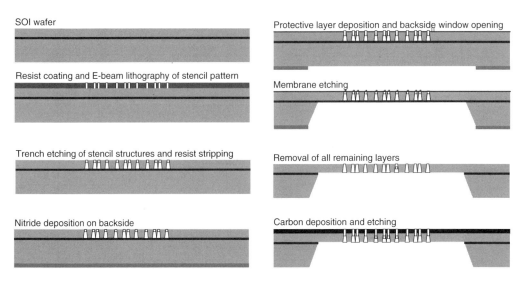

FIGURE 12.5
SOI wafer flow process.

dissociated by the plasma to form a Teflon-like sidewall polymer, which is deposited on the mask and the Si surface. During the following etch step, the polymer film is removed under the ion bombardment mainly at the trench bottom and not at the trench sidewall. Subsequently, fluorine radicals can attack and etch the Si at the bottom of the trench. This directionality increases the vertical etch rate over the lateral etching. Because of the individual parameter adjustment of these two steps, the alternating passivation and etching cycles provide a good control of the anisotropy and stencil profile. A special demand of the ion optics concerns the shape of the trench profile. In order to reduce the influence of ions scattered at the trench sidewalls, a retrograde profile with a sidewall angle larger than 90° is preferred. Scattered ions would blur the aerial image and thus reduce resolution and image contrast. The problem is addressed by designing the openings to recede by a small amount as it is shown in the scanning electron microscopy (SEM) cross-section image of a retrograde trench in Figure 12.6. After the Si trench etch process, the remaining resist layer is stripped, and a PECVD Si_3N_4 layer is deposited on the wafer backside. Within a backside lithography and dry etch step, an open Si window has been defined in the Si_3N_4 layer. This window defines the size of the final Si-membrane. Afterwards a CVD oxide layer is deposited onto the top Si layer and stencil openings, working as protection during the membrane etch process. The membrane is fabricated by wet etching the open Si area on the wafer backside in a two-step wet-chemical etch process up to the SiO_2 etch-stop layer. This buried oxide layer has to be pinhole-free to avoid damage to the membrane layer during the etching.

The first membrane etch step has been carried out in a stainless steel etch cell, where the SOI wafer frontside is mechanically sealed from a KOH etch solution. This step runs to a preliminary Si membrane thickness of 25 μm, which is solid enough for safe removal of the etched wafer. The second step runs outside of the etch cell in a TMAH solution, where the preetched SOI wafer is positioned in a single wafer carrier. This provides stress-free handling without any outer forces during the membrane etching and therefore high reliability and yield from this membrane etch process. The change of the etch chemistry from step one to step two is due to the high selectivity of the TMAH solution to the buried SiO_2 and CVD oxide layers.

Because of the compressive stress of the oxide layers, the membrane becomes wrinkled in the final state of the Si etching [Figure 12.7(a)]. After removing the backside nitride and both oxide layers, the tensile stress of the membrane dominates and stabilizes the whole mask system, resulting in a flat Si-membrane [Figure 12.7(b)].

In the last process step, a carbon protective layer is deposited on top of the membrane, protecting the Si-membrane against ion implantation during exposure. The effect of implantation is unique for IPL, since other charged-particle lithographies use electrons for exposure. Although high-energy electrons can cause local damage to the atomic lattice (e.g. displacements) as well, additional charge is removed by the electrical contacts of the membrane, and no additional atoms are introduced into the lattice. In contrast, ions with their large mass not only create larger damage to the crystal lattice but they are also implanted and are typically located on interstitial lattice positions. This implantation leads to a stress change of the membrane layer and a resulting pattern placement error. For this reason, a protective layer has to be applied, reducing the overall stress change during exposure. In order to fulfill this purpose, a dynamic equilibrium of implantation of He-ions and out-diffusion of recombined He-atoms has to be achieved. Amorphous carbon has been identified as the main candidate for this layer [7]. The carbon is deposited in a direct reactive sputter process with a mixture of argon and nitrogen. With this technique it is possible to deposit carbon layers with low compressive or even tensile stress. The carbon films developed so far nearly meet most of the requirements for protection layers [8,9]: low intrinsic stress, high thermal emissivity, and environmental stability (in nitrogen or argon

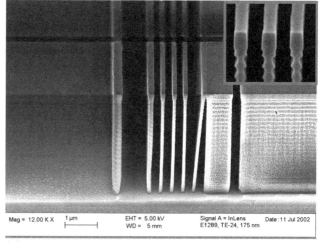

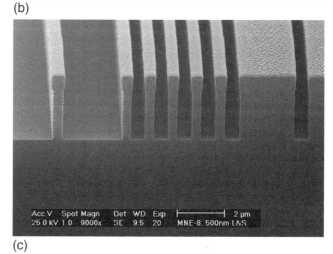

FIGURE 12.6
Retrograde trenches: (a) 175 mm, (b) 250 mm, and (c) trench profile.

Masks for Ion Projection Lithography

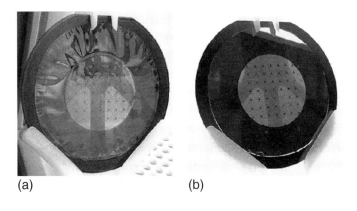

FIGURE 12.7
Stencil mask (a) before and (b) after backside and nitride and oxide layer etch. [29]

atmosphere). Some parameters however, such as stress uniformity, stress repeatability, and the magnitude of the equilibrium dose, require further improvement.

12.3 Pattern Placement and Process-Induced Distortions

Due to the reduced mechanical stability of stencil masks compared to standard optical masks, the issue of pattern placement and stress-related distortions is one of the most critical issues of charged-particle lithography. Since IPL was targeting at higher throughput for mid- to high-volume production, a large pattern area of $50 \times 50\,mm^2$ has been chosen on the mask. This layout is different from electron-based technologies, which typically use small subfields, separated by stabilizing struts of bulk material. While this mask concept increases the mechanical stability, the throughput of these systems is reduced considerably. On the other hand, the large pattern area of IPL masks makes a precise pattern placement control even more challenging. This is even more the case since the total membrane diameter of 126 mm is significantly larger than the actual pattern area. The reason for this large membrane area results from the fact that the IPL mask is an active part of the ion-optical imaging system. Thus, any distortion in the electrical potential due to the bulk wafer ring, results in ion-optical distortions of the image. For this reason, the wafer ring should have a certain minimum distance to the pattern area and the membrane has to be accordingly large.

There are four main contributors to the total image placement (IP) error: placement errors of the e-beam writer, process-induced distortions, pattern-induced distortions, and mounting-induced distortions. The latter are typically accounted for in the tool overlay budget, and they can be of considerable size (several hundred nanometers) if mounting of the mask is not done carefully enough. The placement errors of the e-beam writer result from errors or noise of the beam deflector and will not be discussed in this chapter, since they are not specific to the IPL technology. Depending on the capability of the used e-beam writer, they are in the range of 10 to 15 nm (3sigma error). The importance of the remaining two contributions from process- and pattern-induced distortions is specific to the stencil mask technology and will therefore be discussed in this chapter in more detail. While they also appear in standard optical masks, their contribution to the total placement error budget of these masks is negligible due to the stiffness of the bulk glass substrate, at least if a low-stress absorber material is used.

12.3.1 Process-Induced Distortions

Process-induced distortions result from changes in layer stress or in the stiffness of the whole mask system. The former is relevant when stressed layers, such as the resist layer, are added or removed or when the overall stress of the membrane changes. The latter appear, when the total stiffness of the mask is reduced by removing larger quantities of bulk material during the membrane etch step. By definition, the total placement error is the difference of pattern positions between the e-beam writing step and the finished mask with free-standing membrane and protective layer on top. It is mainly caused by the stress relaxation of the prestressed membrane after removing the bulk Si layer underneath the membrane.

As described earlier, the prestress of the membrane is applied by ion implantation of boron ions and subsequent annealing of the material. During the annealing step, the implanted ions move from interstitial sites to lattice positions and become part of the crystalline lattice. Since their atomic radius is smaller than that of silicon atoms, the disturbed lattice is distorted, resulting in tensile stress. A linear dependence between implantation dose and resulting stress can be observed and is also expected from theory [10].

After implantation, the prestressed Si-top layer bows the wafer edge upwards, since the membrane can gain elastic energy by contraction. The deformation stops when the gained energy is in equilibrium with the elastic energy required for the distortion of the bulk wafer part. When the stiffness of this bulk layer is reduced significantly during membrane etch by removing all bulk material within the membrane area, the total elastic energy can be reduced further by an additional membrane contraction, leading to a larger deformation of the wafer ring. Due to this additional membrane contraction, the bulk wafer ring is tilted inwards. The pattern of the membrane area, however, is radially distorted in the outward direction. This inversion is due to the reduced stiffness of the membrane within the pattern area, resulting from patterning of the open stencils. With a Si-membrane thickness of 3 μm and a membrane stress of 4 MPa, this effect results in pattern distortions of 150 nm 3 sigma. Due to the circular shape of the membrane, these distortions mainly have radial symmetry and can thus be compensated by a magnification correction. However, when SOI wafers with a wafer-flat, i.e., a nonrotational symmetry are used, or if the prestress of the membrane has been nonuniform, a magnification correction is not sufficient.

It could be concluded that a stress-free membrane is the optimum choice in terms of pattern placement. However, a stress-free membrane is mechanically unstable. An inplane stress leads to additional stiffness in vertical direction due to stress-stiffening effects. Therefore, a small tensile stress is required to stabilize the membrane. The resulting placement errors have to be calculated and compensated during e-beam writing. Thus, due to stability reasons it has not been the goal of IPL-stencil mask development to suppress stress-related distortions completely but rather to control, predict, and compensate these distortions with the required precision. As a consequence, the key issue of stress-related distortion control becomes the control of the absolute membrane stress, as well as of the local stress uniformity.

12.3.2 Pattern-Induced Distortions

The process-induced distortions described earlier are due to the stiffness change of the bulk wafer part when the membrane is formed. Additionally, membrane stiffness changes due to the pattern definition, which results in an additional distortion component. The creation of open stencils allows the membrane to distort locally and thus to relax internal stress. Since the pattern density is typically inhomogeneous over the pattern area, the resulting distortions also have a large nonuniform component. As a consequence, these distortions cannot be compensated by a simple radial correction, like a magnification correction and they vary from pattern to pattern. This type of distortion has to be calculated in advance and corrected by a shift of the original design pattern, either during the e-beam writing or in advance in the design.

A numerical method for calculating mechanical distortions is provided by finite element (FE) modeling [11]. This method uses a discrete numerical approach to solve the differential equations determining the distortions. While it can be used to predict process-induced distortions on the macroscopic wafer-scale, the application of this method to pattern-induced distortions is not straightforward. This is due to the pattern complexity and the relevant length scale of several 10 to 100 nm. A discrete FE-mesh with this element size, simulating the whole pattern area would require an incredible amount of computing power. For this reason, alternative methods have to be applied in order to describe the pattern-induced membrane distortions on the wafer-scale correctly.

The most promising approach is to average the local membrane stiffness on an appropriate length scale and to attribute an equivalent stiffness to the corresponding FEs. This equivalent stiffness can be obtained, for example by calculating the distortions of a periodic unit cell of the pattern with a fine mesh and by replacing the unit cell by a single FE with the same mechanical response afterwards. However, this method can only be applied to periodic patterns. In case of nonperiodic patterns, other methods have to be applied like an approximate prediction of the equivalent stiffness by analyzing the void ratio, i.e., the fraction of open areas. The main challenge of equivalent stiffness methods is to find an appropriate length scale over which the mechanical properties are averaged and which describes the resulting distortions with the required precision. This length scale typically depends on the pattern density gradients and is thus pattern-specific. Additionally, effects of anisotropy have to be accounted for and need to be characterized by an appropriate set of parameters [11].

12.3.3 Distortion Control

Since not all distortions can be suppressed completely, methods to control and compensate the residual distortions have to be developed. As the main origin of distortions is internal stress, this stress has to be controlled precisely in terms of absolute values, as well as the local uniformity. Two main contributions determine the final membrane stress: the initial stress of the SOI wafer and the ion implantation for adjusting the membrane stress. While the former is mainly an issue of the SOI manufacturing process [28], the latter is part of the mask fabrication. A final membrane stress between 4 and 5 MPa is the optimum choice for a mechanically stable membrane with still small distortions. In this case, the uniformity over the wafer has to be controlled to better than 0.1 MPa, which results in a pattern distortion of 10 nm.

SOI wafers can be fabricated in different ways. The conventional approach is to bond an oxidized silicon wafer onto another wafer and to thin the bonded wafer by mechanical or chemical methods [12]. More sophisticated approaches use ion implantation of hydrogen atoms to separate both wafers [13]. As an alternative to bonding methods, ion implantation of oxygen can directly be used to create the buried oxide layer [14]. Since the Si-crystal lattice is severely damaged during the implantation process, a subsequent annealing step has to be performed. Afterwards, the thickness of the top silicon layer is increased by epitaxy. The advantage of these nonbonded materials is the potential of an increased stress uniformity compared to bonded wafers. This is due to the fact that local wafer bow, particles, or contamination layers reduce the homogeneity of the bonding process and thus finally of the initial layer stress after annealing and cooling down to room temperature. On the other hand, particles can shield the ion implantation process for nonbonded wafers and thus create pinholes in the buried oxide layer. These pinholes can later damage the membrane layer during the bulk silicon etch step when the membrane is formed. However, the increased stress uniformity of nonbonded wafers is a significant advantage for the application in stencil masks. Depending on the specific fabrication process, the membrane layer can have an initial prestress even before implantation of boron ions [28]. This offset has to be taken into account when adjusting the final membrane stress.

After choosing a material with optimized stress homogeneity, it is important to have a uniform implantation dose over the wafer during the membrane stress adjustment step. Any variation of implanted ion dose would lead to local stress variations and thus finally to error in the distortion prediction. In practice, the required stress uniformity of 0.1 MPa can be achieved after membrane stress adjustment, as shown in Figure 12.8. This figure shows the local stress distribution of the SOI layer of an implanted SOI wafer. The distribution has been measured by local curvature measurements before and after layer removal and by calculating the corresponding stress value with the Stoney-equation.

When the membrane stress has been adjusted by choosing an appropriate implantation dose of boron ions, the resulting process- and pattern-induced distortions have to be compensated. While some of the distortions, such as the radial component, can be compensated by the ion optics, others have to be removed by a shift of the design pattern. The required shift vectors can be calculated by FE modeling of the full wafer geometry. To determine the total pattern shift, the difference of distortions between the e-beam exposure step on the bulk SOI wafer and the final state of the stressed and free-standing membrane, coated with the carbon protective layer, is calculated. An example of the FE model and a calculated distortion pattern of a membrane with small pattern density are shown in Figure 12.9.

With a special preprocessing software the design is shifted accordingly. As described earlier, an equivalent stiffness model has to be used in case of a patterned membrane. Since the initial stress distribution of the membrane layer is put into the model and has a direct influence on the quality of the prediction, this distribution either has to be uniform over the wafer and from wafer to wafer or it has to be measured in advance. The latter is

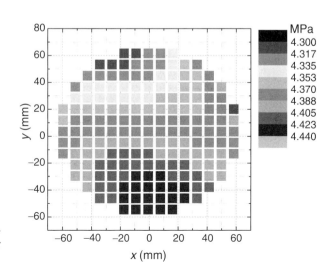

FIGURE 12.8
Local membrane stress distribution, indicating a stress uniformity of 0.1 MPa over the membrane area [28].

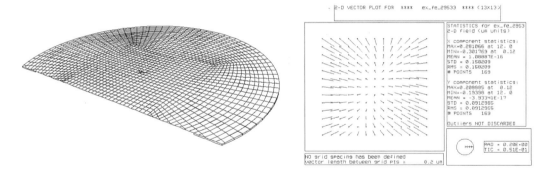

FIGURE 12.9
Cross-section view of a FE model (left), showing the outer bulk wafer ring and the thin membrane. Using this model, process-induced distortions can be calculated (right).

not easily achievable in practice, since most of the accurate stress measuring methods are destructive.

With the method described earlier, the distortions of a patterned membrane can be calculated. It is, however, important to keep the distorted state of the membrane constant over time, i.e., during exposure in the IPL tool. At this point, a significant difference of ions compared to electrons as exposing particles becomes apparent. In contrast to electrons, ions will be implanted into the stencil mask and thus constantly change the membrane stress. For this reason, the protective layer has to protect the underlying membrane against implantation and should additionally not change its layer stress during ion exposure. This can be achieved with an amorphous material like carbon, which forms a dynamical equilibrium between implantation of ions and outgassing of recombined atoms. Typically, this equilibrium is reached after a certain threshold dose. For lower doses, the protective layer stress changes as well. This additional stress change has to be taken into account when calculating the pattern correction. In order to minimize placement errors, the initial stress change, as well as the final equilibrium stress of the protective layer, has to be reproducible. Additionally, for practical reasons the threshold dose should be as small as possible in order to reduce long overhead times. The deposition process and the material properties of the protective layer have to be optimized accordingly.

12.4 Metrology of Stencil Masks

In addition to the front-end processing described earlier, the back-end processes like mask metrology, inspection, repair, and cleaning have to be adapted to the specific properties and requirements of stencil masks as well. Therefore, the following three sections will address these topics. Since the practical work on IPL-stencil masks for semiconductor applications has been terminated at the end of 2001, the results presented in these sections represent a rather early stage of development.

This section addresses two areas of metrology for IPL masks:

1. Metrology for IP
2. Metrology for CD

12.4.1 Tools for Measurement

Today the measurements on conventional masks are made by techniques that involve optics, SEM, and atomic force microscopy (AFM). All these are equally applicable to the measurements on IPL masks.

Among optical tools there are many in the market and new ones are constantly being developed. Some such tools are MuTec2010/LWM for CD measurement and LMS IPRO for IP measurement. LMS IPRO, however, has also been employed for making CD measurements.

SEM, a nonoptical measurements technique, is now being routinely used in the wafer fabrications and can be used for IPL masks, as the IPL masks have adequate electrical conductivity to avoid charging during ion beam exposure.

The AFM tool from Digital Instruments is also a nonoptical technique that can be employed for CD measurement on IPL masks. Figure 12.10 describes the use of such a tool. The system uses a standard tip that is scanned through a rectangular shaped trench giving a convolution of the rectangular signal. The CD value is taken from the measurement signal at the positions indicated by the solid triangles shown in the picture. They correspond to the top edge of the stencil opening. Thus, the shape of the edge of the stencil opening is decisive for the accuracy of the measurement result. Also, since, the shape of the tip has a strong influence on the measurement accuracy and due care must be taken to ensure the integrity of the tip.

The use of AFM is considered critical for measurement especially after membrane etching. The measurement accuracy also depends on the shape of the top of the trenches, but as the corners are very sharp, very accurate results are feasible. With other forms of tip shapes, information on the sidewalls of the opening can also be obtained.

12.4.2 Test Mask

Stencil test masks for the above measurements have been fabricated with a membrane diameter of 126 mm using a 150-mm SOI wafer (Figure 12.11). At the center of the mask is a pattern field of 60 mm × 60 mm filled with square holes of 50 μm × 50 μm separated by 50 μm resulting in 25% open area. A set of test designs consisting of square holes of sizes

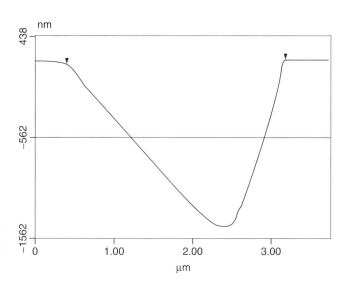

FIGURE 12.10
AFM measurement of a silicon mask–space structure. The CD is given by the distance of the solid triangles. The image shows the convolution of the rectangular opening and the biased measurement tip cone.

Masks for Ion Projection Lithography

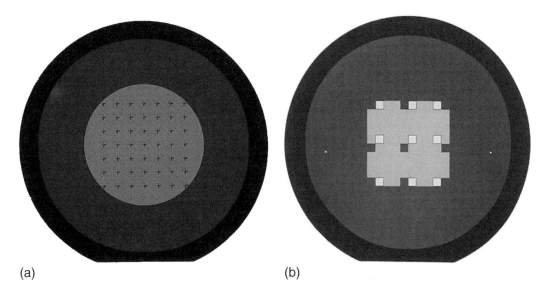

FIGURE 12.11
Test masks: (a) circular; (b) square.

ranging from 0.5 to 2 μm is placed at the four corners of the pattern area. Another set of 5 mm × 5 mm standard designs with line widths down to 0.7 μm is also placed at the centers of the four edges of the pattern area.

In addition, the test mask also consists of 13 × 13 crosses of a length of 10 μm and a line width of 1 μm.

12.4.3 Method of Measurement

It, however, needs to be understood that the conventional measuring instruments are not specifically designed to adapt to unconventional masks like the one under discussion. In order to make measurements with these instruments, the test mask is placed on a quartz plate. A problem encountered in conducting measurement on membrane substrate, which this test mask is made of, was that the airflow within the measuring system to assure temperature stability (22°C) caused the membrane to vibrate. This vibration made it difficult to utilize autofocus option on the machine during any measurement. The problem was resolved by employing an advanced TV focus procedure. The TV system takes a series of pictures in a z-range of 10 μm around the approximate location of the focus plane. As the membrane vibrates, images taken in the same z-position refer to different distances from the stencil membrane surface to the objective plane, which results in a different focal length. This increases the error in the measurement. Each image in Figure 12.12(a) was analyzed by plotting the intensity versus the lateral position (x or y) as shown in Figure 12.12(b). The maximum slope of the intensity profile was then calculated. In addition, the lateral position at 50% of the maximum intensity was also recorded. These values were obtained for all images taken in the chosen z-range. Then, the maximum slopes were displaced as a function of the z-position as shown in Figure 12.13. The focus was defined to be the z-position of maximum slope, so the sharpness peak indicates the focus. After this, the next step was to determine the lateral position of the structure. The lateral positions (left and right edge) of the edges were plotted against z. Then regression

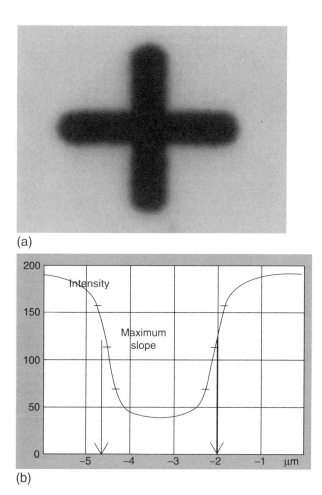

FIGURE 12.12
(a) Test image; (b) intensity profile.

line was calculated for both, and the mid-position between the intersection points of the regression lines with the z-focus was the wanted position value (Figure 12.14).

12.4.4 Image Placement Metrology

IP, like with any conventional mask, is a critical parameter for IPL-stencil masks as well. In earlier days, the IP was controlled by writing send-ahead membrane masks and then making the data correction based on the feedback from the send-ahead mask.

A key factor that influences the IP is the stress on mask membrane that needs to be controlled and minimized. One way to address this is to make the data correction based on the pattern analysis for possible distortion before any writing is carried out.

A correction scheme to minimize the IP placement error is described in a flow diagram shown in Figure 12.15. The mask layout patterns are analyzed by the software and then parameters are calculated, which determine the equivalent stiffness contrasts for the patterned stencil membrane areas. This set of constants is used as inputs for FE calculations. The other set of parameters are defined by the blank specifications. Wafer geometry values (size, thickness, flatness, warp, etc.), layer stresses, and thicknesses for silicon, silicon oxide, silicon nitride, and resist are part of the basic parameters set for the FE calculations. The objective here is to assure tolerances that need not be modeled by FE

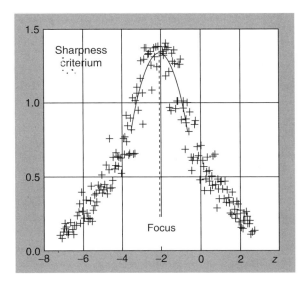

FIGURE 12.13
Maximum slope in dependency of z-position.

analysis. If this is not possible, more basic input data can be used for the calculations, e.g., the warp of the SOI mask wafer blank.

12.4.4.1 Repeatability of IP Measurement

The resulting repeatability of the stencil mask from the earlier mentioned measurements was found to be approximately as good as that of a chrome mask. A plot of the repeatability of the IP measurement is displayed in Figure 12.16. Here, ten complete measurement cycles were executed. The length of the x and y bars located at every measurement structure displays the range in x and y for the individual measurement location.

The maximum 3 sigma at a single measurement site and the 3 sigma of the mean of all measurement sites are plotted. The maximum 3 sigma for x and y directions were 4 nm and 3 nm, whereas the 3-sigma mean for x and y directions were 2 nm for each.

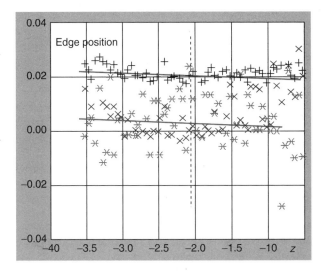

FIGURE 12.14
Edge position versus z.

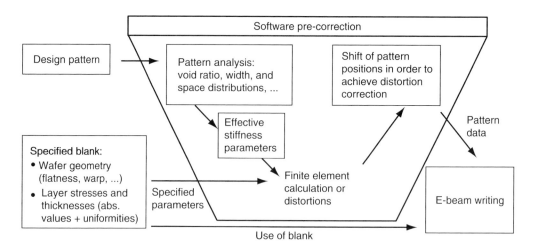

FIGURE 12.15
Concept of software precorrection for placement control.

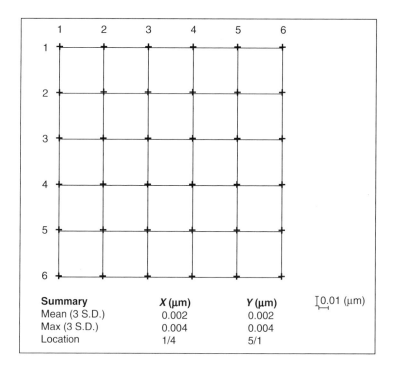

FIGURE 12.16
Repeatability (10 measurements) of placement control.

12.4.4.2 Accuracy of IP Measurements

Accuracy here is defined as the maximum of 3-σ deviation of measurements before and after turning the mask by 90°. The values are calculated after introducing a scaling and angle correction. For a stencil membrane, the result showed the maximum 3-σ accuracy

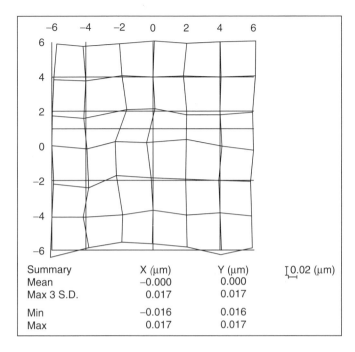

FIGURE 12.17
Accuracy of placement measurement.

as 24 nm, remembering that the spec of the measuring instrument was 25 nm. The results are plotted in Figure 12.17.

12.4.5 CD Metrology

In the case of CD measurement, the MueTec tool has been successfully employed for making measurements in reflective, as well in transmitive, modes. The repeatability in the reflective mode was found to be significantly better than in the transmitive mode. The readings from the two modes were also found to be different.

In transmitive mode, the incoming light encounters high-aspect ratios, and the final results are affected by the information from the whole trench, whereas in the reflective mode, the machine reads only the narrowest end at the top of the silicon opening. These could be the reasons for the difference between the readings from the two modes. Moreover, the changed intensity profile and reduced intensity of the transmitted signal can also be the cause of relatively poor repeatability in the transmitive mode.

12.4.5.1 CD Repeatability

In the case of a 150-nm-thick layer, the 3-σ CD repeatability for reflective measurements with MueTec 2010 was found to be 22 nm along the x direction and 18 nm along the y direction. These readings were taken where the membrane vibration due to the clean room airflow was not quite eliminated. In the absence of any vibration much better results can be obtained [4]. Repeatability measurement with Leica LMS IPRO in the absence of any membrane vibration gave a 3-σ value of 3 nm [15].

12.4.5.2 CD Uniformity

The CD uniformity measurements have also been made with Leica LMS IPRO, where a result of CD uniformity obtained from LMS IPRO is also shown in Figure 12.18.

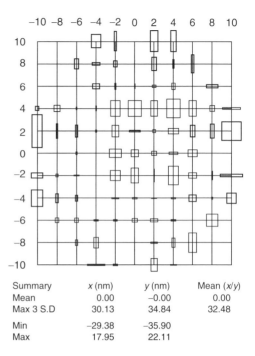

FIGURE 12.18
CD uniformity measured with LMS IPRO at stencil crosses of nominal width 1 μm over a field of 5.5 × 5.5 cm².

Summary	x (nm)	y (nm)	Mean (x/y)
Mean	0.00	−0.00	0.00
Max 3 S.D	30.13	34.84	32.48
Min	−29.38	−35.90	
Max	17.95	22.11	

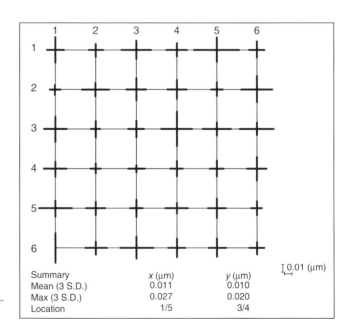

FIGURE 12.19
Repeatability of CD uniformity measurement.

Summary	x (μm)	y (μm)
Mean (3 S.D.)	0.011	0.010
Max (3 S.D.)	0.027	0.020
Location	1/5	3/4

Using the method described earlier in Section 12.4.3, the repeatability of CD uniformity was also measured. The result is shown in Figure 12.19. Here the maximum 3-sigma values were found to be 27 and 20 nm along x and y directions.

This is obviously not as good as for the chrome mask, but this discrepancy could be related to the vibration of the mask in the airflow that was not eliminated during the measurement.

12.5 Defects Inspection Repairs

Defect inspection of IPL-stencil masks can be made with optical, as well as e-beam, inspection tools. Optical inspection tools are the workhorses for current optical mask manufacturing. In most cases, blue or mid-UV light are used to inspect masks for deep UV lithography. This means that the inspection of leading-edge masks is being done in non-actinic lights. The practice of employing non-actinic optical inspection could then also be extended to ion projection stencil masks. It should, however, be kept in mind that imaging by ion projection is different from the properties of optical imaging, and so the defect printability in the two may not be the same. When working with e-beams, both transmitted and secondary electron images can be used for inspecting the masks.

12.5.1 Optical Inspection Tools

Optical inspections have been carried out with KLA-Tencor 351 operating with 488 nm wavelength light. Contact hole and line or space patterns of a stencil mask have been inspected. The inspection capability has been analyzed with a test mask with programmed defects for various defect types and feature. This has resulted in simulation where potential optical defects can be inspected. Results of a regular stencil mask, composed of profiles like gate arrays, contact holes, and fields with square blocks, showed that inspection is possible. Structures with feature sizes of 0.25 to 0.5 μm, as well as with the circular contact holes, have been found to be very critical to inspect (Figure 12.20). To conduct further studies, a special test mask was designed. The mask contains 13 different defect types of various sizes like extension, intrusions, corner defects, undersized and oversized contact holes, misplaced contact holes, CD variations in x or y direction, and pinholes.

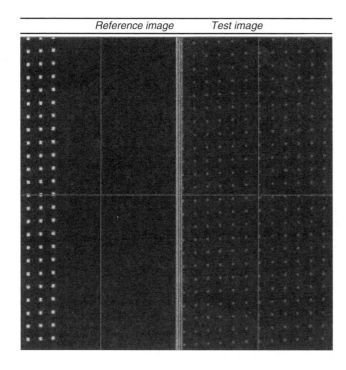

FIGURE 12.20
Contact holes' images of nominal size 0.25 μm.

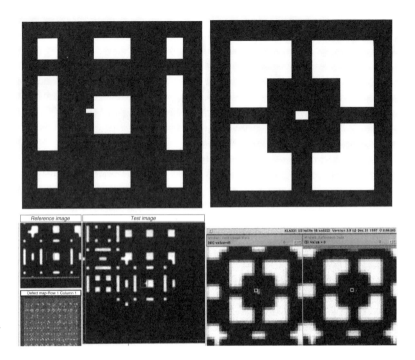

FIGURE 12.21
Design and inspection results of intrusion (left side) and pinhole (right side) with KLA-Tencor 351.

In Figure 12.21, successful inspection results for programmed intrusions and programmed pinholes are displayed. Additional defects on the test masks could also be inspected successfully. In Figure 12.22, an extension in a square contact hole and CD oversizing are also displayed.

12.5.2 E-Beam Inspection Tools

E-beam transmission inspection can be seen as the closest actinic technique among the currently developed defect inspection technologies.

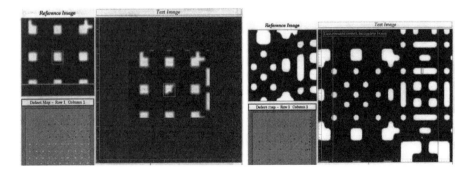

FIGURE 12.22
Inspection results of KLA-Tencor 351 for an extension in square opening (left side) and of an area with oversized CD (right side).

Masks for Ion Projection Lithography

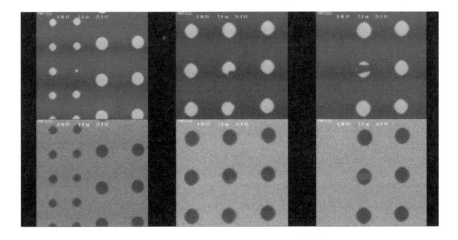

FIGURE 12.23
Inspection with Holon ESPA T-600.

E-Beam inspection has bean carried out with HOLON ESPA T-600 tool [16]. Patterns of different sizes and shapes have been inspected successfully. Figure 12.23 shows that for secondary electron signals, as they are used in high throughput e-beam inspection tools, the contrast is significantly lower for defects situated in the silicon trenches. However, with optimized parameters of inspection, both signals could be used for stencil mask inspection.

12.5.3 Defect Repairs

Focused ion beam (FIB) techniques provide potential solutions for repairing the IPL-stencil masks. These masks can have two kinds of defects. There are opaque defects, where the unwanted materials need to be removed, and then there are clear defects caused by the absence of material, where new material needs to be added or grown. Both types of defects can be repaired with ion beam-induced etching and deposition techniques [17].

In theses methods, a gas is introduced near the defect, which can be part of a chemical reaction induced by the beam. Depending on the type of gas, carbon and silicon are etched, or a high Z-material, e.g., tungsten or platinum, can be deposited.

The working principles of stencil mask pinhole repair and modification of openings are exhibited in Figure 12.24(A). Ion beam-induced deposition can also create side growth, which then can be used to form bridges by moving the ion beam over the stencil opening with proper speed. By employing several deposition and trim processes, complicated pattern inside the stencil mask openings can be achieved as shown in Figures 12.24(B)–(D).

With a 30-keV Ga$^+$ ion beam and current density of 7 mA/cm^2 it is possible to grow the platinum material across a 0.5 to 3.0-μm wide gap with a growth rate of 23 nm/s. Bridges deposited this way required trimming with a low current FIB to mill away overdeposited and resputtered material.

Innovative FIB deposition techniques provide bridges without the need of subsequent rework as demonstrated in Figure 12.25 [17]. The scan area can be advanced along with the leading edge of the bridge. The minimum deposition time per area in this case

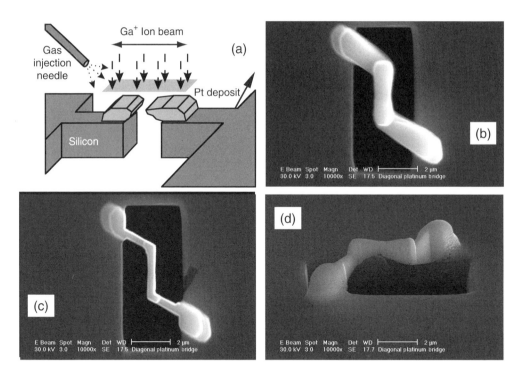

FIGURE 12.24
(a) FIB repair principles of stencil masks; (b) deposition of "zig-zag" platinum bridge with 350 pA Ga$^+$ ion beam; (c) and (d) trimming to 0.4-μm width with 11-pA beam. (IPL White Paper 1999.)

corresponds to a horizontal growth rate of 10 nm/s. Postdeposition trimming was not required for these bridges.

A complete filling of stencil mask openings is not necessary because the low-energy ($\leq$10 keV) ions can be stopped by a layer as thin as 0.1 μm.

IPL pattern transfer is insensitive to out-of-plane material of the FIB-induced depositions. Furthermore, some milling into the membrane material that occurred in the example of Figure 12.24 can be tolerated.

Moreover, there is no "staining problem" when using gallium ions for stencil mask FIB repair.

FIGURE 12.25
Platinum bridge across a slot in 2.75-μm-thick Si membrane, bridge deposited with 9 pA, 30 keV Ga$^+$ ion beam scanning a rectangle that moved across the slot as the bridge was formed. No postdeposition milling was required. (IPL White Paper 1999.)

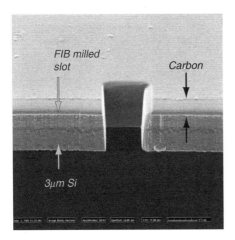 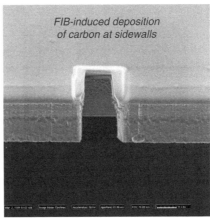

FIGURE 12.26
FIB repair on carbon-coated silicon membrane. (IPL White Paper 1999.)

Another example of ion milling and deposition is shown in Figure 12.26. Here, stencil mask repair was done on carbon-coated silicon membrane using FIB tool from Micrion Corporation. The SEM images show FIB milling of a slot into the C/Si membrane and FIB-induced deposition starting from the slit border.

12.6 Cleaning of IPL Masks

There are two approaches to the availability of clean IPL masks.

The first is the prevention, that means that due care be taken to keep the masks clean to start with. The use of pellicle, however, is not an option due to the very nature of stencil masks.

There are other means, however, to ensure that the mask remains clean throughout its usage on the line. One way is to design the system such that the stencil masks are maintained in a vertical position during their handling with protective shields at both sides. The shields are retracted only during the exposure to allow ions pass through the stencil opening. Moreover, the technique for protecting masks from contamination can also be adopted from the ones used for wafers, e.g., the use of mini-environment and SMIF boxes, etc.

There are, however, a number of techniques to clean a mask should it get contaminated with foreign particles. One aspect of cleaning operation involves the cleaning of silicon surface prior to the deposition of the carbon layer. Here standard silicon wafer cleaning methods, such as Huang or Piranha cleaning, can be applied. Cleaning of the mask after the carbon deposition can also be done but requires special care and close inspection of the carbon layer to ensure no damage is done to the surface.

There are also techniques that involve dry laser cleaning and liquid-film laser cleaning. The mechanism of "dry laser cleaning" involves a fast thermal expansion of material when it absorbs a short laser pulse. Excimer laser pulses have typical pulse duration of 15 to 25 ns. Though the amplitude of this thermal expansion is low, just a few nanometers, the short timescale results in high acceleration, which gives rise to forces large enough to

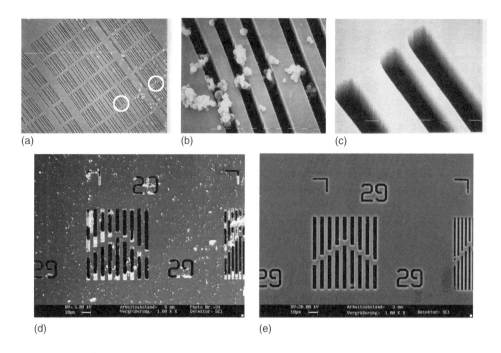

FIGURE 12.27
Laser cleaning of silicon stencil mask contaminated by Al_2O_3 particles with 0.2 to 2.0 μm diameter. (a)–(c) Four KrF-laser (248 nm) pulses. The irradiated spot area was 0.5 × 0.5 mm². Removal of the particles was observed after 4 pulses of dry laser cleaning at 350 mJ/cm². The pulses were applied with intervals of a few seconds. SEM bar = 10 and 1 μm, respectively. (d) and (e) One KrF-laser (2468 nm) pulse of 690 cm². (IPL White Paper 1999.)

overcome the adhesion forces of small particles. Figure 12.27 shows laser cleaning by removal of 200-nm particles [18]. There was a plan to concentrate on even smaller particles and advanced innovative laser cleaning methods that will be suited for particle removal in small-width trenches of carbon coated silicon stencil mask.

In the liquid-film laser cleaning method, a thin liquid film is applied to the surface to be cleaned prior to laser irradiation. This liquid film can be a thin water film that condensed on the sample surface from a flow of humid air. If the laser radiation is strongly absorbed by the sample to be cleaned, as is the case for silicon and excimer laser radiation, then the absorbed radiation will heat a thin surface layer of the sample, from which heat is transferred to the liquid film. This heat transfer is fast enough to result in strong superheating and high pressure within the liquid film. The liquid film will subsequently burst as a jet of steam away from the surface [18]. As an example, Figure 12.28 shows polystyrene spheres (size 500 nm) in trenches within a 3-μm thick silicon stencil mask membrane before laser cleaning: the pattern in this particular mask membrane was not properly etched through the membrane, so that the created pattern yielded trenches with sharp column-like structures at the bottom. The depth of the trenches is estimated as 2 to 2.5 μm. The particulate contamination is seen on the surface, at the sidewalls of the trenches, as well as down to the bottom of the trenches. Liquid-film laser cleaning was performed with one XeCl-laser (308 nm) pulse at 440 mJ/cm². The irradiated area was essentially clean as shown in Figure 12.28. The cleaning procedure removed the 500-nm polystyrene spheres from the surface, from the sidewalls of the trenches, as well as from the rough bottom of the trenches.

The cleaning does not affect the CD or the IPs.

Masks for Ion Projection Lithography 297

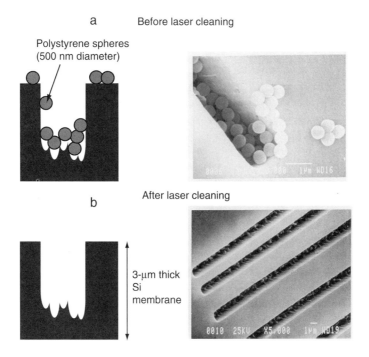

FIGURE 12.28
Liquid-film laser cleaning: (a) Si trench with 2 to 2.5 μm depth contaminated with 500 nm polystyrene spheres; (b) all particles removed by liquid-film laser cleaning with one XeCl (308 nm) laser pulse at 440 cm². (IPL White Paper 1999.)

LMS-IPRO placement measurements have been done on small area (14 mm × 14 mm) stencil masks before and after laser cleaning. In the first laser cleaning, half of the mask membrane field was exposed to 50 XeCl laser pulses at 1 Hz and a fluence of 225 mJ/cm². The second LMS-IPRO measurements did not detect any change in pattern placement. The next laser cleaning on the same membrane was directed to the entire membrane field applying 170 XeCl pulses with a fluence of 300 mJ/cm². The third consecutive measurement with the LMS IPRO did not detect any change in pattern placement.

12.7 Mask Stability and Stitching Errors

Mask stability and the stitching errors are two critical areas of IPL masks that, if not properly addressed, could result in array distortion and feature misplacement.

12.7.1 Mask Stability

The irradiation of the stencil mask with the beam of helium ion can cause significant (several degrees) rise in its temperature resulting in array distortion and feature misplacement. The heat buildup at the mask level must then be counteracted by introducing some sort of mask cooling mechanism. But because of the large and thin stencil mask membrane, it is not feasible to remove the heat builtup by any conductive means. Instead,

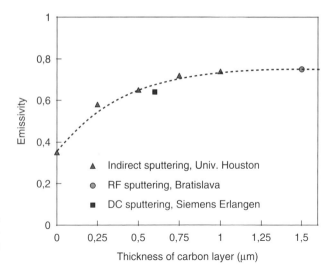

FIGURE 12.29
Influence of carbon layer thickness on C/Si membrane emissivity (measured at 8 to 14 μm wavelength). (IPL White Paper 1999.)

the technique being used to maintain the mask temperature at a constant value (preferably room temperature) is by inducing radiation cooling of masks. Radiation cooling is feasible as the power density input of the ion beam to the stencil masks is only in the order of mW/cm^2 (2 mW/cm^2 was planned for the IPL beta tool, 12 mW/cm^2 for the production tool). Thus, an important parameter for effective radiation cooling is the thermal emissivity of the membrane. Figure 12.29 shows measured thermal emissivity of carbon-coated silicon membranes versus the carbon layer thickness [19]. It shows an emissivity value of ≈0.6 for a carbon layer with 0.5 μm thickness on properly prepared silicon membranes. The radiation cooling was designed for 0.4 stencil mask emissivity. Therefore, thinner carbon layers of 0.1 to 0.2 μm thickness can be used as they are completely stopping 10 keV helium ions.

Figure 12.2 shows the ion beam illumination principles for the IPL stitcher, which was foreseen as the production tool using 200-mm stencil masks. There is a fixed aperture plate such that only one specific mask subfield is illuminated. Aperture plate and stencil mask are at the same temperature (room temperature) and have the same thermal emissivity. The ion beam tends to heat up the region of the aperture plate and the stencil mask membrane, which are illuminated by the ion beam. Cooled lens electrodes tend to cool down the mask area by absorbing the radiation from the mask elements that are at the higher temperature. If the ion beam is turned off, then the mask is irradiated with an external radiation source to maintain its temperature at a constant value.

The radiation cooling has been verified both theoretically and experimentally based on the idea that temperature of the condenser electrodes can be optimized to stabilize the mask temperature [20]. Experimental work on radiation cooling has been carried out at the University of Houston, TX [21]. Carbon-coated silicon membranes were irradiated with a 10-keV helium ion beam with 2.15 mW/cm^2 power density. In proper distance from the C/Si membrane, a temperature controlled cylindrical cone was placed simulating the geometrical configuration of the condenser of the IPL process development tool. With the help of an *in situ* stress measurement system, the stress of the C/Si membrane was measured when changing the temperature of the cylindrical cone. Figure 12.30 shows the experimental result: the stress of the ion-beam-heated mask is equal to that of the original mask at room temperature (25°C) when the cooled cylinder is about 10°C below room temperature. This result is in excellent agreement with the IMS-Vienna analytical model [21].

Masks for Ion Projection Lithography 299

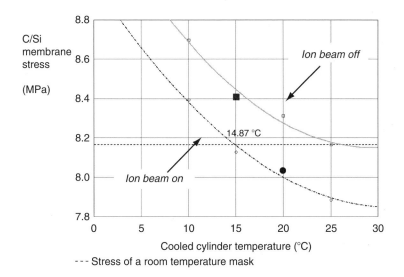

FIGURE 12.30
C/Si membrane stress versus temperature of the cylindrical cone with and without helium ion beam. (IPL White Paper 1999.)

Rigorous FE modeling of stencil mask radiative cooling has been done at the University of Wisconsin [22].

Figure 12.31 shows the assumed configuration and the stencil mask configuration with an IBM Talon mask layout enlarged to 4× reticle size. In this figure "Electrode 1" of the condenser system is close to the source and "Electrode 6" just above the mask. IMS-Vienna has shown by an analytical approach that the uniformity is optimal when cooling electrodes are far away from the mask, i.e., close to the ion source. This is implemented in the FEM simulation: Electrodes 1 and 2 are cooled. Assuming 1.6 mW/cm^2 power load at the stencil mask, the Figure 12.31 shows the calculated temperature contours across the mask membrane. With radiation cooling, the temperature difference is reduced to 0.14 K. This temperature difference translates into 2.3 nm placement error at the stencil mask even without magnification correction.

In an IPL production tool, the power load at the stencil mask was planned up to 12 mW/cm^2. Consequently, the temperature of the cooled electrodes will have to be significantly below the room temperature. Calculations have shown −50°C cooling temperature to be sufficient, which can be engineered without major difficulty. With this radiation cooling, the impact of thermally-induced placement errors can significantly be reduced.

12.7.2 Effect of Pattern Stitching on CDs

When using complementary stencil masks instead of ESE masks, some patterns have to be split in order to solve the donut problem or in some cases where a critical opening is narrow and very long and may suffer strain without any reinforcement. In such cases, masks "A" and "B" of a complementary pair are exposed consecutively, so that the two patterns can barely touch each other. The overlay error of these two complementary masks can result in a CD variation at those positions where the pattern has been split.

Any CD nonuniformity as result of this split can be overcome or minimized by meeting the following conditions:

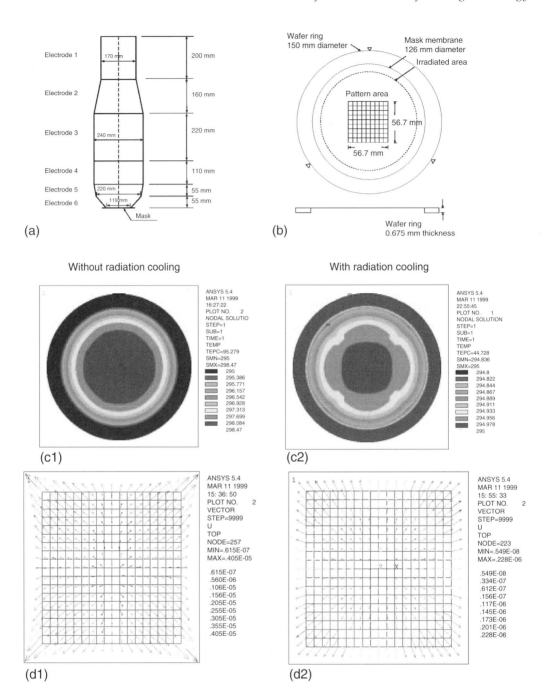

FIGURE 12.31
(a) Schematic of the simulated IPL exposure station, showing the cooled lens electrodes and the mask system, as well as the assumed thermal emissivity and temperature of the different electrodes. (b) Schematic of the IPL-stencil mask used in the FE simulation with modified IBM Tylon mask layout with 56.7 mm mask design field. Percentage void fractions are identified on each subregion. (c) Temperature contours across the mask membrane. (c1) Electrodes 1 and 2 were fixed at 295 K. The maximum temperature differential was 3.47 K. (c2) Electrodes 1 and 2 were cooled at 278.8 K. The temperature was reduced to 0.14 K. (d) IPD vector maps of the membrane pattern area. (d1) Electrodes 1 and 2 were fixed at 295 K. The maximum displacement vector was 40.5 nm. (d2) Electrodes 1 and 2 were cooled at 278.8 K. The maximum displacement vector was 2.3 nm. No magnification corrections have been taken. (IPL White Paper 1999.)

- Both masks A and B of the complementary pair have the same pattern density distribution. Therefore, in-plane distortion (IPD) errors due to inhomogeneous pattern densities are minimized.
- Masks A and B are written on the same e-beam writer in consecutive manner. Thus, systematic errors (except for mounting errors) will be virtually the same in both masks. Consequently, only statistical errors and mounting errors of e-beam writers have to be taken into account.
- While exposing masks A and B in an IPL system, the wafer remains stationed on its stage. Thus, there are no technologically induced wafer distortions to be taken into account.

With regard to the stitching exposure of complementary patterns, a theoretical and experimental study has been carried out at the University of Houston, Texas [23]. Other simulations also have been performed with a software developed by aiss/pdf Solutions [24]. These studies have resulted in close agreement between the two simulation methods.

The stitching of complementary patterns depends on the shape of the stencil mask openings, the ion-optics blur, and resist development. When assuming perfect rectangular stencil mask openings and a zero-blur system, longitudinal displacements between the two complementary patterns A and B would lead either to "blooming" (due the exposure region with double dose) or complete failure. Moreover, sharp staircase line patterns could result from lateral displacements. However, in a realistic IPL system with a ratio of blur or CD of 0.9, there is the possibility to restrict the double-exposure region to an area that is smaller than CD side length. Thus, after the resist development, a line with very small but acceptable change in width can be obtained.

A simulation of "ProBeam3D/IPL" of 100 nm CD stitching is shown in Figure 12.32. This result shows the developed line width versus longitudinal displacement assuming the exposure with 75 keV helium ions, a 60-nm FWHM blur, and a PMMA type resist with $\gamma = 5$ contrast. For all simulations a resist thickness of 400 nm was assumed. As to be expected from the presence of beam blur, there is very little influence if the stencil mask openings are perfectly rectangular or rounded. The resulting slope is about $\Delta CD/\Delta LD = 35\,nm/30\,nm = 1.2$.

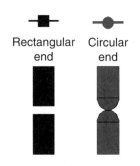
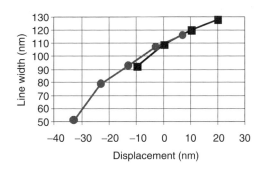

FIGURE 12.32
ProBeam3D/IPL simulation of 100-nm CD stencil mask stitching: influences of stencil mask opening shape and longitudinal offset. (IPL White Paper 1999.)

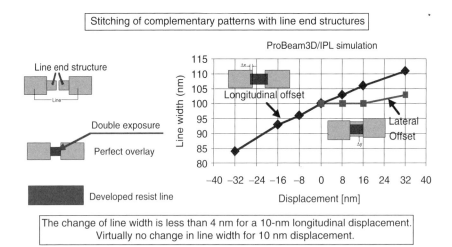

FIGURE 12.33
ProBeam3D/IPL simulation of complementary stencil mask stitching for 100 nm lines: influence of line-end structure. Exposure with 75 keV He^{+} ions, PMMA type resist with contrast of $\gamma = 10$ (IPL White Paper 1999.)

This slope can be decreased with a resist of higher contrast; further decrease of the slope can be realized by using line-end structures.

Figure 12.33 shows ProBeam3D/IPL simulation results for 75-keV helium-ion beam exposure of 100 nm CD lines assuming 60-nm blur, a PMMA type resist with $\gamma = 10$ contrast and stencil mask openings with line-end structures (2× CD width). The figure shows a slope of $\Delta CD/\Delta LD = 20\,nm/50\,nm = 0.4$ for longitudinal displacement. For lateral displacement, there is virtually no influence on CD control. Line-end structure also plays an important role on the CD value at the junction of two complementary patterns as is shown in the figure.

With the results of the experimental and simulation studies it was concluded that a complementary mask overlay error of 9.6 nm leads to a CD variation of 4.3 nm because the ratio between CD variation and longitudinal displacement error can be as low as 0.45. This is the case when the blur or CD ratio is 0.9, and a high contrast resist is used ($\gamma = 10$). This means that even when stitching at critical patterns, it is possible to print within ±10% CD variation.

Mask stability is maintained by keeping its temperature under control. The mask aligner consists of a cooling mechanism that can dissipate any heat gathered by the stencil mask during exposure to the ion beam. When the ion beam is turned off, a switching mechanism reactivates a thermal radiation source to maintain the mask temperature stability.

The thermal emissivity of 2.5-μm-thick silicon membranes has been evaluated (measurements were performed at 8 to 14 μm wavelength). The emissivity at the n-type side of the membrane was 0.3 and at the p-type side 0.2. With 0.5-μm-thick carbon coating, the radiation cooling capability of 3-μm-thick Si membranes is improved considerably to a thermal emissivity value of 0.65 [19].

There is no change of mask temperature as the membrane is subjected to radiation cooling and to heating either by the ion beam or by a thermal radiation (light) source. Turning the ion beam on while reducing the power of the thermal radiation source, the mask is always in a steady-state condition, which is also the case during wafer exposure. As mask membrane and mask frame are at the same temperature, thermal conduction is not involved. The mask temperature is virtually the same for dense pattern (metal) and sparse pattern (contacts) stencil masks.

It is well known that ion implantation into the silicon surface changes the crystal matrix. Thus, after a few exposures, a tensile Si membrane is to get wrinkled. As mentioned earlier, the problem is addressed by sputtering a low-stress carbon layer. Subsequent to deposition, the carbon layer is exposed to a helium ion beam with energy high enough for the ions to penetrate to the C/Si interface but not reaching the Si surface. Through this treatment an amorphous but porous carbon structure (the density is only 60% of compact carbon) is obtained. The carbon layers are stable under storage in nitrogen or argon atmosphere. Exposing such carbon layers with lower ion energy, i.e., stopping the ions within the carbon layer, a stable stress situation is obtained even for very high helium ion beam doses [6–8].

The thermal radiation (light) to the stencil mask can be adjusted in a nonuniform manner in order to correct stencil mask distortions and thus to ensure distortion-free imaging at the wafer (which can be controlled by *in situ* measurements) [4]. Theoretical modeling for this concept has been done, which was presented at the 5th NGL Workshop organized by Internatinal SEMATECH in August 2001.

12.8 Summary and Closing Remarks

Due to the physical properties of ions, IPL has several technical advantages over electron lithographies in terms of NA or depth of focus, field size, and resist sensitivity. There are, however, some technical issues that are specific to the properties of ions, like the implantation into the mask membrane and the resulting stress change during exposure. These effects do not pose a fundamental limit on the applicability of this technology, but they need to be solved before an industrial application of IPL is feasible.

Based on the decision by International SEMATECH to concentrate on the further development of optical lithography and to support EUV as main NGL technology, the work on IPL was stopped towards the end of 2001. Some further work was done in the first months of 2002 on "ion projection direct structuring" (IPDS) for patterned magnetic media [1,25]. IMS-Vienna concentrates further work on large-field charged-particle optics on 200× reduction: with electrons for projection maskless lithography [1,26] and with ions for a projection focused ion multibeam tool [27].

References

1. H. Loeschner, G. Stengl, H. Buschbeck, A. Chalupka, G. Lammer, E. Platzgummer, H. Vonach, P.W.H. de Jager, R. Kaesmaier, A. Ehrmann, S. Hirscher, A. Wolter, A. Dietzel, R. Berger, H. Grimm, B.D. Terris, W.H. Bruenger, G. Gross, O. Fortagne, D. Adam, M. Böhm, H. Eichhorn, R. Springer, J. Butschke, F. Letzkus, P. Ruchhoeft, and J.C. Wolfe, Large-field particle beam optics for projection and proximity printing and for maskless lithography, *JM3—J. Microlithography, Microfabrication Microsystems*, 2 (1; Jan.), 34–48 (2003).
2. R. Kaesmaier and H. Loeschner, *Proc. SPIE*, 3997, 19–33 (2000).
3. S. Eder-Kapl, H. Loeschner, M. Zeininger, O. Kirch, G.P. Patsis, V. Constantoudis, and E. Gogolides, MNE'2003, Cambridge, 22–25 Sept., 2003, t.b.p. in *Microelectronics Eng.*, 2004.
4. E. Haugeneder, A. Chalupka, T. Lammer, H. Loeschner, F.-M. Kamm, T. Struck, A. Ehrmann, R. Kaesmaier, A. Wolter, J. Butschke, M. Irmscher, F. Letzkus, and R. Springer, *Proc. SPIE*, 4764, 23–31 (2002).

5. J. Butschke, A. Ehrmann, B. Höfflinger, M. Irmscher, R. Käsmaier, F. Letzkus, H. Löschner, J. Mathuni, C. Reuter, C. Schomburg, and R. Springer, *Microelectronic Eng.*, 46, 473–476 (1999).
6. F. Letzkus, J. Butschke, B. Höfflinger, M. Irmscher, C. Reuter, R. Springer, A. Ehrmann, and J. Mathuni, *Microelectronic Eng.*, 53, 609–612 (2000).
7. J.R. Wasson, J.L. Torres, H.R. Rampersad, J.C. Wolfe, P. Ruchhoeft, M. Herbordt, and H. Löschner, *J. Vac. Sci. Technol. B.*, 15, 2214–2217 (1997).
8. P. Ruchhoeft, J.C. Wolfe, J. Wasson, J. Torres, H. Wu, H. Nounu, N. Liu, M.D. Morgan, and R.C. Tiberio, *J. Vac. Sci. Technol. B.*, 16, 3599–3601 (1998).
9. P. Hudek, P. Hrkút, M. Držik, I. Kostič, M. Belov, J. Torres, J. Wasson, J.C. Wolfe, A. Degen, I.W. Rangelow, J. Voigt, J. Butschke, F. Letzkus, R. Springer, A. Ehrmann, R. Kaesmaier, K. Kragler, J. Mathuni, E. Haugeneder, and H. Löschner, *J. Vac. Sci. Technol. B.*, 17, 3127–3131 (1999).
10. H. Holloway and S.L. McCarthy, *J. Appl. Phys.*, 73 (1), 103–111 (1993).
11. R. Tejeda, G. Frisque, R. Engelstad, E. Lovell, E. Haugeneder, and H. Löschner, *Microelectronic Eng.*, 46, 485–488 (1999).
12. J.B. Lasky, *Appl. Phys. Lett.*, 48, 48–78 (1986).
13. M. Bruel, *Electronic Lett.*, 31 (14), 1201–1205 (1995).
14. A.J. Auberton-Herve, B. Aspar, and J.L. Pelloie, *Nato Advanced Research Workshop*, Kluwer Academic Publishers, Dordrecht, 1995, pp. 3–14.
15. A. Ehrmann, T. Struck, E. Haugeneder, H. Loeschner, J. Butschke, F. Letzkus, M. Irmscher, and R. Springer, *Proc. SPIE*, 3997, 385–394 (2000).
16. I. Santo, N. Anazawa, T. Okagawa, and S. Matsui, *Symposium on Charged Particle Optics*, Tsukuba, Japan, 29–30 Oct., 1998, 132nd Committee on Electron and Ion Beam Science and Technology, Japan Society for the Promotion of Science, Tokyo, 1998, pp. 119–122 (in Japanese).
17. A.J. DeMarco and J. Melngailis, *J. Vac. Sci. Technol. B.*, 17, 3154–3157 (1999).
18. W. Zapka, R. Lilischkis, and H.P. Zappe, *EMC-1999: The 16th European Mask Conference on Mask Technology for Integrated Circuits and Micro-Components*, Munich, German, 15–16 Nov., 1999 (VDE/VDI), Published in *Proc. SPIE*, 3996, 92–96 (2000).
19. D. Braun, R. Gajic, F. Kuchar, R. Korntner, E. Haugeneder, H. Loeschner, J. Butschke, F. Letzkus, and R. Springer, *J. Vac. Sci. Technol. B.*, 21, 123–126 (2003).
20. H.F. Glavish, G. Stengl, H. Loeschner, A. Chalupka, US Patent 4,916,322.
21. J.L. Torres, H.N. Nounu, J.R. Wasson, J.C. Wolfe, J. Lutz, E. Haugeneder, H. Löschner, G. Stengl, and R. Kaesmaier, *J. Vac. Sci. Technol. B.*, 18, 3207–3209 (2000).
22. B. Kim, R. Engelstad, E. Lovell, A. Chalupka, E. Haugeneder, G. Lammer, H. Löschner, J. Lutz, and G. Stengl, *J. Vac. Sci. Technol. B.*, 16, 3602–3605 (1998).
23. R. Kaesmaier, H. Löschner, G. Stengl, J.C. Wolfe, and P. Ruchhoeft, *J. Vac. Sci. Technol. B.*, 17, 3091–3097 (1999).
24. H. Hartmann, A. Petraschenko, S. Schunk, R. Steinmetz, E. Haugeneder, and H. Loeschner, *Proc. SPIE*, 3996, 105–107 (2000).
25. A. Dietzel, R. Berger, H. Loeschner, E. Platzgummer, G. Stengl, W.H. Bruenger, and F. Letzkus, *Advanced Materials*, 15 (14), 1152–1155 (2003).
26. C. Brandstätter, H. Loeschner, C. Brandstätter, H. Loeschner, G. Stengl, G. Lammer, H. Buschbeck, E. Platzgummer, H.-J. Döring, T. Elster, and O. Fortagne, *SPIE Microlithography — Emerging Lithographic Technologies* VIII, 24–26 Feb., 2004, Santa Clara, California, USA, t.b.p. in *Proc. SPIE*, 5374 (2004).
27. H. Loeschner, E.J. Fantner, R. Korntner, E. Platzgummer, G. Stengl, M. Zeininger, J.E.E. Baglin, R. Berger, W.H. Bruenger, A. Dietzel, M.-I. Baraton, and L. Merhari, *MRS (Materials Research Society) 2002 Fall Meeting, Boston, USA*, 2–6 Dec., 2002, Published in *Proc. MRS*, 739, 3–12 (2003).
28. F.-M. Kamm, A. Ehrmann, H. Schaefer, W. Pamler, R. Kaesmaier, J. Butschke, R. Springer, E. Haugeneder, and H. Loeschner, *Jpn. J. Appl. Phys., Part 1*, 41 (6B), 4146–4149 (2002).
29. J. Butschke, Die SOI Scheibe der Mikroelektronik als neue Prozessbasis für nanostrukturierte Silizium Membranmasken, Thesis (URN: urn:nbn:de:bsz:93-opus-13633; URL: http://elib.uni-stuttgart.de/opus/volltexte/2003/1363/).

13

Mask for Proximity X-Ray Lithography

Masatoshi Oda and Hideo Yoshihara

CONTENTS
13.1 PXL System 305
13.2 X-Ray Mask Structure 306
      13.2.1 Membrane 307
      13.2.2 Absorber 309
13.3 Fabrication 310
      13.3.1 Mask Processes 311
      13.3.2 EB Writing 312
      13.3.3 Dry Etching 312
      13.3.4 Frame Bonding 314
13.4 Defect Inspection and Repair 314
13.5 X-Ray Masks for LSI Fabrication 314
13.6 Summary 315
References 316

Proximity x-ray lithography (PXL) was proposed 30 years ago as a technology to perfectly replicate mask patterns to a wafer using soft x-rays [1]. Although the PXL was confirmed to have sufficient resolution to form patterns below 100 nm soon after its proposal [2], it could not be used in industry for a long time because x-ray sources were too weak and mask fabrication was too difficult. The development of a compact synchrotron radiation (SR) ring [3] gave the industry an x-ray source that could produce x-rays at an intensity strong enough for practical use. The SR ring together with improved mask technology has made PXL the most promising technology for making sub-100-nm patterns. Here, x-ray mask technology is introduced.

13.1 PXL System

A PXL system consists of an x-ray source, a mask, and a stepper, and uses soft x-rays with wavelengths between 0.5 and 1.5 nm. x-rays with shorter wavelengths are not suitable because their higher transparency makes both resist sensitivity and mask contrast too low. On the other hand, x-rays at longer wavelength degrade pattern resolution due to the large diffraction.

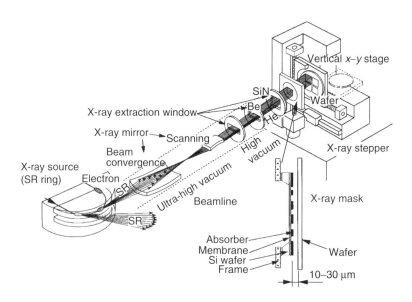

FIGURE 13.1
Schematic of the SR lithography system.

Figure 13.1 shows a PXL system using an SR ring. In this SR lithography system, x-rays from the SR ring are directed to the mask through a beam line. A wafer coated with resist is set behind the mask leaving a small proximity gap of the order of a few ten micrometers. The stepper aligns the wafer.

In PXL, the objective is to replicate to the wafer patterns that are of the same size as the mask patterns. Thus, the mask patterns must be of the right size and all features must be at right locations according to the design layout.

Furthermore, the substrate must be thin enough to allow maximum transmission of the soft x-rays. On the other hand, the absorbers have to be thick enough to stop the x-rays from reaching the wafer. The biggest issue is how to produce absorber patterns with high accuracy on such a thin membrane.

13.2 X-Ray Mask Structure

X-ray masks differ significantly in construction from the photomasks. An x-ray mask consists of absorber patterns on a membrane held by a Si wafer, as shown in Figure 13.2, where the Si wafer is mounted on a frame. The membrane corresponds to the glass substrate of a photomask. The frame is made of materials having high rigidity, such as Pyrex glass or SiC, which is needed so that the masks can be easily and safely handled.

For the membrane to stay flat, it must have tensile stress. The absorber patterns also have stresses, though they are produced unintentionally. These stresses tend to deform the mask as shown in Figure 13.3. The deformation must be very small so that it does not affect the accuracy of the x-ray masks. To keep the Si wafer from being deformed by membrane stress, Si wafers as thick as 2 mm are used. Absorber stress must also be kept small so as not to deform the membrane.

Mask for Proximity X-Ray Lithography

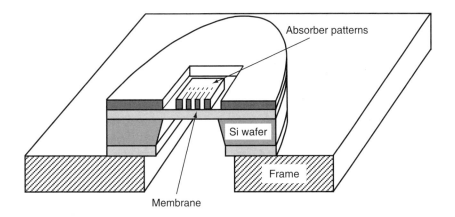

FIGURE 13.2
Schematic of an x-ray mask.

13.2.1 Membrane

The essential membrane requirements are as follows: (a) high transparency to soft x-rays, (b) good smoothness, (c) good flatness, (d) high dimensional stability, (e) high mechanical strength, (f) high chemical durability, (g) high optical transparency, and (h) ease of fabrication. To achieve high x-ray transparency, the membrane material must be made of light elements with small x-ray absorption coefficients. Recently used membrane materials are Si, SiN, SiC, and diamond (Table 13.1). In addition to these materials, organic films, such as Mylar, have also been studied. Organic films, however, suffer from dimensional and thermal stability problems.

The deposition of SiN [4] and SiC [5] is carried out by low-pressure (LP) CVD, whereas the deposition of diamond is typically done by microwave plasma CVD [6]. The stress of a SiN film can be easily controlled by adjusting the temperature and gas flow ratio during low-pressure CVD as shown in Figure 13.4 [4]. Films with low tensile stress are deposited at high temperatures or at a large NH_3 flow rate. The stresses of SiC and diamond can also be controlled by adjusting deposition conditions. SiN is amorphous, so the film surface is smooth after deposition. On the other hand, SiC [7] and diamond [6] are poly-crystals, so

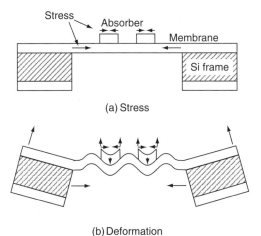

FIGURE 13.3
Membrane and absorber stress, and mask deformation caused by the stress.

TABLE 13.1

Properties of Membrane Materials

	Young's Modulus (GPa)	Thermal Expansion Coefficient (deg^{-1})	Density (g/cm^3)
Si	160	3.7×10^{-6}	2.33
SiN	160	2.1×10^{-6}	3.18
SiC	460	4.6×10^{-6}	3.21
Diamond	1050	3.5×10^{-6}	3.52

the surfaces are rough due to crystal grains. The roughness is removed by mechanical polishing after deposition.

Figure 13.5 shows deformation of SiN, SiC, and diamond membranes caused by Ta absorber stress. As the absorber patterns having compressive stress of about 30 MPa at the upper left spread, the membranes deform. The deformation decreases as the membrane's Young's modulus increases, and the smallest deformation is in the diamond membrane. Young's modulus is thus a very important factor for highly accurate masks.

For highly accurate optical alignment, highly optical transparency is needed. Figure 13.6 shows the transparency of diamond membrane for wavelengths ranging from 400 to 800 nm. The transparency varies periodically with wavelength, which results from interference from light reflected at the surfaces. The transparency can be improved to over 80% at every wavelength between 500 and 800 nm by depositing antireflection material, such as SiO$_2$, on both sides of the membrane.

The membrane must have enough durability for x-rays. When the membrane contains hydrogen atoms from the deposition source gasses, the stress and transparency are varied by x-ray exposure [8]. It has been reported that SiC and diamond with good film quality have good x-ray durability.

Unlike the Si-based materials in Table 13.1, the diamond does not have an x-ray absorption edge near 0.7 nm because it has no Si element. Thus, x-rays having wavelengths shorter than 0.7 nm can be used.

Recently, it was reported that such short-wavelength x-rays make it possible to replicate patterns smaller than a nanometer. The PXLs with short-wavelength x-rays are referred as second-generation PXL [9].

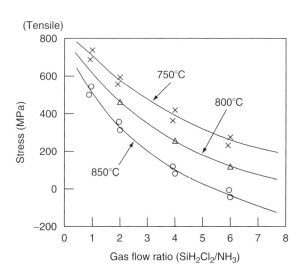

FIGURE 13.4
SiN stress and deposition condition of low-pressure CVD.

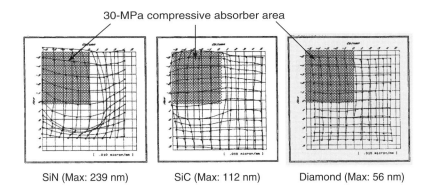

FIGURE 13.5
Membrane material and its deformation caused by absorber stress.

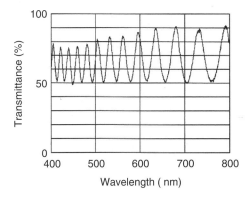

FIGURE 13.6
Optical transmittance of diamond membrane.

13.2.2 Absorber

The required absorber properties are as follows: (a) a high absorption coefficient, (b) high stress controllability, (c) capability of patterning below 100 nm, (d) high chemical and radiation durability, and (e) the ease of fabrication. The high absorption coefficient requirement is met by using high-density heavy metals as absorber. Au has been used as an absorber material since the beginning of x-ray mask development [10] because it has a large x-ray absorption coefficient, and fine patterns can easily be formed. Recently, Ta [11], W [12], and their compounds [13,14] have been used as absorber materials, which can be patterned by dry etching [11–14]. These materials have approximately the same absorption coefficient as gold at x-ray wavelengths ranging from 0.5 to 1.5 nm. Physical properties of absorber materials are summarized in Table 13.2.

Absorber stress must be small so that it does not cause membrane distortion (Figure 13.3). Generally, in films deposited by sputtering, the stress can be controlled by the pressure during deposition. Figure 13.7 shows the dependence of films' (Ta, W, and Re) stress on pressure. For each material, the stress changes from compressive to tensile, as the pressure increases. For Ta, the tensile stress reduces again with pressure and is zero at around 8.5 Pa. The stress of W and Re also comes close to zero when in the high-pressure region. However, stress-free Ta film deposited at high pressure has smaller density than

TABLE 13.2

Properties of Absorber Materials

	Young's Modulus (GPa)	Thermal Expansion Coefficient (deg^{-1})	Density (g/cm^3)
Au	88	1.5×10^{-5}	19.3
Ta	190	6.5×10^{-6}	16.7
W	410	4.6×10^{-6}	19.3
WTi, TaBN	—	—	15–16

that deposited at low pressure. Thus, the low-pressure region is used for depositing Ta absorber films. In Figure 13.7, the gradient of the stress–pressure curve near the stress-free point is smaller for Ta than for Re or W, indicating the stress in Ta film can be controlled more precisely.

To control the stress even more precisely, annealing is performed after deposition. The stress in Ta film shifts to compressive side by annealing as shown in Figure 13.8. Ta films deposited by sputtering have columnar grains. Oxygen diffuses along the grain boundaries from the film surface and causes oxidation in Ta films, making the stress compressive. Stress-free Ta films can be obtained by depositing tensile films after annealing for an appropriate time.

The stress in Ta or W compounds having an amorphous structure is more stable. These films can also be deposited by sputtering, and stress control can be done in the same way as for Ta film. However, contrary to Ta, the stress changes from compressive to tensile by annealing [13].

13.3 Fabrication

The key areas in the fabrication of x-ray masks are: (1) processing (2) e-beam writing, (3) dry etching, and (4) frame bonding, discussed in the following.

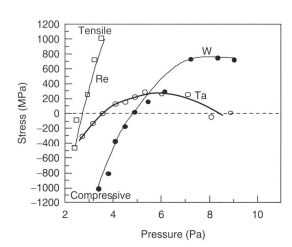

FIGURE 13.7
Film stress and gas pressure during sputtering.

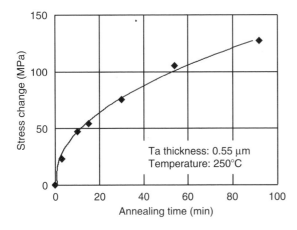

FIGURE 13.8
Stress change of Ta film caused by annealing in air.

13.3.1 Mask Processes

There are two processes for making x-ray masks. One is a membrane process, in which absorber patterning is carried out after back-etching [Figure 13.9(a)]. The other is a wafer process, in which absorber patterning is done before back-etching [Figure 13.9(b)]. In the wafer process, no membrane breakage occurs. However, improving the pattern placement accuracy is difficult because large pattern shifts occurs during back-etching. In the membrane process, masks with high pattern-placement accuracy can be produced, although the apparatus must have special functions for handling wafers with a membrane. In recent days, the membrane process has been mostly favored because the process leads to highly accurate masks.

As shown in Figure 13.9(a), a 2-μm-thick SiC or diamond film is deposited on a 4-in. Si wafer. A Ru film is deposited on the membrane to a thickness of 20 nm by sputtering. Then Ta film is deposited on Ru by ECR-sputtering. Following this, SiO$_2$ is then deposited on Ta by using ECR-plasma CVD technique. The Ru film stabilizes Ta stress [15], and the

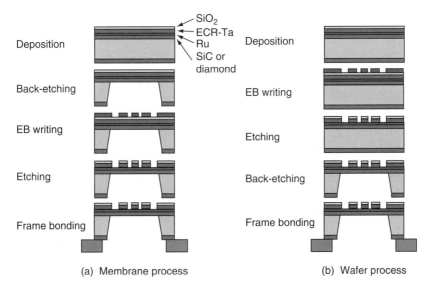

FIGURE 13.9
X-ray mask fabrication process.

SiO$_2$ is used as a mask for Ta etching. After the deposition of these films, back-etching is carried out using KOH solution. Resist patterns are formed on the membrane with an electron beam (EB) writer and the SiO$_2$ film is etched by reactive ion etching. Next, the Ta film is etched using the SiO$_2$ as a mask. Finally, the Si wafer is bonded to a glass frame. In the wafer process in Figure 13.9(b), x-ray masks are produced using similar deposition and patterning techniques.

To produce highly accurate x-ray masks, one can correct pattern position shifts by a method known as previous analysis of distortion and transformation (PAT) of coordinates [16] or product-specific emulation (PSE) [17]. In this method, first, send-ahead masks are made to obtain information about pattern position shifts. Next, working masks are produced by an EB writing process, in which patterns are delineated so as to compensate for the pattern position shifts. If the pattern position shifts are sufficiently reproducible, very highly accurate masks can be made. Nippon Telegraph and Telephone Corporation (NTT) has produced x-ray masks with feature-placement accuracy below 25 nm by using this method [18].

The membrane process has three key operations: EB writing to form resist patterns on the thin membrane with high accuracy, etching of absorber film on the membrane, and the frame bonding.

13.3.2 EB Writing

EB writing must be performed on a thin membrane supporting a heavy metal absorber. Electrons scattering from the absorber strongly affect the formation of resist patterns. Therefore, an EB writer with acceleration voltage of 100 keV was developed, which is significantly higher than the conventional 30 keV systems (Figure 13.10) [19]. This EB writer can form resist patterns with a width of 50 nm. The high resolution comes from the excellent beam sharpness and small forward-scattering in the resist film. Due to the high acceleration voltage, most of the electrons pass through the membrane. The small numbers of electrons that scatter backward in the absorber disperse widely. Therefore, the proximity effect is very small. Using this writer, even complex fine patterns, such as LSI patterns (more complex than 4-Gbits DRAMs), can be formed with a large margin [14].

The mask holder for the EB writer must be designed carefully. The surface of the holder has to be such that electrons passing through the membrane do not scatter backward [20].

13.3.3 Dry Etching

In an x-ray mask, pattern width precision must be as good as pattern placement accuracy. Since the absorber patterns are thicker than 0.3 μm, the etching system must be able to form precise patterns having an aspect ratio larger than 5. There are some dry etching systems for metal etching. Here, electron cyclotron resonance (ECR) ion stream etching [21] is introduced (Figure 13.11). In this system, etching gases are effectively decomposed in the low-pressure chamber by ECR discharge, and ion energy incident on the etched surface is controlled to be below 100 eV. Therefore, large selectivity can be expected. The main etching gas is C$_{12}$ for Ta and Ta compounds, and SF$_6$ for W compounds.

Generally, metal etching depends strongly on the substrate temperature. Therefore, temperature control of the membrane during absorber etching is very important. Helium cooling is very helpful in these etching systems.

Furthermore, in etching such small patterns, we must take into account the microloading effect that means decreased etching rate with decreased pattern widths.

Figure 13.12 shows Ta patterns etched by ECR systems; 70-nm-wide patterns are clearly formed.

Mask for Proximity X-Ray Lithography

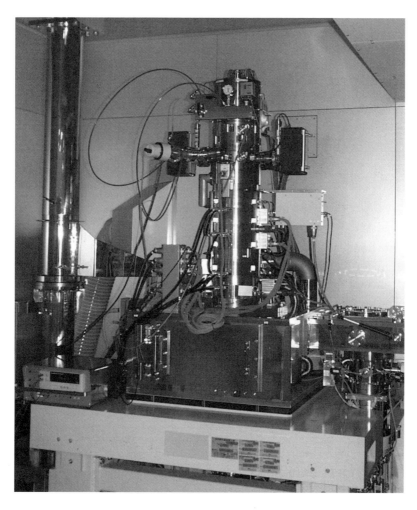

FIGURE 13.10
50-nm-wide resist patterns delineated by an EB writer with an acceleration voltage of 100 kV.

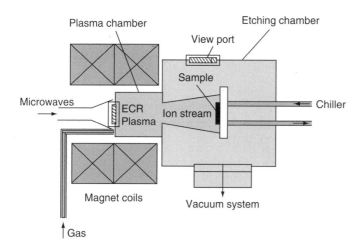

FIGURE 13.11
Schematic of the ECR ion stream etching system.

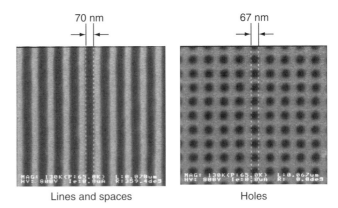

FIGURE 13.12
70-nm-wide Ta absorber patterns.   Lines and spaces   Holes

13.3.4 Frame Bonding

The membrane can suffer deformation from frame bonding. There are two ways to resolve this problem. One is to use a bonding method that does not deform the wafer. For this purpose, the one-point bonding method [22] was developed. In this method, a very small area of a wafer is bonded to a frame, and only that area contacts the frame surface. This is useful for mask processes, in which bonding is the final step. The other way is a process in which the bonding is done before patterning the absorber. In this process, bonding must be very hard so that the wafer does not peel off during the patterning process. Anodic bonding was developed for this purpose [23].

13.4 Defect Inspection and Repair

Defect inspection and repair are crucial to the production of x-ray masks for LSI processes. A scanning electron beam system has been investigated for this purpose. For SiN or diamond membrane, which is an insulator, it is difficult to inspect defects directly because of the charge-up problem. Mask defects are therefore confirmed by inspecting wafers having resist patterns replicating the mask patterns [24]. One can directly inspect defects on SiC membrane because it has low conductivity. To repair defects in x-ray masks, a focused-ion-beam repair system has been studied [25,26]. The system uses a Ga ion beam focused to a spot less than 10 nm in diameter. Transparent defects are repaired by ion-beam-induced deposition of Ta, and opaque defects are repaired by ion milling or gas-assisted etching. Figure 13.13 shows an example of a repair made to Ta absorber mask.

13.5 X-Ray Masks for LSI Fabrication

Many LSIs and devices have been produced using x-ray lithography. IBM developed x-ray masks with 0.5-μm LSI patterns [27] and x-ray masks with 0.25-μm LSI patterns [28]. Mitsubishi Electric developed masks with 1-Gbit DRAM patterns in 1989 [29]. NEC reported LSI masks in 1994 [30]. NTT developed x-ray masks with 200-nm [31] and 100-nm [32] LSI patterns. NTT and Association of Super-Advanced Electronics Technologies

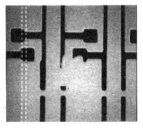
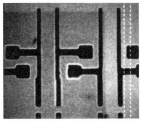

(a) Before repair   (b) After repair

FIGURE 13.13
Ta absorber patterns repaired by a focused ion-beam system.

FIGURE 13.14
X-ray mask.

(ASET) developed masks with 100-nm LSI patterns with position accuracy of 25 nm in 1999 (Figure 13.14) [33]. The usefulness of x-ray lithography has been thoroughly proven by these studies.

13.6 Summary

x-ray masks mainly consist of a thin membrane and absorber patterns. The membranes are made of Si, SiN, SiC, or diamond film and Ta and W, or their compounds are used as the absorbers. Stress control of these films is crucial in mask fabrication. An e-beam writer with high acceleration voltage is necessary for pattern delineation on heavy metals. Highly accurate masks can be fabricated by adopting a membrane process, in which back-etching is carried out before the absorber film is patterned and then by correcting any distortion. A scanning EB system and a focused-ion-beam system enable defect inspection and repair. Highly accurate x-ray masks have been produced, and the usefulness of x-ray lithography has been proven.

References

1. D.L. Spears and H.I. Smith, High-resolution pattern replication using soft x-rays, *Electron. Lett.*, 8, 102–104 (1972).
2. R. Feder, E. Spiller, and J. Topalian, Replication of 0.1-μm geometries with x-ray lithography, *J. Vac. Sci. Technol.*, 12, 1332–1335 (1975).
3. T. Hosokawa, T. Kitayama, T. hayasaka, S. Ido, Y. Uno, A. Shibayama, J. Nakata, and K. Nishimura, NTT superconducting storage ring—Super-ALIS, *Rev. Sci. Instrum.*, 60, 1783–1786 (1989).
4. M. Sekimoto, H. Yoshihara, and T. Ohkubo, Silicon nitride single-layer x-ray mask, *J. Vac. Sci. Technol.*, 21, 1017 (1982).
5. M. Yamada, M. Nakaishi, J. Kudou, T. Eshita, and Y. Furumura, An x-ray mask using Ta and heteroepitaxially grown SiC, *Microelectron. Eng.*, 88, 135–138 (1989).
6. H. Windischmann and G.F. Epps, Properties of diamond membranes for x-ray lithography, *J. Appl. Phys.*, 68, 5665–5673 (1990).
7. K. Yamashiro, M. Sugawara, H. Nagasawa, and Y. Yamaguchi, Smoothing roughness of SiC membrane surface for x-ray masks, *Jpn. J. Appl. Phys.*, 30, 3078–3082 (1991).
8. W.A. Johnson, R.A. Levy, D.J. Resnick, T.E. Saunders, A.W. Yanof, H. Betz, H. Huber, and H. Oertel, Radiation damage effects in boron nitride mask membranes subjected to x-ray exposures, *J. Vac. Sci. Technol. B.*, 5, 257–261 (1987).
9. T. Kitayama, K. Itoga, Y. Watanabe, and S. Uzawa, Proposal for a 50 nm proximity x-ray lithography system and extension to 35 nm by resist material selection, *J. Vac. Sci. Technol. B.*, 18, 2950–2954 (2000).
10. H.I. Smith, D.L. Spears, S.E. Bernacki, x-ray lithography: a complementary technique to electron beam lithography, *J. Vac. Sci. Technol.*, 10, 913–917 (1973).
11. H. Sekimoto, A. Ozawa, T. Ohkubo, and H. Yoshihara, *A High Contrast Submicron x-ray Mask with Ta Absorber Patterns*, Extended Abstracts of the 16th (1984 International) Conference on Solid State Devices and Materials, Kobe, 1984, pp. 23–26.
12. R.R. Kola, G.K. Celler, J. Frackoviak, C.W. Jurgensen, and L.E. Trimble, Stable low-stress tungsten absorber technology for sub-half-micron x-ray lithography, *J. Vac. Sci. Technol. B.*, 9, 3301–3304 (1991).
13. K. Marumoto, H. Yabe, S. Aya, K. Kise, and Y. Matsui, Total evaluation of W–Ti absorber for X-ray mask, *Proc. SPIE*, 2194, 221–230 (1994).
14. S. Tsuboi, Y. Tanaka, T. Iwamoto, H. Sumitani, and Y. Nakayama, Recent progress in 1× x-ray mask technology: feasibility study using ASET-NIST format TaXN x-ray masks with 100 nm rule 4 Gbit dynamic random access memory test patterns, *J. Vac. Sci. Technol. B.*, 19, 2416–2422 (2001).
15. M. Shimada, T. Tsuchizawa, S. Uchiyama, T. Ohkubo, S. Itabashi, I. Okada, T. Ono, and M. Oda, Development of highly accurate x-ray mask with high-density patterns, *Jpn. J. Appl. Phys.*, 38, 7071–7075 (1999).
16. S. Uchiyama, S. Ohki, A. Ozawa, M. Oda, T. Matsuda, and T. Morosawa, Improving x-ray mask pattern placement accuracy by correcting process distortion in electron beam writing, *Jpn. J. Appl. Phys.*, 34, 6743–6747 (1995).

17. D. Pusito, M. Struns, and M. Lawliss, Overlay enhancement with product-specific emulation in electron-beam lithography tools, *J. Vac. Sci. Technol. B.*, 12, 3436–3439 (1994).
18. M. Oda, M. Shimada, T. Tsuchizawa, S. Uchiyama, I. Okada, and H. Yoshihara, Progress in x-ray mask technology at NTT, *J. Vac. Sci. Technol. B.*, 17, 3402–3406 (1999).
19. S. Ohki, T. Watanabe, Y. Takeda, T. Morosawa, K. Saito, T. Kunioka, J. Kato, A. Shimizu, T. Matsuda, S. Tsuboi, H. Aoyama, H. Watanabe, and Y. Nakayama, Patterning performance of EB-X3 x-ray mask writer, *J. Vac. Sci. Technol. B.*, 18, 3084–3088 (2000).
20. K.K. Christenson, R.G. Viswanathan, and F.J. Hohn, x-ray mask fogging by electrons backscattered beneath the membrane, *J. Vac. Sci. Technol. B.*, 8, 1618–1623 (1990).
21. T. Tsuchizawa, H. Iriguchi, C. Takahashi, M. Shimada, S. Uchiyama, and M. Oda, Electron cyclotron resonance plasma etching of α-Ta for x-ray mask absorber using chlorine and fluoride gas mixture, *Jpn. J. Appl. Phys.*, 39, 6914–6918 (2000).
22. M. Oda, T. Ohkubo, and H. Yoshihara, One-point wafer bonding for highly accurate x-ray masks, *Proc. SPIE*, 2512, 152–159 (1995).
23. J. Trube, J. Chlebek, J. Grimm, H.-L. Huber, B. Lochel, and H. Stauch, Investigation of process latitude for quality improvement in x-ray lithography mask fabrication, *J. Vac. Sci. Technol. B.*, 8, 1600–1603 (1990).
24. M. Sekimoto, H. Tsuyozaki, I. Okada, A. Shibayama, and T. Matsuda, x-ray mask inspection using replicated resist patterns, *Jpn. J. Appl. Phys.*, 33, 6913–6918 (1994).
25. I. Okada, Y. Saitoh, T. Ohkubo, M. Sekimoto, and T. Matsuda, Repairing x-ray masks with Ta absorbers using focused ion beams, *Proc. SPIE*, 2512, 172–177 (1995).
26. A. Wangner, J.P. Levin, J.L. Maner, P.G. Blauner, S.J. Kirch, and P. Longo, x-ray mask repair with focused ion beams, *J. Vac. Sci. Technol. B.*, 8, 1557–1564 (1990).
27. R. Viswanathan, R.E. Acosta, D. Seeger, H. Voelker, A. Wilson, I. Babich, J. Maldonado, J. Warlaumont, O. Vladimirsky, F. Hohn, D. Crockatt, and R. Fair, Fully scaled 0.5 μm metal–oxide semiconductor circuits by synchrotron x-ray lithography: mask fabrication and characterization, *J. Vac. Sci. Technol. B.*, 6, 2196–2201 (1988).
28. R. Viswanathan, D. Seeger, A. Bright, T. Bucelot, A. Pomerene, K. Petrillo, P. Blauner, P. Agnello, J. Warlaumont, J. Conway, and D. Patel, Fabrication of high performance 512Kb SRAMs in 0.25 μm CMOS technology using x-ray lithography, *Microelectron. Eng.*, 23, 247–252 (1994).
29. N. Yoshioka, N. Ishio, N. Fujiwara, T. Eimori, Y. Watakabe, K. Kodama, T. Miyachi, and H. Izawa, Fabrication of 1-Mbit DRAMs by using x-ray lithography, *Proc. SPIE*, 1089, 210–218 (1989).
30. K. Fujii, T. Yoshihara, Y. Tanaka, K. Suzuki, T. Nakajima, T. Miyatake, E. Orita, and K. Ito, Applicability test for synchrotron radiation x-ray lithography in 64-Mb dynamic random access memory fabrication processes, *J. Vac. Sci. Technol. B.*, 12, 3949–3953 (1994).
31. K. Deguchi, K. Miyoshi, H. Ban, T. Matsuda, T. Ohno, and Y. Kado, Fabrication of 0.2 μm large scale integrated circuits using synchrotron radiation x-ray lithography, *J. Vac. Sci. Technol. B.*, 13, 3040–3045 (1995).
32. M. Oda, S. Uchiyama, T. Watanabe, K. Komatsu, and T. Matsuda, x-ray mask fabrication technology for 0.1 μm very large scale integrated circuits, *J. Vac. Sci. Technol. B.*, 14, 4366–4370 (1996).
33. H. Aoyama, T. Taguchi, Y. Matsui, M. Fukuda, K. Deguchi, H. Morita, M. Oda, T. Matsuda, F. Kumasaka, Y. Iba, and K. Horiuchi, Overlay evaluation of proximity x-ray lithography in 100 nm device fabrication, *J. Vac. Sci. Technol. B.*, 18, 2961–2965 (2000).

Section V

Mask Processing, Materials, and Pellicles

14

Mask Substrate

Syed A. Rizvi

CONTENTS
14.1 Introduction 321
14.2 Material 322
     14.2.1 Glass 322
     14.2.2 Chrome Film 322
     14.2.3 Molybdenum Silicide ($MoSiO_xN_y$) Film 323
14.3 Effect of Substrate on CD Uniformity and Image Placement 323
References 324

14.1 Introduction

With the evolution in microlithography and mask design the mask substrate also has undergone many transitions.

In the early days of the semiconductor industry, the mask substrate (also called mask blanks) used to be 2 in. × 2 in. The mask size has since then been growing, and at the present the standard size is 6 in. × 6 in. The thickness of the substrate has grown from 0.060 in. to the present 0.25 in. Today a 0.25-in.-thick 6 in. × 6 in. substrate is referred as 6025 plate. The driving force behind this increase in the substrate size has been, among other factors, the large chip size and advancement in the step/scan exposure systems. Larger 9 in. × 9 in. substrates have also been made but their use has not been predominant because of the difficulties and expenses involved in tooling and in modifying many related manufacturing equipment. Today's exposure systems cover a larger distance along one of the two axes that could result in the use of rectangular 6 × 9-in. substrate, but here also due to tooling difficulties the implementation of rectangular substrate was not feasible.

The starting material for substrate is glass plate sputtered with chrome-based film that is then coated with photoresist. The composition of chrome film changes from the bottom to the top. The bottom portion of the films acts as glue in order to ensure a good adherence of chrome to glass surface. The top layer of the film acts as an antireflective (AR) coating to reduce the undesirable reflection that takes place during the exposure cycle.

14.2 Material

14.2.1 Glass

The size of the substrate is not the only factor of importance when considering substrate's structure. The glass plate also has undergone through a series of transformation and forms the basis of all substrates. It plays an important role in the structure and workings of substrates.

In earlier days, the standard material used to be soda lime, which was then replaced by a superior quality material known as White Crown for its reduced defects. Later when the thermal expansion of glass during exposure became an issue the White Crown was replaced with boro-silicate glass, which had a lower coefficient of thermal expansion (CTE). The next improvement was the introduction of fused silica (also known as quartz), since the CTE of fused silica was even lower than that of boro-silicate. In addition to its low CTE, this fused silica material also exhibited better transmission at 365 nm referred as UV wavelength that was the standard in those days. This transparency of the material became more important when the industry moved from 365 to 248 nm and now 193 nm illuminations.

For the upcoming shorter wavelengths of 157 nm, an F2-doped fused silica material with 76% transmission has been introduced [1,2].

Fused silica has been the material for the leading edge technology for quite sometime, but with 193 nm exposure over an extended period the material has been found to exhibit color centers and compaction. The color center formation causes a low level of fluorescence at about 400 nm, and compaction causes a small change in refractive index. However, these changes may affect the optics of the exposure systems but have no detrimental effect on photomask because the energy involve here is very small [3].

Schott-Lithotec, a supplier of mask blank, has also introduced a promising DUV blank known as Zerodur® with a spec of zero thermal expansion to meet the current requirements [4].

There are a number of substrate-related features that affect the CD uniformity that should be <10 nm.

Plate flatness need to be <1.0 μm. The spec on some of the plates has been quoted as low as 0.5 μm.

14.2.2 Chrome Film

At present the absorber on the glass is a compound of chrome consisting of Cr, N_2, O_2, and possibly other elements. The composition of the film varies from the bottom to the top serving different purposes. The bottom layer acts like a glue to improve the adherence of chrome to glass. The top surface acts as an antireflective coating to minimize the undesirable reflection that may take place inside the system.

The top and bottom layers of the film constitute a very small portion of the bulk of the chrome material that acts as the opaque film. A typical thickness of the film is 100 nm with an optical density of 3.0, which amounts to 0.1% (or less) transmission.

As regards to AR, a coating based on a three-layer Fabry-Perot structure with reflectance of <1% has also been reported [5].

Smaller chrome thickness (59–73 nm) for improved performance has been explored and the results are promising [5].

14.2.3 Molybdenum Silicide (MoSiO$_x$N$_y$) Film

Molybdenum silicide (MoSiO$_x$N$_y$), commonly known as MoSi film, was first used to improve the adhesion to fused silica [1]. Now MoSi is the key player in the structure of embedded attenuated phase shift masks (EAPSMs). These EAPSMs will continue to be used for the next few more generations.

The MoSi film is sandwiched between the glass and the chrome film. The MoSi has a tendency to flake and redeposit on the mask during the processing of the substrate and cause some yield loss. There are also issues with the exposure durability and the chemical durability of the cleaning process of the film.

In addition to MoSi, there are other promising candidates, such as TaN/Si$_x$N$_y$, TiN/Si$_x$N$_y$, and CrAlO$_x$N$_y$, that are being looked at [5].

Current status of films: The current status of the chrome and materials for EAPSM film has also been summarized in Table 14.1 [5].

14.3 Effect of Substrate on CD Uniformity and Image Placement

The homogeneity of glass in terms of its optical properties, e.g., refractive index, transmission, and birefringence, has its direct impact on the CD uniformity [5,6]. A material developed by Corning quotes a low birefringence of <1 nm/cm and refractive index homogeneity as <4 ppm.

The inhomogeneity of chrome and other phase-related films on the glass can equally affect the CD uniformity.

TABLE 14.1

Status of Substrate Films [5]

Characteristics	Absorbers for Binary Masks Requirements Fully Met	Absorbers for Binary Masks Requirements Partially Met	Materials for EAPSM Films Requirements Fully Met	Materials for EAPSM Films Requirements Partially Met
Optical density > 3	Yes			
Phase shift of 180 ± 5° at exposure wavelength			Yes	
Reflectivity < 15%		Yes		Yes
Transmission 5–25%				Yes
High conductivity to prevent charging during the e-beam patterning	Yes			
Low roughness, no pinholes, no particles	Yes			Yes
Ease of etch		Yes		Yes
Chemical resistance (no adhesion failure or change in optical properties with cleaning)	Yes			Yes
Uniformity (thickness, transmission, index of refraction)	Yes		Yes	

Today's plates are considerably thicker compared to the earlier ones, but because of their increased size the phenomena of "sag" and "distortion" when mounted on the exposure systems may still occur. Strains suffered by the plates under this condition can directly contribute to the image placement (IP) errors. A deflection of 0.62 μm can give 40 nm IP error [3].

IP is also affected by the CTE of the glass. Even in fused silica, a change of 0.08°C can change the IP by 10% of the allowed tolerance [3]. Plate flatness needs to be <1.0 μm. The spec on some of the plates has been quoted as low as 0.5 μm. Out-of-spec flatness may affect CD uniformity and IPs.

Coated film may cause some stress on the glass, and after the pattern is made some of the stress is released that can bend the glass causing an IP error.

References

1. J.G. Skinner, Photomask Fabrication for Today and Tomorrow, Short Course 122, SPIE Education Service Program, 2001.
2. Asahi Glass Brochure/Website/Photomask Substrate.
3. P. Rai-Choudhury, *Handbook of Microlithography, Micromachining and Microfabrication*, vol. 1, SPIE Press, 1997, pp. 377–474. Ballingham, Washington, USA.
4. Schott Lithotec Brochure: Mask Blanks, IC Advanced Packaging.
5. R. Walton, Photo blanks for advanced lithography, *Solid State Technol.*, October, 2003.
6. B.B Wang, Residual birefringence in photomask substrates, *J. Microlith., Microfab., Miscrosyst.*, 1 (1), 43–48 (2002).

15

Resists for Mask Making

Benjamen Rathsack, David Medeiros, and C. Grant Willson

CONTENTS
15.1 Introduction 325
15.2 Photomask Resist Requirements 325
15.3 Resist Materials 327
    15.3.1 Nonchemically Amplified Resists 327
        15.3.1.1 Positive Tone Resists Based on Chain Scission 327
        15.3.1.2 Positive Tone Resists Based on Dissolution Inhibition 329
        15.3.1.3 Negative Tone Based on Cross-Linking 331
    15.3.2 Chemically Amplified Resists 332
        15.3.2.1 Positive Tone Chemically Amplified Resists 333
        15.3.2.2 Negative Tone Chemically Amplified Resists 334
15.4 Resist Implementation Challenges 335
    15.4.1 Resist–Blank Interactions 335
        15.4.1.1 Nature of Photomask Blanks 335
        15.4.1.2 Resist Footing 335
    15.4.2 Resist Outgassing 336
    15.4.3 Mask Shop Issues 336
References 337

15.1 Introduction

Resist materials used for mask making continue to evolve as the demand for higher resolution and the production of complex assist features introduce new challenges. This chapter provides a brief history of the development of mask-making resists and addresses some of the most relevant challenges faced in patterning of photomasks.

15.2 Photomask Resist Requirements

Photomasks are fabricated using both electron beam (e-beam) and laser lithography processes. E-beam lithography at 50 kV is used for high-resolution photomask production at the 90- and 65-nm nodes [1,2]. High voltage e-beam systems using 100 kV have printed features as small as 20 nm for direct write and imprint applications [3]. An increase in accelerating voltage improves resolution but decreases resist sensitivity [4] and increases resist heating [5,6]. Resist heating also increases with higher beam current, which produces proximity linewidth errors. Resist sensitivity (dose) and beam current issues limit the throughput of the e-beam mask writer systems. E-beam tools like the EBM-4000 (50 kV) using a current density of 20 A/cm^2 need resists with 5 µC/cm^2 dose sensitivity to write a 100-nm node reticle in around 7 h [2]. The transition to high accelerating voltage e-beam systems has driven the evaluation and development of new resist systems that have higher dose sensitivity, lower temperature sensitivity, lower exposure outgassing, long postexposure delay stability, dry etch resistance, and compatibility with chromium substrates.

Laser photomask lithography provides higher throughput (4-h write times) than e-beam by splitting a laser source into multiple beams that write simultaneously. The ALTA tools produced by ETEC Systems use 32 beams to brush an exposure area [7]. The ALTA 3700 laser photomask writer uses a continuous wave, 363.4 nm, argon ion laser for imaging resist features with 0.5-µm dimensions [8]. Nonchemically amplified (NCA) resists based on two-component diazonaphthoquinone (DNQ)-novolak chemistry are used for these systems due to their transparency and bleaching optical properties. Higher resolution laser lithography systems have been made through the reduction in exposure wavelength from 363 to 257 nm [9]. The reduction in wavelength into the deep ultraviolet (DUV) has driven the evaluation of novel, NCA resist materials [10], and chemically amplified resists (CARs) commonly used for 248-nm wafer lithography [11,12]. New DUV laser photomask tools are also developed that function like microsteppers using a micromirror device (spatial light modulator), which regulates pixel exposure using a pulsed 248-nm laser [13,14]. The transition to DUV exposure wavelengths has driven the evaluation of resist materials that have the appropriate optical transparency, postexposure delay stability, dry etch resistance, and compatibility with chromium substrates.

Photoresist imaging processes for both photomask and IC fabrication are similar as shown in Figure 15.1. The photoresist is spin-coated on an antireflective, quartz substrate. The quartz substrate has a 6-in. × 6-in. surface dimension and is 0.25 in. thick. The thin opaque layer on the photomask is typically chromium, which provides a mechanically strong layer that prevents light transmission through the mask. A heterogeneous oxynitride layer is grown on the chromium layer to minimize reflections off of the mask.

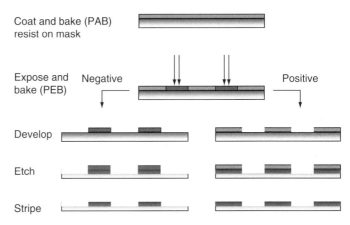

FIGURE 15.1
Positive and negative photoresist processing for photomasks.

Resists for Mask Making

The photomask blanks are coated with resist and then postapplication baked (PAB) to remove solvent. An electron or laser beam directly exposes individual features of the circuit pattern into the photoresist in a serial fashion. It can take many hours to directly print all the features into the photomask where it takes only seconds to expose an image of that photomask using a projection lithography tool.

CARs are postexposure baked (PEB) to drive an acid-based deprotection reaction that renders the films soluble in an aqueous base. Optical resists are also baked after exposure to diffuse the variation in photoactive compound (PAC) concentration in the film caused by standing wave interference effects from reflections at the air-resist and resist-substrate interfaces. Even though the reflections from the chromium surface are small, the high contrast of modern resists demands a PEB to achieve the maximum process latitude [15]. Optical photoresists are most commonly developed with an aqueous base developer (0.26 N tetramethyl ammonium hydroxide or TMAH), while most scission-based e-beam resists are developed using organic solvents. The developed features are etched into the chromium layer using a wet or dry etch (Cl_2/O_2 gas). The remaining photoresist is stripped off the photomask using a sulfuric acid and peroxide mixture or by oxygen ashing. The photomask is typically cleaned multiple times to prevent defects, inspected, and a pellicle that is repaired as necessary is placed over the patterned features to maintain a defect-free imaging plane on the mask.

15.3 Resist Materials

15.3.1 Nonchemically Amplified Resists

NCA resists were used in early photomask lithography applications. NCA resists are described on the basis of their functions in the following section: positive tone resists that undergo chain scission, positive tone resists that convert dissolution inhibitors into soluble species, and negative tone resists based on cross-linking. These resist materials have evolved to provide the resolution, sensitivity, etch resistance, and chemical stability for photomask fabrication.

15.3.1.1 Positive Tone Resists Based on Chain Scission

The majority of early positive tone e-beam resists were based on polymer chain scission. Polymer chain scission reduces the molecular weight of polymer chains upon radiation with e-beam exposure, which increases the solubility of the polymer in organic solvents. The difference in dissolution rate between exposed and unexposed polymers provides the dissolution contrast necessary for imaging. The exposure sensitivity for chain scission resists is quantified by a material parameter, $G(s)$, that describes the number of scissions per unit (100 eV) absorbed dose. Typical scission efficiencies $G(s)$ range from 1.3 for low sensitivity resists like poly(methyl methacrylate) or PMMA to 10 for high sensitivity polymers like poly(butene-1-sulfone) or PBS [16].

One of the earliest positive-tone, scission-based resists is poly(methyl methacrylate), as shown in Figure 15.2. Polymer chain scission appears to be initiated by radiolysis of the main-chain carbon to carbonyl bond [17,18]. The resulting tertiary radical rapidly rearranges to cleave the main chain and form volatile products. The main-chain scissions reduce the molecular weight of the polymer and thereby increase its dissolution rate in an organic developer like methyl-isobutyl ketone: isopropanol (1:3 ratio). The dissolution

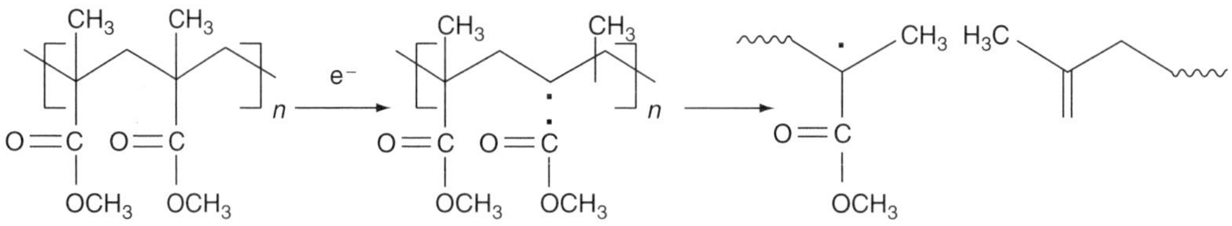

FIGURE 15.2
Poly(methyl methacrylate) undergoes a scission reaction upon e-beam exposure.

properties of PMMA as a function of exposed energy and polymer molecular weight have been studied in effort to increase the resist contrast [19,20].

Researchers have worked since the 1960s to make PMMA analogs with improved sensitivity and lithographic performance. A "terpolymer" analog, consisting of methyl methacrylate, methacrylic acid, and methacrylic anhydride, was found to improve the scission efficiency $G(s)$ from around 1.3 to 4.5 using 10 µC/cm^2 at 10 kV [16]. Furthermore, highly electron-withdrawing groups, such as halogens, have been introduced at the alpha position of the acrylate moiety to aid in the stabilization of main-chain radical that results from radiolysis. EBR-9 made by Toray is one example of a resist developed from halogenated methacrylate homo- and copolymers. Even though PMMA resists have demonstrated high resolution, down to 10 nm using high voltage tools [21], they still have had difficulties meeting the 1–2 µC/cm^2 exposure dose requirements needed for production worthy throughput at 10 kV.

These stringent dose requirements were finally achieved in positive resists based on PBS [22]. PBS has very high sensitivity [$G(s)$ of 10], enabling it to image features under 1 µC/cm^2 at 10 kV. PBS is an alternating copolymer of sulfur dioxide and 1-butene that undergoes main-chain scission between the carbon and sulfur bond as shown in Figure 15.3. The temperature-dependent reaction produces sulfur dioxide and organic by-products [23]. An organic solvent mixture consisting of 30% methyl-propyl ketone and 70% methyl-isoamyl ketone is commonly used to develop the PBS resist. The dissolution rate of PBS is surprisingly humidity dependent.

PBS has been one of the most widely used commercial e-beam resists between 1980s and mid-1990s. PBS has demonstrated good exposure latitude and linearity down to around 500 nm [24,25]. In 1996, PBS coated mask blanks represented 52% of all those shipped from Hoya. Unfortunately, both PBS and PMMA have poor resistance to the halogen-based dry etch process used to etch chromium. Hence, they are limited to wet etch applications that include a significant intrinsic bias. This bias results from undercutting and has driven the need for resists with better dry etch resistance.

In the 1980s, two-component e-beam resists were developed that combined the high sensitivity of the poly(sulfone) with the high etch resistance of novolak polymers. The

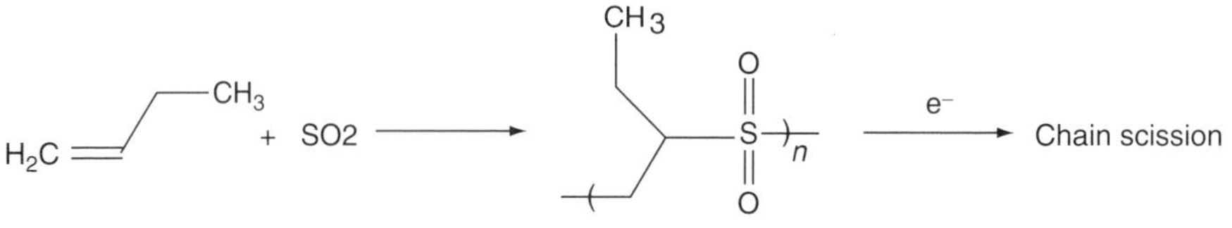

FIGURE 15.3
PBS undergoes a scission reaction upon e-beam exposure.

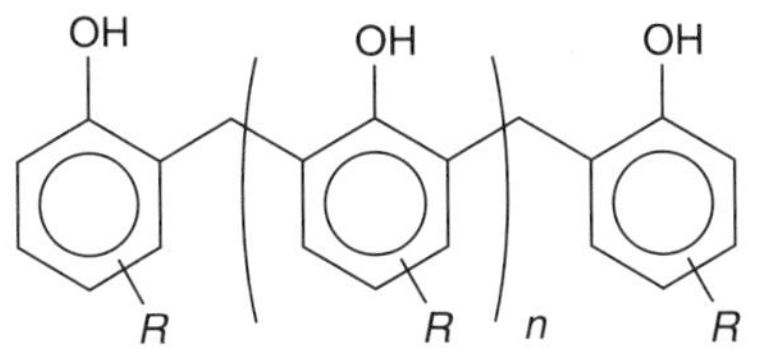

Resists for Mask Making

FIGURE 15.4
Novolak polymer contains cyclic rings that provide higher dry etch resistance. The sensitizer is based on PMPS.

etch resistance of the novolak polymer is driven by the existence of cyclic rings that have a high carbon to hydrogen ratio. Bell Laboratories developed a resist called NPR, which consists of phase-compatible blends of a novolak copolymer and poly(2-methyl-1-pentene) or PMPS [26] as shown by its structure in Figure 15.4. IBM developed a similar platform called the sulfone/novolak system or SNS based on m-cresol novolak and a specialized poly(sulfone) copolymer [27,28]. The inclusion of the phenolic-based novolak polymer enabled the use of aqueous base developer, instead of organic developers and provided excellent etch resistance. The poly(sulfone) copolymer plays a dual role as a dissolution inhibitor and a sensitizer in these systems. The scission of poly(sulfone) produces sulfur dioxide and other volatile organics that no longer inhibit the dissolution of novolak. Even though the use of novolak increases etch resistance, SNS-type resists outgas and require exposure doses that are higher than PBS (5–10 $\mu C/cm^2$).

In the mid-1990s, multi-pass e-beam exposure tools, such as MEBES 4500, were developed to deliver higher total exposure doses providing partial exposure during each pass. This drove the use of resist materials that had better dry etch resistance even though they had lower sensitivity [29]. These multi-pass tools enabled the use of a new positive tone resist based on poly(methyl-chloroacrylate-co-methylstyrene), which provided a balance between sensitivity and etch resistance. Halogenation of the acrylate drives the sensitivity, and the cyclical styrene moiety drives the etch resistance, as shown in Figure 15.5. Dai Nippon developed a commercial resist called ZEP 7000 based on this chemistry that was widely used for photomask fabrication during the 180-nm node. ZEP 7000 requires around 8 $\mu C/cm^2$ at 10 kV and provides dry etch resistance for production photomask fabrication [30].

15.3.1.2 Positive Tone Resists Based on Dissolution Inhibition

Two-component DNQ-novolak resists were widely developed to support optical, projection lithography (435–365 nm wavelengths). These materials found limited use in e-beam applications (AZ-5206) due to low sensitivity, but resists like TOK IP3600 were widely adopted for 363.4-nm laser photomask lithography [31]. DNQ-novolak resists are capable of 500-nm resolution for production reticles and have demonstrated resolution down to 300 nm using high contrast resists [15].

DNQ is a PAC that inhibits the dissolution of novolak polymer resin. Exposure converts the DNQ chromophore compound from a base insoluble compound to a base soluble

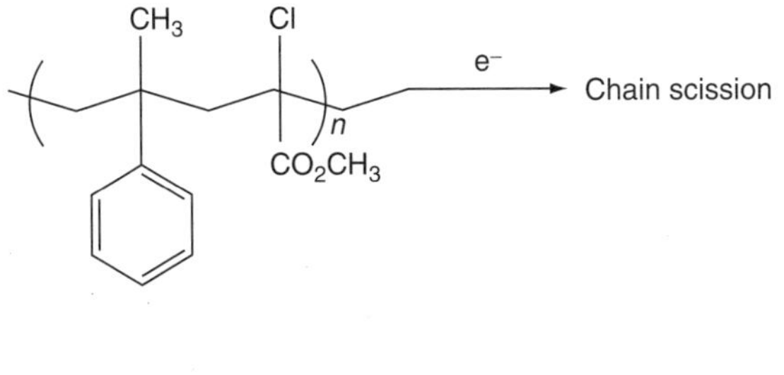

FIGURE 15.5
ZEP 7000 commercialized by Dai Nippon is based on poly(methyl-chloroacrylate-co-methylstyrene), which provides a balance between sensitivity and etch resistance.

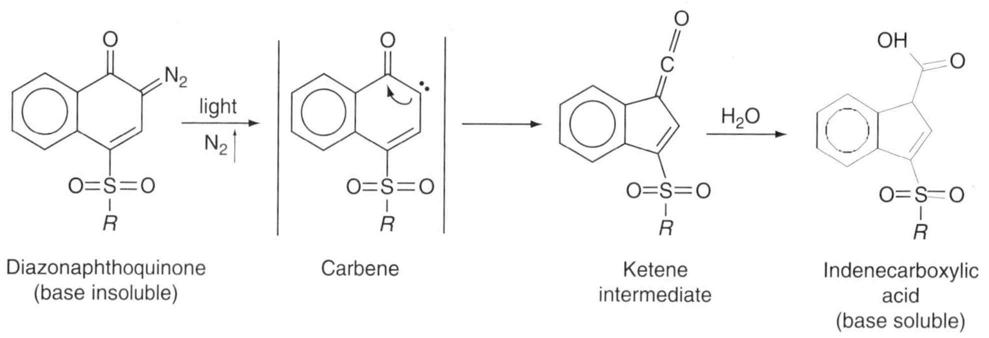

FIGURE 15.6
Photoreaction for a DNQ PAC.

indenecarboxylic acid through the mechanism shown in Figure 15.6 [32]. The carbene is hypothesized to rearrange to form a ketene intermediate, via the Wolff rearrangement. In the presence of water, the ketene forms the base soluble indenecarboxylic acid. The need for water to complete the conversion of DNQ into a base soluble product has led to the use of humidified air in the focus subsystem of production laser writers [33].

Novolak is typically synthesized from a mixture of *meta-* and *para-*cresol, a 35–40% aqueous solution of formaldehyde and an oxalic acid catalyst. The formaldehyde produces methylene linkages in the two *ortho* and one *para* positions on the cresol to form novolak polymer. The efficiency of novolak inhibition has been linked to the interaction of the sulfonate substituents [34] on the DNQ and to the number of *ortho–ortho* bonds in the novolak resin [35]. DNQ is postulated to reduce the probability of deprotonation of hydroxyl groups on novolak due to hydrogen bonding between the sulfonate linkages and hydroxyl groups on the phenolic polymer.

The contrast of DNQ-novolak resists derives from the influence of the unexposed and exposed DNQ molecules on novolak as shown in Figure 15.7. DNQ dramatically decreases the dissolution rate of novolak resin in aqueous base. The carboxylic acid photoproduct of the exposed DNQ actually increases the development rate above the dissolution rate of the pure novolak resin in aqueous base. The changes in the solubility properties switch of DNQ-novolak photoresists with exposure creates a nonlinear development rate response that allows the fabrication of square resist feature cross-sections from a Gaussian-shaped aerial image.

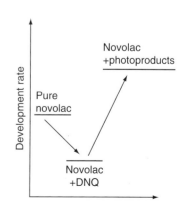

FIGURE 15.7
Dissolution response of DNQ-novolak photoresists with exposure.

Resists for Mask Making 331

PACs have also been developed that bleach upon exposure in the DUV (248–257 nm) and inhibit novolak in a fashion analogous to DNQ. A tri-functional diazopiperidione PAC has been synthesized with sulfonate linkages that enable novolak inhibition and high dissolution contrast upon exposure [36]. A diazopiperidione-novolak resist has produced 500-nm features using a 257-nm laser writer. The resolution of this type of resist was limited by the high absorbance of novolak in the DUV. Research continues on developing polymers that are transparent in the DUV and inhibited by PACs based on diazopiperidione.

15.3.1.3 Negative Tone Based on Cross-Linking

Early negative tone e-beam resists were based on epoxy and styrene-based moieties. A common epoxy-based resist called COP is based on the copolymer of glycidyl methacrylate and ethyl acrylate developed at Bell Laboratories [37]. The epoxy substituent on the COP resist forms inter-chain linkages during a radiation-initiated cross-linking reaction, which makes the resist insoluble. The radiation-initiated reaction propagates through a chain reaction until termination by water or some quencher. The chain reaction propagation leads to high exposure sensitivity. However, reaction propagation continues after exposure has stopped, which generates a CD dependence on the postexposure delay called "dark reaction." COP resists are no longer widely used in production due to resist swelling, poor chromium etch resistance, and dark erosion effects.

Negative e-beam resists have also been developed from polystyrene derivatives, such as poly(chloromethylstyrene). The exposure sensitivity of polystyrene has been increased through the addition of halogen or halomethyl groups in the para-substituted position on the ring as shown in Figure 15.8 [38]. It is proposed that the radiation induces cleavage of the carbon–halogen bond to create a free radical. The free radical induces cross-linking without a propagating chain reaction leading to dark erosion effects. The contrast of styrene-based resists has been increased through the use of high molecular weight polymers at constant dispersity. Even though polystyrene-based resists have demonstrated high contrast and no dark erosion effects, resist swelling due to organic solvent development has restricted production use.

Hydrogen silsesquioxane (HSQ) is an inorganic oxide that has demonstrated utility as a high resolution, negative tone resist. Dow Corning commercialized HSQ as a product called FOx (flowable oxide) primarily for spin-on-glass applications. The chemical structure of HSQ is based on a silicon dioxide network containing reactive SiH bonds on the edges of the molecule as shown in Figure 15.9. Heat [39] and e-beam exposure [40] cleave the SiH bonds enabling the formation of a SiO cross-linked network, which is insoluble in standard TMAH or KOH developers. The solubility difference or dissolution contrast between SiH and cross-linked SiO moieties provides HSQ's function as a negative tone resist.

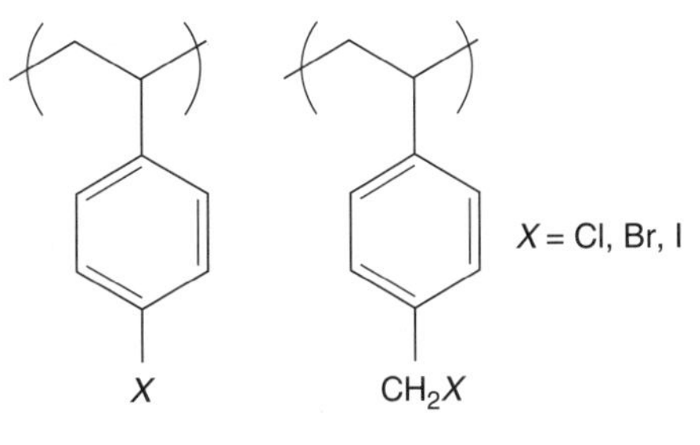

FIGURE 15.8
Halogen *para*-substituted analogs increase the sensitivity of negative polystyrene resists.

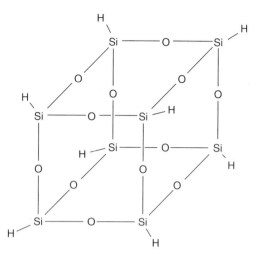

FIGURE 15.9
HSQ is an inorganic oxide that forms an insoluble network or negative tone resist upon exposure (redrawn).

HSQ has demonstrated high resolution and linewidth control without the swelling common to other negative tone, polymeric materials that require organic solvent development. HSQ has been used to print 30-nm structures for imprint lithography masks [41] and 10–30-nm structures for direct write silicon applications [42,43]. In these applications, a thin layer (50–100 nm) of HSQ is coated from methyl isobutyl ketone. HSQ is coated thin due to its high dose requirements in the order of 500–1000 $\mu C/cm^2$ at 50 kV e-beam voltages.

These high dose requirements have driven the use of HSQ in bilayer applications for high voltage (50–100 kV) and low voltage (1 kV) applications [44,45]. Bilayer resist processes consist of a thin imaging layer (HSQ) coated on a thicker transfer layer typically, hard baked novolak resins. The thin top layer is imaged and developed. Then, an O_2 RIE etch is used to transfer the pattern through the underlying novolak-based film. Resist lines have been printed 800 nm high and 40 nm wide (15:1 aspect ratio) using a 140-nm-thick HSQ film coated on 900 nm of novolak resin and exposed at 1000 $\mu C/cm^2$ (50 kV) [44]. The resolution is a tribute to the lithography and the aspect ratio demonstrated by high anisotropic etching.

A similar HSQ bilayer resist scheme has been used for low voltage applications. Typically, new e-beam exposure systems are developed with higher voltages to produce stiffer beams capable of higher resolution. However, multi-beam e-beam systems are developed that utilize low voltages to increase throughput. Low voltages produce much broader e-beam distributions (forward scatter), which have a very thin penetration depth for imaging. This application requires very thin resists. Dense resist lines have been imaged down to around 30 nm using a 24-nm-thick HSQ layer coated on 100- to 400-nm-thick novolak transfer layers (44 $\mu C/cm^2$ at 1 kV) [45]. HSQ is still used only in development, but this inorganic and high etch resistant material has demonstrated high-resolution capabilities.

15.3.2 Chemically Amplified Resists

Increases in circuit density and speed demand higher resolution from both the resists used in device manufacturing and naturally as well from the resists used to print photomasks. Although the standard fourfold reduction employed in convention projection lithography indicates a relaxed resolution-demand for photomask imaging, the low k_1, sub-half wavelength regime the industry is currently operating in requires significant resolution enhancement techniques (RETs), such as optical proximity correction (OPC) and the use of subresolution assist features (SRAFs) [46,47]. The critical dimensions of such assist features are closely approximating those of the device features on the wafer and as such there is ever-increasing demand for very high resolution resists for mask-building applications. While many of the nonamplified resists presented in the previous section can certainly accommodate this high-resolution criterion, they do so only at relatively high doses, thereby extending write times and limiting throughput. Following the trend adopted by device manufacturing since the mid-1980s, mask makers are looking more and more to CARs as a solution to high sensitivity, high resolution photomask imaging.

First proposed in the early 1980s by researchers at IBM [48–50], CARs are based on the radiation-induced generation of a catalytic species, usually strong acid, which brings about multiple chemical transformations that change the solubility of the polymeric matrix of the resist in developer, usually standard aqueous TMAH solution. CARs are available in both positive tone and negative tone versions, and have been developed for use with a variety of imaging wavelengths (257, 248, 193, and 157 nm), as well as with ionizing radiation sources, such as e-beams, x-rays, or soft x-rays (extreme ultraviolet or EUV). However, while there are ample material selections available for auditioning for mask making with laser or e-beam tools, resist companies offer relatively few formulations specifically designed for this application, as the low volume-demand has made it prohibitively expensive to devote substantial research investments towards this endeavor. Nonetheless, a variety of high performance CAR formulations have emerged from adaptations of optical resists that are extending the complexity of masks, which in turn allow for very low k_1 lithography with the state-of-the-art, high NA steppers.

15.3.2.1 Positive Tone Chemically Amplified Resists

Device manufacturing has adopted the use of CARs since the 1980s, as these materials afford the resolution and sensitivity requirements needed for high throughput scaling of microelectronic features. The majority of the commercially available CARs are positive tone in nature. Recently, the mask making industry has focused on the use of these materials. The basis of a positive tone resists is that of the radiation-induced deprotection of an aqueous base soluble that is partially blocked with a labile protecting group. This deprotection induces the solubility of the exposed regions of the resist in developer. The first CARs reported were based on protected poly(p-hydroxystyrene) and were designed to be used with the 248-nm DUV radiation provided by mercury arc lamps or the KrF laser [48–50]. Subsequently, positive tone CARs have been developed for 193-nm (ArF) and 157-nm (F$_2$) radiations as well. For applications with 257-nm lasers, the resists designed for KrF lithography often provide similar performance at this slightly longer wavelength. Interestingly, these are the resists that have also proven to be most suitable for e-beam exposure as well. While the similarity in wavelengths makes the utility with 257-nm lasers somewhat predictable, the usefulness of 248-nm CARs with EBL appears to be largely attributable to the ability of the PHS matrix to provide an efficient source for matrix sensitization while the corresponding acrylates or cyclic olefins used in 193- and 157-nm positive tone CARs undergo a combination of chain scission and recombination, rather than deprotection, and result in difficult to control and/or insensitive materials. What follows is a brief survey of the positive CAR formulations that have been specifically adapted for applications in mask making with either laser or e-beam mask writers.

There are two general classes of positive tone CARs, those that require a PEB to accelerate deprotection and those that do not require PEB. These two classes will be designated as high-activation and low-activation energy resists, respectively, although these classifications are actually an oversimplification as some resists that require PEB are more correctly defined as medium or even low activation energy. The majority of commercially available resist fall into the prior category and are derivatives of materials originally designed for microelectronics patterning with 248-nm photolithography.

Among these are the UV-series available from Shipley Company. These materials, such UV-II HS, UV-6, and UV-110 have been shown to be useful for both e-beam and 257-nm laser mask writing. These systems are based on the environmentally stable chemically amplified photoresist (ESCAP) platform originally described by Ito from IBM [51]. This family or resists comprises a copolymer of hydroxy(styrene) and *t*-butyl acrylate, which results in very high contrast between exposed and unexposed areas, due to the dramatic difference in dissolution rates between the *t*-butyl ester and the corresponding deprotected carboxylic acid. These resists offer, as the name indicates, a significant improvement in stability to airborne contaminants that can lead to postexposure delay effects. This results from their ability to be processed such that they take advantage of the "annealing effect" to minimize the efficacy of amines or other contaminants from penetrating the film surface. One reported example of the use of UV-5 for e-beam radiation demonstrated dense 120-nm features with a 50-keV mask writer [52].

Conversely, some CAR formulations have been designed specifically for e-beam applications. Some of these materials have garnered considerable attention for photomask manufacturing. Among the notable examples in this category are the FEP materials developed by Fuji-Arch, the REAP series from TOK, and the DX family of resists from Clariant. For example, recently the high level of performance of FEP-171 has been reported for 90-nm reticle production with e-beam exposure [53], while process optimization of Clariant's DX1100P was recently reported with DUV laser writing [54].

In the category of the PEB-free resists, one well-documented example is the KRS family of materials developed by IBM. These materials are based on partially protected PHS (ketal blocking groups). They have been reported to offer high resolution coupled with enhanced cross-plate dimensional uniformity attributable to the independence of baking inhomogeneities encountered during processing of quartz mask blanks [54,55]. Furthermore, derivative formulations of KRS materials have developed that offer enhanced throughput via higher sensitivity by formulation optimization [56].

15.3.2.2 Negative Tone Chemically Amplified Resists

Like their NCA analogs, most negative tone CARs are based on cross-linking. The difference is in the underlying mechanism by which this cross-linking occurs and, in turn, the performance attributes that result from this mechanism. Being CARs, these materials are also based on radiation-induced formation of acid and a subsequent acid catalyzed chemical reaction that modifies the dissolution characteristics in the exposed areas. Thus, negative tone CARs offer high sensitivity akin to the positive tone. In these systems, the inherent unexposed polymer film is soluble in aqueous base, and the acid catalyzed reaction enables covalent bond formation between polymer chains and multifunctional cross-linking agents that are formulated with the polymer.

Again, similar to the positive tone CARs, resists first developed for 248-nm photolithography have also found acceptance as e-beam or 257-nm laser resists for mask making. The most widely reported are UVN-30 material from Shipley and NEB Series from Sumitomo. These materials are proprietary formulations based on acid catalyzed cross-linking of poly(*p*-hydroxystyrene) and as such provide sufficient etch resistance to chlorine etching. For example, both high resolution and sensitivity have been reported with NEB-22 by researchers at Motorola and Cornell [57,58]. They offer a high performance alternative to positive tone resists that is particularly attractive for masks with large clear areas, as the write times can be significantly lowered when using this tone of resist. These materials are useful in the manufacturing of complimentary mask sets, where a positive tone resist would otherwise be required to generate a bright-field mask, leading to prolonged write times.

15.4 Resist Implementation Challenges

15.4.1 Resist–Blank Interactions

15.4.1.1 Nature of Photomask Blanks

The composition of a photomask blank plays a role in the resist and etch pattern fidelity. Photomask blanks are made of quartz coated with an opaque film. The opaque film is a heterogeneous layer consisting of chromium oxide, chromium, carbon, and nitrogen [12]. Chromium oxide makes the substrate antireflective to reduce flare in scanners and to partially reduce reflections for mask laser writers. Carbon was originally introduced to control or slow down wet etch rates. However, the recent transition to dry etch processes warrants a reexamination of the concentration of carbon needed for etch control. Nitrogen is used as a carrier gas during the chromium deposition process. The concentration of each component in the opaque film impacts the chemical and physical interactions between a resist and mask blank. Thus, the antireflection layers play an important role in mask patterning.

15.4.1.2 Resist Footing

One of the interactions between the mask blank and resist is resist "footing." Resist footing is commonly attributed to a combination of chemical effects [59] and exposure effects like optical interference from reflections with the substrate [12]. Chemically induced resist footing has increased with the use of CARs. It has been hypothesized that the nitride in the chromium oxynitride film creates a reaction pathway for amine formation at the substrate surface. Amines at the substrate-resist interface can neutralize the acid generated from the exposure of the CAR. Neutralization of the acid prevents polymer deprotection and hence dissolution thereby producing a resist foot in a positive resist. Others have suggested that the high surface energy of the chromium oxynitride induces inhomogeneous dissolution behavior that manifests itself as a footing after development [56].

Resist footing in CARs has been observed on silicon nitride substrates in integrated circuit (IC) manufacturing [60,61]. The IC industry has reduced substrate-based resist footing through the use of organic BARCs and oxide cap deposition. Both methods are designed to provide a barrier between the substrate and resist. The mask industry has demonstrated that a barrier approach using an organic BARC reduced chemical footing on a photomask substrate using e-beam lithography [62]. The implementation of an organic BARC is really gated by the ability to coat defect-free films. The use of an oxide cap on a photomask is challenging, since it will impact the reflectivity properties (flare) of the mask substrate. The impacts of substrate components like nitrogen, as well as of barrier layers, are still evaluated to reduce resist footing. The mask industry also continues to evaluate new CARs that are less sensitive to substrate and environmental amine contamination.

15.4.2 Resist Outgassing

The outgassing of organic materials during the exposure process has come under a great deal of scrutiny in recent years, as volatile by-products can potentially deposit on columns of e-beam systems or coat the optical elements of a laser mask writer. This phenomenon is of particular concern with positive tone CARs, as these systems involve the radiation-induced degradation of photoacid generators and the subsequent deprotection of acid labile groups — two reactions are known to form volatile species, and in some cases, highly reactive species. A variety of studies have been conducted, most notably at MIT-Lincoln Labs [63], which show that in most cases the primary source of detected outgassed species generated during exposure of resists is attributable to the PAG decomposition. Additionally, it has been demonstrated that through careful selection of resist components, including the deprotection groups of the polymeric matrix, resist outgassing can be contained in some cases below the detection limits of the analysis. Among current materials research in this area, activities at the University of Texas are focused on the development of low outgassing systems, such as mass-persistent materials, that are not designed to form volatile species upon deprotection.

15.4.3 Mask Shop Issues

The successful integration of a resist process into mask manufacturing requires adequate blank storage, as well as uniform coating and baking conditions, to provide high linewidth control and low defectivity. Traditionally, resists have been precoated on blanks by a supplier and shipped to mask manufactures for patterning. NCA resists have been widely precoated on blanks due to their long postcoat stability. However, CARs have been more challenging to precoat and store on blanks due to their high sensitivity to environmental contaminants, such as amines.

Mask blank vendors have determined that nitrogen purged storage of precoated CARs is required to meet 10-nm linewidth control specifications over a 4-week period [59]. The most stable CARs show at least 20-nm linewidth changes over a month stored in common packaging without nitrogen purging. The use of precoated resist blanks using CARs will require thorough studies on postcoat delay stability and blank storage protocols to consistently meet stringent linewidth control requirements.

The integration of CARs has challenged mask makers to consider coating on demand. Coating-on-demand processes require resist thickness and bake temperature uniformity control. Uniform resist thickness is achieved through a balance of exhaust flow, spin speed, and air flow control during the rotation of the blank. Bake temperature uniformity is difficult to achieve across the surface of a thick quartz blank, since quartz has poor heat conductive properties. Multizone hot plates have been developed to provide more uniform bake temperatures during the ramp and equilibrium periods of the bake process. The uniformity of the postapplication and postexposure bakes can be tested on unexposed resist films using dark erosion techniques [64]. The bake uniformity is measured indirectly through the resist thickness uniformity after a long develop process. The resist remaining after development is related to the residual casting solvent concentration left in the resist film after the bake.

The integration of CARs challenges mask makers to use more integrated tool sets like cluster tools, as well as SMIF pods, for blank storage and processing. This high level of integration is needed to reduce defectivity and to enable linewidth scaling demonstrated by e-beam and laser beam mask lithography.

Development of advanced resists for mask making represents a serious dilemma for the semiconductor industry. Mask making is an ever-increasing component to the cost of device manufacturing and as such should demand the focused attention of the resist materials development community. However, the volumes of resists that are sold for mask making are so low that it is not possible to make a business case for the recovery of the development costs. Hence, the resist suppliers cannot and should not expend resources on mask resist development. Consequently, this development activity needs to be subsidized by the end users or it is not likely to occur. We are pleased to report that the research consortia, including the Semiconductor Research Corporation (SRC) and SEMATECH, are now funding small research efforts for these activities.

References

1. F. Abboud, K. Baik, V. Chakarian, D.M. Cole, R.L. Dean, M.A. Gesley, H. Gillman, W.C. Moore, M. Mueller, R. Naber, T.H. Newman, R. Puri, F. Raymond, and M. Rougieri, *Proc. SPIE*, 4754, 704–715 (2002).
2. Y. Hattori, M. Kiyoshi, A. Ken-ichi, Y. Takayuki, U. Satoshi, M. Taiga, N. Eiji, N. Shimomura, T. Yamashita, N. Yamada, A. Sakai, H. Honda, T. Shimoyama, K. Nakaso, H. Inoue, Y. Onimaru, K. Makiyama, Y. Ogawa, and T. Takigawa, *Proc. SPIE*, 4754, 696–703 (2002).
3. H. Takemura, H. Ohki, and M. Isobe, *Proc. SPIE*, 4754, 689–695 (2002).
4. T.R. Groves, *J. Vac. Sci. Technol. B.*, 14, 3839–3844 (1996).
5. S. Babin, *J. Vac. Sci. Technol. B.*, 21 (1), 135–140 (2003).
6. S. Babin, *J. Vac. Sci. Technol. B.*, 15 (6), 2209–2213 (1997).
7. C.H. Hamaker and P.D. Buck, *Proc. SPIE*, 3236, 42–54 (1997).
8. C.G. Morgante and C.H. Hamaker, *Proc. SPIE*, 4066, 613–623 (2000).
9. M. Bohan, C.H. Hamaker, and W. Montgomery, *Proc. SPIE*, 4562, 16–37 (2002).
10. B.M. Rathsack, P.I. Tattersall, C.E. Tabery, K. Lou, T.B. Stachowiak, D.R. Medeiros, J.A. Albelo, P.Y. Pirogovsky, D.R. McKean, and C.G. Willson, *Proc. SPIE*, 4345, 543–556 (2001).
11. S.E. Fuller, W. Montgomery, J.A. Albelo, W. Rodrigues, and A.H. Buxbaum, *Proc. SPIE*, 4409, 306–311 (2001).
12. B.M. Rathsack, C.E. Tabery, J.A. Albelo, P.D. Buck, and C.G. Willson, *Proc. SPIE*, 4186, 578–588 (2000).
13. T. Sandstrom, T.I. Fillion, U.B. Ljungblad, and M. Rosling, *Proc. SPIE*, 4409, 270–276 (2001).
14. T. Sandstrom and N. Eriksson, *Proc. SPIE*, 4889, 157–167 (2002).
15. B.M. Rathsack, C.E. Tabery, S.A. Scheer, C.L. Henderson, M. Pochkowski, C. Philbin, P.D. Buck, and C.G. Willson, *Proc. SPIE*, 3678, 1215–1226 (1999).
16. C.G. Willson, in: *Introduction to Microlithography*, Second Edition, American Chemical Society, Washington, D.C., 1994 (Chapter 3).
17. W.M. Moreau, *Semiconductor Lithography Principles, Practices and Materials*, First Edition, Plenum Press, New York, 1988.
18. H. Hiroaka, *Macromolecules*, 9, 359 (1976).
19. A. Uhl, J. Bendig, J. Leistner, U. Jagdhold, L. Bauch, and M. Bottcher, *J. Vac. Sci. Technol. B.*, 16 (6), 2968–2973 (1998).
20. D.G. Hasko, S. Yasin, and A. Mumtaz, *J. Vac. Sci. Technol. B.*, 18 (6), 3441–3444 (2000).
21. W. Chen and H. Ahmed, *J. Vac. Sci. Technol. B.*, 11 (6), 2519–2523 (1993).
22. M.F. Bowden, L.F. Thompson, and J.P. Ballantyne, *J. Vac. Sci. Technol.*, 12 (6), 1294–1296 (1975).
23. J. Brown and J. O'Donnell, *Polymer*, 22, 71 (1981).
24. H. Kobayashi, T. Higuchi, K. Asakawa, and Y. Yokoya, *Proc. SPIE*, 3236, 498–510 (1997).
25. W.P. Shen, J. Marra, and D.V.D. Broeke, *Proc. SPIE*, 2884, 48–67 (1996).
26. M. Bowden, L. Thompson, S. Farenholtz, and F. Doerries, *J. Electrochem. Soc.*, 128, 1304 (1981).
27. Y.Y. Cheng, B.D. Grant, L.A. Pederson, and C.G. Willson, International Business Machines Corporation, U.S. Patent 4,398,001, 1983.

28. D.R. Medeiros, A. Aviram, C.R. Guarnieri, W.S. Huang, R. Kwong, C.K. Magg, A.P. Mahorowala, W.M. Moreau, K.E. Petrillo, and M. Angelopoulos, *IBM J. Res. Dev.*, 45 (5), 639–650 (2001).
29. F. Abboud, R. Dean, J. Doering, W. Eckes, M. Gesley, U. Hofmann, T. Mulera, R. Naber, M. Pastor, W. Phillips, J. Raphael, R. Raymond, and C. Sauer, *Proc. SPIE*, 3096, 116 (1997).
30. C. Constantine, D.J. Johnson, R.J. Westerman, T. Coleman, T. Faure, and L. Dubuque, *Proc. SPIE*, 3236, 94–103 (1997).
31. P.D. Buck, A.H. Buxbaum, T.P. Coleman, and L. Tran, *Proc. SPIE*, 3412, 67–78 (1998).
32. R. Dammel, *Diazonaphthoquinone-based Resists*, vol. TT 11, SPIE Optical Engineering Press, 1993.
33. C.H. Hamaker, G.E. Valetin, J. Martyniuk, B.G. Martinez, M. Pochkowski, and L.D. Hodgson, *Proc. SPIE*, 3873, 49–63 (1999).
34. K. Uenishi, Y. Kawabe, T. Kokubo, S. Slater, and A. Blakeney, *Proc. SPIE*, 1466, 102–116 (1991).
35. C.L. McAdams, L.W. Flanagin, C.L. Henderson, A.R. Pawloski, P. Tsiartas, and C.G. Willson, *Proc. SPIE*, 3333, 1171–1179 (1998).
36. B.M. Rathsack, Photoresist Modeling for 365 nm and 257 nm Laser Photomask Lithography and Multi-analyte Biosensors Indexed through Shape Recognition, Ph.D. dissertation, 2001.
37. C.G. Willson, in: *Introduction to Microlithography*, Second Edition, American Chemical Society, Washington, D.C., 1994 (Chapter 3).
38. Y. Tabata, S. Tagawa, and M. Washio, in: L.F. Thompson, C.G. Willson, and J.M.J. Frechet (Eds.), *Materials for Microlithography*, ACS Symposium Series 266, American Chemical Society, Washington, D.C., 1984, 151–163.
39. Y.K. Siew, G. Sarkar, X. Hu, J. Hui, A. See, and C.T. Chua, *J. Electrochem. Soc.*, 147, 335–339 (2000).
40. H. Namatsu, Y. Takahashi, K. Yamazaki, T. Yamaguchi, M. Nagase, and K. Kurihara, *J. Vac. Sci. Technol. B.*, 16 (1), 69–76 (1998).
41. D.P. Mancini, K.A. Gehoski, E. Ainley, K.J. Nordquist, D.J. Resnick, T.C. Bailey, S.V. Sreenivasan, J.G. Ekerdt, and C.G. Willson, *J. Vac. Sci. Technol. B.*, 20 (6), 2896–2901 (2002).
42. H. Namatsu, Y. Watanabe, K. Yamazaki, T. Yamaguchi, M. Nagase, Y. Ono, A. Fujiwara, and S. Horiguchi, *J. Vac. Sci. Technol. B.*, 21 (1), 1–5 (2003).
43. Falco C.M.J.M. van Delft, *J. Vac. Sci. Technol. B.*, 20 (6), 2932–2936 (2002).
44. Falco C.M.J.M. van Delft, J.P. Weterings, A.K. van Langen-Suurling, and H. Romijn, *J. Vac. Sci. Technol. B.*, 18 (6), 3419–3423 (2000).
45. A. Jamieson, C.G. Willson, Y. Hsu, and A. Brodie, *Proc. SPIE*, 4690, 1171–1179 (2002).
46. J.F. Chen, T.L. Laidig, K.E. Wampler, R.F. Caldwell, A.R. Naderi, and D.J. Van Den Broeke, *Proc. SPIE*, 3236, 382–396 (1997).
47. W. Maurer and C. Freidrich, *Proc. SPIE*, 3546, 232–241 (1998).
48. H. Ito, C.G. Willson, and J.M.J. Frechet, Digest of Technical Papers of 1982 Symposium on VLSI Technology, 1982, pp. 86–87.
49. H. Ito and C.G. Willson, Technical Papers of SPE Regional Technical Conference on Photopolymers, 1982, pp. 331–353.
50. H. Ito and C.G. Willson, US Patent 4,491,628, 1985.
51. W. Conley, B. Brunsvold, F. Buehrer, R. Dellaguardia, D. Dobuzinsky, T. Farrell, H. Ho, A. Katnani, R. Keller, J. Marsh, P. Muller, R. Nunes, H. Ng, J. Oberschmidt, M. Pike, D. Ryan, T. Cottler-Wagner, R. Schulz, H. Ito, D. Hofer, G. Breyta, D. Fenzel-Alexander, G. Wallraff, J. Opitz, J. Thackeray, G. Barclay, J. Cameron, T. Lindsay, M. Cronin, M. Moynihan, S. Nour, J. Georger, M. Mori, P. Hagerty, R. Sinta, and T. Zydowsky, *Proc. SPIE*, 3049, 282–299 (1997).
52. C.M. Falco, J.M. van Delft, and F.G. Holthuysen, *Microelectron. Eng.*, 46, 383 (1999).
53. J. Butschke, D. Beyer, C. Constantine, P. Dress, P. Hudek, M. Irnscher, C. Koepernik, C. Krauss, J. Plumhoff, and P. Voehringer, *Proc. SPIE*, 5256, 334–354 (2003).
54. H.A. Fosshaug, A. Bajramovic, J. Karlsson, K. Xing, A. Rosendahl, A. Dahlberg, C. Bjoernberg, M. Bjuggren, and T. Sandstrom, *Proc. SPIE*, 5256, 355–365 (2003).
55. W.-S. Huang, R.W. Kwong, W.M. Moreau, M. Chace, K.Y. Lee, C.K. Hu, D. Medeiros, and M. Angelopoulos, *Proc. SPIE*, 3678, 1052–1058 (1999).
56. D.R. Medeiros, *J. Photopolym. Sci. Technol.*, 15, 411–416 (2002).
57. D.R. Medeiros, K.E. Petrillo, J. Bucchignano, M. Angelopoulos, W.-S. Huang, W. Li, W. M. Moreau, R. Lang, R.W. Kwong, C. Magg, and B. Ashe, *Proc. SPIE*, 4562, 552–560 (2002).

58. E. Ainley, K. Nordquist, D.J. Resnick, D.W. Carr, and R.C. Tiberio, *Microelectron. Eng.*, 46, 375–378 (1999).
59. M. Hashimoto, H. Kobayashi, and Y. Yokoya, *Proc. SPIE*, 4186, 561–577 (2000).
60. M. Mori, T. Watanabe, K. Adachi, T. Fukushima, K. Uda, and Y. Sato, *Proc. SPIE*, 2724, 131–138 (1996).
61. J. Chun, C. Bok, and K. Baik, *Proc. SPIE*, 2724, 92–99 (1996).
62. M. Hashimoto, F. Ohta, Y. Yokoya, and H. Kobayashi, *Proc. SPIE*, 4409, 312–323 (2001).
63. R.R. Kunz and D.K. Downs, *J. Vac. Sci. Technol. B.*, 17, 3330 (1999).
64. M. Hashimoto, F. Ohta, Y. Yokoya, and H. Kobayashi, *Proc. SPIE*, 4562, 682–693 (2002).

16

Resist Charging and Heating

Min Bai, Dachen Chu, and Fabian Pease*

CONTENTS
16.1 Introduction 341
16.2 Electron Beam and Solid Interactions 343
    16.2.1 Energy and Charge Deposition 343
    16.2.2 Secondary Electron Emission 345
16.3 Resist Charging 345
    16.3.1 The Dynamic Charging Model 345
    16.3.2 Deflection of the Incoming Beam 347
    16.3.3 Experimental Characterization of Charging Effects 349
16.4 Resist Heating 357
    16.4.1 Heating of Resist during the Electron Beam Exposure 357
        16.4.1.1 Temperature Simulations 357
        16.4.1.2 Experimental Verification 359
    16.4.2 The Impact of the Localized Heating on Resists 361
    16.4.3 The Effect of Writing Strategies on Resist Heating 361
    16.4.4 The Impact of Resist Heating on CD Variation 363
    16.4.5 Strategies to Mitigate Resist-Heating Effect 363
16.5 Summary 364
References 364

16.1 Introduction

Almost since 1975 when electron beam lithography (EBL) was introduced for the manufacture of photomasks, resist charging has been attributed to pattern placement errors. Several experiments were reported in which patterns were laid down in a particular order, and the resulting displacements were ascribed to charging of the resist in previously exposed regions causing undesired displacement of the incoming beam. In the early 1990s, the surface potential of just-exposed resists was directly measured; the results indicated that for the conditions of EBL for mask making, the effect was not only much smaller than could cause measurable deflection but also the sign of the surface potential was positive in many cases. Because the results were unexpected and the exposure of the

---

* Fabian Pease contributed the "Introduction".

resist was carried out with a flood rather than with a focused beam, these experiments were not regarded as conclusive. In the late 1990s, the ETEC Corporation and the Semiconductor Research Corporation (SRC) sponsored a project to revisit the charging issue to settle the matter. Shortly thereafter Intel and the SRC sponsored a project into the heating of resist during EBL in mask manufacture. Members of the staff at ETEC had developed some data on the relationship between temperature rise, sensitivity, and displacement of the feature edge. Initially the project was restricted to measuring the thermal properties, particularly thermal conductivity, of the polymeric resist films because a simulator already existed and all that was lacking were good values for thermal conductivity and specific heat. Later the project widened in scope to include direct measurement of temperature rise.

The two students who carried out the work were Min Bai, on the charging problem, and Dachen Chu, on the heating and are the authors of this chapter. They received advice from a wide variety of people and reported their results each quarter to the Mask Advisory Group set up in 1998 by Mark McCord and Neil Berglund and made up of industrial representatives to advise Stanford University Microstructures Group on research related to photomask engineering.

A notable asset was the loan of a Kelvin probe from ETEC for directly measuring the surface potential of resist that had just been exposed with a focused pencil beam; this eliminated the only valid objection to the earlier measurements of Ingino et al. [10]. Within a year Min Bai, with some help from Dan Pickard and Corina Tanasa, had verified the earlier experimental results, i.e., in mask making the surface potential never reaches a value large enough to cause significant displacement of the incoming beam and is often positive. She developed a model to explain this behavior and demonstrated that the beam displacement is indeed negligible unless the resist is so thick that the exposing beam fails to generate a region of induced conductivity that reaches through the thickness of the film. She then carried out a series of experiments to characterize the electron beam (e-beam) induced conductivity for a variety of insulating films and showed that even those films with the lowest induced conductivity would not charge sufficiently to cause appreciable displacement of the incoming beam provided that the film thickness is no greater than the maximum range (normal to the surface) of the incident electrons.

The first phase of the thermal project was to determine the thermal conductivity of polymeric resists in collaboration with Professor Ken Goodson (Department of Mechanical Engineering, Stanford University), who had developed a laser reflectance technique. By measuring the rate of decay of reflectance of a metal film on top of the resist film it was possible to determine the out-of-plane conductivity. To measure the in-plane conductivity, a different technique was employed, also in collaboration with Professor Goodson. One thin film electrode was used to generate heat, and the electrical resistances of nearby sensing electrodes were measured to determine their temperatures. The in-plane and out-of-plane thermal conductivities were indistinguishable for the polymeric resists tested. This was in contrast to earlier experiments on polyimide films, in which the in-plane thermal conductivity could be as much as three times that of out-of-plane. However, it was felt that the relevant measure of thermal conductivity was that during exposure with an e-beam because the free carriers generated might contribute significantly to the total thermal conductivity. So a new set of experiments was designed, in which micron- and submicron-sized thin film thermocouple (TFTC) junctions were used to measure directly the temperature rise during exposure. This work resulted in thermocouple junctions that were, as far as we know, smaller and faster than any others.

In parallel with the experimental program, a modified Green's function approach was developed and brought on line to simulate resist heating. Using the values of thermal conductivity determined in the original set of experiments, the spatial and transient responses of resist temperature to a given beam exposure could rapidly be determined. The thermocouple experiments verified the simulations. Thus, the free electrons generated by electron bombardment appear to have little contribution to total thermal conductivity. This conclusion is also in agreement with calculations based on Min Bai's measurements of e-beam induced conductivity (EBIC).

The modeling was then extended to determine the distribution of temperature rise as a result of writing with different exposure strategies. Depending on the exposure dosage and resist thickness this might be several tens of degree Celsius. How far this affects the placement of the edge of the developed feature depends on the resist material, but CD errors of 0.1 nm/°C have been quoted.

16.2 Electron Beam and Solid Interactions

16.2.1 Energy and Charge Deposition

As the electrons enter the solid, they are scattered by the positively charged nuclei and by the negatively charged atomic electrons. When the primary electron is scattered by a nucleus, the exchange of energy is essentially zero; this is called an elastic scattering event. The primary electron could also ionize the atom by exciting an atomic electron and produce x-ray or Auger electrons, it could collide with a valence electron to generate secondary electron (SE), or it could also interact with the solid lattice and generate phonons. These are all examples of inelastic scattering, which will change both the direction of travel and energy of the primary electron. A sequence of the scattering events continues until the primary electron loses all its energy and becomes thermalized inside the solid or at some point escapes from the sample. Some of the atomic electrons excited by inelastic scattering of the primary electrons have sufficient momentum that they too will contribute to the energy dissipated as they travel through the target material.

As a direct result of the scattering events, both charge and energy are deposited in the exposed region. The Monte Carlo method has been the most popular approach for simulating such interactions between the e-beam and solid targets. By assigning probabilities to the parameters defining the scattering events, we can simulate a large number of electron trajectories, as well as the charge and energy deposition profile in the specimen. The Monte Carlo models have been fully described by Joy [1]. An example of the simulation is shown in Figure 16.1 and Figure 16.2.

Two consequences of inelastic scattering in resist are exposure of the resist and the generation of free electron hole pairs. These free charge carriers diffuse randomly in the sample, some recombine with each other shortly after generation, some separate under the influence of external field or space charge, and those reaching the surface with sufficient energy might overcome the potential barrier and escape from the sample as secondary electrons. Inside the resist, there is a distribution of charge caused by the trapped primary electrons and generated free electron hole pairs. At the surface, there is a thin layer of positive charge caused by the secondary electron emission. This surface charging can divert the trajectories of further incoming electrons and cause errors in beam landing position. For electrons with 20 kV or higher accelerating voltages, the majority of the energy is dissipated as heat in both resist and substrate. The temperature rise in resist changes the resist sensitivity and can lead to variations in the patterned feature sizes. In this chapter, we describe both the surface charging and the heating of the resist.

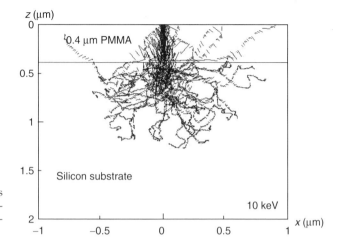

FIGURE 16.1
Monte Carlo simulation of the trajectories of one hundred 10-keV electrons scattered in 400 nm PMMA on the top of silicon substrate.

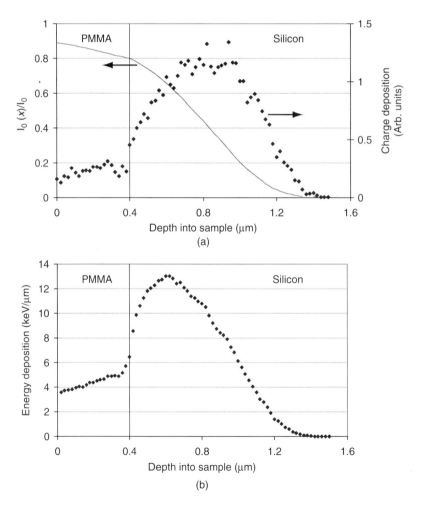

FIGURE 16.2
The Monte Carlo simulation of 10-keV electron scattering in 400 nm PMMA on top of silicon substrate. (a) The charge deposition profile in the specimen and the primary beam current versus depth into the sample (normalized over original primary beam current). (b) The energy deposition profile in the specimen.

16.2.2 Secondary Electron Emission

The process of the secondary electron escaping from the sample surface is called secondary emission. The secondary emission coefficient δ is defined as the ratio of the secondary electron current to the primary beam current. The energy spectrum of all the electrons emitted from the specimen has two main peaks and a number of smaller ones (Figure 16.3). By convention, electrons with energy of 50 eV up to the primary beam energy are classified as backscattered electrons (BSE). Those with energy below 50 eV are classified as secondary electrons [2].

The energy distribution of secondary electrons typically has a peak in the range of 1–5 eV and a half-width ranging from 3 to 15 eV, depending on the material and its surface condition (Figure 16.4) [3]. The secondary electrons are emitted from the shallowest layer on the surface, and the measured mean escape depth is 0.5–2 nm for metals and 10–20 nm for insulators [3]. Because of their low energy, secondary electrons are sensitive to the local spurious field on the specimen surface. The secondary electrons gain different kinetic energies between the surface and the detector if the specimen potential is varied. Positive specimen voltages cause a reduction, while the negative specimen voltages cause an increase in the secondary electron kinetic energy, so that the whole secondary electron spectrum is shifted along the energy axis, which actually sets the fundamental of quantitative voltage measurement by collecting secondary electron currents [4].

16.3 Resist Charging

16.3.1 The Dynamic Charging Model

Surface potential on the specimen after e-beam exposure is the key measure of the charging effect. Ideally, charging induced beam deflection could be predicted from any given spatial distribution of the surface potential. Various theoretical models have been constructed to simulate both the surface potential and the beam deflection.

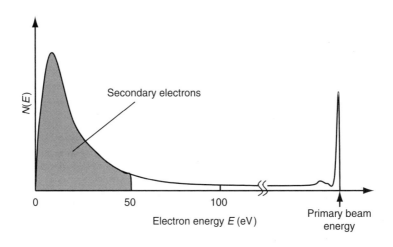

FIGURE 16.3
Typical spectrum of electrons emitted from specimen under e-beam exposure. Secondary electrons are classified as those with energies <50 eV [2].

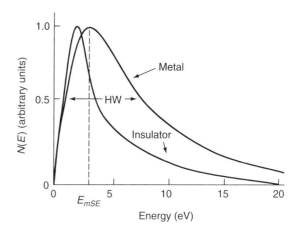

FIGURE 16.4
Relative energy distribution of secondary electrons from metal and insulator [3].

While studying pattern displacement caused by sample charging in EBL, we are most interested in the surface potential after the e-beam has stopped exposing a certain area and moved to adjacent locations on the work piece. However, the charging process is dynamic in nature, and both the space charge distribution and the electric field across the sample evolve with time. Any charging model must take into account the physical interactions between the high-energy electrons and the solid target, as well as the charge transport across the specimen caused by EBIC. Dynamic charging models have been pursued by different authors [5–7]; common among them are several underlying assumptions that define the problem under typical EBL conditions:

1. The front of the specimen is always floating under e-beam irradiation, so it is an open circuit condition.
2. Suppose the net current entering the sample is I_t. According to charge conservation and continuity of current, the current in the exposed sample should be everywhere equal to I_t.
3. The EBIC renders a locally conductive channel to the originally insulating sample and provides a transport path for the trapped charge.

Therefore, the charging process can be described by the following equation:

$$I_t = I_0(x) + g(x)\, E(x,t) \cdot A + \varepsilon_0 \varepsilon_r \frac{\partial E(x,t)}{\partial t} \cdot A \qquad (16.1)$$

where $I_0(x)$ is the current of the primary beam at depth x, $E(x,t)$ the electric field, ε_r dielectric constant, ε_0 the permittivity of the free space, $g(x)$ the EBIC, and A is the scanned area. The net current I_t received by the sample is the primary beam current less than lost through backscattering (I_{BS}) and secondary emission (I_{SE}). The second component in Equation (16.1) stands for the leakage current caused by EBIC, and the third component is the displacement current due to the change of electric field versus time during transient state. When the sample is initially uncharged at time zero, we can solve the linear different equation with initial condition $E(x,t) = 0$ and get

$$E(x,t) = \frac{I_t - I_0(x)}{g(x) \cdot A}\left(1 - e^{-\frac{g(x)t}{\varepsilon_0 \varepsilon_r}}\right) \qquad (16.2)$$

Resist Charging and Heating

When (as is usually the case) the substrate is grounded we obtain

$$V(t) = \int_0^d E(x,t)\,dx \quad (16.3)$$

where d is the specimen thickness.

This model is one-dimensional (vertical) and so assumes that the extent of the scanned area is much larger than the resist thickness. As we shall see in the following, this assumption is valid when determining the deflection of the incoming beam by the surface charges. The lateral straggling of electrons can thus be neglected. We also assume that the EBIC $g(x)$ is proportional to the local power dissipated [8].

Examples of calculated values of surface potential are shown in Figure 16.5 [9]. The results indicate that the surface potential can be positive or negative, and becomes more negative as the resist thickness increases; this is to be expected because a larger fraction of the incident primary electrons is stopped in thicker resist films. As pointed out earlier, secondary electron emission leads to a positive surface potential.

16.3.2 Deflection of the Incoming Beam

In most EBL tools, the spatial boundary condition for solving electrostatic problems is set by two grounded planes (Figure 16.6): the specimen substrate, either the chrome layer on the mask or the backside of the silicon wafer; and the bottom of the final pole piece. The grounded substrate ensures that charges from the e-beam leak away to give a final condition that is steady state, while the grounded final lens restricts the relevant extent of field distribution caused by charging of the specimen. Under such boundary conditions, an image charge model was proposed by Ingino et al. [10] in 1994 to predict the beam deflection caused by a locally charged area in the exposed sample.

The electric field caused by the surface potential would be equivalent to that caused by a sheet of charge on the surface with charge density:

$$\sigma = \frac{\varepsilon_r \varepsilon_0 V}{d} \quad (16.4)$$

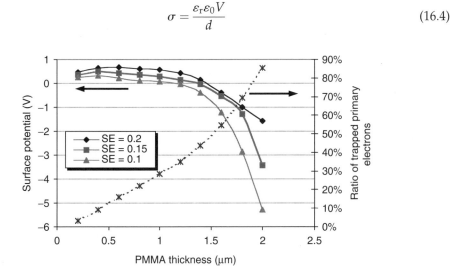

FIGURE 16.5
The simulated surface potential above PMMA resist with different thicknesses on silicon substrate, irradiated by 10-keV e-beam, current density $10^4\,\text{pA}/\text{mm}^2$. The solid lines represent the voltages.

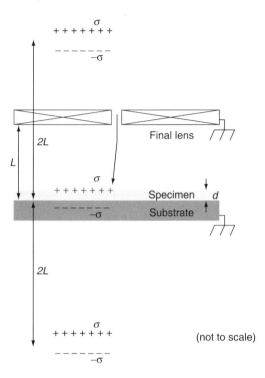

FIGURE 16.6
The image charged model with double ground planes. The sample thickness is d and working distance is L. An infinite number of image dipoles are generated, several of them are shown. The e-beam gets deflected as it comes past the final lens and travels towards the work piece.

where ε_r is the relative dielectric constant of the material and ε_0 is the permittivity of the free space. Due to the existence of two ground planes, a series of images of the surface charge will be formed, so that an infinite number of image dipoles are generated. However, the electric field due to the dipole is proportional to r^{-3}, r is the radial distance between the dipole center and the electron, so the series converges very rapidly. In most cases, the total deflection can be calculated by summing the contributions of only the first-order dipoles.

Figure 16.7 demonstrates a simple case with a charged square distributed symmetric to the X-axis, and the ideal traveling path of the electron is along the Z-axis. The distance between the ideal and the actual landing position is the beam deflection induced by the charge. The trajectory of the electrons can be tracked stepwise as they come out from the grounded pole piece [11]. And Figure 16.8 shows several examples of the simulated beam deflection as a function of the distance from the charged area edge to the ideal beam landing position. With 1 V on a 100 × 100-μm^2 square, a 10-keV beam is deflected by 3.7 nm in the worst case (when the ideal beam landing position is right on the edge of the square). This error increases to about 35 nm when a 1 × 1-mm^2 area is charged, with a 1 × 1-cm^2 pad the beam could be deflected up to 140 nm. As the beam moves farther away from the charged area, the resulting beam displacement falls off almost exponentially.

When the working distance is fixed, the charged area dimension varied, the worst-case deflection of a 10-keV beam caused by 1-V surface potential is shown in Figure 16.9. The beam deflection goes up almost linearly with increasing square size and starts to level off when the dimension becomes comparable to the working distance. If the charged area is fixed but the working distance is varied, the resulting beam deflection does not start to drop off until the working distance becomes comparable to the charged square size (Figure 16.10). This is because the influence of the image charge dipoles does not become significant until the size of the charged area approaches the working distance.

Resist Charging and Heating

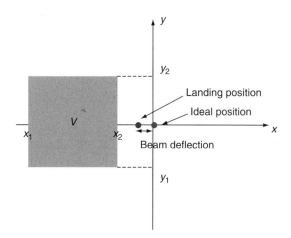

FIGURE 16.7
Top view of the simulated beam deflection configuration. The square is charged to a surface potential V and is symmetric to the X-axis. The distance between the ideal beam position and actual landing position is the beam deflection we try to calculate.

The charging induced beam deflection is proportional to the value of the surface potential while inversely proportional to the primary beam energy. As shown in Figure 16.8, 1 V on the 1 × 1-cm^2 charged pad deflects the 10 keV incoming beam up to 140 nm but only deflects the 50-keV beam up to 28 nm.

16.3.3 Experimental Characterization of Charging Effects

There are a variety of ways to experimentally characterize the pattern displacement induced by sample charging in EBL. However, all the different approaches can be generally classified into two categories, the direct approach and indirect approach. In a direct approach, the pattern distortion and displacement following a particular exposure sequence are measured by analyzing the exposed pattern itself. While in an indirect approach, usually the surface potential on the exposed sample is measured. People have been trying to tackle the charging problem in both ways.

One of the earliest attempts to study the charging effect in EBL was back in the 1990s, carried out by Cummings [12] at Bell Laboratories. They measured charging induced pattern placement error in a fixed spot raster scan direct write EBL system EBES3. Figure 16.11 illustrates the basic strategy of the experiment. A 20-keV e-beam was used to expose the resist in this study. But the thickness of the tri-level resist structures under test were up to 2.62 μm, far larger than the typical resist thickness used for mask making (usually 400 nm). Charging induced beam landing position error up to 0.5 μm was reported, the polarity and magnitude of the resulting surface potential, however, were unspecified.

In 1995, Liu et al. [13] at Stanford introduced a quite unique arrangement to measure surface potential on exposed resist samples (Figure 16.12). The e-beam of a scanning electron microscope is directed parallel and close to the surface of a resist sample that has been exposed by a flood beam. The surface potential on the exposed resist is determined by measuring the lateral deflection of the SEM beam, and the deflection of the SEM beam is measured by viewing the image shift of the grid. Some of the experimental results are as expected, for example, thicker resists or lower electron energy result in more negative charging. The negative surface potential increases nonlinearly with increasing resist thickness and decreasing electron energy. A surprising and disputed result is that thinner resists or higher electron energies can give rise to positive charging. However, the authors were constrained to use a flood beam to expose the resist, while in real mask making conditions a focused beam is always used.

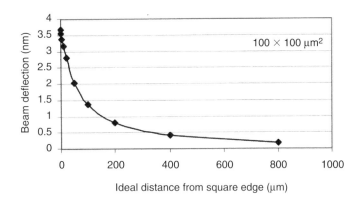
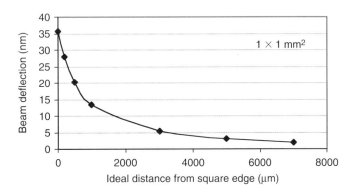
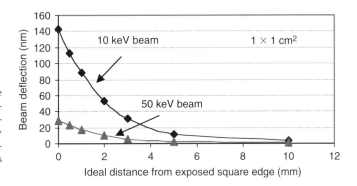

FIGURE 16.8
Beam deflection caused by the charged square with various dimensions (Figure 16.7). The surface potential on the pad is 1 V, primary beam energy is 10 keV, working distance is 14 mm, and sample thickness is 400 nm.

More recently, Bai et al. [14,15] have measured the surface potential on the resist after and/or during e-beam exposure using two independent techniques. First, a Kelvin probe was used to monitor the surface potential immediately following the exposure of the resist by a focused scanning beam (Figure 16.13) [14]. Before exposure, the uncharged sample is positioned under the probe for calibration to null off any DC offset. The sample is then moved under the focused beam and an area measuring several millimeter squares is exposed to the desired dosage. With a calibrated micrometer drive, the exposed region can be moved back underneath the Kelvin probe within a few seconds after exposure to monitor the resulting surface potential and its decay.

The film thickness dependence of surface potential is studied for PBS resist (*Experiment* I, Table 16.1) and the results are presented in Figure 16.14. It is very clear that the change of surface potential versus film thickness in PBS follows the similar general trend as

Resist Charging and Heating 351

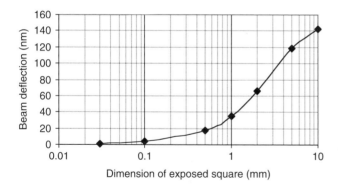

FIGURE 16.9
Beam deflection versus the dimension of exposed square. The surface potential is 1 V, beam energy is 10 keV, and working distance is 5 mm.

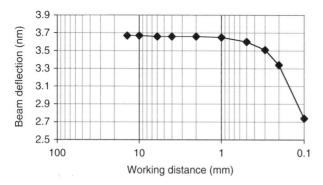

FIGURE 16.10
Beam deflection versus working distance. The surface potential on the 100 × 100 μm² square is 1 V, and beam energy is 10 keV.

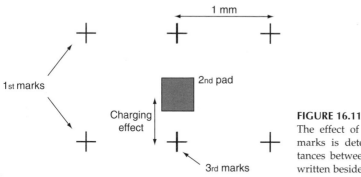

FIGURE 16.11
The effect of the charged large pads on the marks is determined by comparing the distances between the 1st marks and 3rd marks written beside the pads [12].

observed by Liu et al. [13] before. Thinner films and higher primary electron energy lead to more positive surface potentials, and surface potential with different polarities is observed under both 10 and 20 keV exposures. In Liu et al.'s earlier work using flood beam, the authors also studied PBS resist exposed by 10 keV electrons, at comparable dosage. It turns out that the resulting surface potentials from 10 keV focused scanning beam and flood beam exposure agree with each other very closely.

The typical postexposure decay of surface potential on PBS resist is presented in Figure 16.15. Instead of following a simple exponential, the surface potential drops rapidly immediately following exposure, and the decay continuously slows down until the voltage stabilizes after a period of time on the order of an hour or so. The decay behavior indicates the existence of postexposure conductivity in the resist film, which also decays

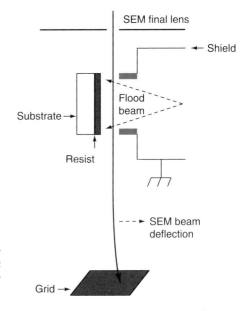

FIGURE 16.12
Experiment to measure directly the surface potential of the resist as a result of electron irradiation by observing the shift of the SEM image of the grid. Resist charging is due to the flood beam [13].

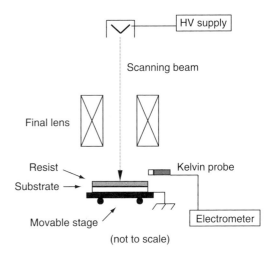

FIGURE 16.13
Experiment for measuring the surface potential above e-beam-exposed resist films using Kelvin probe.

with time. Similar results have also been reported by Yang and Sessler [16] and Weaver et al. [17].

The effect of varying exposure dose on surface potential was also studied. Over the range of dose used in the experiment (from 1 to 4 μC/cm^2 on PBS), the surface potential remains virtually constant at both the thicknesses, 520 and 800 nm (*Experiment* II, Table 16.1; Figure 16.16). The dosage independent phenomenon suggests that there exists significant EBIC during exposure. Without any leakage current caused by EBIC, the charge in resist therefore the surface potential would increase proportionally versus incidental dose. The observed constant surface potentials also indicate that a balance between incoming e-beam current and EBIC induced leakage current has happened before the dose reaches 1 μC/cm^2.

Qualitatively, the experimental results on PBS also agree very well with the modeling results on PMMA. For the surface potential and resist thickness relationship, the experimental data (Figure 16.14) follows the same general trend as the modeling result does

TABLE 16.1

The Different Exposure Conditions in the Kelvin Probe Experiment

Resist	Beam Energy (keV)	Current (nA)	Scanning Area (mm²)	Exposure Time (s)	Dose (μC/cm²)	Thickness (nm)
Experiment I						
PBS	10	1.2	4.2 × 1.4	100	2.0	200–1200
PBS	20	3.0	4.9 × 1.9	125	4.0	200–1200
Experiment II						

Resist	Beam Energy (keV)	Current (nA)	Scanning Area (mm²)	Dose (μC/cm²)	Thickness (nm)
PBS	20	3.0	4.9 × 1.9	1–4	520, 800

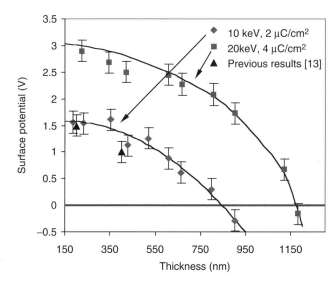

FIGURE 16.14
PBS resist surface potential versus film thickness. The 10 and 20 keV e-beams are used for exposure. At 10 keV, a 4.2 × 1.4 mm² area is scanned by 1.2 nA beam current for 100 s (2 μC/cm²). At 20 keV, a 4.9 × 1.9 mm² area is scanned by 3 nA beam current for 125 s (4 μC/cm²). The results are compared to an earlier measurement made using 10 keV flood beam exposure with similar dosage [13].

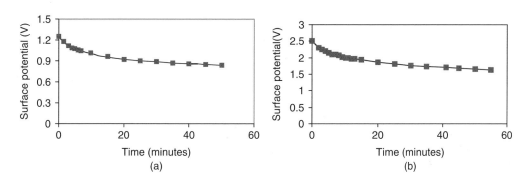

FIGURE 16.15
PBS surface potential postexposure decay. (a) The 10 keV e-beam, 2 μC/cm² exposure on 520 nm resist film. (b) The 20 keV e-beam, 4 μC/cm² exposure on 400 nm resist film.

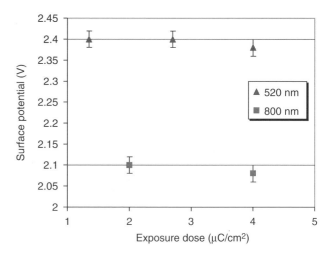

FIGURE 16.16
PBS resist surface potential immediately following exposure versus dose. The e-beam energy is 20 keV, resist thicknesses are 520 and 800 nm.

(Figure 16.5). The dose independent surface potential (Figure 16.16) is consistent with the simulated temporal change of the voltage, which indicates a very rapid saturation of the potential after the exposure. Both the experiment and the modeling imply that the EBIC plays a very important role during the charging process.

To predict charging induced pattern displacements in EBL, perhaps the most important parameter to know is the resulting surface potential after the e-beam has stopped exposing a particular area and started to write in its vicinity. However, to better understand the dynamic charging process, an *in situ* measurement of surface potential while the e-beam is still exposing the specimen becomes desirable. This measurement was then made by a special secondary electron collector to monitor secondary electron current while the specimen is under e-beam exposure [15]. The secondary electron spectrum is shifted along the energy axis as the potential on the specimen is varied [4], so that different portions of the integral spectrum can be detected by the secondary electron collector. From the change of collected secondary electron current, the change of surface potential on the exposed samples can be deduced.

The planar secondary electron collector consists of three parallel components (Figure 16.17). A copper metal piece 25.4 mm in diameter is 12 mm away from the sample holder; a copper grid (64% opening, 250 μm pitch) is stretched by a copper ring and put in between, 6 mm away from both the collector and sample holder. The three-layer structure is supported by four ceramic pillars, which also electrically insulate each layer. The diameter of the grid is 12.7 mm, which gives the collector an acceptance angle of 104°. A 1-mm hole is drilled in the top copper collector and the grid, through which the e-beam is guided to expose the sample. During the measurements, the grid is always held at ground, and the copper collector is slightly positively biased (1.5 V) to increase the collection efficiency. Secondary electrons from the specimen travel through the grid to reach the collector. The secondary electron collector is calibrated and the secondary electron current can be related to the potential on the substrate. PMMA resists on silicon substrate were exposed under different conditions (Table 16.2).

Tests were carried out on PMMA resist films from 0.4 up to 1 μm exposed by 10 and 25 keV e-beams. As an example, Figure 16.18 shows the result for 0.8 μm PMMA film. There is essentially no change of collected secondary electron current following the initial turn-on of the e-beam, indicating virtually no change of surface potential on the exposed samples. As indicated by Monte Carlo simulation [9], the range of a 10-keV e-beam in PMMA is roughly 2.4 and 12.5 μm at 25 keV, far larger than the thickness of PMMA films

Resist Charging and Heating

TABLE 16.2

Different Exposure Conditions on PMMA Resist Films for the *in Situ* Surface Potential Measurements Using the Secondary Electron Collector

Beam Energy (keV)	Beam Current (pA)	Scanning Area (μm²)	Scan Cycle Time (s)	Dosage Per Frame (μC/cm²)	Resist Thickness (μm)
10, 25	75	100 × 40	1.26	1.2	0.4–1
5	80	100 × 40	1.26	1.28	0.8, 0.63, 0.36

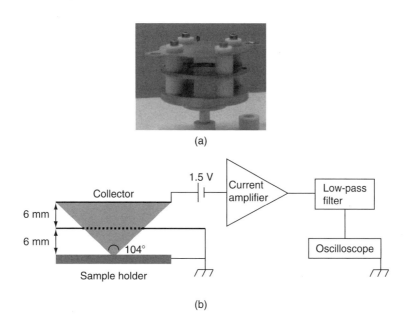

FIGURE 16.17
Experiment of secondary electron collection for *in situ* surface potential measurements during e-beam exposure. (a) The planar secondary electron collector and energy filter. (b) Schematic of experimental setup.

under test. With the majority of primary electrons going to the conducting substrate and trapped electrons leaking through the conductive channel formed by EBIC [18,19], charge left in the exposed films is negligible, leading to the negligible surface potential.

The studies were extended into a much lower beam energy regime, from 3 to 5 keV. As expected, charging becomes much more evident when the resist film thickness approaches the electron range, and a significant fraction of the exposing electrons become trapped in the sample. According to Monte Carlo simulation [9], the largest distances electrons can travel into PMMA film at 5 and 3 keV are 0.7 and 0.35 μm, respectively. Different film thickness was chosen to measure the effects, when the beam is partially or completely absorbed by the resist.

Figure 16.19 shows the results of 5-keV exposure. The secondary electron current collected from 0.63 and 0.8 μm PMMA films increases following the initial scan and starts to saturate after 3–4 scan frames, indicating negative surface potential building up on both samples. However, the collected secondary electron current signal from 0.8 μm PMMA becomes very uneven as charge accumulates and saturates. This unevenness may be caused by the lateral field between the negatively charged exposed regions and the unexposed regions on the sample, diverting some of the secondary electron from

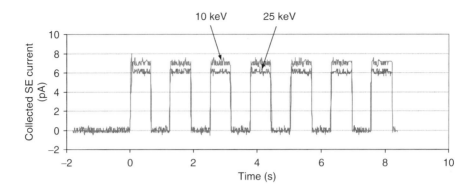

FIGURE 16.18
Collected secondary electron signal while 0.8-μm PMMA resist is exposed by 10 and 25 keV e-beams. Beam current 75 pA, exposure area 100 × 40 μm², scan cycle time 1.26 s, and dosage per frame 1.2 μC/cm². The secondary electron signal is square-wave like because the beam is modulated to scan only half of each frame.

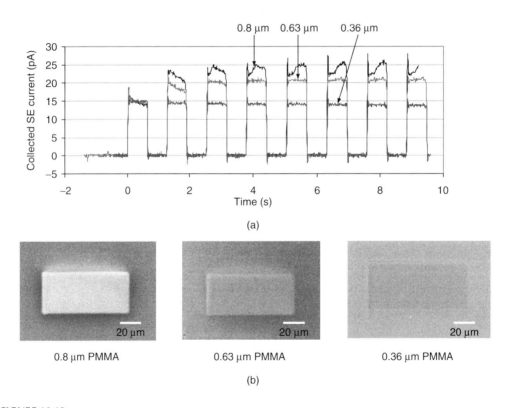

FIGURE 16.19
The 5 keV e-beam exposure of PMMA resist with different thicknesses. Beam current 80 pA, exposure area 100 × 40 μm², scan cycle time 1.26 s, and dosage per frame 1.28 μC/cm². (a) Collected secondary electron current signal. (b) Pictures taken after exposure at lower magnification.

the secondary electron collector. It is also possible that a non-penetrating e-beam can charge the PMMA film to its breakdown voltage [20], and the spatially nonuniform breakdown on the charged area would cause the sloped secondary electron signal. At 0.36-μm PMMA, however, a slight decline of the collected secondary electron current during exposure was observed, which would correspond to a small positive surface potential.

As a qualitative verification of the secondary electron current measurements, some pictures of the charged area after exposure at the same beam energy but at lower magnification were taken, as shown in Figure 16.19. By comparing the brightness contrast between the charged area and uncharged background, one can see the 0.8-μm PMMA film has been charged strongly negatively so that the 100×40-μm^2 pad appears significantly brighter than the background. The contrast is much less on the 0.63-μm film. In the picture of the 0.36-μm film, the charged pad actually looks slightly darker than the unexposed background, indicating a positive potential on the surface.

From the calibration of the secondary electron collector, the surface potential on the PMMA films can be quantitatively deduced according to the secondary electron current measurements. E-beams of 10 and 25 keV caused essentially no charging (<0.2 V) on PMMA resists up to 1-μm thick. This is consistent with the negligible pattern displacement induced by charging observed in previous work under similar experimental conditions [9]. There is a significant rise of surface potential when the thickness of resist becomes comparable to the electron range. For the 5-keV exposure, the estimated surface potential is ~0.3 V on the 0.36-μm PMMA, ~ −2.5 V on the 0.63-μm PMMA, while more than −50 V on the 0.8-μm PMMA. A key result of the surface potential analysis is that it is critical whether or not the film thickness exceeds the electron range in the resist. With a range of ~0.7 μm in PMMA resist, the 5-keV beam has led to a substantial difference in the surface potential on the 0.63 and 0.8 μm films. This observation further demonstrates the importance of EBIC in exposed insulating materials. It is also implied that the free carriers induced by the irradiating beam have a mean free path that is negligible compared to the resist thickness.

16.4 Resist Heating

During e-beam exposure, electrons deposit a large amount of energy into the covering films and substrate of photomasks. Most of the energy is converted to heat and can cause temperature rise in resist during exposure. The resist-heating changes the sensitivity of the resist and affects the accuracy of final patterns. To achieve higher resolution and throughput, higher beam voltages and currents are to be used. This will lead to increased power density and raise the resist-heating temperature close to its adiabatic limit. In this section, we review the theoretical and experimental works of resist heating, the dependency of resist sensitivity on temperature variation, and their effect on critical dimension (CD) errors.

16.4.1 Heating of Resist during the Electron Beam Exposure

16.4.1.1 Temperature Simulations

Accurately modeling the temperature rise of resist during the e-beam exposure is crucial to understand and minimize the resist-heating effect. To calculate the temperature rise of resist during exposure, both numerical and analytical methods have been used. Finite element method (FEM) simulation was used as a major numerical method [21–23]. However, the four-dimensional nature of temperature calculation requires a tremendous amount of computational power therefore limits the practical usage of FEM. Alternatively, a variety of analytical methods have been developed to calculate resist temperature profiles based on Green's function model.

Temperature evolution in resist can be determined by solving the heat transfer equation:

$$\frac{\partial T(x,y,z,t)}{\partial t} = \alpha \nabla^2 T + \frac{1}{\rho C_p} g(x,y,z,t) \tag{16.5}$$

where T is the temperature rise, α the thermal diffusivity, ρ the density, and C_p is the specific heat. The g represents the heat source generated by the interaction between the incoming electrons and solid sample. Typically the heat generation terms were obtained from Monte Carlo simulations [1].

The general solution for the temperature $T(x, y, z, t)$ can be solved by Green's function method [24]:

$$T(x,y,z,t) = \int dt' \iiint dx'\, dy'\, dz' \frac{1}{\rho C_p} G(x,y,z,t,x',y',z',t')\, g(x',y',z',t') \tag{16.6}$$

The coordinates x, y, z, and t represent the temperature field, and the coordinates x', y', z', t' represent the heat source. Green's function $G(x, y, z, t, x', y', z', t')$ represents the temperature at (x, y, z, t) due to an instantaneous point source of unit strength at (x', y', z', t'). Due to the linear nature of the heat transfer equation, the temperature field can be obtained by integrating Green's function over the heat source region.

To solve Green's function, boundary conditions need to be defined. The following assumptions are widely used in resist-heating simulations (the coordinate system is shown in Figure 16.20):

1. Thermal radiation from the resist surface is ignored; the top resist surface at $z = 0$ is assumed to be adiabatic, i.e., $\partial T/\partial z = 0$.
2. The temperature rise is zero when x, y, and z go to infinity (assuming the lateral dimension and the thickness of photomask are much larger than the e-beam exposure area and the electron penetration depth in the substrate).
3. Ideal thermal contact between the interfaces; T and $k(\partial T/\partial z)$ are continuous at interfaces.

Green's function can be decomposed into x, y, and z directions.

$$G(x,y,z,t,x',y',z',t') = G(x,x',t,t')\, G(y,y',t,t')\, G(z,z',t,t')$$

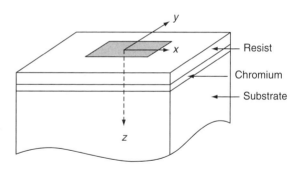

FIGURE 16.20
Coordinates used in modeling. The shaded area represents the exposure area; the exposure center is chosen to be the origin of the coordinate system.

In the x and y directions, the solutions are the Gaussian solutions for free space heat diffusion:

$$G(x,x',t,t') = \frac{\exp(-(x-x')^2/4\alpha(t-t'))}{\sqrt{4\pi\alpha(t-t')}}$$
$$G(y,y',t,t') = \frac{\exp(-(y-y')^2/4\alpha(t-t'))}{\sqrt{4\pi\alpha(t-t')}} \qquad (16.7)$$

The challenging part is to solve Green's function in z direction. To fully simulate the resist heating, the resist layer, chrome layer, and quartz substrate, all need to be taken into account. Solving Green's function in such a nonhomogeneous structure is mathematically challenging and complicated. Different simplifications have been used to facilitate computation. Ralph et al. [25] neglected the resist and chrome layers and assumed that the resist temperature is the same as the surface temperature of the substrate. Abe et al. [26] assumed that the bottom surface of resist is in contact with an ideal heat sink and only considered the heat transfer in the resist layer. Both assumptions simplify the multi-layer material problem into a single material problem. Many authors, for example, Murai et al. [27], Yasuda et al. [28], Groves [29], have used these simplified models in resist-heating simulations. However, these two different models could lead to very different results. For example, under a same beam condition, the maximum temperature increase was calculated as 14 K using Ralph's model and 750 K using Abe's model [26]. Chu et al. [30] developed a general solution of Green's function for a multi-layer structure, which unifies Abe's and Ralf's models. For example, for a $0.2 \times 0.2\,\mu m^2$ square exposed by an e-beam of 50 kV acceleration voltage and 250 A/cm^2 current density for 40 ns, Chu's model predicts a maximum resist temperature rise of 42 K. The material parameters used in the calculation are shown Table 16.3. In the cases of exposing a larger field, a number of subfields are exposed sequentially. Since the earlier exposures affect later ones, resist heating is dependent on the writing strategies and sequences. For instance, resist-heating effects of two different writing strategies are shown in Figure 16.21. The complexity of the general Green's function model increases exponentially as the number of layers increases, making the model more suitable for the structure with two or fewer covering layers on top of substrate. Babin et al. [31,32] developed a software package, TEMPTATION, to calculate resist heating in multi-layer structures with asymptotic analytical formulas, but the mathematical details are not disclosed.

16.4.1.2 Experimental Verification

Although a number of authors have simulated resist heating with different models and techniques, direct resist temperature measurements to verify those models were very limited. This is due to the difficulty in conducting *in situ* temperature measurements with sufficient spatial and temporal resolutions. Babin et al. [33] measured the transient e-beam

TABLE 16.3

The Material Properties Used in Resist-Heating Calculation

	k (W/cm/K)	ρ (g/cm^3)	C_p (J/g/K)
Resist	0.002	1.2	1.5
Chromium	0.63	7.2	0.465
Quartz	0.014	2.2	0.75

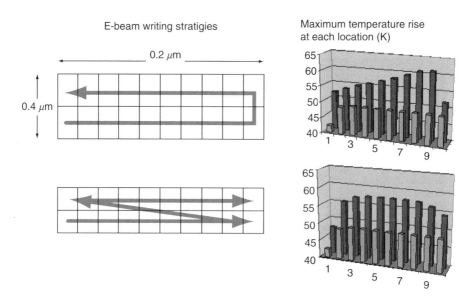

FIGURE 16.21
Maximum temperature rises within each subfield of a 0.4-μm × 2-μm area exposed using two writing strategies. The beam condition is 50 kV voltage, 250 A/cm^2 current, and 10 μC/cm^2 dose and assumes no dwell time between sequential exposures. The left-side figures describe two writing strategies: the top one is serpentine writing and the bottom one is raster writing. The corresponding maximum temperature rises in each exposed square are plotted in the right-side figures.

induced temperature rise on an alumina surface using an YBa$_2$Cu$_3$O$_7$ super-conducting thermometer. But their measurements were conducted at liquid nitrogen temperature. Iranmanesh and Pease [34] used TFTCs to measure steady-state silicon surface temperature profiles under e-beam irradiation. However, the relatively large size of the devices used in both experiments (on the order of micrometer scale) provides inadequate spatial resolution and could distort the temperature field during measurements. Chu et al. [35] fabricated TFTCs with a minimum spatial resolution of 100 nm (Figure 16.22) and used these TFTCs to measure both steady-state and transient resist temperature profiles at room temperature (Figure 16.23). As shown in Figure 16.24, under irradiation by a 15 keV, 150 nA e-beam of 1.7 μm radius for 100 μs, resist heating of 62 and 18 K was measured on quartz substrates and silicon substrates, respectively. Under the same e-beam conditions on quartz substrate, the electron dose of 5 and 15 μC/cm^2 resulted in temperature rises of 25 and 40 K, respectively. These direct temperature measurements show good agreement with the multi-layer Green's function model.

Babin and Kuzmin [36] also used an indirect method to verify the simulation results obtained from the TEMPTATION software. A large area of resist was uniformly exposed and then underdeveloped. Since the resist heating changes the sensitivity of resist, the effective dosage (effective absorbed energy) of exposed areas changes accordingly. The variation of effective dose is related to the variation of residue thickness of underexposed resist. Assuming that all these relationships are linear, the image of residue resist should resemble the profiles of the maximum temperature of the resist during exposure. Babin and Kuzmin [36] obtained good matching between the resist residue thickness profiles and the simulated temperature field profiles. Compared with direct temperature measurements, the residue resist image method has the advantage of obtaining a large field of view of resist surface profile but has the disadvantage of lack of quantitative temperature values.

Resist Charging and Heating 361

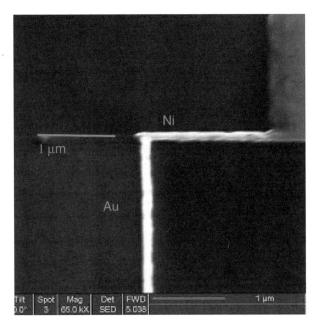

FIGURE 16.22
SEM of Ni/Au TFTC of 100-nm junction size.

16.4.2 The Impact of the Localized Heating on Resists

Different resists respond differently to resist heating. The dependence of resist sensitivity on temperature variation is a key parameter to characterize. Kratschmer and Groves [37] quantified this parameter by measuring the normalized difference in exposure dose required for a full development. Four types of resist, DQN (with a sensitivity of ~25 μC/cm² at 50 kV), XPR (with a sensitivity of ~3 μC/cm² at 50 kV), PBS (with a sensitivity of ~7 μC/cm² at 50 kV), and NXR (with a sensitivity of ~3 μC/cm² at 50 kV), were characterized. The DQN-type resist was found to be most vulnerable and PBS was most immune to resist heating. Yasuda et al. [38] characterized the sensitivity of PMMA versus temperature rise and observed the shape of patterns exposed at different temperatures. For PMMA, remarkable pattern shape distortions appear between 100 and 150 K heating where the glass transition temperature occurs. Recently, Babin [39] characterized the dependency of resist sensitivity on temperature change for a group of resists, including PMMA, PBS, SPR-700, EBR-900, ZEP-7000, and UVII-HS. The main results are shown in the Table 16.4.

16.4.3 The Effect of Writing Strategies on Resist Heating

Babin [40] compared the effects of different writing strategies on resist heating. In his experiments, a variable shape beam system with 50 kV acceleration voltage and 95 A/cm² current density was used to expose SPR-700 and EBR-900 resists (both with a sensitivity of ~40 μC/cm² at 50 kV). In a single-pass exposure with a relatively high (0.02 cm²/s) throughput, a 13.8% effective dose change was observed in raster writing, and 130% effective dose change was observed in vector writing. The heating effect was found to be dramatically reduced when the one-pass exposure was replaced by a four-pass exposure. At the cost of reducing the throughput by a factor of 4, the effective dose change reduced to 1% for raster writing and 16.8% for vector writing.

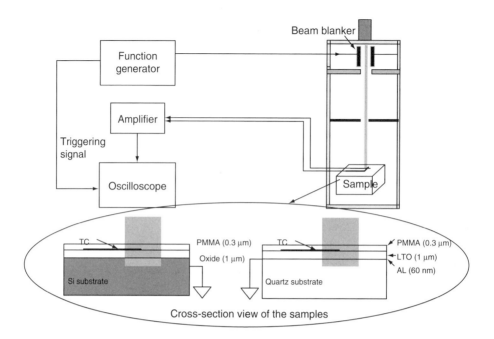

FIGURE 16.23
Schematic of experimental setup of resist-heating measurements using TFTCs. TTL reference signal was used to turn "on" and "off" the e-beam. The resultant TFTC signals were recorded after amplification. TFTCs on both silicon and quartz substrates were used in the measurements. Cross-section views show the detail structure of samples.

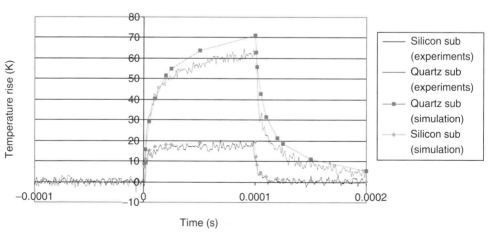

FIGURE 16.24
Experimental and calculated temperature profiles at the bottom surface of resist on both silicon and quartz substrates (1 μm oxide layer exists between the TFTCs and the substrates), irradiated by a 15-kV Gaussian beam. Beam current is 150 nA and the beam radius is 1.7 μm.

TABLE 16.4

Change of Resist Sensitivity (%/°C)

Resist	ΔT Range	dD/dT (%dose/°C)
PMMA	0–42	0.88
PBS	0–30	0.19
	30–42	0.79
EBR-900	0–24	0.27
	24–35	0.5
SPR-700	0–40	0.18
	40–56	0.37
ZEP-7000	0–110	0.25
	110–138	1.2
UVII-HS	0–30	0.125

(Courtesy S. Babin.)

16.4.4 The Impact of Resist Heating on CD Variation

The major concern of resist heating in mask fabrication is CD variation. Sakurai et al. [41] investigated the resist heating induced CD error for ZEP-7000 and two unspecified chemical amplified resists (CAR) under different exposure conditions. The CD errors for CARs were found to be smaller and more stable than CD errors for ZEP-7000. Their experimental results showed that the current density and shot size influenced CD error strongly for ZEP-7000, but weakly for CARs. Kuwahara et al. [42] also observed that a CAR is less sensitive to resist heating when compared with conventional resist. For conventional resist, the maximum temperature rise of resist controls the degree of breakage of the cross-linking. Therefore, the temperature variation within a writing field results in a variation in sensitivity. For CAR, the sensitivity is controlled by a postexposure bake (typically 100–150 K), where the acids created by e-beam irradiation diffuse. As long as the resist-heating temperature is kept lower than the postexposure bake temperature or the dwell time of heating is kept shorter than the time required for acid diffusion, acid diffusion does not take place during the writing and the resist heating would not cause much CD variation.

16.4.5 Strategies to Mitigate Resist-Heating Effect

The following factors contribute to the resist heating:

1. Total energy absorbed in the resist and substrate during exposure
2. The immunity of resist to temperature variations during exposure
3. Writing strategies in photomask fabrication.

Choosing an appropriate resist and writing strategy can minimize resist-heating effects. For resists, since the total energy absorbed in the resist and substrates are proportional to the sensitivity of resist, using a sensitive resist (e.g., requires less dosage) can directly reduce the temperature rise. In addition to the resist sensitivity to exposure dose, the resist immunity to temperature variation is also crucial to reduce the heating effect. CAR was found to be not only sensitive to exposure dose but also more immune to temperature variations. As long as the temperature rise in resist is lower than the postexposure temperature of the CAR (typically 100–150 K), using a CAR is an effective way to minimize resist-heating effect. The resist-heating effect can also be reduced by choosing multi-pass writing strategies. When the interval time between the successive exposures is long enough, the resist temperature rise will be reduced by a factor of the number of writing passes. Kuwahara et al. [42] measured that the CD errors of 2-μm lines and spaces in the writing field were reduced from 22 to 6 nm by changing one-pass writing to four-pass writing.

16.5 Summary

From the results of the charging modeling and experiments, it appears that under typical mask making conditions (e.g., 400 nm of resist on conducting substrate, exposed with >10-keV electrons), resist charging is much less serious in EBL than had been previously reported. Most recently, Dobisz et al. [43] also reported that if more than ~25% of the primary electrons are transmitted through the resist to an underlying conducting layer, the EBIC effects were sufficient to discharge any significant charge buildup. In addition, the EBL tools used for mask fabrication have been moving towards higher beam energy. While as has been pointed out, the beam deflection induced by charging is inversely proportional to the primary beam energy. Therefore, resist charging does not seem to be a serious concern in terms of pattern displacement on masks. However, in other e-beam applications, such as direct writing, charging could still be a significant issue, especially when insulating samples thicker than the primary electron range are involved.

Resist heating is one of the major contributors to CD variation in photomask fabrication. The exposure dosage and thermal immunity of resist determine the line width variation. Typical maximum temperature rise in resist is several tens of degree Celsius and will cause a few nanometers in CD errors. Because heating is an accumulative effect, writing strategies of different e-beam tools also affect the results. In practice, choosing appropriate resist and writing strategies can minimize the resist-heating effect.

References

1. D.C. Joy, *Monte Carlo Modeling for Electron Microscopy and Microanalysis*, Oxford University Press, New York, 1995.
2. John T.L. Thong, *Electron Beam Testing Technology*, Plenum Press, New York, 1993, 39 pp.
3. H. Seiler, Secondary electron emission in the scanning electron microscope, *J. Appl. Phys.*, 54, R1–R8 (1983).
4. E. Menzel and E. Kubalek, Secondary electron detection systems for quantitative voltage measurements, *Scanning*, 5, 151–171 (1983).
5. B. Gross, G.M. Sessler, and J.E. West, Charge dynamics for electron-irradiated polymer-foil electrets, *J. Appl. Phys.*, 45 (7), 2841–2851 (1974).
6. David A. Berkeley, Computer simulation of charge dynamics in electron-irradiated polymer foils, *J. Appl. Phys.*, 50 (5), 3447–3453 (1979).
7. G.M. Sessler, Charge dynamics in irradiated polymers, *IEEE Trans. Electr. Insul.*, 27 (5), 961–973 (1992).
8. W.H. Sullivan and R.L. Ewing, A method for the routine measurement of dielectric photoconductivity, *IEEE Trans. Nucl. Sci.*, NS-18 (6), 310–317 (1971).

9. M. Bai, Insulator Charging in Electron Beam Lithography, Ph.D. dissertation, Stanford University, Stanford, CA, 2003.
10. J. Ingino, G. Owen, C.N. Berglund, R. Browning, and R.F.W. Pease, Workpiece charging in electron beam lithography, *J. Vac. Sci. Technol. B.*, 12 (3), 1367–1371 (1994).
11. J.J. Hwu, Y.-U. Ko, and D.C. Joy, Computer modeling of charging induced electron beam deflection in electron beam lithography, *Proc. SPIE*, 3998, 239–246 (2000).
12. K.D. Cummings, A study of deposited charge from electron beam lithography, *J. Vac. Sci. Technol. B.*, 8 (6), 1786–1788 (1990).
13. W. Liu, J. Ingino, and R.F.W. Pease, Resist charging in electron beam lithography, *J. Vac. Sci. Technol. B.*, 13 (5), 1979–1983 (1995).
14. M. Bai, R.F.W. Pease, C. Tanasa, M.A. McCord, D.S. Pickard, and D. Meisburger, Charging and discharging of electron beam resist films, *J. Vac. Sci. Technol. B.*, 17 (6), 2893–2896 (1999).
15. M. Bai, W.D. Meisburger, and R.F.W. Pease, Transient measurement of resist charging during electron beam exposure, *J. Vac. Sci. Technol. B.*, 21 (1), 106–111 (2003).
16. G.M. Yang and G.M. Sessler, Radiation-induced conductivity in electron-beam irradiated insulating polymer films, *IEEE Trans. Electr. Insul.*, 21, 843–848 (1992).
17. L. Weaver, J.K. Shultis, and R.E. Faw, Analytic solutions of a model for radiation-induced conductivity in insulators, *J. Appl. Phys.*, 48, 2762–2770 (1977).
18. J.J. Hwu and D.C. Joy, A study of electron beam-induced conductivity in resists, *Scanning*, 21, 264–272 (1999).
19. M. Bai, R.F.W. Pease, and D. Meisburger, Electron beam induced conductivity in PMMA and SiO_2 thin films, *J. Vac. Sci. Technol. B.*, 21 (6), 2638–2644 (2003) (to be published).
20. H. Gong and C.K. Ong, Discharging behavior on insulator surfaces in vacuum: a scanning electron microscopy observation, *J. Phys.: Condens. Matter*, 9, 1631–1636 (1997).
21. N.K. Eib and R.J. Kvitek, Thermal distribution and the effect on resist sensitivity in electron-beam direct write, *J. Vac. Sci. Technol.*, 7 (6), 1502–1506 (1989).
22. K. Nakajima and N. Aizaki, Calculation of a proximity resist heating in variably shaped electron beam lithography, *J. Vac. Sci. Technol. B.*, 10 (6), 2784–1788 (1992).
23. A. Wei, W.A. Beckman, R.L. Engelstad, and J.W. Mitchell, *Finite Element Analysis of Localized Heating in Optical Substrates Due to E-beam Patterning*, 20th Annual BACUS Symposium, 2000.
24. M. Ozisik, *Heat Conduction*, 2nd edition, John Wiley & Sons, New York, 1993 (Chpter 6).
25. H. Ralph, G. Duggan, and R.J. Elliott, *Proceeding of the 10th International Conference on Electron and Ion Beam Science and Technology*, 1982, 219 pp.
26. T. Abe, K. Ohta, H. Wada, and T. Takigawa, The electron-beam column for a high-dose and high-voltage electron-beam exposure system EX-7, *J. Vac. Sci. Technol. B.*, 6 (3), 853–857 (1988).
27. F. Murai, S. Okazaki, N. Saito, M. Dan, The effect of acceleration voltage on linewidth control with a variable-shaped electron beam system, *J. Vac. Sci. Technol. B.*, 5 (1), 105–109 (1987).
28. M. Yasuda, H. Kawata, and K. Murata, Resist heating effect in electron beam lithography, *J. Vac. Sci. Technol. B.*, 12 (3), 1362–1366 (1994).
29. T.R. Groves, Theory of beam-induced substrate heating, *J. Vac. Sci. Technol. B.*, (14), 3839–3844 (1996).
30. D. Chu, R.F.W. Pease, and K. Goodson, Modeling resist heating in mask fabrication using a multilayer Green's function approach, *Proc. SPIE*, 4689, 206 (2002).
31. S. Babin, I.Y. Kuzmin, and G. Sergeev, Advanced model for resist heating effect simulation in electron-beam lithography, *Proc. SPIE*, 2884, 520 (1996).
32. S. Babin, I.Y. Kuzmin, and G. Sergeev, Software tool for temperature simulation in electron-beam lithography: TEMPTATION, *Proc. SPIE*, 3236, 464 (1998).
33. S. Babin, M.E. Gaevski, and S.G. Konnikov, Measurement and simulation of temperature dynamics under electron beam, *J. Vac. Sci. Technol. B.*, 19 (1), 153–157 (2001).
34. A. Iranmanesh and R.F.W. Pease, Temperature profiles in solid targets irradiated with finely focused beams, *J. Vac. Sci. Technol. B.*, 1 (2), 739–743 (1983).
35. D. Chu, D.T. Bilir, K.E. Goodson, and R.F.W. Pease, Submicron thermocouple measurements of electron-beam resist heating, *J. Vac. Sci. Technol. B.*, 20 (6), 3044–3046 (2002).
36. S. Babin and I.Y. Kuzmin, Experimental verification of the TEMPTATION (temperature simulation) software tool, *J. Vac. Sci. Technol. B.*, 16 (6), 3241–3247 (1998).

37. E. Kratschmer and T.R. Groves, Resist heating effects in 25 and 50 kV e-beam lithography on glass masks, *J. Vac. Sci. Technol. B.*, 8 (6), 1898–1902 (1990).
38. M. Yasuda, H. kawata, K. Murata, K. Hashimoto, Y. Hirai, and N. Nomura, Resist heating effect in electron beam lithography, *J. Vac. Sci. Technol. B.*, 7 (6), 1362–1366 (1994).
39. S. Babin, Measurement of resist response to heating, *J. Vac. Sci. Technol. B.*, 21, 135–140 (2003).
40. S. Babin, Resist heating with different writing strategies for high-throughput mask making, *Microelectron. Eng.*, (53), 341–344 (2000).
41. H. Sakurai, T. Abe, M. Itoh, A. Kumagae, H. Anze, and I. Higashikawa, Resist heating effect on 50-keV EB mask writing, *Proc. SPIE*, 3748, 126–136 (1999).
42. N. Kuwahara, H. Nakagawa, M. Kurihara, N. Hayashi, H. Sano, E. Maruta, T. Takikawa, and S. Noguchi, Primary evaluation of proximity and resist heating effects observed in high-acceleration voltage e-beam writing for 180-nm-and-beyond rule reticle fabrication, *Proc. SPIE*, 3748, 115–125 (1999).
43. E.A. Dobisz, R. Bass, S.L. Brandow, M. Chen, and W.J. Dressick, Electroless metal discharge layers for electron beam lithography, *Appl. Phys. Lett.*, 82 (3), 478–480 (2003).

17

Mask Processing

Syed A. Rizvi

CONTENTS
17.1 Mask Processing: Introduction 367
17.2 Resists and Developers 368
17.3 Pattern Transfer: Etching of Chrome 369
     17.3.1 Wet Etching of Chrome 370
     17.3.2 Dry Etching of Chrome (or Other Underlying Material) 370
          17.3.2.1 Plasma Reactors 370
          17.3.2.2 Plasma Applications and Processes 370
17.4 Resist Stripping and Cleaning 371
     17.4.1 Wet and Dry Cleaning Processes 371
          17.4.1.1 Cleaning with Wet Processes 372
          17.4.1.2 Cleaning with Dry Processes 372
          17.4.1.3 Cleaning with Semidry Processes 373
     17.4.2 Cleaning of Embedded Attenuated Phase Shift Masks 373
References 374

17.1 Mask Processing: Introduction

The term "mask processing" relates to all steps starting from resist development down to the final cleanup, until the pellicles are mounted on the mask.

For the conventional chrome-on-glass (CoG) structure, the mask processing involves the following basic steps:

1. Developing the mask after exposure to laser or e-beam
2. Pattern transfer: etching the chrome that is now exposed after resist is developed out
3. Stripping the resist
4. Final cleanup

The CoG masks are also referred as binary masks or simply BIMs. In many cases, steps (3) and (4) are carried out in one single operation.

Depending on the type of resists, some bake operations may be required during the processing. Deviation from the above process can occur for masks with more complex structures.

Typically, the CoG mask blanks are coated with chrome and resist film at the supplier's site before they are delivered to the mask shop. At the mask shop the blanks are exposed on laser or e-beam writers and then run through the appropriate developer resulting in the emergence of resist patterns on the plate.

Although the subject of resists and developers has been addressed in Chapter 15, for the sake of continuity a very short review of the topic will be given here.

17.2 Resists and Developers

One of the criteria of resist classification is whether the resist is positive or negative.

Positive resist is the one where its exposed portion is developed out leaving behind the unexposed resist. This step results in the emergence of a pattern of unexposed resist and chrome film that is no longer covered by the resist. In case of a negative resist it is the unexposed portion that is developed out leaving behind the exposed resist. Here also, a pattern of resist and chrome emerges.

While specifics may differ, there is a set of specs that the resist is expected to meet as shown in Tables 17.1 and 17.2 [1,2]. The values in these tables are for generic resist and may vary from the norm as required by individual cases.

Although most mask blanks are received already coated from the supplier, there are times when some of the resist coatings are to be done at the mask shop facility. The procedure is called "spin-coat," where the resist is delivered onto the surface of mask mounted on a chuck that is then spun at a speed of a few thousand revolutions per minute. The chuck when spun at a prescribed speed causes the resist to spread with the formation of a uniform film thickness across the mask surface. A typical spec on resist thickness is 300–450 nm with a uniformity of ± 3 nm.

Before exposure the masks need to be baked to remove the solvents from the resist.

There are two techniques commonly employed for developing masks after exposure.

One technique is the immersion method where the mask is dipped into a tank of developer where the chemical reaction takes place. An advantage of this method is that it can readily be adapted to batch processing where several plates can simultaneously be

TABLE 17.1

Resist Material Properties Requirements [2] (Courtesy of John G. Skinner & Associates.)

Material Property	Requirements
Resist solution shelf life	>6 months
Batch to batch reproducibility	<5% in composition and molecular weight
Resist film coating shelf life	>3 months
Resist thermal properties	Glass transition temperature[a] $T_g > 80°C$; decomposition temperature $T_d > 120°C$
Wet etch Cr chemistry	No degradation or adhesion failure
Dry etch Cr chemistry	Minimum 1:1 selectivity in Cl_2O_2-based plasma[b]
Solubility	Environmentally safe spin coating solvents and aqueous base developers
Strippability	Removal in commercial amine-based stripping solutions or O_2, halogen-based plasma

[a] T_g is the glass transition temperature. It is the temperature at which the resist changes from a glassy amorphous state to rubbery state.
[b] The plasma, used for "dry-etching" the chromium film, may change according to the chromium film composition.

TABLE 17.2

Resist Lithographic Properties Requirements [2] (Courtesy of John G. Skinner & Associates.)

Lithographic Property	Requirements
Sensitivity	
E-beam	2.0 μC/cm² at the rate of 10 keV
Laser	100 mJ/cm² at wavelength 365 nm
Contrast (γ)	>4 for 85° feature wall slope
Resolution	<0.3 μm (for OPC features)

processed. This method, however, has a drawback of contaminating the mask that later requires additional clean up actions.

The other more commonly practiced technique is "spin-develop" with mechanism similar to "spin-coat" mentioned earlier. Here the plate is always immersed in fresh developer at the beginning of each cycle and hence is less prone to contamination.

The uniformity of development across the plate can have an effect on CD uniformity; and hence due to the nature of spin-develop a degree of radial CD variation on the plate is possible.

A number of modifications in spin-develop have been introduced to improve the CD uniformity. Some of such methods employ single spray nozzle, multi-spray nozzle, and puddle development techniques. In the puddle development technique, the plate is slowly spun, while the developer is sprayed from the nozzle (or nozzles).

In all cases the developer is spray-rinsed and spin-dried. Table 17.3 [2,3] shows the effects of various techniques on CD uniformity.

17.3 Pattern Transfer: Etching of Chrome

After the resist has been developed, the next step is to transfer the resist pattern on to the underlying chrome film that now becomes uncovered by the removal of resist. This is done by the etching of the exposed chrome. However, at this point it is important to examine the plate to make sure that the resist is completely developed out. With certain types of resists, an iterative process of CD-measurement and redevelopment may be necessary until the required CD is achieved.

In some cases, traces of resist known as "scum" are left behind in the open windows that can interfere with the etching of chrome. These scums can be removed by a quick exposure to oxygen plasmas. The process is called "descum."

TABLE 17.3

CD Uniformity over 6-in. Mask [2,3]

Dev. Method	3 Sigma (nm)	Radial Error (nm)	Side Error (nm)
Single Spray Nozzle	32	23	12
Dual Spray Nozzle	16	18	10
Puddle	11	7	4
Dip	19	11	8

The etching of chrome has historically been carried out with wet chemicals that are regarded as a part of wet processing. In recent years, due to increased demands on the tolerance of shrinking features, many plasma etch processes known as dry etch processes are introduced onto the manufacturing lines for photomasks.

17.3.1 Wet Etching of Chrome

The processing stations for wet etch can be similar to those of develop stations, namely, immersion tanks or spin stations.

The chemicals commonly used for chrome etch are cerric ammonium nitrate and certain acids that include perchloric acid, acetic acid, nitric acid, and hydrochloric acid [2].

Due to the liquid nature of the chemicals, the wet process tends to be isotropic and causes certain degree of under-cuts. These under-cuts, however, turn out to be helpful in minimizing the effect of slope that the resist profile exhibits towards its edges.

17.3.2 Dry Etching of Chrome (or Other Underlying Material)

As the feature sizes are getting smaller and the tolerance of their size are getting more stringent, the industry is moving towards dry etch, which is more or less of an anisotropic process and requires very little or zero bias in the transferring image form resist to chrome film. Dry etch can also meet the stringent tolerance required by the state-of-the-art designs.

Considering a 100-nm technology node it would seem that the mask feature would be 400 nm. However, when OPCs are involved, the 4:1 rules breakdown, and in such cases mask feature needs to be significantly below 200 nm. At present, 100-nm or even smaller features (on mask) are pursued.

Dry etching involves use of plasma (a mixture of electrons, ions, and various neutral species). In today's vocabulary, dry etch has become synonymous with plasma etch.

17.3.2.1 Plasma Reactors

There are various types of reactors for the creation and application of plasma that are used for the etching of chrome or whatever underlying material is there to be etched. Some of such examples of the etching systems are given in the following:

Ion milling: Here ions are accelerated towards the target, where it is the mechanical impact of the ions rather than any sort of chemical reaction that does the etching of the chrome film.

Reactive ion etching (RIE) *and magnetic enhanced RIE* (MERIE): In this case a reactive species in the plasma chemically reacts with the target to increase etching rate. The composition of chrome etching is $CH_2Cl_2 + O_2$ [2].

Inductive coupled plasma (ICP): This is a low-pressure high-density plasma. It improves CD control and uniformity. ICP is also good for low defect counts.

17.3.2.2 Plasma Applications and Processes

In the case of CoG or binary masks only the chrome film needs to be etched, but with the emergence of phase shift masks new processes are fashioned that can etch other materials also, such as molybdenum silicon (MoSiON). There are also chromeless masks, where features are etched into quartz, which is now being done with plasmas.

In one example of MERIE, the etch parameters for Cr and MoSiON have been cited as the Cl_2/O_2 with gas-assisted etching (GAE).

Mask Processing

For Cr the composition involved is Cl_2/O_2 with GAE, and for MoSiON the composition is CF_4/O_2. The GAE increases the etch selectivity 1.8 times higher than without GAE [4].

There are a number of factors that affect the plasma etching and need to be addressed.

An important factor is chrome loading, that is, the amount of chrome on the mask, it affects several plasma etch responses, e.g., resist selectivity, Cr etch rate, overall CD uniformity, and uniformity within mask [5].

During the dry etch of chrome a certain amount of resist is lost and appears as redeposited polymers and debris on mask surface adding to increased defect counts.

The resist lost can also affect the CD uniformity and etch bias. The phenomena of resist loss are related to poor selectivity. The objective is then to minimize this resist loss by improving the selectivity.

The CD control in uniformity and etch bias show opposite trend lines with standard chemistry of $He/Cl_2/O_2$. Increasing the oxygen flow can improve the uniformity, but it also decreases the selectivity.

In order to overcome the limitations of the two opposite trends, what is needed is to develop a process with improved selectivity to photoresist and with reduced dependence on O_2 flow. There have been chemistries proposed that could provide this benefit.

Hydrogen and carbon containing gases are considered as the promising alternatives.

Several gases proposed are H_2, HCl, and NH_3, and carbon containing gases to promote selectivity are C_2F_6, CCl_4, C_3F_8, CHF_3, CH_4, and CF_4–H_2, etc. [6].

Another work, also on chrome etch, reports on achieving 90-nm features on masks using ICP reactor. Features of this small require a number of process optimization other than just plasma, such as the type of resist, processing, and the writing scheme [7].

Unaxis, another major supplier of plasma etch systems, has its Generation-4 ICP introduced into the market in 2004 that will be addressing the 65-nm technology node [8].

As mentioned earlier, there are phase shift masks that require etching of the quartz to the right depth. Here, after the opening of chrome windows, it is the glass (quartz) that is to be etched. In the case of quartz etching, there is no under-layer that can be used as etch stop. In such cases, the technique is to etch for a predetermined time that can be guaranteed for the desired depth.

The work referred here utilized ICP source with gas composition as CHF_3/CF_4 [9].

17.4 Resist Stripping and Cleaning

Cleaning of photomask starts from the stripping operation where the unwanted resist after chrome etch is to be removed.

However, simply stripping of the resist does not result in perfectly clean mask. There can be defects arising from various sources where some may be as common as water marks whereas others may be subtler and extremely small in size. These defects and other contaminants can adhere to mask surface by either van der Waals force or electrostatic force and can be detected only by sophisticated techniques.

17.4.1 Wet and Dry Cleaning Processes

In general, the cleaning operations fall under the class of wet processes where masks are cleaned with a specific type of solution; or it can fall under the dry process where the mask is exposed to a plasma environment or subjected to high-energy photons for its cleaning. At present, most operations are carried out using wet processes although dry processes are beginning to emerge at many facilities.

17.4.1.1 Cleaning with Wet Processes

The wet process where the mask is subjected to some form of liquid treatment can be further classified as mechanical or purely chemical by its nature.

Example of mechanical treatment is the scrubbing of masks with specially designed brush or sponge, whereas in the other case the mask is immersed into a tank of chemicals where the reactions between the contaminants and chemicals clean the mask.

There are also techniques that employ high-pressure spray cleanup that can be used to take the advantage of mechanical impact of the spray, as well as reactions with the chemicals, to dislodge the contaminants from the mask surfaces.

17.4.1.1.1 Chemistry of Wet Process

Most of the wet processes involve a mixture of H_2SO_4 and H_2O_2 in the ratio of 4:1 used at 90°C, commonly know as Piranha Clean, and is primarily used for removing the resist and heavy organic material. It works as an oxidant and attacks the hydrocarbons [1].

Also, another material used for mask cleaning is a mixture of H_2O, H_2O_2, and NH_4OH in the ratio of 5:1:1 used at room temperature, known as RCA Standard Clean-1 or simply SC-1. This chemistry was designed for removing traces of organic impurities from the mask surface by (1) solvating action of NH_4OH and the oxidation capability of the H_2O_2. The NH_4OH also serves as a complexant for many metallic contaminants. In this case the peroxide in the solution oxidizes the surface, and then the ammonium hydroxide dissolves this oxide. Although this sequential growth and etching of the surface help in the removal of particles, it also results in the undesired micro-roughening of substrate. Recent research has shown that lowering the NH_4OH concentration ratio to 0.01–0.25 greatly reduces the micro-roughening while retaining the particle removal efficiency of the SC-1 [1].

17.4.1.1.2 Application of the Wet Chemistry

Two common practices for the application of chemicals to the mask are:

(a) Immersing the mask into a tank of chemicals
(b) Spraying chemicals on mask while it is spun on a chuck

The use of immersion tanks allows batch processing of masks and has been in practice since the early days of mask making. The spinning processes because of their superior cleaning results are more frequently employed, although this is a single mask processing technique and could result in a lower throughput compared to batch processing.

After wet cleaning, the plates need to be run through final rinse and dried.

Newer techniques like ultrasonic cleaning and megasonic cleaning are becoming quite prevalent. Also, technique of scrubbing the masks with brushes or sponges using "tank" or "spinners" are commonly employed.

Two major issues with chemical cleaning are disposition of used chemicals and redeposition of particles especially in case of immersion cleaning.

17.4.1.2 Cleaning with Dry Processes

Dry cleaning is mainly associated with use of plasma that reacts with the contamination resulting in a by-product, which is then flushed out by the flowing gas. There is also another area of dry processing known as laser-assisted-cleaning, developed by Radiance Services Company [10,11], which uses high-energy photons that can break bonds that hold particles to surface without affecting the surface. Another dry cleaning system utilizes flowing gas that sweeps the particles away from the mask area. Since the process is dry and uses no water or toxic chemicals, the benefits of this technique may include a reduced need for deionized water, chemical handlers, and waste treatment systems in semiconductor facilities.

The gas flow characteristics must be optimized to avoid turbulence otherwise redeposition of particles could occur.

The use of UV radiation has also been known to help strip resist from mask by weakening the bonds of remaining particles after the first strip process.

There is also an on-line *in-situ* dry cleaning process known as plasma mechanical activation and extraction of particle contamination (PLASMAX). This system, jointly developed by Beta Squared and Los Alamos National Labs (LANL), employs resonant spherical harmonic vibration to lift particles from surface, which then suspends, traps, and channels these particles down the vacuum port, thus preventing the particle deposition on the mask surface. The technique is capable of cleaning a mask substrate within 25 s.

This dry cleaning technology can be directly integrated into the exposure system, which can serve as an *in-situ* mask cleaning process, very effective for NGL masks where pellicle cannot be employed.

17.4.1.3 Cleaning with Semi-dry Processes

In addition to the wet and dry cleaning processes there are processes that can be classified as semi-dry process.

In one case, by employing vapor of some liquid (water, isopropanol, and ethanol), a layer of water is deposited on the mask, after which the particle is "hit" with a laser beam. The heated water turns into steam and lifts the particle off the mask surface, and the particle then carried away by a stream of gas flow across the mask surface [12].

17.4.2 Cleaning of Embedded Attenuated Phase Shift Masks

The above paragraphs mainly dealt with the conventional CoG masks, but the structure of some of the advanced masks may differ from the conventional masks and new methods for their cleanup are constantly being developed and implemented. One such example is the case of embedded attenuated phase shift masks (EAPSM), which are MoSiON-based masks. The structure of EAPSM is considerably different from that of the CoG masks. MoSiON can react differently with some of the chemicals used in CoG cleaning. Dry process is more commonly used for MoSiON-based masks because of the ease of plasma etching, but the process also leaves the surface with polymer residues and plasma debris, the removal of which requires due care not to alter the phase and transmitivity of the MoSiON film.

Traditionally, H_2SO_4–H_2O_2 mixture (SPM) followed by NH_4OH–H_2O_2 mixture (APM) has been used for removing photoresists and organic residues from previous processes.

SPM has been seen to cause variations in phase and transmittance compared to other oxidizing chemistries. Hence, ozone-involved chemistries, namely, H_2SO_4–O_3 mixture (SOM) and Ozonated DI Water (DIO3®) from Akrion are introduced to replace SPM (Figure 17.1) [12].

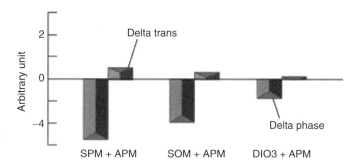

FIGURE 17.1
The effect of different cleaning chemistries on phase and transmittance of EAPSM [11]. (Courtesy of Akrion.)

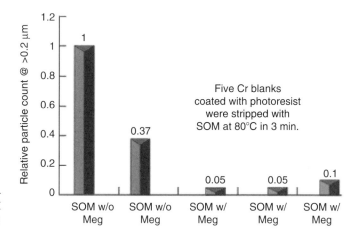

FIGURE 17.2
The effect of megasonic on photoresist removal efficiency of SOM chemistry [11]. (Courtesy of Akrion.)

The efficiencies of SOM and DIO3 for heavy polymer removal, such as stripping photoresists, can be limited. However, by incorporating megasonic technology resist removal has been enhanced as shown in Figure 17.2 [12].

References

1. P. Rai Choudhury, *Handbook of Microlithography, Micromachining and Microfabrication*, vol. 1, SPIE Optical Engineering Press, 1997, pp. 377–474. Bellingham, WA, USA.
2. John G. Skinner, *Photomask Fabrication for Today and Tomorrow*, SPIE Short Course, SC122, 2001, pp. 104–105.
3. Seong-Yong Moon, Won-Tai Ki, Byung-Cheol Cha, Seong-Woon Choi, Hee-Sun Yoon, and Jung-Min Sohn, *19th Annual Symposium on Photomask Technology*, vol. 3873, 1999, pp. 573–576.
4. H. Handa, S. Yamauchi, K. Hosono, and Y. Miyahara, Dry etching technology of Cr films to produce fine pattern reticles under 720 nm with ZEP-7000, in: *19th Annual Symposium of Photomask Technology*, vol. 3873, 1999, pp. 98–106.
5. C. Constantine, R. Westremann, and J. Plumhoff, Plasma etch of binary Cr mask, in: *19th Annual Symposium of Photomask Technology*, vol. 3873, 1999, pp. 93–97.
6. M.J. Buie, B. Stoehr, and Y.C. Huang, Chrome etch for <0.13 micron, in: *21st Annual BACUS Symposium of Photomask Technology*, vol. 4562, 2001, pp. 633–640.
7. M. Mueller, S. Komarov, and K.H. Baik, Dry etching of chrome for photomasks for 100 nm technology using CAR, *Photomask Japan*, 350–360 (2002).
8. Unaxis Website: *http://semiconductors.unaxis.com/en/download/65%20nm%20Dry%20Etch.pdf*.

9. S.A. Anderson, R.B. Anderson, M.J. Buie, M. Chnadrachood, J.S. Clevenger, Y. Lee, N. Sandlin, and J. Ding, Optimization of a 65 nm alternating phase shift quartz etch process, in: *23rd Annual BACUS Symposium on Photomask Technology*, 5256, 66–75 (2003).
10. Semiconductor International Website: *http://www.reed-electronics.com/semiconductor/article/CA163977*.
11. Radiance Services Website: *http://www.radianceprocess.com/rad.html*.
12. R. Novak, I. Kashkoush, and G.S. Chen, Today's binary and EAPSMs need advanced mask cleaning methods, *Solid State Technol.*, February, 45–46 (2004).

18

Mask Materials: Optical Properties

Vladimir Liberman

CONTENTS
18.1 Introduction 377
18.2 Metrology 378
    18.2.1 Bulk Materials 378
    18.2.2 Thin Films on Substrates 379
    18.2.3 Freestanding Thin Pellicle Films 379
    18.2.4 Laser-Based Lifetime Materials Testing 381
18.3 Laser Lifetime Requirements 381
18.4 193- and 248-nm Optical Materials 382
    18.4.1 Fused Silica for Mask Blanks 382
    18.4.2 Pellicles for 193-nm Applications 383
        18.4.2.1 Lifetime Studies 383
        18.4.2.2 Ambient Effects 384
    18.4.3 APSM Materials for 193 and 248 nm 385
18.5 157-nm Lithography Generation 385
    18.5.1 Modified Fused Silica 386
    18.5.2 Pellicle Strategies for 157-nm Lithography 387
        18.5.2.1 Soft Pellicles 387
        18.5.2.2 Hard Pellicles 388
        18.5.2.3 Other Pellicle Solutions 388
    18.5.3 Absorber Materials 389
    18.5.4 Attenuating-Phase Shifting Masks 389
    18.5.5 Contamination and Cleaning 389
Acknowledgments 391
References 391

18.1 Introduction

In the last 15 years, photolithography has moved from utilizing continuous wave (CW) ultraviolet (UV) mercury lamp sources to pulsed excimer lasers, starting with the deep ultraviolet (DUV) generation (248 nm) and recently transitioning to the 193-nm wavelength. The next generation lithographic tools may utilize even the shorter wavelength of 157-nm excimer lasers. This descent into the deep ultraviolet and even the so-called "vacuum" ultraviolet (VUV) region of the spectrum places stringent Xdemands on the optical properties of mask materials. The peak temperatures of irradiated materials are expected to be higher with excimer lasers than those for the CW case because the duty cycle of excimer lasers, even for a multi-kilohertz pulse repetition rate, is in the order of 10^{-5} for a nominal 10 ns pulse width, and hence the peak laser power is quite high. This increase in peak power, coupled with the shorter wavelengths, has placed unprecedented requirements on the durability and purity of optical materials. Indeed, for the 193- and 157-nm lithography generations, lifetime under irradiation is one of the most critical parameters in prescreening choices for mask and pellicle materials.

In this chapter, I will review how mask-related optical materials for the 248–157-nm lithography generations have kept up with the requirements for transparency and durability under laser irradiation. I begin with topics common to all ultraviolet wavelengths, such as metrology considerations and material lifetime expectations. I will then review the state of mask materials for the 248- and 193-nm generations, where remaining challenges are engineering improvements. Then, I will discuss the development of mask blanks, pellicles, absorber materials, and attenuating phase shift mask (APSM) materials for 157-nm lithography. I will also review the challenges posed by photo-induced contamination and cleaning that become particularly pronounced in the VUV range of the spectrum.

18.2 Metrology

In discussing UV metrology relevant to mask materials, I will deal with optical measurements of: (1) bulk material properties which, for example, dominate transmission of 157-nm mask blanks; (2) thin films on substrates, such as absorber materials and APSM materials; and (3) freestanding thin films, such as pellicles. I will briefly discuss relevant issues for each of these measurements. I will also describe strategies for laser-based *in-situ* transmission measurements, required for lifetime material assessment.

18.2.1 Bulk Materials

In evaluating the transmission of mask blanks, one must be mindful of bulk losses, such as bulk absorption, which are intrinsic to the material, as opposed to surface losses, which will be a function of surface finish or surface contamination and are, thus, not intrinsic. In discussing bulk material transmission, it is common to distinguish between internal and external transmission. External transmission is the quantity measured experimentally, which includes surface reflections. Internal transmission can be derived from external transmission (see the following) and describes nonreflective losses in the material. For a sample of length l the internal transmission is

$$T_{\text{int}} = 10^{-(2\beta + \alpha l)} \tag{18.1}$$

where β refers to a single surface loss, including both absorption and scatter, and α is the internal absorption coefficient per centimeter. In Equation (18.1), I ignore the effects of bulk scatter, since they are negligible over a 6-mm mask blank thickness. To obtain the internal transmission as specified in Equation (18.1) from the external transmission, which is actually measured, one can use the following relation:

$$T_{\text{int}} = T_{\text{ext}}(n+1)^4/(16n^2) \qquad (18.2)$$

where n is the refractive index of the material. For fused silica, $n = 1.56$ at 193 nm and $n = 1.51$ at 248 nm. For modified fused silica that is used in 157-nm applications, the index will vary somewhat depending on the amount of fluorine doping but may be taken as 1.67. Thus, working backwards from Equation (18.2) with $T_{\text{int}} = 100\%$, T_{ext} is 87.8%, 90.7%, and 91.9% for 157 nm, 193 nm, and 248 nm wavelengths, respectively. These numbers represent the maximum transmission one can achieve for an uncoated fused silica blank. The actual transmission of the mask blank is evaluated with a spectrophotometer after appropriate surface precleaning to remove contamination.

18.2.2 Thin Films on Substrates

When exploring new APSM or absorber materials, one often finds that optical properties of constituent materials have not been characterized with sufficient accuracy in the ultraviolet region. While transmission/reflection and ellipsometric measurements in the near-ultraviolet can be performed to high accuracy and precision, special challenges arise for metrology below 200 nm [1]. Since both moisture and oxygen are strongly absorbent in that wavelength region, the instruments must either be purged with an inert gas or operated under vacuum. The ambient constraints complicate instrumental design and significantly increase sample transfer and measurement times. All the transmitting windows, bulb envelopes, and photomultiplier windows must be made either of UV grade synthetic fused silica or, for operation below 180 nm, of magnesium fluoride or calcium fluoride. For polarizers operating over a wide wavelength range down to the VUV, Rochon prisms, manufactured of air-spaced magnesium fluoride elements, are used.

In recent years, much progress has been made by metrology equipment companies to address the VUV wavelength region. Robust methods of deriving optical constants of thin films in the VUV region have been demonstrated by both ellipsometric [2,3], as well as reflection/transmission [4], methods. In general, reflection/transmission methods rely on thin film fringe analysis; thus, this technique is most effective for film thicknesses above d_{min} [4]:

$$d_{\text{min}} \approx \lambda_{\text{min}}/2n \qquad (18.3)$$

where λ_{min} is the lowest wavelength of the measured spectrum, and n is the refractive index of the thin film. For $n = 1.5$ and $\lambda_{\text{min}} = 150$ nm, the minimum thickness that can be reliably modeled is ≈ 50 nm. Below this thickness, VUV spectroscopic ellipsometry is a preferred, though considerably more laborious, method of analysis.

18.2.3 Freestanding Thin Pellicle Films

Unsupported soft pellicles (see Section 18.4.2) have thicknesses in the order of a micron. Considering coherent superposition of light over the thickness d of a pellicle, its transmission T_p is given by

$$T_p = \left| \frac{n t_{01}^2 \, e^{-i\delta}}{1 - r_{01}^2 \, e^{-2i\delta}} \right|^2 \qquad (18.4)$$

where

$$n \equiv n_r - ik \tag{18.5}$$

is the complex refractive index,

$$\delta = \frac{2\pi n d}{\lambda} \tag{18.6}$$

$$t_{01} = \frac{2}{n+1} \tag{18.7}$$

$$r_{01} = \frac{1-n}{1+n} \tag{18.8}$$

As a function of wavelength, transmission of a given pellicle will exhibit oscillations with peak positions determined by film thickness and refractive index, and the peak transmission is given by:

$$T_{\text{peak}} = 100 e^{-\alpha d} \tag{18.9}$$

where α is the absorption coefficient defined as

$$\alpha = \frac{4\pi k}{\lambda} \tag{18.10}$$

In measuring peak transmission one has to make sure that the spectrometer spectral resolution, or bandwidth, is set significantly smaller than the fringe spacing; otherwise, the measured transmission profile will be a convolution of the real transmission with the spectrometer bandwidth, resulting in artificially lowered fringe contrast. This effect becomes especially significant in the ultraviolet spectral region (see Figure 18.1).

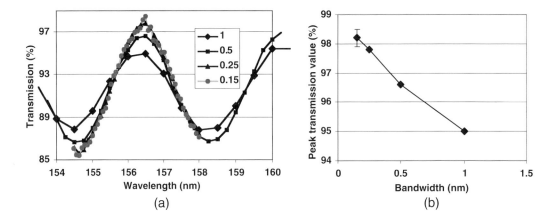

FIGURE 18.1
Effect of spectrometer slit bandwidth on a measurement of a freestanding pellicle film. (a) Transmission spectrum for several bandwidth values. Numbers in the legend refer to bandwidth in nanometers. (b) Peak transmission value as a function of bandwidth.

18.2.4 Laser-Based Lifetime Materials Testing

In addition to initial optical properties, materials must withstand an appropriate lifetime dose of excimer laser irradiation. Lifetime materials testing needs to be done under specified ambient conditions. For 193- and 248-nm lithographies, air is the ambient near the mask, while for 157-nm lithography all tests must be done under nitrogen. Accurate laser-based *in-situ* transmission measurements are indispensable for material lifetime evaluation. The preferred measurement method is laser-based ratiometry (Figure 18.2) [1]. A typical measurement setup involves using a beamsplitter before the sample to reflect part of the incident beam onto a reference detector. The main part of the laser beam is incident on the sample detector, which is positioned directly behind the sample station. First, a sample is retracted out of the beam path and a ratio of the sample to the reference detector is obtained for a 100% measurement. Then, the sample is brought into the transmitted beam, and a sample reading is obtained. The ratio of the sample to the 100% reading yields the transmission of the material under study. For incident laser fluences above $0.05\,\text{mJ}/\text{cm}^2/\text{pulse}$, pyroelectric detectors are well suited to the task of transmission measurements. For fluences much lower than this range, photodiode-based detectors may be more appropriate because of their higher gain.

18.3 Laser Lifetime Requirements

In this section, I will estimate the laser durability requirements for mask materials under laser irradiation. While the calculations were originally done for pellicles, the lifetime target is broadly applicable to a variety of mask-related materials, since it is simply an estimate of the cumulative dose delivered to the pellicle [5].

I assume that pellicle damage is a linear process, therefore, that the correct quantity to specify is cumulative dose. The simplest method for calculating the total dose at the pellicle is to work backwards from the exposure dose E_w required by the resist at the wafer. The dose E_r transmitted through any clear region in the reticle per exposure is

$$E_r = \frac{E_w}{M^2 T} \tag{18.11}$$

where M is the reduction ratio of the projection lens and T is its transmission. To estimate the dose at the pellicle, I make the reasonable assumption that the pellicle is effectively in the far field of the reticle. For a pellicle–reticle separation of 5 mm and a 0.8-NA lens with

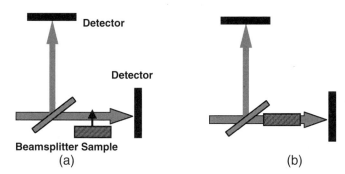

FIGURE 18.2
Schematic representation of *in-situ* laser-based transmission measurement. (a) Measurement configuration for obtaining 100% baseline transmission. (b) Measurement configuration for obtaining sample transmission.

a 4× reduction ratio, reticle features are blurred over a disc approximately 1 mm in radius (5*0.8/4). If we further assume that the spatial variation of patterns on the reticle is primarily much less than 1 mm (i.e., there is little long-range variation in pattern density), the dose delivered to the pellicle per exposure E_p is simply

$$E_p = E_r T_R \qquad (18.12)$$

where T_R is the reticle transmission. Although light diverges between the reticle and the pellicle, the distance is sufficiently short that the spatial extent of the illuminated region is virtually unchanged, and the decrease in energy density is negligible. Finally, the total dose transmitted through the pellicle E_t is the product of the dose per exposure and the total number of fields exposed, or

$$E_t = \frac{FWRE_w T_R}{M^2 T} \qquad (18.13)$$

where F is the number of fields per wafer, W is the total number of wafers printed before repelliclization.

Using Equation (18.13), we can estimate the total dose that a pellicle must be able to withstand. The number of fields F was estimated at about 150 by assuming a 500-mm^2 field and 300-mm wafers (from the International Technology Roadmap for Semiconductors). I assume a relatively bright field reticle with a transmission T_R of 75% as a reasonable upper bound. I estimated the lens transmission T at 30% and a reduction ratio M of 4. The number of wafers that are printed with a given reticle varies substantially between types of products (memory prints many more than microprocessors, which in turn prints many more than ASIC). The current target for E_t of 5 kJ/cm^2 corresponds to a conservative estimate of 10,000 wafers and 20 mJ/cm^2 resist dose. For an estimate of 5000 wafers, and a resist dose of 10 mJ/cm^2, I obtain a total dose of 1.2 kJ/cm^2. Note that these targets vary substantially with input assumptions associated with the product type being printed.

It is also critical to specify the maximum allowable change in transmission over the required lifetime. Targets for previous generations have been 1% [6]. In fact, spatial variation of transmission is of primary concern due to its impact on linewidth control. In practice, this equates to variation with time given the pattern density variation over distance scales greater than the 1-mm radius blur derived. For this reason, 1% remains a reasonable target.

18.4 193- and 248-nm Optical Materials

18.4.1 Fused Silica for Mask Blanks

The optical requirements for fused silica mask blanks are significantly relaxed compared to those for lens materials. Thus, transparency of the material only needs to be insured over a 6-mm thickness. Material lifetime needs to be established over a 5–10-kJ/cm^2 dose at incident pulse fluences of <0.1 mJ/cm^2/pulse. Prior to excimer-laser-based lithography, a great variation in the quality of fused silica existed, with some material damaging quickly under 193-nm laser irradiation even after a few million pulses at fluences of <10 mJ/cm^2/pulse [7]. However, with the advent of high purity UV-grade fused silica with impurity contents below parts per million, achieving consistent transmission of a blank reticle for wavelengths at or above 193 nm has become less of an art but rather a quality control task.

For 248-nm applications, the maximum transmission of a fused silica blank limited only by reflection losses (i.e., 91.9%) is readily realizable. For 193-nm, the losses may be small but are not negligible. Polishing-related surface losses have been measured for a number of samples at $\beta = 0.0015$/surface at 193 nm [8] [see Equation (18.1)], translating into a 0.7% transmission loss for the two surfaces. For unirradiated material, the bulk absorption coefficient α for a number of grades of fused silica is found to be between 0.001 and 0.005/cm for a large majority of samples [8], corresponding to transmission losses of $\approx$0.1–0.7% at 193 nm for a 6-mm thick blank. So, for an uncoated mask blank, an initial transmission of 89–90% is achievable at 193 nm. For 193- and 248-nm laser irradiations at the fluence levels <0.1 mJ/cm^2/pulse, no degradation of high purity excimer-grade fused silica should occur even for doses of tens of kilo-joules per centimeter square. There may be, however, initial laser-induced transient effects, such as bulk bleaching and/or surface cleaning (see also Section 18.5.5). The total cleaning effect for two surfaces has been found to be 1–2% at 193 nm [8]. Transmission bleaching, a bulk phenomenon, may be up to 1% over the reticle thickness, occurring over a dose of 1 kJ/cm^2. By proper selection of raw material, bleaching effects can be minimized.

18.4.2 Pellicles for 193-nm Applications

The use of pellicles to enhance manufacturing yield by protecting a reticle from dust particles has been firmly established in semiconductor industry. For lithography at i-line (365 nm) and g-line (436 nm), nitrocellulose has been widely used as pellicle film. However, this material is not suitable for wavelengths at or below 248 nm because it has neither the initial transparency nor the lifetime under laser irradiation. Therefore, pellicles for the 248- and 193-nm lithographies are typically fluorocarbon derivatives. There are two common types of fluorocarbons used for pellicle films. One of these is Teflon AF from DuPont Corporation, which is a copolymer of tetrafluoroethylene and 2,2-bis(trifluoromethyl)-4,5-difluoro-1,3-dioxole [9]. The other material is CYTOP, manufactured by Asahi Glass, which is based on a perfluorocyclo-oxy-aliphatic monomer unit [10]. While the starting material appears to meet the stringent demands of lithographic applications, such as thickness uniformity and transparency, its degradation when exposed to low-intensity 193-nm irradiation needs to be studied systematically.

18.4.2.1 Lifetime Studies

Previous studies have addressed long-term stability of 248-nm and i-line pellicles, when exposed to 248-nm irradiation, some finding exceptional durability of fluoropolymer pellicles [11–13]. A systematic study of 193-nm pellicle degradation was performed at MIT Lincoln Laboratory [14]. The tests were conducted in a controlled ambient environment, comparing pellicle durability in a nitrogen and an air ambient. The two types of experiments performed were lifetime irradiation tests that attempt to simulate the production environment, and pellicle outgassing studies that are designed to address photochemical changes on a molecular level.

For lifetime irradiation studies, laser irradiation at 400 Hz pulse repetition rate and an incident fluence of about 0.1 mJ/cm^2/pulse were used. In that fluence regime, the lifetime was found to depend only on the total dose, i.e., the number of pulses × fluence per pulse. In the course of irradiation, pellicles were periodically measured *ex-situ* with a UV-visible spectrophotometer. Detailed fringe analysis was employed to extract thickness and optical properties of pellicle material under the laser irradiation. For normal lifetime

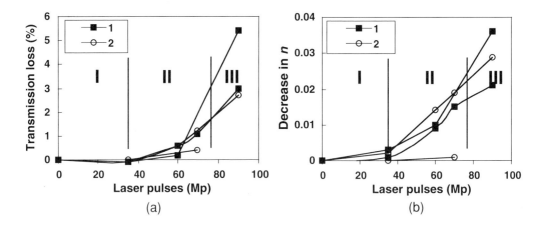

FIGURE 18.3
Results of lifetime tests of several 193-nm pellicles. (a) Laser-induced 193-nm transmission loss. (b) Laser-induced decrease in 193-nm refractive index. "1" refers to pellicles optimized for both 193- and 248-nm use. "2" refers to pellicles optimized just for the 193-nm use. Roman numerals I, II, and III delineate different damage regimes, as defined in the text. Fluence for all exposures is $0.1\,mJ/cm^2/pulse$.

assessment, pellicles were irradiated in flowing air to simulate the mask ambient for 193-nm use. From the measured transmission degradation and modeling, the following regimes of pellicle behavior were observed (see Figure 18.3):

(I) Induction period is about $3\,kJ/cm^2$ incident dose during which no change in transmission occurred.

(II) Initial pellicle degradation, taking place between 3 and $5\,kJ/cm^2$ for most materials, expressed primarily by a decrease in the real part of the refractive index. During this regime, pellicle discoloration could be observed. Pellicles could still be suitable for lithographic use through most of this regime.

(III) Accelerated degradation, for total incident doses above 5–$6\,kJ/cm^2$, expressed by substantial changes in both the refractive index and physical thickness of the pellicle and an increase in absorption at shorter wavelengths.

18.4.2.2 Ambient Effects

To test effects of ambient on pellicle lifetime, researchers compared irradiation under dry nitrogen (<10 ppm O_2) with that in air. Since with oxygen in the ambient irradiation may generate ozone, a high airflow condition was selected so that the ozone concentration was less than 0.15 ppm at all times. Considerably greater degradation for the spot irradiated under N_2 ambient was observed. In order to understand the photochemistry responsible for these markedly different degradation rates, the researchers at MIT Lincoln Laboratory employed a separate gas chromatograph/mass spectrometer-based detection apparatus to measure small amounts of outgassing material evolving from pellicles under 193-nm excimer laser irradiation. Results of these studies and previous research of fluorocarbon photochemistry suggest two alternative pathways for laser-induced changes observed in pellicles:

(1) Irradiation under nitrogen causes formation of radicals, which have long lifetimes (many minutes) in the absence of any quenching reagent, such as oxygen or water (hydrogen). As a result, the radiation-induced radical concentration builds up until radical–radical interactions begin to occur, resulting in

the formation of unsaturated —CF=CF— bonds. The pi-electrons accompanying these bonds then give rise to the increased absorbance in the far ultraviolet.

(2) During irradiation in oxygen, the photo-generated radicals are quickly converted to peroxy groups. When the concentration of peroxy groups becomes high enough, peroxy–peroxy or peroxy–fluorocarbon radical interactions occur, resulting in the formation of organic peroxides, which exhibit lower absorbance than the perfluoroalkenes formed in the absence of oxygen or hydrogen.

18.4.3 APSM Materials for 193 and 248 nm

The use of APSM is a popular resolution enhancing technique. The technology has been applied mostly to printing at the contact levels, but it has been also demonstrated as useful for isolated and dense lines, usually in conjunction with off-axis illumination [15–18]. The geometry of an APSM is similar to that of a binary mask, except the absorber layer is replaced with a film (or stack). This film must meet two fundamental requirements: it transmits a small amount of the incident light, ~5–15%, and it retards the phase of this transmitted light by half a wavelength. As a result, negative interference takes place at the edges of open areas, and the slope of the aerial image is sharpened. Durability of the attenuating films is of a specific concern, since such materials are designed to absorb incident radiation. Furthermore, their optical properties must remain constant not only at the wavelength of use but also at the inspection wavelength, which is typically higher. At 248 nm, APSMs are commonly used in manufacturing, and the material of choice is MoSi [19]. At 193 nm, APSMs are still in the evaluation process. Several material alternatives, such as chromium oxyfluoride [20,21], MoSiON-based layer [22], and Si-based composites [23] have been proposed and tested. It has been found that nonstoichiometric materials are more prone to laser-induced changes through oxidative processes [24]. Minimizing defect densities in the deposited films is important; for example, improved deposition methods or postdeposition annealing show improved durability for chrome oxyfluoride-based films [21].

A novel approach to APSM fabrication for 193- and 248-nm lithographic use involves optical superlattices, comprised out of alternating titanium nitride and silicon nitride layers with thicknesses in the nanometer range [25]. The optical attenuation is controlled by increasing the TiN_x layer thicknesses. The 193-nm radiation durability of these films improves with increasing the number of bilayers from 1 to 10 for the same total thickness. Furthermore, multilayer stacks whose outer layer was TiN_x had better radiation hardness than an equivalent stack with a SiN_x as the outer layer. These trends suggest that intrinsic layer properties rather than defect density (which one would expect to increase with the larger number of interfaces) dominate degradation behavior.

18.5 157-nm Lithography Generation

As we have seen from the discussion earlier for 193-nm materials, both the initial optical properties and lifetimes of mask-related materials are either close to or are already meeting production requirements. However, 157-nm lithography presents a set of unique challenges to the mask community. The number of materials that retain sufficient transparency below 193 nm is significantly limited. For instance, conventional fused silica used for 193-nm applications and soft pellicles for 193-nm have neither the initial transparency nor the lifetime required for use at 157 nm.

18.5.1 Modified Fused Silica

Even the highest quality 193-nm fused silica is not transmissive enough at 157 nm. During the initial exploration of 157 nm as a candidate lithographic wavelength, a considerable effort was spent on identifying potential replacements for the fused silica as a mask material. A number of fluoride materials, such as calcium fluoride or magnesium fluoride, have excellent transparency in that wavelength region. However, the single crystal fluoride materials have coefficients of thermal expansion, which are more than an order of magnitude higher than that of fused silica. This fact raised considerable concern that thermally induced distortions due to either mask writing or wafer exposure would be unacceptably large in printed features [26,27]. Furthermore, if fluoride materials were used for mask blanks, completely new chemical processes would have to be developed by mask blank suppliers and mask houses. So even while the industry sets up a task force to address the manufacture and characterization of full size CaF_2 mask blanks, the search for improvement of fused silica continued.

In fact, a closer look at the absorption edge of fused silica as early as 1978 suggested a possibility of extending its transmission range through potential material modification [28]. The solution came from the optical communications industry, where doping of a silica core fiber with fluorine was used to reduce the refractive index of the material without incurring additional transmission losses. In 1993, Kyote et al. [29] showed that this fluorine doping method was able to significantly extend the transmission range of fused silica to the VUV. In the late 1990s, further progress was made to improve the transparency of modified fused silica by refinement of earlier techniques [30]. The general approach consists of two steps: reduce the hydroxyl content of fused silica since the OH moieties absorb at wavelengths above ~155 nm; and introduce fluorine into the glassy network, in order to titrate all the unpaired electrons on the silicon atoms. Figure 18.4 shows the transmission comparison of conventional UV grade fused silica and modified fused silica [31]. Several modified fused silica samples were evaluated at MIT Lincoln Laboratory, and indeed some met the criteria for low initial absorption, as well as for stable transmission at the pulse counts and fluences listed earlier. The intrinsic absorption coefficient of the modified fused silica can be as low as 0.02/cm (base 10) at 157 nm [31], resulting in a transmission of 85.5% over a 6-mm thick blank assuming no surface losses.

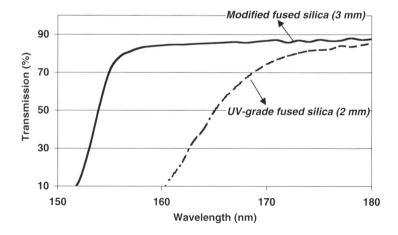

FIGURE 18.4
Transmission comparison of a standard UV-grade fused silica and a modified fused silica samples.

18.5.2 Pellicle Strategies for 157-nm Lithography

Just as the conventional fused silica is unsuitable for 157-nm use, conventional 193-nm fluorocarbon pellicle materials do not have sufficient transparency or the requisite lifetime under laser irradiation. In addition to the resulting change in transmission, outgassing from these pellicles would most likely be unacceptably high for the lens environment. Similar to the concerns of the mask blank community described earlier, a radical departure from the 193-nm pellicle concept would involve a substantial burden to the infrastructure, including pellicle suppliers, lithographic tool manufacturers, and end users. In an earlier industrial poll [32], almost 60% of the participants preferred soft pellicles as a solution for the 157-nm pellicle technology. Unfortunately, as of today, soft pellicles with appropriate laser lifetime have not been engineered. In the following, I will describe the current problems of soft pellicle durability. I will also mention other proposed pellicle solutions, such as hard pellicles or removable pellicles.

18.5.2.1 Soft Pellicles

Conventional fluorocarbon pellicle materials, such as Teflon AF, have transparencies of ≈50%/μm at 157 nm. Irradiation tests performed at MIT Lincoln Laboratory showed that their transmission degrades rapidly for incident laser doses on the order of 10 J/cm^2, leading to eventual pellicle rupture (Figure 18.5). During the last two years, comprehensive studies have been undertaken to develop materials with higher initial transparencies and higher laser durability. One approach for modification of standard Teflon-based fluorocarbons favored promoting alternation in the polymer backbone to avoid long CF_2 and CH_2 runs [33]. Thickness-optimized polymers with transparencies as high as 98% per micron were demonstrated with this approach. Initial laser-based lifetime testing of these materials was promising, suggesting a correlation between lifetime and absorption (Figure 18.6). However, further efforts aimed at increasing material lifetime did not show promise. In a comprehensive set of experiments that examined laser durability as a function of polymer composition, effects of testing ambient that included higher

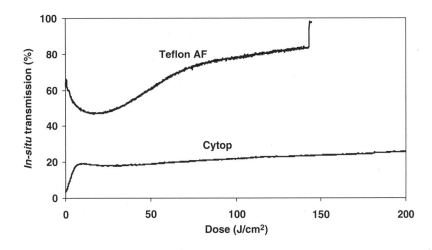

FIGURE 18.5

The 157-nm laser-induced degradation of 1-μm-thick pellicles designed for 193-nm applications. Pellicles were exposed at 500 Hz and 0.1 mJ/cm^2/pulse. The sharp discontinuity in the Teflon AF trace at 150 J/cm^2 indicates bursting of the membrane.

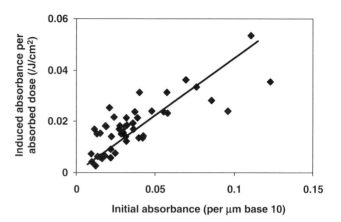

FIGURE 18.6
Induced absorbance per absorbed dose versus initial absorbance for several developmental fluorocarbon thin films for 157-nm pellicle applications. Exposure fluence was $0.1\,mJ/cm^2/pulse$. Solid line is drawn to guide the eye.

concentration of oxygen and moisture and various pretreatments of freestanding films, the longest lifetime obtained was around $5\,J/cm^2$ [5].

Several factors may be limiting soft pellicle lifetimes at 157 nm. Photochemical decomposition rates are determined by a combination of reaction quantum yields and absorbance. Comparative studies of polymer photochemistry at the three lithographic wavelengths did not find large differences in quantum yields for chain scission and crosslinking versus wavelength [34]. It is plausible, therefore, that significant reduction in absorbance can result in reduced degradation at 157 nm. However, changes in polymer chemical structure, such as introduction of reactive hydrogen into the backbone, can open new photochemical channels for polymer degradation. Photochemical hydrogen abstraction reactions leading to formation of absorbing carbon–carbon double bonds have been described in the literature [35]. Additionally, extrinsic effects, such as incorporation of airborne hydrocarbons into the pellicle membrane, can substantially increase absorbance of pellicle material even for low intrinsic absorbance.

18.5.2.2 Hard Pellicles

In this proposed solution, the pellicle is manufactured from a thinned-down fluorinated fused silica to a sub-millimeter thickness [36]. The optical properties and the lifetime under 157-nm irradiation are not a concern for this material, as described in Section 18.5.1. Since the pellicle is thick enough to become an optical element, any distortions due to gravity, mounting, or thermal stresses have to be minimized [37]. Hard pellicles have been a subject of several publications and have been addressed by an industrial task force [32]. As of now, it appears to be a viable solution for the 157-nm lithography, through added complexity of a lithographic tool design and higher projected costs.

18.5.2.3 Other Pellicle Solutions

Other ideas for mask protection have been inspired by next-generation lithographic technologies, such as x-ray or EUV, where the use of conventional pellicles is not possible. For instance, micron-thickness inorganic pellicles would alleviate lifetime issues associated with organic materials while keeping the pellicle thin enough that it does not become a lens element [38]. Such membranes, however, need to be made with very low residual stress in order to avoid breakage. Other particle protection ideas involve thermophoretic methods or the use of removable pellicles [39].

18.5.3 Absorber Materials

Absorber material thicknesses continue to shrink in an effort to minimize the aspect ratio of mask features. At the same time, optical density of 3.0 is considered a minimum safe value for the binary photomasks. Commercial 193- and 248-nm absorber films employ graded structures consisting of chromium and chromium oxynitride layers for control of the reflectivity. There were some concerns that these materials would be too transparent at 157 nm [40]. For instance, the absorption coefficient of bulk chromium drops by about 20% from 193 to 157 nm [41]. However, detailed measurements of 193-nm-based absorbers showed their extendibility to 157 nm [42], at least in equivalent thickness.

18.5.4 Attenuating-Phase Shifting Masks

Much less work has been done to develop 157-nm APSMs as compared to the 193- and 248-nm studies described in Section 18.4.3. A few candidate materials have been reported, such as TaSiO [43], chromium aluminum oxynitride [44], and composite material films consisting of a silicon dioxide host combined with various oxides and nitrides, generally of refractory metals [45]. The limited available data indicate that some of these materials are modified by laser irradiation during use [46]. A bilayer approach utilizing a stack with a thin metal layer for an attenuator and a transparent spin-on-glass layer for phase adjustment has also been demonstrated [47]. This APSM stack offers easy control of attenuation and phase, has good durability, and the approach is extendible to higher wavelengths.

18.5.5 Contamination and Cleaning

There is a shared concern within the lithographic community that special transport and storage procedures must be maintained to prevent the exposure of mask materials to airborne contaminants, especially at the short wavelength of 157 nm. Transmission uniformity requirements on the reticle are $\pm 0.25\%$. For comparison, 1–2 monolayers of physisorbed hydrocarbon or water results in approximately 1% transmission drop per surface at 157 nm [48] (see Figure 18.7). Total transmission nonuniformities on this same order have previously been reported on reticles after shipping [49], presumably from localized variations in surface coverage and contaminant chemistry. A potential alternative to specialized transport procedures is performing a UV-based pretreatment of the reticle prior to its insertion in the scanner beam line. This irradiation at short wavelengths in the presence of oxygen has been found to be very effective in removing hydrocarbon and water-based contaminants [50,51]. It is an all-dry process that can also be performed "off-line" in a separate compartment to prevent potential recontamination of lens elements in the projection optics. The reaction mechanism for material removal has been studied before [50–52] and has been found to be between UV-induced radicals formed in the polymer film and gas phase oxygen molecules reacting with the surface, leading to the production of volatile products. Photo-induced ozone formed in the gas phase significantly enhances the reaction.

Detailed studies have been performed at MIT Lincoln Laboratory on the removal of hydrocarbon contaminants from reticle surfaces by irradiating them with UV lamps in an oxygen-containing ambient [53]. Immediately after irradiation, the samples were load-locked into a vacuum UV spectrometer. In one set of experiments, 157-nm mask blanks were shipped with no special handling or storage procedures. It was found that water readily desorbed from modified fused silica once placed within a dry nitrogen

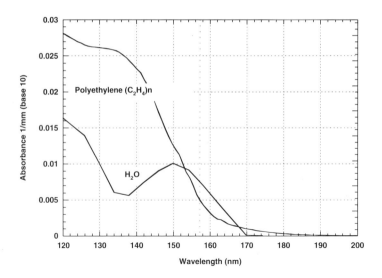

FIGURE 18.7
Absorbance per nanometer of a thin layer of polyethylene and condensed water. (Data from Ref. [48].)

environment, but that hydrocarbon residues tended to remain on the surface until cleaned with the lamp. After lamp-based cleaning of these reticles, a transmission recovery on the order of 1% at 157 nm was found to occur.

The second set of experiments provided a controlled level of precontaminant using highly conjugated hydrocarbon films. These materials provide a "worst case" scenario as far as the chemistry of removable hydrocarbon contaminants is concerned. Clearing doses are shown in Figure 18.8(A) for ~10 nm films at oxygen concentrations ranging from 0.1% to 10%. The removal rate is independent of the oxygen concentration within ~15%, and accordingly with the ozone concentration [54]. Figure 18.8(B) compares the removal rates at two different wavelengths: with a 172-nm lamp delivering 2 mW/cm^2 and with a mercury bulb with an output of 30 mW/cm^2 total power, delivering 1.5 mW/cm^2 into the 185 nm

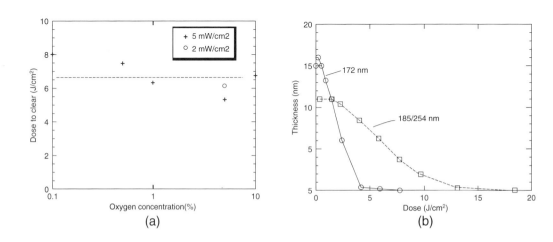

FIGURE 18.8
Clearing dose required to remove a 10-nm film of hydrocarbon (a) using the 172-nm bulb as a function of oxygen concentration. The estimated normalized film thickness as a function of incident dose at the surface is shown in (b) for 1% oxygen and two wavelengths.

line and the rest into the 254 nm line. The results suggest that 185 nm is significantly more efficient compared to 254 nm at generating radicals in these graphitized films.

Finally, changes have been observed in the reflectivity and surface roughness of chrome-based absorber films irradiated in high levels of oxygen for extended periods of time. This effect may place a constraint on the total exposure time and oxygen levels used in lamp cleaning of reticles, due to an increase in flare and possible complications during mask inspection.

Acknowledgments

I would like to thank my colleagues at MIT Lincoln Laboratory for many useful discussions and insights. I would also like to thank our many industrial collaborators for providing samples for studies and for discussions of the data. This work was sponsored by Collaborative Research and Development Agreements between MIT Lincoln Laboratory and SEMATECH and between MIT Lincoln Laboratory and Intel Corporation. Opinions, interpretations, conclusions, and recommendations are those of the authors and are not necessarily endorsed by the United States Government.

References

1. V. Liberman and M. Rothschild, Ultraviolet and vacuum ultraviolet sources and materials for lithography, in: P. Misra and M.A. Dubinskii (eds.), *Ultraviolet Spectroscopy and UV Lasers*, Marcel Dekker, New York, 2002, pp. 1–33.
2. J.N. Hilfiker, B. Sing, R.A. Synowicki, and C.L. Bungay, Optical characterization in the vacuum ultraviolet with variable angle spectroscopic ellipsometry: 157 nm and below, *Proc. SPIE*, 3998, 390–398 (2000).
3. P. Boher, J.P. Piel, P. Evrard, and J.P. Stehle, New purged UV spectroscopic ellipsometer to characterize thin films and multilayers at 157 nm, *Proc. SPIE*, 3998, 379–389 (2000).
4. V. Liberman, T.M. Bloomstein, and M. Rothschild, Determination of optical properties of thin films and surfaces in 157-nm lithography, *Proc. SPIE*, 3998, 480–491 (2000).
5. A. Grenville, V. Liberman, M. Rothschild, J.H.C. Sedlacek, R.H. French, R.C. Wheland, X. Zhang, and J. Gordon, Behavior of candidate organic pellicle materials under 157 nm laser irradiation, *Proc. SPIE*, 4691, 1644–1653 (2002).
6. D. Cote, K. Andresen, D. Cronin, H. Harrold, M. Himel, J. Kane, J. Lyons, L. Markoya, C. Mason, D. McCafferty, M. McCarthy, G. O'Connor, H. Sewell, and D. Williamson, Micrascan III-performance of a third generation, catadioptric step and scan lithographic tool, *Proc. SPIE*, 3051, 806–816 (1997).
7. J.H.C. Sedlacek and M. Rothschild, Optical materials for use with excimer lasers, *Proc. SPIE*, 1835, 80–88 (1993).
8. V. Liberman, M. Rothschild, J.H.C. Sedlacek, R.S. Uttaro, A. Grenville, A.K. Bates, and C. Van Peski, Excimer-laser-induced degradation of fused silica and calcium fluoride for 193-nm lithographic applications, *Opt. Lett.*, 24, 58–60 (1999).
9. P.R. Resnick, The preparation and properties of a new family of amorphous fluoropolymers: Teflon AF, *Proc. MRS*, 167, 105–110 (1990).
10. Y. Fukumitsu M. Kohno. Dust cover superior in transparency for photomask reticle use and process for producing same. U.S. Patent Number 4,657,805.

11. M. Kashiwagi, H. Matsuzaki, and N. Nakayama, Development of deep-UV and excimer pellicle (membrane longevity), *Proc. SPIE*, 2793, 372–386 (1996).
12. W.N. Partlo and W.G. Oldham, Transmission measurements of pellicles for deep-UV lithography, *IEEE Trans. Semicon. Manufact.*, 4, 128–133 (1991).
13. P. Yan and H. Gaw, Lifetime and transmittance stability of pellicles of I-line and KrF excimer laser lithography, in: *Proceedings of the KTI Microelectronics Seminar*, 1989, p. 261.
14. V. Liberman, R.R. Kunz, M. Rothschild, J.H.C. Sedlacek, R.S. Uttaro, A. Grenville, A.K. Bates, and C. Van Peski, Damage testing of pellicles for 193-nm lithography, *Proc. SPIE*, 3334, 480–495 (1998).
15. M. Fritze, P.W. Wyatt, D.K. Astolfi, P. Davis, A.V. Curtis, D.M. Preble, S.G. Cann, S. Deneault, D. Chan, J.C. Shaw, N.T. Sullivan, R. Brandom, and M.E. Mastovich, Application of attenuated phase-shift masks to sub-0.18-μm logic patterns, *Proc. SPIE*, 4000, 1179–1192 (2001).
16. R.J. Socha, W.E. Conley, X. Shi, M.V. Dusa, J.S. Petersen, F. Chen, K. Wampler, T. Laidig, and R. Caldwell, Resolution enhancement with high-transmission attenuating phase-shift masks, *Proc. SPIE*, 3748, 290–314 (1999).
17. Y.-M. Ham, S.-M. Kim, S.-J. Kim, S.-M. Bae, Y.-D. Kim, and K.-H. Baik, Sub-120-nm technology compatibility of attenuated phase-shift mask in KrF and ArF lithography, *Proc. SPIE*, 4186, 359–371 (2001).
18. C.-M. Wang, S.-J. Lin, C.-H. Lin, Y.-C. Ku, and A. Yen, Printing 0.13-μm contact holes using 193-nm attenuated phase-shifting masks, *Proc. SPIE*, 4186, 275–286 (2001).
19. R. Jonckheere, K. Ronse, O. Popa, and L. Van den Hove, Molybdenum silicide based attenuated phase-shift masks, *J. Vac. Sci. Technol. B.*, 12, 3765–3772 (1994).
20. K. Nakazawa, T. Matsuo, T. Onodera, H. Morimoto, H. Mohri, C. Hatsuta, and N. Hayashi, CrO_xF_y as a material for attenuated phase-shift masks in ArF lithography, *Proc. SPIE*, 4066, 682–687 (2000).
21. K. Mikami, H. Mohri, H. Miyashita, N. Hayashi, and H. Sano, Development and evaluation of chromium-based attenuated phase shift masks for DUV exposure, *Proc. SPIE*, 2512, 333–342 (1995).
22. S. Kanai, S. Kawada, A. Isao, T. Sasaki, K. Maetoko, and N. Yoshioka, Development of a MoSi-based bilayer HT-PSM blank for ArF lithography, *Proc. SPIE*, 4186, 846–852 (2001).
23. S.J. Chey, C.R. Guarnieri, K. Babich, K.R. Pope, D.L. Goldfarb, M. Angelopoulos, K.C. Racette, M.S. Hibbs, M.L. Gibson, and K.R. Kimmel, Novel Si-based composite thin films for 193/157-nm attenuated phase-shift mask (APSM) applications, *Proc. SPIE*, 4346, 798–805 (2001).
24. B.W. Smith, L. Zavyalova, A. Bourov, S. Butt, and C. Fonseca, Investigation into excimer laser radiation damage of deep ultraviolet optical phase masking films, *J. Vac. Sci. Technol. B.*, 15, 2444–2447 (1997).
25. P.F. Carcia, R.H. French, G. Reynolds, G. Hughes, C.C. Torardi, M.H. Reilly, M. Lemon, C.R. Miao, D.J. Jones, L. Wilson, and L. Dieu, Optical superlattices as phase-shift masks for microlithography, *Proc. SPIE*, 3790, 23–35 (1999).
26. T.M. Bloomstein, M. Rothschild, R.R. Kunz, D.E. Hardy, R.B. Goodman, and S.T. Palmacci, Critical issues in 157 nm lithography, *J. Vac. Sci. Technol. B.*, 16, 3154–3157 (1998).
27. J. Chang, A. Abdo, B. Kim, T. Bloomstein, R. Engelstad, E. Lovell, W. Beckman, and J. Mitchell, Thermomechanical distortions of advanced optical reticles during exposure, *Proc. SPIE*, 3676, 756–767 (1999).
28. P. Kaminow, B.G. Bagley, and C.G. Olson, Measurements of the absorption edge in fused silica, *Appl. Phys. Lett.*, 32, 98–99 (1978).
29. M. Kyote, Y. Ohoga, S. Ishikawa, and Y. Ishiguro, Characterization of fluorine-doped silica glasses, *J. Mat. Sci.*, 28, 2738–2744 (1993).
30. L. A. Moore C. Smith, Vaccum ultraviolet transmitting silicon oxyfcuoride Cithography glass. U.S. Patent Number 6,242,136.
31. V. Liberman, T.M. Bloomstein, M. Rothschild, J.H.C. Sedlacek, R.S. Uttaro, A.K. Bates, C. Van Peski, and K. Orvek, Materials issues for optical components and photomasks in 157-nm lithography, *J. Vac. Sci. Technol. B.*, 17, 3273–3279 (1999).
32. J. Cullins and E. Muzio, 157-nm photomask handling and infrastructure: requirements and feasibility. *Proc. SPIE*, 4346, 52–60 (2001).

33. Roger H. French, Robert C. Wheland, Weiming Qiu, M.F. Lemon, Edward Zhang, Joseph Gordon, Viacheslav A. Petrov, Victor F. Cherstkov, and Nina I. Delaygina, Novel hydrofluorocarbon polymers for use as pellicles in 157 nm semiconductor photolithography, *J. Fluorine Chem.* (to be published). Vol. 122 (2003) pp. 63–80.
34. T.H. Fedynyshyn, R.R. Kunz, R.F. Sinta, R.B Goodman, and S.P. Doran, Polymer photochemistry at three advanced optical wavelengths, in: *Forefront of Lithographic Materials Research, Proceedings of International Conference on Photopolymers*, vol. 12, 2000, pp. 3–16.
35. R.P. Wayne, *Principles and Applications of Photochemistry*, Oxford University Press, Oxford, 1988, pp. 142–145.
36. J. Miyazaki, T. Itani, and H. Morimoto, Requirements for reticle and reticle material for 157 nm lithography: requirements for hard pellicle, *Proc. SPIE*, 4186, 415–422 (2001).
37. E.P. Cotte, A.Y. Abdo, R.L. Engelstad, and E. Lovell, Dynamic studies of hard pellicle response during exposure scanning, *J. Vac. Sci. Technol. B.*, 20, 2995–2999 (2002).
38. J.R. Maldonado, S. Cordes, J. Leavey, R. Acosta, F. Doany, M. Angelopoulos, and C. Waskiewicz, Pellicles for X-ray lithography masks, *Proc. SPIE*, 3331, 245–254 (1998).
39. P. Mangat and S. Hector, Review of progress in extreme ultraviolet lithography masks, *J. Vac. Sci. Technol. B.*, 19, 2612–2616 (2001).
40. B.W. Smith, A. Bourov, M. Lassiter, and M. Cangemi, Masking materials for 157 nm lithography, *Proc. SPIE*, 3873, 412–420 (1999).
41. E.D. Palik (ed.), *Handbook of Optical Constants of Solids*, vol. II, Academic Press, San Diego, 1991, pp. 374–384.
42. D.A. Harrison, J.C. Lam, G.G. Li, A.R. Forouhi, and G. Dao, Modeling of optical constants of materials comprising photolithographic masks in the VUV, *Proc. SPIE*, 3873, 844–852 (1999).
43. O. Yamabe, K. Watanabe, and T. Itani, Evaluation of high-transmittance attenuated phase shifting mask for 157-nm lithography, *Jpn. J. Appl. Phys.*, 41, 4042–4045 (2002).
44. S. Kim, E. Choi, H. Kim, J. Son, and K. No, Simulation and characterization of silicon oxynitrofluoride films as a phase-shift mask material for 157-nm optical lithography, *Proc. SPIE*, 4691, 1696–1702 (2002).
45. B.W. Smith, A. Bourov, L. Zavyalova, and M. Cangemi, Design and development of thin film materials for 157-nm and VUV wavelengths: APSM, binary masking, and optical coatings applications, *Proc. SPIE*, 3676, 350–359 (1999).
46. V. Liberman, M. Rothschild, N.N. Efremow Jr., S.T. Palmacci, J.H.C. Sedlacek, and A. Grenville, Long-term laser durability testing of optical coatings and thin films for 157-nm lithography, *Proc. SPIE*, 4691, 568–575 (2002).
47. V. Liberman, M. Rothschild, S.J. Spector, K.E. Krohn, S.C. Cann, and S. Hien, Attenuating phase shifting mask at 157 nm, *Proc. SPIE*, 4691, 561–567 (2002). pp. 957–988.
48. E.D. Palik (ed.), *Handbook of Optical Constants of Solids*, vol. II, Academic Press, San Diego, 1991.
49. J. Zheng, R. Kuse, A. Ramamoorthy, G. Dao, and F. Lo, Impact of surface contamination on transmittance of modified fused silica for 157 nm lithography, *Proc. SPIE*, 4186, 767 (2000).
50. J.R. Vig and J.W. Le Bus, UV/ozone cleaning of surfaces, *IEEE Trans. Parts, Hybrids, Packaging*, PHP-12, 365 (1976).
51. M.E. Frink, M.A. Folkman, and L.A. Darnton, Evaluation of the ultraviolet/ozone technique for on-orbit removal of photolyzed molecular contamination from optical surfaces, *Proc. SPIE*, 1754, 286–294 (1992).
52. B. Ranby and J.F. Rabek, Photodegradation, *Photo-oxidation and Photostabilization of Polymers*, John Wiley and Sons, New York, 1975.
53. T.M. Bloomstein, V. Liberman, M. Rothschild, N.N. Efremow Jr., D.E. Hardy, and S.T. Palmacci, UV cleaning of contaminated 157-nm reticles, *Proc. SPIE*, 4346, 669–675 (2001).
54. P. Warneck, *Chemistry of the Atmosphere*, Academic Press, New York, 1988, pp. 100–102.

19

Pellicles

Yung-Tsai Yen, Ching-Bore Wang, and Richard Heuser

CONTENTS

- 19.1 History 396
- 19.2 Overview of Pellicles 396
  - 19.2.1 Introduction 396
  - 19.2.2 The Use of a Pellicle 396
- 19.3 Optical Requirements 396
  - 19.3.1 Transmission Versus Film Thickness 396
  - 19.3.2 Particle Size Versus Frame Height 398
  - 19.3.3 Focus Change Versus Film Thickness 399
- 19.4 Anatomy of a Pellicle 399
  - 19.4.1 Film 400
    - 19.4.1.1 Manufacturing Process 400
    - 19.4.1.2 Transmission and Material 400
    - 19.4.1.3 AR Coating 402
  - 19.4.2 Frame 402
    - 19.4.2.1 Frame Coating 402
    - 19.4.2.2 Vent Hole and Filter 403
  - 19.4.3 Mounting Adhesive or Gasket 403
  - 19.4.4 Backside Cover 404
- 19.5 Inspection 404
  - 19.5.1 Inspection of Film Transmission and its Uniformity 404
  - 19.5.2 Inspection of Film Particles 405
  - 19.5.3 Inspection of Frame 405
  - 19.5.4 Inspection of Adhesive 406
- 19.6 Handling and Environment 406
  - 19.6.1 Controlling Static Charge 406
  - 19.6.2 Mounting 407
  - 19.6.3 Cleaning 407
- 19.7 Long-Term Stability of a Pellicle on a Photomask 407
  - 19.7.1 Outgassing and Crystallization 407
  - 19.7.2 Material Stability 408
- 19.8 The Future of Pellicles 409
  - 19.8.1 Soft Pellicle 409
  - 19.8.2 Hard Pellicle 409
  - 19.8.3 Removable Pellicle or Cover 409
  - 19.8.4 No Pellicle 409
- References 410

19.1 History

The term "pellicle" is used to mean "film," "thin film," or "membrane." Beginning in the 1960s, thin film stretched on a metal frame, also known as a pellicle, was used as a beam splitter for optical instruments. It has been used in a number of instruments to split a beam of light without causing an optical path shift due to its small film thickness. In 1978, Shea and Wojcik [1] at IBM patented a process to use the pellicle as a dust cover to protect a photomask or reticle (hence all will be called "photomask" in the rest of this chapter). In this chapter, the word "pellicle" will be used only to mean a "thin film dust cover to protect a photomask."

19.2 Overview of Pellicles

19.2.1 Introduction

A pellicle is used primarily for two purposes — to increase die yield and reduce overall photomask handling, i.e., cleaning and inspection. It is a thin film stretched on a frame used to protect a photomask from particle contamination. Today, the pellicle has become an integral component in the manufacturing process for most IC manufacturers, and high-resolution projection photolithography systems use it in the manufacturing of thin film magnetic reading heads, LCD flat panels, micro-electromechanical system (MEMS), etc.

19.2.2 The Use of a Pellicle

During the printing process, the image of any particle on the pellicle film will be out of focus on the wafer plane and therefore has only a blurry shadow, which has a minimal effect on the photomask's image on a wafer. Without pellicle protection, a photomask can easily get a particle on its surface and form a distorted image on the wafer, creating a defect on a chip. Before the time when pellicles were used, a photomask would require daily cleaning and inspection. Consequently, the photomask would easily become contaminated from the environment or damaged from the cleaning process, resulting in a low die yield and high replacement costs.

The use of a pellicle in an optical projection system is illustrated in Figure 19.1. Once a pellicle is properly attached on the photomask, the surface that is covered by the pellicle is free from future outside particle contamination. The original quality of the photomask can therefore be preserved. Now only a brief inspection of the pellicle film and photomask surface is required to insure the quality of the photomask.

19.3 Optical Requirements

19.3.1 Transmission Versus Film Thickness

A pellicle needs a good light transmission and long-term transmission stability. The transmission depends on the film thickness, film material, and any antireflective coating

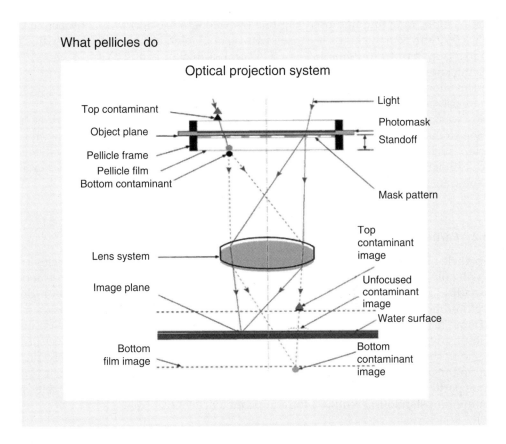

FIGURE 19.1
The use of a pellicle.

(ARC) on the film. The Transmission stability depends on the material used, the light wavelength, and intensity used for the optical system.

The transmission of a thin film is dependent on the film thickness, light wavelength, incident angle, and light absorption of the film. A normal incident light on a nonabsorptive thin film, the maximum transmission of the film happens when the film thickness is an integer multiple of an optical half wavelength, i.e., the half wavelength of the light divided by the refractive index. That is $k\lambda/2n$, where $k = 0, 1, 2, \ldots$, λ is the wavelength, and n is the refractive index.

The transmission minimum happens when the film thickness is at the optical quarter wavelength from a transmission maximum. That is $k\lambda/2n + \lambda/4n$. The minimum transmission is $(1 - 4((n-1)/(n+1))^2)$. For $n = 1.5$, the minimum transmission is 0.84. From a transmission spectrum of a thin film, one can calculate the thickness and the refractive index.

For a projection aligner or a wafer stepper using multiple wavelengths, the film thickness selection is dependent on the light source intensity and photoresist sensitivity at different wavelengths. The film thickness is usually chosen to have a stable transmission over some thickness range. For a multiple wavelength system, the transmission versus film thickness shows a "beat" pattern and stabilizes to a constant transmission when the thickness is high. Ronald S. Hershel published a good introduction paper on pellicles in 1981 [2].

For example, for a thin film used at g-line (wavelength 436 nm), with a refractive index of 1.5, a 0.72-μm film thickness is chosen with a small tolerance of ± 0.01 μm, while for a broadband system from g-line to i-line (365 nm), a thickness of 2.85 μm is chosen with large thickness tolerance of ± 0.2 μm. It has to be noted that for a broadband system because photoresist sensitivity is wavelength dependent, each system with different photoresist values can have a different optimal thickness.

ARC was introduced to improve the transmission and reduce the sensitivity of transmission to the film thickness variation. At the beginning, vacuum deposition with metal fluoride, such as calcium fluoride, was used. ARC with fluoropolymer was first used by Micro Lithography, Inc. (MLI) in 1984. Multiple-layer ARC was also introduced later [3].

Some transmission curves of different films are shown in Figure 19.3.

19.3.2 Particle Size Versus Frame Height

The pellicle film is kept at a fair distance from the photomask, so that any particles on the film will only give a blurry shadow on the wafer. However, if the particle is large enough, the shadow can reduce enough intensity of the light on the wafer to cause a defect. Therefore, the particle size and the distance between the film and photomask surface, i.e., frame height or standoff, need to be considered. The standoff is dependent on the maximum expected particle size and allowable light intensity reduction in the process and should be proportional to particle size and reduction ratio, and inversely proportional to the numerical aperture and partial coherence of the illumination system. For an opaque particle in a single lens system, the following calculation can be used to determine the minimum standoff D required for an opaque particle with diameter P:

$$D = P(M/\text{NA}/\sigma)\left[R^{-0.5} - 1\right]/2$$

where M is the reduction ratio, R is the percentage of intensity reduction, NA is the numerical aperture of the projection optics on the wafer, and σ is the partial coherence of the illumination system. The particle image diameter is P/M, and the distance from the particle image to the photomask image on the wafer is D/M^2. The diameter of the shadow is $(2\text{NA} \cdot D/M^2 + P/M)$. Examples of the reduction ratios and numerical aperture values are shown below:

- First Perkin Elmer wafer aligner, 1973: $M = 1$, NA $= 0.167$
- First GCA stepper, 1976: $M = 10$, NA $= 0.28$
- Most current wafer stepper, 2003: $M = 4$, NA $= 0.6$

Assuming the partial coherence σ from the illumination system is 1, the following examples show the different calculations used in determining standoff with different reductions in intensity on the wafer plan:

- For $R = 4\%$ reduction in intensity: $D = (M/\text{NA}) \times P \times 2$,
- For $R = 1\%$ reduction in intensity: $D = (M/\text{NA}) \times P \times 4.5$,
- For $P = 0.1\,\text{mm} = 100\,\mu\text{m}$, the calculated standoff is presented in Table 19.1.

For particles on the glass surface, the apparent distance from particles to the pattern surface is D/n, where D is the thickness of the glass and n is the refractive index of the glass.

TABLE 19.1
Standoff for Different Wafer Steppers and Aligners

	M	NA	M/NA	D @ R = 4% (mm)	D @ R = 1% (mm)
Perkin Elmer	1	0.167	6.0	1.2	2.7
GCA	10	0.28	36	7.2	16
4:1	4	0.6	7	1.4	3.2

For a real system, the minimum standoff is also dependent on the partial coherence of the illumination system, wavelength of the light, and size of the pattern, i.e., the diffraction pattern. Pei-Yang Yan et al. [4] published a paper on the printability of pellicle defects for deep UV (DUV) lithography in 1992.

19.3.3 Focus Change Versus Film Thickness

In addition, the film thickness of the pellicle on the pattern side can change the depth-of-focus by $t/n/M^2$, where n is the refractive index of the pellicle film and t is the thickness of the film. For a 1:1 broadband projection wafer aligner using a pellicle with 2.85 μm film thickness and refractive index of 1.5, there is a 1.90-μm depth-of-focus correction, while for a 4:1 stepper the depth-of-focus change is only $t/24$. For a film thickness of 1 μm, this is a shift of only 0.04 μm.

19.4 Anatomy of a Pellicle

There are several key features to a pellicle that allow it to properly perform its function. A typical pellicle is shown in Figure 19.2.

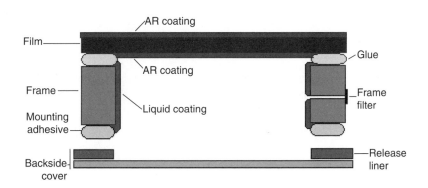

FIGURE 19.2
Cross-section of a Micro Lithography, Inc. (MLI) pellicle.

19.4.1 Film

The Film in a pellicle provides a physical barrier to prevent outside contamination, i.e., particles or vapor outgassing, from reaching the photomask surface. At the same time, because it is thin it provides an optical path with minimum focus and transmission distortion.

19.4.1.1 Manufacturing Process

Dip-coating, chemical vapor deposition, and spin-coating have been used to create pellicle film. Currently, most pellicle film is produced by the spin-coating process [5,6]. In 2003, a pellicle as big as 582 mm × 348 mm for LCD photomask was produced with spin-coating process. The pellicle film can also be coated with antireflective materials to give it suitable antireflective properties. The ARC process can be done by spin-coating or vacuum deposition with low refractive index materials. Fluoropolymers, which are used for ARC or for DUV film, create a low-energy surface and can make it easier to remove particles from the pellicle surface.

19.4.1.2 Transmission and Material

The Transmission depends on the film thickness, type of ARC, and light absorption of film material and the light wavelength used by the wafer aligner or wafer stepper. Nitrocellulose was the film material initially used and can be used for g-line (436 nm) or i-line (365 nm) wafer steppers and wideband projection wafer aligners. However, this material begins to absorb just below 350 nm and cannot be used below 350 nm. Cellulose esters, such as cellulose acetate and cellulose acetate butyrate, have good transmission above 300 nm while amorphous per-fluoropolymer materials, such as Teflon AF® (DuPont™) or Cytop® (Asahi Glass Co. Ltd.), can be used for 248-nm and 193-nm steppers. Examples of the different types of film and their transmission curves are shown in Table 19.2 and Figure 19.3, respectively.

The film material must have the proper uniformity, mechanical strength, optical transmission, and cleanliness to allow continuous replication of the photomask image onto the wafer surface. Specifically, a few necessary characteristics are as follows:

(1) *Transmission uniformity*: As most film is generated from spin-coating, uniformity is from the center of the pellicle film to the edge.

TABLE 19.2

MLI's Film Specifications

Material[a]	Part No.	Thickness (μm)	Double-Sided AR Coating	Transmission	Wavelength (nm)
NC	100	2.85	No	91% (avg.)	350–450
	102	2.85	Yes	97% (avg.)	350–450
	105[b]	2.85	Yes	99.5% (min.)	380–420
	122[c]	1.40	Yes	99% (min.)	365, 436
CE	201[c]	1.40	Yes	99% (min.)	365, 436
FC	603[d]	0.81	No	99% (min.)	248, 365, 436
	703[d]	0.54	No	99% (min.)	193, 248, 365

[a] NC: nitrocellulose; CE: cellulose ester; FC: fluorocarbon polymer.
[b] MLI U.S. Patent # 4,759,990.
[c] MLI U.S. Patent # 5,339,197.
[d] MLI U.S. Patent # 5,772,817.

Pellicles

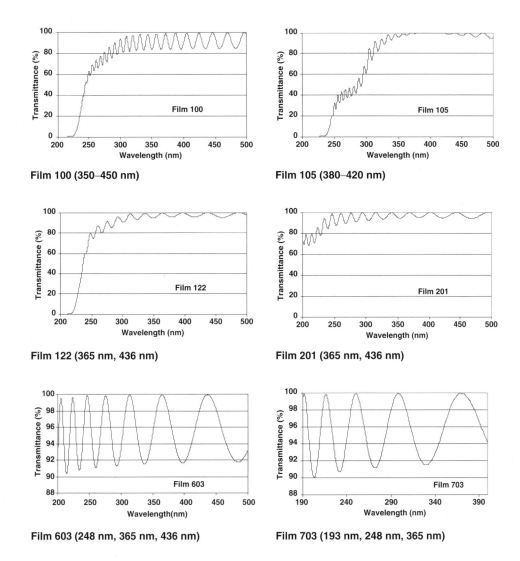

FIGURE 19.3
Typical transmission curves.

(2) *Mechanical strength*: The film and glue adhesion must be able to withstand certain air pressure from a nitrogen or air blow-off gun with a 2-mm or larger opening at all angles. For fluoropolymers used for DUV pellicle film or ARC, it is very difficult to find a suitable glue to bond the film to the frame due to the film's low surface energy. Therefore, the glue developed for this purpose can sometimes show only a marginal strength and limited lifetime of adhesion strength. Adhesion strength, i.e., adhesion of glue on the frame, should be checked with each vendor's pellicle.

(3) *Usage life*: The pellicle lifetime can vary greatly, depending on pellicle materials and the light source of the wafer stepper or wafer aligner, i.e., light source wavelength, intensity and light filter used. All material components in a pellicle are subject to UV light degradation, oxidation degradation, and outgassing and should therefore be considered as having a limited lifetime.

19.4.1.3 AR Coating

ARC on a pellicle can improve the transmission and its uniformity over the entire pellicle. The ARC on a pellicle also makes the transmission less sensitive to the thickness variations of the base film. ARC can be deposited with inorganic material, such as calcium fluoride, or spin-coated with a fluoropolymer. Fluoropolymer ARCs have an additional advantage of creating a low-energy surface, and therefore it is easier to blow off particles from the pellicle surface compared with most inorganic ARCs. Because DUV pellicles for 248 and 193 nm use perfluoropolymer, no ARC is currently used for these pellicles.

19.4.2 Frame

The frame is used to support the film and to be bonded on a photomask. It must be mechanically rigid, flat, stable, create no contamination, and easy for inspection. The material is typically of a black, anodized aluminum alloy.

19.4.2.1 Frame Coating

The current frame manufacturing process creates a pellicle frame that has a very rough, irregular surface. Coating is therefore used to give the pellicle frame a smoother surface. Hidden particles on the irregular surface are sealed by the coating, while the coating allows for easier detection of particles on the frame surface. Coating is often used in conjunction with an adhesive or liquid-like material to catch possible airborne particles. Without coating on the pellicle frame, particles can potentially hide in frame crevices and eventually fall on the pellicle film and/or photomask. It is important that the coatings have no adverse affects on the pellicle. For example, the liquid coatings must be UV resistant and have minimal outgassing to avoid any transmission loss of the film and condensation or crystallization on the photomask surface, respectively. Figure 19.4 and Figure 19.5 show frame surfaces of an MLI pellicle without coating and with coating.

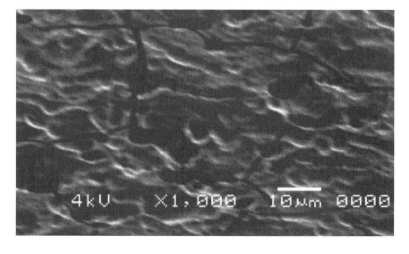

FIGURE 19.4
Scanning electron microscope picture of a frame without coating.

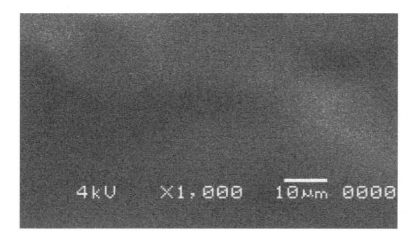

FIGURE 19.5
Scanning electron microscope picture of a frame with coating.

19.4.2.2 Vent Hole and Filter

During air shipment, a pelliclized photomask is subject to significant air pressure differentials, causing the volume of air under the pellicle film to expand or contract [7]. This can cause the film to damage or even break. Consequently, a vent hole, i.e., breath hole, in the pellicle frame was developed to equalize the air pressure differentials inside and outside the pellicle film during air shipment. A vent hole with a cap screw was first used by Intel in the early 1980s for shipment of an Ultratech wafer stepper reticle from one factory to another at different altitudes. Then filter on the vent hole was introduced by Kasunori Imamura in a patent [8].

Although a vent hole with a filter can equalize the pressure from inside to outside of a frame, the hole itself is always a potential place for hidden particles even when the wall of the hole is coated with a pressure-sensitive adhesive. Therefore, unless it is necessary to ship by air or use at different altitudes, it is not recommended to use a vent hole with filter.

In addition, with the introduction of a single-layer cast pressure-sensitive adhesive, which has a tight seal on a photomask or a reticle, a fast mounting process can trap some air and cause a bulge of the film. A frame with a vent hole with filter can eliminate this problem.

With the concern of environmental outgassing and photomask container outgassing, especially for DUV photolithography, one has to get enough outgassing data in deciding whether a vent hole with a filter is right for the process because the filter can connect the inside of a pellicle to possible outside vapor contamination.

19.4.3 Mounting Adhesive or Gasket

The mounting adhesive or gasket is used to bond the pellicle to the photomask and is preapplied on the frame with a release liner. There are two different types of adhesives that have been used — carrier adhesive and noncarrier adhesive. Their pictures are shown in Figure 19.6. An adhesive with a carrier is a double-sided coated pressure-sensitive acrylic or rubber adhesive with a polyurethane foam, vinyl foam, or solid carrier. Foam adhesives were used widely in the early stages of pellicle manufacturing. Adhesives without a carrier can be applied to the frame from a one-layer transfer tape or

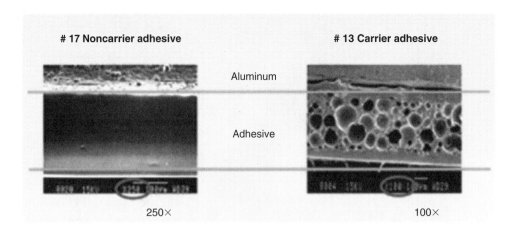

FIGURE 19.6
MLI #17 noncarrier adhesive versus #13 carrier adhesive.

cast in place on the frame from hot melt, UV-cured or emulsion pressure-sensitive adhesives. Both types of adhesives have a thickness that varies anywhere from about 0.10 to 0.80 mm.

19.4.4 Backside Cover

The backside cover was first introduced by MLI in 1982 and is now widely used in the field. The purpose of a backside cover is to seal the inside of the pellicle against airborne particles during transportation. This cover is then removed before the pellicle is applied to the photomask.

19.5 Inspection

Because of the pellicle's structure, only the film can be inspected automatically. Other components have to be visually inspected. Defects should be categorized, i.e., killing defect, functional defect, or cosmetic defect, and appropriate actions should be taken. Functional defects should be well defined between vendor and user, while cosmetic defects should be compared with a sample standard. To properly inspect a pellicle, attention must be given to several components — transmission and uniformity of the pellicle film, particle contamination on the film, frame, adhesive, and overall integrity of the pellicle.

19.5.1 Inspection of Film Transmission and its Uniformity

To inspect transmission, a spectrophotometer is used to measure pellicle film transmission and its uniformity. Consideration must be given to the background stray light noise, resolution, i.e., bandwidth of the slit, and the angle of the measuring incident light from the spectrophotometer to the pellicle. Although some pellicles might occasionally have been out of specification from some pellicle suppliers, transmission generally presents no problem because each pellicle's transmission is individually measured.

Transmission uniformity is typically inspected with a monochromatic light, usually a green mercury light or a helium–neon light. With the aid of a monochromatic light, detection of nonuniform spots on the film can easily be found, such as those from the spin-coating process. Similarly, a laser reflection inspection machine has been used by MLI to inspect the coating uniformity of uncoated wafer stepper film. Uncoated film thickness can easily be measured by calculations from the transmission or reflection spectra of a spectrophotometer or any commercial film thickness-measuring machine.

19.5.2 Inspection of Film Particles

Particle standards should be used for inspection calibration. Specifically, 1-, 0.5-, and 0.3-μm standards on pellicle surface should be used to calibrate any inspection tool. Even for visual inspection, these standards should regularly be shown to operators to insure accurate inspection.

Currently, there are three methods of inspecting film contamination — human-eye, laser scan, and video camera inspections. The human eye is quite sensitive. MLI has found that with proper lighting a particle as small as 0.3 μm can be detected (calibrated with standard polystyrene beads on pellicle film and verified by a laser scan machine). However, human inspection has a nonquantitative nature. There are several factors involved in this inspection process, such as operator eye sensitivity, ability to focus, inspection angle and position, film distance from the eye, incident and background light intensity, and the eye's pupil response to background light. The ability to detect a particle below 1 μm can be very different between operators.

Even inspection with machines can give irreproducible results. With laser inspection of film particles, the best reproducibility is approximately $\pm$ 20%. This limitation is primarily controlled by the scan line overlap and scan line stability, which is affected by the vibration limit of the scanning optics. Laser scan can detect particles smaller than 0.3 μm.

The video camera can easily detect particles smaller than 0.3 μm with proper illumination. With current high-speed computer video capture and processing, this is the best method to detect particles on pellicle film.

All the three inspection methods have their limitations. Human eye, laser scan, and video camera detections are the detections of scattered light, not the real particle size. The real particle size can be substantially larger than the size its scattering light appears, sometimes as large as 10 times that of the detected particle size. In addition, laser detection is limited to about 2–3 mm from the frame's edge, depending on the laser scanning angle, intensity distribution of the laser spot, height of the frame, and light scattering of the frame edge.

Comparing human eye, laser scan and video inspections, the advantage of the laser and video inspections is that they are mechanical processes that generate more reproducible results, whereas human inspection under strong light scattering presents an unpleasant condition to work. However, human eye inspection does have the desired sensitivity, fast inspection speed, ability to inspect up to the frame's edge, and even the ability to determine which side of the film the particle lies on, at least for larger particles.

19.5.3 Inspection of Frame

A frame is usually machined, sandblasted, and black anodized. Currently, human-eye inspection is used under a projector light to inspect such pellicle frames. However, an operator cannot distinguish particles of 3 μm or even larger from frame irregularities, which came from machining marks, a rough surface from sandblasting, or a porous anodized surface with etch pits. Therefore, it is very difficult to differentiate between a frame surface irregularity and particle contamination. In addition, the strong background scattered light from the frame makes the task even more difficult.

Inspection of frame particles under a microscope has been attempted but has not been successful. Seen under a microscope, the irregularity of the frame will become even more obvious and more difficult to distinguish from particles. Therefore, surface coating of the frame is necessary to give the frame a smooth surface and make automated and human inspection reasonably possible.

19.5.4 Inspection of Adhesive

Foam adhesives previously used had holes larger than 50 μm on the sidewall. The irregularity of the foam material made it almost impossible to detect small particles on such an adhesive. Only a rigorous cleaning procedure was performed to ensure cleanliness. As the illustration in Figure 19.6 shows, cast-in-place, noncarrier adhesives allow the surface to be much smoother, making it easier to differentiate between particles and adhesive irregularities. However, there is still difficulty detecting particles on the edge of the frame and adhesive.

19.6 Handling and Environment

A pellicle has to be made clean in a clean room area, inspected, stored and shipped in a clean box and bag, mounted on the photomask, stored and used for a long time in a photomask box. It is expected that all these steps would not create any more particles or contamination. Actually, many steps of handling and inspection can degrade the cleanliness of the pellicle. Usually, many different sizes and shapes of pellicles have to be handled in one area, making automation a difficult task and human handling the only solution. Even holding a pellicle for inspection is a difficult task and can be a source of contamination. The handling of a pellicle is not a small task. Standardization of pellicle sizes might be a good idea to solve some of our handling problems.

19.6.1 Controlling Static Charge

The importance of controlling static charge during pellicle handling cannot be overemphasized. Given that a pellicle is plastic, e.g., film, release liner and backside cover, it is easy to generate static charge during handling. For example, peeling the release liner or backside cover from the mounting adhesive can generate 2000 V of electricity, and if not neutralized quickly, it can almost instantly attract particles from the surrounding environment onto the pellicle. Also the packaging materials — such as shipping boxes, bags, and storage containers — are all plastic. Handling these materials in the production line can easily generate electrostatic charge and attract particles to the working area. Therefore, it is crucial to control the electrostatic discharge (ESD) at the working area. Under appropriate airflow conditions, proper antistatic equipment should be used to reduce ESD and allow for at least 10 s for a charging plate monitor to decay from 1000 to 100 V and thus minimize contamination from electrostatic attraction. Without a clean and proper ESD environment, contamination can be generated and lead to a defective photomask, causing it to fail during incoming inspection, outbound inspection, or even after repeated uses. Proper airflow is also needed at the mounting machine and working table to insure cleanliness.

19.6.2 Mounting

Putting a pellicle onto a photomask is called mounting. Ideally, a mounting machine with automatic handling and automatic inspection should do the job. In reality, only the film of a pellicle can be automatically inspected. Mounting operations, peeling the release liner or the backside cover from a pellicle can create static charge and contamination. Therefore, a simple solution would be to use a mounting machine that can easily be kept clean and to support it by operator-assisted visual inspection.

Even force has to be used in mounting the pellicle on a photomask. The mounting fixture can damage the pellicle frame edge and get contaminated. Mounting tools have to be kept clean all the time because cross-contamination is possible. It is recommended that a strong antistatic environment with good surface airflow on the machine be used. Although mounting accuracy is typically obtained, proper communication between pellicle manufacturer and equipment designer is necessary for many sizes of pellicles. Yen [9] has invented the process of putting each pellicle on a mounting plate and shipping the whole package to the customer. The customers do not have to touch the pellicle directly in handling and can put the whole package into a simple mounting machine after an inspection of the pellicle.

19.6.3 Cleaning

If particles are generated onto the pellicle film, blowing may be used in an attempt to remove these particles. Specifically, a filtered, deionized air or nitrogen gun with a needlepoint blower is preferred. Blowing is only effective in removing large particles and might generate some small particles and contaminate the environment in the clean area if not used carefully.

Cleaning mounting adhesive residue on a photomask after removal of the pellicle can be a challenging job in an area using multiple vendors because different vendors use different mounting adhesives. Different vendors may supply different cleaning methods or solutions.

19.7 Long-Term Stability of a Pellicle on a Photomask

The long-term stability requirement of a pellicle on a photomask depends on the expected usage lifetime of the photomask and the sensitivity of the lithography process. Most of us would like the lifetime of pellicle on a photomask to be infinite. Unfortunately, most of the materials in a pellicle and photomask should be considered to have only a limited lifetime.

19.7.1 Outgassing and Crystallization

The long-term usage and storage of a pellicle on a photomask provides a real challenge for pellicle design and material selection. Ideally, a pellicle on a photomask will seal off all outside particle and vapor contamination. Unfortunately, the organic components in a pellicle can lead to outgassing and they may contaminate even the area it is supposed to protect. The most susceptible outgassing components of the pellicle are the mounting adhesive, inside coating and glue, all being a mixture of polymers and low molecular weight organic compounds, such as residue solvent, plasticizer, antioxidant, and UV stabilizer. The chrome surface on the photomask is known as an active surface, which can absorb many organic chemicals on it. Unprotected chrome surfaces made from sputtering can absorb material from the photomask container outgassing and make it impossible to coat a positive photoresist on its surface after only 1 day. Outgassing and crystallization can also come from the box or storage container and easily contaminate the area that is not under pellicle protection. Environmental outgassing or vapor can also contaminate the pellicle, reduce its transmission, or form crystals even underneath the pellicle [10]. Outgassing or vapor can easily get into the pellicle-protected area through the vent hole with a filter or through the film directly when the film is thin and permeability to the vapor is high.

For example, in 1986, crystals of 2,5-di(*t*-amyl)quinone were found on a pellicle protected photomask surface. In one extreme case, even within a day of mounting pellicle on a photomask, crystals could be found on the chrome pattern edge, making the photomask useless. The origin of the crystals was identified by us as coming from the outgassing of antioxidant stabilizer of 3M447 — a rubber-type double-sided pressure-sensitive tape used for pellicle mounting adhesive [11]. This antioxidant stabilizer was 2,5-di (*t*-amyl) hydroquinone. Recently, in 2001, a crystal was formed on the pattern surface and identified as a quinone. The quinone comes from the dimerization of the antioxidant as a stabilizer in the hot melt pressure-sensitive adhesive.

An outgassing test should be performed for any new pellicle qualification. In addition, a lifetime test for particle generation on the photomask under the pellicle should be completed in a real production environment as well.

19.7.2 Material Stability

All the materials used for a pellicle have to be subjected to some photodegradation from UV light and oxidation degradation from air and outgassing. In addition, film thickness can change due to the residue solvent evaporation out of the film, polymer molecular rearrangement, and humidity change. The film for a pellicle is always chosen to resist UV light radiation, which may degrade the film transmission or mechanical strength. However, other components, such as film glue, mounting adhesive, or frame coating, are do not necessarily have to go through a rigorous checking. For example, one of our early mounting adhesives, i.e., a polyethylene vinyl acetate hot melt pressure-sensitive adhesives neither nor without antioxidant, lost its pressure sensitivity in 6 months. In our laboratory, a silicone adhesive used to mount the film to the frame lost its adhesion on a perfluoropolymer film only after a few days after being fully cured. Some mounting adhesives which use hot melt pressure-sensitive adhesive, such as poly-styrene-ethylene-butylene-styrene (SEBS), poly-styrene-isoprene-styrene (SIS), or poly-butylene polymer, were neither DUV stable nor oxidation stable without an antioxidant.

As discussed above, the material of the container, i.e., the box, used for pellicle and photomask has to be screened for outgassing because some of them do contain possible outgassing low molecular organic compounds in the plastics.

Depending on the functional requirement, not all the degradations will result in the defect generation for a photomask with a pellicle. However, depending on the usage lifetime, sometimes it may take a long time to find out if a pellicle is suitable for the process.

19.8 The Future of Pellicles

The introduction of 157 nm for next-generation optical lithography has created a need for new pellicle materials optimized for that wavelength. Currently, there are four strategies being considered.

19.8.1 Soft Pellicle

First, new fluorocarbon-based polymers have to be developed that are transparent, damage-resistant, and that possess mechanical properties to enable their preparation in very thin pellicle form [12]. Initial results from Lincoln Labs at Massachusetts Institute of Technology demonstrated that commercial fluoropolymers used for pellicles at 248- and 193-nm wavelengths, such as Teflon AF® and Cytop®, rapidly burst under irradiation with 157 nm light because they lost sufficient mechanical integrity. Consequently, an extensive program was initiated to develop and screen novel fluoropolymer candidates with the desired properties for 157-nm lithography. Although some polymers did show promising transmission, their lifetime is still insufficient due to photochemical darkening. Research is still needed to solve this fundamental problem.

19.8.2 Hard Pellicle

A hard pellicle [13,14] is simply a thin, quartz glass on a frame. Fluorinated fused silica has a sufficient lifetime for the 157-nm process. Although thickness control and thickness uniformity are a challenge, a good parallelism, i.e., thickness uniformity, has been achieved. However, even thin fused silica is several hundred times thicker than a soft film pellicle. With a typical thickness of 800 μm, a hard pellicle would act as an additional optical element and impact the imaging and overlay performances. Developers have achieved good optical homogeneity and surface finish for hard pellicles, but improvements to the pellicle mounting and adhesive thickness control are required to keep the pellicle bending low in order to avoid significant optical distortion. The purge of the pellicle cavity with inert gas is also a topic to be studied. The 157-nm lithography process is under very low humidity, which also raises an ESD problem. A circular pellicle with a circular photomask should be used to minimize any distortion.

19.8.3 Removable Pellicle or Cover

The third strategy is to use the soft pellicle as a photomask cover only during transportation and storage. The pellicle would then be removed before exposure and remounted after exposure [15]. There are three mounting options that are under consideration — adhesive, magnetic, and a modified reticle carrier. Although the process can be easily proven in a research line, the long-term contamination control and inspection will still be a challenge, and the process will have to be proven in the very costly production line.

19.8.4 No Pellicle

The last approach is a pellicle-less solution. However, after 20 years of using pellicles the "no pellicle" proposal will be a challenge because the feature size of IC is much smaller than before — 0.1 μm versus 4 μm. In addition, the photomask is easier to become contaminated because the feature size is now much smaller. A reliable, constant inspection feedback system is, therefore, necessary for the success of this method.

At this time of writing, it seems that the hard pellicle will produce enough lifetime and contamination-free protection for a 157-nm reticle. The production of a clean, defect-free hard pellicle and mounting without distortion or repetitive distortion still has some challenges to overcome.

References

1. Vincent Shea and Walter J. Wojcik, U.S. Patent 4,131,363, 1978.
2. Ronald S. Hershel, Pellicle protection of integrated circuit masks, *Proc. SPIE*, 275, Semiconductor Micro Lithography VI (1981).
3. Yung-Tsai Yen, U.S. Patent 4,759,990, 1988.
4. Pei-Yang Yan, Michael S. Yeung, and Henry T. Gaw, Printability of pellicle defects in DUV 0.5 µm lithography, *Proc. SPIE*, 1604, 106–117 (1992).
5. Ray Winn, U.S. Patent 4,378,953, 1983.
6. Ray Winn, U.S. Patent 4,536,240, 1985.
7. Robert W. Murphy and Rick Boyd, The effect of pressure differentials on pelliclized photomasks, *Proc. SPIE*, 2322, 187–201 (1994).
8. Kasunori Imamura, U.S. Patent 4,833,051, 1989.
9. Yung-Tsai Yen, U.S. Patent 5,168,993, 1992.
10. Naofumi Inoue, Hiroaki Nakagawa, Masahiro Kondou, and Masanori Kitajima, Pellicle vs. influence of clean room environments, *Proc. SPIE*, 2512, 60–73 (1995).
11. Chris Yen and C.B. Wang, Potential Particle Problem from an Adhesive, MLI Technical Publication, 1986.
12. Roger H. French, Rober C. Wheland, Weiming Qiu, M.F. Lemon, Gregory S. Blackman, Xun Zhang, Joe Gordon, Vladimir Liberman, A. Grenville, Roderick R. Kunz, and Mordechai Rothschild, 157-nm pellicles: polymer design for transparency and lifetime, *Proc. SPIE*, 4691, 576–583 (2002).
13. Emily Y. Shu, Fu-Chang Lo, Florence O. Eschbach, Eric P. Cotte, Roxann L. Engelstad, Edward G. Lovell, Kaname Okada, and Shinya Kikugawa, Hard pellicle study for 157-nm lithography, *Proc. SPIE*, 4754, 557–568 (2002).
14. Kaname Okada, K. Ootsuka, I. Ishikawa, Yoshiaki Ikuta, H. Kojima, T. Kawahara, T. Minematsu, H. Mishiro, Shinya Kikugawa, and Y. Sasuga, Development of hard pellicle for 157 nm, *Proc. SPIE*, 4754, 569–577 (2002).
15. Andy Ma, Arun Ramamoorthy, Barry Lieberman, C.B. Wang, Q.R. Bih, Kevin Duong, and Corbin Imai, Removable Pellicle, in: *Sematech Pellicle Risk Assessment Workshop*, September 27, 2001.

Section VI

Mask Metrology, Inspection, Evaluation, and Repairs

20
Photomask Feature Metrology

James Potzick

CONTENTS
20.1 Introduction 413
20.2 The Feature Edge 414
20.3 Costs and Benefits of Mask Feature Metrology 415
20.4 Measurement Uncertainty and Traceability 420
20.5 Terminology 421
20.6 Parametric and Correlated Errors 423
20.7 Differential Uncertainty 425
20.8 The "True Value" of a Photomask Linewidth—Neolithography 427
20.9 Some General Notes on Linewidth Metrology 429
20.10 Conclusion 430
Acknowledgments 431
References 431
Bibliography 432

20.1 Introduction

This chapter discusses some general issues with regard to measurement of the size and placement of the features on a photomask. The size is often called the linewidth or CD, and the placement with respect to another feature is often called the pitch. Figure 20.1 illustrates the difference, showing the cross-section of two parallel chrome lines on a quartz substrate.

The concepts discussed here are not specific to any one kind of instrument but apply especially to scanning electron microscopes (SEMs) and optical microscopes in transmission and reflection, to scanning probe microscopes (SPMs), and to scatterometers. These concepts also apply to binary chrome, attenuated phase shift, hard phase shift, and chromeless features, as well as subresolution assist features, both 2-dimensional (lines and spaces) and 3-dimensional (contact holes, etc.) features. All linewidth measurements are assumed to be averaged over a specified length segment to average the higher spatial frequencies of line edge roughness. Binary chrome features will be used in examples and illustrations for simplicity.

Since all linewidth and placement measurements derive from the location of a feature's edges, this chapter starts with a discussion of the geometric definition of a line's edge and its position. Then the econometric rationale for feature metrology and measurement uncertainty evaluation are presented, followed by practical definitions for the terms

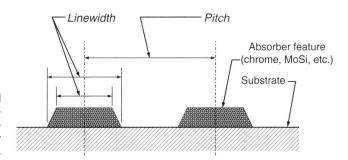

FIGURE 20.1
Cross-section view of two parallel chrome lines on a quartz substrate illustrating the difference between linewidth and pitch. Note the linewidth for the case shown is not uniquely defined.

relevant to uncertainty calculations. Then some notes on parametric errors and the correlated errors often found in comparing measurements at different sites, followed by the neolithography model of integrating metrology and modeling into photomask design and wafer exposure process optimization, and finally some general notes on the relation between the mask metrology process, the wafer manufacturing process, and real mask features.

20.2 The Feature Edge

As intimated in Figure 20.1, vertical and flat chrome feature edges are rarely found on real photomasks. High resolution SEM and AFM images show most features to look more like the "real feature" in Figure 20.2, with poorly defined edges. This raises the all-important question: Where are the "edges" that define the linewidth?

The first rule of metrology is to define the measurand. An attempt to deal with this issue can be found in SEMI Standard P35 *Terminology for Microlithography Metrology* [1]. Here the real chrome feature with its complex and irregular geometry is represented by a simplified feature model (or object model), Figure 20.2, with a well-defined linewidth, center, etc. This representation is used in feature metrology to define the edges and in subsequent application of the metrology results, for example, wafer exposure modeling.

Figure 20.2 shows only two of the many possible choices for a linewidth feature model. The first model represents the line with a rectangular cross-section, whose width is unambiguous. The second uses a trapezoidal cross-section, which represents that line's actual shape better, but now the sidewall angles and height must be specified and the linewidth is no longer unambiguous but must be defined relative to the trapezoid. There might be some reason, determined by the application of the measurement data, to choose

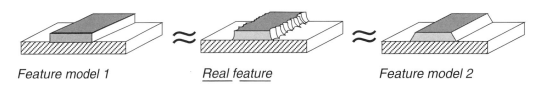

FIGURE 20.2
The complex chrome feature can be represented by a feature model with well-defined linewidth.

Photomask Feature Metrology

the width at the top or bottom or half way up, but often this choice will be completely arbitrary.

The feature model and its measurement data can be used in a wafer exposure image model to predict this mask feature's performance in printing wafers (proximity effects, defect printability, etc.). The advantage of *Feature model 2* is that this feature model better represents the actual feature shape and will result in more accurate modeling results. The disadvantage of *Feature model 2* is that "the linewidth" may now depend on the height above the substrate.

SEMI Standard P35 further suggests that geometric complexity can be traded for measurement uncertainty by drawing "bounding boxes" around the feature, or at least around its edges, as in Figure 20.3. The *line edge bounding box* is meant to encompass the line's "edge," so that there is sufficiently high probability that the edge is inside the box for the purpose intended by the measurement. There will be a probability distribution for the edge's position inside the bounding box, with an expectation value and a variance. For the ideal case of a smooth straight-line with constant rectangular cross-section, the inner, outer, and mean bounding boxes are identical, and the line edge bounding box has zero width. See SEMI Standard P35 [1] for details of this approach.

To be conservative one might, for example, choose a line edge bounding box for a feature model, so that for any conceivable purpose everyone would agree that the edge is inside this box. To cover all conceivable applications of this feature, there can be no presumption of where this edge is within the box. This results in a rectangular probability distribution for the edge position, with expectation value at the center of the box. In practice, one might exclude chrome asperities from the box, as in Figure 20.3, if they are deemed not relevant to any function of the feature.

20.3 Costs and Benefits of Mask Feature Metrology

Every measurement of a feature's size (linewidth or CD) or placement on a photomask is made for a reason. Usually, the measurement leads to a decision, often involving a business transaction. A mask manufacturer may measure a few critical features to decide whether and how to adjust the write, develop, or etch process; or may measure some or all features to determine if they meet customer specifications in order to decide whether

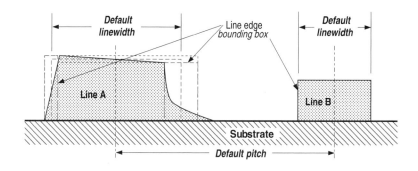

FIGURE 20.3
A feature model may be enclosed in "bounding boxes" to encompass limits on the positions of its edges and center.

or not to ship the mask. A mask user may measure some features to decide whether to accept or reject the mask.

Every such measurement contains unknown errors, however. Since the errors are unknown (else they would have been removed), they are best characterized by probability distributions. Thus, a measurement result is a probability distribution of likely values for the measurand, with a mean and a variance. The mean is the expected value (or best estimate) of the measurand, and the square root of the variance is the standard measurement uncertainty. The net measurement error is the sum of the individual errors, and the variance of its probability distribution is the sum of the variances for the individual errors (assuming they are uncorrelated). See Section 20.6: "Parametric and correlated errors."

There is a probability $p_1 < 1$ that a feature that measures in tolerance is actually in tolerance (with probability $1 - p_1$ that it is not), and a probability p_2 that a feature that measures out of tolerance is actually out of tolerance (Figure 20.4). Suppose a mask is measured for compliance with specifications prior to being shipped to the customer. The mask is judged to be "in specification," or "out of specification" based on the measurement result. There is a cost c_{12} of shipping an out-of-tolerance mask, and a lower cost c_{21} of scrapping an in-tolerance mask. The cost of scrapping an out-of-tolerance mask is smaller yet, and the cost of shipping an in-tolerance mask is zero.

In general, c_{ij} is the cost of action i if the part is actually in measurement result category j (e.g., in specification or out of specification) and p_j is the probability that the part is actually in category j, with the normalizing constraint $\Sigma p_j = 1$. Then the *expected* cost of action i is [2]:

$$C_i = \Sigma c_{ij} p_j$$

See Table 20.1 as an illustration of the 2-action example above. There is an additional cost c_0 of simply making the measurement, regardless of the outcome. For many products, the cost of shipping bad product c_{12} can be very high compared to the others.

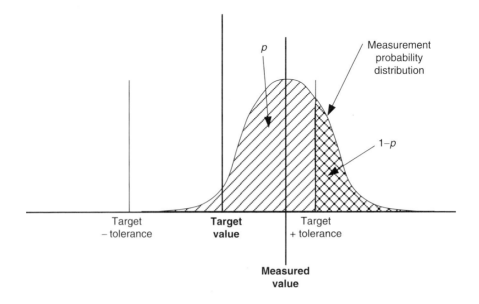

FIGURE 20.4
The specified value (target) and tolerance for a photomask feature, and the probability distribution of a measurement result. The probability part in tolerance is p, out of tolerance $(1 - p)$.

TABLE 20.1

A Measurement Indicates with Probability p that a Feature Meets Specifications. Then There is a Probability $1 - p$ that the Feature is out of Specification. This Table Shows the Expected Cost of Shipping or Scrapping the Mask, Depending on p. This Table is Readily Generalized to More Complex Cases

Action	Cost of Action if Mask is Actually in Spec	Cost of Action if Mask is Actually out of Spec	Expected Cost of Action, C_i
Ship Mask	c_{11}	c_{12}	$p\, c_{11} + (1-p)\, c_{12}$
Scrap Mask	c_{21}	c_{22}	$p\, c_{21} + (1-p)\, c_{22}$

Then the net expected cost of metrology is

$$\text{net expected cost} = C_i + c_0$$

The costs of various possible actions c_{ij} can be estimated from business economic considerations, and the probabilities p_j from evaluating the measurement uncertainty.

The expanded measurement uncertainty U is directly proportional to the standard deviation of the measurement probability distribution in Figure 20.4:

$$U = k\sqrt{\text{var(measurement)}}$$

where k is a constant (the "coverage factor") chosen from the t-table to represent the desired confidence interval, and var(x) is the variance of the probability distribution (normal or otherwise) of x. At the extremes of measurement uncertainty, as

$$U \to \infty, p \to 0 \quad \text{and} \quad C_1 \to c_{12}, C_2 \to c_{22}$$

and as $U \to 0, p \to$ (1 or 0) depending on the measurement result, and

$$C_1 \to (c_{11} \text{ or } c_{12}), \quad C2 \to (c_{21} \text{ or } c_{22})$$

That is, if $U = \infty$ (no measurements and no prior knowledge of the process), the probability of the feature's falling within the finite tolerance interval is 0 (no knowledge, but this is not to say it is impossible), but if $U = 0$, the feature is clearly in tolerance or it is not (no uncertainty). If the part measures to be its target value but $U > 3 \cdot tolerance$, then the probability that the part is actually in tolerance is less than 1/2 (because less of the area under the probability curve in Figure 20.4 lies between the $-tolerance$ and $+tolerance$ limits than outside those limits) and one cannot safely conclude the part meets its specifications. In practice, uncertainty of the value of the measurand is always finite because of prior knowledge about the measurand (there are usually practical bounds) and the process, even without a measurement. This fact justifies measuring a statistical sample of product and inferring the uncertainty of the remaining measurands.

The measurement cost is related to U because the uncertainty can often be reduced by spending more resources on the metrology, e.g., by increasing the number of repeat measurements, more carefully controlling the metrology environment, purchasing more sophisticated and expensive equipment, improving operator training, etc. Thus, the metrology cost can be approximated by

$$c_0 \propto 1/U$$

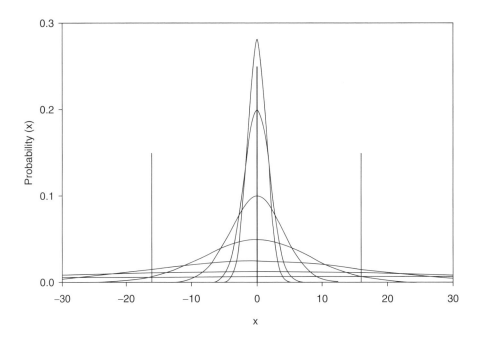

FIGURE 20.5
Probability distributions of measurement results for different possible measurement uncertainties (2.8, 4, 8, 16, 32, 64, and 128 nm, 2σ), for a feature meeting its target value (Example 1). The two vertical bars represent an example tolerance of ± 16 nm.

The measurement uncertainty can be regarded as an independent variable, directly related to the cost of metrology.

In a manufacturing environment, the value of the measurement data must exceed the cost of making the measurement. The objective of mask metrology is to determine which action will minimize the total expected cost ($C_i + c_0$) for the action chosen. Since a decision based on this measurement may have serious economic consequences, it is important that these costs and probabilities (thus measurement uncertainty) be acknowledged and evaluated.

Two examples will be given. Assume $c_0 = 1/U$ in relative cost units, where U is the expanded uncertainty of the measurement ($k = 2$), and that $c_{11} = 0$ and $c_{12} = 1$. These numbers can be scaled to fit particular situations.

In example 1, a mask feature measures exactly to specification, with the measurement results distributed as one of the curves in Figure 20.5. The different curves represent possible measurement uncertainties. Then the expected cost of shipping this part is shown in Figure 20.6 for different customer specified tolerances. For any specified tolerance, there is an optimum measurement uncertainty that minimizes the expected cost. In this case, for a tolerance of 16 nm (heavy curve in Figure 20.6), the minimum expected cost occurs at a measurement uncertainty of 14 nm, 2σ. As the measurement uncertainty increases, the likelihood that the feature is in tolerance diminishes, and the expected cost approaches c_{12}.

In example 2, a mask feature is 16 nm larger than its specification (Figure 20.7). The corresponding expected cost curves are shown in Figure 20.8. If the tolerance is 16 nm, the feature is borderline; the expected costs for larger tolerances (feature is in specification) appear in the lower part of Figure 20.8.

Since the cost factors c_{ij} can differ widely from each other, it is important to understand and control the uncertainty of these measurements.

Photomask Feature Metrology 419

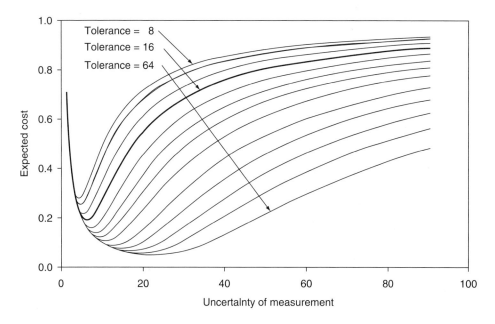

FIGURE 20.6
Expected cost of shipping this apparently good mask in Example 1, for various *specified* tolerances (8 nm, 9.5, 11.3, 13.5, 16,..., to 64 nm). Each tolerance curve has a minimum cost at a different measurement uncertainty.

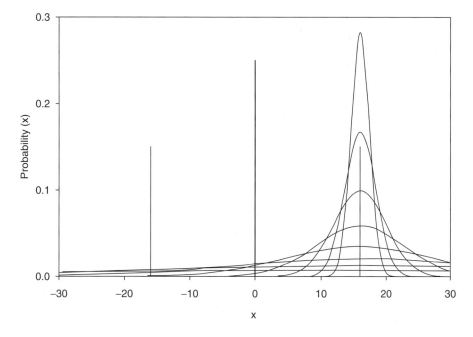

FIGURE 20.7
Probability distributions of measurement results for the same possible measurement uncertainties shown in Figure 20.5, with the measured value offset by 16 nm from its specification (Example 2).

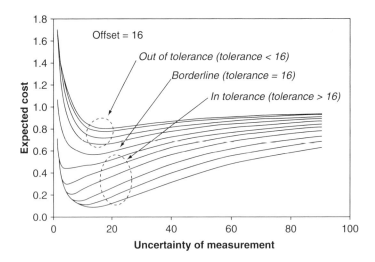

FIGURE 20.8
Expected cost of shipping for various specified tolerances (Example 2). The borderline case, where the tolerance equals the offset, is disconnected from the others.

20.4 Measurement Uncertainty and Traceability

Measurement uncertainty is defined by the International Organization for Standardization ISO [3] as a parameter, associated with the result of a measurement that characterizes the dispersion of the values that could reasonably be attributed to the measurand.

Numerically, it is the square root of the sum of the variances of the probability distributions of all the possible (assumed independent) measurement errors (both random and systematic) multiplied by a stated factor chosen to represent the desired confidence interval, as described in ANSI/NCSL Z540-2-1997b [4], which is essentially the same as the ISO document *Guide to the Expression of Uncertainty in Measurement* [5].

A measurement procedure on an object results in an indicated value $I(a)$ for a measurand a and an unknown error $\varepsilon(a)$:

$$a = I(a) + \varepsilon(a)$$

Since $\varepsilon(a)$ is unknown and a second measurement will generally yield a different $I(a)$, each of these terms has a variance and an expectation value. For $I(a)$ these can be found by repeating the measurement n times. Then its expectation value and variance are

$$\langle I(a) \rangle = \frac{1}{n} \sum_{i=1}^{n} I_i(a), \quad \mathrm{var}(I(a)) = \frac{1}{n-1} \sum_{i=1}^{n} (I_i(a) - \langle I(a) \rangle)^2$$

The error term $\varepsilon(a)$ is the sum of many errors ε_j from various sources; by nature their values are unknown, and only their probability distributions can be known or estimated. If an error ε_j were known exactly, it would have been removed. Since the size of the error is not known, its probability distribution represents all that is known about it. This probability distribution is based on all of the available information on the possible values of ε_j.

If a variable x (or ε) has a normalized probability distribution $p(x)$

$$\int_{-\infty}^{\infty} p(x)\mathrm{d}x = 1$$

then its expectation value and variance are given by

$$\langle x \rangle = \int_{-\infty}^{\infty} xp(x)\mathrm{d}x, \quad \mathrm{var}(x) = \int_{-\infty}^{\infty} (x - \langle x \rangle)^2 p(x)\mathrm{d}x$$

Since all known errors have been removed,

$$\langle \varepsilon(a) \rangle = 0$$

Then

$$\langle a \rangle = \langle I(a) \rangle$$

and

$$u^2(a) = \mathrm{var}(a) = \mathrm{var}(I(a)) + \mathrm{var}(\varepsilon(a))$$

$\sqrt{\mathrm{var}(I(a))}$ is the standard deviation of the measurement results and is the measurement repeatability. $\langle I(a) \rangle$ is the best estimate of the true value of a, and $u(a)$ is its uncertainty. The probability distribution in Figure 20.4 is usually Gaussian (because of the central limit theorem) with standard deviation $u(a)$.

Perhaps the greatest contribution of ISO's *Guide to the Expression of Uncertainty in Measurement* is the recognition that systematic errors have probability distributions just like the random errors (although usually continuous) and contribute to the measurement uncertainty in the same way.

Multiplying $u(a)$ by the coverage factor k, taken from the t-table for the desired confidence interval,

$$U(a) = k\, u(a)$$

results in an interval $\pm U(a)$ about the measurement result $\langle I(a) \rangle$ that has a 95% ($k = 2$) or 99% ($k = 3$) probability of containing the true value.

$U(a)$ is called the expanded uncertainty, and $u(a)$ is the standard uncertainty. Since the true value a is not known (else why measure?) and all measurements of continuous values have unknown errors, this probabilistic interpretation is the best that can be done, combining all relevant knowledge about the measurand.

20.5 Terminology

When evaluating or comparing measurements it is important that the metrology terms used have well defined and commonly understood meanings. The following definitions of metrology terms are taken from the ISO publication *International Vocabulary of Basic and General Terms in Metrology* [3] and are accepted by national measurement laboratories around the world:

- *Error* (of measurement): Result of a measurement minus a true value of the measurand.
- *Random error*: Result of a measurement minus the mean that would result from an infinite number of measurements of the same measurand carried out under repeatability conditions.
- *Systematic error*: Mean that would result from an infinite number of measurements of the same measurand carried out under repeatability conditions minus a true value of the measurand.

The magnitude and sign of a measurement error are unknown; otherwise it would have been removed. Knowledge of an error is represented by a probability distribution. Essentially, the measurement uncertainty represents the combined widths of the probability distributions of all of the possible errors [5].

- *True value* (of a quantity): Value consistent with the definition of a given particular quantity.
- *Traceability*: Property of the result of a measurement or the value of a standard whereby it can be related to stated references, usually national or international standards, through an unbroken chain of comparisons all having stated uncertainties.

A linewidth standard from a national measurement laboratory is traceable to the definition of the meter [6]:

- *Meter*: The length of the path traveled by light in vacuum during the time interval of 1/299,792,458 of a second.
- *Second*: The duration of 9,192,631,770 periods of the radiation corresponding to the transition between the two hyperfine levels of the ground state of the cesium-133 atom.

The realization of the meter by the cesium clock is the ultimate length standard, unambiguously defined and the same for everyone under all conditions. It is a natural standard, not an artifact, and is unconditionally stable over very long periods of time, is internationally accepted, and is universally accessible. Should a traceable length standard become lost or damaged, its replacement will be traceable to the same reference and thus directly related to the lost standard.

A measurement is traceable if the measurand has been compared, through measurements, to a specified reference standard (such as an artifact standard or the definition of the meter) with documented measurement uncertainty. While this uncertainty may be small or large, the uncertainty of an untraceable measurement is not known. Therefore, an untraceable measurement commands little user confidence and provides no information on the probabilities p_j.

In addition to helping to identify the action with lowest expected cost, there are other reasons measurement traceability may be desirable. Traceability to a *national standard* or to a *defined base unit* (e.g., the definition of the meter) may be needed:

- If a part's function is dimensionally dependent
- For a high confidence level in long-term stability
- For comparing experiment with theory
- For consistent measurements between distant manufacturing sites
- For comparing products from different manufacturers

- For comparing measurements in a way that is mutually acceptable to all parties involved in a transaction
- For resolving differences between buyer and seller
- For ensuring compliance with legal requirements (such as by government agencies or ISO 9000)
- For resolving differences between different measurement techniques

Traceability to an *in-house artifact* may be adequate for some manufacturing purposes, such as process monitoring, if the artifact is sufficiently stable.

20.6 Parametric and Correlated Errors

Measuring linewidths accurately can be difficult even on binary photomasks. See Section 20.8: "The 'true value' of a photomask linewidth — neolithography."

When an optical, electron, or scanning probe microscope is used to measure the linewidth of a feature, it forms a scaled image of the linewidth object measured. A scatterometer forms a Fourier intensity "image." These images differ from the object because of diffraction or electron scattering and other effects, but only the image can be measured directly, not the object. The measurement process consists of comparing the image of the linewidth object to the modeled image of a similar theoretical linewidth object or to the real image of an artifact linewidth standard, using the instrument's calibrated scale [7].

Linewidth is intrinsic to the object, independent of the method of measurement. But we can measure only an object's image in an instrument [8], and the image depends both on object and tool parameters $\{P_i\} \equiv P_1, P_2, P_3, \ldots, P_N$ (e.g., chrome thickness, chrome complex index of refraction n and k, edge roughness; illumination wavelength, objective lens NA, etc.; see Table 20.2 for more examples). The tool parameters are not intrinsic to the object, but still affect the measured image. Consequently, we must relate the image to the object through an imaging model [9], which predicts the instrument's image of the object for specific parametric conditions and identifies the locations of the object's edges in the image. (Ideally one would apply the inverse model to the image to derive a description of the object, but this can be very difficult and the inverse model may not have a unique solution.) The instrument image must be modeled (or the difference in images if two measurements are to be compared) in order to identify the locations of the object's edges, and the image scale calibrated to measure their separation.

A similar imaging process prints wafers from the mask, but the exposure tool's parameters are usually different from the metrology tool's parameters (the object parameters, of course, are identical). Both imaging processes can be modeled to predict the wafer image from the metrology image, mitigating the expense of printing test wafers [10,11].

The measurement error $\varepsilon(a)$ can usually be expressed in terms of errors in the measurement parameters $\{\delta P_i\}$,

$$\varepsilon(a) = f(\{\delta P_i\})$$

where $f(\{P_i\})$ is the measurement process model, and δP_i is an error in the parameter P_i. In general, if $y = f(\{\delta P_i\})$, then

$$\text{var}(y) = \sum_{i=1}^{N} \left(\frac{\partial f}{\partial P_i}\right)^2 \text{var}(\delta P_i) + 2 \sum_{i=1}^{N-1} \sum_{j=i+1}^{N} r(\delta P_i, \delta P_j) \frac{\partial f}{\partial P_i} \frac{\partial f}{\partial P_j} \sqrt{\text{var}(\delta P_i)\, \text{var}(\delta P_j)} + \cdots$$

TABLE 20.2

A List of Possible Optical Instrument Parameters

Object Terms	*Instrument Terms*
Chrome edge runout	Tool-induced shift
Chrome n	Scale factor calibration
Chrome k	Substrate temperature
Substrate thickness	Substrate temperature variation
Chrome thickness	Air temperature, pressure, RH
Feature proximity	Illumination wavelength
	Illumination NA
	Objective NA
	Sampling aperture
Alignment Terms	Data noise filter
Specimen cosine alignment	Proximity effects
	Photometer linearity
Defocus	Image modeling parameters
Illumination alignment	Interferometer resolution
	Photometer resolution
	Laser wavelength uncertainty
	Laser polarization mixing
	Abbè error
	Optical image distortion
	CCD linearity (x, y, intensity)
	Image processing algorithms

where $r(\delta P_i, \delta P_j)$ is the correlation coefficient between these two parameters, and $\cdots$ means higher order terms. In general, $-1 < r(x_i, x_j) < +1$, $r(x_i, x_j) = r(x_j, x_i)$, $r(x_i, x_i) = 1$, and $r(x_i, -x_i) = -1$. If x_i and x_j are uncorrelated, then $r(x_i, x_j) = 0$. One might call δP_i a parametric error, and $(\partial f/\partial P_i)\sqrt{\mathrm{var}(\delta P_i)}$ the corresponding parametric uncertainty component. One might think of $\{P_i\}$ as a vector **P** in "parameter space" and of the parameter error $\{\delta P_i\}$ probability distributions as a cloud $\delta \mathbf{P}$ around the point **P**.

Correlated parametric errors can sometimes be beneficial when comparing measurements on the same object at two sites by squeezing the cloud in some directions.

If, during a measurement, errors in or perturbations of the measurement parameters are independent of each other, then $r(\delta P_i, \delta P_j) = 0$ for $i \neq j$ and

$$u^2(a) = \mathrm{var}(I(a)) + \sum \left(\frac{\partial f}{\partial P_i}\right)^2 \mathrm{var}(\delta P_i)$$

because $I(a)$ is random and $r(I(a), \varepsilon(a)) = 0$.

The imaging model and its input parameters can have errors, which lead to linewidth measurement errors. Very often in photomask metrology the parameters are nearly independent, $r(\delta P_i, \delta P_j) \approx 0$, and the corresponding linewidth a parametric uncertainty components are found by modeling perturbations to imaging parameters P_i to find $\partial a/\partial P_i$, estimating the parameter uncertainties $u(P_i)$, and determining the parametric uncertainty components $u(P_i)\,\partial a/\partial P_i$.

A traceable linewidth measurement can be costly, even with a traceable linewidth standard, because the standard and the object may not match in all parameters that affect the measurement process. Those parameters are properties of the objects (other than the measurand), in which there may otherwise be little interest. See Ref. [8] for a discussion about tuning the measurement uncertainty for economical manufacturing process control.

20.7 Differential Uncertainty

If measurements are to be compared with each other, they must be traceable to a common standard. For example, a mask may be measured prior to shipping to the customer, and the customer may measure the mask prior to acceptance. These measurements should agree with each other, so both buyer and seller can agree that the mask meets its specifications.

If a mask supplier and his customer both measure a feature on the same mask, they will in general obtain different results. What is the uncertainty of this difference? What is the likelihood they will agree on whether that feature meets its specification?

The measurement result at site A is

$$a = I(a) + \varepsilon(a)$$

and at site B

$$b = I(b) + \varepsilon(b)$$

with

$$\varepsilon(a) = f(\{\delta P_i\})$$

and

$$\varepsilon(b) = g(\{\delta Q_i\})$$

Then

$$u^2(a-b) = \text{var}(a) + \text{var}(b) - 2r(a,b)\sqrt{[\text{var}(a)\ \text{var}(b)]}$$

Assume the parameters during one measurement are independent of each other, as is often the case:

$$r(P_i, P_j) \approx 0 \text{ and } r(Q_i, Q_j) \approx 0 \quad \text{for } i \neq j$$

Also $r(I(a), I(b)) = 0$ because they are both random. Then

$$\text{var}(a) = \text{var}(I(a)) + \sum \left(\frac{\partial f}{\partial P_i}\right)^2 \text{var}(\delta P_i), \quad \text{var}(b) = \text{var}(I(b)) + \sum \left(\frac{\partial g}{\partial Q_i}\right)^2 \text{var}(\delta Q_i)$$

and

$$u^2(a-b) = \text{var}(a-b) = \text{var}(a) + \text{var}(b) + \sum \left(\frac{\partial f}{\partial P_i}\right)^2 \text{var}(\delta P_i) + \sum \left(\frac{\partial g}{\partial Q_i}\right)^2 \text{var}(\delta Q_i)$$

$$-2 \sum_i \sum_j r(P_i, Q_j) \frac{\partial f}{\partial P_i} \frac{\partial g}{\partial Q_j} \sqrt{\text{var}(\delta P_i) \text{var}(\delta Q_j)}$$

$$\times \sum \text{var}(\delta R_i) \left(\frac{\partial f}{\partial R_i} - \frac{\partial g}{\partial R_i}\right)^2$$

There can be three types of parameter, P_i, Q_j, and R_k, where the R_k are all of those P's and Q's, which are identical at sites A and B. For those common mode terms where $P_i = Q_j \equiv R_i$, $r(P_i, Q_j) = r(R_i, R_i) = 1$, the parametric terms (those containing Σ) become

The general case: All of the object parameters are common mode because the same object is measured at both sites. With common-mode parameters $P_k = Q_k \equiv R_k$, $r(P_i, Q_j) = 0$ for $i, j \neq k$ by the assumption above, and the general expression for $u^2(a - b)$ becomes

$$u^2(a - b) = \text{var}(I(a)) + \text{var}(I(b)) + \sum_{i \neq k} \left(\frac{\partial f}{\partial P_i}\right)^2 \text{var}(\delta P_i) + \sum_{i \neq k} \left(\frac{\partial g}{\partial Q_i}\right)^2 \text{var}(\delta Q_i)$$

$$+ \sum_{k} \text{var}(\delta R_k) \left(\frac{\partial f}{\partial R_k} - \frac{\partial g}{\partial R_k}\right)^2$$

Special case 1: If all $P_i = Q_j \equiv R_i$, then there are no P's or Q's left, and

$$u^2(a - b) = \text{var}(I(a)) + \text{var}(I(b)) + \sum \text{var}(\delta R_k) \left(\frac{\partial f}{\partial R_k} - \frac{\partial g}{\partial R_k}\right)^2$$

An example would be the use of a transmission optical microscope at site A and a reflection mode optical microscope at site B ($f \neq g$), but same wavelength, NA, etc.

Special case 2: If $f = g$ (same kind of instrument), then $(\partial f / \partial R_i) - (\partial g / \partial R_i) = 0$ and

$$u^2(a - b) = \text{var}(I(a)) + \text{var}(I(b)) + \sum_{i \neq k} \left(\frac{\partial f}{\partial P_i}\right)^2 \text{var}(\delta P_i) + \sum_{i \neq k} \left(\frac{\partial g}{\partial Q_i}\right)^2 \text{var}(\delta Q_i)$$

An example would be use of transmission optical microscopes at both sites A and B, but with different wavelengths, NAs, etc.

Special case 3: If $f = g$ and all $P_i = Q_j$ (same instrument parameters and same object parameters), then

$$u^2(a - b) = \text{var}(I(a)) + \text{var}(I(b))$$

An example would be use of SEMs at both sites A and B with the same beam energy, effective beam diameter, detector sensitivity and linearity, specimen charging control, Abbé error, etc. Under these conditions the uncertainty in the difference of measurements at the two sites is only the combined uncertainties due to their repeatabilities. Since the instruments are identical ($f = g$ and all $P_i = Q_j$) and the object is the same at both sites, the systematic errors are all common mode and do not affect this difference measurement.

Obviously, if one site uses an optical microscope and the other site uses an SEM, then clearly $f \neq g$ and few instrument parameters can be common, and the general case with its many terms must be used. Therefore, it is difficult to obtain consistent agreement between optical and SEM measurements.

In each of these special cases,

$$u^2(a - b) \leq \text{var}(a) + \text{var}(b)$$

which is less than it would have been had the measurements been totally uncorrelated. For all cases, the expanded uncertainty is

$$U(a - b) = k\sqrt{\text{var}(a - b)}$$

and, since the true value of $a - b$ is zero, there is a 5% chance (if $k = 2$) that $|a - b| > U(a - b)$. Consequently, sites A and B might disagree on whether the feature meets its specification,

Photomask Feature Metrology

particularly if $|a - b| > 2 \cdot tolerance$. It is important that parties to such a transaction understand the role of measurement uncertainty here in order to resolve such disagreements and minimize potential rework costs.

20.8 The "True Value" of a Photomask Linewidth—Neolithography

Recall the ISO definition of true value, "a value consistent" consistent with the definition of a given particular quantity.

What is the definition of photomask linewidth? It depends on the use to which the measurement data will be put. Ultimately, the true value of the photomask linewidth produces the observed feature size on the printed wafer. This definition, however, is not always useful because this linewidth is not intrinsic to the mask but depends on wafer exposure, development, and etch conditions. For this reason, a definition based on the actual geometry of the chrome line is usually preferred. If the 3-dimensional chrome features had straight, vertical, and flat edges, this would not be a problem, but they definitely do not.

However, an even better definition—which produces the *desired* feature size on the printed wafer — can be realized by integrating mask metrology into the lithography process design.

A lithography process optimization loop is shown in Figure 20.9, whose lower half is a "virtual wafer fabrication," or a suite of linked software products designed to simulate the various lithography subprocesses (the upper half of the figure is the real fabrication). The process designer can adjust the process parameters, such as exposure, defocus, post exposure bake, develop time, etc. (either manually or automatically), by printing as many virtual wafers as necessary. Since the simulation software may not be perfect, virtual fabrication optimization provides good initial values for the lithography parameters, but printing a few real test wafers may still be necessary.

Integrating the mask design and metrology into this process results in the neolithography [10,11] scheme shown in Figure 20.10. The virtual fabrication is the top left block, and

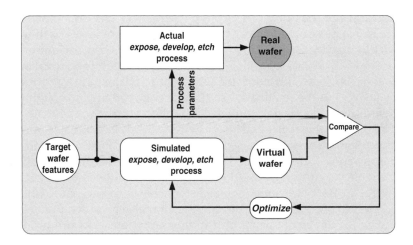

FIGURE 20.9
Lithography process optimization via the "virtual fabrication."

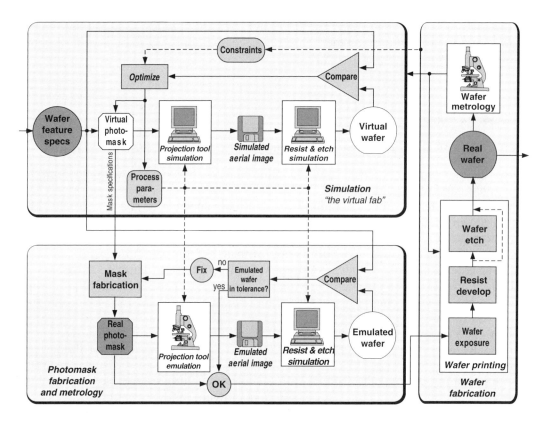

FIGURE 20.10
Neolithography, the integration of photomask design and metrology into lithography process optimization.

the photomask parameters, feature sizes and placements, assist features, phase shifters, etc., as well as the lithography process parameters, are established here.

Both sets of parameters are passed to the photomask fabrication and metrology block, lower left. Here the real mask is fabricated and measured. The preferred metrology tool, however, is exposure aerial image emulation — a transmission optical microscope whose wavelength, polarization, objective NA, coherence parameter, and illumination apodization are adjusted to match those of the exposure tool. Only the magnification is different, and the 3-dimensional (through focus) aerial image of the photomask's critical features is measured. A photomask linewidth standard is not needed here, only accurate scale calibration.

Ideally this aerial image would match the simulated aerial image in the virtual fabrication block directly above, but the real mask has chrome edge runout and roughness, printing errors, and defects, which were not simulated. The effects on the wafer of these differences can be seen by applying this real aerial image to the same resist and etch simulator used in the virtual fabrication. This results in an emulated wafer that can be compared directly with the wafer specifications. Defect printability, mask error enhancement factor (MEEF), and other proximity effects can be assessed directly in the emulated wafer. If some of the emulated features are out of tolerance, it may be possible to adjust some of the lithography process parameters to bring the mask into specification instead of scrapping it.

Note that many of the components of neolithography are software products — very inexpensive to acquire and use compared to their hardware counterparts. It behooves users of these products to urge their suppliers to improve the accuracy and interoperability of lithography process simulators.

20.9 Some General Notes on Linewidth Metrology

A metrology process can be represented by the operation

$$\text{process model} \oplus \text{feature model} \rightarrow \text{output model}$$

In this case, the process model represents the metrology process, and the *feature model* is like those in Figure 20.2. These models are abstractions of the complex realities they represent, a simplification usually required in order to make the modeling tractable and the measurement practical. The output model of this metrology process is the feature model with the metrology results attached, including the associated measurement uncertainty. Additional measurement uncertainty arises from inevitable differences between both the process and feature models and their respective realities. The *measurement error* is the difference between the measurement result and the unknown true value, and the *measurement uncertainty* is expressed as a confidence interval representing the variance of the measurement errors. The measurement uncertainty includes components from model infidelity in addition to scale calibration, repeatability, environmental factors, etc. A confidence interval of 95% (or $k = 2$ for normally distributed errors) is used in the examples, in accordance with international custom. That is, the probability that the true value of the measurand lies within the range (*measurement result* $\pm$ *expanded measurement uncertainty*) is 95%.

A manufacturing process can be represented in a similar manner. In particular, if that process is wafer exposure, then the same feature model for the photomask features can be used for both the mask metrology and exposure processes:

$$\text{exposure model} \oplus \text{photomask feature model} \rightarrow \text{wafer feature model}$$

Errors and uncertainties in the photomask feature model propagate through the exposure model to become manufacturing errors — differences between a wafer feature's size or placement and its target value — and corresponding manufacturing uncertainties (the error variances). In analogy with measurement uncertainty, tolerances on wafer features encompass mask measurement uncertainties, including differences between the models and their respective realities, as well as the effects of tolerances for exposure parameters and photomask features. The MEEF and other optical proximity effects are good examples of the wafer *exposure model* operating on photomask feature size and placement variations to produce nonlinear variations in wafer feature size and placement under some conditions.

Real microlithographic features often have irregular shapes and rough edges; it is neither possible nor necessary to know the exact shape of a feature to be measured. The purpose of the *feature bounding boxes* is to account for such edge details as top-to-bottom runout and along-the-line irregularities that are often observed. In such cases, the bounding boxes help define the measurand. To the extent that such details are not known, not relevant, or too complex to be considered, the bounding boxes represent the feature with a simpler geometry and mix these disregarded details into the measurement uncertainty. For the ideal line with known edge geometry and no edge irregularities, the line edge bounding box can have zero width. The bounding box approach simplifies metrology issues for the quasi-thin-film features often encountered in microlithography.

The definition of a measurand can depend on the purpose for which a measurement is made, and the measurement error depends on the definition of the measurand. It is up to the user to specify or define the measurand in a way that suits his present purpose and in an unambiguous way. Otherwise interpretation of the measurement may result in error, and the measurement uncertainty may be meaningless or impossible to ascertain. In other words, the true values of feature edge positions, centerline, centroid, and linewidth can depend on the purpose to which the corresponding measurement results are put.

The probability distribution and expectation value for the position of an edge within the line edge bounding box are determined as described in *ANSI Z540-2* [4]. Default values for these assume that the edge is equally likely to be anywhere inside the line edge bounding box. In that case the expectation value of the line edge location is the center of the line edge bounding box, and the edge position uncertainty (at the 95% confidence level) is 0.577 · width of line edge bounding box. The corresponding linewidth measurement uncertainty component is 0.816 · width of line edge bounding box if the right and left edge location uncertainties are uncorrelated, and 1.154 · width of line edge bounding box if they are mirror-image correlated (as is often approximately the case) [5].

In most cases, the width or centroid or edge positions of the bounding box are measured from its image in a metrology tool; inferring the width of the bounding box from this image usually requires modeling of the image-forming process. The bounding box should be constructed so that its image in the metrology tool can be modeled with the modeling tools available. If the image is not modeled accurately, additional measurement uncertainty will accrue [8].

20.10 Conclusion

Assigning a single number to a photomask linewidth implies vertical and smooth edges on the etched metal lines. High-resolution images of photomask lines reveal that this is rarely true, obfuscating the meaning of the term "linewidth." A practical solution is to represent each edge as a probability distribution within an edge bounding box and include the combined variances for the two edges in the linewidth measurement uncertainty. If the edges show runout or undercut, then the edge probability distribution correlation must be taken into account.

There are costs associated with mask metrology, but in a well-designed process the benefits outweigh the costs. In fact it may be possible to calculate an optimum level of resources to devote to mask metrology in a production environment, but this requires an understanding of the measurement uncertainty.

The measurement uncertainty is expressed in terms of a confidence interval about the measurement mean with a stated probability of containing the true value of the measurand. In particular, the expanded measurement uncertainty is the square root of the sum of the variances of the probability distributions of all the possible measurement errors (both random and systematic), taking into account possible correlations, multiplied by a stated factor chosen to represent the desired confidence interval.

Those correlations can sometimes reduce the measurement uncertainty, for example, when using a photomask linewidth standard to evaluate the linearity of a linewidth metrology tool. A practical approach may be to model the measurement process, evaluate the effects of parametric uncertainties by perturbing the parameters in the model, estimate the uncertainties of these parameters in the measurement system, and combine these results. In comparing measurements (at the supplier's and customer's sites, or at different sites on a mask) some of these parameters may have correlated effects, reducing the uncertainty of the comparison.

Many of the problems with chrome edge definition, defect printability, proximity effects, etc., can be obviated by measuring mask feature performance instead of mask feature geometry. The neolithography design model places a virtual wafer fabrication — a collection of interoperable process simulation applications, in a feedback loop — on the desk of the lithography process designer. He uses this tool to design a mask (including OPC and phase shifters, as required) and to set the wafer printing parameters (with mask and parameter tolerances), which will produce the desired patterns on the wafer, balancing product performance with process latitude.

The mask fabrication shop also has this tool; they both use the same process models and lithography process parameters. The mask shop fabricates the mask, measures critical features, and uses this data with the virtual fabrication to predict mask performance (defect printability, MEEF, etc.). The mask shop then determines if the mask will perform as required, which, if any, features to repair, if an exposure parameter adjustment will bring the mask performance into specification, etc.

Accurate and comprehensive process models appear to be essential ingredients for overcoming the economic problems of making and measuring masks of ever increasing complexity.

Acknowledgments

Thanks to Drs. Tyler Estler (NIST) and Robert Larrabee (NIST, retired) for many helpful discussions on the ideas in this chapter.

References

1. SEMI Standard P35, *Terminology for Microlithography Metrology*, SEMI International Standards, 3081 Zanker Rd., San Jose, California 95134.
2. D.V. Lindley, *Making Decisions*, second ed., John Wiley & Sons, New York, 1985.
3. *International Vocabulary of Basic and General Terms in Metrology*, ISO, 1993, 60 pp., ISBN 92-67-01075-1.
4. *U.S. Guide to the Expression of Uncertainty in Measurement*, ANSI/NCSL standard Z540-2-1997b.
5. *Guide to the Expression of Uncertainty in Measurement*, ISO, 110 pp., ISBN 92-67-10188-9, 1995.
6. B.N. Taylor, *The International System of Units* (SI), NIST Special Publication 330, 1991.
7. J. Potzick, Accuracy in integrated circuit dimensional measurements, in: *Handbook of Critical Dimension Metrology and Process Control*, vol. TR52, SPIE, Bellinhgam, Washington, September 1993, pp. 120–132 (Chapter 3).
8. J. Potzick, The problem with submicrometer linewidth standards, and a proposed solution, in: *Proceedings of SPIE 26th International Symposium on Microlithography*, vol. 4344-20, 2001.
9. D. Nyyssonen, R. Larrabee, Submicrometer linewidth metrology in the optical microscope, J. *Res. National Bureau Standards*, 92 (3), (1987).
10. J. Potzick, Photomask metrology in the era of neolithography, in: *Proceedings of the 17th Annual BACUS/SPIE Symposium on Photomask Technology and Management*, vol. 3236, Redwood City, California, September 1997, pp. 284–292.
11. J. Potzick, The neolithography consortium, in: *Proceedings of SPIE 25th International Symposium on Microlithography*, vol. 3998-54, 2000.

Bibliography

1. J. Potzick, J.M. Pedulla, M. Stocker, Updated NIST photomask linewidth standard, in: *Proceedings of SPIE 28th International Symposium on Microlithography*, vol. 5038–34, 2003.
2. A. Starikov, et al., Applications of image diagnostics to metrology quality assurance and process control, in: *Proceedings of SPIE 28th International Symposium on Microlithography*, vol. 5042–39, 2003.
3. J. Potzick, Measurement uncertainty and noise in nanometrology, in: *Proceedings of the International Symposium on Laser Metrology for Precision Measurement and Inspection in Industry*, 1999, pp. 5–12 to 5–18, Florianopolis, Brazil.
4. J. Potzick, Accuracy differences among photomask metrology tools and why they matter, in: *Proceedings of the 18th Annual BACUS Symposium on Photomask Technology and Management*, Redwood City, California, SPIE vol. 3546–37, September 1998, pp. 340–348.
5. J. Potzick, Accuracy and traceability in dimensional measurements, in: *Proceedings of SPIE 23nd International Symposium on Microlithography*, vol. 3332–57, 1998, pp. 471–479.
6. J. Potzick, *Antireflecting-Chromium Linewidth Standard, SRM 473, for Calibration of Optical Microscope Linewidth Measuring Systems*, NIST Special Publication SP-260-129, 1997.
7. R. Silver, J. Potzick, and J. Hu, Metrology with the ultraviolet scanning transmission microscope, in: *Proceedings of the SPIE Symposium on Microlithography*, vol. 2439–46, Santa Clara, California, 1995, pp. 437–445.
8. J. Potzick, Noise averaging and measurement resolution (or a little noise is a good thing), *Rev. Sci. Instrum.*, 70 (4), 2038–2040 (1999).
9. J. Potzick, New NIST-certified small scale pitch standard, in: *1997 Measurement Science Conference*, Pasadena, California, 1997.
10. J. Potzick, Re-evaluation of the accuracy of NIST photomask linewidth standards, in: *Proceedings of the SPIE Symposium on Microlithography*, vol. 2439–20, Santa Clara, California, 1995, pp. 232–242.
11. D. Nyyssonen, in: *Spatial Coherence: The Key to Accurate Optical Metrology*, SPIE Vol. 194, Applications of Optical Coherence, 1979, pp. 34–44.

21
Optical Critical Dimension Metrology

Ray J. Hoobler

CONTENTS
21.1 Introduction 433
21.2 Methodology 435
    21.2.1 Scatterometry 435
    21.2.2 Spectroscopic Ellipsometry 436
    21.2.3 Normal-Incidence Spectroscopic Reflectance (Nonpolarized) 438
    21.2.4 Normal-Incidence Spectroscopic Ellipsometry (Polarized Reflectance) .. 438
21.3 Rigorous Coupled-Wave Analysis 440
    21.3.1 Theory Overview 440
    21.3.2 RCWA Implementation 440
        21.3.2.1 Real-Time Regression Analysis 441
        21.3.2.2 Libraries 441
21.4 Applications 441
    21.4.1 After Development Inspection of Binary Mask 443
    21.4.2 After Etch Inspection 444
        21.4.2.1 Chrome-Grating Structures 444
        21.4.2.2 Quartz Grating Structures 446
        21.4.2.3 Alternating Aperture Phase Shift Masks 446
        21.4.2.4 Advanced Process Control 453
21.5 Conclusions 454
Acknowledgments 454
References 454

21.1 Introduction

Current trends in semiconductor fabrication indicate that chip level feature sizes will decrease below 0.1 µm (100 nm) in the very near future. While traditional techniques based on electron microscopy exist for routinely measuring submicron dimensions, these techniques will eventually face the same limitations of optical microscopy when the features of interest are of the same dimension as the wavelength of electromagnetic radiation used for imaging. In response to this need, a new interest in diffraction-based optical methods has arisen. These techniques do not directly image a feature of interest, but, rather, rely on implementation of the rigorous coupled-wave analysis (RCWA) for periodic grating structures, which are included as test patterns on the sample.

The determination of important, critical dimensions via optical techniques is appealing for several reasons:

- The sample is exposed to only visible light; the method is nondestructive to 193-nm photoresist or subject to charging effects.
- The technique is capable of measuring the critical dimensions of grating structures down to approximately 40 nm.
- Minimal facilities are required for installation (no high vacuum, cooling, or shielding of electromagnetic fields [EMFs]).
- Like optical thin film metrology, optical critical dimension (OCD) technology can be integrated into process tools enabling advanced process control (APC).

Unlike imaging technologies that can measure a variety of features, optical methods incorporate test structures in the form of gratings. A simple grating structure is shown in Figure 21.1; features, such as rounding, notching, and footing, can also be included in the model if desired.

The typical sample requirements for OCD technology are as follows. Target structures are limited by the illumination and should have a minimum dimension of 50 µm × 50 µm with a nominal pitch of 180 nm or greater. The step height for a photoresist structure should be >200 nm and have a linewidth >40 nm. The lines should extend to edges of the test site. In some respects, the test site requirements for OCD measurements on mask are not as stringent as those at the wafer level. The minimum dimensions above for wafer level measurements are dictated by how much space is designed into the open space between devices. At the mask level, these test sites will be a factor of 4 larger. Minimum linewidths and pitch values are also scaled, thus for 100-nm wafer level lines, the typical mask measurement is 400 nm. One might argue that given these larger dimensions, imagining methods will still dominate and the need for optical diffraction techniques is minimal. However, the great advantage of optical methods is providing profile information that can greatly assist process development. Obtaining cross-section information by traditional imaging methods is a time-consuming and destructive process.

In the following section, the various optical methods will be discussed. Section 21.3 will outline the RCWA used in calculating the linewidths and profiles of the grating structures used for test sites. Section 21.4 will present theoretical and experimental results in the application of normal-incidence spectroscopic ellipsometry (SE) to measurements on masks.

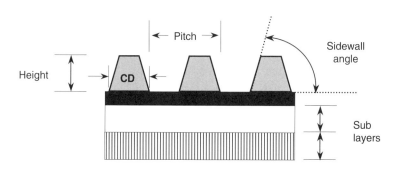

FIGURE 21.1
Grating parameters to be determined by OCD measurements.

21.2 Methodology

While many similarities exist between current optical technologies, different methodologies have distinct characteristics that can influence OCD measurements; these methodologies can be categorized as scatterometry, SE, normal-incidence spectroscopic reflectance and normal-incidence SE. Each of these methods requires light to be diffracted (or "scattered") from a periodic grating structure designed to provide information on CD features inside the active components of the IC. Unlike scanning electron microscopes, these optical techniques are indirect and require the user to model the diffracted light based on the optical properties and structure of the grating. These models employ complex analysis routines based on theories, such as the RCWA, which provides a method for calculating the diffraction of electromagnetic waves by periodic grating structures. Depending on the experimental method, analysis can be done in "real time" using standard nonlinear regression techniques to optimize the modeled spectrum to the experimental data or by searching a library of pregenerated spectra.

21.2.1 Scatterometry

In scatterometry, a polarized beam of monochromatic light is reflected from a grating structure at an incident angle θ. The diffracted light is dependent on the grating structure, and information can be obtained by examining how the light is diffracted from the 0th order beam into the higher-order beams or by measuring the intensity of the 0th order reflection as a function of incident angle (Figure 21.2 and Figure 21.3). This later method is commonly referred to as angular 2θ scatterometry.

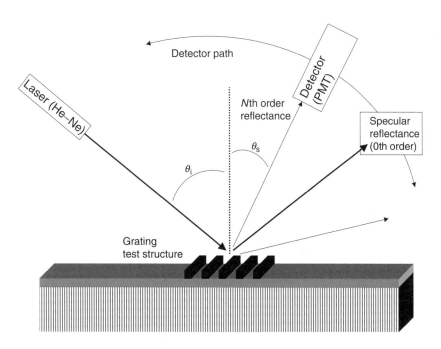

FIGURE 21.2
Scatterometry.

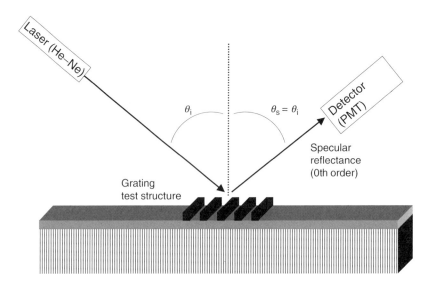

FIGURE 21.3
2θ scatterometry.

Angular 2θ scatterometry has several appealing features; the most obvious is the simplicity of the experimental design: a monochromatic light source (most commonly a helium–neon laser) and a detector. Angular 2θ scatterometry has proven effective at determining linewidths and line/space profiles for critical dimensions, as well as providing information on profile changes [1]. The technique can also offer sub-nanometer precision. Angular 2θ scatterometry has not shown the sensitivity to provide detailed profile information, and small pitches will require shorter wavelengths than currently provided from a simple helium–neon laser [2]. Finally, as with all oblique angle methods, RCWA methods are computationally intense and usually require the user to generate a library, which the experimental data can be compared against; however, once a library is in place, the time to find the best match is relatively fast. Additional optical components can be added to the system, allowing polarization and phase information to be obtained (single wavelength ellipsometry), which can increase sensitivity at the expense of the instrument simplicity.

While this methodology has been established, spectroscopic methods dominate in the optical CD metrology arena. This is undoubtedly due to the added complexity of automating multi-angle measurements and the apparent lack of profile sensitivity [2]. SE has also become a standard tool for thin film measurements, and in many cases existing equipment based on SE can be used. For APC applications, the system has the potential for a large footprint due to both the mechanical complexity and the necessity of an X/Y stage. As with all oblique angle measurements, positioning in the z-direction (focus) is critical.

21.2.2 Spectroscopic Ellipsometry

Researchers and process engineers routinely rely on SE in the determination of optical properties of thin films, and the technique has become the standard by which all optical methods are compared. The advantages of SE arise from the ability to monitor the change in polarization state as a function of wavelength. If light of a known polarization is reflected from the surface of a substrate or thin film, the magnitude of the amplitude for the s- and p-waves (as defined by the surface) will be altered and experience a phase shift. This is expressed in the fundamental equation of ellipsometry as

$$\rho = \tan \Psi e^{i\Delta}$$

where the ratio of the magnitude for the s- and p-waves is proportional to $\tan \Psi$ and phase information is given by Δ. Ψ and Δ are the parameters determined by ellipsometric measurements [3]. By monitoring these parameters over a large number of wavelengths, one can eliminate the need for multi-angle measurements for most thin-film measurements. SE is routinely used to monitor film thickness and optical properties in IC fabrication. A schematic of the typical experimental layout is shown in Figure 21.4.

Extension of SE to OCD measurements has proceeded rapidly due to the presence of existing hardware. By using RCWA methods, researchers have shown the ability to model profile changes over a wide range of conditions, and the model results have shown good correlation to CD-SEM [4]. Precision is also subnanometer. The current analysis methods are divided between library methods and real-time regression analysis. While library-based systems offer relatively high speed, the time required to create a new library can be long (6–12 h), depending on the computer used. Small changes to the film stack (addition of a single layer) may require the regeneration of the library. Recently, the application of real-time regression analysis has also been introduced. As expected, this requires a computer with a large number of processors, which adds to the expense and complexity of the system.

While the hardware is identical to that used in thin film metrology, focus and position of the grating structure relative to the incident light are critical. As with angular 2θ scatterometry, the system has the potential for a large footprint and the necessity of an X/Y stage. The number of experimental parameters that must be accounted for—angle of incidence, angular spread, and wafer tilt—can make tool-to-tool matching challenging.

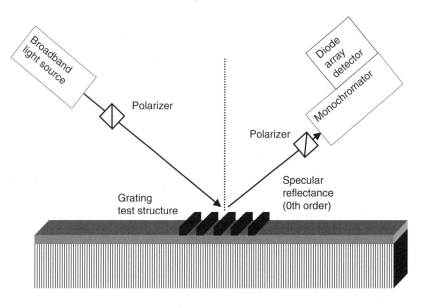

FIGURE 21.4
Spectroscopic ellipsometry.

Finally, since recent research has indicated that oblige angle (approximately 65–75°) measurements can have strong correlations to CD height and width; the application of near-normal incidence SE measurements can have advantages for analysis of periodic grating structures [5].

21.2.3 Normal-Incidence Spectroscopic Reflectance (Nonpolarized)

Spectroscopic reflectance has become the most popular method for thin-film measurements primarily due to its speed. Spectroscopic reflectance can be used to reliably monitor film thickness for materials where the optical properties are known (usually determined from SE measurements). Reflectometers offer simple mechanical design that can have a footprint only slightly larger than the wafer itself making the systems ideal for inline monitoring and APC. While the technology is well established for measurements on isotropic thin films, the lack of a polarized light source means any data gathered from an anisotropic grating structure will be a linear combination of transverse electric (TE) and transverse magnetic (TM) components (p- and s-waves) of the observed reflectance. The inability to separate these components will most likely hinder any practical application to monitor critical dimensions without using additional information from CD-SEM measurements [6].

21.2.4 Normal-Incidence Spectroscopic Ellipsometry (Polarized Reflectance)

Unlike standard reflectance techniques, addition of a polarizer in the optical path allows for the separation of TE and TM modes for the reflected light (Figure 21.5); if an anisotropic sample is found in the grating structure, this can provide the means for determining linewidths and analyzing complex profiles [7]. The measured reflectance is a complex function of the angle, φ, between the polarizer transmission axis and the sample grating lines given by

$$R(\varphi) = R_{TE} \cos^4(\varphi) + R_{TM} \sin^4(\varphi) + \sqrt{R_{TE} + R_{TM}} \cos(\Delta) \sin^2(\varphi) \cos^2(\varphi)$$

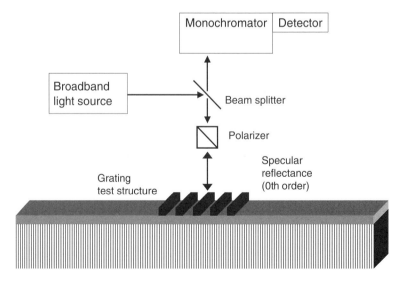

FIGURE 21.5
Normal-incidence spectroscopic ellipsometry.

where R_{TE} and R_{TM} is the reflectances with the polarizer parallel and perpendicular to the grating lines, respectively. Measurement at an arbitrary third angle allows the determination of Δ (or $\cos(\Delta)$), providing the ability to make the equivalent of an ellipsometric measurement.

In this configuration, the polarization of the incident light is set with respect to the periodic grating structure on the sample allowing several different modes of data acquisition. TE mode where the electric field of the incident beam polarization is parallel to the grating lines; TM mode where the electric field of the incident beam polarization is perpendicular to the grating lines. The selection of measurement mode is an important consideration and depends on the sample material, and feature size. For analysis of complex structures, phase information can be obtained by acquiring reflectance data at a polarization of 45° with respect to the grating structure.

The normal-incidence spectroscopic ellipsometer maintains much of the simplicity in mechanical design found in a standard reflectometer. The additional polarizer may be fixed or rotating; however, rotation of the polarizer is preferred over movement of the sample in most applications. Addition of the polarizing element has no effect on the footprint making the system amenable for integration, inline monitoring, and APC. As with standard reflectometers, tool-to-tool matching is straightforward and depends on a relatively simple optical alignment and the initial wavelength calibration of the detector. A schematic of optical layout and the NANOmetrics 9010M OCD system are shown in Figure 21.6.

While analysis can utilize pregenerated libraries as found in most SE applications, the use of normal incidence greatly reduces the complexity of the RCWA calculations to the point that high-end desktop computers are sufficient for most tasks and the analysis can be done in real-time. Unlike libraries, where a large set of possible answers are stored, real-time analysis requires the scientist or engineer to input nominal starting values. The use of real-time regression analysis also provides a great deal of flexibility in making modifications when processes are updated and for the research and development of new processes.

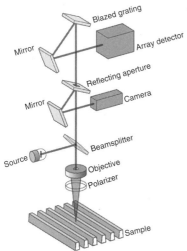
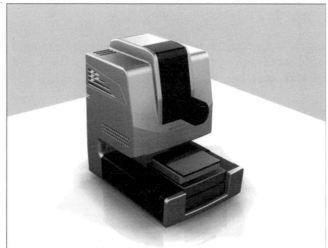

FIGURE 21.6
Schematic for a normal-incidence spectroscopic ellipsometer and picture of the corresponding product (NANOmetrics, Inc.).

21.3 Rigorous Coupled-Wave Analysis

The RCWA method provides an exact method for calculating the diffraction of electromagnetic waves by periodic grating structures. Details can be found in the paper by Moharam et al. [8]; a brief overview will be provided here.

21.3.1 Theory Overview

Analogous to optical measurements of thin films, OCD measurements require an accurate model to be generated. Parameters of this model can be varied, and the calculated spectra saved in a database or the parameters can be varied using a regression analysis until the modeled spectrum matches the sample spectrum. In this formulation, the grating layer is sliced into a stack of layers, and the complex dielectric function (ε) is expanded in a Fourier series of $2N + 1$ harmonics along the x direction of the grating region with grating pitch D:

$$\varepsilon(x) = \sum_{j=-N}^{+N} \varepsilon_j \cdot \exp\left(i\frac{2\pi j}{D}x\right)$$

In the limit of N approaching infinity, the solution is exact. In practice, the number of harmonics required is reduced to a number determined by the material forming the grating structure. This expansion is visually depicted in Figure 21.7.

From this Fourier expansion, Maxwell's equations can be solved for the electric field in each layer, matched via E-M boundary conditions and the spectrum calculated.

21.3.2 RCWA Implementation

The implementation of the RCWA analysis for critical dimensions has proceeded along two lines: libraries and real-time regression. While fundamentally different in their approach for process monitoring, each is based on the same theories put forth in RCWA. As discussed in the following, the large computational requirements for performing RCWA calculations placed on oblique angle measurements usually demanded the use of libraries; however, calculations can be carried out in real-time using multiprocessor systems.

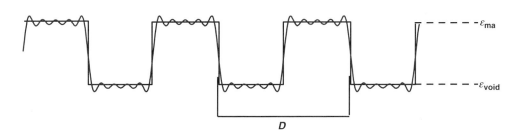

FIGURE 21.7
Representation of a grating and its Fourier reconstruction of order N.

21.3.2.1 Real-Time Regression Analysis

For thin film optical metrology, all analyses are done in real-time. That is, a spectrum is obtained and specific parameters (e.g., thickness, refractive index, etc.) are varied using nonlinear regression algorithms to find the best fit between the experimental and model spectra. The computational requirements for these calculations are minimal, and optical instruments for measuring thin film have been available since the mid-1970s. In contrast, the theory surrounding RCWA has been developed over the last 20 years but required significant computational resources. Even today, oblique angle, SE still requires 16–32 CPUs to perform calculations on the timescale of the measurement (5–10 s). Normal-incidence SE greatly reduces the computational requirements allowing real-time analysis using high-end desktop computers (>2 GHz, two processors). The main advantage of real-time calculations is the ease in which changes can be made to existing measurement recipes where layers can be added/deleted/changed in the optical model.

21.3.2.2 Libraries

The most appealing feature of libraries is their speed. Once a library is generated, analysis will be limited by the measurement time itself, since the time required to look-up the best spectrum should be <1 s. The use of libraries was also driven by the desire to avoid the expense and space requirements for multi-processor systems required for oblique angle SE measurements. A main limitation for libraries is the rigid framework from which they must be developed and implemented. Even for a simple photoresist grating structure, the number of spectra, which must be generated, can be quite large. For example, for a target CD of 400 nm, the range of CD values may be 350–800 nm. For 1-nm resolution, 450 spectra must be generated. Variation of the sidewall angle may cover 60–95° and require a resolution of 1° (35 spectra). Thickness information must also be included for the photoresist (150–500 nm) and chrome oxide (150–300 nm): 350 and 150 spectra, respectively. Together, this corresponds to a database of 8.3×10^8 spectra. These assumptions are probably over simplified, since there has been no effort to include detailed profile information (rounding, notching, etc.) that may be required to fully characterize the observed spectrum. Database generation is highly dependent on the computational facilities available, but is reported to be 6–12 h for dedicated multiprocessor workstations.

21.4 Applications

The use of OCD technology to measure critical dimensions in production environments is relatively new, and the application of OCD measurements to masks is still being evaluated. However, the possibility of obtaining profile information from a spectroscopic technique has generated a great deal of interest given the difficulty and destructive nature of obtaining cross-sectional information from SEM techniques. OCD measurements typically fall into two categories: after development inspection (ADI) where the grating structures are composted of photoresist and after etch inspection (AEI) where the grating structures are chrome, chrome quartz, or quartz. In the following sections, the application of normal-incidence SE to these grating structures will be considered. OCD has been extensively tested for across wafer CD measurements. Figure 21.8 illustrates the excellent agreement between calculated profiles and profiles from X-SEM. Even subtle features, such as footing (photoresist), notching (poly-silicon gate oxide structure), and silicon trench isolation, can be determined. OCD measurements have also shown excellent correlation with CD-SEM measurements. While traditional

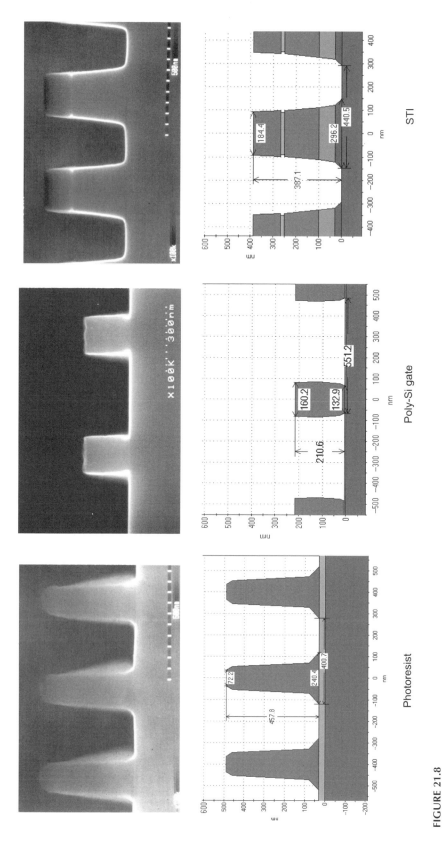

FIGURE 21.8
Comparison of X-SEM and profiles determined using OCD metrology (NANOmetrics, Inc.).

Optical Critical Dimension Metrology

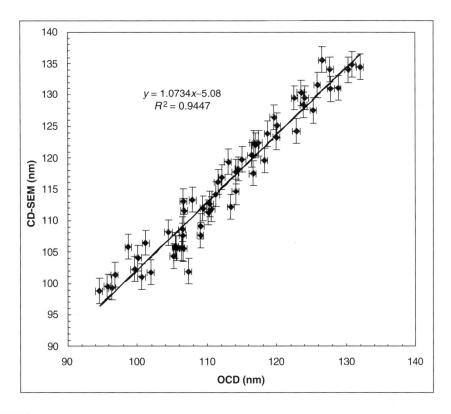

FIGURE 21.9
CD-SEM, OCD correlation plot. CD-SEM repeatability was estimated to be ~4 nm. OCD repeatability ~1 nm (CD-SEM data courtesy of ASML, OCD data NANOmetrics, Inc.).

CD-SEM measurements have repeatability on the order of several nanometers, OCD measurements have a repeatability that is typically subnanometer. It is important to remember that CD-SEM measurements are highly localized, while optical methods average over a much larger number of lines. Typical linewidth variations require an average of 20 CD-SEM measurements over the grating target to generate high-quality correlation plots. A CD-SEM, OCD correlation plot is shown in Figure 21.9. In this example, four CD-SEM measurements were averaged and used to estimate the uncertainty. Additional correlation studies are available in references [7,10].

21.4.1 After Development Inspection of Binary Mask

The measurement of photoresist grating structures using normal-incidence SE can provide rapid evaluation of mask writer performance. A typical after development inspection (ADI) photoresist grating structure is shown in Figure 21.10. The grating structure consists of patterned photoresist on chrome oxide/chrome film on quartz. Unlike CD metrology at the wafer level where lines with sub-100-nm CD have been generated, the smallest CDs at mask level are typically 200 nm. With current technology focused on printing 90–110-nm features at wafer level, 400-nm CDs at the mask level are common. The ADI photoresist structures can have significant variation depending on the mask writing process. Figure 21.11 shows a well-defined grating structure (a) and a poorly defined grating structure where the photoresist material has not been totally cleared (b). A comparison of two spectra shows the distinct spectral changes due to the grating structures.

FIGURE 21.10
ADI structure for OCD analysis.

Examples of the experimental R_{TE} and R_{TM} data and the calculated model results for a typical grating structure are shown in Figure 21.12. The good spectral fit obtained an indication of the model accuracy. The corresponding profile is shown in Figure 21.13(A). This profile was confirmed by cross-sectional SEM (X-SEM). In many cases, a model based on a modified trapezoid is sufficient to model the experimental data; the bottom CD, top CD, and sidewall angle are determined during the real-time regression analysis. An example for a 0.3-μm site for photoresist on chrome structures is included in Figure 21.13(B). The repeatability of these measurements is very good with the standard deviation from the mean determined by making 30 measurements sequentially is 0.5 nm or less.

The sensitivity of the normal-incidence SE can be seen in Figure 21.14, where the predicted spectral changes for varying linewidth with a constant 1:2 linewidth to pitch ratio are shown. Definitive spectral changes show how the variations in linewidth allow for unambiguous determination of the critical dimension under study.

21.4.2 After Etch Inspection

ADI allows the process engineer to evaluate the patterning process before the etch process. AEI allows the process engineer a chance to evaluate the tool's performance. For photomask, the AEI has typically relied on top-down CD-SEM for characterization with little or no profile information. This is partly due to the difficulty in obtaining cross-sections. In the past, when the etch process was confined to the chrome layer of the photomask, the profile information was not expected to be significant; however, the need to produce smaller features at the wafer level has lead to the development of phase-shift mask where the quartz substrate is also etched to a specified depth. Unlike wafer fabrication, where "stop" layers can be included in the production, the etch process for photomasks is controlled by time alone.

21.4.2.1 Chrome-Grating Structures

The analysis of chrome-grating structures by OCD techniques is only in its infancy; however, a typical structure for chrome grating on quartz can be simulated to observe the sensitivity to changes in CD. Figure 21.15 shows the calculated reflectance spectra for a chrome grating on a quartz substrate. The pitch was 1600 nm and share changes in the TE and TM spectra as the CD is varied. The simulation shows good sensitivity to linewidth changes over a fairly narrow range of values. Figure 21.16 shows a similar simulation for a 400-nm grating structure with a pitch of 800 nm. Again, the spectral changes are significant; however, the reflectivity for this type of sample is extremely low, less than 20% at most wavelengths.

Optical Critical Dimension Metrology

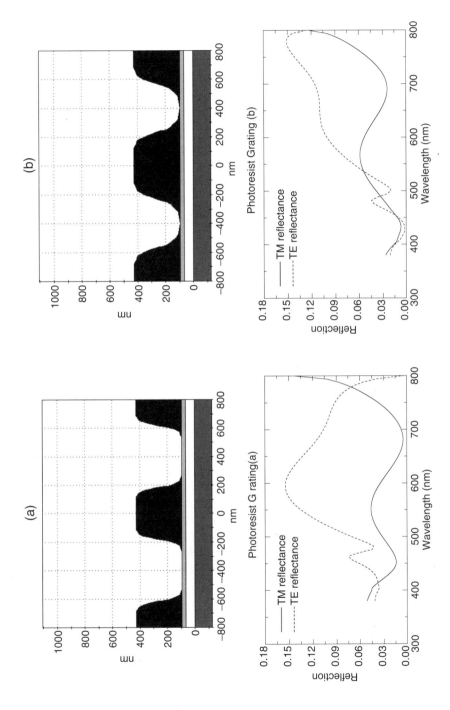

FIGURE 21.11
Calculated spectra for a well-resolved (a) and poorly resolved (b) grating.

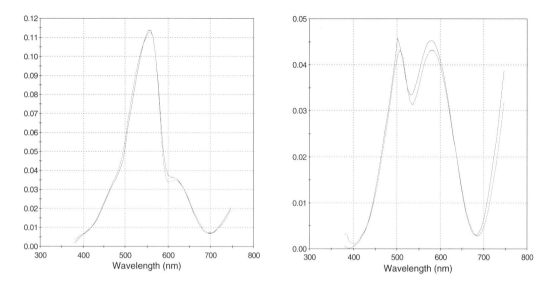

FIGURE 21.12
Experimental and model reflectance spectra of TE and TM data. The line/space ratio for the grating structure is 500 nm/500 nm.

21.4.2.2 Quartz Grating Structures

While chrome-grating structures will most likely continue to be monitored by CD-SEM, the ability of OCD techniques to provide profile/depth information will extend the applications to etched structures where information is not obtainable with traditional CD-SEM. In many respects, these structures are similar to those generated in the dielectric etch step seen in wafer fabrication. Traditional SE and normal-incidence SE has proven effective at determining the profile of trench structures seen in dual-damascene processes [9,10]. Simulations for an etched quartz grating on quartz shows the overall reflectivity to be quite low (Figure 21.17); however, the spectral changes as a function of CD are readily apparent, and the spectra show several sharp features that aid in the analysis. The spectra were calculated with a pitch of 1600 nm. Additionally, spectral changes as a function of etch depth for the quartz-grating structure also show a continuous variation (Figure 21.18). Spectral changes for 400-nm lines with a pitch of 800 nm are shown in Figure 21.19 and Figure 21.20.

Preliminary investigations have shown sensitivity to chrome/quartz structures and the analysis of etched quartz structures is ongoing (NANOmetrics, Inc., Internal Communication, 2002).

21.4.2.3 Alternating Aperture Phase Shift Masks

Development of etched quartz structures has led to the development of alternating aperture phase shift masks (altPSM), which provide a reticle-based, resolution enhancement technology for producing sub-100-nm devices [11]. These masks provide one possible solution for overcoming limitations of diffraction effects seen when printing features smaller than the illumination wavelength of the lithography exposure tool.

The basic structure of an altPSM and the resolution enhancement over a traditional binary mask are shown in Figure 21.21. The inability of the binary mask to resolve closely spaced structures is due to the constructive interference of the light originating from a

Optical Critical Dimension Metrology

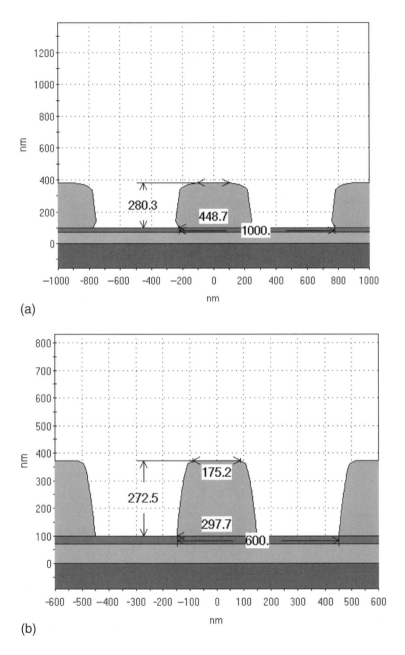

FIGURE 21.13
Calculated profiles for photoresist gratings, 500 nm (a) and 300 nm (b).

coherent source. The etched quartz aperture results in a phase shift for the light compared to the unetched aperture. This phase shift is given by

$$\Delta\varphi = 2\pi d(n-1)/\lambda$$

where $\Delta\phi$ is the phase shift, d is the depth of the etched trench, n is the index of refraction at the illumination wavelength, λ. For a 180° phase shift, an etch depth of 243.6 nm is required for 248-nm processes ($n = 1.509$). However, even if the target depth is achieved,

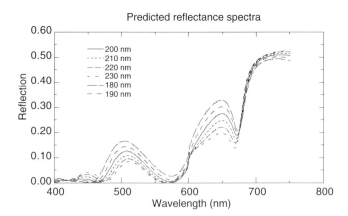

FIGURE 21.14
Calculated reflectance spectrum (RTE) showing variation as a function of linewidth.

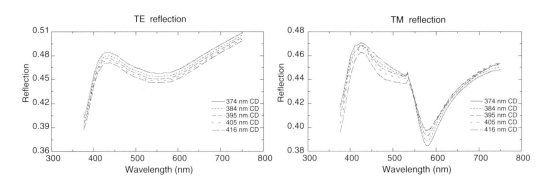

FIGURE 21.15
Calculated reflectance spectra for a chrome grating, showing variation as a function of linewidth (pitch: 1600 nm).

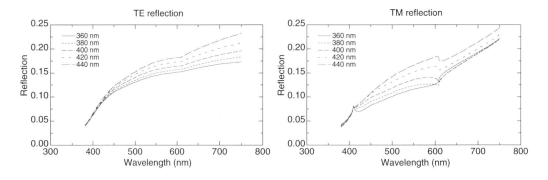

FIGURE 21.16
Calculated reflectance spectra for a chrome grating, showing variation as a function of linewidth (pitch: 800 nm).

topography effects from the structures will scatter light and result in an image imbalance. Several recent articles address possible solutions; two of which are shown in Figure 21.22 [12,13].

Electromagnetic field (EMF) simulations have shown the ability to correct the inherent image imbalance, and each of the proposed etch structures poses considerable challenges to standard metrology. While CD-SEM can readily address bias measurements, under-

Optical Critical Dimension Metrology 449

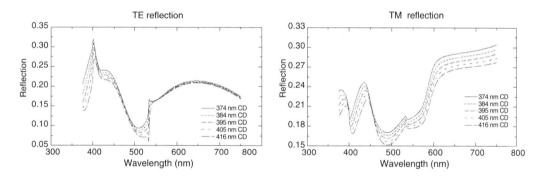

FIGURE 21.17
Calculated reflectance spectra for a quartz grating structure showing variation as a function of linewidth (pitch: 1600 nm).

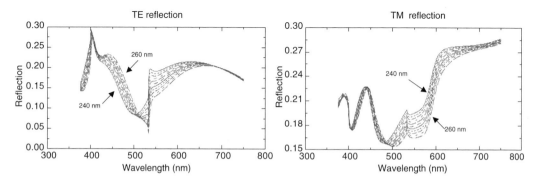

FIGURE 21.18
Calculated reflectance spectra for a quartz grating structure showing variation as a function of etch depth (linewidth: 400 nm; pitch: 1600 nm).

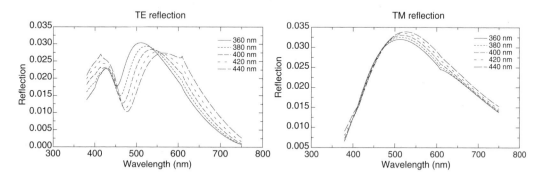

FIGURE 21.19
Calculated reflectance spectra for a quartz grating structure showing variation as a function of linewidth (pitch: 800 nm).

cutting can only be obtained from cross-sectional images; a combination of bias and undercut may prove the most effective solution to image imbalance. OCD metrology has the potential to measure all of these parameters simultaneously. The differences in the calculated spectra based on bias, undercut, and depth for a modeled altPSM (Figure 21.23) are shown in Figures 21.24–21.26, respectively. The simulated structure incorporates a

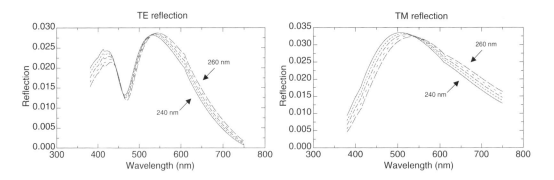

FIGURE 21.20
Calculated reflectance spectra for a quartz grating structure showing variation as a function of etch depth (linewidth: 400 nm; pitch: 800 nm).

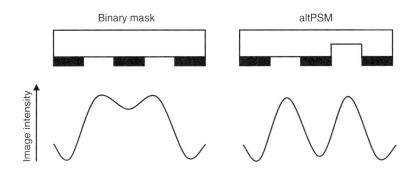

FIGURE 21.21
Reticle-based resolution enhancement using an altPSM ensures light passing through adjacent apertures are 180° out of phase.

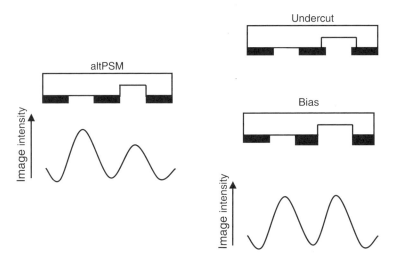

FIGURE 21.22
Image imbalance and possible modifications to the trench structure.

Optical Critical Dimension Metrology 451

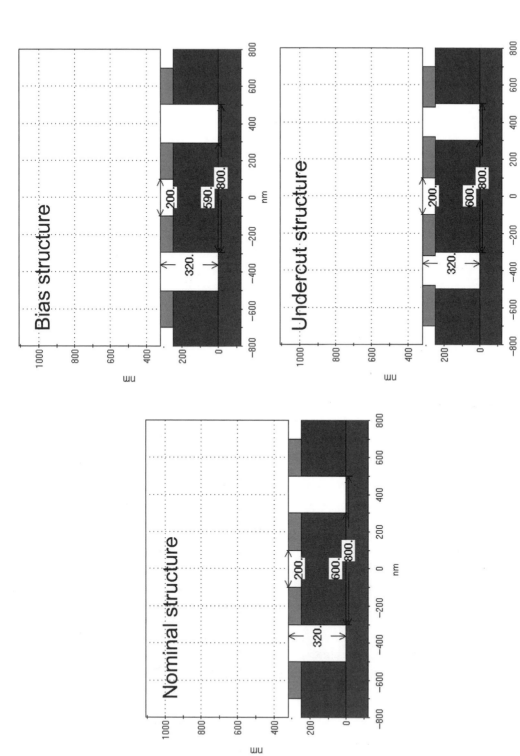

FIGURE 21.23
Test structure used for bias, undercut and depth simulations.

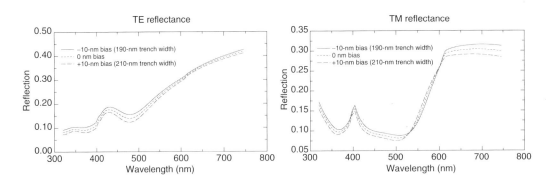

FIGURE 21.24
Calculated reflectance spectra showing changes due to etch bias structure.

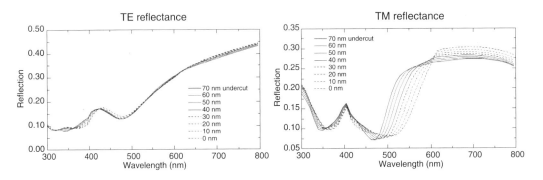

FIGURE 21.25
Calculated reflectance spectra showing changes due to undercut of the chrome absorber.

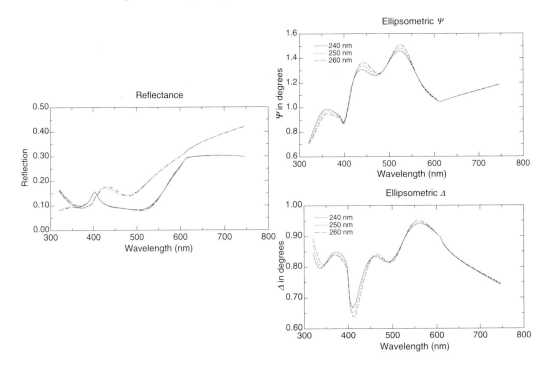

FIGURE 21.26
Calculated reflectance spectra and NISE spectra showing changes due to quartz depth variations for the altPSM.

Optical Critical Dimension Metrology

70-nm chrome absorber with 250-nm etched quartz structures. Figure 21.24 shows the change in TE and TM reflectance for a $\pm$ 10-nm bias. The change in spectral intensity changes smoothly for both TE and TM. This is in contrast to changes in the reflectance spectra due to an undercut of the absorber, Figure 21.25. Here, the changes are quite dramatic for the TM reflectance spectrum but very subtle for the TE mode. Thus, it should be possible to determine both parameters simultaneously with minimal parameter correlation in the RCWA analysis. Changes in the calculated TE and TM reflectance spectra are quite small; however, because the third angle measurement is possible, the phase difference between the TE and TM modes can be readily calculated. By monitoring the changes in the ellipsometric parameters Ψ and Δ, a complete analysis of the altPSM structure should be possible (Figure 21.26).

21.4.2.4 Advanced Process Control

With a constant push for yield enhancement, APC has driven metrology manufactures to provide new sensors and tools, as well as adopt standards for integration [14]. OCD technologies can be implemented for both ADI and after etch inspection (AEI) processes. The general schematic is shown in Figure 21.27. For ADI, OCD can provide monitoring and feedback for mask writer performance using specific targets. Depending on the available inputs, measurement results can then be used to tune the mask writer for the next mask. An integrated measurement could also provide immediate fault control if the measurement falls outside a specified range reducing the possibility of multiple masks misprocessed while waiting for off-line analysis. The measurement time required for OCD is minimal (4–10 s/site) and requires no special sample preparation. This work would be analogous to ADI measurements performed at the wafer level in photolithography process control [15].

For AEI, the ability to control the etch depth will be critical for advanced phase shift mask. Again, the ability to make measurements directly after the etch process will provide immediate feedback and enable process control. While analogous to dielectric etch processes, engineers working on dielectric etch processes can obtain cross-sectional

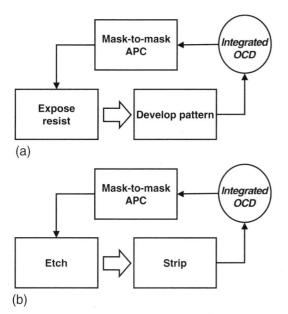

FIGURE 21.27
Simple APC control for ADI (a) and AEI (b) processes.

images, if necessary, with minimal difficulty. OCD technology can offer a unique perspective for mask makers who must otherwise rely on various forms of surface profiling to obtain depth information.

21.5 Conclusions

The current OCD technologies provide a new platform for measuring across wafer CD variation and etch depth. While the applications have focused on a number of processes related to IC fabrication, many of the same benefits are obtainable in mask fabrication. The OCD platform provides a unique tool for nondestructive profiling and will let mask makers to gain valuable information about their process using optical spectroscopy. Information obtained via RCWA calculations can be used to obtain top and bottom CD linewidths, as well as profile information, such as sidewall angle, footing, or undercuts. This information can be obtained with relative ease and provide process engineers with the ability to carefully control critical steps in the mask fabrication process, offering greater CD uniformity across the mask.

Acknowledgments

I would like to thank the Dr. Edwin Boltich, Dr. Weidong Yang, Dr. Milad Tabet, Ms. Ebru Apak, and Dr. Jiangtao Hu and the Applications group at NANOmetrics for their many helpful discussions and particularly Mr. Nagesh Avadhany for his encouragement to do this work. Additionally, I would like to thank, Dr. Turgut Sahin of Applied Materials for the discussion of AEI for quartz grating structures.

References

1. C. Raymond, M. Murnane, S. Prins, S.S.H. Naqvi, et al., Multiparameter grating metrology using optical scatterometry, *J. Vac. Sci. Technol. B.*, 15, 361–368 (1997).
2. Yiping Xu and Ibrahim Abdulhalim, Spectroscopic Scatterometer System, U.S. Patent #6,483,580: 2002.
3. Harland G. Tompkins and William A. McGahn, *Spectroscopic Ellipsometry and Reflectometry: A User's Guide*, John Wiley, New York, 1999, pp. 20–21.
4. Xinhui Niu, Nickhil Jakatdar, Junwei Bao, and Costas J. Spanos, Specular spectroscopic scatterometry, *IEEE Trans. Semicond. Manuf.*, 14, 97–111 (2001).
5. Hsu-Ting Huang, Wei Kong, and Fred Lewis Terry, Jr., Normal-incidence spectroscopic ellipsometry for critical dimension monitoring, *Appl. Phys. Lett.*, 78 3983–3985 (2001).
6. Yuya Toyoshima, Isao Kawata, Yasutsugu Usami, et al., Complementary use of scatterometry and SEM for photoresist profile and CD determination, *Proc. SPIE*, 4689, 196–205 (2002).
7. Weidong Yang, Roger Lowe-Webb, Rahul Korlahalli, et al., Line-profile and critical dimension measurements using a normal incidence optical metrology system, in: *Proc, IEEE/SEMI Adv. Semicond. Manuf. Conference*, 2002, 119–124.

8. M.G. Moharam, Drew A. Pommet, Eric B. Grann, and T.K. Gaylord, Stable implementation of the rigorous coupled-wave analysis for surface-relief gratings: enhanced transmittance matrix approach, J. Opt. Soc. Am. A., 12 (5), 1068–1076 (1995).
9. Vladimir A. Ukraintsev, Mak Kulkarni, Christopher Baum, et al., Spectral scatterometry for 2D trench metrology of low-K dual-damascene interconnect, Proc. SPIE, 4689, 189–195 (2002).
10. Ray Hoobler, Rahul Korlahalli, Deepak Shivadprassad, et al., Monitoring dielectric etch: application of optical critical dimension technology to dual damascene structures, in: *Advanced Process Control/Advanced Equipment Control Symposium XIV*, 2002.
11. P. Rhyins, M. Fritze, D. Chan, C. Carney, B.A. Blachowicz, M. Viera, and C. Mack, *Characterization of Quartz Etched PSM Masks for KrF Lithography at the 100 nm Node*, 21st Annual BACUS Symp. Photomask Technology, 3–5 October 2001, SPIE, vol. 4562, pp. 486–495.
12. David J. Gerold, John S. Petersen, and Marc D. Levenson, Multiple pitch transmission and phase analysis of six types of strong phase-shifting masks, Proc. SPIE, 4346, 72 (2001).
13. Effects of altPSM Design on Image Imbalance for 65 nm, *http://www.e-insite.net/semiconductor/index.asp?layout = articlePrint &articleID = CA273367*, February 2003.
14. James Moyne and Joe White, Existing and envisioned control environment for semiconductor manufacturing, in: James Moyne, Enrique del Castillo, and Arnon Max Hurwitz (eds.), *Run-to-Run Control in Semiconductor Manufacturing*, CRC Press, New York, 2001, pp. 115–124.
15. J.M. Holden, T. Gubiotti, W.A. McGaham, M.V. Dusa, and T. Kiers, Normal-incidence spectroscopic ellipsometry and polarized reflectometry for measurement and control of photoresist critical dimension [4689-133], Proc. SPIE, 4689, 1110–1121 (2002).

22

Photomask Critical Dimension Metrology in the Scanning Electron Microscope[*],[**]

Michael T. Postek

CONTENTS
- 22.1 Introduction 458
- 22.2 Fundamental SEM Architecture 460
  - 22.2.1 Accelerating Voltage/Landing Energy 460
- 22.3 SEM Signals 462
  - 22.3.1 Electron Range 463
  - 22.3.2 Secondary Electrons 463
    - 22.3.2.1 SE Signal Components 465
    - 22.3.2.2 Collection of Secondary Electrons 466
  - 22.3.3 Backscattered Electrons 466
    - 22.3.3.1 Collection of Backscattered Electrons 467
- 22.4 CD-SEM Metrology 467
  - 22.4.1 Nondestructive SEM Inspection and Metrology 467
    - 22.4.1.1 Sample Charging 468
    - 22.4.1.2 Low Accelerating Voltage 472
  - 22.4.2 Linewidth/Critical Dimension Measurement 474
  - 22.4.3 Particle Metrology 475
  - 22.4.4 Line Edge Roughness 475
  - 22.4.5 Automated CD-SEM Characteristics 476
    - 22.4.5.1 Instrument Reproducibility 476
    - 22.4.5.2 CD-SEM Accuracy 476
    - 22.4.5.3 Charging and Contamination 477
    - 22.4.5.4 System Performance Matching 477
    - 22.4.5.5 Pattern Recognition/Stage Navigation Accuracy 477
    - 22.4.5.6 Throughput 477
    - 22.4.5.7 Instrumentation Outputs 477
- 22.5 Instrument and Interaction Modeling 478
  - 22.5.1 Modeling of the SEM Signal 478
    - 22.5.1.1 Accurate SEM Modeling 479

---

[*] Contribution of the National Institute of Standards and Technology. This work was supported in part by the National Semiconductor Metrology Program at the National Institute of Standards and Technology; not subject to copyright.

[**] Certain commercial equipment is identified in this chapter to adequately describe the experimental procedure. Such identification does not imply recommendation or endorsement by the National Institute of Standards and Technology, nor does it imply that the equipment identified is necessarily the best available for the purpose.

			22.5.1.2 Comparison Metrologia/MONSEL Codes 480

                  22.5.1.3 Simulating the Effects of Charging.. 480
        22.5.2 Inverse (Simulation) Modeling.. 480
        22.5.3 SE Versus BSE Metrology.. 481
        22.5.4 Modeling and Metrology... 482
        22.5.5 Shape Control in the Lithography Process... 483
        22.5.6 Model-Based Metrology... 483
        22.5.7 CD Measurement Intercomparison.. 484
22.6 Instrument Calibration... 485
        22.6.1 Magnification Certification.. 485
                22.6.1.1 Definition and Calibration of Magnification............................ 485
        22.6.2 Linewidth Standard... 487
22.7 CD-SEM Instrument Performance 488
        22.7.1 The Sharpness Concept... 488
                22.7.1.1 Reference Material RM 8091 490
                22.7.1.2 Performance Monitoring... 490
        22.7.2 Increase in Apparent Beam-Width in CD-SEMs 491
        22.7.3 Contamination Monitoring.. 492
22.8 High Pressure/Environmental SEM... 492
22.9 Telepresence Microscopy 493
22.10 Conclusion 494
Acknowledgments... 494
References 495

22.1 Introduction

Critical dimension (CD) control begins at the photomask. Therefore, photomask metrology is a principal enabler for the development and manufacturing of current and future generations of semiconductor devices. With the potential of 100, 65, and 45 nm or even smaller linewidths and high aspect ratio structures, the scanning electron microscope (SEM) remains an important tool, which is extensively used in many phases of semiconductor manufacturing throughout the world. The SEM provides higher resolution analysis and inspection than is possible by current techniques using the optical microscope and higher throughputs than scanned probe techniques. Furthermore, the SEM offers a wide variety of analytical modes, each of them contributing unique information regarding the physical, chemical, and electrical properties of a particular specimen, device, or circuit [1]. Due to recent developments, scientists and engineers are finding and putting into practice new, very accurate and fast SEM-based measuring methods in the research and production of photomasks.

Photomask dimensional metrology, especially that associated with the SEM, has not evolved as rapidly as the metrology for integrated circuit and resist features on wafers. This has largely been due to:

1. the distinct emphasis placed on the value of wafer production as opposed to mask production;
2. the fact that far fewer photomask metrology and inspection instruments are needed in production applications;

3. the fact that photomask metrology technology significantly leverages wafer metrology technology improvements;
4. the distinct technological advantages afforded by the 4× or 5× reduction used in the optical steppers and scanners of the lithography process;
5. the fact that there was previously a lesser need to account for the real three-dimensionality of the mask structures [2].

Where photomasks are concerned, many of the issues challenging wafer dimensional metrology at 1× are reduced by a factor of 4 or 5 and thus have been swept aside — temporarily. This is rapidly changing with the introduction of advanced masks with optical proximity correction and phase shifting features used in 100 nm and smaller circuit generations. The *International Technology Roadmap for Semiconductors* (ITRS) 2001 edition states, "mask linewidth controllability fails to meet the requirements of the chip-makers." [3]. Fortunately, photomask metrology generally directly benefits from the advances made for wafer metrology, so many issues can be readily resolved.

Electron beam-based photomask metrology has been around for many years, but has not been as extensively utilized in the production metrology of photomasks and thus, has not been as hot a topic for industrial research and development. In fact, finding a *dedicated* scanned electron beam photomask metrology instrument today is not possible, since most, if not all, mask inspection is done on modified wafer metrology instruments. In 1984, Postek [4,5] presented some early work on SEM metrology of photomasks. At that time, all phases of SEM-based semiconductor metrology were in their rudimentary stages. This was recently updated [2], and some fundamental comparisons were drawn, as shown in Table 22.1. But, as discussed in the following, there are still unique issues to be solved in this form of dimensional metrology.

Optical instrumentation for the determination of photomask precision and accuracy has remained in key metrology positions, since the first linewidth standards developed at the National Bureau of Standards through the work of Dr. Diana Nyyssonen and others (see the review by Nyyssonen and Larrabee, 1987 [6]). Dr. Nyyssonen and her collaborators

TABLE 22.1

Comparison of Selected Characteristics of SEM Photomask Metrology Instruments Past and Present (from Ref. [4])

Characteristic	1985 Instrument	Current Instrument
Instrumentation	Modified laboratory SEM	Modified dedicated CD-SEM
Lens technology	Flat, 45° or 60° pinhole final lens	Extended field and other
Automation	Nonexistent	Common
Electronics	Analog/digital	Fully digital
Image averaging	Rudimentary	Sophisticated and built-in
Scan rate	Slow scan	TV rate
Electron source	Lanthanum hexaboride	Field emission
Linewidth measurement	Rudimentary	Model and library-based
Throughput	Poor (manual)	Vastly improved
Charging	Problem	Acceptable solutions
Sample size	Broken mask/full mask with difficulty	Full mask
Sample contamination	Marginal	Improved
Signal-to-noise ratio (S/N)	Poor	Greatly improved
Cost	$150K	$2.5M+

pointed out, in a series of papers, that the fact that the wavelength of the commonly used visible light in the optical measuring tools was comparable to the feature sizes of interest led to serious metrology limitations. She discussed the need to improve the instrumentation and to mathematically model the effects of diffraction in the image and thereby develop a meaningful criterion of which point on the image corresponds to the edge of the feature whose dimensions were measured [7–9]. Nyyssonen developed such procedures and a model that was used for the calibration and issuance of NIST's first photomask linewidth standards [10,11]. Perhaps the main message of this work was that submicrometer optical metrology was more difficult than commonly envisioned at the time, and that many factors came into play that were often overlooked, ignored, or inadequately treated in many practical applications. With the continuing drive of the semiconductor industry towards sub-100-nm dimensions, this attitude had to change if the anticipated future needs for decreased uncertainty were to be met. The principles developed by Nyyssonen and these coworkers are used today with the next generation of photomask standards [12].

Electron beam-based photomask metrology has been around for many years but has not been as extensively utilized in the production metrology of photomasks as optics. The issues regarding SEM metrology are as complex as those for optical metrology and were reviewed by Postek and Vladár [13], Postek and Larrabee [14], and Postek and Joy [15].

22.2 Fundamental SEM Architecture

All SEMs, whether they are general-purpose laboratory instruments or specialized instruments for inspection and dimensional measurements of integrated circuit structures, function essentially the same. The SEM is named in this way because, in it, a finely focused beam of electrons is moved or scanned from point-to-point over the specimen surface in a precise, generally square or rectangular pattern called a raster pattern (Figure 22.1). The primary beam electrons originate from an electron source and are accelerated towards the specimen by a voltage usually between 100 V (0.1 kV) and 30,000 V (30 kV). For semiconductor production wafer applications, an automated critical dimension SEM (CD-SEM) tool typically operates between 400 and 1000 V. Even lower accelerating voltage applications are being developed (see Sections 22.2.1 and 22.4.1.2). The electron beam travels down the column where it undergoes an electron optical demagnification by one or more condenser lenses. This demagnification reduces the diameter of the electron beam from as large as several micrometers to nanometer dimensions. Depending on the application, magnification range, resolution required, and specimen nature, the operator optimizes the image by choosing the proper accelerating voltage and amount of condenser lens demagnification. Generally, depending on operating conditions and type of instrument, the electron beam diameter is few nanometers in diameter when it impinges on the sample.

22.2.1 Accelerating Voltage/Landing Energy

A clarification needs to be made between the terms accelerating voltage and landing energy because of changes in the modern SEM designs. Ultimately what is of interest is the energy of the electron striking the sample. Historically, accelerating voltage was equal

Photomask CD Metrology in SEM 461

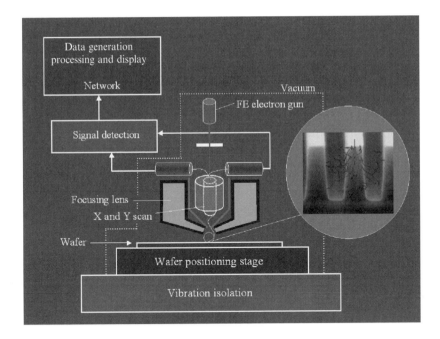

FIGURE 22.1
Schematic drawing of the SEM column. (Courtesy of Dr. Andras Vladár.)

to the landing energy, but today there can be a difference between the two terms. In this chapter, they will be used synonymously unless otherwise differentiated. In SEMs whether they are laboratory or CD instruments, the electrons drawn from the filament are driven down the column by a potential applied between the filament region of the electron gun and the anode. The anode is kept at ground potential. In earlier instruments, the accelerating voltage was varied at the gun and the energy of the electrons impacting the sample was essentially the same as the accelerating voltage (minus any bias voltage applied to some electron guns). A 10-kV accelerating voltage resulted in an electron achieving a landing energy of essentially 10 keV. As accelerating voltages were reduced from 30 to 10 and 1 kV, and lower, the brightness, signal-to-noise ratio and other factors suffered. One way of dealing with this problem is to use high accelerating voltage in the gun region and the upper portions of the column, and then decelerate the electrons to the desired level either in the final lens or at the sample by applying a bias potential. For instance, the electron may be set in motion at the gun area by a 4-kV energy, achieve 4 keV in the column, but will arrive with a 1-keV landing energy due to a 3-kV bias applied to the specimen. In this case, the brightness, resolution, etc., are improved, but the landing energy of the electron is different from the accelerating voltage applied to the filament.

Scanning coils in the microscope column precisely deflect the electron beam in a raster pattern controlled by a digital or analog X and Y scan generators. This deflection is synchronized with deflection of the visual and, when provided, photographic cathode ray tubes (CRTs), so there is a point-by-point visual representation of the signal generated by the specimen as it is scanned. Each point, as a composite, forms the image. Unlike the optical microscope, there is no "image" of the sample formed anywhere in the column. The smaller the area scanned in the raster pattern relative to the fixed size of the display CRT (or other calibrated entity, such as the pixel — see Section 22.6.1.1), the higher the magnification. Alternatively, the smaller the area on the sample represented by one pixel the higher the effective magnification. The proper calibration of either the raster pattern (i.e., magnification calibration) or pixel size is essential [16]. The SEM is capable of extreme high magnification and resolution depending on the instrument design [17,18]. In comparison, many optical microscopes, depending on their illumination source, have their best resolution limited by the effects of diffraction to about 0.25–0.5 μm.

Figure 22.1 describes essentially the design of a "top-down" type metrology instrument. This type of instrument is optimized for throughput and is typically the type of instrument modified for photomask inspection and metrology. In this type of instrument, there is no tilting of the sample possible. Other instrument designs incorporating tilt are also available (especially in the laboratory-type and defect review designs) thus enabling the observation of sidewall structure, cross section information, and optimization for x-ray collection. Some of the newer instruments also provide the ability to obtain three-dimensional stereoscopic information.

One of the major characteristics of the SEM, in contrast to the optical microscope, is its great depth-of-field, which can be 100–500 times greater than that of an optical microscope depending on the SEM instrument conditions employed. This characteristic allows the SEM to produce completely in-focus micrographs of relatively rough surfaces even at high magnifications. This is usually not an issue with relatively thin photomask structures but even with a large depth-of-field, some high aspect ratio semiconductor structures are larger than the depth-of-field afforded by this instrument. This is especially true where some of the newer instruments are concerned. "Confocal-like" imaging has been implemented in some of the newer instruments to overcome this problem by taking several images at different focal points, then merging all the images in the computer. The SEM is also parfocal; it can be critically focused and the astigmatism corrected at a magnification higher than that needed for work; then the magnification can be reduced without a loss of instrument sharpness.

The proper operation of a SEM requires maintaining the electron microscope column under high vacuum, since electrons cannot travel for any appreciable distance in air. The requirement for vacuum does tend to limit the type of sample that can be viewed in the SEM. The mean free path (MFP) or the average distance an electron travels before it encounters an air or gas molecule must be longer than the SEM column. Therefore, depending on the type of instrument design, ion pumps, diffusion pumps, or turbomolecular pumps are utilized to achieve the level of vacuum needed for the particular instrument design. An alternative design incorporating higher pressure in the specimen environment has been utilized in laboratory applications but has only recently been applied to photomask metrology (see Section 22.8). This type of instrument may have greater application in SEM photomask metrology in the very near future.

22.3 SEM Signals

The interaction of an energetic electron beam with a solid results in a variety of potential "signals" generated from a finite interaction region of the sample [1]. A signal, as defined here, is something that can be collected, used, measured, or displayed on the SEM. The most commonly used of the SEM signals are the secondary and backscattered electron (BSE) signals. The distribution and general intensity of these two signal types are shown in Figure 22.2. The electron beam can enter into the sample and form an interaction region, from which the signal can originate. The size of the interaction region is directly related to the accelerating voltage of the primary electron beam, the sample composition, and the sample geometry (discussed in later sections). Those signals that are produced

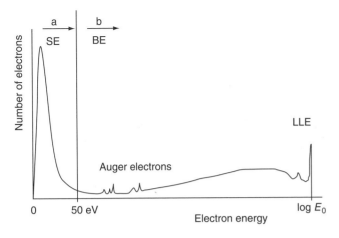

FIGURE 22.2 Distribution and intensity of some of the typical SEM signal types. The arrows denote the energy ranges of (a) SE signal at and below 50 eV and (b) BSE above 50 eV electron signal.

within the interaction region and leave the sample surface can be potentially used for imaging, if the instrument is properly equipped to collect, display, and utilize them.

22.3.1 Electron Range

The primary electron beam can enter into the sample for some distance, even at low accelerating voltages. Thus, it is important to understand and define this interaction volume. The maximum range of electrons can be approximated in several ways [19]. Unfortunately, due to a lack of fundamental understanding of the basic physics underlying the interaction of an electron beam in a sample (especially at low accelerating voltages), there are no equations that accurately predict electron travel in a given sample [20]. Over the past several years, new work has improved this situation a great deal. It is very important to have an understanding of the electron beam penetration into a sample to understand if the electron beam will penetrate into or through the chromium. One straightforward expression derived by Kanaya and Okayama [21] has been reported to be one of the more accurate presently available for approximating the range in low atomic weight elements and at low accelerating voltages. Coupling this with Monte Carlo modeling (see Section 22.5.) provides a wealth of information about a sample and its interactions with the electron beam. The calculated trajectories of electrons in a photoresist layer are shown for 5 keV and 800 eV (Figure 22.3). The electron ranges shown approximate the boundaries of the electron trajectories as if the material was a continuous layer. In the case of a multifaceted structure, such as the one illustrated with three lines, the electrons follow far more complex paths than shown in the higher accelerating voltage case. The electrons penetrate (and leave and potentially penetrate again) the surfaces in every direction making the signal formation process more difficult to accurately model.

22.3.2 Secondary Electrons

The most commonly collected signal in the CD-SEM is the secondary electron (SE); most micrographs readily associated with the SEM are mainly (but, not exclusively) composed of SE (Figure 22.4). Some of the SEs, as discussed earlier, are generated by the primary electron beam within about the first few nanometers of the specimen surface, their escape depth varying with the accelerating voltage and the atomic number of the specimen. Typically, this depth ranges from 2–10 nm for metals to 5–50 nm for nonconductors. SEs

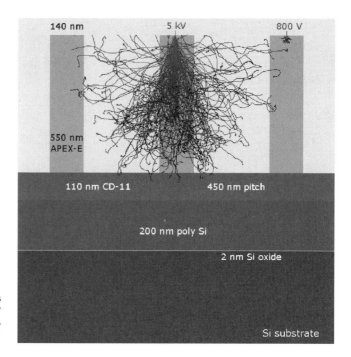

FIGURE 22.3
Monte Carlo electron trajectory plots for high 5 kV (left) and low 800 V (right) accelerating voltages. (Courtesy of Dr. Andras Vladár.)

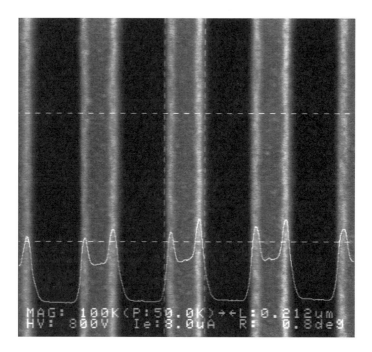

FIGURE 22.4
SE image of photoresist.

also result from BSEs as they leave the specimen surface or collide with inside surfaces of the specimen chamber [22]. The number of SEs emitted from a particular sample relates to the SE coefficient of the materials comprising the sample and other factors, such as surface contamination [23–27].

The SEs are arbitrarily defined as those electrons generated at the sample that have between 1 and 50 eV of energy. The SEs are the most commonly detected for low accelerating voltage inspection due to their relative ease of collection and since their signal is much stronger than any of the other types of electrons available for collection. However, due to the low energy of the SEs, they cannot escape from very deeply in the sample and thus their information content is generally thought to be surface specific. Consequently, the information carried by SEs potentially contains the high-resolution sample information of interest for metrology.

Unfortunately, the reality of the situation is not that straightforward. SEs do not originate only from the location of the primary electron beam. The nature of the imaging mechanism, instrument, and detector all have a direct effect on the SE image observed. This is an extremely important point because this means that the "secondary" electron signal, usually collected in the SEM, is a composite of a number of signal mechanisms. It has been calculated that the number of remotely generated electrons (i.e., energy less than 50 eV) is much larger than those generated from the primary electron beam interaction by a factor possibly greater than 3 [28]. It should be clearly understood that because of electron scatter in a sample, SEs can and do originate from points other than the point of impact of the primary electron beam [29–31].

22.3.2.1 SE Signal Components

There are four potential locations from which the electrons composing the SE image can be generated [22]. The SE signal is composed not only of those SEs generated from initial interaction of the electron beam as it enters the sample (SE-1) but also from SEs generated by the escape of elastically and inelastically scattered BSEs when they leave the sample surface (SE-2). The BSEs can have multiple interactions with other structures on the sample or other internal instrument components and generate more SEs (SE-3). Stray SEs (SE-4) coming from the electron optical column itself may also enter the detector (Figure 22.5). BSEs can also be collected as components of the SE image if their trajectory

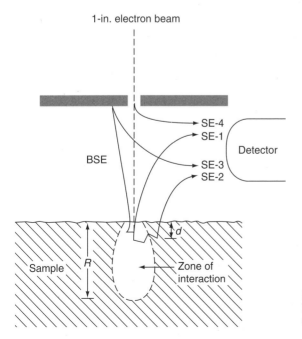

FIGURE 22.5
Diagrammatic description of the possible four derivations of SEs in a typical laboratory SEM.

falls within the solid angle of collection of the electron detector. This is one reason why the modeling of the SE signal is so complex.

Peters [30,31] measured the components of the SE signal from gold crystals. He found that, depending on the sample viewed, for the total SE image the contribution of the SE-2 is approximately 30%, and the contribution to the image of the SE-3 electrons is approximately 60% as compared to approximately 10% of the image contributed by the SE-1 derived signal. The standard SE detector does not discriminate between these variously generated electrons, and thus the collected and measured SE signal is composed of a combination of all of these signal-forming mechanisms. For metrology, the difficulties in interpreting this composite signal can lead to interpretation errors. These errors can be highly variable, and they have a strong dependence on sample composition, sample geometry, and to a lesser or greater extent (depending on instrument design) on other physical factors, such as an instrument's internal geometry that induces anomalies in the detector collection field (i.e., stage motion). Furthermore, since this signal is highly variable and often instrument specific, it is extremely difficult to model.

22.3.2.2 Collection of Secondary Electrons

In most SEMs, SEs are generally collected by the use of a scintillator-type detector of the original design of Everhart and Thornley [32] often referred to as an E/T detector or a modification of that design. Other detector types, including the microchannel-plate electron detector [33,34], are also possible. Due to the low energy of the SE signal, any local electric or magnetic fields easily influence the electron paths; therefore, this detector is equipped with a positively biased collector to attract the SEs. The collection efficiency of an SE detector relates directly to its position, potential, and the field distribution at the sample. Detectors that have a location at some off-axis angle, as in many laboratory instruments designed to accept detectors for x-ray microanalysis, show asymmetry of detection with respect to the orientation of feature edges, i.e., the right and left edges are different for a vertical narrow structure (e.g., resist line). In these cases, it is not possible to achieve symmetrical video profiles. For metrology, the symmetry of the video profile is very important. Asymmetry of the image is a diagnostic that indicates some sort of misalignment whether it is a specimen, a detector, or a column. Furthermore, it is not easily determined if the video asymmetry demonstrated is derived from the position of the detector, from other problems introduced by the instrument's electronics, by column misalignments, by specimen/electron beam interactions, specimen asymmetries, or by a random summing of all possible problems.

22.3.3 Backscattered Electrons

BSEs are those electrons that have undergone either elastic or inelastic collisions with the sample atoms and are emitted with an energy that is larger than 50 eV. A significant fraction of the BSE signal is composed of electrons close to the incident beam energy. This means that a 30-keV primary beam electron can produce a BSE of 24–30 keV (as well as SEs). A 1-keV primary electron beam can produce close to 1 keV BSEs that can be collected and imaged or can interact further with the sample and specimen chamber. The measured BSE yield varies with the sample, detector geometry, and chemical composition of the specimen but is relatively independent of the accelerating voltage above about 5 kV. Because of their high energy, BSEs are directional in their trajectories and are not easily influenced by applied electrostatic fields. Line-of-sight BSE striking the ET detector contributes to all SE micrographs.

BSEs have a high energy relative to SEs, and thus they are not affected as greatly by surface charging. Thus, optimization of collection using sample tilt and collector bias can often enable observation of uncoated, otherwise charging samples. However, it should be cautioned that the charging although not observed in BSE imaging has not been dissipated as discussed later in this chapter.

22.3.3.1 Collection of Backscattered Electrons

BSEs emerge from the sample surface in every direction, but the number of electrons in any certain region of the hemisphere is not equal. BSEs, because of their higher energies, have straight-line trajectories; consequently they must be detected by placing a detector in a location that intercepts their path. This may be accomplished by the use of a solid-state diode detector [35], a microchannel-plate electron detector [36–39], or a scintillator detector positioned for this purpose [40]. The size and position of the detector affect the image, and thus affect any measurements made from it. Therefore, the particular characteristics of the detector and its location must be taken into account when analyzing the observed BSE signal for any application.

BSEs can also be collected through the use of energy filtering detectors or low-loss detectors [41,42]. Energy filtration has the advantage of detecting those electrons that have undergone fewer sample interactions (low-loss) and thus have entered and interacted with the sample to a lesser degree (i.e., over a smaller volume of the specimen) and thus carries higher resolution information. This type of detector has been used successfully at low accelerating voltages, although it does suffer from signal-to-noise limitations. The energy-filtering detector holds promise in assisting in understanding the generation of the signal measured in the CD-SEM. Many of the input parameters are well known for this detector; therefore, electron-beam interaction modeling becomes more manageable. This is especially helpful in the development of accurate standards for CD metrology. Other types of electron detectors are available in the SEM, and these detectors have been reviewed by Postek [43,44]; the reader is directed to these publications for further information.

22.4 CD-SEM Metrology

In 1987, when a review of SEM metrology was done [22], the predominant electron sources in use were the thermionic emission type cathodes, especially tungsten and lanthanum hexaboride (LaB_6). The SEM columns were also much less sophisticated at that time. CD-SEM metrology was in its infancy, and these instruments were essentially only the modified laboratory instruments. In later reviews [43,44], many major changes and improvements in the design of SEMs were introduced, especially the predominance of field emission cathodes and new improved lens technology. The reader is referred to those publications for details regarding those improvements.

22.4.1 Nondestructive SEM Inspection and Metrology

The majority of CD metrology of photomasks is currently done under "nondestructive" SEM conditions. Nondestructive inspection in an SEM implies that the specimen is not altered before insertion into the SEM, and that the inspection in the SEM itself does not modify further use or functions of the sample. The techniques used in "nondestructive" operation of the SEM, as used in this context, have been used in practice for only about the past decade and a half. Historically, scanning electron microscopy was done at relatively high (typically 20–30 kV) accelerating voltages in order to obtain both the best signal-to-noise ratio and image resolution. At high accelerating voltages, nonconducting samples require a coating of gold or a similar material to provide conduction to ground for the electrons and to improve the SE signal generation from the sample. Further, early instruments were designed to accept only relatively small samples so that a large sample, such as a photomask or wafer, typical of the semiconductor industry, needed to be broken prior to inspection. This was not cost-effective, since for accurate process monitoring it was necessary to sacrifice a mask or several rather expensive wafers during each processing run. As wafers and masks became larger and more complex, this became even more undesirable. Therefore, in recent years this procedure for production CD inspection and metrology has been abandoned. Modern on-line inspection during the production process of semiconductor devices is designed to be nondestructive, which requires that the specimen be viewed in the SEM without a coating and totally intact. This required a substantial overhaul of the fundamental SEM design, thus driving the incorporation of field emission sources for improved low accelerating performance, large chamber capability, improved lens designs, clean pumping systems, and digital frame storage. Therefore, the semiconductor industry is, and has been, the primary driver for many of the modern improvements to the SEM.

Photomasks are generally measured and inspected in modified wafer metrology tools, as stated in Section 22.1. Defects in the photomask, either random or repeating, are sources of yield loss and must be found. Defects may occur in the glass, photoresist, or chrome and appear as pinholes, bridges, glass fractures, protrusions, solvent spots, intrusions, or even as missing geometrical features. Currently, *in situ* repair processes are being implemented.

22.4.1.1 Sample Charging

Sample charging is one of the largest impediments to accurate photomask metrology in the SEM. The ITRS states that charging in the SEM is a difficult metrology challenge. Charging was shown to be a significant detriment to the metrology of photomasks in an extensive work by Postek [4,5]. Images in Figure 22.6 are showing the effect of charging on a photomask sample from that early work. Figure 22.6 (left) is a sample of an uncoated binary photomask at 0° of tilt, and Figure 22.6 (right) shows the same sample at 45° of tilt. In a laboratory-type SEM where this work was initially done, tilting of the sample minimizes the effect of charging because it facilitates the collection of the SEs. However, most CD-SEMs are designed to view the photomasks normal to the electron beam and not tilted. Therefore charging is a real possibility.

The same effect is graphically shown in Figure 22.7 (left) for a photomask normal to the electron beam at several accelerating voltages and in Figure 22.7 (right) for a sample having 45° of tilt. The line profiles show that as a sample is tilted, the effect of charging can be minimized or eliminated at lower accelerating voltages, and that the accelerating voltage can be pushed to higher levels as the tilt is increased.

In the 1984 paper, Postek postulated that the specimen charging effect not only caused a deceleration of the electron beam that would change its landing energy, it could also deflect the beam. If this takes place, the landing of the electron beam can be in a location different from the calibrated position, thus inducing measurement error. Gross sample charging can result in a "mirror microscopy" effect. In this case, the electron beam is scanned throughout the specimen chamber, even imaging the final lens polepiece, apertures, and other instrument components. Subtle charging, deflecting the beam a

Photomask CD Metrology in SEM

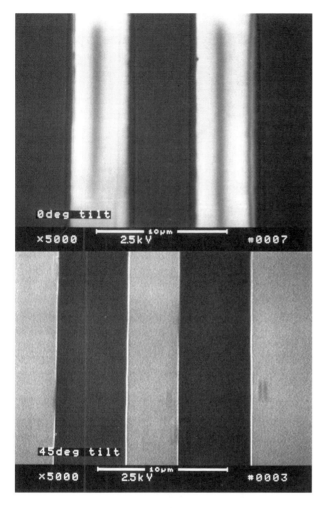

FIGURE 22.6
(Top) Example of specimen charging viewed on a photomask viewed normal to the electron beam in an early lanthanum hexaboride SEM. (Bottom) Example of suppression of specimen charging viewed on a photomask when the mask is viewed at 45° of tilt in an early lanthanum hexaboride SEM.

few pixels is a more insidious problem because it is often not detected. This was explored in a modeling study by Davidson and Sullivan [45]. In a series of Monte Carlo simulations, the effects of a build-up of various potentials on the structure of interest were studied, and it was shown that the beam could be deflected by several nanometers. Thus, a significant nonreproducible measurement error could result. Because of these issues, the ITRS has stated, "alternative paths should be sought."

Since electron beam photomask metrology is typically accomplished with modified wafer CD-SEMs, higher accelerating voltages (greater than about 2 kV) are not usually available. Also, tilt stages are commonly not present in these instruments because they are optimized for high throughput. Therefore, low or very low accelerating voltage imaging is employed in order to avoid charge build-up. In some ways, these are compromises that *may* not always provide the best answer. A typical image from one of these instruments is shown in Figure 22.8.

22.4.1.1.1 Charge Elimination

Charging must be overcome in order to obtain any meaningful data from the SEM. Gross charging can readily distort the image, and subtle charging can deflect the beam leading to measurement errors. There are essentially four possible approaches to deal with charging on photomasks as follows.

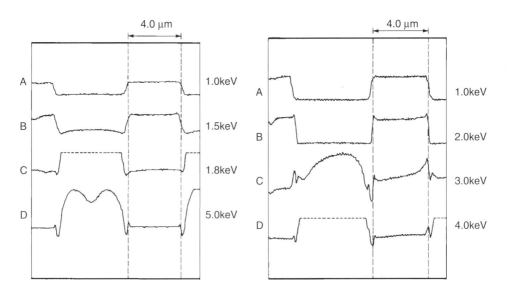

FIGURE 22.7
(Left) Graphical representation of the potential build-up of charge on a photomask sample viewed normal to the electron beam in a lanthanum hexaboride SEM. The dotted line in C depicts charging beyond the range of the system. (Right) Graphical representation of the potential build-up of charge on a photomask sample viewed at 45° of tilt to the electron. Notice in comparison to the previous figure that the uncontrolled charge build-up occurs at a much lower accelerating voltage. The dotted line in D depicts charging beyond the range of the system.

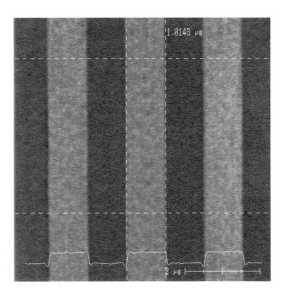

FIGURE 22.8
Image and measurement from a field emission CD-SEM modified for photomask inspection taken at 600 V landing energy.

22.4.1.1.1.1 Sample Coating

The traditional manner to overcome charging is to make the sample conductive by coating it with a metal, such as gold. This is the typical approach that has been taken in laboratory-type inspection of all types of insulating samples. This of course is not only destructive to the sample but also is an undesirable approach to the problem, since gold is a fast diffuser into silicon, and any gold in a fabrication is deemed unacceptable. Where

Photomask CD Metrology in SEM 471

photomasks are concerned, coating also destroys contrast because the coating covers the entire surface leading to contrast derived only from slight topographic difference in the structure.

22.4.1.1.1.2 Charge Balance
The most common approach to neutralizing charging is adjustment of the accelerating voltage or landing energy to approach points where the incoming electron beam and the total signal leaving the sample are balanced. Charge balance can typically be achieved depending on an optimization of a number of factors, such as tilt, extraction field, scanning speed, and final lens configuration [Figures 22.9(b) and (d)]. Low accelerating voltage and achieving charge balance is discussed further in Section 22.4.1.2.

22.4.1.1.1.3 Backscattered Electron Imaging
One technique for reduction of charging effects in the image has always been to collect the high-energy BSEs. In some newer in-lens instruments, the SEs can be filtered out, and the BSE signal can be collected [Figures 22.9(a) and (c)]. This does not neutralize the charge

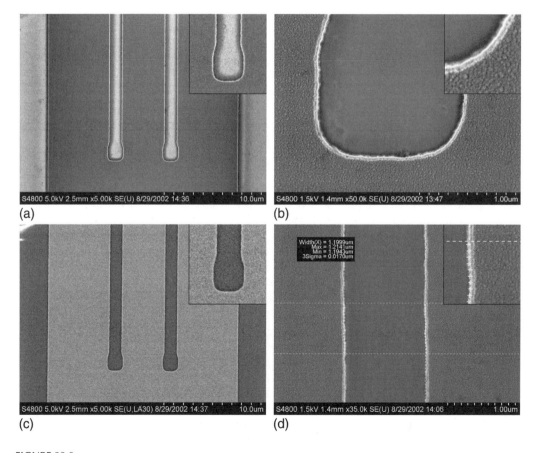

FIGURE 22.9
Photomask inspected at 5 and 1.5 kV in a state-of-the art standard "laboratory" high-resolution field emission SEM capable of working with samples up to 200 mm in diameter. SE image (a) shows charge build-up. Energy-filtered image (c) demonstrates the reduction in sample charging. Low accelerating voltage images (b) and (d) show high-resolution details of the chromium and a measurement of LER. Insets are selected areas at higher digital magnifications. (Courtesy of Hitachi High Technology.)

per se; it only provides an image having charge minimization. The detrimental effects of charging upon the metrology may still occur.

22.4.1.1.1.3.1 Low Loss Electron Imaging. Low loss imaging, as introduced in Section 22.3.3, is a subset of BSE imaging where the electrons are energy filtered in such a manner that only those that have minimally interacted with the sample are collected. These are the low loss electrons. These electrons have been demonstrated to have great surface sensitivity and reduced apparent charging [46,47]. Overall sample charging is not eliminated, and beam deflection by surface charging can still occur.

22.4.1.1.1.4 High-Pressure Microscopy
An alternative metrology technique that has not been fully explored until now has been the employment of high-pressure or environmental microscopy. This methodology employs a gaseous environment to neutralize the charge. For various technical reasons, high-pressure microscopy has mostly been employed on specimens of biological nature and not on many semiconductor samples. Although potentially desirable for charge neutralization, this methodology has not been seriously employed in photomask or wafer metrology until just recently. High-pressure microscopy offers advantages of the possible application of higher accelerating voltages and different contrast mechanisms. This is a new application of this technology to this area, but it shows great promise in the inspection, imaging, and metrology of the photomasks. This methodology is discussed further in Section 22.8.

22.4.1.2 Low Accelerating Voltage

Low acceleration voltage operation (i.e., low landing energies) for production and fabrication of photomasks remains of great interest to the semiconductor industry [48–51]. At low accelerating voltages (from 200 V to 2.5 kV), it is possible to inspect photomasks and in-process wafers in a nondestructive manner.

On-line inspection in an SEM of photomasks does not have one of the major concerns as with wafers. That is the potential that high-energy electrons interacting with the sample can also damage sensitive devices [52–54]. Photomasks have been successfully viewed in the SEM at higher accelerating voltages, but sample charging remains an issue that has to be minimized. Also, where photomasks are concerned, damage can be induced either by deposited charge lifting and damaging the chromium or by deposited contamination, which could change the optical characteristics of the mask. However, more modern testing and review of these concerns need to be done. Low accelerating voltage inspection is also thought to eliminate, or at least to minimize, these concerns. Low accelerating voltage operation is generally defined as work below 2.5 keV — generally within a range of about 0.2–1.2 keV. Further advantages derived from operating the SEM at low accelerating voltages are that the electrons impinging on the surface of the sample have less energy; therefore they penetrate into the sample a shorter distance. The electrons also have a higher cross section (probability) for the production of SEs near the surface where they can more readily escape and be collected. Thus, in this context, nondestructive evaluation requires that the sample is not to be broken, and that it is to be viewed in an instrument at an accelerating voltage below the point where electron beam damage will become a problem. Higher voltage inspection and metrology of photomasks are discussed in Section 22.8.

For low accelerating voltage operation, it is imperative to keep the primary electron beam accelerating voltage at the minimum practical values. In this way, ideal beam energy usually occurs where the SE emission volume results in a near zero SE build-up on the surface of the sample (see Section 22.4.1.2.1). This ideal beam energy can vary from sample to sample (and from location to location) depending on the incident current, nature of the substrate, or the type and thickness of the photoresist. Variation of only 100 V in accelerating voltage or of a couple of degrees of specimen tilt may change a uselessly charging image to one relaying useful sample information. The conductivity in several photoresists was recently studied by Hwu and Joy [55], thus underscoring the capricious nature of this type of sample.

22.4.1.2.1 Total Electron Emission

The behavior of the total electrons emitted (δ) from a sample per unit beam electron is shown in Figure 22.10. This graph is extremely significant to nondestructive low accelerating voltage operation. The principles demonstrated here are crucial to the successful imaging of insulating specimens, such as shown in Figure 22.11. The points where the total emission curve crosses unity (i.e., E_1 and E_2) as shown in fig 22.10 are the points where there is, in principle, no net electrical charging of the sample (i.e., emitted electrons

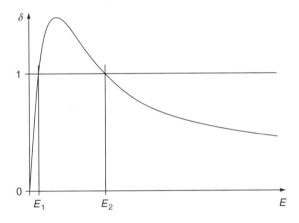

FIGURE 22.10
Typical total electron emission curve for nondestructive SEM metrology and inspection. The E_1 and E_2 points denote the points where no charging is expected to occur on the sample.

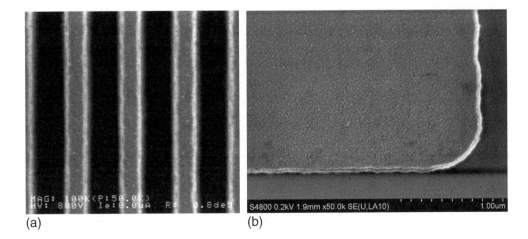

FIGURE 22.11
(a) Low accelerating voltage image of uncoated photoresist. (b) Chromium structure of the photomask viewed at very low landing energy. (Courtesy of Hitachi High Technology.)

equal incident electrons). During irradiation with the electron beam, an insulating sample, such as photoresist or quartz, can collect beam electrons and develop a negative charge. This may result in a reduction in the primary electron beam energy incident on the sample. If the primary electron beam energy of the electrons impinging on the sample is 2.4 keV and the particular sample has an E_2 point at 2 keV, then the sample will develop a negative charge to about −0.4 keV to reduce the effective incident energy to 2 keV and bring the yield to unity. This charging can have detrimental effects on the electron beam and degrade the observed image. If the primary electron beam energy is chosen between E_1 and E_2, there will be more electrons emitted than are incident in the primary beam, and the sample will charge positively. Positive charging is not as detrimental as negative charging, since positive charging is thought to be only limited to a few volts. However, positive charging does present a barrier to the continued emission of the low-energy SEs. This reduction in the escape of the SEs limits the surface potential but reduces signal as these electrons are now lost to the detector. The closer the operating point is to the unity yield points E_1 and E_2, the less the charging effects. Each material component of the specimen being observed has its own total emitted electron/beam energy curve, and so it is possible that, in order to eliminate sample charging, a compromise must be made to adjust the voltage for all materials. For most materials, an accelerating voltage in the range of about 0.2–1.2 keV is sufficient to reduce charging and minimize device damage. Specimen tilt also has an effect on the total electron emission, and it has been reported that increasing tilt shifts the E_2 point to higher accelerating voltages [56,57]. This is a very complex signal formation mechanism because the number of detected electrons depends not only on the landing energy of the primary electrons but also on the number and trajectories of the emitted electrons that are strongly influenced by the local electromagnetic fields.

22.4.2 Linewidth/Critical Dimension Measurement

Linewidths and other CDs must be accurately controlled to ensure that the manufactured integrated circuit performance matches design specifications. This control begins at the photomask. However, traditional light-optical methods for the linewidth measurement of VLSI and ULSI geometries are approaching an inability to attain the accuracy or precision necessary for inspection of these masks, and a growing number of manufacturers are moving to the SEM. Since present photomask fabrication methods employ resist exposure radiation of very short wavelength, it follows that testing and measuring the fabricated structures would involve similar short wavelength optics and high resolution.

Two measurements critical to the semiconductor industry are "linewidth" and "pitch or displacement" (Figure 22.12). Linewidth is the size of an individual structure along a particular axis, and pitch or displacement is the measurement of the separation between the same positions on two or more nearly identical structures [2,4,14,58,59].

Unlike the optical microscope, the range in SEM magnification can continuously span more than 4 orders of magnitude. All SEM linewidth measurement systems rely on the accuracy of this magnification, computed from numerous internal instrument operational factors, including working distance and acceleration voltage. Although acceptable for most applications, the magnification accuracy of many typical SEMs may be inadequate for critical measurement work, since the long-term magnification may vary with time. For critical, reproducible linewidth measurement, all sources of magnification instability must be minimized to achieve long-term precision and measurement accuracy. Other sources of error in image formation and linewidth measurement have been outlined by Jensen [58], Jensen and Swyt [59], and Postek [18] and must be corrected and monitored before the SEM can make accurate, reproducible measurements.

Photomask CD Metrology in SEM

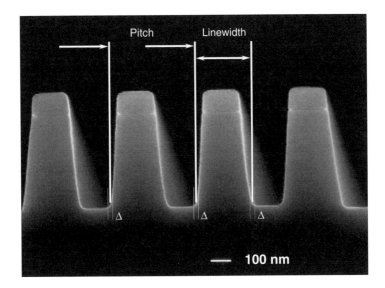

FIGURE 22.12
Pitch and linewidth comparison.

22.4.3 Particle Metrology

Particle metrology and characterization is important to photomask fabrication. Particles are a significant problem for all semiconductor manufacturing [60,61]. Particles are induced by the processing steps and equipment, as well as by the inspection process itself. The SEM has numerous moving parts. Each can generate particles through wear mechanisms. As the mask is transferred into and out of the system, particles can be generated from contact with the mask transfer machinery. Movement of the wafer into and out of the vacuum causes some degree of turbulence, which can mobilize particles, possibly depositing them on the wafer surface. Particles can also be formed by temperature and pressure changes during the sample exchange process leading to water vapor condensation, droplet formation, and liquid-phase chemical reactions. Clearly, the size of the specimen, as well as the size of the particles, must also be considered in such a specification in order to make it meaningful to a specific process. Reduction of particle generation is also important to the performance of the instrument, since a charged particle landing on a sensitive portion of the instrument can rapidly compromise the resolution of the SEM, especially at low accelerating voltages.

22.4.4 Line Edge Roughness

CD control begins at the photomask, and line edge roughness (LER) is a measure of localized roughness along a line, which cannot be averaged out. It is something that must be measured both on the mask and later on the product to understand the transfer function between the two. The *International Technology Roadmap for Semiconductors* (ITRS) 2001 edition defines "LER Control" as the local linewidth variation (3σ total, all frequency components included, for both edges) evaluated along a distance equal to four technology nodes (for wafer metrology). Measurement of the roughness with a precision to tolerance ratio of 20% is required. No specification has been defined for photomask metrology. The ITRS calls for very precise LER measurements at the wafer. At this time, there are no known solutions for measurements of LER with the required precision.

The requirements for LER measurements are different and more stringent than those for linewidth measurements. The latter are about a general shape and width of the line; LER measurements are about the (small) departures from the general shape and width. Linewidth measurements can be (and generally are) done faster using a few averaged line scans. LER measurements require many individual line scans. In linewidth measurements, a certain amount of averaging along the length of the line is desired, while for LER measurements this is not an option. Collection of the right amount and right kind of information in both cases is essential, but for image-based LER measurements the requirements make acceptable throughput more difficult to achieve.

There are several different roughness measurement methods. Most new CD-SEMs have some types implemented, and IC manufacturers already have a variety of ways of measuring some types of roughness. Nevertheless, there is no consensus on the definitions, no common measurement procedures, and no agreement on the details; therefore very different results may emerge. These methods are still a subset of linewidth variation measurements that do not account for the finer details of LER, and important information can go unnoticed.

22.4.5 Automated CD-SEM Characteristics

The Advanced Metrology Advisory Group (AMAG) comprised of representatives from the International SEMATECH (ISMT) consortium member companies; the National Institute of Standards and Technology (NIST); and ISMT personnel, joined to develop a unified specification for an advanced SEM-CD measurement instrument [62]. This was initially developed for wafer inspection instrumentation; however, any improvements in that technology are readily applied to photomask inspection and metrology. Therefore, this specification has direct applicability to SEM photomask metrology as well. This specification was deemed necessary because it was felt that no single CD measurement instrument (other than the SEM) or technology would provide process engineers, in the near future, with the tools that they require providing lithographic and etching CD measurement/control for sub-180-nm manufacturing technology. The consensus among AMAG metrologists was that CD-SEMs needed improvement in many areas of performance. The specification addressed each of these critical areas with recommendations for improvement and a testing criterion for each. This specification was designed to be a "living document" in that as the instruments improved the specifications must also be improved. The critical areas that were targeted for improvement in that document were as follows.

22.4.5.1 Instrument Reproducibility

Confidence in an instrument's ability to repeat a given measurement over a defined period is imperative to semiconductor production. The terms reproducibility and repeatability are defined in general terms in ISO documentation [63]. The new SEMI document E89-0999 expanded on these definitions and includes the term precision [64]. The various components of reproducibility are useful in the interpretation and comparison of semiconductor fabrication process tolerance.

22.4.5.2 CD-SEM Accuracy

The semiconductor industry does not yet have traceable linewidth standards relevant to the kinds of features encountered in VLSI fabrication. A great deal of focused work is progressing in that area and is discussed in a later section of this chapter. Achieving accuracy requires that important attributes necessary in a measurement system must be evaluated [65]. Currently, some of the measurable entities include beam steering accuracy, linearity and sensitivity testing of working algorithms, analysis of instrument sharpness, and the apparent beam-width (ABW).

22.4.5.3 Charging and Contamination

Contamination and charging are two of the most important problems remaining in SEM-based IC metrology. While charging starts to show up as soon as the electron beam hits the sample and contamination tends to build up somewhat more slowly, they act together to change the number, trajectory, and energy of the electrons arriving into the detector, and to make it difficult to make precise measurements. It is very difficult to measure contamination and charging independently. New approaches to dealing with sample charging are discussed in Section 22.8.

22.4.5.4 System Performance Matching

System matching refers to measurement output across several machines. The matching of instrumentation typically applies to all machines that have the same hardware by virtue of the fact that they have the same model number. Matching between manufacturers and different models is deemed desirable, but due to instrument design variations this is seen as unattainable at this time. Matching error is a component of reproducibility, and within the ISO terminology it is the measurement uncertainty arising from changing measurement tools. The matching specification targeted is <1.5 nm difference in the mean measurement between two tools for the 180-nm generation CD-SEMs, or 2.1 nm range of the means across more than two tools.

22.4.5.5 Pattern Recognition/Stage Navigation Accuracy

Pattern recognition capture rate is characterized as a function of pattern size and shape characteristics, layer contrast, and charging, and must average >97% on production layers. Errors need to be typed and logged, so that they are available for an analysis of pattern recognition failures. Stage accuracy and repeatability for both local and long-range moves must be measured for each of 5-μm, 100-μm, and "full–range" across-wafer stage movements. CD-SEMs must be able to measure features that are 100-μm from the nearest pattern recognition target.

22.4.5.6 Throughput

Throughput is an important driver in semiconductor metrology. The throughput CD-SEM specification is designed to test the high-speed sorting of production wafers by a CD-SEM. Throughput must be evaluated under the same conditions as the testing of precision, contamination and charging, linearity and matching, which is using the same algorithm and SEM configuration, and the same wafers.

22.4.5.7 Instrumentation Outputs

CD control at 100 nm and below demands sophisticated engineering and SEM diagnostics. There are a number of outputs that metrologists require from an advanced tool in addition to the output CD measurement number itself. These include raw line scan output, total electron dose, signal smoothing parameters, detector efficiency, signal-to-noise ratio, pattern recognition error log, and others outlined in the text of the AMAG document.

Once this AMAG specification was circulated to the manufacturers, rapid improvements in many of the targeted areas occurred resulting in drastically improved CD-SEM instrumentation. Over the years, application of the metrics developed in the specification has shown a progressive improvement in the performance of the instrumentation.

22.5 Instrument and Interaction Modeling

It is well understood that the incident electron beam enters into and interacts directly with the sample as it is scanned (Figure 22.3). This results in a variety of potential signals generated from an interaction region whose size is related to the accelerating voltage of the electron beam and the sample composition. The details of this interaction are discussed in Section 22.3. For historical and practical reasons, the two major signals commonly used in SEM imaging and metrology are divided into two major groups: BSEs and SEs. Transmitted electrons have also been utilized for specific metrology purposes of specialized masks [66,67]. However, it must be understood that this distinction between SE and BSE becomes extremely arbitrary, especially at low beam energies. Other commonly used signals include the collection and analysis of the x-rays, Auger electrons, transmitted electrons, cathodoluminescence (light), and absorbed electrons; but they will not be discussed here but can be found elsewhere [68–70].

22.5.1 Modeling of the SEM Signal

The appearance of a scanning electron micrograph is such that its interpretation seems simple. However, it is clear that the interaction of electrons with a solid is an extremely complex subject. Each electron may scatter several thousand times before escaping or losing its energy, and a billion or more electrons per second may hit the sample. Statistical techniques are appropriate means for attempting to mathematically model the interactions. The most adaptable tool, now, is the so-called Monte Carlo simulation technique. In this technique, the interactions are modeled and the trajectories of individual electrons are tracked through the sample and substrate (Figure 22.3). Because many different scattering events may occur and because there is no *a priori* reason to choose one over another, algorithms involving chance-governed random numbers are used to select the sequence of interactions followed by any electron (hence the name, *Monte Carlo*). By repeating this process for a sufficiently large number of incident electrons (usually 1000 or more), the effect of the interactions is averaged, thus giving a useful idea of the way in which electrons will behave in the solid. The Monte Carlo modeling techniques were initially applied to x-ray analysis in order to understand the generation of this signal. Today the Monte Carlo technique is being applied to the modeling of the entire signal generating mechanisms of the SEM.

The Monte Carlo modeling technique provides many benefits to the understanding of the SEM image. Using this technique, each electron is individually followed; everything about it (its position, energy, direction of travel) is known at all times. Therefore, it is possible to take into account the sample geometry, the position and size of detectors, and other relevant experimental parameters. The computer required for these Monte Carlo simulations is modest and so, even current high-performance desktop personal computers can produce useful data in a reasonable time. In its simplest form [71–77], the Monte Carlo simulation computes the BSE signal. Since this requires the program to follow and account for only those electrons that have energy higher than 50 eV, it is relatively fast. More time-consuming simulations that calculate SE signal generation, as well as the BSE signal takes somewhat longer depending on the input parameters and the detail of the required data.

By further dividing the electrons based on their energy and direction of travel as they leave the sample, the effect of the detection geometry and detector efficiency on the signal profile can be studied. While the information regarding the BSEs is a valuable first step under most practical conditions, it is the SE signal that is most often used for imaging and metrology in the SEM and recent work to improve this model is being done [73,75]. Simulating the SE image is a more difficult problem because many more electrons must be computed and followed. While this is possible in the simplest cases, it is a more difficult and time-consuming approach when complex sample geometry is involved. The importance of being able to model signal profiles for a given sample geometry is that it provides a quantitative way of examining the effect of various experimental variables (such as beam energy, probe diameter, choice of signal used, etc.) on the profile produced. This also gives a way of assessing how to deal with these profiles and determine a criterion of line edge detection for given edge geometry and thus, better calculate the linewidth. However, now efficient, more accurate Monte Carlo techniques are available but are still under development and as such have some limitations.

22.5.1.1 Accurate SEM Modeling

Accurate SEM metrology requires the development of an appropriate, well-tested computer model. Early Monte Carlo models for metrology were based on the pioneering work of Drs. David Joy, Dale Newbury, and Robert Myklebust and others [78]. More recently, a Monte Carlo computer code specifically designed for CD metrology has been developed at NIST [73,75] and is undergoing continual development and improvements. With this program, the location of an edge in a patterned silicon target has been determined to an error below 6 nm from comparisons between computed and measured BSE and SE signals in an SEM. The MONSEL (Monte Carlo for SE) series of Monte Carlo computer codes are based on first-principles physics [73]. The code simulates the BSE, secondary, and transmitted electron signals from complex targets in the SEM. Measurements have been made on a specially made target composed of a 1-μm step in a silicon substrate in a high-resolution SEM [79]. By overlaying the measured data with the simulation (which predicts the expected signal for a given target geometry), it is possible to determine the position of a measured feature in the target to a low level of uncertainty [80]. This work proved that it is possible to obtain agreement between theoretical models and controlled experiments.

An example of a comparison between measured and simulated data is shown in Figure 22.13 for the SE signal for a 1-keV electron beam energy near the edge of the step in the silicon target [80]. The modeled edge starts at the zero position in the figure and extends to 17 nm because of the measured approximately 1° wall slope. The solid line shows the simulation result for the signal around the edge, which agrees well in shape with the experimental data, given by the dashed lines. Without the simulation, one would not be able to determine the edge location accurately and be forced to use a "rule of thumb." All one could conclude is that the edge occurred somewhere within the region of increased signal. With the simulation, one can determine the edge location within <3 nm, which is a reduction in edge uncertainty by at least a factor of 4 due to the modeling. The results of Figure 22.13 were produced by the Monte Carlo code named MONSEL-II, which is a

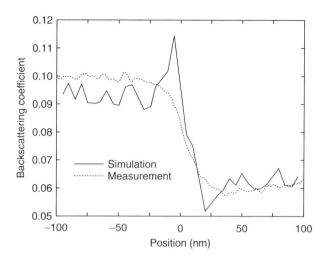

FIGURE 22.13
Monte Carlo modeled data and experimental data overlain as a comparison.

variation for two-dimensional targets. An extension of this code, named MONSEL-III, has been written to compute three-dimensional targets. All of these codes are available from NIST [81]. This model has not been extensively applied to photomask metrology at this time, but NIST and ISMT are currently working together to accomplish that task.

22.5.1.2 Comparison Metrologia/MONSEL Codes

The NIST MONSEL modeling code is primarily a research tool, and it was designed for researchers. Metrologia (SPECTEL Research) is a commercially available modeling program employing a Monte Carlo computer code [82]. Collaboration between NIST and developers of Metrologia enabled the testing and comparison of the two programs in order to determine the level of agreement. A number of modeling experiments were run with both codes. The operations and results of the two codes are now mutually compatible and in general agreement.

22.5.1.3 Simulating the Effects of Charging

A photomask is an insulator with generally isolated chromium structures. Therefore, even the conductive chromium structures can build up a charge, since they are not grounded. The build-up of a positive or negative charge on such an insulating specimen remains a problem and if possible should be avoided. Charging can affect the electron beam and thus the measurements. Accurate metrology in the SEM requires that an accurate charging model be developed or the charging be eliminated. Ko and coworkers [83–85] have quantitatively investigated the effects of charging utilizing Monte Carlo modeling. Specimen and instrumentation variations make charging difficult to reproduce, and thus study in the quantitative manner is necessary for accurate metrology. The deflection of the electron beam by charge build-up was also studied by Davidson and Sullivan [45], and a preliminary charging model was developed and published. Studies of the induced local electrical fields were also done by Grella et al. [86]. An alternative approach to the charging issue is discussed in Section 22.8.

22.5.2 Inverse (Simulation) Modeling

Monte Carlo modeling of a defined structure is a very valuable metrology tool. Taking those data that have been derived by the model and forming an image analog (inverse

Photomask CD Metrology in SEM

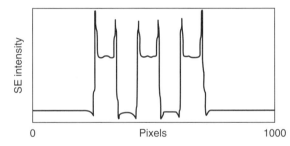

FIGURE 22.14
Monte Carlo modeled line scan of the SEs of photoresist.

modeling) is a powerful tool as well. The image analog is valuable because all the measurement parameters are fully known. This image can then be used to test metrology algorithms and compare measurement instruments. Figure 22.14 is a Monte Carlo modeled line scan of a 1-μm silicon line with nearly vertical edges. The pixel spacing is 1.15 nm. This was modeled using the NIST MONSEL code. Once these data were obtained, various beam diameters were convoluted to add beam diameter effects. The next step converts the line scan into an SEM image analog as shown in Figure 22.15. Once noise and alphanumerics have been added, the image appears similar to any SEM image. Compare the simulated Figure 22.15 with actual image of Figure 22.11a. The emulated image is a standard image file and thus can be imported into any commercial measurement programs or input to other metrology instruments. This form of modeling has been used to test various metrology algorithms, including the ABW algorithm (see Section 5.3).

22.5.3 SE Versus BSE Metrology

A photomask has metallic structures, and thus secondary or BSEs are produced in sufficient quantities for effective electron collection. Closer scrutiny and understanding

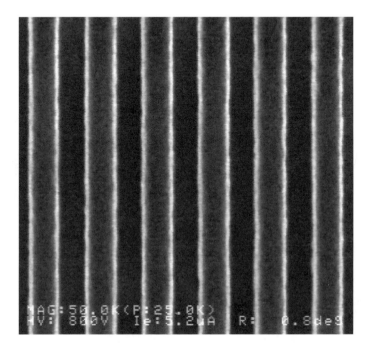

FIGURE 22.15
Completed simulated SEM image based on the same data shown in Figure 22.14.

regarding the differences between the secondary and BSE measurements must be done. The majority of the current production CD-SEMs measure the SE image because of the larger signal afforded by that mode of operation. High signal enhances throughput. In the past, some tools used BSE image-based measurements, but the method has lost favor mainly due to the poorer signal-to-noise ratio signal that led to slower measurements and reduced throughput. Measurement of the BSE signal offers some distinct advantages. The major advantage is found in the modeling of the signal. It is much easier to model the BSE image, since the BSE trajectories are reasonably well understood. It was found [88] that there is an apparent difference in the width of the resist lines between the secondary and BSE image measurements in the low accelerating voltage SEM. It has been observed and documented that there are differences in the measurements between the SE and the BSE images in laboratory and production line instruments as well [88,89]. These differences are identical to the results published earlier in a similar experiment using the microchannel-plate electron detector [33,34]. Figure 22.16 (left) is an SE image and Figure 22.16 (right) is a BSE image. In this example, there is a measurement difference of 17 nm between the two modes of electron detection. SE image taken with highly tilted (but not cross-sectioned) samples revealed that the walls in the imaged structures are sloped. This could account for some of the differences between the imaging modes; however, scanning probe microscope line scans demonstrated a more vertical wall profile in the measurement region.

The discrepancy between the SE image and the BSE image is not explainable by the currently available electron beam interaction models. Well-characterized conductive samples are needed to exclude the charging effects produced by the sample. There are three components to this experiment: instrument, sample, and operator. The operator component is eliminated by automation, and the sample issues can be excluded by proper sample selection, thus leaving the instrument effects to be studied.

22.5.4 Modeling and Metrology

The semiconductor industry for both photomasks and wafers requires fully automatic size and shape measurements of very small, three-dimensional features. These features are currently 100 nm and smaller in size, and these measurements must be done in seconds with an accuracy and precision approaching atomic levels. One main problem

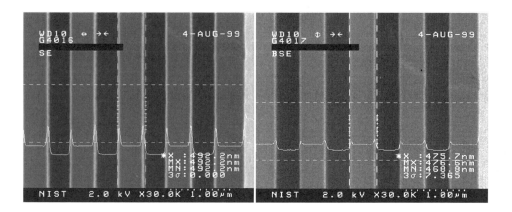

FIGURE 22.16
SE and BSE image comparison of uncoated photoresist. (Left) SE image demonstrating a width measurement of 492.2 nm. (Right) BSE image showing a width measurement of 475.7 nm.

TABLE 22.2
Comparison of Measurement Algorithms

Algorithm	Space Width (nm)	Linewidth (nm)
Peak	109.52	91.15
Threshold	91.65	110.6
Regression	75.63	125.9
Sigmoid	92.95	110.52

is that the measurements are done with primitive edge criteria (regression algorithm, threshold crossing, etc.) and are certain, not necessarily justified presumptions and beliefs. Table 22.2 shows the results of the application of several algorithms to the measurement of a simulated line image as described earlier. A simulated image is extremely valuable in this measurement because all the input parameters to the simulated image are fully known; hence the pitch, linewidth, and space width are known. A similar discrepancy among width measurements was demonstrated experimentally in the SEM Interlaboratory Study, and attempts were made to explain the differences. To accurately determine where the measurement of width should be made on the intensity profile, an accurate model is required.

The images and line scans taken with the CD-SEM contain much more information than is generally used [92]. Modeling the possible cases can help to draw correct conclusions and makes possible to use more accurate, customized measuring algorithms.

22.5.5 Shape Control in the Lithography Process

The lithography process is expected to create the photomask features (chromium lines, clear spaces, etc.) within certain ideal shapes with designed sizes and tolerance. The overall goal is to find a reliable link, some kind of transfer function between the designed and developed resist features and the final etched chromium structures. Care and concern for the control of shape in the photomask fabrication process is imperative, so that the chromium features are fabricated with the desired sizes and tolerances.

22.5.6 Model-Based Metrology

Model–based metrology is a new concept, which will ultimately combine a number of currently developing areas into a single approach. The major application of this concept has been in wafer metrology, but it is just as applicable to photomask metrology. Initial application of model-based metrology to current production CD-SEM instrumentation has resulted in the demonstration of a 3× improvement in measurement precision [91]. The main component of model-based metrology is a database of measured video waveforms from production samples and library of modeled waveforms [92–94]. Any waveform coming from a new structure is compared to a number of waveforms or line scans in the library. The result can be extremely accurate; not just the width of the line but the top corner rounding the wall angle and even height of the resist line are correctly reported. Figure 22.17 shows a cross-section of the photoresist line and the computed line shape superimposed. The shape of the line was calculated from the top-down CD-SEM image of the same line.

In addition to the modeled structure, modeling of the signal path, including the electronics and signal processing of the SEMs, remains an essential area of research and

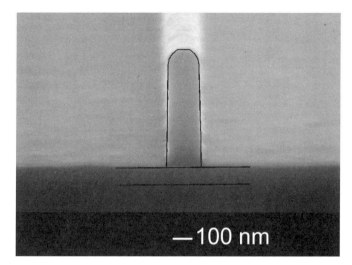

FIGURE 22.17
Cross-section of a photoresist line with the structure calculated form top-down view and modeled data.

is critical to the successful development of an integrated and accurate model. Recent work by Dr. Andras Vladár and others at NIST investigated the performance of CD-SEM instrumentation and have demonstrated that the metrology resolution may be several times better than the resolution by incorporating the modeling [91]. The model-based metrology is evolving rapidly and is being implemented in industrial CD metrology instrumentation but has yet to be fully applied to photomask metrology.

22.5.7 CD Measurement Intercomparison

From the above discussions it becomes clear that inferring a photomask line's width from its SEM image requires assumptions about how the instrument interacts with the sample to produce the image, and how, quantitatively, the resulting apparent edge positions differ from the true ones. For the initial steps to understand this "probe–sample" interaction quantitatively and to verify models of instrument behavior, Villarrubia et al. [95] employed highly idealized samples, fabricated in single crystal silicon [96]. The lines were electrically isolated from the underlying wafer by a 200-nm thick oxide to permit electrical critical dimension (ECD) measurements. Table 22.3 summarizes the results of that work* for three measurement techniques: SEM, atomic force microscopy (AFM), and ECD. An uncertainty budget for each metrology technique was developed according to NIST and ISO guidelines [97,98], which listed the major components contributing to the measurement uncertainty. Listing these components in an honest manner provides a valuable tool for determination

TABLE 22.3

Linewidth Results on Single-Crystal Silicon Sample

Technique	Width (nm)	3σ Uncertainty (nm)
SEM	447	7
AFM	449	16
ECD	438	53

* The numbers in Table 22.3 reflect the total uncertainty of the measurements. Certain components of it (e.g. precision) can be negligible, but others contribute to the large uncertainty. The reader is encouraged to review this chapter in order to understand how these numbers were obtained.

of opportunities for improving the measurement accuracy. In the SEM, the scale (i.e., magnification) component accounted for about one-half of the overall uncertainty. This is an area where improvements can be made. This work provided a confidence in the ability to determine edges and thus provide a meaningful linewidth measurement. It is important to continue to test our understanding of the metrology for samples that approximate as closely as possible the samples of greatest industrial interest.

22.6 Instrument Calibration

Accuracy of measurements and precision of measurements are two separate and distinct concepts [2,99,100]. Process engineers want accurate dimensional measurements, but accuracy is an elusive concept that everyone would like to deal with by simply calibrating their measurement system by using a standard developed and certified at the National Institute of Standards and Technology (NIST). Accurate feature-size measurements require accurate determination of the position of both the left and right edges of the feature being measured. The determination of edge location presents difficulties for all current measurement techniques because of the reasons discussed in earlier sections. Linewidth or CD measurement is a left-edge-to-right-edge measurement (or converse). Therefore, an error in absolute edge position in the microscopic image of an amount ΔL will give rise to an additive error in linewidth of $2\Delta L$. Without an ability to know the location of the edges with good certainty, practically useful measurement accuracy cannot be claimed. For accurate SEM metrology to take place, suitable models as discussed above must be developed, verified, and used.

Recently, the need has been identified for three different standards for SEM photomask metrology. The first standard is for the accurate certification of the magnification of a nondestructive SEM metrology instrument, the second standard is for the determination of the instrument sharpness, and the third is an accurate linewidth measurement standard.

22.6.1 Magnification Certification

Currently, the only certified magnification standard available for the accurate calibration of the magnification of an SEM from NIST is SRM 484. SRM 484 is composed of thin gold lines separated by layers of nickel providing a series of pitch structures ranging from nominally 1 to 50 μm [101]. Newer versions have a 0.5-μm nominal minimum pitch. This standard is still very useful for many SEM applications. During 1991–1992 an interlaboratory study was held using a prototype of the new low accelerating voltage SEM magnification standard, Standard Reference Material (SRM) 2090. This standard was initially fabricated [102,103] and released as a prototype Reference Material (RM 8090) [104]. This RM was rapidly depleted from stock, and a second batch of the artifacts is currently fabricated. As this RM/SRM has undergone significant design improvements since its first issuance, it will soon be released under a different number (Figure 22.18).

22.6.1.1 Definition and Calibration of Magnification

In typical scanning electron microscopy, the definition of magnification is essentially the ratio of the area scanned on the specimen by the electron beam to that displayed on the photographic CRT or other calibrated entity, such as the pixel. The size of the photographic CRT is fixed; therefore by changing the size of the area scanned on

486 *Handbook of Photomask Manufacturing Technology*

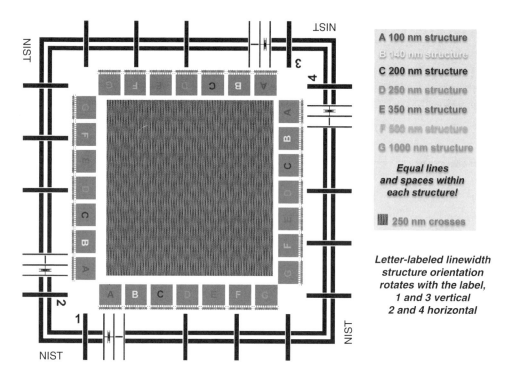

FIGURE 22.18
New design of the SEM low-accelerating voltage SEM standard redesigned from the original RM 8090.

the sample, the magnification is either increased or decreased. Today, where SEM metrology instruments are concerned, the goal is not necessarily to calibrate the magnification as previously defined and discussed, but to calibrate the size of the pixel in both the X and Y directions of the digital measurement system. Since the digital storage system is common to the imaging, the "magnification" therefore is also calibrated. It should be noted that because of the aspect ratio of the SEM display screen, the number of pixels in X may differ from the number in Y, but the size of the pixel must be equal in both X and Y. This is an important concept because in order for a sample to be measured correctly in both X and Y the pixel must be square. The concept of pixel calibration and magnification is essentially identical, and pitch measurements can be used to adjust either. Adjustment of the calibration of the magnification should not be done using a width measurement until an accurate model (as discussed earlier) is available. This is because width measurements are especially sensitive to electron beam/specimen interaction effects. This factor cannot be ignored or calibrated away. Fortunately, this factor can be minimized by the use of a pitch-type magnification calibration sample, such as SRM 484, RM 8090, or the new standard when issued (Figure 22.18). A NIST photomask standard (SRM 474, SRM 2059) can also be used for a magnification calibration as well; however, charging of the artifact must be overcome (as discussed earlier). Other commercial magnification calibration standards can also be used. These standards must also be based on the measurement of "pitch." As described earlier, pitch is the distance from the edge of one portion of the sample to a similar edge some distance away from that first edge. In a pitch standard that distance is certified, and it is to that certified value that the magnification calibration of the SEM is set. Under these conditions, the beam scans a calibrated field width in X and Y. That field width is divided by the number of pixels making up the measurement system, thus defining the measurement unit or the pixel size. The larger the number of pixels available (512, 1024, 2048, etc.)

the finer the measurement "ruler." If we consider two lines separated by some distance, the measurement of the distance from the leading edge of the first line to the leading edge of the second line defines the pitch. Many systematic errors included in the measurement of the pitch are equal on both of the leading edges of the structures being measured; these errors, including the effect of the specimen beam interaction, therefore cancel. This form of measurement is therefore self-compensating. The major criteria for this to be a successful measurement is that the two edges measured must be similar in all ways. SEM pixel/magnification calibration can be easily calibrated to a pitch.

The measurement of a width of a line (as discussed earlier) is complicated in that many of the errors (vibration, electron beam interaction effects, etc.) are now additive. Therefore, errors from both edges are included in the measurement and can only be removed through modeling. The SEM magnification should not be calibrated to a width measurement, since these errors vary from specimen to specimen due to the differing electron beam/sample interaction effects. Effectively, with this type of measurement, we do not know the accurate location of an edge in the video image, and more importantly it changes with instrument conditions. Postek et al. in the Interlaboratory Study, demonstrated, that the width measurement of a 0.2-μm nominal linewidth ranged substantially among the participants. Calibration based on a width measurement requires the development of an electron beam modeling, as described previously. This is the ultimate goal of the SEM metrology project at NIST. The methodologies developed within this project have been shown to be successful for special samples, such as x-ray [68,69] and scattering with angular limitation in projection electron beam lithography [105,106] (SCALPEL) masks measured in the SEM and in the linewidth correlation study.

22.6.2 Linewidth Standard

During the past several years, three significant areas directly related to the issuance of an SEM linewidth standard relevant to semiconductor wafer production have significantly improved. Application of these same principles could result in the development of a photomask linewidth standard. Collaborative work at ISMT is currently in process to fill that need. The first area is modeling. Substantial improvements in the modeling of the electron beam–solid-state interaction have occurred. ISMT support in the modeling area has been crucial to the progress that has been made. The ISMT co-sponsoring with NIST of several electron beam/instrument interaction workshops at the SCANNING International meetings over the past several years has provided a forum that for the first time drew model builders from all over the world. This has resulted in significant and more rapid progress in the area of electron beam interaction modeling. The NIST MONSEL computer codes have been significantly improved and experimental verification of the modeling have produced excellent results on certain well-defined structures.

Second, confidence in the model has been fostered by comparison to commercial code through a NIST/SPECTEL research model comparison also fostered by ISMT. This forward-looking project facilitated the third component that was a partially ISMT-funded linewidth correlation project discussed earlier [96]. For the first time, three metrology methods were carefully applied to a given, well-characterized structure and more importantly an uncertainty of the measurement process was thoroughly assessed.

A prototype linewidth test pattern was recently developed and placed on the AMAG test wafer set [107]. The wafer set will be composed of semiconductor process specific materials and it is designed to be measured with several metrology techniques, including the NIST length scale interferometer [108]. The AMAG test wafer will serve as the prototype for a traceable integrated circuit production specific linewidth standard; a similar photomask standard using similar metrology philosophy is also in process.

22.7 CD-SEM Instrument Performance

In industrial applications, such as semiconductor production, users of automated SEM metrology instruments would like to have these instruments function without human intervention for long periods of time, and to have some simple criterion (or indication) of when they need servicing or other attention. Self–testing is now beginning to be incorporated into commercial instruments to verify that the instrument is performing at a satisfactory performance level. Therefore, the realization of the need for the development of a procedure for periodic performance testing has finally been realized. A number of potential parameters can be monitored and some of them have been reviewed by Joy [109] and Allgair et al. [62]. Two measures of instrument performance have been suggested for incorporation into future metrology tools. These are a measure of sharpness and a measure of the ABW.

22.7.1 The Sharpness Concept

SEMs are utilized in the inspection and metrology of photomasks, as well as semiconductor production. These instruments are approaching full automation. Once a human operator is no longer monitoring the instrument's performance and multiple instruments are used interchangeably, an objective diagnostic procedure must be implemented to ensure data and measurement fidelity and instrument matching. The correct setting of the sharpness and the knowledge of its value are very important for these production line instruments. A degradation of the sharpness of the image of a suitable test object can serve as, of perhaps several, indicators of the need for maintenance. Postek and Vladár [110] first published a procedure based on this sharpness principle, and it has subsequently been refined into a user-friendly stand-alone analysis system [111,112]. This concept was based on the objective characterization of the two-dimensional Fourier transform of the SEM image of a test object for this purpose and the development of appropriate analytical algorithms for characterizing sharpness. The major idea put forth in those papers was that an instrument can be objectively tested in an automated manner and the solution provided was one approach to the problem. A third paper outlining in the public domain a statistical measure, known as the multivariate kurtosis, was also proposed to measure the sharpness of SEM images [113,114]. In that study, using a then state-of-the-art automated CD-SEM the need for instrument performance monitoring was clearly demonstrated by taking a series of linewidth measurements of photoresist lines. The initial measurement of the lines with this instrument resulted in an average width measurement of 247.7 nm (1σ standard deviation of 5.62 nm). Under the conditions studied, the instrument was found to be functioning below its ultimate performance and produced less sharp images, and hence wider than expected lines. The performance of the instrument was checked and improved after changing the final lens aperture and correctly adjusting the electron optical column. The same photoresist lines were measured again and their width was 238.1 nm (1σ standard deviation of 4.37 nm) on the same sample. The nearly 10-nm difference in the measurement resulted from the performance difference of the same instrument and not the product. This is important to the production engineers, since unnecessary reworking of the product may have resulted. With smaller and smaller CDs on the horizon, the correct setting of the SEM becomes indispensable and periodic performance monitoring is very important.

It is known that the low frequency changes in the video signal contain information about the large features and the high frequency ones carry information of finer details. When an SEM image has fine details at a given magnification, namely, there are more

Photomask CD Metrology in SEM 489

high-frequency changes in it, we say they are sharper. A procedure based on the Fourier transform technique on SEM images can analyze this change. Other procedures based on the Fourier transform technique can also be found [115–117]. Since an SEM image is composed of a two-dimensional array of data, the two-dimensional Fourier transform generates a two-dimensional frequency distribution. Based on the computed frequency spectra of selected SEM images, it can be observed that when an SEM image is visibly sharper than a second image, the high spatial frequency components of the first image are larger than that of the second. This was developed into a software program developed in collaboration between NIST and SPECTEL Research called SEM Monitor (Figure 22.19). The important point is not the particular technique of how the SEM image is analyzed for sharpness, but the fact that it can and should be analyzed. Currently, the public domain programs are available from NIST, or the SEM Monitor program is available commercially through SPECTEL Research. Some of the CD-SEM manufacturers have already begun to incorporate the analysis procedure concept in the more recent CD-SEM models.

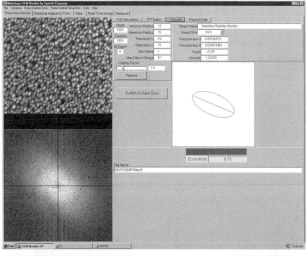

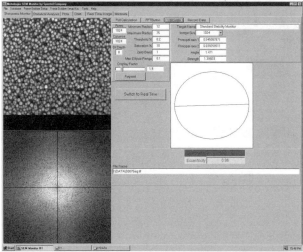

FIGURE 22.19
SEM monitor analysis comparison of RM 8091. (Top) Image demonstrating a degree of astigmatism and (bottom) well-focused image.

22.7.1.1 Reference Material RM 8091

Reference material RM 8091 is one type of specimen in a class of samples appropriate for testing the sharpness of the SEM [118]. RM 8091 is a specimen identified as "grass" in the semiconductor manufacturing vocabulary. Grass is a result of preferential masking during the reactive ion etching. Grass is fully compatible with state-of-the-art integrated circuit technology. This RM can be used at either high accelerating voltage or low accelerating voltage (Figure 22.20). In the current version, the standard is available as a diced square sample (capable of being mounted on a specimen stub) for insertion into either a laboratory or wafer inspection SEM. 200- or 150-mm special (drop-in) wafers with recessed areas for mounted chip-size samples also available. These can easily be loaded as any other wafer.

22.7.1.2 Performance Monitoring

The sharpness technique can be used to check and optimize two basic parameters of the primary electron beam, the focus and the astigmatism. Furthermore, this method makes it possible to regularly check the performance of the SEM in a quantitative, objective form. The short time required to use this method makes it possible that it can be performed regularly before new measurements take place. To be able to get objective, quantitative data about the resolution performance of the SEM is important, especially where high-resolution imaging or accurate linewidth metrology is important. The Fourier method, image analysis in the frequency domain, summarizes all the transitions of the video signal constituting the whole image, not just one or some lines in given directions. This improves the sensitivity (signal-to-noise ratio) and gives the focus and astigmatism information at once. The best solution would be if this and other image processing and analysis functions were incorporated as a built-in capability for all research and industrial SEMs.

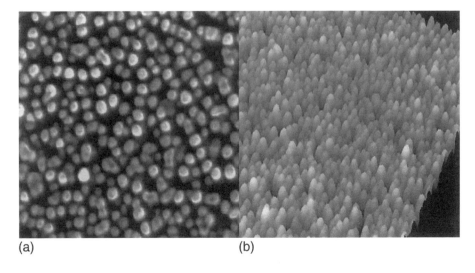

(a) (b)

FIGURE 22.20
Etched grass (a) SEM image of a silicon sample, "grass" that is a result of preferential masking during the reactive ion etching. (b) An AFM image that illustrates the three-dimensional structure of the grass sample (field width = 180 nm).

22.7.2 Increase in Apparent Beam-Width in CD-SEMs

Along with the NIST sharpness approach, the measurement of ABW is one other potential diagnostic procedure for the periodic determination of the performance of on-line CD-SEM measurement tools. ABW is a quantitative measure of the sum of all the factors contributing to the apparent electron beam size as it scans across the edge of the sample (Figure 22.21). The measurement of ABW-like sharpness provides a single comparable number. Archie et al. [119] have reviewed this concept and using experiments and Monte Carlo modeling demonstrated the value of this procedure.

Measurement instruments in the production environment have demonstrated that the ABW is greater than expected for the performance level of a given instrument. Under a given set of conditions, an instrument that is potentially capable of demonstrating 4-nm "resolution" often will demonstrate 20-nm or greater beam-width on production samples. It would be expected that, due to electron beam interaction effects, the guaranteed resolution and ABW would not be identical, but they should be much closer than the 5× (or greater) difference being observed. This phenomenon does not appear to be isolated to any particular instrument type or manufacturer. This presents a serious problem for metrology at 100 nm and below. The measurement of ABW in advanced metrology tools has been recommended in the advanced SEM specification developed by the Advanced Metrology Advisory Group.

The increase in apparent beam diameter is a function of a number of factors, including beam diameter, wall angle, sample charging, sample heating, vibration, or the image capturing process. Based on the current knowledge of the ABW situation, one possibility is that sample charging is apparently affecting the electron beam as it scans the sample. The electron beam is potentially being deflected as the beam scans the dynamically charging and discharging sample. The sample then appears to be "moving" as the image is stored in the system. The image capturing process is averaging what appears to be a moving sample, and thus the edges become enlarged. Another possibility is an environmental effect — vibration. Vibration would have a similar detrimental effect on the image by increasing the measurement. This can only be tested with a fully conductive sample.

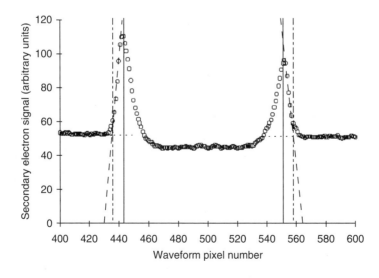

FIGURE 22.21
ABW analysis of the left and the right edges. [From C. Archie, J. Lowney, and M.T. Postek, *Proc. SPIE*, 3677, 669–686 (1999).] With permission.

One problem associated with this issue is that when a pitch is used as a sanity check on instrument calibration, that measurement will be correct due to the self-compensation characteristics of the measurement. However, subsequent linewidth measurements may be detrimentally affected by this effect because of the additive nature of the errors in this type of measurement. Thus, erroneous conclusions may be potentially drawn from the data.

22.7.3 Contamination Monitoring

The deposition of contamination on the surface of any specimen in the SEM is a pervasive problem. The low surface roughness of RM 2091 makes this standard useful in the determination of specimen contamination deposition. Since this standard is susceptible to the effects of contamination, care must be taken to always operate the instrument on a clean area and not dwell too long on any particular area. For this reason, RM 8091 is also a good sample of the measurement of contamination.

22.8 High Pressure/Environmental SEM

New developments in SEM design have not been restricted only to electron sources and lens designs. Concepts in the management of the vacuum have also evolved. Not all applications of the SEM require high vacuum in the specimen chamber, and many samples are damaged or distorted during the specimen preparation processes. An alternative technique to low accelerating voltage SEM for semiconductor metrology and inspection of photomask that minimizes, if not eliminates the charging is environmental or high pressure SEM [2,120]. This application has been reviewed by Postek et al. [121]. High pressure/environmental SEM was originally proposed early in the development of SEM, has slowly developed and has been most recently utilized to obtain previously unattainable data in biological, food, and chemical science applications. The application of environmental microscopy to production semiconductor metrology is new because of the need for the technological combination and implementation of high resolution, high signal field emission technology in conjunction with large chamber and sample transfer capabilities to the environmental microscope technology. This overall combination of technology has not been available until just recently.

High pressure SEM methodology employs a gaseous environment surrounding the sample to help neutralize the charge. Typically the gas used for photomask inspection is water vapor (although other gasses can be used). A typical SEM operates with a sample chamber pressure of about 6.7×10^{-3} Pa (5×10^{-5} Torr). For high-pressure microscopy work, the chamber pressure is allowed to rise to the realm of about 20–160 Pa by the injection of the water vapor (as compared to atmospheric pressure of 101,325 Pa). These operating conditions can be magnitudes different than current standard SEM operating parameters. High-pressure microscopy offers the advantage and possible application of higher landing energies or accelerating voltages, different contrast mechanisms, and charge neutralization [118]. Higher landing energies means that higher resolution imaging is possible than at the lower accelerating voltages. But, beam penetration is increased. This methodology employs a gaseous environment to help diminish the charge build-up that occurs under irradiation with the electron beam. Although potentially very desirable for the charge reduction [122,123], for various technical reasons, this method-

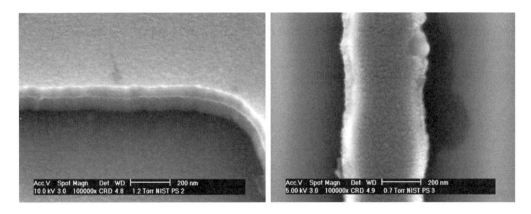

FIGURE 22.22
High-pressure high-resolution SEM images: clearly resolved fine structures, profile, edge and surface roughness, and surface contamination. (Courtesy of FEI Company.)

ology has not been seriously employed in semiconductor inspection or metrology until just recently [118,121]. This is a relatively new application of this technology to this area and much still needs to be learned. But, this technology shows great promise in the inspection, imaging, and metrology of photomasks in a charge-free operational mode. It has been reported that even at high accelerating voltage, injection of air of as little as 20 Pa (0.15 Torr) into the specimen chamber can reduce the charging potential of an insulator at the surface by as much as an order of magnitude [120]. In addition, this methodology affords a path that minimizes, if not eliminates, the need for charge modeling that is needed for higher accuracy measurements. The modeling of charging is exceptionally difficult, since each sample, instrument, and operating mode can respond to charging in different ways. Therefore, this methodology shows great potential, if the optimal balance can be achieved in a reproducible manner. Further research is currently underway to understand the ways to optimize these operating conditions. The reader is directed to the work of Danilatos [122,123] and Postek et al. [2] for further information. Environmental SEM has the potential of solving the charging problem associated with measurement of semiconductor structures (Figure 22.22). Some technical complications do exist in the application of this technology. Currently, no application of low-vacuum scanning electron microscopy to CD metrology has occurred in the production environment; however, this methodology holds some promise for the future.

22.9 Telepresence Microscopy

Telepresence microscopy is an application of the currently available telecommunications technology to long-distance scientific endeavors [124]. Long distance is a relative concept. This can refer to collaboration across the country or from one distributed location within a single company to another. Telepresence is currently applied to electron microscopy in several locations where unique analytical facilities (such as those at NIST) can be used via Internet connection. Potentially this can provide tremendous savings to a company where asset sharing can now be rapidly and effectively used or remote unique facilities can be reached without the requirement of expensive and time-consuming travel. This also has tremendous potential for the photomask or wafer fabrication facility since the engineer can monitor the process remotely without having to enter the clean room. NIST, Texas Instruments, and Argonne National Laboratory worked together to develop a microscopy collaboratory testbed to demonstrate the value of telepresence microscopy in the industrial environment. This testbed shows the value of this technology for technology transfer to organizations having distributed manufacturing facilities, such as Texas Instruments and between organizations, such as NIST and Texas Instruments [125].

22.10 Conclusion

Metrology will remain a principal enabler for the development and manufacturing of future generations of semiconductor devices. With the potential of much less than 100-nm linewidths and high aspect ratio structures, the SEM remains an important tool that is extensively used in many phases of semiconductor manufacturing throughout the world. The SEM still provides higher resolution, *localized* analysis, and inspection than that afforded by current techniques using the optical microscope and higher throughputs than scanned probe techniques. The current instruments must improve accuracy and sensitivity to those characteristics of the photomask that matter to the lithography process. Accurate metrology with this instrument requires the development and availability of traceable standards. Today, magnification (line scale) calibration artifacts traceable to SI unit of length are available for the SEM, and traceable width standards are in the near future.

The first commercial SEM was developed in the late 1960s. This instrument has become a major research tool for many applications, providing a wealth of information not available by any other means. The SEM was introduced into the semiconductor production environment as a CD measurement instrument in the mid-to-late 1980s, and this instrument has undergone a significant evolution in recent years. This instrument now holds a leading role in modern manufacturing. The evolution is not ended with the improvements provided by newer technologies, such as modeling, and the potential afforded by improved electron sources. This tool will continue to be the primary CD measurement instrument for years to come.

Acknowledgments

The author would like to thank and acknowledge the excellent collaboration and technical support provided by Trisha Rice, Ralph Knowles, Ed Griffith, and others at FEI Company in obtaining the high pressure/environmental micrographs from the MDA 600; and Sarah White, Hideo Naito, Robert Gordon, and others at Hitachi High Technology in obtaining the high resolution micrographs from the new S-4800. He would like to thank Marylyn Bennett, Bill Banke (IBM), and Bhanwar Singh (AMD) for supplying the photomasks, the Office of Microelectronics Programs for their support during this research, and Dr. Robert Larrabee, Dr. Andras Vladár, and Mr. Samuel Jones for their technical comments, assistance, and figures used in this chapter.

References

1. M.T. Postek, in: J. Orloff (ed.), *Handbook of Charged Particle Optics* (ed Jon Orloff), CRC Press, New York. 1997, pp. 363–399.
2. M.T. Postek, A.E. Vladár, and M.H. Bennett, *SPIE 22nd BACUS Symposium on Photomask Technology*, vol. 4489, 2002, pp. 293–308.
3. Semiconductor Industry Association, *International Technology Roadmap for Semiconductors*, 2001 edition, http://public.itrs.net.
4. M.T. Postek, *Proc. SPIE*, 480, 109–118 (1984).
5. M.T. Postek (1997), *Scanning Electron Microscopy/Handbook of Charged Particle Optics* (ed Jon Orloff), CRC Press, New York. 1997, pp. 1065–1074.
6. D. Nyyssonen and R.D. Larrabee, *NBS J. Res.*, 92 (3), 187–204 (1987).
7. D. Nyyssonen, *Appl. Opt.*, 16, 2223–2230 (1977).
8. D. Nyyssonen, *Proc. SPIE*, 194, 34–44 (1979).
9. W.M. Bullis and D. Nyyssonen, in: N.G. Einspruch (ed.), *VLSI Electronics: Micro-structure Science*, vol. 3, Academic Press, New York, 1982, pp. 301–346 (Chapter 7).
10. D. Swyt, *Proc. SPIE*, 129, 98–105 (1978).
11. M.C. Croarkin and R.N. Varner, NIST Technical Note 1164, National Bureau of Standards, Gaithersburg, MD, 1982.
12. J. Potzick, J.M. Pedulla, and M. Stocker, *SPIE 22nd BACUS Symposium on Photomask Technology*, vol. 4489, 342–348 (2002).
13. M.T. Postek and A.E. Vladár, in: Alain Diebold (ed.), *Handbook of Silicon Semiconductor Metrology*, Marcel Dekker, New York, 2000, pp. 295–333 (Chapter 14).
14. M.T. Postek and R.D. Larrabee, in: S. Mahajan and L. Kimmerling (eds.), *Concise Encyclopedia of Semiconducting Materials and Related Technologies*, Pergamon Press, New York, 1992, pp. 176–184.
15. M.T. Postek and D.C. Joy, *NBS J. Res.*, 92 (3), 205–228 (1987).
16. M.T. Postek, A.E. Vladár, S.N. Jones, and W.J. Keery, *NIST J. Res.*, 98 (4), 447–467 (1993).
17. M.T. Postek, in. K. Monahan (ed.), *SPIE Crit. Rev.*, 52, 46–90 (1994).
18. M.T. Postek, *NIST J. Res.*, 99 (5), 641–671 (1994).
19. S.G. Utterback, Non distructive Submicron dimensional metrology using the scanning electrone microscope. Review of progress in NDE, La Jolla, California, August 1986, pp. 1141–1151.
20. D.C. Joy, *Inst. Phys. Conf.*, Ser. No. 90, EMAG, 1987, pp. 175–180 (Chapter 7).
21. K. Kanaya and S. Okayama, *J. Phys. D.*, 5, 43–58 (1972).
22. M.T. Postek and D.C. Joy, *NBS J. Res.*, 92 (3), 205–228 (1987).
23. L. Reimer, *Scanning Electron Microscopy/1979/II*, SEM, Inc., 1979, pp. 111–124.
24. L. Reimer, *Scanning*, 1, 3–16 (1977).
25. L. Reimer, Scanning Electron Microscopy 1982/SEM Inc/Chicago III, pp. 299–310.
26. L. Reimer, *Electron Beam Interactions with Solids*, 1984, 299–310.
27. L. Reimer, *Scanning Electron Microscopy. Physics of Image Formation and Microanalysis*, Springer-Verlag, New York, 1985.
28. H. Seiler, *Z. Angew. Phys.*, 22 (3), 249–263 (1967).
29. H. Drescher, L. Reimer, and H. Seidel, *Z. F. Angew. Physik*, 29, 331–336 (1970).
30. K.-R. Peters, *Scanning Electron Microscopy/1982/IV*, SEM, Inc., 1982, pp. 1359–1372.
31. K.-R. Peters, *Scanning Electron Microscopy/1985/IV*, SEM, Inc., 1985, pp. 1519–1544.
32. T.E. Everhart and R.F.M. Thornley, *J. Sci. Instrum.*, 37, 246–248 (1960).
33. M.T. Postek, W.J. Keery, and N.V. Frederick, *Rev. Sci. Instrum.*, 61 (6), 1648–1657 (1990).
34. M.T. Postek, W.J. Keery, and N.V. Frederick, Development of a Low-Profile Microchannel-Plate Electron Detector System for SEM Imaging and Metrology. *Scanning*, 12, I–27–28 (1990).
35. S. Kimoto and H. Hashimoto, in: T.D. McKinley, K.F.J. Heinrich, and D.B. Wittry (eds.), *The Electron Microscope*, Proc. Symp., Washington, 1964, John Wiley, New York, 1966, 480–489.
36. M.T. Postek, W.J. Keery, and N.V. Frederick, Development of a Low-Profile Microchannel-Plate Electron Detector System for SEM Imaging and Metrology. *Scanning*, 12, I–27–28 (1990).

37. M.T. Postek, W.J. Keery, and N.V. Frederick, *EMSA Proceedings*, 1990, pp. 378-379.
38. P.E. Russell, *Electron Optical Systems*, SEM, Inc., 1990, pp. 197–200.
39. P.E. Russell and J.F. Mancuso, *J. Microsc.*, 140, (1985) 323–330.
40. V.N.E. Robinson, *J. Phys. E: Sci. Instrum.*, 7, 650–652 (1974).
41. O.C. Wells, *Appl. Phys. Lett.*, 19 (7), 232–235 (1979).
42. O.C. Wells, *Appl. Phys. Lett.*, 49 (13), 764–766 (1986).
43. M.T. Postek, *SPIE Crit. Rev.*, 52, 46–90 (1994).
44. M.T. Postek, *NIST J. Res.*, 99 (5), 641–671 (1994).
45. M.P. Davidson and N. Sullivan, *Proc. SPIE*, 3050, 226–242 (1997).
46. M.T. Postek, A.E. Vladár, O.C. Wells, and J.L. Lowney, *Scanning*, 23 (5), 298–304 (2001).
47. O.C. Wells, M. Mc-Glashan-Powell, A.E. Vladár, and M.T. Postek, *Scanning*, 23 (6), 366–371 (2001).
48. T Ahmed, S.-R. Chen, H.M. Naguib, T.A. Brunner, and S.M. Stuber, *Proc. SPIE*, 775, 80–88 (1987).
49. M.H. Bennett, *SPIE Crit. Rev.*, 52, 189–229 (1993).
50. M.H. Bennett and G.E. Fuller, *Microbeam Analysis*, 649–652 (1986).
51. F. Robb, Proc. SPIE, 775, 89–97 (1987).
52. P.K. Bhattacharya, S.K. Jones, and A. Reisman, *Proc. SPIE*, 1087, 9–16 (1989).
53. W.J. Keery, K.O. Leedy and K.F. Galloway, *Scanning Electron Microscopy/1976/IV*, IITRI Research Institute, 1976, pp. 507–514.
54. A. Reisman, C. Merz,, J. Maldonado, and W. Molzen, *J. Electrochem. Soc.*, 131, 1404–1409 (1984).
55. J.J. Hwu and D.C. Joy, *Scanning*, 21, 264–272 (1999).
56. M.T. Postek, *Scanning Electron Microscopy/1984/III*, IITRI, 1985, pp. 1065–1074.
57. M.T. Postek, *Review of Progress in NDE*, 6 (b), 1327–1338 (1987).
58. S. Jensen, *Microbeam Analysis*, San Francisco Press, San Francisco, CA, 1980, pp. 77–84.
59. S. Jensen and D. Swyt, *Scanning Electron Microscopy I*, 393–406 (1980).
60. M.H. Bennett and G.E. Fuller, *Microbeam Analysis*, 649–652 (1986).
61. M.H. Bennett, *SPIE Crit. Rev.*, 52, 189–229 (1993).
62. J. Allgair, C. Archie, G. Banke, H. Bogardus, J. Griffith, H. Marchman, M.T. Postek, L. Saraf, J. Schlessenger, B. Singh, N. Sullivan, L. Trimble, A.E. Vladár, and A. Yanof, *Proc. SPIE*, 3332, 138–150 (1998).
63. International Organization for Standardization 1993, International Vocabulary of Basic and General Terms in Metrology – ISO, Geneva, Switzerland, ISBN 92–67–01075–1, 1993, 60 pp.
64. SEMI, Document E89-0999 – Guide for Measurement System Capability Analysis, 1999.
65. W. Banke and C. Archie, *Proc. SPIE*, 3677, 291–308 (1999).
66. M.T. Postek, J.R. Lowney, A.E. Vladár, W.J. Keery, E. Marx, and R.D. Larrabee, *NIST J. Res.*, 98 (4), 415–445 (1993).
67. M.T. Postek, J.R. Lowney, A.E. Vladár, W.J. Keery, E. Marx, and R. Larrabee, Proc. SPIE, 1924, 435–449 (1993).
68. M.T. Postek, K. Howard, A. Johnson, K. Mc Michael, *Scanning Electron Microscopy – A Students Handbook*, Ladd Research Industries, Burlington, Vermont, 1980, 305 pp.
69. O.C. Wells, *Scanning Electron Microscopy*, McGraw Hill, New York, 1974, 421 pp.
70. J.I. Goldstein, D.E. Newbury, P. Echlin, D.C. Joy, C. Fiori, and E. Lifshin, *Scanning Electron Microscopy and X-ray Microanalysis*, Plenum Press, New York, 1981, 673 pp.
71. G.G. Hembree, S.W. Jensen, and J.F. Marchiando, *Microbeam Analysis*, San Francisco Press, San Francisco, CA, 1981, 123–126.
72. D.F. Kyser, in: J.J. Hren, J.I. Goldstein, and D.C. Joy, *Introduction to Analytical Electron Microscopy*, Plenum Press, New York, 1979, pp. 199–221.
73. J.R. Lowney, *Scanning Microscopy*, 10 (3), 667–678 (1996).
74. J.R. Lowney, M.T. Postek, and A.E. Vladár, *Proc. SPIE*, 2196, 85–96 (1994).
75. E. Di Fabrizio, L. Grella, M. Gentill, M. Baciocchi, L. Mastrogiacomo, and R. Maggiora, *J. Vac. Sci. Technol. B.*, 13 (2), 321–326 (1995).
76. E. Di Fabrizio, I. Luciani, L. Grella, M. Gentilli, M. Baciocchi, M. Gentili, L. Mastrogiacomo, and R. Maggiora, *J. Vac. Sci. Technol. B.*, 10 (6), 2443–2447 (1995).

77. E. Di Fabrizio, L. Grella, M. Gentill, M. Baciocchi, L. Mastrogiacomo, and R. Maggiora, *J. Vac. Sci. Technol. B.*, 13 (2), 321–326 (1995).
78. D.C. Joy,. *Monte Carlo Modeling for Electron Microscopy and Microanalysis*, Oxford University Press, New York, 1995, 216 pp.
79. M.T. Postek, A.E. Vladár, G.W. Banke, and T.W. Reilly, in: Edgar Etz (ed.), MAS Proceedings, 1995, pp. 339–340.
80. J.R. Lowney, M.T. Postek, and A.E. Vladár, in: Edgar Etz (ed.), MAS Proceedings, 1995, pp. 343–344.
81. Contact Dr. Jeremiah R. Lowney at the National Institute of Standards and Technology.
82. M.P. Davidson, *Proc. SPIE*, 2439, 334–344 (1998).
83. Y.-U. Ko and M.-S. Chung, *Proc. SPIE*, 3677, 650–660 (1999).
84. Y.-U. Ko, S.W. Kim, and M.-S. Chung, *Scanning*, 20, 447–455 (1998).
85. Y.-U. Ko and M.-S. Chung, *Scanning*, 20, 549–555 (1998).
86. L. Grella, E. DiFabrizio, M. Gentili, M. Basiocchi, L. Mastrogiacomo, and R. Maggiora, *J. Vac. Sci. Technol. B.*, 12 (6), 3555–3560 (1994).
87. M.T. Postek, *Rev. Sci. Instrum.*, 61 (12), 3750–3754 (1990).
88. M.T. Postek, W.J. Keery, and R.D. Larrabee, *Scanning*, 10, 10–18 (1988).
89. N. Sullivan, Personal communication.
90. J. McIntosh, B. Kane, J. Bindell, and C. Vartuli, *Proc. SPIE*, 3332 (1999).
91. J.S. Villarrubia, A.E. Vladár, J.R. Lowney, and M.T. Postek, A scanning electron microscope analog of scatterometry. *Proc. SPIE*, 4689, 304–312 (2002).
92. M.P. Davidson and A.E. Vladár, *Proc. SPIE*, 3677, 640–649 (1999).
93. A.E. Vladár and M.T. Postek, New way of handling dimensional measurement results for integrated circuit technology, *Proc. SPIE*, 5038 (2003) (in press).
94. J. Villarrubia, A.E. Vladár, and M.T. Postek, simulation study of repeatability and bias in the CD-SEM. Proceedings, *Proc. SPIE*, 5038 (2003) (in press).
95. J.S. Villarrubia, R. Dixson, S. Jones, J.R. Lowney, M.T. Postek, R.A. Allen, and M.W. Cresswell, *Proc. SPIE*, 3677 (1999) 587–598.
96. R.A. Allen, N. Ghoshtagore, M.W. Cresswell, L.W. Linholm, and J.J. Sniegowski, *Proc. SPIE*, 3332, 124–131 (1997).
97. B. Taylor and Kuyatt, NIST Technical Note 1297, 1994.
98. International Organization for Standardization 1997, Guide to the Expression of Uncertainty in Measurement (corrected and reprinted 1995), This document is also available as a U.S. National Standard NCSL Z540-2-1997.
99. R.D. Larrabee and M.T. Postek, *SPIE Crit. Rev.*, CR52, 2–25 (1993).
100. R.D. Larrabee and M.T. Postek, . Solid–State Elec. 36(5):673–684, 1993.
101. J. Fu, T.V. Vorburger, and D.B. Ballard, *Proc. SPIE*, 2725, 608–614 (1998).
102. B.L. Newell, M.T. Postek, and J.P. van der Ziel, *J. Vac. Sci. Technol. B.*, 13 (6), 2671–2675 (1995).
103. B.L. Newell, M.T. Postek, and J.P. van der Ziel, *Proc. SPIE*, 2460, 143–149 (1995).
104. M.T. Postek and R. Gettings, Office of Standard Reference Materials Program NIST, 1995, 6 pp.
105. R.C. Farrow, M.T. Postek, W.J. Keery, S.N. Jones, J.R. Lowney, M. Blakey, L. Fetter, L.C. Hopkins, H.A. Huggins, J.A. Liddle, A.E. Novembre, and M. Peabody, *J. Vac. Sci. Technol. B.*, 15 (6), 2167–2172 (1997).
106. J.A. Liddle, M.I. Blakey, T. Saunders, R.C. Farrow, L.A. Fetter, C.S. Knurek, A.E. Novembre, M.L. Peabody, D.L. Windt, and M.T. Postek, *J. Vac. Sci. Technol. B.*, 15 (6), 2197–2203 (1997).
107. M.T. Postek, A.E. Vladár, and J. Villarrubia, Is a production critical scanning electron microscope linewidth standard possible? *Proc. SPIE* 3988: 42–56 (2000) (in press).
108. J.S. Beers and W.B. Penzes, *J. Res. Natl. Inst. Stand. Technol.*, 104, 225–252 (1999).
109. D.C. Joy, *Proc. SPIE*, 3332, 102–109 (1997).
110. M.T. Postek and A.E. Vladár, *Proc. SPIE*, 2725, 504–514 (1996).
111. M.T. Postek and A.E. Vladár, *Scanning*, 20, 1–9 (1998).
112. A.E. Vladár, M.T. Postek, and M.P. Davidson, *Scanning*, 20, 24–34 (1998).
113. N.-F. Zhang, M.T. Postek, and R.D. Larrabee, *Proc. SPIE*, 3050, 375–387 (1997).

114. N.-F. Zhang, M.T. Postek, and R.D. Larrabee, Image sharpness measurement in scanning electron microscopy. Part 3. Kurtosis, *Scanning* 21: 256–262 (1999).
115. K.H. Ong, J.C.H. Phang, J.T.L. Thong, *Scanning*, 19, 553–563 (1998).
116. K.H. Ong, J.C.H. Phang, J.T.L. Thong, *Scanning*, 20, 357–368 (1998).
117. H. Martin, P. Perret, C. Desplat, and P. Reisse, *Proc. SPIE*, 2439, 310–318 (1998).
118. M.T. Postek and A.E. Vladár, *Proc. SPIE*, 2000 (in press).
119. C. Archie, J. Lowney, and M.T. Postek, *Proc. SPIE*, 3677, 669–686 (1999).
120. D.C. Joy, *Proc. SPIE*, 4689, 1–10 (2002).
121. M.T. Postek, A.E. Vladár, T. Rice, and R. Knowles, Potentials for high pressure environmental SEM microscopy for photomask dimensional metrology. *Proc. SPIE*, 5038, 315–329 (2003) (in press).
122. G.D. Danilatos, *Adv. Electronics Electron Phys.*, 71, 109–250 (1998).
123. G.D. Danilatos, *Microsc. Res. Tech.*. 25, 354–361 (1993).
124. M.T. Postek, M.H. Bennett, and N.J. Zaluzec, *Proc. SPIE*, 3677, 599–610 (1999).
125. NIST Telepresence videotape/CD is available through the Office of Public and Business Affairs, Gaithersburg, MD 20899.

23
Geometrical Characterization of Masks Using SPM

Sylvain Muckenhirn and A. Meyyappan

CONTENTS

23.1 Introduction 500
23.2 Scanning Principles 500
    23.2.1 Scanning Sensors 500
        23.2.1.1 Sensor Types 500
    23.2.2 Scanner Calibration (X, Y, Z Using VLSI Standards) 501
    23.2.3 Scanning Modes 501
        23.2.3.1 Contact Mode 502
        23.2.3.2 Alternative Contact (Also Known as Intermittent Contact or Tapping Mode) 502
        23.2.3.3 Pixel Mode 502
        23.2.3.4 Noncontact Mode 503
    23.2.4 Scanning Algorithms 503
        23.2.4.1 One-Dimensional Scan 503
        23.2.4.2 Adaptive Scan 504
        23.2.4.3 Two-Dimensional Scan 504
23.3 Tips 504
    23.3.1 Tip–Sample Interaction 505
        23.3.1.1 Tip Radii and Cone Angle Effect 507
        23.3.1.2 Tip Shape Effect 508
        23.3.1.3 Tip Stiffness Effect 513
    23.3.2 Tip Calibration 514
        23.3.2.1 Calibration Structures 514
    23.3.3 Tip Shape Removal or Tip De-Convolution 516
23.4 Mask Applications 517
    23.4.1 Profile 520
    23.4.2 Defect 521
    23.4.3 Linewidth Uniformity 522
    23.4.4 Sidewall Roughness 524
    23.4.5 Line Roughness, Line Edge Roughness 524
    23.4.6 Shape Characterization 525
    23.4.7 Depth, Step Height 525
    23.4.8 Roughness 525
    23.4.9 Width 526
23.5 AFM Selection and Test Methodology 528
Acknowledgments 529
References 529

23.1 Introduction

As the feature dimensions get smaller and circuits get more complex, the demand for comprehensive measurements of reticule geometries increases. The 3D characterization of binary masks and phase shift masks (PSM) is required to foresee and guarantee the quality of the transferred image. Scanning probe microscopy (SPM) is ideally suited for making these characterizations. It quantitatively profiles lines and trenches in three dimensions. SPM is a nondestructive technique, allowing for the preservation of the integrity of the mask. The technique is insensitive to differences in the intrinsic characteristics of the materials (chromium on quartz, resist on conductive or nonconductive layers).

The most widely known forms of SPMs are scanning tunneling microscopy (STM) and atomic force microscopy (AFM). STM was developed by Binnig and Rohrer in 1982 [1]. Later in 1986, Binnig et al. [2] developed the AFM to overcome the limitations of STM, which could image only conducting materials. In this chapter, we shall also address instruments, such as the High Resolution Profiler of KLA-Tencor, Stylus Nano Profilometer of FEI Company, as well as various AFMs made by several manufacturers, including Veeco Instruments for a more inclusive discussion on the possibilities of probe microscopes. The illustrations in this chapter are mostly from AFM even though there are some examples from other tools.

In the following sections, we present scanning modes, scanning algorithms, tips, tip–sample interactions, as well as practical examples of profile angle, width, sidewall roughness, line roughness, line edge roughness and linewidth uniformity, and critical dimension measurements. Also, an example of test methodology used to verify the adequacy of the SPM to mask characterization is presented.

23.2 Scanning Principles

SPMs are used primarily to study the surface properties of materials [3,4]. In our case, the sample to be scanned is a mask. All SPMs contain a probe tip, a sensor that accurately locates the vertical position of the tip, a feedback system that controls the vertical position of the tip, and a piezoelectric scanner that moves the tip relative to the sample in a raster pattern. A computer system drives the scanner, measures the data, and converts it into an image.

The tip is linked to a cantilever. Light, electrons or electrodes are linked to the cantilever and monitor its spatial location. The attraction or repulsion between the tip and the surface causes the cantilever to deflect. The deflection is read as a change in voltage.

Scanning the tip across the surface generates the image as shown in Figure 23.1.

23.2.1 Scanning Sensors

SPMS use various techniques to detect the position of the probe tip, most of these are optical techniques, but there are others that use STM, capacitance, strain gauge, and other techniques.

23.2.1.1 Sensor Types

The most commonly used method to detect the position of the probe tip is to employ an optical sensor. In the most common scheme, a laser beam bounces of the back of the cantilever onto a position sensitive photodetector [5,6]. As the cantilever bends, the

Geometrical Characterization of Masks Using SPM

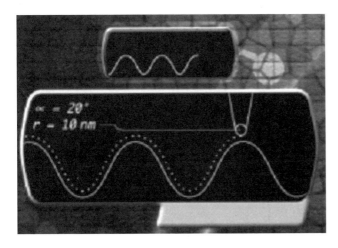

FIGURE 23.1
Surface scanned by a tip (bottom) and resulting scan line (top).

position of the laser beam on the detector shifts. The detector can measure displacements of light as small as 1 nm. There is also mechanical amplification because of the distance between the detector and the cantilever the laser beam travels. Hence, the system can detect sub-nanometer vertical movement of the probe tip.

There are other techniques for detecting the movement of the tip, such as heterodyne interferometer used with STM tip [7–9], or by employing a piezoelectric material as a cantilever such that it can detect its own deflection electrically [10]. One other technique is to use the cantilever as part of a capacitor, and the change in capacitance will be a measure of the position of the tip [11].

23.2.2 Scanner Calibration (*X, Y, Z* using VLSI Standards)

All SPMs use piezoelectric scanners to position the stage under the probe. The electronics drive the scanner in a raster fashion. These scanners have the characteristics of the piezoelectric materials like nonlinearities and hysteresis, but most manufacturers have built-in corrections for their optimum operations [12,13]. There are standards and reference structures that can be used for calibrating and checking the calibration of the scanners.

VLSI and others standards, Inc. and others manufacture a variety of standards and reference samples that can be used to calibrate the *X, Y,* and *Z* scanners. The manufacturer of the tool may also supply some of these standards.

The surface topography standards are NIST certified in all three dimensions. These traceable standards are available in various heights (18–180 nm) and various pitches (1.8–20 μm). The surface topography references have a uniform pitch of 3 or 10 μm pitch with various nominal heights from 18 to 180 nm. These references are not NIST certified.

The step height standards are features etched directly into 25 mm × 25 mm × 3 mm quartz substrates. These come in heights varying from 8 nm to 1.8 μm and are NIST certified. Moreover, in addition to NIST there are other manufacturers who make standards and reference samples.

These standards can be used to calibrate the scanners of the tool.

23.2.3 Scanning Modes

Irrespective of the sensor used, SPM suppliers are advertising various scanning modes describing the physical level of interaction between the tip and substrate. Forces of interaction, as shown in Figure 23.2, range from micro- to pico-Newton. Scanning modes are referred as contact mode, alternative or intermittent contact mode, and noncontact mode.

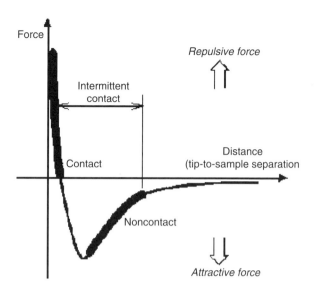

FIGURE 23.2
Inter-atomic force versus distance curve.

23.2.3.1 Contact Mode

In contact mode, the probe and the sample surface are in soft contact. A typical sequence can be described as approaching the surface, contacting the surface, scanning the surface, retracting away from the surface as shown in Figure 23.3. The force of interaction is in the range of 10^{-8} to 10^{-6} N [14].

23.2.3.2 Alternative Contact (Also Known as Intermittent Contact or Tapping Mode)

In alternative or intermittent contact mode, popularly known as tapping mode, the tip is intermittently in contact with the surface of the sample. A typical sequence can be described as approaching the surface, contacting the surface, scanning the surface with alternate contact/retract of small vertical amplitude and high frequency, retracting away from the surface. This scheme is shown in Figure 23.4(a). Interaction forces: 10^{-9} to 10^{-7} N [15].

23.2.3.3 Pixel Mode

One extreme variant of the intermittent contact mode is the pixel mode where alternate contract/retract is of higher amplitude and smaller frequency, and where contact with the surface is performed without any displacement of the tip, scanning is performed pixel by pixel while tip is away from the surface [16,17]. This is shown in Figure 23.4(b) Interactions forces: 10^{-7} to 10^{-6} N.

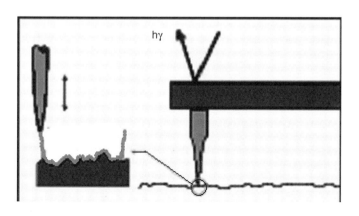

FIGURE 23.3
Contact mode.

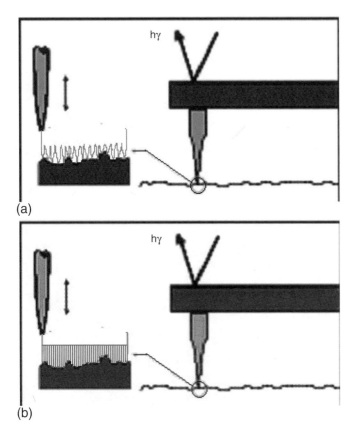

FIGURE 23.4
(a) Alternative-intermittent contact mode; (b) pixel mode.

23.2.3.4 Noncontact Mode

In noncontact mode, the probe is not in touch with the sample but is maintained at a small distance away from the sample by maintaining the attractive force at constant level. A typical sequence can be described as approaching the surface, sensing the surface, scanning over the surface at a defined flying height with small vertical amplitude, retracting away from the surface (Figure 23.5). Here the interaction forces are of the order of 10^{-12} to 10^{-10} N [7].

23.2.4 Scanning Algorithms

In addition to the mode of operation, there are various algorithms for scanning the samples depending on the structure to be imaged. These vary from 1D mode to adaptive mode to 2D mode. These algorithms decide on the way the tip interacts with and hence obtains the information on the surface being imaged. Normally, the scan direction is X in the XY plane and the variation in the direction normal to the XY plane (Z) is displayed in an XYZ plot.

23.2.4.1 One-Dimensional Scan

One-dimensional scan is typically used for acquiring data in an XY grid over a surface to report surface roughness or step height (Z) using a cone tip. The step increment in the lateral directions (X and Y) is maintained constant over the area of data collection. This scheme is shown in Figure 23.6.

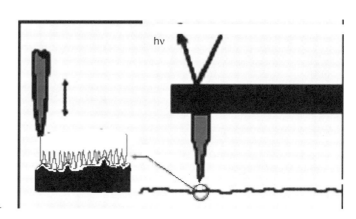

FIGURE 23.5
Noncontact mode.

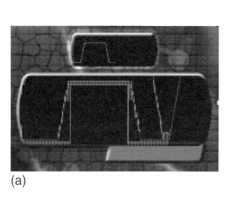
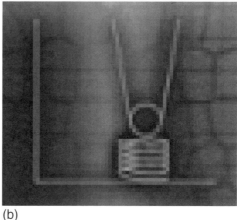

(a)   (b)

FIGURE 23.6
(a) X step equal along the scan line, additive in scan direction; (b) point of interaction between the tip and the sample.

23.2.4.2 Adaptive Scan

Adaptive scan is typically used for acquiring step height, using a cylindrical tip where the information close to the sidewall is important. This scheme is shown in Figure 23.7.

23.2.4.3 Two-Dimensional Scan

Two-dimensional scan is typically used for acquiring width (critical dimension) and sidewall information using a boot tip as shown in Figure 23.8. This involves sensing the surface in the scan direction (X), as well as in the vertical direction (Z).

23.3 Tips

Tips are manufactured in various ways (MEMS technology, Focus ion beam etching, etc). Tips could be made out of mono-crystalline silicon, silicon dioxide, silicon nitride, diamond, carbon nanotube, and other materials. They can be doped, conductive, non-conductive, metal coated, diamond-like carbon coated, and so on. They can be made in

Geometrical Characterization of Masks Using SPM

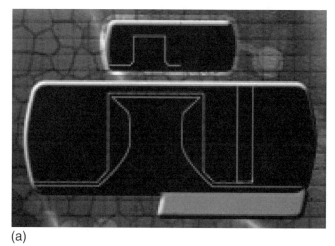

FIGURE 23.7
(a) X step adjusted to Z variations, additive or null in scan direction; (b) point of interaction between the tip and the sample.

various shapes. Some of these tips are shown in Figure 23.9 and Figure 23.10. Typical tip and cantilever specifications are shown in Table 23.1.

23.3.1 Tip–Sample Interaction

As the probe tip scans the sample surface, the resulting image is the convolution of the tip shape with the features of the surface. It is essential to understand the interaction of the probe tip with the surface in order to arrive at characteristics imaged by the SPM [18].

TABLE 23.1

Typical Tip and Cantilever Specifications

Material	Single Crystal Silicon
Tip height	$>7\,\mu m$
Apex or corner radius	$<10\,nm$
Cantilever length	$125\,\mu m$
Cantilever width	$35\,\mu m$
Cantilever thickness	$4\,\mu m$
Cantilever stiffness	$35\,N/m$
Cantilever resonance frequency	$347\,kHz$

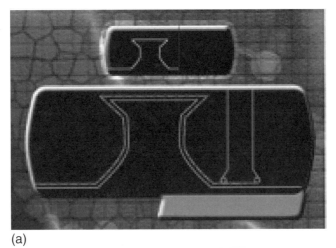

(a)

(b)

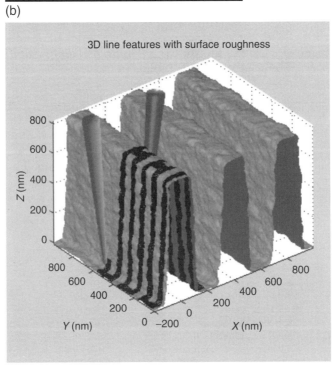

(c)

Geometrical Characterization of Masks Using SPM

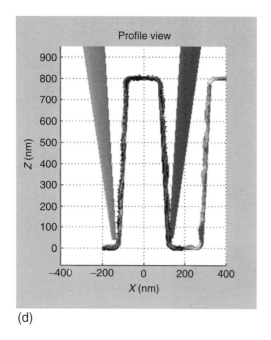
(d)

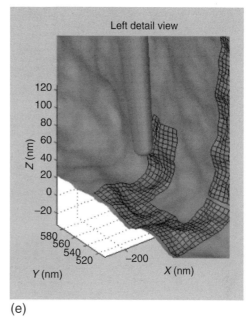
(e)

FIGURE 23.8
(a) X step adjusted to Z variations, additive, null or subtractive in scan direction; (b) points of interaction between the tip and the sample; (c) X step adjusted to Z variations, additive, null or subtractive in scan direction (Xidex Corporation); (d) points of interaction between the tip and the sample (Xidex Corporation); (e, left detail) point of interaction between the tip and the sample (Xidex Corporation).

23.3.1.1 Tip Radii and Cone Angle Effect
23.3.1.1.1 Cone Tips

A probe with a 30-nm tip radius, shown in Figure 23.11(a), finely reproduces the gently rolling surface but has difficulty defining the fine structure shown in Figure 23.11(b); this requires the use of a probe with 10-nm tip radius as shown in Figure 23.11(c).

A probe with a 60° tip cone angle [Figure 23.11(d)] does not reproduce the real amplitude of the rolling surface, whereas a probe with a 20° tip cone angle [Figure 23.11(e)] allows high fidelity imaging of the surface. A probe with 20° cone angle tip cannot image the 90° sidewall structure [Figure 23.11(f)].

Note that various combinations of tip shape and scanning algorithm can return a false angle of the structure measured [Figures 23.11(g) and (h)]. An overhang structure will be characterized up to the overhang capability of the tip using a boot tip and 2D scanning algorithm [Figure 23.11(i)].

A corollary to these facts is that the tip radius and tip cone angle must stay stable to allow resolution to stay identical within the scan image or from image to image. Horizontal cross-section of the tip and its combination with the direction of the scanning algorithm can have an effect on the correctness of the measurement as well [Figures 23.11(j)–(l)].

23.3.1.1.2 Cylindrical Tips, Boot Tips, Tilted Cone Tips

A tilted cone tip is usually a small angle cone tip with a mechanism allowing a tilted approach to the surface. A parallel sidewall tip is defined as having parallel sidewalls and circle, square or rectangular cross-section at least up to the height of the structure measured. A boot tip is defined as presenting an overhang part towards the bottom of

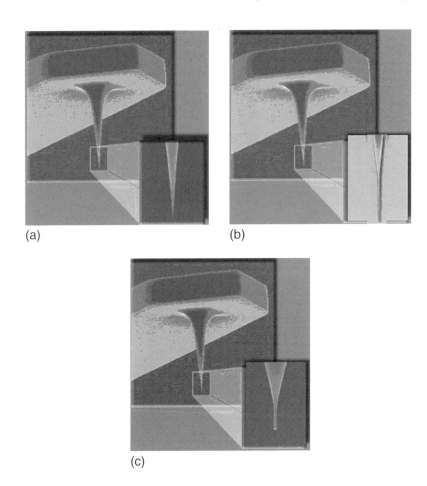

FIGURE 23.9
(a) Cone tip (Team Nanotec); (b) cylindrical tips (Team Nanotec); (c) boot tips (Team Nanotec).

the tip, within the height of the structure measured. In all these cases, the finite tip radius will affect the measurement.

The sketches in Figure 23.12 explain the impact of the radii of these tips on the measurement. There is a blind area beneath the tip and any detail of the measured structure beneath the edge of tip curvature and height of flight will not be detected.

In semiconductor application, it is essential to understand that this blind area is representative of the area of uncertainty allowed by the measurement when characterizing the effect of the foot at the bottom of the structure measured. This foot blindness is also influenced by the local slope of the very structure measured.

Obviously, any lack of symmetry (tip with different tip corner radii, left and right; structures with different local slopes, left and right) will affect the end result. Integrity of the tip must be maintained throughout the whole process of approaching the surface, scanning, and retracting. Any tip modification during any of these phases will undermine both reproducibility and accuracy of the measurement.

23.3.1.2 Tip Shape Effect

The general tip shape will have an impact on the representation of the structure measured. Subtracting the tip shape is often difficult and subject to the quality of the

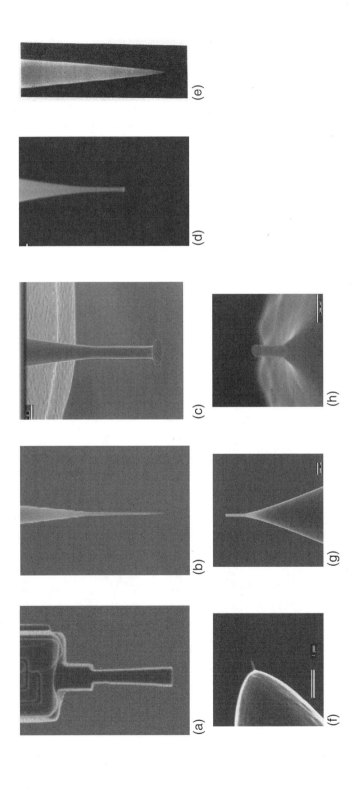

FIGURE 23.10
(a) 80-nm wide, cylindrical, pixel mode, FIB, diamond tip (FEI company); (b) <40-nm wide, cylindrical, noncontact mode, MEMS, silicon tip (Team Nanotec); (c) 1.5-µm wide, large overhang, cylindrical, noncontact mode, MEMS, silicon/silicon nitride tip (Team Nanotec); (d) <50-nm wide, boot, noncontact mode, MEMS, silicon tip (Team Nanotec); (e) <10° cone angle tip, <10 nm apex radius, noncontact mode, MEMS, silicon tip (Team Nanotec); (f) nanotube grown on AFM tip (Xidex Corporation); (g) <100-nm wide, round tip, noncontact mode, MEMS, silicon tip (Team Nanotec, Veeco Instruments); (h) <100-nm wide, round tip, noncontact mode, MEMS, silicon tip (Team Nanotec, Veeco Instruments).

510   *Handbook of Photomask Manufacturing Technology*

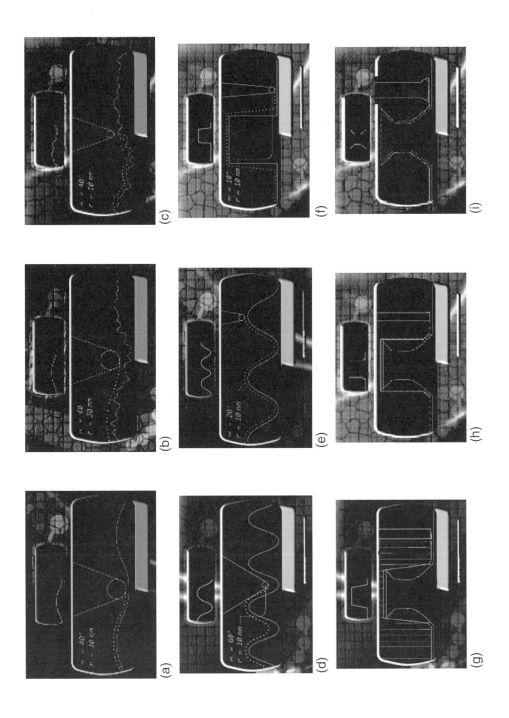

Geometrical Characterization of Masks Using SPM 511

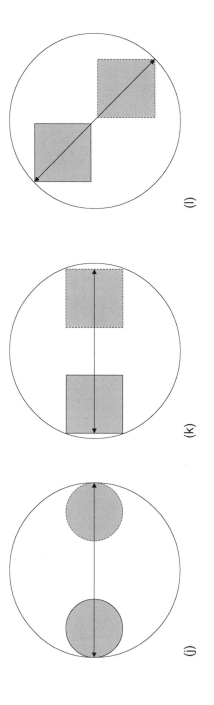

FIGURE 23.11
(a) Image produced by a 30-nm radius tip of a gently rolling surface; (b) image produced by a 30-nm radius tip of a rough surface; (c) image produced by a 10-nm radius tip of a rough surface; (d) image produced by a 60° angle cone tip of a rolling surface; (e) image produced by a 20° angle cone tip of a rolling surface; (f) image produced by a 20° angle cone tip of a 90° sidewall structure; (g) image produced by a parallel sidewall tip of an overhang structure using 1D scanning algorithm; (h) image produced by a parallel sidewall tip of an overhang structure using an adaptive scanning algorithm; (i) image produced by a boot tip of an overhang structure using 2D scanning algorithm; (j) diameter measured by a round tip with X scanning direction; (k) diameter measured by a square tip with X scanning direction of displacement; (l) diameter measured by a square tip with a diagonal scanning direction.

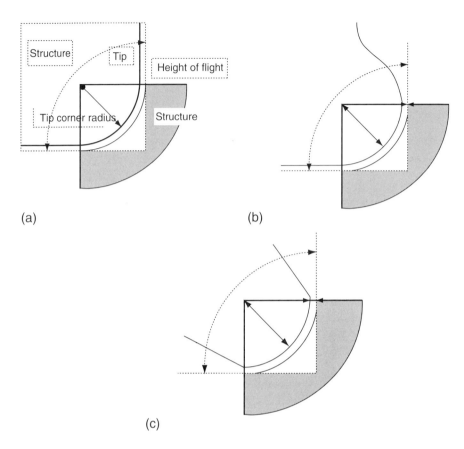

FIGURE 23.12
(a) Cylindrical tip; (b) boot tip; (c) tilted cone tip.

characterization of this tip shape. For metrology purposes, it is good however to keep in mind some simple impacts of the tip shape that do not necessitate a full high fidelity characterization of the tip.

As defined in the previous section, the rounding of the corner has an impact on the final metrology result. When local slope of the structure can be measured, the tip corner radius impact can be estimated and subsequently corrected. But when local slope of the structure cannot be evaluated (top corner of the structure where two unknown shapes, tip corner and structure corner, combine to give a string of pixels, and bottom of the structure where tip corner radius creates blindness), it is wise to set up an analysis algorithm so as to avoid any measurement extraction in these areas (Figure 23.13).

We can calculate the top and bottom metrology exclusion area height as the higher of the tip corner radius or the structure corner radius. In both cases, the minimum value of the metrology exclusion area would be the effective tip corner radius. Practically, as both the tip corner radius and the structure corner radius are often unknown, a careful analysis of the scan lines allows us to set up top and bottom metrology exclusion areas outside the rounding part of the scan lines.

Geometrical Characterization of Masks Using SPM

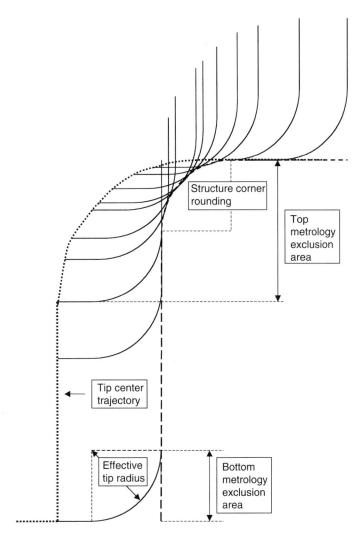

FIGURE 23.13
Metrology exclusion area.

23.3.1.3 Tip Stiffness Effect

Lateral tip stiffness is very often an overlooked parameter. This parameter has an important effect on the validity of the measurement of profile and width. AFMs are in effect monitoring the position of the backside of the cantilever. The assumption is that there is a constant difference between the position of the backside of the cantilever and any spatial point of the tip. If, for any reason, this difference varies (tip wearing, corner rounding), it is assumed that recalibration will allow correction of the problem on an ongoing basis.

However, if, for mechanical reasons, the lateral stiffness of the tip is less than the surface reactive forces induced by the sensor threshold sensitivity, the spatial relationship between the points in the tip and the back of the cantilever will be modified in an elastic

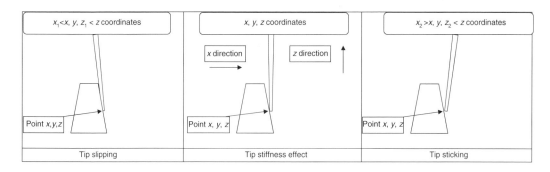

FIGURE 23.14
Tip stiffness effect versus pedestal shape, bending of the tip can occur at any height along the tip length; (left) tip slipping; (right) tip sticking.

way. This will result in reporting of a false measurement points, either by earlier contact with the surface (sticking of the tip due to attractive forces bending the tip towards the surface) or slippage of the tip along the surface before reaching sensor threshold sensitivity (Figure 23.14).

23.3.2 Tip Calibration

23.3.2.1 Calibration Structures

There are different structures that can be used for calibration of probe tips, namely silicon nano edge (SNE), flared silicon ridge (FSR), vertical parallel structure (VPS), etc. There are variations on these structures as they have evolved over the years to meet the stringent requirements of the scanning probe metrology. The manufacturers of SPMs generally include information on probe tip calibration in their manuals. What follows here is a procedure on how a probe tip may be calibrated.

23.3.2.1.1 Tip Width Determination Using Silicon Nano Edge

The following procedure can be used to calibrate the tip using the SNE (Figure 23.15). It is written to help users understand the interaction between the tip and the nano edge, and compute the tip width from the scanned data.

Scan the nano edge in a clean area. Setting of the scan must encompass an x size big enough to scan both sides of the structure. Number of scan lines must allow statistical averaging of the measurement. The analysis procedure can be carried out manually or automatically. So it will be a good idea to save the nano edge scan image. We need to compute the parameters: distance from top to measure; width subtraction for future reference. We can use the following method of computation.

Figure 23.20 shows a single scan across the nano edge; D is the distance from the top of the nano edge to the measurement location; L is the width of the scanned image at that location; W is one-half of the subtraction required to find the tip width. It is indicated in the figure. We can write an expression for W as

$$W = (D - r)\tan\theta + r$$

Geometrical Characterization of Masks Using SPM

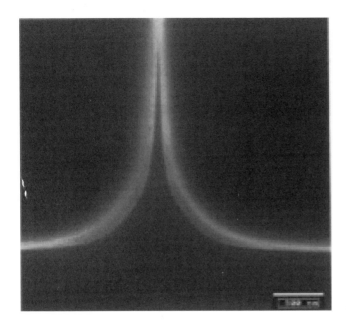

FIGURE 23.15
Improved SNE (Team Nanotec).

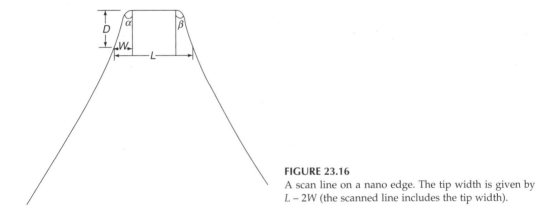

FIGURE 23.16
A scan line on a nano edge. The tip width is given by L − 2W (the scanned line includes the tip width).

where D can be set to 80 nm, r = 7.5-nm (radius of the SNE), and θ is the average angle computed from the left and right side slopes $(90° - (\theta \; \theta_2)) / 2$. For example, if the left and right side slopes are both 70°, $\theta = 20°$. $\theta = (\alpha + \beta) / 2$ in Figure 23.16.

With the computation of W we are in a position to determine the effective tip width as

$$\text{tip width} = L - 2W$$

This is only the effective tip width and not the real tip width because we have not included the height of flight in our calculations. We are interested in using the effective tip width to reach precision or accuracy.

However, while precision can be reached based on assumptions used above for the radius of the SNE and identical height of flight over various materials, accuracy cannot be

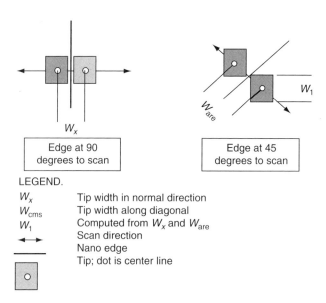

FIGURE 23.17
Tip width at different angles.

claimed as the 7.5-nm value (TEM evaluation) used for definition of the edge radius is approximate, and as height of flight from material to material can vary versus the scanning mode used.

While the above-described procedure allows you to characterize tip width in the direction of scanning, rotating scanning direction, or structure by 45° angle will allow you to characterize diagonal size of the tip (Figure 23.17).

23.3.2.1.2 Flared Silicon Ridge and the Overhang

FSR is usually used to further characterize a tip (Figure 23.18). The use of the FSR allows determination of more tip shape parameters useful to validate structure measurement (Figure 23.19).

By combining results from SNE and FSR tip calibration, we are able to reconstruct the tip to a close approximation [19]. Assumptions of the radius of the corner of the FSR must be accepted to further assess radius of the corner of the tip (Figure 23.20).

23.3.2.1.3 Vertical Parallel Structure

A simplified tip characterization can be used for production purpose where the use of the overhang capability of the tip is not required. Precision using VPS (Figure 23.21) is high due to the excellent line-width uniformity and the low sidewall and line roughness. This structure also can be used routinely to characterize sidewall slopes of cone tips.

23.3.3 Tip Shape Removal or Tip De-convolution

Various algorithms are used for tip de-convolution, or tip shape removal. For metrology purpose, the width of the tip in the scanning direction is sometimes simply subtracted (line) or added (trench) during the width calculation, and removed from the image as a

Geometrical Characterization of Masks Using SPM

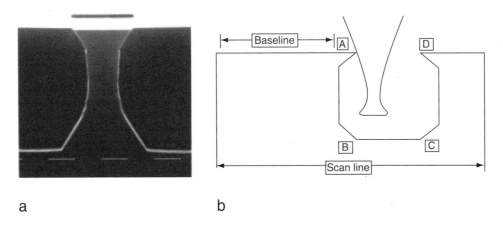

FIGURE 23.18
(a) FSR (Team Nanotec); (b) relationship boot tip/structure on FSR.

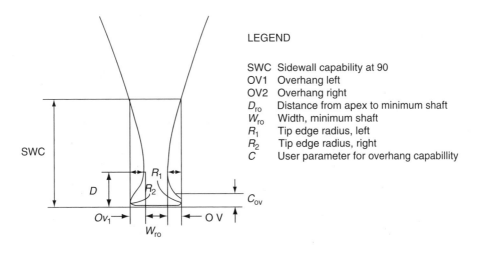

FIGURE 23.19
Tip shape parameters extracted from FSR scanning.

segment (Figure 23.22). Some algorithms are more sophisticated, and they will allow a more complete tip shape removal [20]. One of these tip-shape removal techniques is illustrated in Figure 23.23.

23.4 Mask Applications

Mask metrology presents various challenges to AFM. It is not unusual to find that process-induced electrical charges on the surface compete with the normal interaction

518 Handbook of Photomask Manufacturing Technology

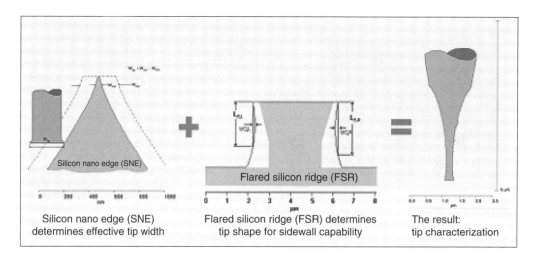

FIGURE 23.20
Tip characterization.

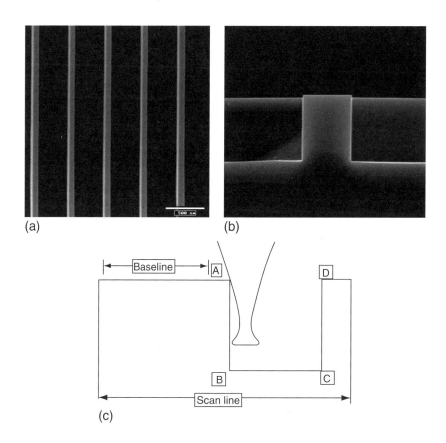

FIGURE 23.21
(a) 100-nm width VPS lines (Team Nanotec); (b) vertical parallel line cross-section (Team Nanotec); (c) relationship boot tip/structure on vertical parallel line.

Geometrical Characterization of Masks Using SPM 519

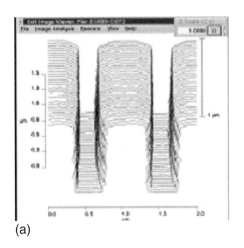 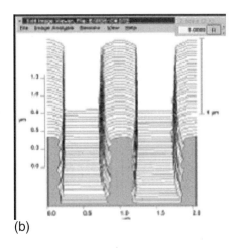

FIGURE 23.22
(a) Raw image before tip width removal (boot tip, noncontact, 2D mode); (b) final image after tip width removal (boot tip, noncontact, 2D mode).

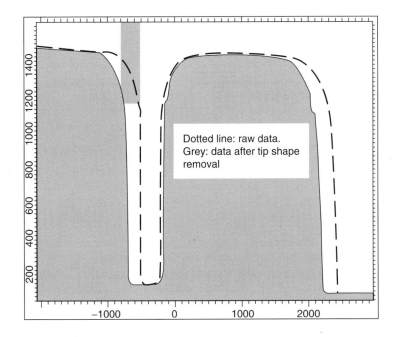

FIGURE 23.23
Tip shape removal (cylindrical tip, intermittent contact, pixel mode) (FEI Company).

forces (Figure 23.24), as well as capillary forces due to insufficient cleaning/drying of the surface. Mask plate temperature must also be stable during the measurements. Choice of scanning mode, discharge apparatus, and careful processing will allow overcoming these obstacles.

While high voltage will create obvious distortions, smaller voltages can still impact reliability of the measurements (Figure 23.25).

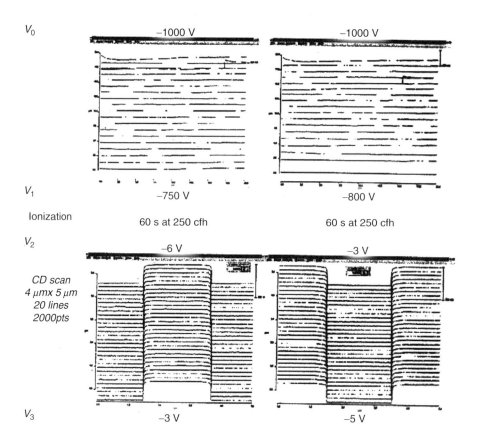

FIGURE 23.24
Upper scans obtained before de-ionization; lower scans obtained after de-ionization (boot tip, noncontact, 2D algorithm).

23.4.1 Profile

Characterizing the profile (cumulative cross-sections along the Y direction) of a structure can prove challenging if the sidewalls are not sheer and smooth. Two main parameters will influence this characterization: tip shape and scanning algorithm. Up to now, only the combination boot tip — 2D algorithm has been proven to allow faithful representation of sidewalls up to tip shape limitations. New sensors are however under development, which may allow in pushing the envelope a bit further (caliper sensor using two tilted cone tips).

Wet etch mask profiles are very often impacted by the difference of etch rate between the top CrO_2 antireflective layer and the bulk Cr layer, while dry tech profiles have proven easier to characterize (Figure 23.26).

To avoid parasitic phase shift of the light, the most inner point of the quartz wall must be recessed from the vertical projection of the outer effective thickness of the metal structure (where exposure light transmission is blocked). Therefore, it is of interest to characterize quartz etch profile [21,22] (Figure 23.27).

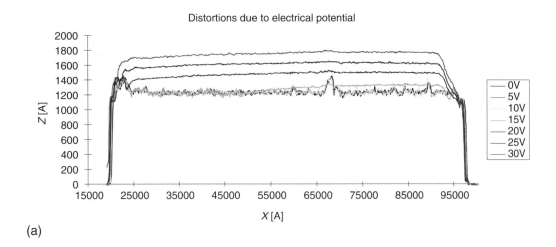

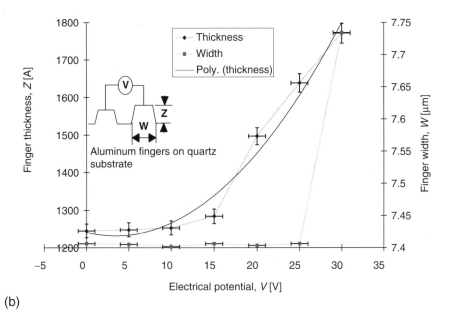

FIGURE 23.25
(a) Profile versus residual voltage (boot tip, noncontact, 2D algorithm); (b) thickness and width versus residual voltage (boot tip, noncontact, 2D algorithm).

23.4.2 Defect

Metal or quartz defects can be characterized by AFM. An interesting application is the characterization of quartz bump defect that would locally modify light phase during exposure. When mask defect coordinates can be linked accurately to a repair tool, monitoring of the etching process allows repair of defects, even though invisible to the repair tool (Figure 23.28).

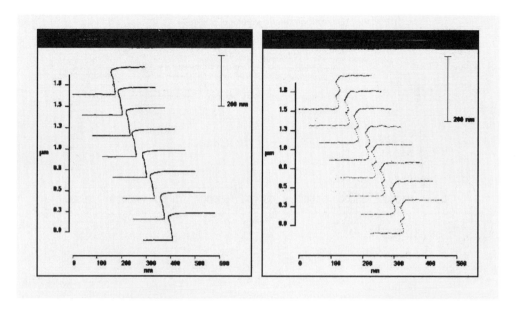

FIGURE 23.26
Dry etch (left) and wet etch (right) mask chromium profile (boot tip, noncontact, 2D algorithm).

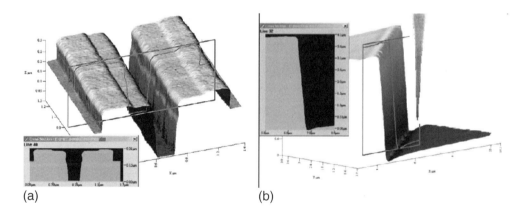

FIGURE 23.27
(a) Phase shift quartz etch trench (boot tip, noncontact, 2D algorithm); (b) phase shift quartz etch trench foot characterization (tilted cone tip, noncontact, adaptive algorithm).

23.4.3 Linewidth Uniformity

The discrete acquisition of multiple scan lines along the Y direction of the structure allows one to evaluate the uniformity of the linewidth [23] with the high resolution of the AFM system: The example below shows five successive scans (acquisitions of 20 scan lines performed at the same position of the structure to be measured along a 5-μm Y size). Thickness, top width, middle width, and bottom width are reported. Bottom line of the chart is lower specification limit (LSL). Middle line of the chart is upper specification limit (USL). One can clearly see that the linewidth variation is higher than the USL minus LSL range (Figure 23.29).

Geometrical Characterization of Masks Using SPM 523

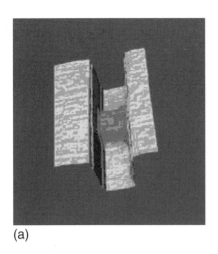
(a)

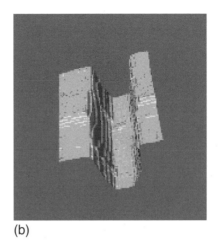
(b)

FIGURE 23.28
(a) Before repair (cylindrical tip, intermittent contact pixel mode, 1D algorithm) (FEI Company); (b) after ion beam repair (cylindrical tip, intermittent contact Pixel mode, 1D algorithm) (FEI Company).

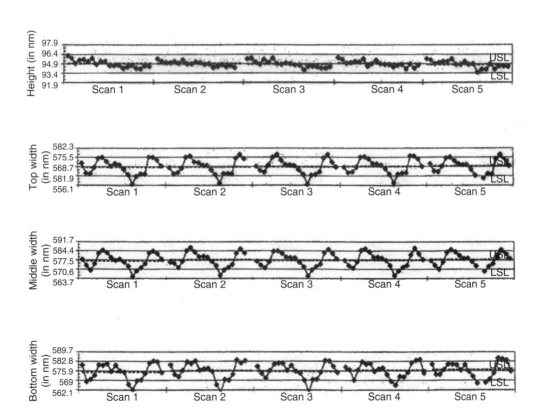

FIGURE 23.29
Line width uniformity and height uniformity on a chromium mask line (boot tip, noncontact, 2D algorithm).

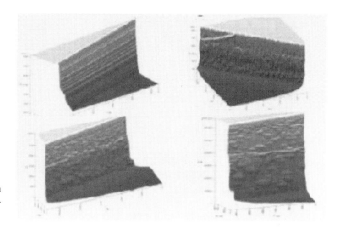

FIGURE 23.30
Sidewall roughness versus quartz etch process (boot tip, noncontact, 2D algorithm).

23.4.4 Sidewall Roughness

While bottom roughness of a phase shift trench has a potential impact on the transmission of the mask material, it is also important to adjust the quartz etch process to ensure that sidewall roughness will not create any parasitic effect. The example included (Figure 23.30) shows various mixtures of quartz etch process inducing various sidewall roughness [24].

23.4.5 Line Roughness, Line Edge Roughness

Unlike line-width uniformity that is a measurement of width variation along a structure, line roughness focuses on one side of the structure. Line roughness reports a variation in the position of a sidewall along a horizontal slice defined by the height threshold in percentage (Figure 23.31). When this percentage is at 100% (top of the structure), the returned number is called line edge roughness [24].

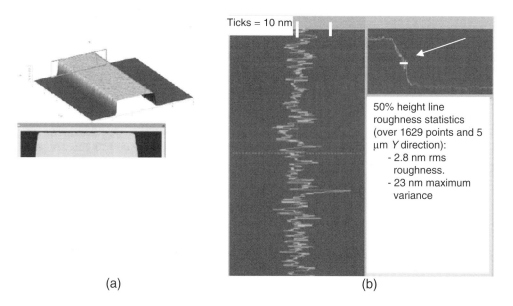

(a) (b)

FIGURE 23.31
(a) Chromium mask, dry etched (boot tip, noncontact, 2D algorithm); (b) Analysis of line roughness of the left side of the structure.

Geometrical Characterization of Masks Using SPM 525

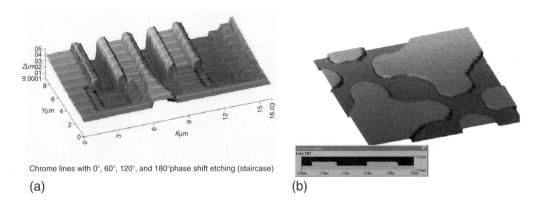

(a) Chrome lines with 0°, 60°, 120°, and 180°phase shift etching (staircase)   (b)

FIGURE 23.32
(a) Staircase shape characterization (boot tip, noncontact, 2D algorithm); (b) OPC shape characterization (boot tip, noncontact, 2D algorithm).

23.4.6 Shape Characterization

A fully characterized tip and a comprehensive tip shape removal algorithm should allow shape characterization (Figure 23.32).

23.4.7 Depth, Step Height

AFM have proven extremely useful in depth monitoring [25]. This is of importance in metrology of PSMs. Results of <0.5 nm 3 σ reproducibility have been reported (Figure 23.33).

23.4.8 Roughness

AFM is the chosen instrument for surface metrology. The recent improvements in AFM technology have enabled better accuracy in measurements and resolutions up to 100th of

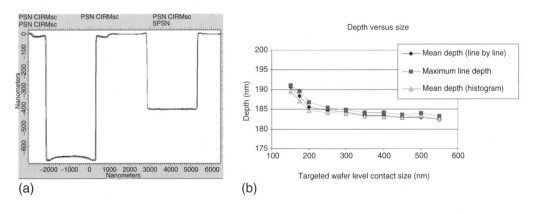

FIGURE 23.33
(a) Depth characterization of a PSM area (cylindrical tip, intermittent contact pixel mode, 1D algorithm); (b) characterization of structure size effect on final depth of the structure (parallel sidewall tip, noncontact, adaptive algorithm).

nanometer are possible. The new AFMs have a lateral resolution of <1 nm and a noise floor <0.05 nm.

AFMs measure the vertical height (z) as a function of position (x, y). Depending on the scan size and the number of samples per scan, the resolution can be varied. The information collected by AFM is the true surface modified by the transfer function that consists of the effects of tip geometry, modification of the sample caused by the tip, and the instrument drifts. AFM measurements at ambient conditions have been shown to measure true roughness. The parameters that can be obtained for micro-roughness from AFM are average roughness (Ra), root mean square roughness (Rq), and peak to valley (PV). We can define these as follows:

$$\text{Ra} = (1/N^2 \Sigma\Sigma |z(x_i, y_j) - z_{av}|)$$

$$\text{Rq} = [(1/N^2 \Sigma\Sigma (z(x_i, y_j) - z_{av})^2)]^{1/2}$$

$$\text{PV} = \max |z(x_i, y_i)| - \min |z(x_i, y_i)|$$

These measurements give information about the height of the surface as a function of location but do not give an overall impression of the surface, such as the variation over the entire surface. To remove this ambiguity information such as the power spectral density could be used. This parameter will give the spatial information required for eliminating the ambiguity whether the roughness is due to a singular irregularity or the spatial variation (Figure 23.34).

23.4.9 Width

AFM allows a precise measurement of the physical width of a feature when tip wearing is under control [19]. Accuracy can be reached through careful tip width calibration. Correlations to TEM have been demonstrated better than 1 nm. Tip width error determination can be calculated as follows:

$$3\sigma_{\text{tip width}} = 3 * [\{(\sigma_{\text{cal.structure}})^2 + (\sigma_{\text{static width}})^2\}/\text{number of scan lines}]^{1/2}$$

Two-dimensional scanning algorithm and boot tip have been demonstrated as the best way to characterize width as it offers the capability to follow sidewall profile variations and vary the X step size.

Static precision (50 scan lines performed on the structure to measure with a y range equal to $0\,\mu$) is characterized. Adjustment of various scanning parameters (clamp, X step, Z/X ratio, set point, etc.) allowed determining the best combination leading to the smallest error induced by the couple tool/sample.

Static precision can be calculated as follows:

$$3\sigma_{\text{static width}} = 3 * [\{(\sigma_{\text{top}})^2 + (\sigma_{\text{middle}})^2 + (\sigma_{\text{bottom}})^2\}/3]^{1/2}$$

$$3\sigma_{\text{static height}} = 3 * \sigma_{\text{height}}$$

$$3\sigma_{\text{static angle}} = 3 * [\{(\sigma_{\text{left}})^2 + (\sigma_{\text{right}})^2\}/2]^{1/2}$$

Characterization of the line-width uniformity is useful to set up the measurement recipe. Feature-width uniformity is characterized by scanning the feature-to-measure with a small y step and a high number of scan lines (at least 50 scan lines). Feature-width uniformity is extracted through statistical formula from feature-width-measurement uniformity by

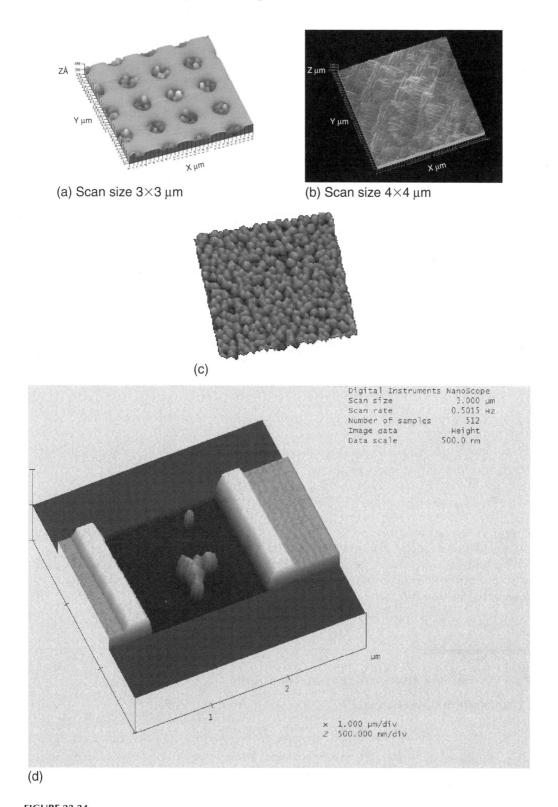

FIGURE 23.34
(a) Tungsten plug (cone tip, noncontact mode, 1D algorithm); (b) Epitaxial silicon (cone tip, tapping mode, 1D algorithm); (c) hemispherical silicon grain (cone tip, noncontact mode, 1D algorithm); (d) roughness measurements on PSM (cone tip, tapping mode, 1D algorithm) (Veeco Instruments).

taking into account the static precision previously determined. Feature-width uniformity is a statistically based characterization as the y step is smaller than the tip size in the direction orthogonal to that of scanning (Y size). This Y size will impact the angle measurements identically. The same procedure can be used to determine the $\sigma_{\text{cal.structure}}$. Knowledge of tip width is not needed for these exercises.

Feature-width uniformity for depth measurement performed with a boot tip will not give a true image of the uniformity of the feature, as the parallelepiped shape of the bottom of the tip will tend to smooth the surface.

Feature-width uniformity can be calculated as follows:

$$3\sigma_{\text{feature width}} = 3 * \{[1/3 * \{(\sigma_{\text{top}})^2 + (\sigma_{\text{middle}})^2 + (\sigma_{\text{bottom}})^2\}] - \sigma_{\text{static-width}})^2\}^{1/2}$$

$$3\sigma_{\text{feature height}} = 3 * \{(\sigma_{\text{height}})^2 - (\sigma_{\text{static height}})^2\}^{1/2}$$

$$3\sigma_{\text{feature angle}} = 3 * \{[[1/2 * \{(\sigma_{\text{left}})^2 + (\sigma_{\text{right}})^2\}] - (\sigma_{\text{static angle}})^2\}^{1/2}$$

The number of scan lines is determined by the dynamic precision required and the feature-width uniformity. Feature uniformity contribution to dynamic precision decreases as the number of scan lines increases.

The number of scan lines must be chosen to reach a compromise between throughput and precision to minimize the impact of the feature-width uniformity characterized. This impact can be calculated as follows:

$$3\sigma_{\text{static measurement error contribution}} = 3 * [\{(\sigma_{\text{feature width}})^2 + (\sigma_{\text{static width}})^2\}/\text{number of scan lines}]^{1/2}$$

Combining $3\sigma_{\text{static measurement error contribution}}$ and $3\sigma_{\text{tip width}}$ allows us to compute a minimum value for the $3\sigma_{\text{dynamic measurement precision}}$:

$$3\sigma_{\text{dynamic measurement precision}} = 3 * \{(\sigma_{\text{tip-width}})^2 + (\sigma_{\text{static measurement error contribution}})^2\}^{1/2}$$

A typical throughput to target is <1 min/site. A typical short-term precision to target (within 100 sites on one mask measured within 100 min) is <2 nm 3σ.

The long-term precision targeted can be calculated over the 1000 sites measured. A short-term exercise can be run once in the morning (with tip number 1) and once overnight (with tip number 2) for 5 days.

23.5 AFM Selection and Test Methodology

In summary, the choice of your AFM is linked to your application:

- Surface roughness <10 Å rms: tapping mode, cone tip, 1D scanning algorithm
- Surface roughness >10 Å rms: noncontact mode, cone tip, 1D scanning algorithm
- Step height: noncontact mode, cylindrical tip, adaptive scanning algorithm, or intermittent contact pixel mode, cylindrical tip, 1D algorithm
- Profile, sidewall roughness, line roughness, line edge roughness, width, and line-width uniformity: noncontact mode, boot tip, 2D scanning algorithm

The above list is indicative as AFM systems, sensors and algorithms are evolving quickly with the increased interest in their capabilities. Whatever your application, the test methodology must be designed to characterize possible tip wearing. These data are important to set up re-calibration frequency, tip change frequency, and to avoid unknown drift in measurements.

Acknowledgments

The authors would like to thank the following for their valuable help in writing this chapter: Marty Klos (Rave LLC), Kelvin Walch (Ash semi-services), Mike Young (KLA-Tencor), Vladimir Ukraintsiev (TI), Johann Greschner (Team Nanotec), Yves Martin (IBM), Troy Morrisson (FEI company), Paul Mac Clure (Xidex Corporation), Vladimir Mancevsky (Xidex Corporation), Kirk Miller (Veeco), Guy Vachet (CNET) Philippe Cochet (Zygo). The authors would also like to thank the following institutions: FEI Company, Veeco Instruments, and Zygo Corporation.

References

1. G. Binnig and H. Rohrer, Scanning tunneling microscopy, *Helv. Phys. Acta*, 55, 726 (1982).
2. G. Binnig, C.F. Quate, and Ch. Gerber, Atomic force microscope, *Phys. Rev. Lett.*, 56, 930 (1986).
3. C.F. Quate, The AFM as a tool for surface imaging, *Surf. Sci.*, 299/230, 980–995 (1994).
4. H.K. Wickramasinghe, Scanned probe microscopes, *Sci. Amer.* (October), 98–105 (1989).
5. G. Meyer and N.M. Amer, Novel optical approach to AFM, *Appl. Phys. Lett.*, 53, 1045 (1988).
6. S. Alexander, L. Hellemans, O. Marti, J. Schneir, V. Eling, P.K. Hansma, M. Longuire, and J. Gurly, *J. Appl. Phys.*, 65, 164 (1989).
7. Y. Martin, C.C. Williams, and H.K. Wickramasinghe, Atomic force microscope — force mapping and profiling on a sub 100 Å scale, *J. Appl. Phys.*, 61, 4723 (1987).
8. D. Royer, E. Dieulesaint, and Y. Martin, Improved version of a polarized beam heterodyne interferometer, in: B.R. McAvoy (ed.), *Proc. IEEE Ultrasonics Symp.*, San Francisco, vol. 432, IEEE, New York, 1985.
9. G. Binnig, C.F. Quate, and Ch. Gerber, Atomic force microscope, *Phys. Rev. Lett.*, 56, 930 (1986).
10. M. Tortonese, R.C. Barrett, and C.F. Quate, *Appl. Phys. Lett.*, 62, 834 (1993).
11. T. Goddenhenrich, U. Lemke, U. Hartmann, and C. Heider, Force microscope with capacitance displacement detection, *J. Vac. Sci. Technol.*, A6, 383 (1988).
12. R.C. Barrettm and C.F. Quate, Optical scan-correction system applied to atomic force microscopy, *Rev. Sci. Instrum.*, 62, 1393 (1991).
13. J.E. Griffith, G.L. Miller, and C.A. Green, A scanning tunneling microscope with a capacitance-based position monitor, *J. Vac. Sci. Technol.*, B8, 2023–2027 (1990).
14. A.L. Weisenhorn, P.K. Hansma, T.R. Albrecht, and C.F. Quate, Forces in atomic force microscopy in air and water, *Appl. Phys. Lett.*, 54, 2651 (1989).
15. D. Vie, H.G. Hansma, C.B. Prater, J. Massie, L. Fukunaga, J. Gurley, and V. Elings, Tapping mode atomic force microscope in liquids, *Appl. Phys. Lett.*, 64, 1738 (1994).
16. A. Mathai and M. Oyumi, Profiling high-aspect ratio features for post-etch metrology, *Yield Management Solutions*, Autumn, 30 (1999).
17. Product information, www.feico.com.
18. G.S. Pingali and R. Jain, Surface recovery in scanning probe microscopy, *Proc. SPIE*, 1823, 151–162 (1992).

19. S. Muckenhirn and A. Meyyappan, Critical dimension atomic force microscope (CD-AFM) measurement of masks, *Proc. SPIE*, 3332, 642–653 (1998).
20. J.S. Villarrubia, Algorithms for scanned probe microscope image simulation, surface reconstruction, and tip estimation, *J. Res. Natl. Inst. Stand. Technol.*, 102 (4), 425–453 (1997).
21. A. Meyyappan, M. Klos, and S. Muckenhirn, Foot (bottom corner) measurement of a structure with SPM, *Proc. SPIE*, 4344, 733–738 (2001).
22. S. Muckenhirn, A. Meyyappan, K. Walch, M.J. Maslow, G.N. Vandenberghe, and J. van Wingerden, SPM characterization of anomalies in phase-shift mask and their effect on wafer features, *Proc. SPIE*, 4344, 188–199 (2001).
23. A. Meyyappan and S. Muckenhirn, Photoresist focus exposure matrix (FEM) measurements using critical-dimension atomic force microscopy (CD AFM), *Proc. SPIE*, 3332, 631–641 (1998).
24. K. Walch, A. Meyyappan, S. Muckenhirn, and J. Margail, Measurement of sidewall, line, and line-edge roughness with scanning probe microscopy, *Proc. SPIE*, 4344, 726–732 (2001).
25. V.C. Jaiprakash, M.E. Lagus, A. Meyyappan, and S. Muckenhirn, High-aspect-ratio depth determination using non-destructive AFM, *Proc. SPIE*, 3677, 10–17 (1999).

24
Metrology of Image Placement

Michael T. Takac

CONTENTS

Introduction 531
24.1 General Overview of $x-y$ Metrology 532
24.2 Coordinate System Traceability Path 535
24.3 Analysis 537
    24.3.1 Regressions 537
    24.3.2 Fourier Transforms 541
24.4 Sampling 546
    24.4.1 Characterization 546
    24.4.2 Dispositioning 546
        24.4.2.1 Many Point Alignment 546
        24.4.2.2 n-Point Alignment 546
        24.4.2.3 Systematic Inclusion 547
        24.4.2.4 Adaptive Metrology 547
        24.4.2.5 Dispositioning as a Function of Application 547
References 548

Introduction

This chapter is intended to provide an understanding of the basic role of $x-y$ image placement metrology in the mask industry.

The ideal objective in mask manufacturing is to render any design without error in zero time and at an affordable cost. Most ideal objectives guide science down a path of innovation in concert with its observational enhancements needed for exploration and verification. Of the observational platform within the mask industry, one metrology tool measures pattern placement in the $x-y$ plane. $x-y$ metrology made its mask industry début during the 1970s in company-captive mask facilities. In the mid-1970s, Boller & Chivens, a company in Pasadena (CA), introduced the linear-dimensional analyzer (LDA) capable of measuring x and y coordinates of features on the mask. The system employed finely graduated glass rulers along the stage x- and y-axes for feature placement measurements. The system repeatability was on the order of a micron. During the 1980s Nikon, Inc., commercially offered a line of $x-y$ metrology tools, also during the 1990s to the present; has dominated the industry with its line of products. During this period, $x-y$ metrology improved from 1000.0 nm down to 3.0 nm 3σ for long-term repeatability.

The mask industry utilizes pattern generators to render features on substrates for the purpose of projection or contact printing. A pattern generator relies on an internal coordinate system as a reference for feature placement within the mask plane. The x–y metrology tools, as the ones mentioned above, also have their own internal coordinate system for the measurement of feature placement on a mask. The ideal feature placement scenario unfolds when the coordinate systems of the pattern generator and x–y metrology converge. Matching coordinate systems is an important objective; however, it is shadowed by the task of matching the geometry of the metrology tool's coordinate system with the theoretical coordinates used during mask pattern design. Therefore, the primary role of x–y metrology functions as a catalyst to facilitate the traceability path for matching the pattern generator to the theoretical design coordinate system; in addition, x–y metrology plays a similar role at the wafer level.

A secondary role for x–y metrology, the most frequently utilized, is in product dispositioning and tool characterization.

The following section addresses a general overview of x–y metrology, the traceability path to the design coordinate, analysis, and sampling.

24.1 General Overview of x–y Metrology

Advancements in x–y metrology lead to using the interaction of light as a replacement for the graduated rulers along the stage x- and y-axes.

Over the last three centuries, scientists have studied the interaction of light with matter and of light to itself. For example, in 1879, Albert Abraham Michelson devised a celebrated experiment (Figure 24.1) known as the Michelson–Morley experiment [1], using the interaction of light to measure the ether wind. Later Fabry and Perot made further contributions to centuries of related research leading to the culmination of what is known today as the interferometer.

Figure 24.2 is a diagram of a present day interferometer used in x–y metrology system [2]. Interferometers are found in precision systems throughout the mask industry. Both the mask pattern generator and x–y metrology rely on stage interferometers to accurately position the mask.

In addition to the interferometer performance, an equally critical x–y metrology function is the capability of resolving and detecting features. A recent improvement to x–y metrology includes a transmitted light illumination for the optical head adding precision and capacity, for the transmitted mode masks, and providing the option to concurrently measure line-width during image placement (Figure 24.3).

FIGURE 24.1
Michelson–Morley experiment.

Metrology of Image Placement

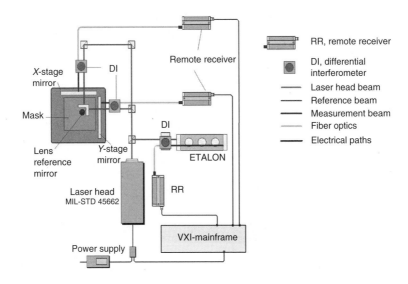

FIGURE 24.2
Stage interferometer of $x-y$ metrology system.

Figure 24.4 shows a photographic view of the stage interferometer and optical head resting on an isolation-table to minimize vibrations.

Figure 24.5 shows a delivery system that is used to fetch mask plates from a magazine to the metrology stage. The delivery system minimizes environmental changes when transferring mask plates to and from the stage.

The $x-y$ metrology tool is enclosed in an environmental chamber to minimize temperature and humidity changes shown in Figure 24.6. Also shown is the control console where the metrologist interacts with the system.

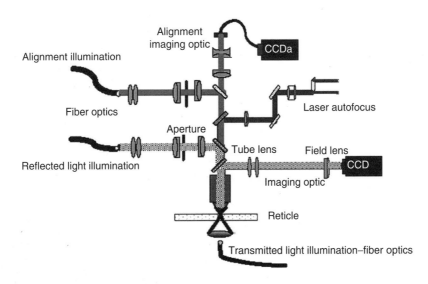

FIGURE 24.3
Optical head providing reflective and transmitted light illumination.

FIGURE 24.4
Stage interferometer and optical head.

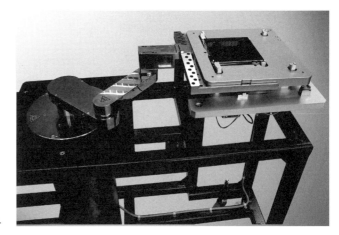

FIGURE 24.5
Mask handler.

FIGURE 24.6
$x-y$ metrology system.

24.2 Coordinate System Traceability Path

The technology of $x-y$ metrology is at the forefront of nanotechnology; this in itself creates a dilemma for standard organizations to keep pace with a fast-moving industry that usually acquires this technology first. The return on investment for $x-y$ metrology hinges on the quality of calibration.

Conventionally, most measurement instrument setup procedures include an industry "standard artifact" for system calibration. A calibration is guaranteed providing there exists a traceability path to a standard scale and a traceability path to the geometry of which the measurement tool was designed to operate. The ideal "standard artifact" should include both a traceability path to scale and geometry. This section covers a general overview to calibration in the event that a standard artifact or a setup procedure does not include a geometric traceability path.

A quote from Lord and Wilson [3]: "In his inaugural address at the University of Erlangen (1872), Felix Klein [4] introduced a unifying principle of profound importance for the understanding of what is meant by 'geometry' in its widest sense. Each kind of geometry is associated with a group of motions, or one-to-one mappings of the space onto itself, that leave intact the geometrical properties of figures in the space. And conversely, the specification of a group of one-to-one mappings of a space onto itself determines a geometry."

Felix Klein's inaugural address is referred to as the Erlangen Program. A variation of it establishes the foundation to self-calibration, whose primary function is the mapping of a physical tool to its intended geometry. Throughout the history of precision engineering, self-calibration methods have come in a rich variety, some of which are summarized in the research works of Evan et al. [5] and Raugh [6].

The intended geometry for $x-y$ metrology in the mask industry is Euclidian having a Cartesian coordinate mapping. In Euclidean geometry, self-calibration is dependent on the concept of *congruence*. Two figures are congruent in Euclidean space when the distance between any two points in one figure is equal to the distance between corresponding points in the other. The correspondence between two congruent figures is an example of a one-to-one mapping between two sets of points. A one-to-one mapping of the Euclidean space onto itself that preserves distances is called an isometry. Examples of Euclidean space isometries are rotations, translations, and reflections. All Euclidean isometries constitute a group, called the group of motions of the Euclidean space, or the *Euclidean group* [3].

One may state the axiom that a calibrated metrology tool having a unity transfer function, containing a one-to-one mapping between input and output throughout its working space, has a set of isometries belonging to the Euclidean group.

Applying the reciprocity principle with a set of specified rigid artifact motions (rotations, translations, or reflections) within the metrology working space, and modifications to the tool's transfer function approaching shape congruency within the artifact motion set maintains a self-calibration.

The subtle difference between calibration and self-calibration is scale. The process of achieving self-consistency within the artifact motion set does not preserve scale. Scale is a relative term and only the shape-preserving properties are maintained. Therefore, a scale standard is needed for the self-calibrated system. The outcome of this exercise results in a calibrated system with accuracy to its geometry approaching the metrology system repeatability, and the scale approaching the accepted industry precision.

Figure 24.7 represents a one-dimensional self-calibration example using transitivity to achieve congruency [7].

FIGURE 24.7
Congruency via views of the same uncalibrated rigid artifact (A) measured at two different locations on an uncalibrated tool (T).

Transitivity is an axiom; where if some initial item is equal to the second and the second is equal to a third, then the initial item is also equal to the third. Since the artifact is assumed rigid, we can use transitivity to assert that the measured length on tool T of L_1 in view 1 must equal to the measured length of L_1 in view 2. If the lengths are not equal, then record the difference in c_2. Adding c_2 to L_1 in view 2 makes its length equal to L_1 in view 1. The c_2 is the calibration mapping that mathematically converts the tool's graduation at T_2 to measure the same as T_1. The process continues by adding c_2 to the length of L_2 in view 1, followed by comparing the L_2's between views 1 and 2, and saving the difference in c_3, and so on. By adding the c_n vector to the tool's graduations (T_n), mathematically makes the distance between all tool graduations equal to the distance between the first two graduations (T_0 to T_1). Equally spaced graduations contain a one-to-one mapping, having shape-preserving properties, hence a self-calibration. A self-calibration presents the artifact's true shape with relative scale and orientation.

In Figure 24.7, the uncalibrated artifact became the agent to obtain artifact view congruency via transitivity. On the other hand, all tool graduation pairs inherited the initial pair's length (T_0 to T_1) in a transitive manner. The initial graduation pair is end points to a line segment fragment, a fragment of the tool that became the reference for all graduation pairs. Such a fragment can also construct the two-dimensional lattice shown in Figure 24.8.

The two pivot points (X) in Figure 24.8 represent the fragment's endpoints. The pivot points are generated by three artifact views having two views with a 90° rotation relative to the first and one translated by one graduation unit in x for this example [8]. Starting with the two pivot points (A and B in Figure 24.9), observe the lattice formation when alternating between the two 90° motions.

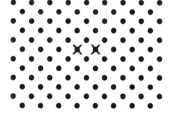

FIGURE 24.8
Two 90° rotations about two different pivot points (represented by "X") give rise to a discrete fixed-point lattice.

FIGURE 24.9
Lattice formation. (Left) The fragment between A and B is rotated 90° about A, forming the point at B'. B could also get to B' by a 180° motion about C. C is a composite pivot point generated by the sum of the two 90° motions. Rotating 90° about B forms the points (A', B'', and C') shown on the right. The continuation of alternating between the two 90° motions forms the lattice in Figure 24.8.

 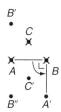

Metrology of Image Placement

This fragment metaphor is a tool for concept building; nevertheless, the lattice is a requirement for two-dimensional calibration. Methods attempting a self-calibration based on rotations about a single pivot point cannot guarantee a calibration; a single pivot point is incapable of generating the required lattice.

The $x-y$ metrology series employs a self-calibration method called "combined corrections." When "combined corrections" is used with different plate sizes on the above lattice prerequisite is maintained [9]. Prior to the $x-y$ metrology series, tool reversal methods consisting of rotations about a single pivot point were used; unfortunately, those methods could not guarantee a calibration.

In addition to "combined corrections" there are a number of self-calibration solutions [10,11] for $x-y$ metrology and, perhaps, additional ones are under development.

24.3 Analysis

Data collected from $x-y$ metrology preserves the spatial location of each measurement. The notion of having data associated with location gives rise to a family of analyses that exploits the informational asset imbedded in location. The subsequent sections cover two analyses' methodologies that apply location as a prime factor in the study of lithographic spatial components.

24.3.1 Regressions

An application of regression analysis on $x-y$ metrology data is a spatial study of any relationship that may have some physical meaning for process control. Figure 24.10 presents $x-y$ metrology data in the form of a positional error vector (PEV) representing an image placement relative to its intended design location.

A two-dimensional PEV field expression by a power series of the form:

$$\text{PEV} = c_0 f_0 + c_1 f_1 + \cdots + c_m f_m + \text{residue} \tag{24.1}$$

where f represents the linearly independent variables and c the coefficients. The coefficients are usually unknown. A method of *least squares* [12] solves the coefficients needed to examine data and perhaps draw meaningful conclusions about any dependencies. Components not represented by the coefficients are generally classified as residues.

Rewriting the general form of Equation (24.1), where f is a matrix of the two-dimensional (X, Y) linearly independent variables as shown in Equation (24.2):

$$\text{PEV} = \sum_{i=0}^{n} \sum_{j=0}^{i} c_k X^{i-j} Y^j + \text{residue}, \text{ where } k = j + \sum_{m=0}^{i} m \tag{24.2}$$

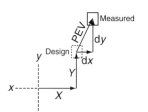

FIGURE 24.10
PEV is the distance vector of a measured feature relative to its design location.

In Equation (24.2), n represents the regression order, where X and Y are the feature's design and c's are the unknown coefficients. Equations (24.3a) and (24.3b) demonstrate an example of the PEV's first-order dx and dy components:

$$dx = a_0 + a_1 X + a_2 Y + \text{residue} \qquad (24.3a)$$
$$dy = b_0 + b_1 X + b_2 Y + \text{residue} \qquad (24.3b)$$

A second-order example is shown in Equations (24.4a) and (24.4b):

$$dx = a_0 + a_1 X + a_2 Y + a_3 X^2 + a_4 XY + a_5 Y^2 + \text{residue} \qquad (24.4a)$$
$$dy = b_0 + b_1 X + b_2 Y + b_3 X^2 + b_4 XY + b_5 Y^2 + \text{residue} \qquad (24.4b)$$

Let d represent the PEV data, f represent the matrix of linearly independent variables, and c the unknown coefficients as in:

$$cf = d \qquad (24.5)$$

Solving for c:

$$c = \left((f^T f)^{-1} f^T d \right)^T \qquad (24.6)$$

where superscript T denotes matrix transpose. Also computed are the vectors of residues (rms).

The c coefficients a_0 through a_n for the dx equation and b_0 through b_n for the dy equation render some of the following common lithographic shapes supporting physical interpretation shown in Figure 24.11.

Figure 24.11 contains the shapes found in the first-order coefficients of Equations (24.3a) and (24.3b). *Translation* is a distance vector shift of the field to a coordinate system that is parallel to the previous co-ordinate system. *Scale* is a proportional factor of one unit in terms of another. *Shear* is a nonorthogonal component relative to right angles. *Rotation* is an angular displacement of the field to a coordinate system that shares a common pivot point with the previous Figure 24.12 shows some common higher-order shapes. The shapes in Figure 24.11 and Figure 24.12 are only a subset from Equation (24.2).

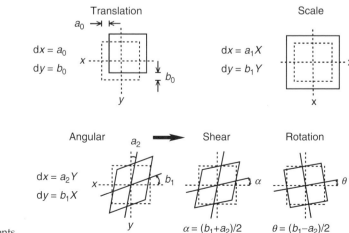

FIGURE 24.11
First-order components.

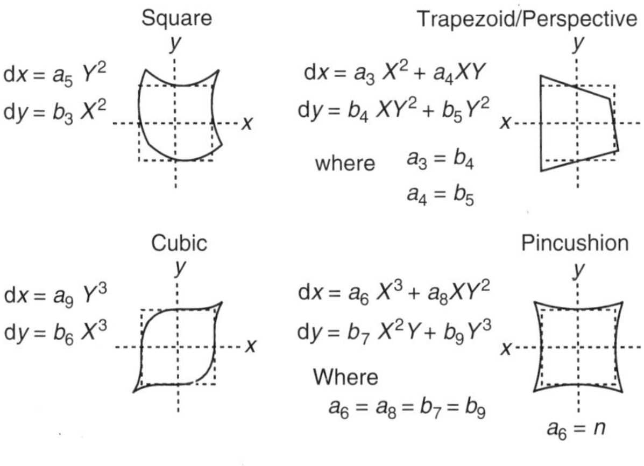

Metrology of Image Placement

Square
$dx = a_5 Y^2$
$dy = b_3 X^2$

Trapezoid/Perspective
$dx = a_3 X^2 + a_4 XY$
$dy = b_4 XY^2 + b_5 Y^2$
where $a_3 = b_4$
$a_4 = b_5$

Cubic
$dx = a_9 Y^3$
$dy = b_6 X^3$

Pincushion
$dx = a_6 X^3 + a_8 XY^2$
$dy = b_7 X^2 Y + b_9 Y^3$
Where
$a_6 = a_8 = b_7 = b_9$
$a_6 = n$

FIGURE 24.12
Second- and third-order components.

Figure 24.13 illustrates a second-order regression example on the $x-y$ metrology data in light gray. In Figure 24.13, the deltas between the second-order fit and data are the residues. The residues are the remaining components not included in the regression coefficients. Usually residues are assumed random normal; however, in cases where residues include nonrandom events, an empirical probability [13] should be considered.

The empirical probability is simply the cumulative function of the data distribution. That is, the smallest data point has a rank-order of $r = 1$, the next largest data value has a rank of 2, and so on. The largest sample value has a rank of n, where n is the sample size and each observation is the rank divided by the sample size (r/n).

The distribution and empirical probability curves shown in Figure 24.14 were taken from mask image placement data containing scale, shear, trapezoid, and random components commonly found in mask lithography. The 3σ value on the distribution is 25.4 nm. However, the 0.997 empirical probability happens to be 30.0 nm. Note that the shape of the distribution is nonnormal.

The empirical probability is calculated from the data, rather than imposing a parametric model on the data. Using a normal model to calculate $n\sigma$ values on lithographic data may cause incorrect results. The normal model is based on the assumption of independent random data. A lithographic PEV field usually contains systematic (nonrandom) data in addition to random data. To treat the field as if it contains only random data may result in incorrect $n\sigma$ values when nonrandom data are present. The empirical probability of 0.997

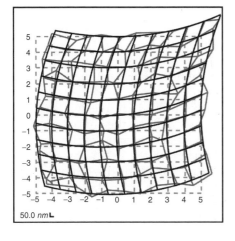

FIGURE 24.13
A second-order regression (dark curve) on $x-y$ metrology data in gray.

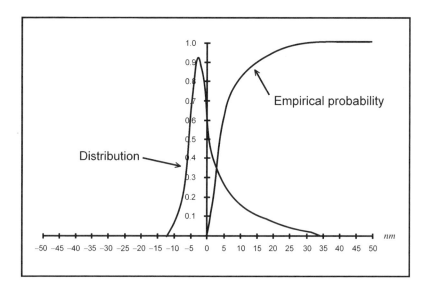

FIGURE 24.14
Empirical probability.

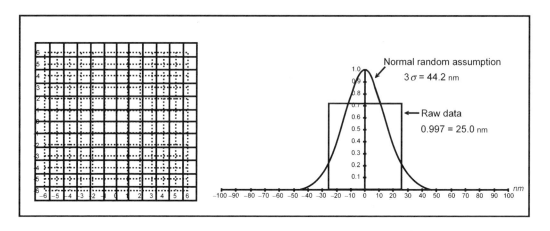

FIGURE 24.15
Comparing random normal assumptions to empirical probability on magnification.

about the mean is, in some cases, similar to a 3σ value. That is, if the data are purely random normal, the 0.997 empirical probability will converge to a 3σ.

The example shown in Figure 24.15 contains a population of PEVs having a magnification component (a common lithographic component).

A 3σ calculation on the magnification data shown in Figure 24.15 results in a 3σ value of 44.2 nm. A 3σ of 44.2 nm implies the raw data population follows the normal distribution shown in Figure 24.15. The square distribution is the actual raw data distribution and from the curve, not one data point exceeds 26.0 nm (25.0 nm @ $0.997p$ < 44.2 nm @ 3σ).

Multiple events having random normal distributions are added in a root mean square (rms) operation. In Figure 24.16, the two wave components have a 3σ of 44.5 nm each. The summation of the assumed random distributions results in 62.9 nm 3σ. The result distribution of the actual vector addition turns out to be zero.

Metrology of Image Placement

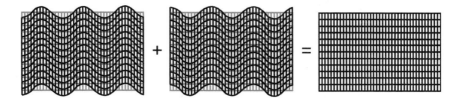

FIGURE 24.16
Vector addition of a wave to its inverse.

As the mask industry moves closer to its ideal objective, traditional practices relying on statistical assumptions need to be reevaluated.

Regression analysis is a viable tool for the study of lithographic systematic components; however, limits to regression analysis occur when examining higher frequency events. In the next analysis section, we shall review analyzing $x-y$ metrology data in the frequency domain.

24.3.2 Fourier Transforms

Data collected from $x-y$ metrology consisting of uniform discrete measurements of mask features can exploit an analysis tool used in many fields of study. The discrete Fourier transform (DFT) [14] and its inverse comes from the Fourier integral of a two-dimensional continuous function of $I(x,y)$ given by Equation (24.7):

$$F(u, v) = \int_{-\infty}^{\infty} \int_{-\infty}^{\infty} I(x, y) e^{-2\pi i (ux+vy)} dx\, dy \qquad (24.7)$$

and its associated inverse transform by Equation (24.8)

$$I(x, y) = \int_{-\infty}^{\infty} \int_{-\infty}^{\infty} F(u, v) e^{2\pi i (ux+vy)} du\, dv \qquad (24.8)$$

The above is the two-dimensional extension of the well-known one-dimensional Fourier transform pair dealing with a continuous function $I(x,y)$. The Fourier transform provides a mathematical model for many of the seemingly complex image placement shapes found in lithographic systems. It is also useful in many fields of study where the frequency content, or spectrum, provides information that enables one to draw conclusions about the data's origin or characteristics; in addition, the transform is deterministic. That is, there is no loss of information in the Fourier coefficients as compared to the potential loss of information found in regression coefficients. In other words, there are no residues in a Fourier solution.

Transforming the above continuous function to a discrete frequency domain defines the PEV's dx and dy components in Equation (24.9)

$$\begin{aligned} dx(x, y) &= \frac{1}{N} \sum_{u=0}^{N-1} \sum_{v=0}^{N-1} Fx(u, v) e^{\frac{2\pi i}{N}(ux+vy)} \\ dy(x, y) &= \frac{1}{N} \sum_{u=0}^{N-1} \sum_{v=0}^{N-1} Fy(u, v) e^{\frac{2\pi i}{N}(ux+vy)} \end{aligned} \qquad (24.9)$$

The $F(u,v)$ notation represents the Fourier coefficient at some u,v location corresponding to a graduation marker on the frequency plane where the translational component (zero frequency) is at the center, having the highest frequencies at the periphery.

A condition of the Fourier transform is that the x and y components of $I(x,y)$ range from minus to plus infinity. The PEV field must meet the continuous requirement or discontinuities may arise at the grid's boundary creating undesirable spatial frequencies. To meet the continuous requirement on a finite field, Equations (24.7) and (24.8) are actually the cosine transform that is a simple outgrowth of the Fourier transform. The cosine transform meets the continuous requirement by representing a periodic symmetric version of the spatial field as shown in Figure 24.17.

The dotted perimeter of the image to the left in Figure 24.17 represents the outermost column row set on an $N \times N$ grid. The image to the right is a field showing the cosine periodicity of the left image. The dark line boundaries in the right image are a pitch/2 distance from the dotted lines. This separation gives a pitch distance between the sampling intervals across the periodic boundaries. Notice the $2N \times 2N$ image is periodic when traveling continuously clockwise or counterclockwise about the center.

The metrology informational content is a function of the sampling interval (pitch) that determines how accurately a set of discrete measurements represents the original field. The following describes the conversion of a single row (k) of PEVs to the frequency domain. Given the following y shape (dashed curve) results in a set of discrete dy PEV vectors as in Figure 24.18.

The curve in Figure 24.18 is known as a *sinusoid* and is written as Equation (24.10):

$$dy(k) = A \cos(\omega k + \phi) \tag{24.10}$$

Fourier claims that any waveform (the y displacement in our example) could be represented by an infinite series of sinusoidal functions. The same is true for any x displacement along a single column. Furthermore, we can extend it to all rows and columns representing a two-dimensional waveform.

In Equation (24.10), variable A is the *amplitude* or size of the dy vector, ω is the *angular velocity*, and ϕ is the starting point in relation to the origin ($k = 0$). The starting point ϕ is also known as the *phase* of the sinusoid.

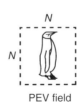
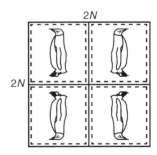

FIGURE 24.17
A nonperiodic field converted to a periodic field about the center.

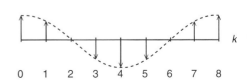

FIGURE 24.18
Single row having a sinusoidal PEV shape.

Metrology of Image Placement 543

Using Euler's formula to write $dy(k)$ as in Equation (24.11):

$$\begin{aligned} dy(k) &= Ae^{i(\omega k + \phi)} \\ &= A(\cos(\omega k + \phi) + i\sin(\omega k + \phi)) \\ &= a + ib \end{aligned} \tag{24.11}$$

where a is real, and ib is imaginary

Equation (24.11) is called a *phasor* and is interpreted as a *dot* that moves around a circle of radius A in the complex plane. In this example, Equation (24.10) contains the real part of a phasor with amplitude A having a counterclockwise ω velocity. Starting at the zero sample on the spatial plane a trace of the phasor's real axis is projected as the dot travels on the circumference in the complex plane shown in Figure 24.19.

In Figure 24.19, the complex plane is shown next to the spatial plane, and the phasor starts on the right side of the real axis by setting $k = \phi = 0$. In Figure 24.20, the phasor traces the dy PEV as sampled in the spatial plane.

When k equals one, the phasor moves to a location on the circle that projects a magnitude on the real axis that is equal to the observed PEV at graduation marker labeled one. The phasor continues traveling around the circle tracing the sampled points.

Alternatively, the example's sinusoid could have been the "sin" function resulting in a magnitude A along the imaginary b axis. The same dy sinusoid shape would result having an appearance of a shift right or left relative to the "cos" function. The real and imaginary components together constitute different waveform shapes.

In Figure 24.20, the phasor traced a spatial component; conversely, the inverse phasor can render a spatial PEV shape. In Figure 24.21, the inverse phasor constructs a translational PEV component when the angular velocity is zero.

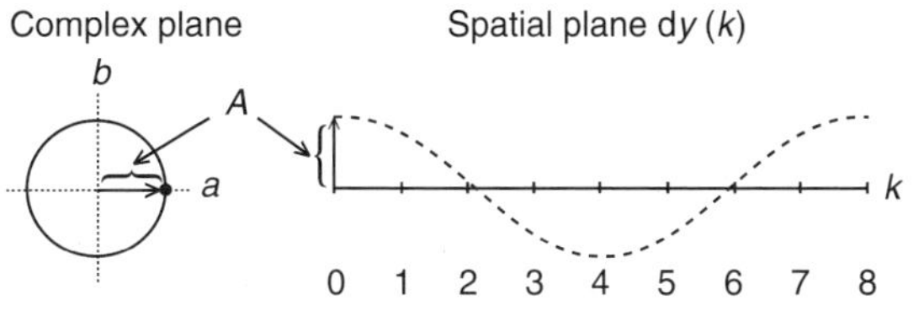

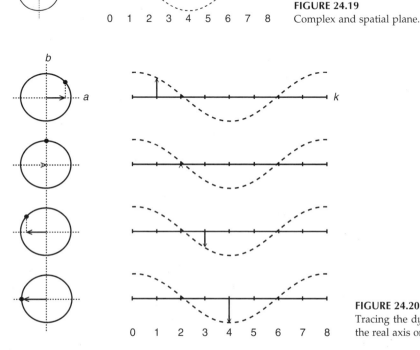

FIGURE 24.19
Complex and spatial plane.

FIGURE 24.20
Tracing the dy spatial magnitude on the real axis on the complex plane.

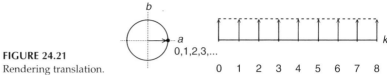

FIGURE 24.21
Rendering translation.

With a zero angular velocity, the d*y* PEVs result in a translation of magnitude *A*. On the other hand, Figure 24.22 shows the phasor constructing a PEV shape at its maximum angular velocity of π.

The maximum velocity represents a zigzag error, often found in stage subsystems. The maximum velocity is limited to the pitch unit between the graduation markers. That is, the greatest change in the shortest distance a PEV with magnitude *A* is from $+A$ to $-A$ in one pitch unit. This maximum velocity equates [Equation (24.12)] to what is known as the *Nyquist frequency* [15]:

$$\text{frequency} = \omega/(2\pi T) \text{cycles/unit} \tag{24.12}$$

where $|\omega| \leq \pi$ and the highest frequency represented by discrete sampling is $1/2T$ cycles/unit, where *T* is the physical unit corresponding to a sample interval or the feature spatial pitch distance. The frequency $1/2T$ cycles/unit is called the *Nyquist frequency* after Harry Nyquist who studied telegraph transmission in the 1920s.

For higher frequencies between graduations, more sample points are needed.

Since the phasor dot's path of motion is circular, its velocity is expressed as an angular velocity in radians per unit.

In Figure 24.23, a speedometer representation displays the phasor's positive angular velocity counterclockwise and clockwise for negative velocity. The velocity range is from zero to $\pm \pi$.

The speedometer's circumference is marked in divisions that equal the number of graduation markers on row *k*. The circle is sliced at π and stretched to form a straight-line as in Figure 24.24.

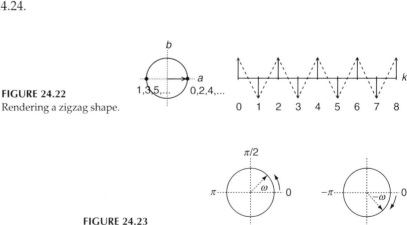

FIGURE 24.22
Rendering a zigzag shape.

FIGURE 24.23
Phasor angular velocity.

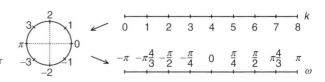

FIGURE 24.24
Spatial graduation markers to angular velocity notation.

Metrology of Image Placement 545

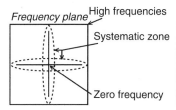
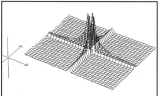

FIGURE 24.25
The lithographic systematic zone in the frequency domain.

Continuing the above scenario on all rows and columns forms a two-dimensional frequency plane.

Each graduation location on the frequency plane represents the angular velocity, and the coefficients at a graduation are the spatial magnitude (in nanometers for example) of the real and imaginary components of Equation (24.11) for the x- and y-axes. Equation (24.12) converts angular velocity to frequency.

The DFT analysis can aid in isolating lithographic systematic events (LSEs). In the frequency plane, the LSEs are normally concentrated along the center rows and columns as shown in Figure 24.25.

However, there are exceptions where the LSE may have patterns throughout the frequency plane. Generally, these patterns are easily recognized by their magnitude relative to the background. Random noise is typically in the background spread uniformly across the plane at relatively low magnitude compared to the LSE. The right view in Figure 24.25 displays the spatial power spectrum in the frequency domain showing a distinct systematic field along the center rows and columns.

Figure 24.26 is an analysis using the inverse Fourier transform as a filter attempting to decouple systematic from random events. The IDFT filter window (upper right window containing the inverse discrete Fourier transform) is operating on the coefficients from the DFT window along the center rows and columns to construct the systematic window (lower center). The residue is the difference between the view and systematic windows.

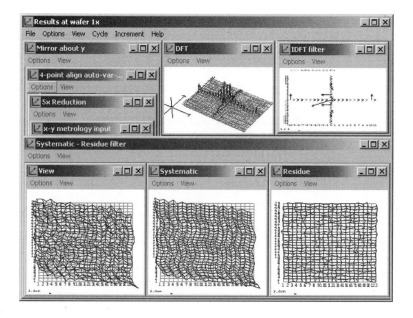

FIGURE 24.26
A Fourier filter decoupling systematic from random events.

24.4 Sampling

The ideal metrology system should sample all features from the aerial image of the entire mask in a fraction of the time that it took to write the mask. Perhaps in a future handbook, discussions will ensure about such a system, but today sampling is a function of frequency, time, economics, and objectives.

Frequency is the resolution of the sample space, the higher the spatial frequency the larger the sample size and greater the information content.

Time is how long it takes to complete a measurement task. Depending on the spatial frequency a task may take days to complete. Under long running tasks, the metrology tool may drift where registration measurements are needed to check and compensate for drift during data analysis.

Economics is the value added from the information of the sample.

Objectives balance the above three to optimize the sampling plan for a given task.

The subject of sampling as it relates to $x-y$ metrology is larger in scope, and within the context of this writing, the focus will be on a subset of these strategies.

24.4.1 Characterization

Sampling strategies for tool characterization or exploratory work often require large volumes of data and typically favor the Nyquist sampling criteria. One can exploit a rich set of tools when treating the datum as a two-dimensional waveform (see Section 24.3.2). Masks having evenly spaced features in an array, such as memory technology, are optimally suited for this strategy.

24.4.2 Dispositioning

Dispositioning poses a classic problem in sampling. In a manufacturing environment a binary decision must be made on the quality of a mask. The number of features on a mask can be in the order of the ratio of the size of the written area divided by the minimum linewidth. This number could easily exceed 10^{10}. Current $x-y$ metrology technology is not capable of measuring as many features to provide an absolute certainty in a binary decision. The following sections outline some binary dispositioning methods for $x-y$ metrology.

The measured features are usually indirect metrology targets or part of the active device element. A typical sample size may range from 15 to 200 plus measurements. Dispositioning may be a combination of limits set on the mean, n-sigma, maximum, or range.

24.4.2.1 Many Point Alignment

Many point alignment uses a first-order regression (see Section 24.3.1) that fits the measured data to the design grid by removing translation and rotation. In some cases, the scale and shear are also removed.

24.4.2.2 n-Point Alignment

The n-point alignment is similar to the many point alignment by removing translation and rotation via n-points relative to the design grid. The n-points are usually the stepper alignment targets used during the projection printing of the mask pattern onto the wafer. In some cases, the scale may be adjusted via the n-points to simulate an auto-variable-magnification feature found in steppers.

24.4.2.3 Systematic Inclusion

The systematic inclusion is a method that may incorporate any of the above alignment strategies with the inclusion of systematic events convoluted with random assumptions predicting the image placement behavior on the unsampled population [16]. The sampling strategy must be uniform across the working area of the mask, avoiding any targets outside this area.

The systematic coefficients may come from a pattern generator characterization repository or from the measured data using a regression to represent a systematic field at a given resolution.

24.4.2.4 Adaptive Metrology

Adaptive metrology [17] incorporates a sampling strategy that is a function of the data from the previous sample. As exemplified in Figure 24.27, the first sample is a uniform low frequency scan across the mask. The next iteration samples the mask at a higher frequency about a region surrounding the largest vector from the previous sample. The process continues up to the Nyquist frequency.

24.4.2.5 Dispositioning as a Function of Application

This strategy employs a philosophy of dispositioning the mask as a function of its application.

For example, in Figure 24.28, mask B at 5× is aligned to mask A at 4× and the customer uses four box-in-box overlay targets in his overlay process control. The request is to overlay mask B relative to mask A using the same four box-in-box targets with auto-variable-magnification and to disposition a critical feature in all repeating device elements of mask B relative to mask A evaluating the results at the wafer 1× level. Any device

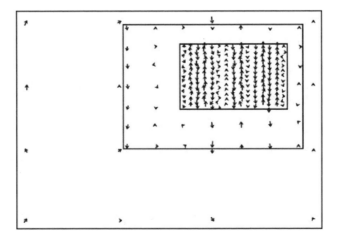

FIGURE 24.27
Adaptive metrology maintains a higher frequency sample about a maximum vector from a previous lower frequency sample.

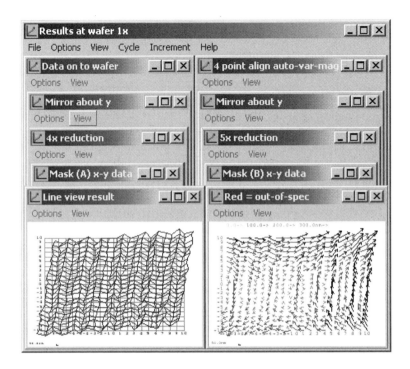

FIGURE 24.28
Dispositioning as a function of application an overlay example of mask B aligned to mask A.

overlay vector exceeding a given value is considered a failure for the latest produced mask B. If the mask passed, forward the field translation and rotational information relative to the four box-in-box targets for wafer overlay process control.

This methodology has a large number of combinations for mask dispositioning and may only be supported by captive mask facilities.

References

1. A.A. Michelson and E.W. Morley, On the relative motion of the earth and the luminiferous ether, *Am. J. Sci.*, 203 (November) (1887).
2. G. Schlueter, K.-D. Roeth, C. Blaesing-Banger, and M. Ferber, Next generation mask metrology tool, in: *Photomask and Next-Generation Lithography Mask Technology IX*, SPIE vol. 4754, 2002.
3. E.A. Lord and C.B. Wilson, *The Mathematical Description of Shape and Form*, Ellis Horwood Limited, Chichester, UK, 1984, pp. 13–37.
4. F. Klein, Vergleichende Betrachtungen uber neuere geometrische Forschungen, *Math. Annalen*, 43 (1903).
5. C.J. Evans, R.J. Hocken, and W. Tyler Estler, Self-calibration: reversal, redundancy, error separation, and absolute testing, *Annals CIRP*, 4 (5/2/1996) (1996).
6. M.R. Raugh Two-dimensional stage self-calibration: role of symmetry and invariant sets of points, *J. Vac. Sci. Technol. B.*, 15 (6; Nov./Dec.) (1997).
7. M.R. Raugh, *Self-consistency and Transitivity in Self-calibration Procedures*, Computer systems laboratory, Stanford University, 1991.

8. M.T. Takac, Jun Ye, M.R. Raugh, R. Fabin Pease, C. Neil Berglund, and G. Owen, Obtaining a physical two-dimensional Cartesian reference, *Am. Vac. Soc. B.*, 15 (6; Nov./Dec.) (1997).
9. M.T. Takac and J. Whittey, Stage Cartesian self-calibration: a second method, *SPIE BACUS*, 3546–3549 (1998).
10. M.R. Raugh, *Self-calibration of Interferometer Stages: Mathematical Techniques for Deriving Lattice Algorithms for Nanotechnology*, Stanford Library, ARITH-TR-02-01, March 2002.
11. Jun Ye, M.T. Takac, C. Neil Berglund, G. Owen, and R. Fabian Pease, An exact algorithm for self-calibration of two-dimensional precision lithography stages, *Precision Eng.*, 20, 16–32 (1997).
12. N.R. Draper and H. Smith, *Applied Regression Analysis*, John Wiley & Sons, New York, QA278.2.D7, 1981, pp. 218–221.
13. J. Weintraub, Introduction to probability plotting, *Microelectronics Manufacturing Technol.*, April (1991).
14. M.D. Levine, *Vision in Man and Machine*, McGraw-Hill, New York, 1985, pp. 260–266.
15. H. Nyquist, Certain topics in telegraph transmission theory, *Trans. AIEE*, 47, 617–644 (1928).
16. T.R. Groves, Statistics of pattern placement errors in lithography, *Am. Vac. Soc. B.*, 9 (6; Nov./Dec.) (1991).
17. Weidong Wang, *Adaptive Metrology and Mask Inspection*, Stanford Electronics Laboratories, SRC Contract No. MC-515, December, 1997.

25

Optical Thin-Film Metrology for Photomask Applications

Ebru Apak

CONTENTS
25.1 Introduction 551
25.2 Optical Techniques 552
    25.2.1 Reflection of Light 552
        25.2.1.1 Interaction of Light with a Single Interface 553
        25.2.1.2 Fresnel's Equations 553
        25.2.1.3 Interaction of Light with Multiple Interfaces 554
    25.2.2 Spectroscopic Ellipsometry 555
    25.2.3 Spectroscopic Reflectance 556
25.3 Optical Properties of Materials 556
    25.3.1 Optical Models 557
        25.3.1.1 Cauchy Model 557
        25.3.1.2 Lorentz Oscillator Model 559
        25.3.1.3 Other Models 560
25.4 Modeling of Photomask Thin Films 560
    25.4.1 Modeling Procedure 560
    25.4.2 Application Examples of Photomask Thin-Film Measurements 566
        25.4.2.1 Application 1 566
        25.4.2.2 Characterization of Thin-Film Layers 567
        25.4.2.3 Application 2 569
        25.4.2.4 Measurement Results 573
25.5 Conclusion 575
Acknowledgments 575
References 575

25.1 Introduction

The main drive for film thickness measurements comes from the desire in the microelectronics industry to achieve planarity across the wafer/mask regardless of the size and density of the patterns. Another important reason is that in order for the devices to function properly, the film layers have to be of a specific thickness. Currently, spectroscopic ellipsometry and spectroscopic reflectometry are the two primary techniques used for thin film metrology in mask fabrication. These measurement techniques allow the user not only to determine film thickness but also to obtain information regarding the optical properties of the materials. After characterization of the material, its thickness and index can be determined accurately. Unlike direct measurements, such as a profilometer, which subjects the sample surface to scratching by its stylus, optical methods have the advantage of being nondestructive. Technological advances in the developments of detectors, light sources, and powerful desktop computers all contribute to improve optical measurements and faster calculation results. This chapter is intended to provide insight into the details of film thickness measurements of photomasks using spectroscopic ellipsometry and spectroscopic reflectometry. The reader is first introduced to the theory of ellipsometry and reflectometry, then to the optical models used in determining film thickness. Finally, the modeling of two photomask examples is given.

25.2 Optical Techniques

Spectroscopic ellipsometry and spectroscopic reflectometry are optical methods that are nondestructive and indirect measurement techniques. Nondestructive from the fact that there is no scratching or damage to the sample surface, and indirect in that ellipsometers and reflectometers do not measure thickness and optical properties directly but measure amplitude change (Ψ), phase shift (Δ), and reflectance ($\Re$), from which thickness and optical constants are extracted by a theoretical model. Techniques, such as surface profiling (profilometer), use a diamond tipped stylus and measure the film thickness by moving the sample underneath the stylus between a coated and an uncoated regions. This method is based on the vertical movement of the stylus caused by surface topology. It can scratch and damage the surface. Additionally, the vibrations caused by stage movement and other noise in the environment all add up to measurement tilt errors. These errors are eliminated with optical techniques.

In order to understand the techniques of ellipsometry and reflectometry, it is necessary to study the properties of light and its interaction with interfaces.

25.2.1 Reflection of Light

Light is an electromagnetic wave, which can be described by Maxwell's equations. A complete description of electromagnetic theory is beyond the scope of this chapter. However, a brief overview is provided here. An electromagnetic wave has four field vectors; the electric field vector E, the magnetic field vector H, the electric-displacement density D, and the magnetic flux density B. To describe the polarization states of light, the electric field vector E is chosen due to the fact that in the course of interaction of light with matter, electrons experience a much greater force exerted by the electric field of the light wave compared to the magnetic field [1]. The magnetic and electric field vectors are perpendicular to each other and are not independent, both have a magnitude, which is a function of time and position. These two field vectors and the direction of propagation are orthogonal; the electric field vector and the direction of propagation are sufficient to define a plane wave. In three dimensions the solution to the wave equation using Maxwell's equations is:

$$\bar{E}(\bar{r}, t) = \bar{E}_0 \exp\left(\frac{-j2\pi\tilde{N}}{\lambda} \bar{q} \cdot \bar{r}\right) \exp(-j\omega t) \tag{25.1}$$

Optical Thin-Film Metrology

where $\bar{E}_0$ is the amplitude of the wave, $\bar{r}$ is the position vector, $\bar{q}$ is the unit vector along the direction of propagation, t is time, ω is the angular frequency, and $\tilde{N}$ is the complex index of refraction. The complex index of refraction is defined as $\tilde{N} = n + ik$, where n is the index of refraction and k is the extinction coefficient.

For more detailed understanding of the electromagnetic theory the reader is recommended to consult an introductory text on electromagnetic theory [2].

25.2.1.1 Interaction of Light with a Single Interface

Figure 25.1 describes Snell's law, which in its general form is:

$$\tilde{N}_1 \sin \theta_1 = \tilde{N}_2 \sin \theta_2 \qquad (25.2)$$

where $\tilde{N}_1$ and $\tilde{N}_2$ are complex index of refraction for the ambient and the film, respectively. The incident light arrives at the sample surface at an angle of θ_1 and is reflected off the surface at θ_3, while some portion of it is transmitted from the interface at an angle of θ_2. For the electric fields in both media to obey Maxwell's equations and boundary conditions, the incident, transmitted, and the reflected waves must all lie in the same plane, the plane of incidence. In Figure 25.2, the p- and s-directions are shown by two orthogonal vectors and indicate the corresponding polarization states. The p-direction lies in the plane of incidence, whereas s-direction is perpendicular to the plane of incidence. E^p and E^s refer to the amplitudes of the electric field parallel and perpendicular to the plane of incidence, respectively.

25.2.1.2 Fresnel's Equations

Fresnel reflection coefficients (r) describe the ratio of the amplitudes between reflected wave and incident wave. These parameters are derived from Maxwell's equations with

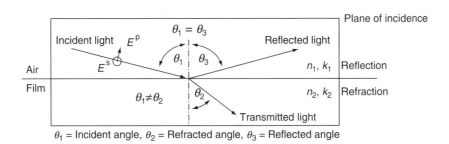

FIGURE 25.1
Propagation of light in the presence of a single interface.

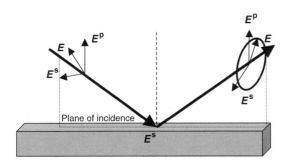

FIGURE 25.2
p and s polarization states of light in the plane of incidence. Linearly polarized incident beam converted to elliptically polarized reflected beam.

the condition that there is no discontinuity of the tangential components of the magnetic and the electric field at the interface [3]:

$$r^P = \frac{E_r^P}{E_i^P} = \frac{\tilde{n}_2 \cos\theta_1 - \tilde{n}_1 \cos\theta_2}{\tilde{n}_2 \cos\theta_1 + \tilde{n}_1 \cos\theta_2} \tag{25.3}$$

$$r^s = \frac{E_r^s}{E_i^s} = \frac{\tilde{n}_1 \cos\theta_1 - \tilde{n}_2 \cos\theta_2}{\tilde{n}_1 \cos\theta_1 + \tilde{n}_2 \cos\theta_2} \tag{25.4}$$

E_r is the amplitude of the reflected wave, E_i is the amplitude of the incident wave, p is the parallel, and s is perpendicular to the plane of incidence.

25.2.1.3 Interaction of Light with Multiple Interfaces

It is helpful to recall that interference phenomenon can be explained by Young's two-slit experiment. In this experiment, Young proved the wave-like behavior of light by letting a beam of light pass through a thin sheet with two slits and fall onto a screen. The waves radiated from each slit formed a pattern of bright spots and dark regions on the screen. The bright areas corresponded to the constructive interference of waves, creating maxima, whereas dark regions were the result of destructive interference creating minima. If the path difference between the two waves is an integer number of wavelengths, then constructive interference occurs, if half-integer then destructive interference occurs. Multiple layers with different index of refraction cause interference patterns depending on the constructive or destructive interference formed by the light reflected from the ambient–film interface and from the film–substrate interface.

In the presence of a second interface, some of the transmitted wave from the first interface upon arriving to the second interface is reflected off of the surface and the rest is transmitted. The reflected light from the second interface is headed towards the first interface and is reflected and transmitted again as shown in Figure 25.3. This phenomenon of reflection and transmission continues, and each partial wave created will be out of phase with a prior wave by the phase factor $e^{-j2\beta}$. The following are the Fresnel's equations for multiple interfaces [3]:

$$r_{12}^P = \frac{\tilde{N}_2 \cos\phi_1 - \tilde{N}_1 \cos\phi_2}{\tilde{N}_2 \cos\phi_1 + \tilde{N}_1 \cos\phi_2} \tag{25.5}$$

$$r_{23}^s = \frac{\tilde{N}_2 \cos\phi_2 - \tilde{N}_3 \cos\phi_3}{\tilde{N}_2 \cos\phi_2 + \tilde{N}_3 \cos\phi_3} \tag{25.6}$$

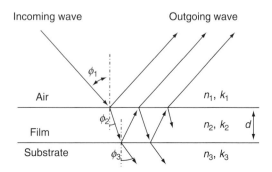

FIGURE 25.3
Propagation of light through different media.

Optical Thin-Film Metrology

$$r_{23}^P = \frac{\tilde{N}_3 \cos\phi_2 - \tilde{N}_2 \cos\phi_3}{\tilde{N}_3 \cos\phi_2 + \tilde{N}_2 \cos\phi_3} \qquad (25.7)$$

$$r_{12}^s = \frac{\tilde{N}_1 \cos\phi_1 - \tilde{N}_2 \cos\phi_2}{\tilde{N}_1 \cos\phi_1 + \tilde{N}_2 \cos\phi_2} \qquad (25.8)$$

The addition of reflected and transmitted waves creates an infinite geometric series from which total the reflection coefficient (R) is obtained. The total reflection coefficient is defined as the ratio of amplitude of total reflected wave to the amplitude of incident wave for p- and s-waves.

$$R^P = \frac{r_{12}^P + r_{23}^P \exp(-j2\beta)}{1 + r_{12}^P r_{23}^P \exp(-j2\beta)} \qquad (25.9)$$

$$R^s = \frac{r_{12}^s + r_{23}^s \exp(-j2\beta)}{1 + r_{12}^s r_{23}^s \exp(-j2\beta)} \qquad (25.10)$$

where β is phase change in the wave as it travels from top of the film to the bottom, and d is the film thickness:

$$\beta = 2\pi \left(\frac{d}{\lambda}\right) \tilde{n}_2 \cos\phi_2 \qquad (25.11)$$

Equations (25.5)–(25.11) contain the necessary quantities that are needed in order to calculate the film thickness d.

25.2.2 Spectroscopic Ellipsometry

Ellipsometry measures the changes in the polarization state of light reflected off the surface of a sample. These changes are monitored in the measured quantities Ψ and Δ, which relate to changes in the amplitude and the phase of the polarization states, respectively. Using two measured quantities provides extra information, which increases the accuracy and the sensitivity of the ellipsometer. Since the Fresnel's coefficients for the p- and s-waves are different, there already exists a phase shift between these two components, and upon reflection and refraction the phase relation between them will be changed:

$$\Delta = \delta_{(E^P - E^s)_{\text{before}}} - \delta_{(E^P - E^s)_{\text{after}}} \qquad (25.12)$$

where δ is the phase difference between the p- and s-waves, and Δ is the phase shift upon reflection.

There is also an amplitude reduction for both the s- and p-waves. $|R^P|$ and $|R^s|$ are the magnitudes of the amplitude reduction. The ratio of these two quantities is defined as:

$$\tan\psi = \frac{|R^P|}{|R^s|} \qquad (25.13)$$

where $\tan\psi$ is a real number. Also, a complex ratio of total reflection coefficients is defined:

$$\rho = \frac{R^p}{R^s} \quad (25.14)$$

$$\rho = \tan \psi \, e^{j\Delta} \quad (25.15)$$

From Equations (25.14) and (25.15) the fundamental equation of ellipsometry is obtained:

$$\frac{R^p}{R^s} = \tan(\Psi) e^{i\Delta} \quad (25.16)$$

In measuring complex films with multiple layers and very thin films, ellipsometry is the preferred method for characterization as it provides information about the phase and the amplitude changes of the p- and the s-waves. For example, in the presence of a thin film the intensity of the reflected light from the sample surface does not change much, whereas there is a noticeable change in Δ.

25.2.3 Spectroscopic Reflectance

Reflectance measurements can provide fast measurement results within small areas and work well for films that have a thickness greater than 35 Å. Reflectance is defined as the ratio of the intensity of the reflected light to the intensity of the incident light:

$$\mathfrak{R} = \frac{I_r}{I_i} \quad (25.17)$$

where I_i is the incident light intensity, and I_r is the reflected light intensity.

A reflectometer measures only the reflected light intensity, I_r. To determine I_i, one measures a reference sample with a known reflectance $\mathfrak{R}_0$. Then the reflectometer determines sample reflectance as:

$$\mathfrak{R} = I_r \frac{\mathfrak{R}_0}{I_{r0}} \quad (25.18)$$

The reflectance $\mathfrak{R}$ then depends on how accurately I_{r0} and $\mathfrak{R}_0$ are determined. It is critical to properly and regularly reference the reflectometer.

Since reflectance $\mathfrak{R}$ is defined as the square of the magnitude of total reflection coefficient R, the fundamental equation for reflectance becomes

$$\mathfrak{R} = |R^p|^2 = |R^s|^2 \quad (25.19)$$

From the above equation it is seen that there is no difference between the p- and the s-waves when measurements are done at normal incidence [3].

25.3 Optical Properties of Materials

For proper characterization, optical constants of materials have to be obtained over the spectral range of interest. Optical constants determine how a material will respond to excitation by an electromagnetic field (light) at a given frequency. They are expressed as a complex dielectric function, $\tilde{\varepsilon} = \varepsilon_1 + \varepsilon_2$ or as a complex index of refraction $\tilde{N} = n + ik = \sqrt{\tilde{\varepsilon}}$.

Optical Thin-Film Metrology

The refractive index n is the ratio of the speed of light in free space to the speed of light in a medium: $n = c/v$. The extinction coefficient k specifies how quickly the light is absorbed as the light intensity decreases during its travel through the material. The relation between extinction coefficient k and absorption coefficient α is:

$$k = \alpha\lambda/4\pi \tag{25.20}$$

In an absorbing medium as the incident wave travels through a distance of z, its amplitude decays exponentially by the equation:

$$I(z) = I_0 e^{-\alpha z} \tag{25.21}$$

After a certain distance called the penetration depth D_P, the intensity decreases to $1/e$ of its original value. The penetration depth is given by:

$$D_P = \lambda/4\pi k \tag{25.22}$$

For example, at a wavelength of 6328 Å, the D_P for silicon is approximately 3.1 mm, and for aluminum approximately 73 Å [3].

Unlike the name suggests, the optical constants are not constant values and vary as a function of photon energy or wavelength. This dependence on the wavelength is called dispersion and states that the material will exhibit different indices at different wavelengths. Many different dispersion models have been developed, and an appropriate dispersion model for a given material must be used to ensure accurate results.

The optical constants n and k arise from light-matter interactions, such as interband and intraband absorptions [4]. Interband absorption occurs when an electron in a bound state absorbs a photon and makes it jump to a higher energy level. This type of optical absorption is seen mostly for semiconductors, dielectric materials, and for metals, which also exhibit strong free carrier absorption. The other type of absorption, intraband absorption, occurs when an electron absorbs a photon and moves to a different energy state within the same band.

The index of refraction and extinction coefficient are not independent but related by Kramers–Kronig relations. These equations relate the real part to the imaginary part of the complex dielectric function or complex index of refraction through a form of Hilbert transforms [5]. If the extinction coefficient is known, then by using Kramers–Kronig relation the index of refraction can be calculated. In the regions where the index of refraction shows strong dispersion the absorption is also expected to be strong, and for regions of weak dispersion the absorption will also be low. In general, oscillator models are Kramers–Kronig consistent, while empirical models, such as the Cauchy equation, are not.

25.3.1 Optical Models

There is a wide variety of dispersion models developed for the ever-increasing number of material types. The most commonly used ones are the Cauchy model, Lorentz oscillator model, and Tauc–Lorentz oscillator model. In addition to these, many equipment and software manufacturers have proprietary models.

25.3.1.1 Cauchy Model

In the visible range, materials, such as dielectrics, show no absorption, i.e., they are transparent. Unlike conductors, dielectrics cannot support an induced current; hence, the waves travel through the material with very little light loss. In this region where the material is transparent, the refractive index decreases as the wavelength increases and the extinction coefficient k is essentially zero. In the visible wavelength range, the dispersion behavior of dielectrics can be expressed by a Cauchy function, which is an empirical relation and given by:

$$n(\lambda) = A_n + \frac{B_n}{\lambda^2} + \frac{C_n}{\lambda^4} \qquad (25.23)$$

where A_n, B_n, C_n are the Cauchy coefficients. The Cauchy relation assumes the extinction coefficient k to be zero at all wavelengths and is not Kramers–Kronig consistent. For materials with a large band gap, the resonant frequency ω_0 lies in the UV range. In this region, these materials demonstrate a strong absorption, and with decreasing wavelength the refractive index increases and then decreases. In the UV range where the extinction coefficient is not necessarily zero, the Cauchy model by itself is not sufficient to describe the dispersion of dielectrics. An absorption tail, such as the Urbach absorption, as shown in Equation (25.21) can be modeled into the Cauchy layer:

$$k(\lambda) = A_k \, e^{B_k(1.24(1/\lambda - 1/C_k))} \qquad (25.24)$$

where A_k is the absorption amplitude, B_k is the broadening, C_k is the absorption band edge fixed at the lowest wavelength of the measured data, and λ is the wavelength of light in microns.

Figures 25.4(a) and (b) show physically reasonable dispersion relations using the Cauchy relation; however, for Figure 25.4(b) one should model an absorption in the UV

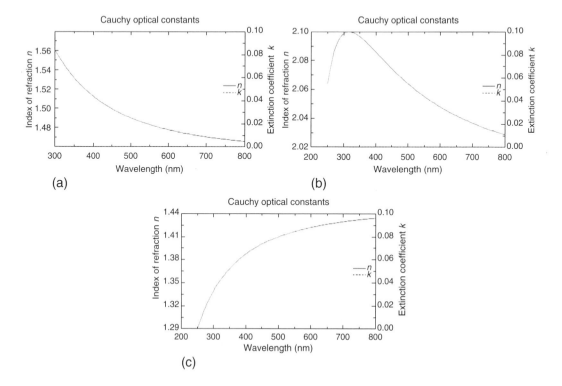

FIGURE 25.4

(a) Cauchy behavior showing index of refraction n versus wavelength, fitting to a Cauchy model. (b) Cauchy behavior fitting to a Cauchy model, however, needs an absorption tail in the UV. (c) Non-Cauchy behavior.

Optical Thin-Film Metrology

where the peak in the index of refraction curve is observed. The peak in the refractive index can be due to the strong scattering that may occur in the presence of multiple phases or grain boundaries, as well as dopants, that absorb light [5]. The most important point to note when modeling dielectrics using a Cauchy model is that the index of refraction should never increase with increasing wavelength when the extinction coefficient is zero, Figure 25.4(c).

25.3.1.2 Lorentz Oscillator Model

Dispersion behavior of metals can be described by a Lorentz oscillator model, which assumes absorption(s) has a Lorentz line shape. The Lorentz oscillator model is analogous to the response that a mass attached on a spring will give. The model is based on the assumption that the motion of electrons bound to the nucleus in the presence of an electric field produces the same response that the motion of a mass on a spring does when subjected to a force [4]. In the case of a harmonically driven mass, the model includes a term containing the acceleration of the mass, a term for viscous damping, the driving force, and a term for the restoring force originating from Hooke's Law, which includes the resonant frequency ω_0. For the Lorentz oscillator model, the mass is analogous to the electron, which is treated as a particle. The Lorentz oscillator model is formulated as:

$$\tilde{\varepsilon}(E) = \left[\tilde{N}(E)\right]^2 = \varepsilon(\infty) + \sum_{i=1}^{n} \frac{A_i}{E_i^2 - E^2 - iB_iE} \tag{25.25}$$

where $\varepsilon(\infty)$ is the value of the real part of the dielectric function at large photon energies, E is the photon energy in electron-volts, $\tilde{\varepsilon}$ is the complex dielectric function expressed as a function of the photon energy E, $\tilde{N} = n + ik$ is the complex index of refraction as a function of photon energy, n (in the summation sign) is the number of oscillators, A_i is the amplitude of the ith oscillator, B_i is the broadening of the ith oscillator, and E_i is the center energy of the ith oscillator [4]. Typically metals require 2–7 oscillators. Figures 25.5(a) and (b) are generated by using a single and a two-oscillator Lorentz models, respectively. The Lorentz oscillator model is Kramers–Kronig consistent.

Another form of the Lorentz oscillator model is the Drude model, where the electrons are free rather than being bound to the nucleus. This model has a single oscillator with

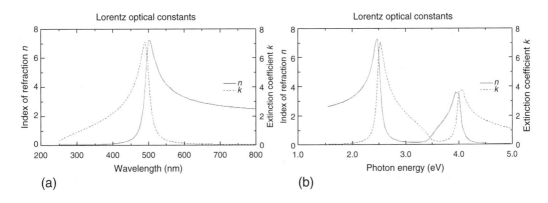

FIGURE 25.5
(a) Optical constants n and k versus wavelength fitting to a single oscillator Lorentz model, Kramers–Kronig consistent. (b) Optical constants n and k versus wavelength fitting to a two-oscillator Lorentz model, Kramers–Kronig consistent.

zero energy and is useful for modeling metals, conducting oxides and doped semiconductors [4].

25.3.1.3 Other Models

In the visible range, some semiconductors can be modeled using the Cauchy relation with the Urbach absorption. However, most of the time, especially in the UV, semiconductors require more sophisticated models. One of the models used for semiconductors, which is also Kramers–Kronig consistent, was developed at J.A. Woollam Co., Inc. The downside of this model is that one has to have some experience as it contains many fit parameters, which may correlate with each other. The parametric model is a more sophisticated version of the Lorentz oscillator model, which allows a higher number of oscillators to be used. This model is also used for other materials besides semiconductors, such as nitrides and oxynitrides.

Besides the above models, there are many more model types, effective medium approximation (EMA), alloy models, models for graded layers, etc. For example, the optical constants of the film layer can be a mixture of optical constants of two known materials. In this case, an EMA model is used to describe the optical constants of the film in interest. Spin-on-glasses, which can be defined as a mixture of silicon dioxide and voids, and oxynitrides, a mixture of oxide and nitride can be modeled using an EMA model. For a polysilicon film an alloy model can be used, which contains amorphous silicon fraction and determines the degree of crystallinity. Alloy models contain the fit parameter alloy fraction to describe the varying compositions of the films. Besides polysilicon, it can be used for a ternary compound semiconductor, such as $Al_xGa_{1-x}As$. For graded films, which have varying optical constants from top to the bottom of the film, graded models are mostly used. Graded models divide the film into homogeneous layers also known as slices and each of the slices has slightly different optical constants.

25.4 Modeling of Photomask Thin Films

Most binary photomasks have a common film stack: the substrate material is quartz, with 700–800 Å of chrome and 200–300 Å of chrome oxide. For most thin film applications, the top most layer is a photoresist. There can be more layers on and underneath the resist, such as BARC, ARC, etc. (Figure 25.6). In almost all unpatterned photomasks, the chrome layer is thick enough to be treated as the substrate. The penetration depth, D_P of chrome at a wavelength of 633 nm is approximately 116 Å, the thickness the intensity of light has dropped to 37% of its original value. At a thickness of four times the depth of penetration, the light intensity has already dropped to 2% of its original value. Figures 25.7(a) and (b) show the calculated Ψ and Δ values of thin chrome film on glass with a thickness range of 100–800 Å. As the thickness increases especially above 500 Å, it is seen that the spectral changes for 500 and 800 Å films are nearly identical. At a thickness of approximately 800 Å, chrome is opaque.

25.4.1 Modeling Procedure

For the film analysis done in this chapter J.A. Woollam's WVASE32 analysis software package is used. More information about the software can be obtained by contacting the

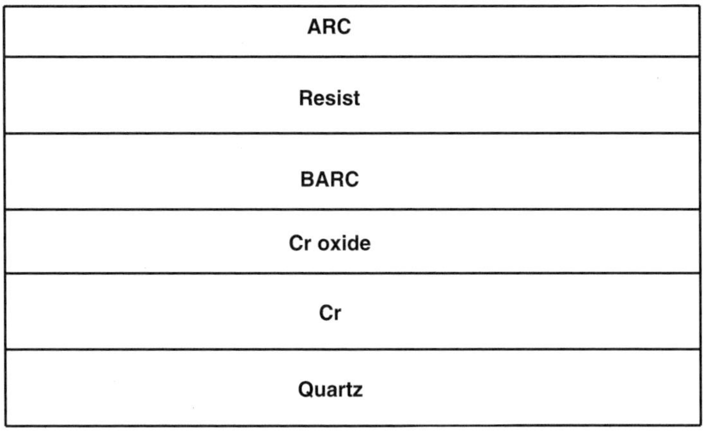

Optical Thin-Film Metrology

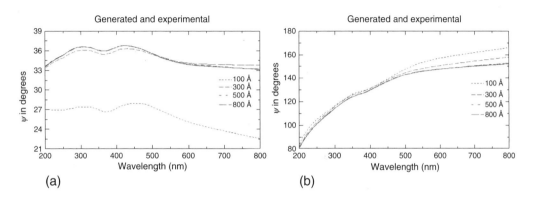

FIGURE 25.6
Film stack of a photomask sample.

FIGURE 25.7
(a) Simulated spectra, Ψ versus wavelength of thin chrome on glass, 100–800 Å. (b) Simulated spectra, Δ versus wavelength of thin chrome on glass, 100–800 Å.

J.A. Woollam Co., Inc., through the company website at www.jawoollam.com. The steps taken in modeling can be conveniently shown in a flowchart, Figure 25.8. In order for the metrology tool to make thickness measurements, an optical model must be built. The first step in thin film analysis is to collect data from the sample, which is normally done with an ellipsometer. However, it is also possible to use a reflectometer or a combination of both. After that, a model is constructed where the materials' optical constants and thicknesses are varied, so that the calculated (generated) data are tried to fit to the experimental data. Finally, the goodness of fit will determine the validity of the model. This section attempts to explain in detail the various steps for building a robust optical model for thin film metrology.

First, data, such as ellipsometry or reflectometry, or combined scan as a function of wavelength are collected for proper characterization of the films. It is important to remember that the optical tools, such as reflectometers and ellipsometers, do not measure the unknown parameters, thickness and optical constants directly, rather they measure reflected beam intensities and polarization states of light. This statement points to the fact that Ψ, Δ, and reflectance are measured quantities whereas thickness, refractive index, and extinction coefficient are calculated quantities. In order for the tool to measure

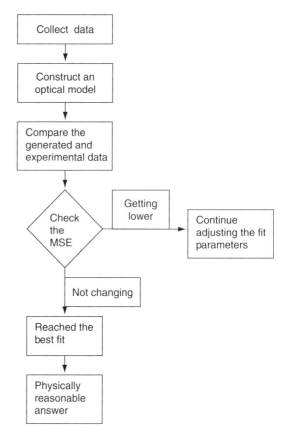

FIGURE 25.8
A flowchart for modeling procedure.

thickness and index, a mathematical model must be incorporated into the measurement tool's software, so the spectroscopic information can be used in determining the physical properties of interest based on a predefined model.

The model is constructed according to the film stack information as determined by the operator incorporating the various kinds of models stated in the previous section. For example, the chrome layer can be modeled using a Lorentz oscillator model, for photoresist layers a Cauchy model can be used.

After a model is constructed, the operator can calculate the SE and/or SR spectrum for their sample. By varying the fit parameters, i.e., thickness, Cauchy parameters, etc., of the various models used for each layer the calculated spectra can be made to "fit" the experimental spectra. A nonlinear regression analysis, such as the Levenberg–Marquardt multivariate regression algorithm, automates this process.

In order to reach a conclusion about the quality of the fit a maximum likelihood estimator called the mean-squared error (MSE) is used. MSE is a good indicator of how well the generated data (modeled data) fit the measured data (experimental data), and is given by the following equation [2]:

$$\text{Fit Value} = \text{MSE} = \frac{1}{N-M} \sum_{i=1}^{N} \left(\frac{y_i - y(x_i)}{\sigma_i} \right)^2 \quad (25.26)$$

where N is the number of Ψ and Δ pairs, M is the number of parameters in the model, the term in parenthesis is the difference between the calculated and measured values, and σ_i is the standard deviation associated with the term in parenthesis. If the modeled data and

Optical Thin-Film Metrology

the measured data match each other well, then the MSE takes a low value, and if the MSE is still high, then the process of adjusting the fit parameters should continue.

The final step consists of the establishment that the analysis has led to yield the best solution being both unique and realistic, and the fit parameters are not strongly correlated. If there is a parameter correlation, the model fit can still be very good, but in reality the model would converge to a wrong solution. For any indirect measurement the operator must always determine whether the calculated values are physically reasonable. To demonstrate this statement, a film stack consisting of multi-layers were modeled, Figures 25.9(a), (b), and Figure 25.10. These figures show good model fits to the experimental data; however, by looking at the optical constants of the film layers, it is seen that the solution is not realistic. The optical constants in Figure 25.11 were obtained by using a Cauchy model, but the dispersion relation deviates from Cauchy-like behavior. Also the nitride layer present in this film stack shows nonrealistic optical constants, Figure 25.12, due to the sharp drop in the extinction coefficient. Figures 25.13(a), (b), and Figure 25.14 show the same data modeled using the above model. The fit is as good as the previous fit and hence would not give any indication about the accuracy of the model. Only when the optical constants are considered, it is seen that the layer modeled using the Cauchy model exhibits the expected dispersion curve, Figure 25.15, and also the optical constants of the nitride layer, Figure 25.16, demonstrate a more realistic and smoother dispersion. This more accurate representation of the data was obtained by avoiding correlation among the

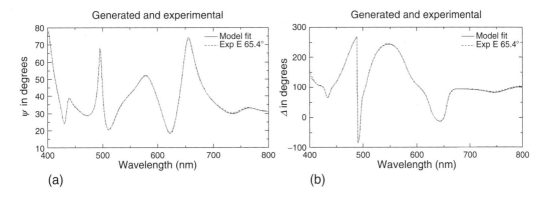

FIGURE 25.9
(a) Model fit to the ellipsometric data; Ψ values versus wavelength. (b) Model fit to the ellipsometric data; Δ values versus wavelength.

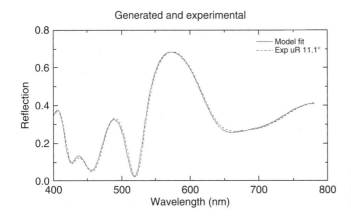

FIGURE 25.10
Model fit to the reflectometer data; reflection versus wavelength.

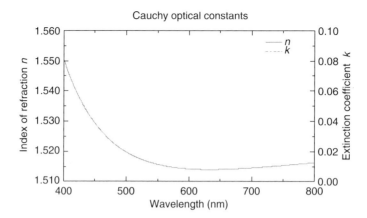

FIGURE 25.11
Non-Cauchy behavior obtained as a result of the above modeling.

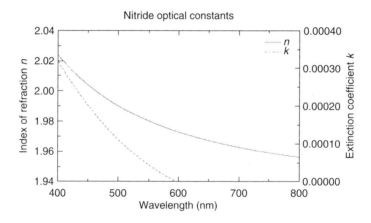

FIGURE 25.12
Nitride optical constants obtained from the above modeling, not real.

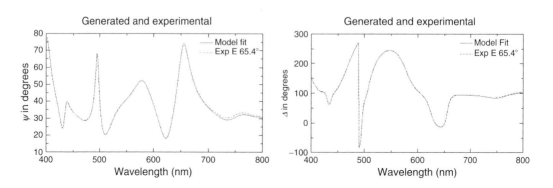

FIGURE 25.13
(a) Model fit to the ellipsometric data; Ψ values versus wavelength. (b) Model fit to the ellipsometric data; Δ values versus wavelength.

Optical Thin-Film Metrology

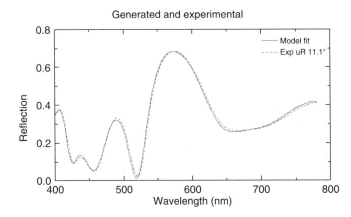

FIGURE 25.14
Model fit to the reflectometer data; reflection versus wavelength.

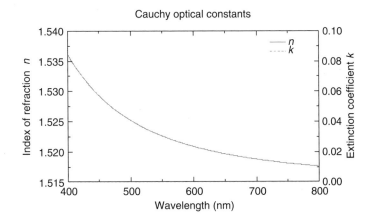

FIGURE 25.15
Cauchy behavior obtained from the above model.

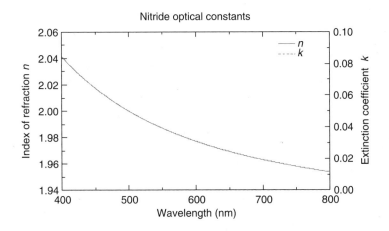

FIGURE 25.16
Nitride optical constants obtained from the above modeling.

fit parameters, i.e., not including as many parameters in the analysis. Hence, it can be said that not only the quality of the fit of the model is sufficient to determine the validity of the model but also optical constants need to be evaluated to see if they are yielding realistic dispersion relations.

As with any nonlinear regression, the Levenberg–Marquart multivariate regression algorithm can yield a wrong solution if it settles in a local minimum. To prevent this, a "global fit" can be performed on one or more of the model parameters, where a number of search points are defined in the parameter space [4]. Global fitting iterates the guess values at the specified limits at equal increments to seek out the lowest MSE, and hopefully the global minimum.

25.4.2 Application Examples of Photomask Thin Film Measurements

For good lithographic performance, film uniformity is very important. The film thickness measurements on photomasks enable the mask developer to monitor the film thickness across the mask, i.e., uniformity of the mask. A uniform mask can provide better control of feature sizes, i.e., less variation in CD. Below are application examples for thin film measurements of the photomask. The application requirements for the samples are to characterize the film layers and optical constants, and measure the film thickness.

25.4.2.1 Application 1

The first example is a typical photomask sample with the following film stack: photoresist/chrome oxide/chrome/quartz. The chrome is thick enough, ~1000 Å, to be treated as the substrate of the film stack. The measurement of the film with unknown optical constants consists of three parts: data collection, modeling of the data, and measurement on the production metrology tool.

25.4.2.1.1 Data Collection

As always the starting point is to collect data from the sample, which was done by performing a combined scan of spectroscopic ellipsometry and spectroscopic reflectometry on a single spot using a NANOmetrics' tool, Nanospec 8000 XSE. The Nanospec 8000 XSE incorporates an ellipsometer and a reflectometer into a single tool, Figure 25.17. The ellipsometer is a J.A. Woollam M-44 VIS rotating analyzer ellipsometer (RAE). The M-44 ellipsometer uses white light, dispersed onto a 44 silicon detector array measuring 44 wavelengths from 430 to 750 nm. The angle of incidence (AOI) for the ellipsometer is approximately 65°. The spectroscopic reflectometer uses normal incidence and two light sources, a halogen lamp and a UV deuterium lamp. It measures in the range of 200–800 nm. The combined scan of ellipsometry and reflectometry is helpful in model development of multifilm stacks with unknown optical constants; reflectometer information is simply not sufficient for accurate determination of optical constants due to possible parameter correlations in the model. The quantities obtained from an ellipsometer scan; Ψ and Δ contain information about the sample, which is essential in the characterization of the unknown optical constants. The Ψ contains the amplitude change after light is reflected from a surface, and Δ includes the phase shift, Equations (25.12)–(25.15).

25.4.2.1.2 Modeling of the Experimental Data

After the data collection a model was built. The reflectance and ellipsometer data were fit simultaneously to the same optical model. The modeled wavelength range for both data sets was 400–800 nm to best match the wavelength range of the M-44 ellipsometer. Additionally, since there is a photoresist layer in the mask, measuring it in the UV can

Optical Thin-Film Metrology

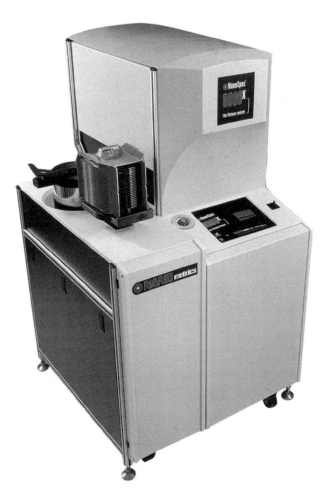

FIGURE 25.17
Nanospec 8000 XSE.

alter its optical properties. The model consisted of a Cauchy model for the resist material, an oscillator type model for chrome oxide and a Lorentz oscillator model for the chrome layer. After fitting the parameters associated with each film layer, the thicknesses of the layers were found to be around 5100 and 280 Å for the photoresist and chrome oxide, respectively, Figure 25.18. Figures 25.19(a), (b), and Figure 25.20 show the ellipsometer and reflectometer experimental data fitting to the model created above, respectively. As it can be seen from the model fits, the experimental and modeled data match well.

25.4.2.2 Characterization of Thin Film Layers

25.4.2.2.1 Photoresist

The processing conditions of photoresists affect the optical constants; therefore, each photoresist film is characterized before its thickness is measured. For this particular example, a Cauchy model was used to describe the dispersion behavior of the photoresist. As stated in Equations (25.20) and (25.21), this is an empirical relationship. A_n is a

2	Resist	5144.4Å
1	Cr$_2$O$_3$	281.41Å
0	Cr	1 mm

FIGURE 25.18
Modeling window, application 1.

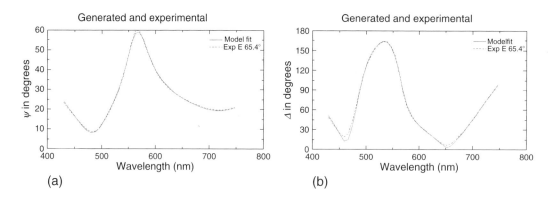

FIGURE 25.19
(a) Model fit to the ellipsometric data; Ψ values versus wavelength, application 1. (b) Model fit to the ellipsometric data; Δ values versus wavelength, application 1.

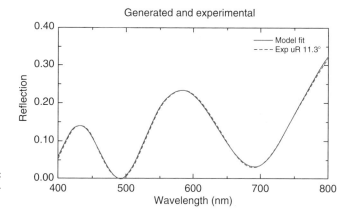

FIGURE 25.20
Model fit to the reflectometer data; reflection versus wavelength, application 1.

constant at long wavelengths, B_n specifies the curvature, and C_n controls the spectrum at shorter wavelengths [2]. After specifying initial values representative of photoresist, A_n, B_n, and C_n were adjusted using a nonlinear regression analysis. An absorption tail was also included which provided a better fit. The important point in the use of the Cauchy model is that the material has to be either transparent or possess a small absorption tail that can be described accurately using the Urbach absorption, Equation (25.21). Hence, in the wavelength range used, the Cauchy model was capable of determining the optical constants for this photoresist layer. The optical constants of this photoresist are shown in Figure 25.21.

25.4.2.2.2 Chrome Oxide

This layer is usually a very thin layer with a thickness varying between 200 and 300 Å. For modeling the chrome oxide layer, an oscillator type model with a single oscillator was used. This model type is Kramers–Kronig consistent. The optical constants are shown in Figure 25.22.

25.4.2.2.3 Chrome

The optical constants of metallic films are very sensitive to deposition conditions. Using the already published values (book values) of optical constants for metallic films is not going to produce good model fits and accurate thickness values. In order to model the

Optical Thin-Film Metrology

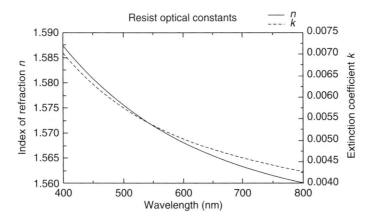

FIGURE 25.21
Optical constants of resist layer, application 1.

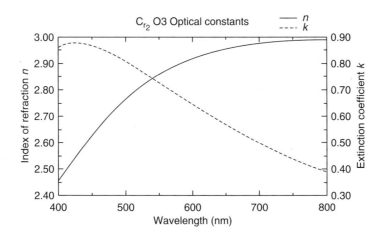

FIGURE 25.22
Optical constants of chrome oxide layer, application 1.

optical constants accurately, a Lorentz oscillator model was used. The model consisted of seven oscillators and the amplitude, broadening and energy parameters of the oscillators were fit. The optical constants of chrome are shown in Figure 25.23.

25.4.2.3 Application 2

In the second application a more complex film stack is analyzed. The photomask film stack is as follows: top anti-reflective/resist/bottom anti-reflective/chrome oxide/chrome/quartz. The chrome is approximately 1000 Å, which makes it thick enough to be the substrate of the film stack. As before, the same steps will be followed for the analysis of this film stack.

25.4.2.3.1 Data Collection

The first step is to collect data from the sample. This was done by performing a spectroscopic ellipsometric scan using variable angle spectroscopic ellipsometry, VASE

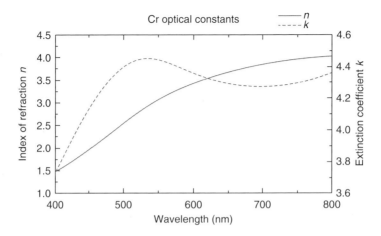

FIGURE 25.23
Optical constants of chrome layer, application 1.

(J.A. Woollam Co.). VASE measurements are powerful for multilayer film stacks as they provide information at multiple AOI over a wide range of wavelengths. By doing so, Ψ and Δ values are collected at every wavelength/angle combination. In this case 65°, 70°, and 75° were used as the AOI, and the scan range was 300–800 nm. The choice of using these angles were primarily due to the fact that most materials have their pseudo-Brewster angles around 65°. The AOI if near the pseudo-Brewster angle provides Δ values that are around 90° and are most sensitive to layer thickness and optical constants.

The VASE ellipsometer is a RAE, with a fixed polarizer, a monochromator placed before the sample, a beam chopper at the end of the monochromator, and a solid-state detector. VASE measurements assume ideal conditions, such as the beam entering the detector, is polarized completely, the interfaces between the films are parallel and flat, zero angular spread and bandwidth and uniform film. For nonideal situations, these options can be defined as fit parameters in the model and included in the nonlinear regression [4].

25.4.2.3.2 Modeling of the Experimental Data

After data collection a model was built. The model consisted of a Cauchy model for the resist material, and oscillator models to characterize the top anti-reflective layer (TARC), bottom anti-reflective layer (BARC), the chrome, and the chrome oxide layer. After fitting the experimental data to the model, the following values were found for thickness; 500 Å of top anti-reflective, 4700 Å of resist, 800 Å of bottom anti-reflective, and 270 Å of chrome oxide, Figure 25.24. Figures 25.25(a) and (b) show the spectra obtained at variable angles showing the model and data fits. The generated and the experimental data fit very well to each other.

25.4.2.3.3 Top Anti-Reflective Layer

The top layer in this sample is a clear-coated substance with an index of refraction around 1.34 at 633 nm. It is spun on the resist and acts as a top layer anti-reflective coating, which suppresses the reflections in the photoresist, and hence improves the image quality and reduces the linewidth variation. An oscillator type model was used to model this layer. A Cauchy layer could have been used if only visible wavelength range was modeled, but the use of the oscillator type model provided better description of the optical constants

Optical Thin-Film Metrology

4	Top anti-reflective	507.16 Å
3	Resist	4699.3 Å
2	Bottom anti-reflective	789.53 Å
1	Cr2O3	274.09 Å
0	Cr	1 mm

FIGURE 25.24
Model window, application 2.

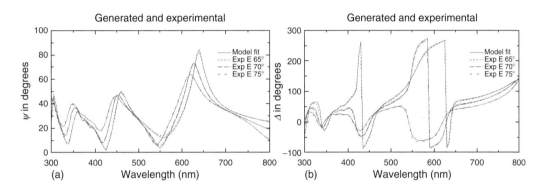

FIGURE 25.25
(a) Model fit to the ellipsometric data; Ψ values versus wavelength, application 2. (b) Model fit to the ellipsometric data; Δ values versus wavelength, application 2.

below 400 nm where absorption occurs, Figure 25.26. This oscillator type model is Kramers–Kronig consistent.

25.4.2.3.4 Bottom Anti-Reflective Layer

The use of bottom anti-reflective coating (BARC) has many benefits; the most important one is that it absorbs most of the light in order to prevent it from being reflected back by the substrate to the resist layer, it can minimize standing waves and improve image contrast, and planarize the surface before resist is deposited. For the BARC film layer, a

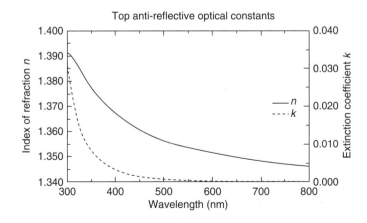

FIGURE 25.26
Optical constants of top anti-reflective layer TARC, application 2.

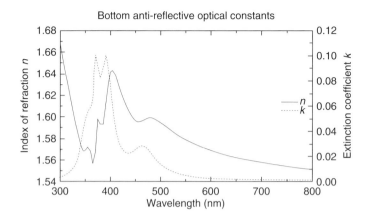

FIGURE 25.27
Optical constants of bottom anti-reflective layer BARC, application 2.

Lorentz oscillator model with seven oscillators was used. The dispersion relation of this film was found to be complex as it exhibited great amount of detail especially between 340 and 410 nm, Figure 25.27. The use of a Cauchy model would have been sufficient only for the transparent region; however, below 430 nm it is not able to describe the complex dispersion relation. In this region the material shows sharp absorption peaks. The Lorentz oscillator model is also Kramers–Kronig consistent. Thickness and parameters, such as broadening, amplitude, and energy, were included in the analysis, and attention was paid to avoid strong correlation between fit parameters.

25.4.2.3.5 Photoresist, Chrome Oxide, Chrome

These layers were modeled in a similar way to the previous example. A Cauchy model was used for the resist and oscillator type models for the chrome oxide and for the chrome. The optical constants of these layers can be found in Figures 25.28–25.30.

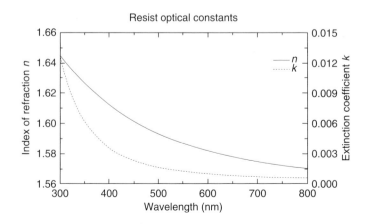

FIGURE 25.28
Optical constants of photoresist layer, application 2.

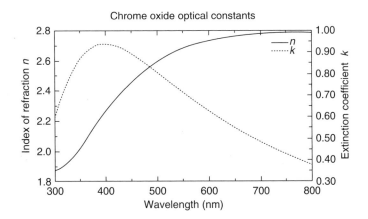

FIGURE 25.29
Optical constants of chrome oxide layer, application 2.

25.4.2.4 Measurement Results

After the sample characterization and model development, both samples were measured on a Nanospec 6100, Figure 25.31. Nanospec 6100 is a noncontact spectroscopic reflectometer tool with a linear array head and vertical light illumination. There are two light sources, a halogen lamp and a UV deuterium lamp. The measurement range included the visible range, 480–800 nm and utilized a 4× lens. The repeatability of both measurements was good with a standard deviation of less than 0.5 Å. This repeatability is an indication of an accurate and a robust optical model. Another measurement consisted of measuring 49 points on the mask in order to check the film uniformity of the mask. The edges of the mask were found to be thinner than the center area as it exhibited values lower than the mean, Figure 25.32.

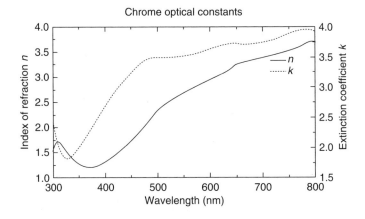

FIGURE 25.30
Optical constants of chrome layer, application 2.

FIGURE 25.31
Nanospec 6100.

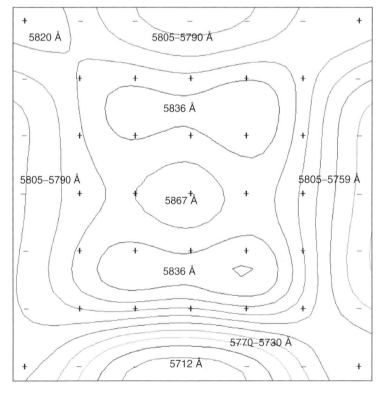

Nanospec/AFT — contour map

FIGURE 25.32
Thickness map of the mask surface.

25.5 Conclusion

Ellipsometry and reflectometry are the two main nondestructive optical techniques used in thin film measurements. Ellipsometry measurements are typically used to characterize new materials and build optical models for complex film stacks. The advantages of reflectometry, such as speed, cost efficiency, and ability, to measure smaller features compared to ellipsometry make reflectometry the most common metrology production tool. Characterization of all samples starts by determining the optical constants n and k of the material and building an appropriate model. The use of correct dispersion models for each material type is essential in obtaining the right optical constants. As in the example of characterizing a photomask, consisting of a resist layer on chrome oxide on chrome, a Cauchy model can be used for the photoresist, and oscillator models for chrome and chrome oxide. If present, additional layers, such as ARC, also require characterizing by using appropriate dispersion models. An important consideration is to find a solution unique to the experimental data. This unique solution will yield the best model fit to the experimental data and the fit parameters associated with each model. In this way, accurate and stable thickness values can be obtained from the metrology tool (Figure 25.32).

Acknowledgments

I would like to express my thanks to Ray Hoobler and Milad Tabet of the Applications Group at NANOmetrics for their guidance and support.

References

1. R.M.A Azzam and N.M. Bashara, *Ellipsometry and Polarized Light*, Elsevier Science B.V., Amsterdam, 1987, pp. 1–7.
2. Max Born and Emil Wolf, *Principles of Optics*, 7th ed., Cambridge University Press, London, 1999, pp. 1–74.
3. H.G. Tompkins and W.A. McGahan, *Spectroscopic Ellipsometry and Reflectometry*, John Wiley & Sons, New York, 1999, pp. 10, 19–21, 93–95, 195, 212–224.
4. J.A. Woollam's Manual, *Guide to Using WVASE*, pp. 2.2–2.58, 7.17, 12.8–12.45.
5. J.H. Simmons and K.S. Potter, *Optical Materials*, Academic Press, New York, 2000, pp. 85–91.

26

Phase Measurement Tools for PSM

Hal Kusunose

CONTENTS
26.1 Introduction 577
26.2 Direct Measurement of Phase-Shift and Transmittance at Exposure Wavelength 577
26.3 Phase-Shift Measurement 578
26.4 Transmittance Measurement 578
26.5 MPM Series 579
26.6 Interferometer Optics 579
26.7 Fringe Scan 582
26.8 Transmittance Measurement 583
26.9 Dependence of Phase-Shift Measurement Value on Pattern Size 584
26.10 Automatic Operation Function 586
26.11 Issue of Long-Term Stability 586
26.12 Summary 586
References 587

26.1 Introduction

Shortening exposure wavelength is an effective method to delineate a finer pattern in lithography. However, from an economical standpoint, it is desirable to postpone the change of the exposure wavelength and to extend the life of the already existing lithography tools. Phase-shift mask (PSM) technology is an already accomplished method for this purpose. Many types of PSMs have been already developed. Attenuated-type embedded phase-shift masks (EPSM) are mainly used for hole patterns, and alternating phase-shift masks (APSM) or chrome-less type masks are used for interconnection.

26.2 Direct Measurement of Phase-Shift and Transmittance at Exposure Wavelength

A PSM is a type of mask that controls the phase of transmitted light. Phase-shift must be controlled correctly because a PSM that has a phase-shift deviating from the designed value gives not only lower-than-expected resolution value, but also causes such harmful effects as focus offset, variation in pattern sizes, etc. Fabrication of PSMs and verification of phase-shift require a phase-shift measurement tool, and it is desirable that this tool be capable of measuring both phase-shift and transmittance with high precision because it is necessary to control both phase-shift and transmittance simultaneously, especially for EPSM. In EPSM production, phase-shift and transmittance of EPSM are virtually dependent on a shifter layer deposition process, and mask blanks makers definitely need a phase-shift and transmittance measurement tool. There is always a possibility that optical properties and film thicknesses change during pattern formation and the cleaning process. Thus, mask shops also need phase-shift and transmittance measurement.

26.3 Phase-Shift Measurement

There are two possible methods for phase-shift measurement. One is to measure directly phase-shift and transmittance by the light of the same wavelength as that of wafer exposure. The other is to obtain phase-shift and transmittance by calculation after the phase-shift measurement by the light of a wavelength different from that of the wafer exposure. In the case of EPSM, film structure is not uniform, and there are cases where the inner structure is very complicated or where multi-layers exist when the layer is strictly viewed. The phase-shift can be induced, not only by shifter layer but also by slight etching of the quartz pattern itself. In these cases, there is a possibility that an error may be produced in the phase-shift calculated by conversion, using refractive indices of the shifter material at measurement and exposure wavelengths. A measurement result obtained by the light with another wavelength may be different from one obtained by exposure light. It is widely believed that the usage of mask transmission light with the same wavelength as that of exposure light is very important to avoid this error.

26.4 Transmittance Measurement

Transmittance has conventionally been measured using a spectrophotometer in the mask blank production process. In this measurement method, the transmittance, which is the result of the combination of both substrate and film, has been calculated by referencing the transmittance of air (air reference transmittance). On the other hand, the definition of transmittance value should be the relative transmittance of the shifter area by referencing that of the opening area, which is quartz reference transmittance. Exposure light intensity is controlled, so that the delineated linewidth of the resist pattern on a wafer may become the designed value in the lithography process. For this reason, the number of cases in which the definition of transmittance for EPSM is changed to the quartz reference is increasing.

Meanwhile, conversion of the air reference to the quartz reference can be performed assuming that the transmittance of the substrate remains constant. However, error can take place during the wafer exposure process, the cause of which is that the transmittance of the patterned area may virtually drop when the etched surface on the substrate becomes coarse during the pattern etching process. In this case, the scattered light portion in the light transmitted through the patterned opening increases, resulting in the decrease of the intensity of the light that transmits through the lens pupil of the exposure tool. This kind of phenomenon suggests that even APSM shifter patterns generated by etching substrates can undergo change in effective shifter transmittance. It might be necessary to measure the transmittance of the shifter area by quartz reference using a measuring tool that has a lens pupil with the same degree of NA as that of the wafer exposure tool in a strict sense.

26.5 MPM Series

MPM series tools are the only kinds of tools commercially available in the world. MPM series tools are capable of directly measuring the phase-shift and transmittance of the PSM by the same light wavelength as that of the wafer exposure tools. MPM100 [1,3], the tool with measuring wavelengths of 436 and 365 nm, MPM248 [2] with 248 nm, and MPM193 with 193 nm, have been available and are now used by many mask blanks makers and mask makers for process development and quality control. The MPM series is also being used for incoming inspection of masks before applying masks to wafer fabrications. Figure 26.1 shows the external view of MPM248.

PSM is also the fundamental technique for F2 lithography, and MPM157, the most advanced model in the MPM series, with a measurement wavelength of 157 nm, was completed and is being released commercially. There are some tools available from other companies that provide phase-shift by complicated calculation. However, the accuracy of phase-shifts by these tools is not established yet. So these tools must be calibrated with MPM, and they are not used for quality assurance applications. In the measurement of phase-shift and transmittance in this field, MPM series tools are now *de facto* standard equipment.

26.6 Interferometer Optics

All the models in the MPM series employ a common structure, which is a compact Mach–Zehnder-type shearing interferometer that is placed behind the objective lens. Figure 26.2 shows the optical system of MPM248 as an example. The light source for illumination is a Hg–Xe lamp for MPM100 and MPM248. A deuterium lamp is used for MPM193. The mask to be measured is illuminated by a light of the same wavelength as that of the wafer exposure and the illumination light is obtained by selecting the light from the emission spectrum by using an interference filter and a spectroscopic prism. Transmitted light mask pattern images are enlarged by an objective lens and laterally shifted by the Mach–Zehnder type shearing interferometer, and then images are projected and overlapped by a pinhole and a camera. In the case of phase-shift measurement, the images from the phase-shift pattern area and from the non-shift pattern area are made to overlap at the image plane that is also the photo-detecting position. In such image overlapping, light intensity at the photo-detector is determined by phase-shift that is the sum of phase-shift between light transmitted through a shifter patterned area and light transmitted through a non-shifter patterned area, and intrinsic phase-shift in an interferometer. For the instruments that utilize two beams interference, fluctuation of optical length at split path in an interferometer can cause significant deterioration of the measurement precision.

FIGURE 26.1
A view of MPM248 system.

Disturbance of airflow and temperature fluctuation mainly affects the measurement accuracy. MPM has the Mach–Zehnder-type shearing interferometer behind the objective lens, and this configuration makes the light paths near the mask common paths. The result is that the split path, which is especially sensitive to precision, is kept only on the inside of the interferometer. Furthermore, the interferometer structure is made of SiC ceramic that has a very small thermal expansion coefficient and high mechanical stiffness, which brings about very stable measurement even under temperature fluctuation and mechanical vibration. All the models in the MPM series guarantee phase-shift measurement repeatability of $3\sigma = 0.5°$ for short-term measurement. Figure 26.3 shows interferometer stability versus ambient temperature drift after the lamp is turned on. The horizontal axis in this figure is elapsed time, the vertical axis on the right is ambient temperature, and the

Phase Measurement Tools for PSM

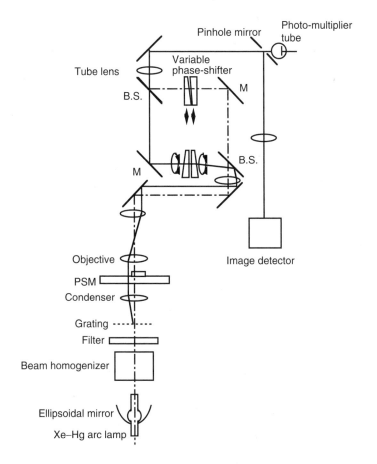

FIGURE 26.2
Interferometer microscope optics.

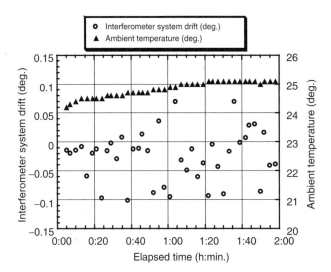

FIGURE 26.3
Interferometer system drift.

vertical axis on the left is the averaged phase-shift after measuring five times the phase-shift on the area where phase-shift does not exist. This measurement result shows that interferometer stability stays within $\pm 0.1°$ against the temperature fluctuation of $1°C$, meaning that this tool is very stable against temperature drift.

26.7 Fringe Scan

The basic operation for measurement of phase-shift and transmittance employs a technique called fringe scan, which modulates one of the optical lengths within the interferometer. The specific fringe scan method employed for the MPM series is to measure the interference signal intensity when an optical wedge (variable phase-shifter) that is placed on one side of the split beam in the interferometer is moved in the perpendicular direction against the optical axis. The light path length of one of the beams is modulated by one fringe. Figure 26.4 shows the various signal waveforms that result when fringe scan is performed at different places on a EPSM. In order to measure the phase-shift value on a PSM, phase difference between the light beams, one transmitted through the phase-shift pattern image portion and the other through the non-shift pattern image portion, is obtained first by performing fringe scan. Next, the phase-shift (intrinsic phase) generated by the difference of the light path length of two beams inside the interferometer is eliminated from the phase-shift value mentioned above. Phase-shift generated inside of the interferometer is usually obtained by performing fringe scan using the area on the mask where the phase-shift is zero. Fringe scanning at the mask area where the phase-shift is zero is not necessary for every measurement. The phase-shift measurement procedure in the MPM series employs the cancellation of the phase-shift generated inside the interferometer. It requires two phase-shift measurements. One is obtained by measuring the phase-shift on the overlapped image by placing a phase-shift pattern image on the left and a non-phase-shift pattern on the right. The other measurement is obtained by an opposite positioning, placing a non-phase-shift pattern image on the left and a phase-shift pattern image on the right. The resulting phase-shift difference is two times larger than that of a phase-shift generated in the actual phase-shifter, so MPM obtains the accurate phase-shift by dividing this number by 2. Figure 26.5 demonstrates this method. Phase-shift obtained by the method described here corresponds to the average value of two measurements, and short-term repeatability becomes high, accordingly. The time

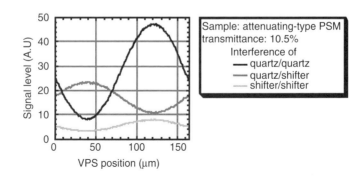

FIGURE 26.4
Fringe scan signal.

Phase Measurement Tools for PSM

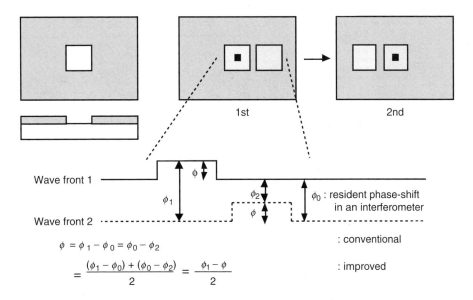

FIGURE 26.5
Improved phase measurement sequence.

required for measurement of either phase-shift or transmittance is about 30 s, which meets the measurement speed requirement for general applications.

26.8 Transmittance Measurement

Transmittance by referencing quartz is defined as the ratio of the transmitted light intensities of two lights, one transmitted through the shifter area and the other transmitted through the non-shifter area. Generally speaking, there are many cases when the transmittance of a small area is influenced by the peripheral patterns. This phenomenon is called the Schwarzschild–Villiger effect [4], and the origin of this phenomenon is stray light in the optical system. Figure 26.6 shows an example of this phenomenon. MPM has made it possible to effectively reduce this stray light by combining low coherent light irradiation and white light interferometer [5]. Unwanted stray lights that are scattered or reflected on the lens surface go through a light path longer than the normal light path. As a result, the optical path length difference between that of normal light and that of stray light becomes larger than the coherent length, and the stray light loses its coherency toward normal light. The stray light becomes a constant component in the interference signal obtained by performing a fringe scan under this situation. In other words, MPM measures the transmittance by the ratio of two signal amplitudes of interference signals both from shifter/shifter (EPSM layer area) and quartz/quartz (pattern opening area), as can be seen in Figure 26.4. This method eliminates a stray light effect by utilizing low temporal coherency. In contrast with the method of simply comparing the light intensities, MPM has made it possible to measure transmittance close to the pattern edge, namely, transmittance of a small pattern. Figure 26.7 shows three kinds of methods to measure transmittance: 1) the method to simply compare the transmitted light brightness; 2) the method to utilize the amplitude of the interference signal without lateral shift; and

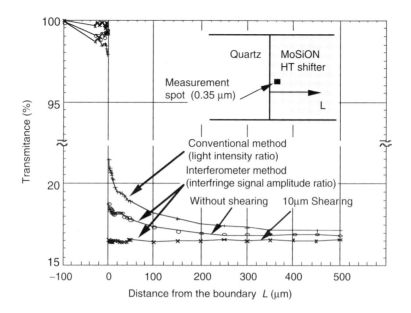

FIGURE 26.6
Transmittance near boundary between an opening and attenuating phase-shifter pattern.

3) the method to utilize the amplitude of the interference signal with lateral shift. The measurement positions are in close vicinity of the pattern edge. This figure clearly shows that the measurement value near the pattern edge is changed when there is influence by stray light. However, the effect of stray light is most effectively eliminated when an interferometer is used and additionally when the image is laterally shifted. Such an illumination condition improves the lateral resolution of transmittance measurement. In fact, the MPM series is designed in such way that a light beam consisting of a light component of spatially coherent light illuminates a mask in the lateral direction, and interference occurs when an image is displaced by a specific distance in the horizontal direction by a shearing interferometer. As an alternative method, it has been reported that reduction of the S-V effect can be achieved without using an interferometer by making the illumination area small enough. However, transmittance measurement error occurs because detected light decreases due to focus error, and this could be a problem in such cases. The MPM series is free from the effect of focusing error because incident light illuminates a large area on the mask. Based on this feature, all the models in the MPM series guarantee transmittance measurement repeatability of $3\sigma = 0.2\%$.

26.9 Dependence of Phase-Shift Measurement Value on Pattern Size

The accuracy of the phase-shift measurement value for small patterns depends on the focusing error. Figure 26.7 shows the relationship between focus height and phase-shift measurement value. This phenomenon corresponds, in reverse manner, to the instance in which focus offset occurs in the wafer exposure process when a PSM with its phase-shift deviates from 180°. The phase-shift measurement value changes relative to the focus

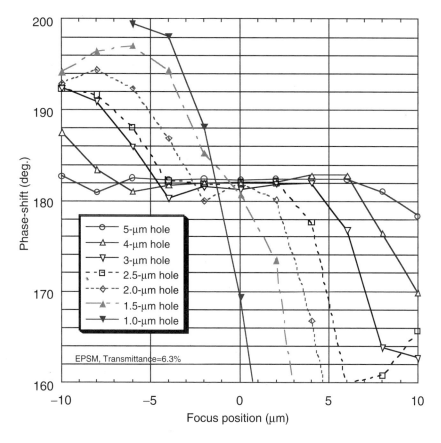

FIGURE 26.7
Focus latitude of phase-shift measurement with MPM248.

point deviation, when patterns with a size near the resolution limit are measured. Thus, when measuring the phase-shift of the actual patterns that are used in wafer exposure, the distance between the mask and objective lens must be strictly controlled. At the same time, for the phase-shift measurement of such a small pattern, the measurement result is directly affected not only by the effect of focus deviation but also by the optical aberration of the lens, since the proportion of higher-order diffraction light is large. An aberration, which is a phase-shift between the 0-order diffraction and the higher order of diffracted light in a small pattern, superposes the phase-shift of the phase-shift pattern itself. It is impossible to separate this superposition. Focus deviation also can be regarded as a kind of wave front aberration, and it is virtually impossible to measure the phase-shift of the patterns with an actual exposure pattern size correctly. Meanwhile, there is a region where the phase-shift measurement value does not depend on focus height near the focus point in a relatively large pattern region, as can be seen in Figure 26.7. This means that the problem of measurement deviation dependent on the focus does not occur when this region is utilized. For this reason, phase-shift measurement result for EPSM is guaranteed by using monitor patterns of about 7.5 μm, which is large enough to avoid this measurement deviation. A phase-shift difference between exposure pattern and monitor pattern could occur due to a micro-loading effect during the etching process, leading to an error factor. It is necessary to optimize the etching process by confirming the dependence of pattern size on etching depths using AFM to know the difference of the etching rate quantitatively, so that correction of the micro-loading effect becomes possible by the management of target value offsets against the monitor pattern.

26.10 Automatic Operation Function

MPM is equipped with an automatic operation function utilizing the automatic recognition capability of a simple pattern edge. It provides continuous and unattended measurement of phase-shift and transmittance over the entire mask area. Teaching of measurement points is performed either by memorizing the operator's measuring procedure and subsequently measuring based on that memorized data, or by inputting the measurement point coordinates from a data file.

26.11 Issue of Long-Term Stability

Differing from those optical microscopes that simply provide observation of optical image, for those optical measurement tools like MPM that output measured numerical numbers, such as phase-shift and transmittance, the effect of contamination on optical systems is recognized as a measurement value change that is directly readable. Thus, the effects of chemical contamination in the atmosphere where tools are installed must be very clearly recognized as a problem or issue of long-term measurement repeatability. Especially in those conventional clean rooms where countermeasures to chemical contamination are not applied, chemical substances such as siloxane, Di-Octyl Phthalate (DOP), and hydrocarbons in the atmosphere are deposited by the irradiation of DUV light, and therefore frequent cleaning of optical systems and parts becomes necessary. To avoid this, it is important to keep the atmosphere around the optical system chemically clean and to install tools such as MPM248 and MPM193 that use DUV light in a chemically clean booth employing activated carbon filter. Unfortunately, the effect of these anticontamination arrangements is not good enough and Lasertec is carrying out the development of a purge mechanism using mechanical parts with minimum out-gas and chemical filters for limited space. The area around the interferometer is especially sensitive to contamination, and it is necessary to make both temperature fluctuation and turbulence as small as possible to achieve good short-term measurement repeatability. Our current goal is to make cleaning of the interferometer area unnecessary even when tools are installed in conventional clean rooms.

26.12 Summary

This chapter describes principles of phase-shift and transmittance measurement with MPM. There are many unique technologies employed in the system, such as the combination of low temporal coherent light illumination with an interferometer and periodically and spatially coherent illumination. They are effective in improving lateral resolution and accuracy of the measurements. MPM series tools have been released with all wavelengths commonly used in photolithography, and play a very important role in meeting the requirements of PSM applications.

References

1. H. Fujita, H. Sano, H. Kusunose, H. Takizawa, K. Miyazaki, N. Awamura, T. Ode, and D. Awamura, Performance of i-/g-line phase-shift measurement system MPM100, *Proc. SPIE*, 2793, 497–512 (1996).
2. H. Kusunose, N. Awamura, H. Takizawa, K. Miyazaki, T. Ode, and D. Awamura, Direct phase-shift measurement with transmitted deep-UV illumination, *Proc. SPIE*, 2793, 251–260 (1996).
3. H. Kusunose, A. Nakae, J. Miyazaki, N. Yoshioka, H. Morimoto, K. Murayama, and K. Tsukamoto, Phase measurement system with transmitted UV light for phase-shifting mask inspection, *Proc. SPIE*, 2254, 294–301 (1994).
4. K. Schwarzschild and W. Villiger, *Astroyhys. J.*, 23 (1906).
5. H. Takizawa, H. Kusunose, N. Awamura, T. Ode, and D. Awamura, Transmittance measurement with interferometer system, *Proc. SPIE*, 2793, 489–496 (1996).

27

Mask Inspection: Theories and Principles

Anja Rosenbusch and Shirley Hemar

CONTENTS
27.1 Introduction 589
27.2 New Challenges: Sub-Wavelength Lithography and Defect Printability 590
27.3 Mask Defect Types 591
    27.3.1 Types of Hard Defects 591
    27.3.2 Different Types of PSM Defects 593
    27.3.3 Minimum Defect Requirements 593
27.4 Inspection of Defects 594
    27.4.1 Basic Principles of Mask Inspection 594
    27.4.2 Aerial Imaging versus Mask Imaging 596
    27.4.3 Aerial Image-Based Mask Inspection 597
Acknowledgments 598

27.1 Introduction

Photolithographic masks are the stencils or templates that are used to replicate the integrated circuit patterns of VLSI devices on silicon wafers. Masks consist of a patterned chrome film over a quartz substrate, or when more advanced, phase masks are produced with MoSi or etched quartz. The desired pattern is transferred to the semiconductor wafer by projecting the mask image at 4:1 or 5:1 reduction onto a photosensitive resist coating with a "step and scan" lithography tool (stepper), developing the resist, and processing the film through the resultant pattern.

    A single mask can be used in the production of hundreds or even thousands of wafers (each wafer containing from tens to hundreds of dies, each of which is processed to become a fully functional device); therefore, an undetected error or defect in the mask can cause significant loss in yield. To minimize this loss, all reticles are inspected for quality control several times during their manufacturing process. Currently, there are three inspection methods being used in the industry: die-to-database inspection, where the mask image is compared to the design data; die-to-die inspection, where the images of nominally identical dies within a mask are compared to each other; and contamination inspection, where the mask is checked for non-pattern-related defects, e.g., particles. When defects are found, the mask is repaired (if possible) or cleaned and reinspected. Conventional mask inspection systems employ short wavelengths and high magnification optics to detect defects down to 60 nm in size.

27.2 New Challenges: Sub-Wavelength Lithography and Defect Printability

As the design rules shrink, designers must turn to sub-wavelength lithography that involves printing features as small as 65 nm (or even smaller) with 193-nm illumination, working below the diffraction limit; wherein the optics cannot faithfully reproduce the desired pattern without the application of resolution enhancement techniques (RETs). RETs include optical proximity correction (OPC), e.g., assist features that are sub-wavelength non-printable features added to enhance the contrast, and phase shifting masks (PSM) where the light from adjacent features undergoes phase inversion to increase contrast and enhance resolution. Manufacturing high-quality RET masks becomes increasingly difficult. That makes making masks fabrication as one of the primary challenges in the 65-nm technology node.

An ambiguous question in mask making is when to call a mask anomaly — pattern or substrate related — an actual defect. Due to the cost awareness and the fact that mask repair holds risk of damage not every anomaly is classified as defect. The current trend is to classify defects on a mask based on their predicted impact onto the wafer. A defect is classified as printable if it causes a CD variation larger than a certain specification (usually ranging from 6–10%).

The actual printability of sub-wavelength defects is a complex issue: some may not print at all causing no effect on yield, while others may produce large pattern errors, effectively killing the device. The relative dimensional scaling between a mask defect and the printed defect is termed mask error enhancement factor (MEEF), calculated as:

$$\text{MEEF} = \frac{\Delta CD_{wafer} \times 4}{\Delta CD_{reticle}}$$

where the factor of 4 accounts for the nominal 4:1 stepper reduction ratio, and ΔCD refers to the change in critical dimension (CD), e.g., the width of a line or the diameter of a contact. In high MEEF pattern densities (>1), a small defect on the critical features of the reticle might be enlarged when transferred to the wafer and cause a printable defect. Detecting such a defect is crucial for guarantying "defect-free" masks.

Figure 27.1 presents two different scenarios. The first example consists of a set of three smaller figures belonging to the left half of Figure 27.1. This set addresses the case of a defective contact. The defect is located at the corner of the contact as shown in the CAD design in the left figure of the set. The mask image (middle image of the set) shows the defect clearly developed at the bottom right corner. The wafer CD SEM image (the right figure from the set) shows the impact of the corner defect. Here the contact itself is slightly

FIGURE 27.1
Defects and their impact on the actual result on wafer: The final wafer result depends on the defect type and the feature density around the defect location. On the left half of the set of three figures the result due to localized defect is shown. It affects only the defect contact. On the right half, consisting of three figures, a high-MEEF area is displayed. Here an isolated pinhole causes two neighboring contacts to bridge.

Mask Inspection: Theories and Principles

distorted at the bottom right. The contacts located next to the defective one are not or only slightly affected. The scenario on the set of right half shows a contact array with an isolated pinhole located in-between two neighboring contacts. The wafer result (right figure of this set) actually shows a bridge between the two contacts resulting from the pinhole defect. The impact of the pinhole is severe. The wafer/die might have to be scraped.

27.3 Mask Defect Types

Defects can be found on four different surfaces of the mask: the patterned area, the backside of the mask, and the front and backside of the pellicle. The defects of most interest to mask maker and mask user are the defects on the patterned surface, as they are in focus while under stepper exposure and will most likely be transferred onto the wafer. Figure 27.2 shows examples of defects on the four different surfaces. Defect type 1a is a chrome defect at the patterned surface. Defect type 1b is a defect on the quartz surface of the mask. A particle or additional chrome falls into this category. Defect type 1c shows a transmission defect in the quartz surface, which is caused by the different quartz thicknesses or a thin residue of a solvent such as cleaning solution. Defect type 2 represents a defect at the backside of the quartz. The defect specifications for this type of defect are usually more relaxed, as they should not affect wafer-printing results. Defect type 3 shows a defect at the front side of the pellicle. Defects on the backside of the pellicle are illustrated in number 4. The defect requirements for defects types 3 and 4 are less aggressive as the pellicle is not in focus during lithography exposure; hence, defects on these surfaces do not affect wafer results. Nevertheless, particles on the back side of the pellicle may fall on the pattern during the lifetime of the mask thus detecting them is still important.

Mask defects are commonly divided into two different categories: hard and soft defects. A defect is called a hard defect if it is not possible to remove it by a cleaning process. Added or missing features in the chrome, phase shifter, or absorber area fall into this category. Pindots, as well as pinholes are called hard defects as well. A defect is called soft defect if it is possible to remove it by a cleaning process. Particles, stains, contaminations, such as crystals, and residue materials are soft defects.

27.3.1 Types of Hard Defects

Figure 27.3 illustrates some of the most common hard defects such as chrome extensions (1), clear intrusions (2), and corner defects (3 and 4). In addition to that missing or added features are called hard defects as well.

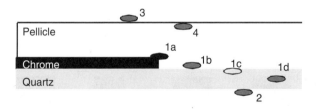

FIGURE 27.2
Defects on the four different surfaces of a mask. Defect types 1 (a, b, c, and, d) are very critical as they are in focus while being exposed. These defects are quite likely to print on the wafer. Defect type 2 located on the backside of the glass and the defects of types 3 and 4 on the pellicle surfaces are less likely to print.

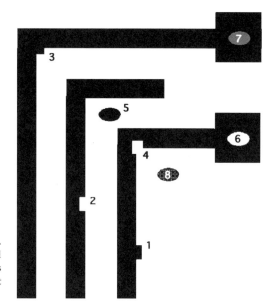

FIGURE 27.3
Classical hard defects illustrated on a binary mask. Defect types are chrome extensions on edges (1) and in corners (3), clear intrusions on edges (2) and corners (4), as well as chrome (5), clear (6) and semitransparent (7,8) spots known as pinholes and pindots.

Defects in the clear quartz area of the mask, such as pindots (5), scratches, and bubbles, are classified as hard defects as well. Defects in the opaque areas of the mask, such as pinholes (6), are in this category of defects. In addition to that there are defects related to the transmission of the mask. Defects like transmission defects in the opaque (7) or semitransparent defects on clear (8) are counted as hard defects.

In addition to these types of defects, any kind of feature mis-sizing and misplacements are called hard defects. Figure 27.4 shows some more examples. Feature misplacement (1) can be caused by errors in the original data preparation of a mask, as well as by mask writer errors. Feature mis-sizing is another hard defect type. A feature can be mis-sized in either direction, x or y (2). Missing and additional features are classified as hard defects (3).

Another type of hard defects is global CD and/or quality change over the entire mask. Edge roughness is one example, as well as global CD uniformity changes due to heating or etching effects. As these changes usually occur gradually over the span of the mask area, state-of-the-art mask inspection hardly detects them.

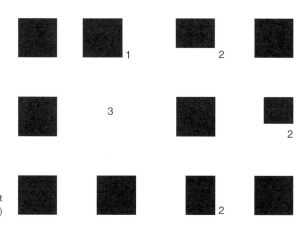

FIGURE 27.4
Examples of hard defects: feature misplacement (1) feature mis-sizing (2) and missing feature. (3)

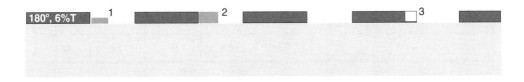

FIGURE 27.5
PSM defect types on an EAPSM: partially added shifter material (1), additional shifter (2), and missing shifter material (3).

27.3.2 Different Types of PSM Defects

Phase shifting masks (PSMs) suffer from the conventional hard defects as described in the previous section. In addition to that, this type of mask is affected by phase-specific defects. Figure 27.5 presents some of the more common phase shifting defects in embedded attenuated PSM (EPSM). A mask might have additional phase material (types 1 and 2) or missing shifter material (type 3). In an embedded attenuated PSM, the absorber material is slightly transmitive; hence, a shifter defect is also transmitive. In other words, its lithography behavior is different from a conventional opaque defect. This presents additional challenge to mask inspection systems, as the system has to support a different rejection mechanism and specification for this kind of defects.

Another type of masks is alternating PSM. In alternating PSM (APSM), an additional type of defect occurs due to the manufacturing process involved in generating an APSM. In an APSM, the shifter area is generated, for example, by an additional etch step. This step might produce defect as shown in Figure 27.6. Defect type 1 is an additional partial shifter in the phase area. This defect might change the phase behavior of the mask. Type 2 presents additional full shifter; hence, the target CD (generated by the interference of non-shifter and shifter areas) might not be guaranteed by the mask. Types 3 and 4 are more conventional. Additional absorber is located in the shifter or non-shifter area. The inspection support of these types of mask is today still the biggest challenge for mask inspection. Phase errors might only be seen at the stepper exposure wavelength. As the inspection wavelength might differ from the stepper exposure wavelength, the inspection system might not be able to detect a phase defect.

FIGURE 27.6
PSM defect types on APSM: additional partial shifter (1), additional full shifter (2), and additional chrome (3 and 4).

27.3.3 Minimum Defect Requirements

At each generation of semiconductor lithography a minimum defect size, based on minimum gate width, has been defined. These are listed in the ITRS roadmap. Figure 27.7 shows the minimum defect requirement as defined in the latest International Technology Roadmap for Semiconductor (ITRS) 2003. In addition, more and more often the minimum defect size criteria are being extended by defect capture criteria based on

Year	2003	2004	2005	2006	2007	2008
Minimum defect size (nm)	80	72	64	56	52	45.6

FIGURE 27.7
Minimum defect size criteria as defined in ITRS 2003.

printability. Defects are classified based on their impact on CD variation in the aerial image or wafer result.

27.4 Inspection of Defects

In this section, basic principles of mask inspection are explained. Mask makers have to provide defect-free masks; hence masks must meet the minimum defect criteria as defined by their customer. It is a usual practice for a mask to be inspected with sensitivity settings below the actual sensitivity requirement in order to have a safety margin against eventual inspection system flaws.

27.4.1 Basic Principles of Mask Inspection

Mask inspection is designed to verify the integrity of a mask. It could be compared to an insurance policy. It ensures that the mask will meet customer requirements regarding defectability. In simpler words, mask inspection verifies that the light is transmitted through a mask as defined in the mask layout. There are two different methods to ensure that.

The first method compares the printed mask features to their actual designs. This method is called die-to-database inspection. Assuming that the mask making process, which transfers the mask design layout into actual mask (chrome) features, had no errors, a die-to-database inspection will detect all differences between the database and the mask itself. If a difference violates the defect criteria given, it will then be classified as a defect. Differences that are detected during an inspection, but do not violate the defect criteria, are usually called nuisance. One of the system requirements for mask inspection systems is to keep the number of these nuisances as small as possible (a usual number is 20). Otherwise defect review becomes too time-intensive.

Figure 27.8 shows the die-to-database comparison of a gate feature. On the left is the database image. The feature is well defined with straight edges and sharp corners.

Database pixels   Image pixels

FIGURE 27.8
Die-to-database inspection compares the feature on the actual mask to the database. This inspection method is applicable to all types of masks.

Mask Inspection: Theories and Principles

The image on the right side shows the binarized image of the actual mask feature. The challenge of die-to-database inspection is to identify real mask defects distinguishing them from systematic errors, such as edge roughness (etch step), or corner rounding (mask writer) introduced by the mask manufacturing itself.

If the mask has more than one die, a second method, known as die-to-die inspection method can be applied. This method assumes that all dies of a mask are similar. Die-to-die inspection between all dies will identify differences between different dies. The drawback of this method is that it cannot identify a systematic error that can occur in all dies, such as additional feature or a little extension as shown in the bottom of Figure 27.9.

Mask inspection is used more than once during the mask manufacturing process. Most commonly a mask inspection (die-to-database) is performed after the mask etch step, in order to identify all mask repair locations. Then another standard inspection is performed after cleanup and pellicle to qualify the mask for customer shipment. As mask inspection is used with relatively high frequency, one of the major aspects of tool selection is efficiency.

The efficiency of a system is based on several aspects, such as detection sensitivity, capture rate, false alarm rate, inspection throughput, and comprehensiveness of the application provided. One major factor is the algorithm efficiency and quality on which the difference (defect) detection is based. The design of these algorithms is based on many system-specific criteria, such as inspection wavelength, image capture mechanism, pixel size and data transfer, and conversion (for die-to-database) rate. In addition, mask-related factors such as mask reflectivity and transmission, mask quality aspects such as uniformity and linearity (especially for die-to-database), edge roughness and pattern fidelity need to be taken into consideration. The introduction of RETs like OPC and PSM increases the complexity of mask inspection drastically. The treatment of subresolution features complicates mask inspection as they usually are in the same intensity range as defects or nuisance.

Another aspect of inspection efficiency is the cost. Inspection costs are based on many factors. The most important ones are data conversion and preparation times (which become even more crucial with OPC and PSM due to data feature explosion or multi-layer processing), number of inspections necessary, time of each inspection (throughout), defect review, classification and disposition time. Mask inspection time is mainly driven by the pixel size used. The smaller the pixel size is, the longer are data preparation and conversion times. In conventional mask inspection systems, smaller pixel sizes are necessary to obtain better detection sensitivity. With smaller pixel size, the probability of nuisance (or false alarms) becomes higher too. Nuisance defects are differences in the mask, which are below the defined mask defect specification requirement. Higher nuisance rate impacts especially review and classification times, hence might impact the inspection costs.

Besides guaranteeing that a mask arrives defect free at a customer, mask inspection also serves as process control tool in the mask shop itself. Mask process issues might manifest themselves by increased etch roughness or particle count, which can be caught by conventional mask inspection using mask imaging.

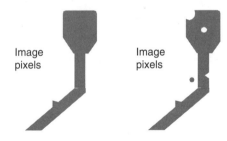

FIGURE 27.9
Die-to-die inspection compares the dies of a multi-die reticle. Differences between the different dies of a mask are detected. Systematic errors like the small extension shown in the bottom of the features cannot be identified by die-to-die inspection.

27.4.2 Aerial Imaging versus Mask Imaging

The mask aerial image is the light intensity distribution at the wafer plane, as produced by the stepper illumination source and projection optics. While the aerial image does not predict the final pattern in the developed photoresist, it does faithfully reproduce all the optical physics, including RET and MEEF. Therefore, an aerial imaging inspection technology will alert the operator of defects, which actually print, ignoring those that may be present on the mask but have no impact on the final result.

Many reticle defects, such as extrusions, protrusions and proximity pinholes, and pindots manifest themselves on the wafer as CD variation and contribute to CD nonuniformity. Figure 27.10 depicts an example of a bump defect between two lines in an alternating PSM. On the wafer this defect produces a CD variation in the lines and the bump itself is not printed at all. Figure 27.11 shows an example of a clear protrusion at the corner of a contact hole. Once again, the defect causes a CD variation on the wafer, as well as shape asymmetry in the contact hole, but the defect itself is not resolved or printed.

Figure 27.12 illustrates the case of the actual MEEF by comparing the reticle and aerial images of a contact defect in an attenuated PSM. The defect that happens to be a dark intrusion at the lower edge of a contact can barely be seen by conventional means. The error however becomes apparent by the CD variation estimated from image cross sections, measuring a 33% CD variation in the aerial image versus a 12% CD variation in the reticle image, implying a MEEF factor of 2.75. The actual CD variation on the wafer printed from this reticle was 35%, consistent with the aerial image measurement.

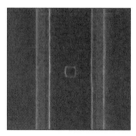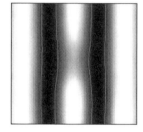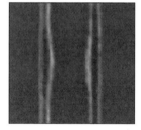

FIGURE 27.10
Line and space defect. Reading from the left to right: defect on the reticle (SEM image); an aerial image of the reticle, and finally the last picture on the right shows the image "as printed" on the wafer (SEM image). The phase bump between two lines causes a CD variation in the lines, as predicted by the aerial image.

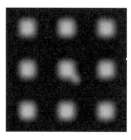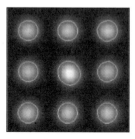

FIGURE 27.11
Contact defect. From left to right: defect on the reticle (high-magnification optical image); the aerial image of the reticle; and finally the image "as printed" on the wafer (SEM image). The clear protrusion from the contact corner causes a CD variation and asymmetry in the contact on the wafer, as predicted by the aerial image.

Mask Inspection: Theories and Principles

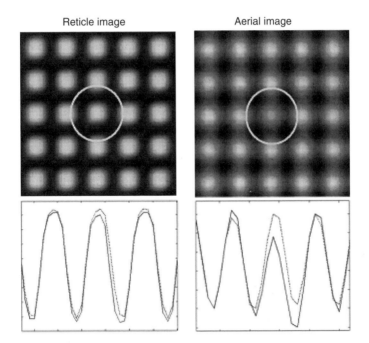

FIGURE 27.12
Edge intrusion defect (bottom edge of contact), MEEF > 1. On the left is an optical image of the defect on the reticle and on the right is the aerial image of the same defect. The plots show a vertical profile cross-section of the defect area, where the reference is shown by the dotted lines and the defect is in solid lines. The CD variation of the defective contact in the reticle is estimated to be 12%, whereas on the aerial image the CD variation is 33%, implying a MEEF of ~2.75.

These examples clearly illustrate how the defect in the aerial image can be much more representative of the transferred pattern than the image of the defect on the reticle, indicating that the aerial image of the mask provides more useful information on the quality of the mask, compared to the magnified mask image.

All this leads to one of the basic problems of current solutions in mask defect detection that is the discrepancy between those defects found by the inspection systems and those of interest to the user. Advanced reticles using RETs are built on physical optical effects to enhance image quality and reduce printable feature size. Sub-wavelength feature sizes, especially in dense patterns, may result in high MEEF values. Inspection systems that do not illuminate the reticle at the stepper wavelength and do not reproduce the physics of the optical path might not be able to account for these effects in the course of inspection.

27.4.3 Aerial Image-Based Mask Inspection

Figure 27.13 presents a schematic comparison of the optical path of a stepper to that of a proposed aerial image-based inspection system.

The aerial image-based mask inspection system inspects the aerial image of the reticle for defects in the aerial image plane. The tool emulates the stepper optics, while looking at a small area of the reticle with high magnification to enable defect detection.

The numerical aperture and illumination settings (σ values) of the stepper can be set on the inspection tool, and various off-axis illumination schemes (e.g., annular, quadrupole) can be applied. The illumination used is an excimer pulse laser, which has the same wavelength and employs the same technology used on steppers. This ensures that

598 Handbook of Photomask Manufacturing Technology

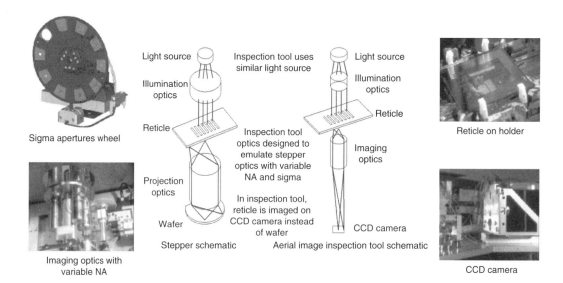

FIGURE 27.13
Schematic comparison of the optical path of a stepper to a proposed aerial image-based inspection system. The inspection system uses a similar illumination source as the stepper typically does — an excimer pulse laser at the same wavelength. The optics of the inspection system are designed to emulate the stepper optical path, including varying NA and Sigma settings. The main difference between the stepper and the inspection system is that while in the stepper the reticle image is reduced and imaged on a resist-coated wafer, in the inspection system the reticle image is magnified and imaged on a CCD camera.

all wavelength-related phenomena, such as phase shift and OPC, are inherently taken into consideration. The inspection concept is a classical die-to-die inspection model, where the identical dies of a reticle are compared with each other to look for defects. The inspection scheme is straightforward:

- Laser light is transmitted through the reticle, creating an aerial image of the reticle by optics, which emulates the stepper optics (in hardware).
- A 2-D CCD camera grabs the aerial images of the two compared dies.
- The images are sent to an image-processing module where algorithms are applied for finding inconsistencies between the two images.

Since the inspection is at-wavelength and using aerial imaging, issues, such as OPC, MEEF, PSM (alternating or attenuated), are taken into consideration without overhead or special effort. The reticle is inspected and qualified in a single step. The aerial images can be later used during the review stage to perform resist modeling simulation and process window analysis.

Acknowledgments

The authors would like to thank SEMATECH and Taiwan Semiconductor Manufacturing Company, Ltd., for providing mask and wafer CD images used in this presentation.

28

Tool for Inspecting Masks: Lasertec MD 2500

Makoto Yonezawa and Takayoshi Matsuyama

CONTENTS
28.1 Introduction 599
28.2 Development Background.. 599
28.3 System Outlines and Features 600
28.4 Application 601
28.5 Technology... 602
    28.5.1 Optics.. 602
    28.5.2 Stage System 603
    28.5.3 Defect Detection Performance 603
    28.5.4 Inspection Time 603
    28.5.5 Usability.. 603
    28.5.6 Autoloader 604
    28.5.7 Clean Unit 604
28.6 Reliability 604
28.7 Safety... 605
28.8 Conclusions.. 605
Reference 605

28.1 Introduction

Photomasks/reticles are used in the exposure process of semiconductor manufacturing, where a circuit pattern is transferred on wafers with a stepper. Photomasks/reticles are required to assure both defect-free quality and quick delivery. To meet these requirements, Lasertec has developed a new photomask/reticle inspection system MD2500, the newest model of this series. This chapter describes the outline and features of the new system.

28.2 Development Background

Masks should possess high quality and zero defects and support quick delivery synchronizing to the progress of semiconductor manufacturing processes. Consequently, a mask defect inspection system is required to have both high defect detection capability and performance fast enough to increase productivity, shortening delivery time in mask manufacturing. In addition, the recent finer design rules of semiconductors have resulted in complicated mask structure and higher-level requirement. As a result, requirements for pattern defect inspection are getting severe year after year. As a solution for improvement of sensitivity in pattern defect detection, some modification and development, including employment of a shorter wavelength inspection light source, are in progress. Also, for the purpose of improving yield, resist pattern inspection after development, which was not performed earlier, is now increasing its importance.

There are other factors affecting productivity and delivery in mask manufacturing. They include the system operating rates and contamination of masks by foreign particles during mask manufacturing process. Considering these factors in the development of the new system MD2500, Lasertec has successfully established highly stable operationality and superb cleanliness of the inspection system.

Additionally, to satisfy various requirements for mask defect inspection systems, Lasertec has considerably improved the designs of the optics, stage control mechanism, and mask transfer system using the technical expertise accumulated through the constant development of Lasertec MD series models.

28.3 System Outlines and Features

MD2500 is the newest model of the Lasertec photomask/reticle defect inspection system, MD series. Figure 28.1 shows the external appearance of MD2500.

FIGURE 28.1
A view of the mask inspection system MD2500.

Tool for Inspecting Masks: Lasertec MD 2500

TABLE 28.1

Machine Specification of the Mask Inspection System MD2500

Item	Specification
Sensitivity	0.20 μm (ADI)
Minimum pattern size	0.375 μm
Image resolution (pixel size)	0.125 × 0.250 μm
Scan time	18 min (without autoloader handling time) 100 mm × 100 mm (6-in. single mask)
Inspection method	Die-to-die/cell-to-cell (cell shift)/mask to mask
Mask type	Binary/alt-PSM/att-PSM (half-tone, tri- tone)/resist
Lens separation	31.0–304.8 mm
Objective lens working distance	7 mm
Macro view	Yes
X–Y stage stroke	314.8 (X) × 314.8 (Y) mm
Stage accuracy	±0.50 μm
Autoloader	Yes (built-in)
Cassette type	Two from SMIF/Canon/Nikon/manual loading table
Cleanliness	Class 1
Footprint	(W) 4000 × (D) 3600 × (H) 2400 mm (including maintenance area)

This new system is intended for photomasks/reticles used for 100-nm node devices. The system provides a defect detection sensitivity of as high as 0.20 μm (after-development inspection, ADI) and inspection time of 18 min (per 100 × 100 mm), which is less than one fifth the time by Lasertec conventional systems (100 min). MD2500 supports two defect inspection methods: (1) die-to-die inspection that compares adjacent chips, and (2) cell-to-cell (cell shift) inspection that compares adjacent patterns (cells) of the same shape and same size [1]. The specifications of the model are listed in Table 28.1.

The optics allow particle inspection on a pattern surface and resist pattern inspection after development (ADI), which leads to improvement of quality and yield in mask manufacturing. Also, the optics here consists of a line confocal imaging structure, resulting in the increased sensitivity and considerable damage reduction on resist at the same time.

The system employs a clean unit satisfying cleanliness of Class 1. Also, the mask transfer system of the autoloader is equipped with a multiarticular robot of cleanliness Class 1, so that it can prevent particles from contaminating masks under inspection. The autoloader, equipped as standard, operates a multiarticular robot for flexibly supporting mask cases for steppers, in addition to SMIF pods and such. Figure 28.2 shows the transfer mechanism mounted on MD2500, a manual transfer attached on the left side and, an SMIF pod on the right side.

28.4 Application

The system can provide three types of inspection: resist pattern inspection (ADI), etching pattern inspection before resist stripping (after etch inspection, AEI), and finished mask inspection after resist stripping. Thus, a single MD2500 can support pattern defect inspection in each mask manufacturing process.

Since ADI allows resist pattern conditions to be inspected before etching, we can find a fatal defect (killer defect) in mask patterns at an earlier stage. The ADI is especially

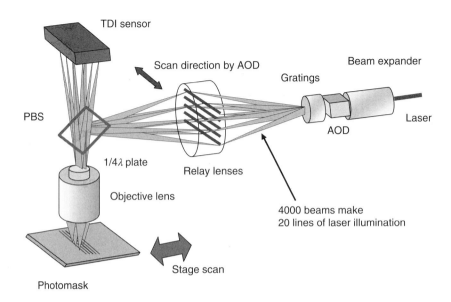

FIGURE 28.2
Fast laser scanning method for line confocal imaging.

effective for identifying those factors that affect quality and yield, which might not be found in pattern inspection after etching and subsequent resist separation. In this respect, ADI makes an important contribution to the improvement of mask quality.

28.5 Technology

28.5.1 Optics

MD2500 optics employ line confocal imaging, so that it can generate images of higher resolution. This feature gives the system high sensitivity capability to inspect masks.

The reflectance inspection light source is an argon ion laser of 488-nm wavelength. A reflected light should have such a wavelength that minimizes damage on resist during resist pattern inspection on masks. At the same time, a wavelength should be short enough to detect minute defects. As a result, a wavelength of 488 nm is regarded optimum for reflected light at present, this wavelength value being derived from expertise on Lasertec blanks inspection systems (MAGICS series). Reflected light is separated into 4000 beams by an acousto-optic deflector (AOD) and a diffraction grating, constituting high-speed multibeam confocal optics (Figure 28.2). With these optical systems, the system provides reduced amount of heat per unit area on a mask and considerably decreases damage on resist (by about 30%) compared to the Lasertec conventional system (MD2100).

The photodetector is equipped with Lasertec original time- delay-and-integration (TDI) image sensor that has optimized sensitivity spectrum at ultraviolet (UV) wavelength region, resulting in higher sensitivity of the system. The pixel size is $0.125 \times 0.250\,\mu m$ after converting to a size on a mask. Both TDI image sensor and moving stage are used to perform line confocal imaging, where the image of a mask pattern can be obtained with high resolution and high S/N ratio, resulting in the accomplishment of high defect detection sensitivity of the MD2500.

MD2500 is equipped with two objective lenses, each at the right side and the left side. The distance between two lenses can be varied between 31.0 and 304.8 mm.

28.5.2 Stage System

The inspection stage consists of an air slider driven by a linear motor. To realize high accuracy location control, the inspection stage is controlled by a laser interferometer. Thanks to these features, the inspection system realizes an overall location accuracy of $\pm$ 0.5 μm or less. The stage stroke is 314.8 × 314.8 mm.

The inspection stage is mounted on an active air vibration-free platform. This platform functions as a vibration absorber against vibration from the floor where the inspection system is installed, as well as the vibration from the autoloader or moving stage for inspection, which contributes to the realization of high sensitivity.

28.5.3 Defect Detection Performance

MD2500 obtains mask pattern images with high resolution and high S/N ratio through the usage of the image sensor whose sensitivity is optimized to UV light employed and also the line confocal imaging optics. As a result, the system improves its defect detection sensitivity to as high as 0.20 μm. Also, the system has the capability of inspecting masks with pattern sizes of as small as 0.375 μm.

With regard to the cell shift defect inspection method, adjacent patterns of the same shape and the same size are inspected on a single screen of one objective lens, which minimizes the signal noise during comparative inspection. As a result, the cell shift defect inspection method has a higher defect detection capability than the die-to-die inspection method [1]. The cell shift inspection method is very effective for masks having repetitive patterns.

28.5.4 Inspection Time

MD2500 completes the scanning over a 100 × 100-mm area in 18 min, which is less than one-fifth the time of our conventional system (100 min). Such a short inspection time is accomplished by the newly employed optics, the stage that moves smoothly with high accuracy and defect detection circuits, which are increased in speed by 10× compared to our conventional defect detection circuits, and processing the signal in parallel at a high speed. As a result, this system has the capability of inspecting three masks per hour, including mask transfer time.

Until now, users have never been satisfied with inspection time. In the course of the MD2500 development, Lasertec has been realizing that the inspection time is as important as defect detection capability, and has succeeded in attaining ultrahigh speed inspection of as fast as 18 min.

28.5.5 Usability

Since the defect inspection system comes equipped with a low-scale macro view for entire mask observation, an operator can easily find the current observation point by an objective lens focused on the mask, and move the mask on the stage to the desired point easily. This macroview function improves operation efficiency in setup of inspection conditions (inspection area) and workability for observation/verification of defects.

Lasertec MD series systems provide capability of constantly displaying pattern images captured by the right and left object lenses. Display color of the image that reflects the area where light is bright portion of the mask is set initially to be red for one objective lens and green for another, respectively. In a superimposed image of right and left lenses, the bright portion without any defect is displayed in yellow, and a bright portion with a defect is in either red or green. This color variation on the superimposed image enables operator to easily locate the defect. Normally, a pattern is displayed on the screen at a magnification ratio of 750 and can be expanded by up to 6×, that is, at a magnification ratio of 4500.

28.5.6 Autoloader

The transfer system of the autoloader employs a 6-axis polyarticular robot of cleanliness Class 1, so that it can prevent particle contamination of a mask under inspection and provide flexible handling, such as rotation, of various mask cases and masks.

The applicable mask cases are an SMIF pod, a stepper case (Nikon/Canon), and a manual transfer stage. Users can select any two of them to mount on the system.

In the conventional autoloader supplied as optional accessory, rotation and inversion of masks were handled by independent robots corresponding to each movement, which resulted in complicated maintenance. MD2500, however, uses the polyarticular robot to perform all kinds of tasks, which makes maintenance easier, and repair/adjustment time shorter. At the same time, the number of handling times by the robot is reduced, leading to a lower risk that the mask may be contaminated by foreign materials or particles. Combined with various interlock sensors, MD2500 provides reliable mask handling.

28.5.7 Clean Unit

The main unit of the inspection system and autoloader are installed in a clean chamber of cleanliness Class 1. In the chamber, the airflow speed is adjustable with the maximum airflow speed of 0.35 m/s, and sufficient positive pressure and ventilatory volume are secured. The clean unit is also equipped with chemical filters.

28.6 Reliability

Since system uptime strongly affects productivity and delivery in mask manufacturing, system downtime due to troubles or other causes should be minimized. Reliable system operation is as important as system performance. MD2500 provides the mean time between failures (MTBF) of 1500 h or more, mean time to repair (MTTR) of 4 h or less (after arrival at the site), and scheduled downtime of 4 h or less per month. These figures mean that MD2500 is a highly stable and highly reliable inspection system.

28.7 Safety

With regard to safety, MD2500 conforms to SEMI S2-0200, SEMI S8-0701, and CE marking as standard. Since the clean robot is installed together with the main unit of the inspection system in the clean unit, the system can transfer masks in clean conditions while the operator remains in a safe environment.

28.8 Conclusions

In addition to high sensitivity and high accuracy, user demands for inspection systems include shortening of inspection time and more diversified inspections. As masks become finer, more advanced technology is required to support those leading edge masks. To satisfy user demands, Lasertec promises to continue research and development for next generation inspection systems with such flexible responsiveness to the user's need, aiming at higher operationality, higher reliability, and more solid safety while improving basic performance characteristics of the systems.

Reference

1. Y. Morikawa, et al., Performance of cell-shift defect inspection technique, *Photomask and X-ray Mask Technology IV*, 1997, p. 404.

29

Tools for Mask Image Evaluation

Axel Zibold

CONTENTS
29.1 Introduction 607
      29.1.1 Reticle Enhancement Techniques 608
      29.1.2 Further Need for Mask Quality Assurance 609
29.2 Aerial Image Measurement Technique 609
29.3 Aerial Image Analysis 612
29.4 Typical Applications for AIMS™ 615
      29.4.1 Defect Analysis 615
      29.4.2 Pre- and Postrepair Qualification 616
      29.4.3 AIMS™ as "Phase Indicator" 619
      29.4.4 Process Optimization Based on Aerial Images 622
      29.4.5 New Applications in the Extreme Ultraviolet Arena 625
      29.4.6 AIMS™ in IC Manufacturing 625
29.5 Summary 625
Acknowledgments 626
References 626

29.1 Introduction

The rapid growth of the semiconductor industry occurred as a result of patterning more and more devices on smaller areas and thereby increasing the density of the integrated circuits (ICs). As this affects all the manufacturing steps, a strong challenge was placed on the micro-lithography, which allows the fabrication of these silicon chips layer by layer using various templates, or photomasks. These are durable, high-precision plates, which contain the microscopic image of each layer of an IC. In the past, photomasks were simply binary made out of patterns of chrome-on-glass substrate. Exposure tools, both wafer steppers and scanners, are critical to transfer the two-dimensional patterns of various photomasks or reticles repeatedly onto photosensitive resist, which is spun onto the wafer to create the structures layer by layer in several process steps, known as pattern transfer. On the wafer the patterns are reduced in size compared to the mask, typically a 4:1 reduction. The photomask image to be printed on the wafer is known as the aerial image. As technology has advanced, so smaller and smaller sizes of aerial images have to be printed on the resist of the wafer. Dense lines and spaces or contact holes are permanently shrinking as we head to the next generation technology node.

TABLE 29.1

Parameters Affecting the Feature Size [13]

CD = $k_1 \lambda / NA$	CD, smallest feature size to be printed λ, wavelength of exposure tool	NA, numerical aperture of exposure tool k_1, process related factor

The photomask as an optical element in the exposure tool plays a critical role in the printing characteristics of features on the photoresist. On one side, the smallest feature size that can be achieved is driven by the performance of the exposure tool (Table 29.1). The smaller the exposure wavelength and the larger the wafer side numerical aperture (NA) of the exposure tool optics, the smaller the feature sizes that can be obtained. This can be seen by wavelength changes from visible light illumination to shorter wavelengths, such as i-line (365 nm), 248 or 193 nm, and by NAs rising to 0.85. Further shrinking is expected from optical lithography as 157 nm and extreme ultraviolet (EUV; 13.4 nm) or electron imaging as examples of next generation lithography (NGL) solutions. On the other side, it has become more and more obvious that mask enhancement techniques push the feature size to become ever smaller; it is the regime where k_1 becomes <1. A landmark occurred in 1994 with the crossover of feature sizes and exposure wavelength [1]. Subsequent to this the mask has assumed a key role in the enabling of smaller linewidths at any given wavelength. At this time, reticle enhancement techniques also started to play a significant role. It became clear that the reticle is acting not just as a replica of the pattern, but also as an important optical element in the whole set up of the exposure tool.

29.1.1 Reticle Enhancement Techniques

Reticle enhancement techniques help to overcome the diffraction limit of the optics; and therefore smaller feature sizes can be obtained for any given wavelength and NA of the exposure tool. Various techniques have been developed, which can be combined in the overall printing process. Different kinds of off-axis illumination, from annular and dipole to quadrupole are in use. Additional sub-resolution features, like assist features, are added on the reticle. Such adjustments to the main features are called optical proximity corrections (OPCs), and help to correct lines or contacts that would otherwise print larger than the specified tolerance. Other OPC features like serifs help to print contact holes, for example, to a more square shape, while hammerheads are used to correct for line end shortening and/or corner rounding.

A further method to overcome the diffraction limit of the optics and enable the printing of smaller feature sizes is phase-shifting masks (PSMs). Such photomasks or reticles affect both the amplitude and phase of the light. Most common are attenuated phase-shift masks (att.PSM), on which clear areas without structure and partially transmitting areas are adjacent to one another. These are known as weak shifters. Alternatively, we have different phase-shift techniques. Also important to mention are the so-called strong shifters, where areas with 0° and 180° phase shift alternately. Further phase shift techniques like chromeless PSM, CPL, are upcoming, too. In general, the use of PSMs in the exposure tool impacts the printing characteristics and allows the printing of smaller features. Additionally the use of PSMs also is driving the illumination conditions in the exposure tools. The degree of coherence of light (σ) needs to be lowered significantly for some of them to reach the desired illumination to print a small feature size.

The printing process and requirements on masks become even more complex in cases where off-axis illumination, phase-shift technology, and OPCs are all in use to obtain smallest feature size, which avoids the need to change to smaller wavelength on the exposure tool.

29.1.2 Further Need for Mask Quality Assurance

It is an obvious consequence that mask evaluation becomes much more important with the increasing complexity of mask design. Defect free masks have to be used in the printing process to ensure that no repeating defects occur on the wafer. Therefore, the printing characteristics of critical mask features, defects, and repairs have to be understood. A rapid method of determining the printing characteristics of a mask without the need to do real wafer prints on resist and followed by wafer SEM measurements is required to speed up the readiness for production, lower costs, and increase yield. Since there is a very large variety of different techniques, such as binary, OPC, phase-shift masks, and different illumination conditions, a system is required that does not need the extensive use of computer programs and mathematical algorithms. Such a solution was created by IBM and Carl Zeiss companies and is known as the aerial image measurement system or AIMS™, which optical emulates the image produced by an exposure tool onto a resist layer [2–6] (Table 29.2). A rapid evaluation of the exposure and depth-of-focus characteristics of real masks can be performed prior to resist validation [2,4]. The optical simulation or, more precisely, emulation is especially strong on real defects where the nature of the defect and interactions to the pattern are unknown. In the case of PSMs, which increasingly dominate for critical layers, the depth-of-focus emulation is an essential feature to obtain reliable results for mask qualification.

Figure 29.1 shows the evolution of the Carl Zeiss AIMS™ tools for higher automation and shorter wavelengths. At the BACUS 2002 conference, first printing results measured on a 157-nm alpha tool were presented [7]. Due to wavelength in the vacuum ultraviolet (VUV) spectrum, major efforts had been required for the optics, as well as for the purging of the complete beam path [8–10]. Now new developments are started to address aerial image measurements for the upcoming immersion lithography which is expected to replace 157 nm and push out EUV. With EUV being discussed as one of the NGL solutions, an aerial imaging system, which operates at these wavelengths, has also been brought into discussion. As EUV imaging is a paradigm shift from transmitive to reflective optics in ultra high vacuum, a significant risk in development of such a system is identified, which can be seen addressed in a number of potential configuration solutions that have been studied up to date [11].

29.2 Aerial Image Measurement Technique

The optical base elements of an aerial image measurement system are an illumination unit and imaging unit. The former contains changeable and adjustable parts to realize the

TABLE 29.2

Characteristics for Exposure Tool Emulation, (*) for Zeiss AIMS™ Fab Systems for 248 nm, as Well as 193 nm

	Exposure Tool	AIMS™
Wavelength	λ	λ
Degree of coherence of light	σ	σ
Illumination type	On-axis/off-axis	On-axis/off-axis
Optics	Reduction M:1	Magnification 150× (*)
Numerical aperture	NA	$NA_{AIMS} = NA/M * 150$
Typical field of view	25 mm × 33 mm	20 μm × 20 μm up to 60 μm × 60 μm

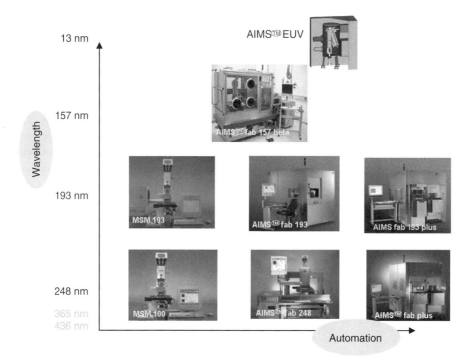

FIGURE 29.1
The Carl Zeiss AIMS™ tools are adapted to a higher degree of automation, and smaller wavelengths starting out from MSM100 (bottom left), an engineering tool, over AIMS™ fab, with automated tool setting, to AIMS™ fab plus (bottom right), where automated mask handling and SMIF are integrated. The product AIMS™ EUV is under consideration and a preliminary design is shown.

illumination type and setting of the degree of coherence of light, i.e., the adjustment of sigma (σ). The imaging part contains changeable pinholes to realize the stepper equivalent setting of NA. Both, the σ and the NA have to cover a wide range of values and be able to adjust to different exposure tool settings with minimal effort at one and the same system. Typical values are $0.25 < \sigma < 1$ and $0.3 < NA < 0.9$, independent of wavelength [12]. Figure 29.2 shows the schematic beam path of the Carl Zeiss AIMS™ fab system for 248- and 365-nm exposure tool emulation. The optical setup is a highly sophisticated derivative of the Carl Zeiss microlithography simulation microscope (MSM)-based optics [2]. For 248 nm and longer wavelengths, a Xe–Hg arc source, a cold mirror for dumping excess IR radiation into a heat sink, and a collector are used for illumination. Main components of the illumination unit are wavelength filters, attenuation filters, sigma aperture diaphragm, optical zoom system, field stop, and condenser lens. The wavelength filters allow the selection of 248, 365, or 436 nm for illumination. Band path is typically less than 10 nm full width half maximum (FWHM). Using a zoom system the adjustment of coherence of light over a wide range can be achieved without the need to exchange the aperture stop. A condenser lens focuses the illumination on the small region of interest on the mask. The imaging unit contains a measurement objective, tube lens, postmagnifying optics, image side aperture diaphragm, and a Bertram lens. The image capture is performed with a UV-light sensitive CCD camera, while cooling of the camera allows low signal-to-noise ratio. An intensity resolution of 12 bits per pixel provides 4096 grey levels allowing sufficient intensity discrimination. On the AIMS™ fab system, a total magnification of 150× is made between mask plane and CCD camera image plane, which contrasts with exposure tools that employ reduction optics. Typically a capture of the object of interest is made in a field of view of less

Tools for Mask Image Evaluation

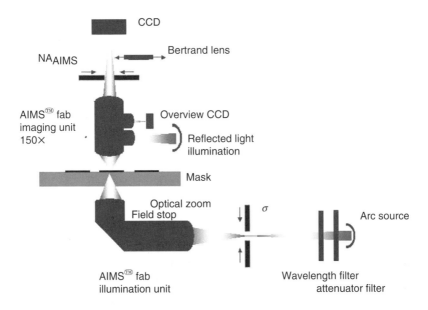

FIGURE 29.2
Schematic view of the optics of Carl Zeiss AIMS™ fab platform.

than 60 μm × 60 μm (standard 20 μm × 20 μm). The mask is placed on a stage that allows the handling of different mask sizes and thickness. The measurement objective is mounted on a turret, which moves parallel to the optical axis and can move to the various focus levels. The choice of a long working distance measurement objective enables the investigation of masks with a pellicle mounted. Additional imaging capability for the purpose of overview is given by reflective light illumination, as well as by low magnification transmitive light. Larger field-sizes can be imaged to navigate to the region of interest.

To realize the stepper emulation, the image side aperture and the sigma aperture diaphragm have to be in conjugated planes. It means when the Bertrand lens is switched into the beam path, then both aperture diaphragms can be imaged simultaneously in focus and viewed in quick succession. On an AIMS™ fab, a typical value of the state-of-the-art measurement lens is 10×/0.23. In this special case, the NA value of the objective NA = 0.23 determines the maximum value of the exposure tool lens, which can be emulated. For a 4× mask it is $NA_{max} = 0.23 * 4 = 0.92$. The σ-value is determined by the ratio of illumination side aperture to imaging side aperture. Depending on the selected NA value, the zoom moves into a specific position to achieve the selected σ-value. The alignment of the respective components is done automatically. In the last few years, it has also become common to use off-axis illumination techniques on the exposure tools to improve the printing capability of smaller feature sizes. Such off-axis illumination patterns can also be incorporated into the AIMS™ tools. Indeed any type of appropriately shaped aperture can be inserted into sigma aperture position and selected automatically. Figure 29.3 shows typical off-axis illumination patterns, which are in use and acquired on AIMS™ tools.

For shorter wavelength systems, i.e., 193 and 157 nm, wavelength-specific excimer lasers and beam homogenizers are used as illuminators. Band path values are < 1 nm for 193 nm and 10 pm for 157 nm (FWHM). Therefore, the beam homogenizer is needed to reduce the speckles in the laser beam and ensure similar illumination uniformity as the longer wavelength systems using the arc source. The lasers are fully integrated, and light emission is only possible during image acquisition. The full beam path of the laser-based systems is encapsulated, and the respective AIMS™ tools correspond to laser safety class 1 systems.

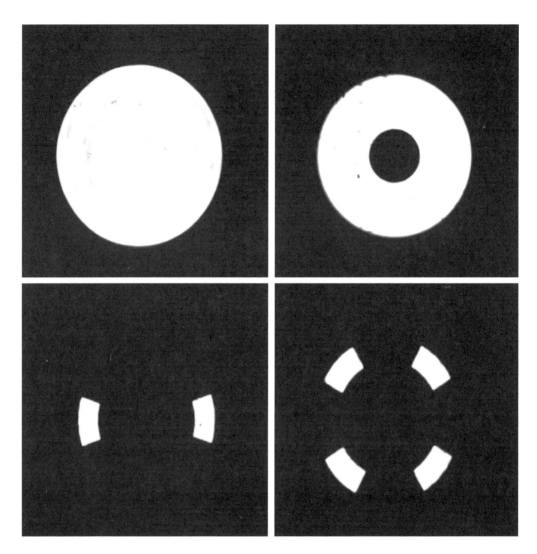

FIGURE 29.3
Different types of illumination aperture σ, which can be viewed with the Bertrand lens moved into the beam. Moving clockwise the picture shows on-axis illumination of a pupil, annular 2:3 pupil, Quasar pupil and Disar pupil image.

All Carl Zeiss AIMS™ tools utilize the same software platform and are operated in the same way. Typically, single image capture times range between 100 ms and 1 s depending on the transmittance of the mask patterns and the lifetime of the source or purge condition of the excimer laser. Therefore, the stack of through-focus aerial images can be obtained within a few seconds.

29.3 Aerial Image Analysis

Aerial images acquired with the AIMS™ tools are recorded either as single images or as a through focus series (TFS) providing a stack of images. A common approach for TFS is to

Tools for Mask Image Evaluation 613

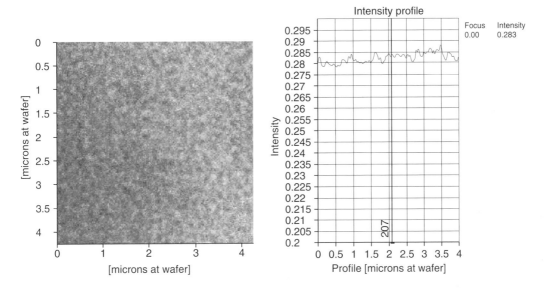

FIGURE 29.4
Reference image, unnormalized, taken at a clear area of a mask (left) and profile plot to illustrate field homogeneity (right).

acquire an odd number (usually 7) of images in equidistant focal steps, having one image in best focus position and equivalent extra- and intra-focal images covering the range of depth-of-focus. In actual use a reference image is measured at a clear mask region. All measured images are normalized with this reference image [4]. The normalization eliminates any system dependent properties, such as field uniformity and image acquisition time, and allows a quantitative analysis of the mask properties for the given stepper settings. Figure 29.4 shows an example of a clear image, which is used for normalization. The profile plot that is selected shows the field distribution given by the alignment of the optical system, the values being averaged perpendicular between the two lines. These values are pixel counts of the CCD camera divided by maximum intensity value of 4096, which is the number of available grey levels. In Figure 29.5 the aerial image is shown of a 5-μm wide isolated line on the mask. The image of the mask feature is normalized by the reference taken on a clear area of the exact same mask. The intensity values shown in the profile plot are arbitrary units and refer to a transmission value through a clear area on the mask, as long as the measurement point is far away from edges or other features, which interfere and lead to disturbing effects. The strongest modulated profile curve shows the best focus values. At best focus a strong effect of the edges can be seen due to interference and intensity values greater than unity are obtained. In the middle of the bright line the intensity is unity. The other profile curves refer to the extra- and intra-focal planes. From the plot it can be seen that a step size of 0.2 μm on the wafer level was used on the AIMS™ tool. In total a focal range of 1.2 μm of an exposure tool was covered by the TFS. This corresponds to a step size of 0.2 μm $*4^2 = 3.2$ μm on the mask level and is the actual step size used on the AIMS™ tool.

On Carl Zeiss AIMS™ tools, both the MSM series and the AIMS™ fab series, stacks of aerial images from through focus measurements are the input for the further image analysis. Algorithms and display routines are used to extract a variety of different image information. This allows the rapid mask or reticle evaluation under exposure tool equivalent optical settings without the need of wafer prints on resist.

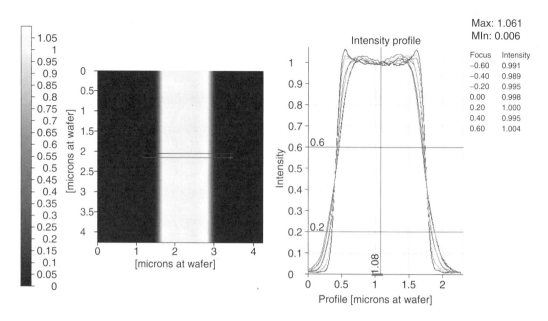

FIGURE 29.5
Normalized image of a 5-μm wide line on a mask ($\lambda = 193\,\text{nm}$, $NA = 0.68$, $\sigma = 0.6$, $M = 4$) and profile plot.

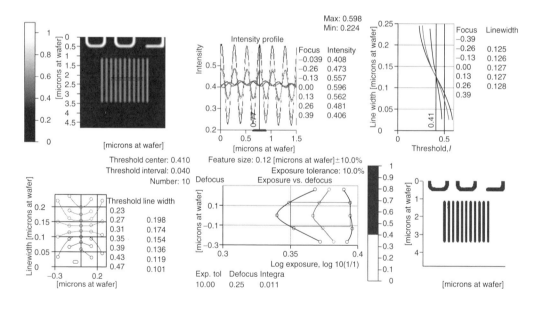

FIGURE 29.6
Dense lines and spaces of 1 μm 1:1 on mask level. The stepper equivalent settings are $\lambda = 193\,\text{nm}$, $NA = 0.7$, $\sigma = 0.6$, and $M = 4$.

Figure 29.6 shows a field of view from dense lines and spaces of pitch 1 μm 1:1 and a variety of standard plots, which provide valuable information about the printability of mask defects [2,4]. A TFS was taken with seven images. The aerial image represents the best focus measurement (upper left window). All images were normalized by taking a clear reference. A quick prediction of the resist behavior can be obtained by the contour plot (lower right window). In a simple threshold model, it is assumed that the expected resist linewidth depending on resist properties and exposure dose is given by selection of a threshold value, in this case the selected threshold value is 0.41. The intensity profile plot (upper middle window) allows the evaluation of peak intensities of bright features and comparison to one another. The cursor shows one feature, which is selected for further evaluation. The linewidth versus threshold plot provides the opportunity to analyze the predicted linewidth for a selected threshold value for different focal layers. The "pivot"-point, where the predicted linewidth does not depend on the focal position, can be clearly seen. In our case, it is at the threshold of 0.41 and the predicted linewidth prints to 127 nm at the wafer level. For further analysis, the so-called Bossung plot can be displayed to extract information about the exposure latitude (lower left window) [13]. Each curve in the plot displays the linewidth as a function of the focus at a fixed threshold level. It is similar to linewidth versus defocus behavior at a fixed dose. The generalization is shown in the exposure–defocus (ED) window (lower middle window). Here the plot shows the maximum allowable focus variation at a given specified exposure tolerance. The mask feature is supposed to print to 120-nm linewidth with a feature size tolerance of ±10%. The rectangular box shows simultaneously possible changes in focus and dose. For an exposure dose tolerance set to 10%, an allowable focus variation of 0.25 μm on wafer level is permitted.

29.4 Typical Applications for AIMS™

29.4.1 Defect Analysis

Masks are used as templates in the wafer fabrication to print hundreds and thousands of ICs on wafers. Should a defect occur on the mask, then it will print as a repeater on every exposure and a significant yield loss will result. Therefore, a mask supplier, which is shipping its product to a wafer fabrication, has to ensure that the mask is defect free in the printing process. Defects are identified by inspection systems, which operate in either a die-to-die or die-to-database mode. Typical defects on a mask can be additional chrome structures, such as chrome bridges, or missing chrome, such as pinholes. However, there are others, which include glass or quartz damage, quartz bumps, and partially transparent films, that result in more or less critical transmission loss and therefore CD error [14].

A quick qualitative view about the severity of a defect can be shown by a contour plot as described above or in the paper [15]. However, the most appropriate judgments for printability of defects or repairs are based either on transmission (normalized intensity) loss or on CD error [16,17]. These are defined as follows:

$$\text{transmission loss} = 100\% * |(T_{\_\text{reference}}(\%) - T_{\_\text{defect}}(\%))/T_{\_\text{reference}}(\%)|$$
$$\text{CD error} = 100\% * |(CD_{\_\text{reference}} - CD_{\_\text{defect}})/CD_{\_\text{reference}}|$$

Figure 29.7 shows an example for lines and spaces of chrome-on-glass, and the centerline exhibits a chrome bridge defect. A profile plot is selected, and in the data window only the best focus curve is displayed. An image analysis with a horizontal slice shows a transmission loss of about 41%. Figure 29.8 shows the same measurement as Figure 29.7. However, in this case a vertical slice was used to determine the transmission loss. A comparable value of 41% is calculated, which results from the maximum peak intensity

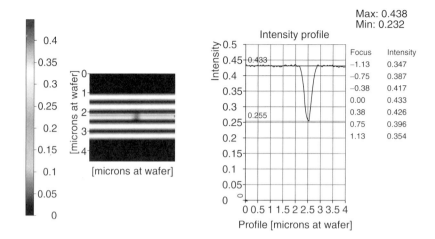

FIGURE 29.7
Defect analysis with profile plot of horizontal slice.

being used in the vicinity as a reference. Typically, several peaks will be averaged as a reference value. Such a high transmission loss is a very clear indication of a defect and the requirement for a repair.

An alternative way to determine the severity of the defect is the CD error. Very often it is a preferred method because in lithography a threshold model, rather than peak intensity, is the more realistic scenario. In this case, operators use the reference feature to determine a threshold value for data evaluation. Such a reference peak is indicated in Figure 29.8 with the left vertical line shown in the intensity versus profile plot. A target wafer CD value can then be used to determine the threshold value. This threshold value is kept fixed for both the reference and the defect, before a comparison of CD values is made. In the example given, 300 nm at the wafer level is selected as a target value, which results in threshold 0.2. For this threshold value of 0.2, the defect prints at 189 nm, which corresponds to a CD error of 37%. Again a clear indication for a printing defect under the given exposure tool settings.

The transmission loss and CD error type of analysis are also adequate for local phase defects, quartz defects, as well as missing chrome defects and others.

29.4.2 Pre- and Postrepair Qualification

In the mask manufacturing process, it is usual that some of the masks will require a repair, otherwise the delivery of a mask of zero defects cannot be guaranteed. Different repair techniques are available, which include laser, focused ion beam, or a micromachining technique. With decreasing feature sizes and more complex reticle enhancement techniques, including phase shift, masks' repair becomes critical, and their repair presents severe problems. A more detailed characterization about the severity of any defect is required to minimize any risk and choose the most appropriate repair technique. Further cost-intensive and time-consuming repairs must be eliminated. Figure 29.9 gives an example of partially repaired phase defect and its quick visual evaluation of the printability by using a contour map.

A solid criterion had to be found to provide sufficient quality assurance of the mask. It can only be given by the exact knowledge of its printability, including defects and the repairs. This is an essential reason to implement AIMS™ technology into the mask manufacturing process [16,18]. This actinic emulation under exposure conditions pro-

Tools for Mask Image Evaluation

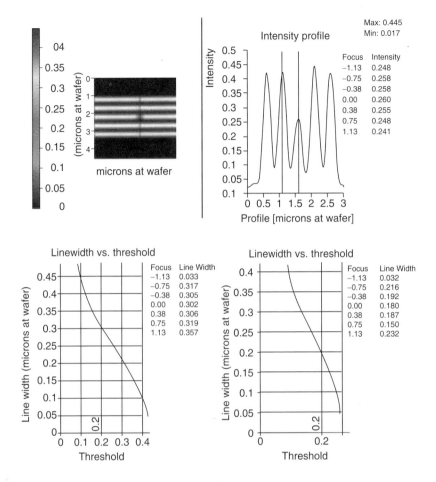

FIGURE 29.8
Defect analysis with profile plot of vertical slice and CD evaluation of reference feature and defect.

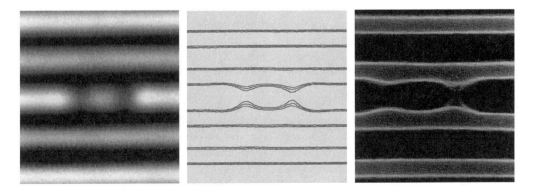

FIGURE 29.9
Comparison of aerial image (left), contour plot (middle) plot, and SEM image of the printed feature on the wafer (right). In the middle a partially repaired phase defect can be seen. The contour map provides a quick visual evaluation of printability after the repair.

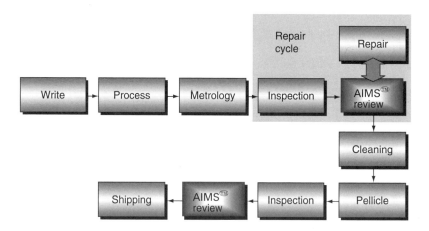

FIGURE 29.10
Schematic flow of mask fabrication process. It shows common uses for AIMS™ tools in the process. Defects and repairs can be visited due to the information provided by the inspection systems to which the AIMS™ can be linked.

vides a method to unambiguously assess, with respect to printability checks, any repaired region. The emulation can be done for any type of binary masks, dense patterns, OPC-enhanced features, and for phase effects at exactly the exposure conditions, which are used later on in the wafer fabrication. The wide range of imaging side NA and illumination (σ, type) settings on the respective AIMS™ tool allows for adjustment to match the different demands of the various wafer fabrication end users.

Figure 29.10 shows a schematic process flow that is common for a mask shop. The main application of the AIMS™ tool is its use in the repair cycle. The mask is run through a defect inspection system to find defects based on the sensitivity of the system. A list of defects is created, and a classification of the type of defects is provided. However, once detected and classified on the inspection system, the operator still needs to understand any implications on printability of the defect. The AIMS™ can be linked via a software interface to the inspection system; and the defect listing, alignment marks, and mask coordinates can be transferred. In some mask shops, additional mask evaluation tools are linked together into a cluster [19,20]. So the mask can be loaded on the AIMS™ tool, and based on this information any defect can be revisited with the unit set to the exposure conditions under which the mask will be used in the wafer fabrication. Based on the result of the through-focus measurement of the defect, the operator can then judge if the defect will print or not and decide whether a repair has to be performed. In cases where only nonactinic defect review is made, it can happen that any misjudgment may lead to an unnecessary, cost-intensive repair. We only need to repair things that print. Based on the AIMS™ measurements, a mask with a non-printing defect can be directly processed with cleaning and further steps. In cases where the operator judges the defect will print, the mask has to be rejected or can be repaired. After the mask is repaired, the operator has to assess whether the repair was successful. Again the criteria are the printability of the repaired area. After defect repair, the mask is loaded on the AIMS™ tool and reviewed under exactly the same exposure conditions as before the repair. With such quality assurance the mask can be passed along the process flow with confidence. Due to this printability check, the mask repair can even be performed in steps and the mask cycled between the AIMS™ and the repair tools. In this way it can be ensured that any defect is not "over repaired" beyond the necessary requirement for printability at the wafer level, thereby reducing the risk of the repair operation. Today in most of the major mask shops, the closing and shortening of the repair cycle is well established for masks of different exposure wavelengths and various nodes.

Typical values for successful repairs are transmission loss values in a range of less than 5%, which indicates that one can tolerate a certain incompleteness of the repair process [21–23]. The approach to use CD error to assess successful repairs is done in practically the same way. Typical values, which verify successful repairs, change from $\pm 10\%$ range to $\pm 5\%$ with smaller nodes [17]. If the normalized intensity profile is compared to the lithography process, as discussed above, then it seems also evident that the CD error is the more appropriate technique to use as a judgment for the printability of really small defects or repairs. Not the maximum peak intensity of the incident light but rather the width of the peak at the threshold value determining the feature size is a closer approximation.

In many cases for pre- and postrepair disposition it is sufficient to evaluate only the best focus curve of the aerial image. For repair verification of PSMs with OPC, a more appropriate method is the comparison of the ED window of the repaired pattern and the ED window of a reference pattern [17]. Figure 29.11 (left side) shows the comparison of an ED-window of a reference pattern (see contact hole, right box) and a defect (see contact hole, left box). The measurement is performed on an attenuated PSM, emulated with $\lambda = 248$ nm, $NA = 0.57$, $\sigma = 0.385$, $M = 4$ with 250-nm feature size at wafer level. On the right side of Figure 29.11, the defect is repaired. In the ED plots, the reference pattern is represented by the solid line, the defect/repair with a dashed line. It can be seen, before the repair of the defect no overlap is given and after the repair of the defect the process window is shifted and now overlaps with the reference pattern.

Over the last years successful verification of repairs and the techniques for 248- and 193-nm lithography masks was proven by using AIMS™ systems, which includes many comparisons to real wafer prints with steppers. Here are some literature references that provide further discussions — for opaque chrome mask defects [22,24], opaque defect repair of MoSi attenuated PSM [16,18,25–28], clear defects on attenuated PSM [16,26,27], quartz bump defects on attenuated and alternating PSM [21,25,29,30], quartz trench damage [22], and aggressive OPC and EPSM contacts [31].

29.4.3 AIMS™ as "Phase Indicator"

Phase errors for PSM masks have an effect on the printability of the mask features depending on the settings of the exposure tools. Defects can occur with both, alternating or attenuated PSMs, which are difficult to repair. The CD of the printing feature can be affected strongly by such phase errors. Optical imbalance between different phase regions can cause distortions of the aerial image and violate the mask specification.

For optimizing PSM manufacturing it is required to control the phase angle. From the AIMS™ a direct phase value cannot be retrieved; however, the AIMS™ can be used as a "phase indicator" [32]. The through-focus measurements reveal for different phase angles different out-of-focus behaviors for the intra-focal or extra-focal measurements. Various attempts have been shown for alternating PSMs [2,32–35] and attenuated PSMs [2] to extract phase values from the linewidth behavior as a function of defocus. The distortions of the Bossung plot curves reveal clearly the impact of a phase error. The work is confirmed by comparison to other measurement equipment and real wafer prints [32,33–36].

Figure 29.12 shows an example of a phase defect on a line and space pattern. The left upper window shows the SEM image, the right upper window the aerial image at best focus taken on the AIMS™ 248-nm tool. The arrows indicate where the Bossung plots were taken. In the nondefective, "good" area symmetric curves can be seen (left lower

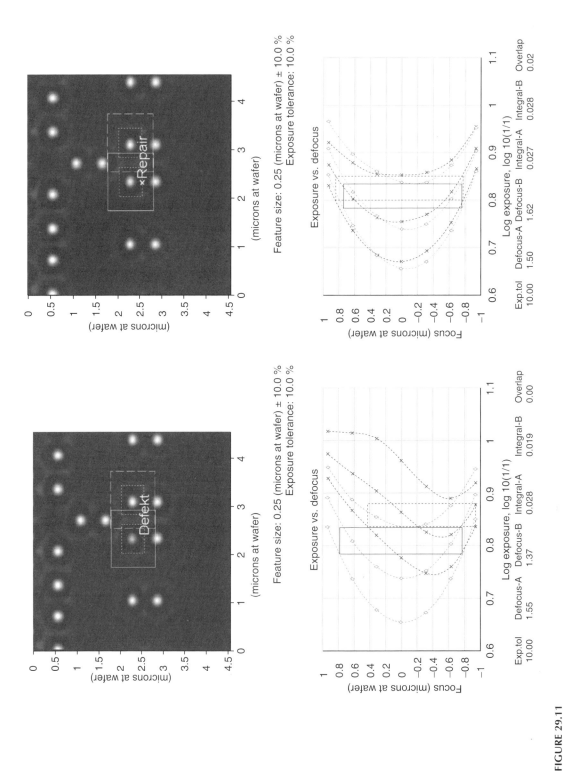

FIGURE 29.11
Comparison of ED windows of the defect and neighboring good feature (left) and of the same defect after repair to the same neighboring good feature (right). The two dashed boxes in the ED window show the changes before and after repair. Measurements are taken on a PSM with $\lambda = 248\,nm$, $NA = 0.57$, $\sigma = 0.385$, and $M = 4$.

Tools for Mask Image Evaluation

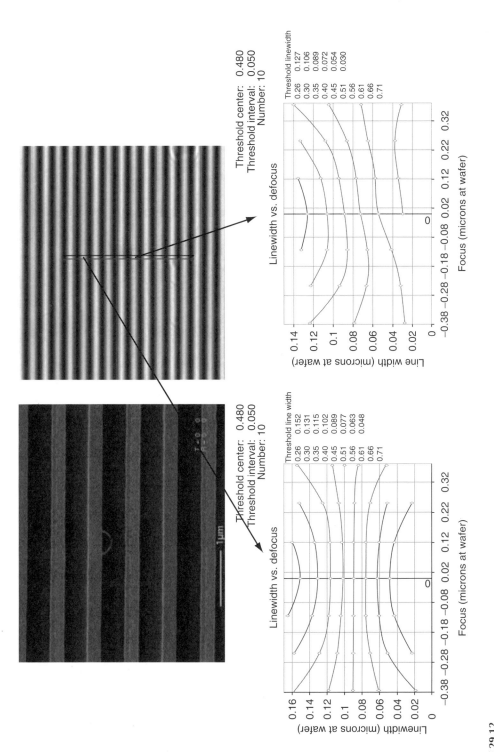

FIGURE 29.12
SEM image, aerial image, and Bossung plots for defective and non-defective area. The phase defect causes a tilt of the Bossung curves. Measurement conditions are $\lambda = 193\,\text{nm}$, $NA = 0.7$, $\sigma = 0.3$, and $M = 4$.

window) and the Bossung plot of the area with the defect shows a tilt of the curves, which can be used for further phase evaluation [37]. The nondefective area shows a flat curve where the linewidth is independent from the defocus, and which corresponds to the isofocal point. Such an isofocal point cannot be found for the phase defect anymore. The stronger the tilt of the Bossung curves the larger the phase error will be [35].

29.4.4 Process Optimization Based on Aerial Images

In addition to defect and repair analysis, the aerial images obtained from an AIMS™ tool can be used to analyze the print capability of mask features for specific exposure tool settings. For example, from the aerial image, the contrast of mask features that are very close to the smallest feature size to be printed, for a specific node, can be extracted. Figure 29.13 shows the aerial image for 280-nm lines and spaces mask pattern of a halftone PSM. The profile plot shows very good feature homogeneity. The contrast as one measure of image quality [13] can be calculated by averaged maximum (Int_{max}) and minimum intensity (Int_{min}) values based on

$$\text{contrast} = Int_{max} - Int_{min}/(Int_{max} + Int_{min})$$

and results in a value of about 0.5.

Beyond that the aerial imaging measurement technique can be used to explore effects and interactions of mask type and resolution enhancement techniques. Different off-axis illuminations, such as annular or quasar, can be tested for mask features, which are either corrected with or without OPC. Figure 29.14 shows the comparison of aerial images taken with the AIMS™ on a MoSi PSM with 18% attenuation. The illumination was either

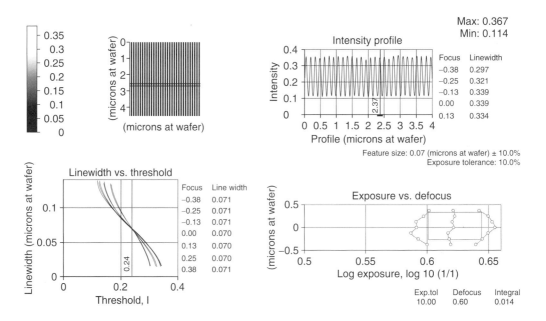

FIGURE 29.13
Dense lines and spaces of 280 nm 1:1 on mask level. The stepper equivalent settings are $\lambda = 193$ nm, $NA = 0.8$, $\sigma = 0.9$, off-axis Disar 69%, and $M = 4$. The linewidth of the selected peak can print to 70 nm on wafer level. For 70 nm $\pm 10\%$ and 10% exposure tolerance a defocus range of 0.6 μm was identified.

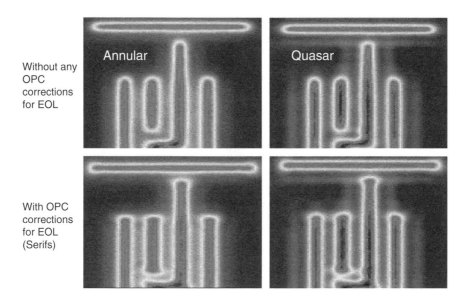

FIGURE 29.14
End of line AIMS® aerial images with 0.7 NA. Lines are 130 nm and spaces 260 nm. Mask is 18% attPSM, $\lambda = 248$ nm ($\sigma_{outer}/\sigma_{inner} = 0.85/0.55$), and $M = 4$ [38,39].

annular or quasar ($\sigma_{outer}/\sigma_{inner} = 0.85/0.55$), the image was with and without OPC correction [38]. Differences are seen clearly in the feature shape, size, and image contrast, for example, quasar illumination produces a higher image contrast. A metric that is applicable for all types of feature is exposure latitude, which allows the determination of defocus tolerance for optical lithography [13]. Exposure latitude corresponds to the maximum amount of exposure variation before the printed pattern falls out of specification. With aerial image measurements it can be extracted from the analysis plots of the through-focus images given in the Carl Zeiss AIMS® software. From TFS of aerial images as presented in Figure 29.14 exposure latitude has been extracted, and various results are displayed in Figure 29.15. Differences in exposure latitude can be seen for different illumination types of the same mask (18% attPSM) and the same NA (=0.7; left window). Annular illumination provides a larger focus latitude. Differences in exposure latitude can be also be seen for lines and space pitches 1:1.5, 1:2 with differing OPC for the same illumination type, same NA = 0.63 and same mask (6% attPSM). Small but different OPC treatments, as well as non-OPC, have a large effect on the exposure and focus latitude [38,39]. The trends in the above investigation correlate also very well with real wafer prints [38–40], and it can be stated with confidence that the AIMS® technique can be used for decisions about medium or aggressive OPC.

Recently, for low k_1 lithography PSM mask layouts with pattern shapes that look completely different from the pattern to which they print have been shown by Han [41]. A decision from the contour of the mask layout can only with difficultly be translated to a wafer level contour. Square contours on the mask print rounded lines on the wafer, heavy OPC on the mask with oval shaped "holes" print to very different looking cells on the wafer level. In these cases, the AIMS® technique is proposed as the best technique to evaluate the contours on the wafer level and verify the desired wafer print and lithographic process. The AIMS® can also be used in the development process of such type of mask features.

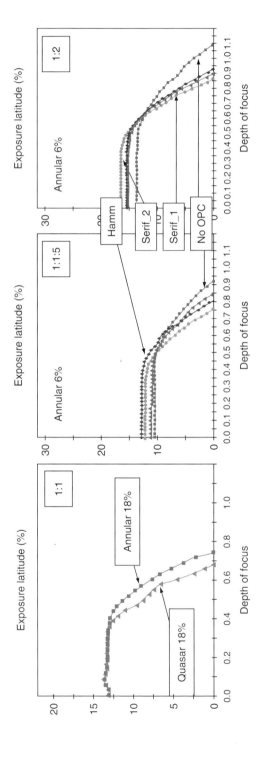

FIGURE 29.15
Exposure latitude extracted from aerial images of AIMS™ 248-nm tool. Differences in exposure latitude can be seen for different illumination types of the same mask (18% attPSM), same NA and for lines and space pitches 1:1.5, 1:2, as well as different OPC, for the same illumination type and same mask (6% attPSM) [38,39].

29.4.5 New Applications in the Extreme Ultraviolet Arena

So far all discussion about AIMS™ applications have been related to patterned masks, or defects and repairs on patterned masks in the wavelength range from 365 nm down to 157 nm. With the attention of the industry turning to the EUV regime, an AIMS™ tool using EUV (13.4 nm) wavelength is being considered for investigations of both patterned and mask blanks [11]. The EUV mask blank is not only a substrate but also has multilayers on top. Substrate defects, which are buried below, are manifested as phase defects and can lead to yield loss. Only actinic AIMS™ measurements at EUV wavelength can help to assess the severity of such a blank defect by assessing printability at the wafer level.

29.4.6 AIMS™ in IC Manufacturing

With smaller nodes, where design rules are below 100 nm and low k_1 factors are in use in the lithography, smaller defects on reticles are becoming a serious problem; and it is expected that smaller defects will occur in increasing numbers in the wafer fabrication [42]. Likely defects on the mask are particles, crystal growth, electrostatic damage, irradiation damage due to the high energy along with the short exposure wavelength, and they may cause as repeaters catastrophic yield loss. With increased numbers of defects and higher complexity of masks, the understanding of such defects can help to decide whether the mask needs to be shipped back to the mask shop, has no influence on the wafer print or can even be used with a tolerable CD error. Since most of these defects act not fully light blocking, they act like phase defects. A rapid method to do the defect disposition without the need for real wafer prints is the AIMS™, where the printability can be determined with actinic measurements under the same exposure tool conditions.

In a further step more defect repair efficiency can be obtained if the mask shop and the wafer fabrication correlate AIMS™ emulation results on both sides [17].

For ongoing reticle qualification, the AIMS™ can monitor the wafer printability of features throughout the lifetime of the reticle. For example, growing defects can be assessed and classified for printability/nonprintability. An early warning from the AIMS™ tool can prohibit significant yield loss caused by wafer printing with defects repeated on each die.

As the AIMS™ is powerful for the predictability of critical issue like line-end, proximity, overlapping ED windows, RET feature, it can shorten and close the loop between IC design and optical lithography decisions in selecting OPC strategies and exposure tool settings [38,39]. The flexibility of varying the illumination type and degree of coherence of light, the NA and the rapid through-focus image acquisition provide a highly efficient process development, and the need to run real wafer prints in the fabrication can be minimized significantly.

29.5 Summary

AIMS™ technology allows the optical emulation of an exposure tool by adjustment of wavelength, wafer side NA, degree of coherence of light (σ), and mask reduction factor. The printability at the wafer level can be obtained rapidly without the need to print real wafers. Defects, repairs, and critical features can be evaluated for different types of mask, such as binary, OPC, and phase shift, independently of both their complexity and interaction of the features in the printing process. The AIMS™ allows the assessment of defects before and after repair. Moreover, the impact of OPC and illumination conditions, as well as phase effects, for optimizing mask development and production can be understood.

This technology is a very good method to assess the capability of lithographic processes and can be used to predict many issues prevalent in low k_1 lithography.

Acknowledgments

I would like to thank many of colleagues from Carl Zeiss Company for helpful discussions concerning AIMS technology during the course of the last years, especially Wolfgang Harnisch, Thomas Engel, Peter Schaeffer, Yuji Kobiyama, and Andrew Ridley. Special thanks related to this chapter I would like to give to Mark Joyner from Metron. For data contributions I would like to thank Wolfgang Dettmann from Infineon, Clare Wakefield from Photronics, and Mircea Dusa from ASML.

References

1. H.J. Levinson, *Principle of Lithography*, SPIE, 2001, p. 257.
2. R.A. Budd, D.B. Dove, J.L. Staples, R.M. Martino, R.A. Furguson, and J.T. Weed, Development and application of a new tool for lithographic mask evaluation, the stepper equivalent aerial image measurement system, AIMS, *IBM J. Res. Develop.*, 41 (1997).
3. AIMS®, TM is a trademark of Carl Zeiss.
4. R.A. Budd, J. Staples, and D.B. Dove, A new tool for phase shift mask evaluation, the stepper equivalent aerial image measurement system AIMS, *Proc. SPIE*, 2087 (1993).
5. R.A. Budd, D.B. Dove, J.L. Staples, H. Nasse, and W. Ulrich, A new mask evaluation tool, the microlithography simulation microscope aerial image measurement system, *Proc. SPIE*, 2197 (1994).
6. R. Martino, R. Furguson, R. Budd, J. Staples, L. Liebmann, A. Molless, D. Dove, and J. Weed, Application of the aerial image measurement system AIMS to the analysis of binary mask imaging and resolution enhancement techniques, *Proc. SPIE*, 2197 (1994).
7. K. Eisner, P. Kuschnerus, J.P. Urbach, C. Schilz, T. Engel, A. Zibold, T. Yasui, and I. Higashikawa, Aerial image measurement system for 157 nm lithography, *Proc. SPIE*, 4889 (2002).
8. P. Kuschnerus, T. Engel, W. Harnisch, C. Hertfelder, A. Zibold, J.-P. Urbach, C. Schilz, and K. Eisner, Performance of the aerial image measurement system for 157 nm lithography, in: *Proc. 19th EMC Conference*, 2003.
9. P. Kuschnerus, T. Engel, A. Zibold, C. Hertfelder, T. Yasui, I. Higashikawa, C.M. Schilz, and A. Semmler, Actinic aerial image measurement tool for 157 nm lithography, *Proc. SPIE*, 5130 (2003).
10. T. Yasui, I. Higashikawa, P. Kuschnerus, T. Engel, A. Zibold, Y. Kobiyama, J.P. Urbach, C. Schilz, and A. Semmler, Actinic aerial image measurement tool for 157 nm mask qualification, *Proc. SPIE*, 5130 (2003).
11. A. Barty, J.S. Taylor, R. Hudyma, E. Spiller, D.W.Sweeney, G. Shelden, and J.P. Urbach, Aerial image microscopes for the inspection of defects in EUV masks, *Proc. SPIE*, 4889 (2002).
12. Typical values obtained on a Carl Zeiss AIMS® fab based system. Stepsize is quasi-continuous $\Delta = 0.01$.
13. A.K.-K. Wong, *Resolution Enhancement Techniques in Optical Lithography*, SPIE Press, vol. TT47, 1999.
14. N. Kachwala and K. Eisner, Defect inspection and repair reticle (DIRRT) design for the 100 nm and sub-100 nm technology nodes, in: *Proceedings of 18th EMC Conference*, 2002.

15. A.C. Rudack, L. Levit, and A. Williams, Mask damage by electrostatic discharge: a reticle printability evaluation, *Proc. SPIE*, 4691 (2002).
16. F. Gans, J. Marion, S. Kolpoth, and R. Pforr, Printability and repair techniques for DUV photomasks, *Proc. SPIE*, 3236 (1997).
17. W. Chou, T. Chen, W. Tseng, P. Huang, C.C. Tseng, M. Chung, D. Wang, and N. Huang, Characterization of repair to KrF 300 mm wafer printability for 0.13 um design rule with attenuated phase shifting mask, *Proc. SPIE*, 4889 (2002).
18. S. Kubo, K. Hiruta, H. Morimoto, A. Yasaka, R. Hagiwara, T. Adachi, Y. Morikawa, K. Iwase, and N. Hayashi, Advanced FIB mask repair technology for ArF lithography (2), *Proc. SPIE*, 3873 (2000).
19. K. Peter, Quality assessment of advanced photomasks using the Q-CAP cluster tool, in: *Proceedings of the 17th EMC Conference*, 2000.
20. D. Bald, S. Munir, B. Lieberman, W. Howard, and C. Mack, PRIMADONNA: a system for automated defect disposition of production masks using wafer lithography simulation, *Proc. SPIE*, 4889 (2002).
21. C. Friedrich, M. Verbeek, L. Mader, C. Crell, R. Pforr, and U.A. Griesinger, Defect printability and repair of alternating phase shift masks, *Proc. SPIE*, 3236 (1997).
22. M. Schmidt, P. Flangigan, and D. Thibault, Photomask repair using an advanced laser based repair system (MARS2), *Proc. SPIE*, 4889 (2002).
23. S. Fan, M. Hsu, A. Tseng, J.F. Chen, D. Van Den Broeke, H. Lei, S. Hsu, and X. Shi, Phase defect repair for the chromeless phase lithography (CPL) mask, *Proc. SPIE*, 4889 (2002).
24. J.C. Morgan, Focused ion beam mask repair, *Solid State Technol.*, March (1998).
25. D.C. Ferranti, J.C. Morgan, and B. Thompson, Advances in focused ion beam repair of opaque defects, *Proc. SPIE*, 2194 (1997).
26. R. Hagiwara et al., Advanced FIB mask repair technology for 100 nm/ArF lithography, *Proc. SPIE*, 4889 (2002).
27. H.W.P. Koops, K. Edinger, J. Bihr, V. Boegli, and J. Greiser, Electron beam mask repair with induced reactions, in: *Proceedings of the 19th EMC Conference*, 2003.
28. H. Kobayashi, M. Ushida, and K. Ueno, Photomask blanks quality and functionality improvement challenges for the 130 nm node and beyond, in: *Proceedings of the 17th EMC Conference*, 2000.
29. B. LoBianco, R. White, and T. Nawrocki, Use of nanomachining for 100 nm mask repair, *Proc. SPIE*, 4889 (2002).
30. B. LoBianco, R. White, and T. Nawrocki, Use of nanomachining for 100 nm mask repair, in: *Proceedings of the 19th EMC Conference*, 2003.
31. Y. Borodovsky, R. Schenker, G. Allen, E. Teijnil, D. Hwang, F.C. Lo, V. Singh, R. Gleason, J. Brandenburg, and R. Bigwood, Lithography strategic for 65 nm node, *Proc. SPIE*, 4754 (2002).
32. U.A. Griesinger, L. Mader, A. Semmler, W. Dettmann, C. Noelscher, and R. Pforr, Balancing of alternating phase shifting masks for practical application: modelling and experimental verification, *Proc. SPIE*, 4186 (2000).
33. Y. Morikawa, Y. Totsu, M. Nashiguchi, M. Hoga, N. Hoyashi, L. Pang, and G.T. Luk-Pat, Study of defect printability analysis on alternating phase shifting masks for 193 nm lithography, *Proc. SPIE*, 4889 (2002).
34. U.A. Griesinger, R. Pforr, J. Knobloch, and C. Friedrich, Transmission and balancing of alternating phase shifting masks (5×) — theoretical and experimental results, *Proc. SPIE*, 3873 (1999).
35. U.A. Griesinger, W. Dettmann, M. Hennig, J. Heumann, R. Köhle, R. Ludwig, M. Verbeek, and M. Zarrabian, Alternating phase shifting masks: phase determination and impact of quartz defects — theoretical and experimental result, *Proc. SPIE*, 4754 (2002).
36. N. Ishiwata, T. Kobayashi, T. Yamamoto, H. Hasegawa, and S. Asai, Fabrication process of alternating phase shift mask for practical use, *Proc. SPIE*, 4066 (2000).
37. R. Köhle, W. Dettmann, and M. Verbeek, Fourier analysis of AIMS images for mask characterization, *Proc. SPIE*, 5130 (2003).
38. M. Dusa, J. van Praagh, A. Ridley, and B. So, Study of mask aerial images to predict CD proximity and line-end shortening of resist patterns, *Proc. SPIE*, 4764 (2002).
39. M. Dusa, J. van Praagh, and A. Ridley, A method for evaluating RETs for advanced masks, *Solid State Technol.*, October (2001).

40. V. Philipsen and R. Jonckheere, A printability study for phase shift masks at 193 nm lithography, in: *Proceedings of 19th EMC Conference*, 2003.
41. W.-S. Han, Lithography technology trend for DRAM devicdes, *Proc. SPIE*, 4754 (2002).
42. V. Philipsen, R. Jonckhere, S. Kohlpoth, C. Friedrich, and A. Torres, Printability of hard and soft defects in 193 nm lithography, *Proc. SPIE*, 4764 (2002).

30
Mask Repair

Randall Lee

CONTENTS
30.1 Defects and Mask Repair 629
30.2 Defect Types 630
30.3 Defect Classifications 631
30.4 Repair Qualification 632
30.5 Repair Tool Commonalities 632
30.6 Laser Defect Repair 633
      30.6.1 Theory of Operation 633
30.7 Focused Ion Beam Mask Repair 635
      30.7.1 Theory of Operation 636
30.8 AFM Nanomachining Mask Repair 640
      30.8.1 Theory of Operation 640
30.9 Electron Beam Mask Repair 641
      30.9.1 Theory of Operation 641
30.10 Next Generation Lithography Mask Repair Challenges 644
      30.10.1 Stencil Masks 644
      30.10.2 Extreme Ultraviolet Lithography Masks 644
References 645

30.1 Defects and Mask Repair

The purpose of the photomask reticle is to faithfully reproduce the desired pattern with correct dimensions in photoresist on the wafer surface when inserted into the stepper optics and exposed on the wafer. Mask defects cause errors in the transferred pattern by diffracting, deflecting, or absorbing photons during exposure. In earlier days, 1× photomasks for full field aligners could have a certain number of killer defects, unrepairable defects that would result in loss of that die. This seems unthinkable now, but then fewer layers and generally smaller die sizes (on smaller wafers) could still provide acceptable wafer yield. With the advent of the wafer stepper, however, reticles with only 1–10 dies per exposure are typical. Reticles must be perfect; any defects must be repaired to prevent unacceptable yield loss. (Whether all defects found will print is another matter [1] and will not be covered here. This chapter only addresses repair of hard defects.)

Mask repair improves photomask manufacturer yields and performance to schedule. Both metrics are critical to maintaining profitability and the customer base. In this chapter, defect types are reviewed first. Repair processes are targeted to the specific type of defect, and the reader must understand the nomenclature. Subsystems common to all repair tools and major repair technologies are described and contrasted next. Last, new lithography technologies and their mask challenges are briefly reviewed.

30.2 Defect Types

During mask inspection, operators classify defects into different categories. Routing of the mask to the next manufacturing step depends on the number and type of defects found.

Any material on the substrate where it is not supposed to be constitutes an opaque defect. Opaque defects block light from going through the mask. Any absorber material missing from where it is supposed to be is a clear defect. Clear defects allow light to go through the mask to be imaged on the wafer, exposing photoresist in wrong places (Figure 30.1).

Soft defects are generally defined to be any defect not of the substrate or absorber material. There are numerous types of soft defects: particles from the environment or

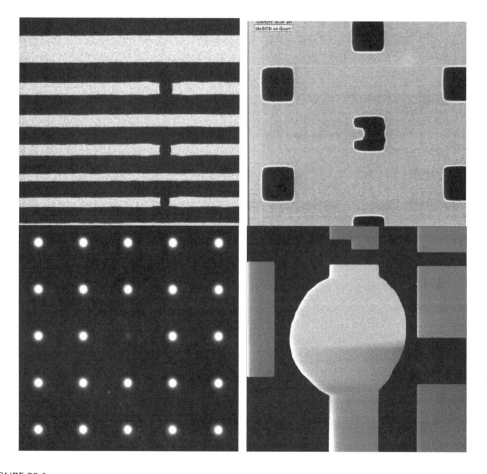

FIGURE 30.1
Images of different types of mask defects. Clockwise from top left: programed chrome bridge defects; contact edge defect; undersized contact; killer "blob" defect. (Courtesy FEI Company.)

process equipment, leftover photoresist residue from a strip process, chemical residue from incomplete removal during a wet process, etc. Soft defects can even arise on masks with pellicles from particles that loosen from the frame and end up on the mask, or cracking of volatile compounds from outgassing of the pellicle adhesive onto the mask surface [2]. Mask manufacturers remove soft defects with dedicated cleaning equipment and processes; this topic will not be covered here, since it is not a repair process *per se*. (However, mask makers may attempt a "repair" if any particles remain after cleaning rather than scrap the mask.)

Alternating aperture phase shift masks (AAPSM) are also known as hard shifter masks. The concept is that by etching the fused quartz substrate in specific areas, phase differences caused by path length differences will create destructive interference of incident light at the wafer surface. By careful placement and sizing of the shifters, contrast is improved at edges, and therefore geometries smaller than the wavelength of light can be imaged in a given stepper [3]. Opaque geometries can be imaged on masks completely without absorbers by correct placement and etching of the substrate geometries (chromeless phase shift mask, or CPSM). Any surface height deviation on the AAPSM substrate that results in image errors on the wafer is a hard shifter defect. "Quartz bumps" jut above the desired surface height; "quartz pits" sink below the surface (Figure 30.2). Both can cause aerial image errors at the wafer and must be repaired when they do.

Other types of typical defects are catastrophic in nature, such as plate scratches (front or backside) or glass defects. They usually result in immediate scrapping of the plate.

30.3 Defect Classifications

Any deviation in an edge of a geometry from the correct position is an edge or "attached" defect. They are usually small in size; large ones become "blob" defects if they are bigger than a couple of microns in extent or larger in comparison to the geometry. They can occur on straight or curved edges. Opaque defects that span the clear area between two absorber geometries are bridge defects. Repair of these defects involves trimming excess

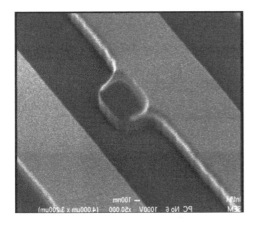 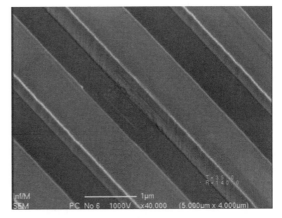

FIGURE 30.2
SEM images of a programed AAPSM quartz defect and after repair. (Courtesy RAVE, LLC.)

absorber back to the line edge for opaque edge defects (extrusions), or adding repair material to fill in missing absorber (intrusions) out to the correct edge.

Isolated defects stand alone. Opaque isolated defects are called "chrome spots" or "pindots." Mask makers usually call clear isolated defects "pinholes."

Generally large in size, "blob" or "killer" defects can obscure complete geometries or fall across several features and are the most difficult type of defect to repair. "Blobs" usually require feature reconstruction.

30.4 Repair Qualification

Qualification of repairs usually requires one or two of the following three steps: reinspection on the defect inspection tool to qualify the whole mask again; measurement of the geometry for edge or geometry placement at the repair site; and transmission metrology, such as normalized peak intensity, and through-focus measurements of the repair site on a specialized tool. Plate reinspection is the most common form of qualifying repairs. Most mask shops inspect leading edge masks at the highest sensitivity and with wavelengths different than the exposure tool, however, and so can find and classify repaired areas as defects even if they would not actually print. Transmission metrology tools can prove to the mask user that the repair would not print and are often used for qualifying repairs on these costly masks.

30.5 Repair Tool Commonalities

Mask repair systems share four common features: a stage, defect navigation software, viewing optics, and a method of carrying out the repair.

Stages come in a wide variety of types but share common requirements. They must hold the reticle in place without slippage during stage motion. They must be accurate enough to place a defect area within a field of view (after mask alignment). And they must keep the site stable with little or no drift in the defect location during the repair process. Laser interferometers help achieve stability and accuracy on many of today's stages. Many of the latest repair tools and stages attain Class 1 cleanliness as well and can have front-end robots for single reticle SMIF pods for the latest high-end mask shop facilities.

The defect repair tool has to navigate to the defect site. All repair tools have software graphic user interfaces (GUIs) to download inspection files specific to the plate, and most can sort through defect classifications so the mask repair engineer or operator can select the defect of choice. The GUI must therefore download the correct inspection file (manual input or plate identification through a bar code reader or RFID tag on the SMIF pod), fracture the inspection file into the internal format used by the tool, locate the alignment points on the mask (standard types or plate-specific), display the defect number and classification, and then go to the selected defect.

Once the tool navigates to the expected defect location, optics generate an image of the area for finding the defect in the field of view. Optics can be based on standard light microscope technology, or on charged particle technology, such as scanning electron beams or focused ion beams (FIBs). Whatever the technology, it has to have sufficient

Mask Repair

resolution to see the defect to be repaired. Many tools now can import image files of the defect site as stored in the MIS database from the inspection tool for referencing, which can greatly aid in locating the smallest defects.

Methods to effect the repair differentiate the capabilities of mask repair tool technologies. The next sections cover these technologies.

30.6 Laser Defect Repair

Mask makers have used laser repair tools since the 1970s to remove opaque defects. They have proven to be highly reliable, easy to use and have high throughput [4]. Such attributes have contributed to their popularity over the years, and they appear on the equipment list of every mask shop in the world.

30.6.1 Theory of Operation

Lasers emit beams of coherent light either continuously or in pulses. Mask repair systems use pulsed lasers. These light pulses may have low average power, but their instantaneous power can be quite high. Pulse durations cover the ranges of nano-, pico- and femtosecond scales. The wavelength of the laser light must be such that it is absorbed by the target material.

For nanosecond and picosecond lasers, the absorber heats up, melts and evaporates from the laser pulse energy. The process can be quite violent on the micron scale: chrome can essentially explode from the surface. Melted "splatter" can deposit nearby, creating more defects, and the substrate surface itself can sustain damage, reducing light transmission or creating a phase defect [5] (Figure 30.3). Thermal diffusion from the target sight to surrounding areas limits repair spatial resolution to about 0.5 microns.

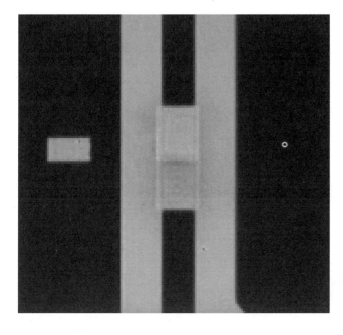

FIGURE 30.3
Optical image of laser absorber repair with chrome debris and substrate damage. (Courtesy SPIE.)

Most types of laser repair systems are imaged aperture tools. That is, the laser illuminates a motorized aperture assembly in the optical chain. That aperture image is demagnified and imaged at the mask level for opaque defect ablation. The tool provides a visual indication of the size of the laser repair box and location on the substrate for the operator. Users can vary the size and shape of the aperture to localize the irradiated area to the defect. The defect is then "zapped" by triggering a laser pulse. Laser power must be controlled tightly to reduce or eliminate splatter or quartz damage.

Femtosecond lasers have a different mechanism for absorber removal [6]. The "light pulse directly excites a large fraction of the valence electrons to antibonding states, causing the material to enter the vapor phase before the electrons transfer their energy to phonons (heat)." The target material transitions directly from solid to vapor with no thermal effects on the adjacent absorber. Substrate damage is minimal. Without thermal effects, repair spatial resolution is mostly a function of the laser spot size at the surface. Repair quality of imaged aperture femtosecond laser tools is better than nano- and picosecond lasers but still limited by diffraction of laser light at the aperture. A newly-announced tool utilizes a scanned Gaussian-shaped laser beam for defect removal and shape reconstruction [7] (Figure 30.4). Repeated laser pulses (while the stage is scanned beneath it) remove small volumes of chrome with every pulse until all of the absorber has been removed in the defect region. Selectivity of chrome to the substrate is very high, so that the substrate can be repeatedly scanned without noticeable degradation. Lines have been drilled in chrome down to 80 nm in width with this tool, though the edges were not as smooth as with larger repairs (Figure 30.5).

Certain tools offer laser deposition for clear repair. A gas is adsorbed locally on the surface. Laser energy heats up the substrate and gas, decomposing the gas and leaving a film behind on the irradiated surface. Laser clear repair has not gained widespread market acceptance due to film qualities, size, and placement issues.

Advantages:

- Opaque defect repair is very fast with very high throughput
- Atmospheric pressure operation
- Historically high reliability
- Easy to learn and operate
- Laser tools (if supplied with a long working distance objective) can repair opaque defects found after pellicles are applied to the mask

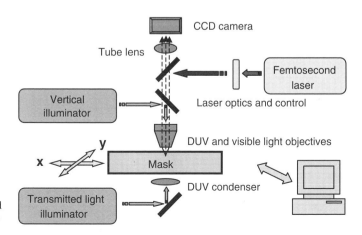

FIGURE 30.4
Block diagram of a new femtosecond laser repair system. (Courtesy SPIE.)

Mask Repair

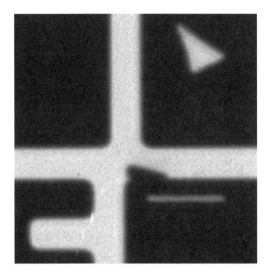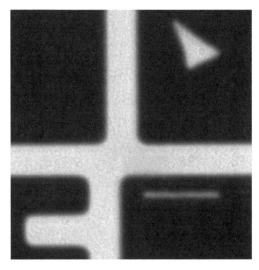

FIGURE 30.5
Transmitted light before/after images of laser repairs with femtosecond laser tool. The defect is approximately 250 nm high. The narrow horizontal line below the defect and the triangle were also made by the laser repair tool. (Courtesy SPIE.)

Disadvantages:

- Clear repair has not been a strong, proven application for laser tools.
- Chrome "splatter" and metal "pullback" from the melting and ablation process can yield more defects or a nonyielding repair, at least for most imaged aperture laser repair tools.
- Substrate damage can also be an issue. (This is not an issue for the femtosecond laser tool described earlier.)
- Light optics cannot match the inherent superior resolution of charged particle systems for imaging small defects, or for minimum size geometry reconstruction.

30.7 Focused Ion Beam Mask Repair

Companies first developed FIB systems for mask repair in the early 1980s, and they have been another workhorse in the mask shop stable ever since. While not as ubiquitous as laser tools, every single high-end commercial mask shop, and captive shop has an FIB for repairing masks. They have better spatial resolution for defect imaging and repair than laser tools. That coupled with the ability to raster-specific patterns with the beam provides the mask shop with ways to fix "blob" defects. This section will review only the most common type of FIB repair tool that uses a gallium source, though researchers have developed many types of source species (silicon, gas field ion sources, indium, etc.).

30.7.1 Theory of Operation

A focused gallium ion beam constitutes the probe that both results in image formation and work on the substrate surface. The source and column together form the focused gallium ion beam and work in a high vacuum environment.

One type of liquid metal ion source (LMIS) consists of a liquid gallium coating on a sharpened, treated metal needle, typically tungsten. Electronics supply a method of heating the source needle to reliquify the source (if necessary) or burn off surface contaminants that inhibit proper source operation. A very high electric field is placed near the source tip. That field literally pulls the liquid gallium away from the tip in a balance between electrostatic forces and surface tension, eventually forming a geometric shape called a Taylor cone. As the field strength is increased further, gallium atoms are ionized and pulled off the tip of the Taylor cone. The total flow of ions from the tip (extraction current) is tightly controlled via a feedback loop for beam stability. The source assembly floats at an acceleration voltage of typically 20–30 kV, so the gallium ions are accelerated down into the focusing optics of the column. Liquid metal flowing down on the needle surface from a reservoir above replenishes the gallium in the tip that is ionized and sent down the column.

Electrostatic lenses in the focusing column serve to collect and focus the ion beam into a spot on the surface. An aperture defines the beam current and spot size. Usually multiple apertures on a strip provide user-selectable beams for imaging and mask repair at that point. An octupole or two provides beam stigmation correction and beam scanning at the substrate surface, and high speed blanking plates deflect the beam off to the side when the beam is not scanning or during scan retraces to prevent substrate damage (Figure 30.6).

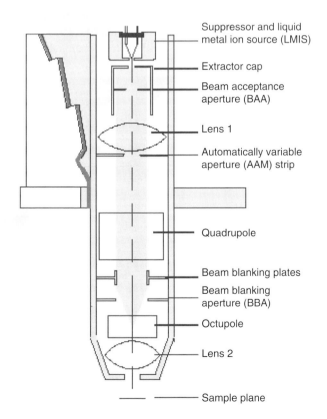

FIGURE 30.6
Schematic of FIB column. (Courtesy FEI Company.)

The beam profile is a Gaussian with extended beam tails due to column aberrations. Gallium ions are relatively heavy; when one strikes the target surface, it produces localized sputtering. The energy dispersed by the ion ejects secondary atoms, ions, and electrons from the upper 20 nm of the target surface. The number and type of secondary particles depend on incident ion energy and sputter yield of the target material, crystal orientation, local topography, and local voltage conditions. The ion itself implants into the substrate. In a mask, implanted gallium ions in a clear area absorb photons, reducing transmission [8]. (This is called "gallium staining.") The higher the implant dose, the lower the transmission. Gallium also absorbs more strongly as the illumination wavelength shortens.

Scanning electronics precisely control where the beam hits at a pixel, how long it dwells there, and how far it moves to the next pixel. A variety of detectors are available: microchannel plates, Everhart–Thornley, secondary ion mass spectrometers (SIMS), and others. Detector bias voltages determine whether secondary electrons or ions are collected for analysis or image generation. Since the very act of generating an image involves scanning the ion beam and looking at sputtered by-products, images must be acquired with a low ion dose to avoid excessive damage and gallium staining.

The implanted ions create another problem: they are big and relatively immobile positive charges and therefore very high positive electric fields can build at the substrate surface. Positive charging can prevent secondary electrons from escaping for reduced signal at the detector, deflect the incoming positive gallium beam causing unstable beam placement, and even cause catastrophic electrostatic discharge (ESD) events on the mask surface. The most popular method of controlling charging in FIB systems involves spraying low-energy electrons onto the target surface at the incident point of the ion beam. These electrons suppress or reduce the surface electric field by recombination with the implanted ions or by other means.

When specific gases are injected into the local area where the ion beam is scanned, chemical, as well as physical, effects come into play. Gas molecules adsorbed on the surface can dissociate from the energy of the impinging ion (collision cascade model). In some cases, material will be deposited on the surface. Tungsten, carbon, platinum, and gold films (among others) have been deposited in such a fashion from precursor gases. The depositions are never pure; gallium and components from the precursor gas are present. However, the depositions have proven to be quite rugged and opaque. FIB clear repair uses carbon as the deposition film of choice. For other gases, the target material and the dissociated gas combine to form a volatile by-product that is pumped away by the vacuum system (gas-assisted etching, or GAE). In this way, local redeposition is reduced or eliminated, and the sputter yield is improved. (It could be considered a type of local reactive ion etch, with good selectivity of etch of the target material to the surrounding materials.) The higher etch yield reduces the amount of gallium dose required to clear an opaque repair, and therefore leads to reduced gallium staining and better transmission.

FIB tools repair opaque defects by sputtering the film away, usually in the presence of a gas to enhance removal and selectivity [9]. MoSiON films for embedded (or attenuated) phase shift masks [10] (EPSMs) show very high etch enhancement (a factor of 10–20) and high selectivity to the substrate with typical process gases, and very little substrate damage beneath the completed repairs [11]. Chrome, however, is very difficult to remove due to its crystalline form on the substrate and the fact that there are very few volatile chrome compounds. Etch enhancement may range from 2× and 5×, at best. Thus physical sputtering is a major component of chrome removal. Sputtering rates increase at sharp topography edges. What this means is that during chrome removal, substrate

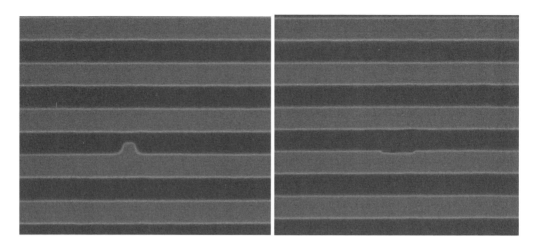

FIGURE 30.7
Before and after FIB images of FIB chrome repair. (Courtesy FEI Company.)

damage is increased around the periphery of a defect and is known as "riverbed." Controlling quartz damage during FIB opaque repair is critical to success (Figure 30.7). As mentioned before, redeposition of sputtered material in the local area must also be minimized.

Carbon films have long been used to fix clear defects in FIB tools. The films display excellent opacity and ruggedness in all cleaning processes, and FIBs are the preferred method for binary mask clear repair (Figure 30.8). These films do not have the same index of refraction as MoSiON films of EPSM masks. Users can either match phase or transmission of EPSM films but not both at this time. Most users opt to match transmission.

"Blob" defects can be repaired with FIB tools. Software overlays video images of defective and good mask areas to create a bitmap repair map. The operator then overlays

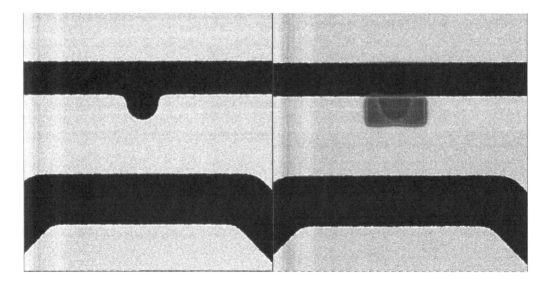

FIGURE 30.8
Before and after FIB images of FIB clear repair. (Courtesy FEI Company.)

Mask Repair

the repair map on the defect and starts the repair. The beam raster scans over only the repair area to fix these killer defects. This is known as pattern copy repair.

Sidewall profiles of both opaque and clear repairs are not vertical due to the Gaussian shape of the beam and beam tails. Gallium staining and quartz damage reduce transmission under an opaque defect repair. A repaired bridge defect may have its edges aligned perfectly with the adjacent lines, has good transmission in the middle of the clear area, and yet fails an aerial image measurement from effects of sidewall profiles, staining, and quartz damage. FIB tools offer biasing, where the edge might be moved into the chrome slightly to allow more light to be transmitted in the clear area. This will allow it to pass AIMS measurement, but the bias should not be so large that it is picked up as an edge defect during a regular defect inspection step.

FIB tool manufacturers are researching CPSM and AAPSM mask repair strategies. For quartz bump repair, defect height maps combined with variable dose milling shows promising results [12–14] (Figure 30.9). Investigators are also examining FIB dielectric deposition for quartz pit repairs [15].

Advantages:

- Excellent MoSiON opaque defect removal
- Excellent binary mask clear repair process
- Good spatial resolution for imaging and repairs
- Good placement specifications for edges and pattern copy repairs

Disadvantages:

- Gallium staining during imaging and repair reduces transmission in those areas
- Carbon films for clear repairs can match transmission of EPSM masks but not phase
- Lower throughput than laser tools
- Substrate damage due to the sputtering process
- Substrate charging affects overall repair quality, requiring mitigation techniques

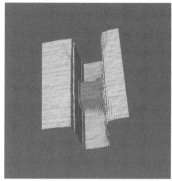

FIGURE 30.9
Left to right: FIB image of an AAPSM quartz defect, and surface profile scans before and after the repair. (Courtesy FEI Company.)

30.8 AFM Nanomachining Mask Repair

Atomic force microscope (AFM) use proliferated in the 1990s in many areas of materials and semiconductor research. A simplified view of how it operates is that a sharp tip is set vibrating at a specific frequency. It is lowered to a surface. The vibration frequency changes due to interaction of atomic forces as the tip gets close to the surface. A feedback loop keeps the tip vibrating at a specific frequency by minutely raising and lowering the tip assembly as it is scanned over a surface. A laser interferometer system bounced off the top surface of the tip senses the vertical height of the tip as it is being scanned, and that information generates a 3-dimensional image of the surface after software analysis. AFM manufacturers and researchers have developed many types of tip shapes and materials for specific uses (Figure 30.10). For some tools and applications, the tip actually contacts the surface for measurements (non-vibrating mode). Tools can now measure critical dimensions, and sidewall and reentrant angles and profiles.

30.8.1 Theory of Operation

In the mask repair application, the AFM both measures the defect and abrades it through controlled, direct contact of a tip to the surface [16,17]. Special tip designs can machine down from the top surface, or cut in from the side on defects. The tool can measure progress of repairs for real-time feedback on the repair process. The tool can image and remove opaque defects of chrome and MoSiON down very small sizes. Missing clear features have been reconstructed in absorbers [18]. It has also been used to repair quartz defects and bumps on APSM masks (Figure 30.2 and Figure 30.11). Cryogenic carbon dioxide blown over the work area removes generated debris to keep the mask clean. Clear repair has not been demonstrated in literature by this tool at this time.

Advantages:

- Able to see and repair the smallest opaque defects
- APSM quartz bump repair
- Good edge placement
- Feedback during the repair process provides minimal substrate damage
- Material type independent

Disadvantages:

- Lower throughput, especially for larger defects
- Frequent tip changes from abrasion wear and tear
- No additive repair process

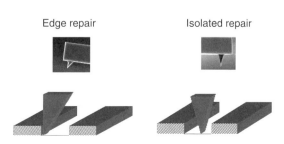

FIGURE 30.10
Special AFM tips used for nanomachining masks, and their type of application. (Courtesy RAVE, LLC.)

Mask Repair 641

FIGURE 30.11
Before and after SEM and aerial image metrology tool images of a quartz defect repair using AFM nanomachining. (Courtesy RAVE, LLC.)

30.9 Electron Beam Mask Repair

Researchers have used scanning electron microscopes for over 40 years. Continual improvements in electron sources and optics have kept these tools at the leading edge of image resolution and elemental analysis. Successful FIB gas processes recently sparked similar interest in gas chemistry work with e-beam tools. Proof of concept experiments confirmed that electron beams can perform some basic mask repair tasks, and developmental tools are now available on the market for investigations in mask repair applications. At least three companies have announced upcoming products for e-beam mask repair.

30.9.1 Theory of Operation

An electron beam constitutes the probe that forms the image and performs work on the substrate surface. Beam formation is different than FIB optics, but beam shaping and scanning and similar to FIB tools.

The two most common types of electron sources are heated filaments and field emission sources. Heated filaments "boil" off electrons from a hot surface (usually tungsten). Field emission sources use high electric fields to lower work functions that in turn allow electrons to escape the emitter surface more easily. Freed electrons from both types of sources are collected and accelerated down the column to the target. Electromagnetic and/or electrostatic lenses focus the beam to the desired shape and size as the electrons accelerate from the source to the substrate down the column. Spot current is usually defined by apertures along the beam path. Deflection coils or plates move the beam around at the substrate plane. Beam blanking subsystems move the beam off the beam path to a blanking aperture in the column, effectively turning the beam off (and on) at the substrate surface.

The standard electron beam for mask repair applications has a Gaussian profile. Electrons striking the surface either implant inside the substrate, drain off to ground through conductive paths, or escape as backscattered electrons (essentially rebounding off atomic nuclei and leaving the substrate surface). Energy from momentum transfer of the impinging electrons also creates secondary electrons that escape from the substrate. Both secondary and backscattered electrons can be collected together or separately and used for image formation though they have different contrast mechanisms. Electrons that implant in the fused quartz substrate or end up on isolated absorber islands (like clear field contacts) can rapidly build up a high charge that can deflect the electron beam from its intended position. Charging can also affect secondary and backscattered electron trajectories and collection, resulting in poor imaging. Some schemes that alleviate charging, such as ESEM™ detectors used in variable pressure SEMs, are commercially available for imaging and analysis tools, though none have been adapted for a mask repair tool yet. Incidental carbon deposition or hydrocarbon "cracking" results from the scanning beam decomposing free hydrocarbons in the vacuum chamber absorbed at the target surface [19,20] (Figure 30.12). This film absorbs light and must be minimized or eliminated for mask repair tools. Cold traps or ESEM™ detectors have helped this problem.

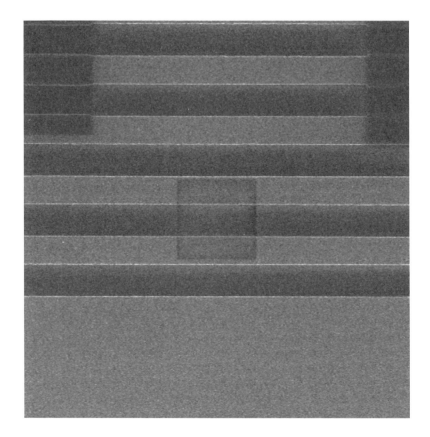

FIGURE 30.12
The center square is hydrocarbon contamination from high magnification SEM imaging; the darker areas in the upper left and right are "landmarks" from FIB imaging. (Courtesy SPIE.)

Mask Repair

FIGURE 30.13
SEM images of types of e-beam gas chemistry work. MoSi EPSM absorber etch is on left, and platinum metal deposition is on right. (Courtesy FEI Company.)

Many companies are actively pursuing e-beam chemistry research for a variety of applications, including mask repair [21]. Gases injected into the vacuum chamber and adsorbed on a substrate surface can be dissociated by the impinging electron beam, though the energy transfer by individual electrons at the gas has a much lower overall average value. Since there is no physical work, such as sputtering (as in an FIB repair tool) at the substrate surface, the surface reaction becomes almost purely chemical from interaction of the dissociated gas components and the target materials (Figure 30.13). Not all chemistries that work with for FIB applications will work in e-beam applications for that reason. Also, backscattered electrons and electrons in the tail of the Gaussian profile may cause accidental deposition or etch outside of the scanned area (sometimes called "overspray").

Some companies offer research tools for e-beam mask repair investigation [22]. Researchers performed seminal works in e-beam etch for tantalum nitride (TaN) and molybdenum silicide (MoSiON) absorbers, and metal deposition for clear repair applications. Deposits have proved resistant to chemical attack from cleaning cycles but must still be qualified for physical ruggedness from mechanical cleaning processes and examined for any issues with reflectivity. Chrome etch remains elusive for e-beam tools, however, as etch rates revealed to date are too slow for production mask repair. Companies promise commercial repair tools for the future and research proceeds apace.

Advantages:

- Damageless imaging for locating defects and setting up repairs.
- Very high possible resolution for locating small defects.
- High selectivity for gas processes due to purely chemical reaction.

Disadvantages:

- Lower yields for gas processes than FIB repair processes.
- Hydrocarbon "cracking" or staining of scanned surfaces.
- Charging effects from implanted electrons.
- No commercial production-qualified tools are offered.

30.10 Next Generation Lithography Mask Repair Challenges

Optical photomasks comprise over 99% of the current market. Experts agree that optical lithography will someday reach its end, perhaps at the 32-nm node, but no one can say that with any certainty. Several technologies are competing to win the title of next generation lithography (NGL). No attempt will be made here to either describe all of the candidate technologies or debate their relative merits; discussion will center on brief descriptions of mask construction for two mask types and implications for future mask repair.

30.10.1 Stencil Masks

Both electron projection lithography and ion projection lithography use stencil masks to expose patterns on resist [23]. The stencil mask substrate is usually a thin membrane of silicon, silicon nitride, or some other material with a few microns thick. A mask writer exposes the pattern in photoresist on the substrate, and after a develop step, the pattern is etched completely through the substrate (hence the name "stencil"). During wafer exposure in the stepper, electrons or ions striking the substrate are absorbed or deflected out of the optical path, and those flying through the etched areas are focused on the wafer below to expose the pattern on the photoresist there.

These masks are very fragile and require special handling. During repair, sidewall angles must be near 90°, since the absorber is several microns thick. Clear defect (absence of absorber) repairs must attach to existing absorber sidewalls, bridge over gaps, have high lateral resolution, be thick enough to absorb the intended particle, and yet rugged enough to withstand cleaning processes. All of the discussed repair tools have difficulties with these many requirements. Research to date shows encouraging results for opaque defect repair.

30.10.2 Extreme Ultraviolet Lithography Masks

EUV technology is unique in that the entire optical chain is comprised of reflective elements, including the mask. Ultra low coefficient of thermal expansion material maintains pattern spatial integrity as the mask substrate heats up during exposure, and the industry appears to have selected the 6-in. × 0.250-in. size as the standard, so the same installed base of mask making tool sets can be used for manufacturing. Forty alternating layers of molybdenum (Mo) and silicon (Si) in layers less than 5 Å thick reflect about two thirds of the incoming EUV photons at 13.4 nm wavelength, which arrive approximately 6° off perpendicular [24,25]. Both Mo and Si oxidize, which ruins the mirror reflectivity, so a capping layer protects them. Tantalum nitride (TaN) is one of the candidate absorber materials, which is patterned on top of the capping layer [26].

Defects can occur in either the absorber or the multilayer mirror. Research continues for repairing mirror defects, both on the surface and buried at the bottom of the mirror stack [27]. Proof of concept work uses e-beams for heating and collapsing the mirror layers above the buried defect to minimize phase errors. For surface mirror defects, research centers on very low-energy broad FIBs to sputter off surface amplitude defects. The resulting relatively large shallow crater in the mirror appears to have minimal effect on reflectivity, though a capping layer must be deposited over the crater to prevent oxidation of the exposed multilayers.

Absorber defects have their own issues. Mask repair of these substrates must remove their opaque defects cleanly with excellent resolution and with no damage to the underlying mirror. Successful repair tools must have very high resolution and capability to repair in high aspect ratios. Commercial FIB repair tools may have problems with gallium ion implantation mixing up the underlying multilayer mirrors, though lower beam energies coupled with an appropriately thick capping layer may overcome this issue [28]. E-beam mask repair research looks very promising for this new technology as it does have high resolution, is nondamaging, and a gas process has been demonstrated for TaN etch.

The future technology of choice for mask repair remains unknown, as does the choice for lithography. Mask repair is and will continue to be a critical part of all mask manufacturing. The recent new entrants of mechanical and e-beam tools for mask repair have reinvigorated research in this field, and all technologies provide the potential for exciting, important discoveries in the future.

References

1. J.N. Wiley and J.A. Reynolds, Device yield and reliability by specification of mask defects, *Solid State Technol.*, July (1993).
2. K. Bhattacharyya, W.W. Volk [KLA-Tencor Corp. (USA)], B.J. Grenon [Grenon Consulting Inc. (USA)], D. Brown, and J. Ayala [IBM Microelectronics Div. (USA)], Investigation of reticle defect formation at DUV lithography, *Proc. SPIE*, 4889, 478–487 (2002).
3. M.D. Levenson, N.S. Viswanathan, and R.A. Simpson, Improving resolution in photolithography with a phase shifting mask, *IEEE Trans. Elec. Dev.*, ED-29, 1828–1836 (1982).
4. Y. Yoshino, Y. Morishige, S. Watanabe, Y. Kyusho, A. Ueda, T. Haneda, and M. Oomiya [NEC Corp. (Japan)], High accuracy laser mask repair system LM700A, *Proc. SPIE*, 4186, 663–669 (2000).
5. P. Yan, Q. Qian, J. McCall, J. Langston, Y. Ger, J. Cho, and B. Hainsey, Effect of laser mask repair induced residue and quartz damage in sub-half micron DUV wafer process, *Proc. SPIE*, 2621, 158 (1995).
6. A. Wagner, R.A. Haight, and P. Longo [IBM Thomas J. Watson Research Center (USA)], MARS2: an advanced femtosecond laser mask repair tool, *Proc. SPIE*, 4889, 457–468 (2002).
7. M.R. Schmidt, P. Flanigan, and D. Thibault [IBM Microelectronics Div. (USA)], Photomask repair using an advanced laser-based repair system (MARS2), *Proc. SPIE*, 4889, 1023–1032 (2002).
8. J. Morgan and T.B. Morrison, *Solid State Technol.*, 43 (7), 195–201.
9. R. Hagiwara, A. Yasaka, K. Aita, O. Takaoka, Y. Koyama, T. Kozakai, T. Doi, M. Muramatsu, K. Suzuki, Y. Sugiyama, H. Sawaragi, M. Okabe, S. Shinohara, M. Hasuada, T. Adachi [Seiko Instruments, Inc. (Japan)], Y. Morikawa, M. Nishiguchi, Y. Sato, N. Hayashi [DaiNippon Printing Co., Ltd. (Japan)], T. Ozawa, Y. Tanaka, and N. Yoshioka [Semiconductor Leading Edge Technologies, Inc. (Japan)], Advanced FIB mask repair technology for 100 nm/ArF lithography, *Proc. SPIE*, 4889, 1056–1064 (2002).
10. B.J. Lin, The attenuated phase shift mask, *Solid State Technol.*, January, 43–47 (1992).
11. C. Marotta, J. Lessing, J. Marshman [FEI Company (USA)], and M. Ramstein [Infineon Technologies AG (Germany)], Repair and imaging of 193 nm MoSiON phase shift photomasks, *SPIE* 4562, 1161–1171 (2002).
12. D. Kakuta, I. Kagami, T. Komizo, and H. Ohnuma [Sony Electronics, Inc. (Japan)], Quantitative evaluation of focused ion beam repair for quartz bump defect of alternating phase-shift masks, *Proc. SPIE*, 4562, 753–761 (2001).
13. J.X. Chen, J. Riddick, M. Lamantia, A. Zerrade, R.K. Henderson, G.P. Hughes [Dupont Photomasks, Inc. (USA)], C.E. Tabery, K.A. Phan, C.A. Spence [Advanced Micro Devices, Inc. (USA)], A.A. Winder, B.A. Stanton, E.A. De.arosa [Micron Technology, Inc. (USA)], J.G. Maltabes,

C.E. Philbin, L.C. Litt [Motorola (USA)], A. Vacca, and S. Pomeroy [KLA-Tencor Corp. (USA)], ArF (193 nm) alternating aperture PSM quartz defect repair and printability for 100 nm node, *Proc. SPIE*, 4562, 786–797 (2001).
14. S. Fan [Toppan Chunghwa Electronics Co., Ltd. (Taiwan)], M. Hsu [ASML MaskTools, Inc. (USA)], A. Tseng [Toppan Chunghwa Electronics Co., Ltd. (Taiwan)], J.F. Chen, D.J. Van Den Broeke [ASML MaskTools, Inc. (USA)], H. Lei [United Microelectronic Corp. (Taiwan)], S. Hsu, and X. Shi [ASML MaskTools, Inc. (USA)], Phase defect repair for the chromeless phase lithography (CPL) mask, *Proc. SPIE*, 4889, 221–231 (2002).
15. H.D. Wanzenboeck [Vienna Univ. Technology (Austria)], M. Verbeek, W. Maurer [Infineon Technologies AG (Germany)], and E. Bertagnolli [Vienna Univ. Technology (Austria)], FIB-based local deposition of dielectrics for phase-shift mask modification, *Proc. SPIE*, 4186, 148–157 (2000).
16. M.R. Laurance [RAVE LLC (USA)], Subtractive defect repair via nanomachining, *Proc. SPIE*, 4186, 670–673 (2000).
17. B. LoBianco, R. White, and T. Nawrocki [RAVE, LLC (USA)], Use of nanomachining for 100 nm mask repair, *Proc. SPIE*, 4889, 909–921 (2002).
18. D. Brinkley [Intel Corp. (USA)], R. Bozak, B. Chiu [RAVE LLC (USA)], C. Ly, V. Tolani [Intel Corp. (USA)], and R. White [RAVE LLC (USA)], Investigation of nanomachining as a technique for geometry reconstruction, *Proc. SPIE*, 4889, 232–240 (2002).
19. B.S. Kasprowicz [Photronics, Inc. (USA)], M. Ananth, and C.-Y. Wang [KLA-Tencor Corp. (USA)], Investigating inspectability and printability of contamination deposited during SEM analysis, *Proc. SPIE*, 4186, 654–662 (2000).
20. C.M. Schilz [Infineon Technologies AG (Germany)], K. Eisner, S. Hien [International SEMATECH (USA)], T. Schleussner, R. Ludwig, and A. Semmler [Infineon Technologies AG (Germany)], Influence of e-beam-induced contamination on the printability of resist structures at 157 nm exposure, *Proc. SPIE*, 4562, 297–306 (2001).
21. V.A. Boegli, H.W.P. Koops, M. Budach, K. Edinger, O. Hoinkis, B. Weyrauch, R. Becker, R. Schmidt, A. Kaya, A. Reinhardt, S. Bräuer [NaWoTec GmbH (USA)], H. Honold, J. Bihr, J. Greiser, and M. Eisenmann [LEO Elektronenmikroskopie GmbH (Germany)], Electron beam-induced processes and their applicability to mask repair, *Proc. SPIE*, 4889, 283–292 (2002).
22. H.W. Koops, K. Edinger, V. Boegli [NaWoTec GmbH (Germany)], J. Bihr, and J. Greiser [LEO Elektronenmikroskopie GmbH (Germany)], Electron beam mask repair with induced reactions, in: *19th European Mask Conference on Mask Technology for Integrated Circuits and Micro-Components*, 2003.
23. K. Suzuki, T. Fujiwara, K. Hada, N. Hirayanagi, S. Kawata, K. Morita, K. Okamoto, T. Okino, S. Shimizu, T. Yahiro, and H. Yamamoto [Nikon Corporation (Japan)], Nikon EB Stepper: the latest development status, *Proc. SPIE*, 4343, 10 (2001).
24. S.D. Hector, EUVL masks: requirements and potential solutions, emerging lithographic technologies VI, *Proc. SPIE*, 4688, 134–149 (2002).
25. SEMI International Standard P38, Specification for absorbing film stacks and multilayers on extreme ultraviolet lithography mask blanks, Semiconductor Equipment and Materials International, San Jose, CA, USA.
26. P.Y. Yan, G. Zhang, A. Ma, and T. Liang [Intel Corporation (USA)], TaN and Cr EUV mask fabrication and characterization, *Proc. SPIE*, 4343, 42 (2002).
27. Y. Deng [Univ. California/Berkeley (USA)], B. LaFontaine [Advanced Micro Devices Corp. (USA)], and A.R. Neureuther [Univ. California/Berkeley (USA)], Performance of repaired defects and attPSM in EUV multilayer masks, *Proc. SPIE*, 4889, 418–425 (2002).
28. Ted Liang, Alan Stivers, Richard Livengood, Pei-Yang Yan, Guojing Zhang, Fu-Chang Lo, Progress in extreme ultraviolet mask repair using a focused ion beam, *J. Vac. Sci. Technol. B.*, 18 (6), 3216–3220 (2000).

Section VII

Modeling and Simulation

31
Modeling and Simulation

Andreas Erdmann

CONTENTS

31.1 Introduction into Lithography Simulation 650
    31.1.1 Aerial and Resist Image Formation 651
    31.1.2 Physical/Chemical Modifications during the Processing of the Photoresist 654
    31.1.3 General Simulation Strategies 655
31.2 Rigorous Modeling of Light Diffraction from Masks 656
    31.2.1 Motivation 656
    31.2.2 General Remarks on Rigorous Electromagnetic Field Solvers 658
    31.2.3 Finite-Difference Time-Domain Method 660
        31.2.3.1 A Short History of FDTD 660
        31.2.3.2 The FDTD Algorithm 660
    31.2.4 Coupling with Standard Lithography Simulations and Performance Considerations 667
    31.2.5 Field Decomposition Techniques and other Enhancements 669
31.3 Application of Simulation to Different Mask Types 670
    31.3.1 Binary Masks 670
    31.3.2 Phase Shift Masks 672
    31.3.3 EUV-Masks 675
        31.3.3.1 Efficient Modeling of EUV-Masks 676
        31.3.3.2 Mask-Induced Imaging Artifacts in EUV Lithography 679
31.4 Simulation of Mask Defects 681
    31.4.1 Standard Defect Printability Considerations 681
    31.4.2 Rigorous EMF Simulation of Defects on Optical Masks 683
    31.4.3 EUV Multilayer Defects 685
31.5 Summary 686
Acknowledgments 687
References 688

Modeling and simulation have become indispensable tools for the understanding and optimization of lithographic processes and for the development of new process technology. Aerial image simulation is used to evaluate the imaging of designed photomasks and to explore the impact of optical parameters, such as numerical aperture (NA), partial coherence σ, and defocus, and wave aberrations on the imaging performance of a projection stepper or scanner. Other simulation approaches are used to describe the impact of the photoresist thickness, of the post-exposure bake (PEB) temperature and of the development characteristics of the photoresist on the total process performance. Lithography simulation is also used as an effective learning tool for new process engineers, scientists, and managers.

Until recently, most of the simulation approaches considered the mask as an infinitely thin object with a transmission T, which is directly derived from the design. For binary masks, the dark chromium features would produce a transmission $T \approx 0$, while the bright clear areas would give $T = 1$. This assumption of an infinitely thin mask is often referred to as the Kirchhoff approach. Unfortunately, the Kirchhoff approach becomes inaccurate when considering many advanced mask technologies, such as optical proximity correction (OPC), phase shift masks (PSM), and masks for extreme ultraviolet (EUV) lithography. This chapter will introduce advanced modeling approaches for modern mask technologies. Rigorous electromagnetic field (EMF) simulation is used for an accurate description of the light diffraction from various types of masks. The advanced modeling approaches can be used to explore the impact of the mask geometry/topography on the imaging and on the full process performance. Several important mask topography-related imaging artifacts will be discussed in detail.

To understand the importance of the advanced mask modeling approaches in a general simulation framework, this chapter will start with a brief review of the most important models and methods, which are used in lithography simulation. Rigorous EMF simulation of light diffraction from masks will be introduced in Section 31.2. Most of the commercially available simulation programs are based on finite-difference time-domain (FDTD) algorithms. Several examples will demonstrate the operation of FDTD and its integration into a lithography simulator. Section 31.2 will conclude with a discussion of field/domain decomposition techniques and other approaches that are used to increase the efficiency of advanced mask modeling. Section 31.3 will discuss mask topography-related imaging artifacts, such as sizing errors, OPC implications, intensity imbalancing, placement errors, and asymmetric process windows (PWs). Special emphasis is put on imaging artifacts in EUV masks. Section 31.4 addresses the defect printability as another important aspect of mask simulation. This includes chromium and phase defects on optical masks, and defects in EUV mask blanks.

31.1 Introduction into Lithography Simulation

The purpose of this introductory section is to provide the reader with a general background of lithography simulation and to understand the mask-specific modeling issues, which are explained in the following sections within a general simulation framework. In general, a simulator for optical or EUV lithography consists of models that describe the projection of a mask into a photoresist on top of a silicon wafer, the interaction of the projected light with the photoresist, and the physical/chemical processes during PEB and development. For an in-depth discussion of the physical/chemical models and numerical algorithms, which are used in lithography simulation, the reader is referred to specific literature and references therein [1–4].

Specifically, this section will introduce the basic ideas for the computation of the intensity distribution in the image plane of a projection system. The resulting aerial image can be used for a first evaluation of the process performance. Some peculiarities, resulting from the projection of the image into a photoresist on top of a wafer-stack will be discussed. The second part of this section gives a brief review of the chemical/physical modifications of the photoresist during the different process steps. Finally, the general structure of a lithography simulator will be introduced.

31.1.1 Aerial and Resist Image Formation

The projection imaging of a mask with a spatially modulated transmission T is one of the most critical steps in semiconductor fabrication processes. A principle sketch of a lithographic projection stepper/scanner is shown in Figure 31.1. The imaging system consists of an illumination optics (light source, condenser) and a projection optics (projection lens, aperture stop). The condenser system is designed to ensure a homogeneous illumination of the mask. The projection system images the mask into the image plane close to the wafer surface. The projection lens transmits only a certain angular range of the light, which is diffracted by the mask, whose sine is bounded by the numerical aperture NA of the projection system.

To compute the intensity of the projected light, the so-called aerial image, several assumptions are made:

- The mask is considered to be an infinitely thin object with a complex valued transmission $T(x, y)$, where x and y specify Cartesian coordinates in the mask plane. The value of T is directly derived from the mask design. In other words, the absolute value of T is close to 0 in chromium covered areas, and $T = 1$ in the transparent areas of a binary mask. The described assumption will be referred as the Kirchhoff approach. In Section 31.2, a more accurate model for the description of the mask will be introduced.

- The projection lens is characterized by a complex valued pupil function $P(\theta_x, \theta_y)$. The P describes the transfer of plane waves through the projection lens. The direction of the plane waves is specified by the angles θ_x, and θ_y with respect to the optical axis of the system. The magnitude of P is 0 outside the numerical aperture NA of the projection lens: $P(\theta_x, \theta_y) = 0$ for $(\sin^2 \theta_x + \sin^2 \theta_y) > NA^2$. In the case, that all plane wave components inside the numerical aperture $(\sin^2 \theta_x + \sin^2 \theta_y) < NA^2$ are transferred without any amplitude or phase distortion ($P = 1$), one speaks of diffraction limited imaging. Misalignments, design, and manufacturing imperfections result in a phase delay ϕ of the plane waves inside the projection lens, which depends on the propagation angles θ_x, θ_y: $P = \exp[-i\ \phi(\theta_x, \theta_y)]$.

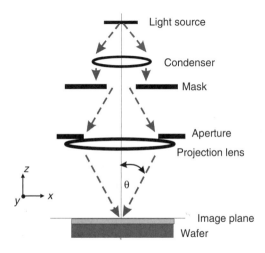

FIGURE 31.1
Principal sketch of an optical projection system used in lithography steppers or scanners.

The described phase effects correspond to wave aberration of the projection system. These wave aberrations result in focus shifts, placement errors, and other imaging artifacts. Additionally, pupil apodization and residual variations of the pupil transmission may result in a dependency of the magnitude of P from the angles θ_x and θ_y.

- In standard lithographic projection systems, the light source can be considered to consist of single source points, which emit mutually incoherent light. The condenser lens transfers the light of a single source point o into a plane wave with the direction θ_x°, θ_y°. The directions for the illumination of the mask are limited by the numerical aperture of the condenser NA_C: $(\sin^2 \theta_x^\circ + \sin^2 \theta_y^\circ) < NA_C^2$. The ratio between the numerical apertures of the condenser and of the projection lens defines the partial coherence parameter σ of the system: $\sigma = NA_C/NA$. The described source concept can also be used for the specification of more advanced illumination schemes, such as annular or multipole illuminations.

With the additional assumption that polarization effects can be neglected in the image formation, the aerial image $I(x, y)$ of the mask with a transmission $T(x, y)$ is obtained by the following formulas:

$$s(f_x, f_y) = F[T(x, y)]$$
$$a(x, y, q_x, q_y) = F^{-1}[P(f_x, f_y) \cdot s(f_x - q_x, f_y - q_y)]$$
$$I(x, y) = \iint_{\text{source}} a(x, y, q_x, q_y) \cdot a^*(x, y, q_x, q_y) \cdot dq_x dq_y$$

(31.1)

The first equation describes the diffraction of a monochromatic plane wave (wavelength λ) from a mask with a complex transmission $T(x, y)$. The diffraction spectrum s specifies the angular distribution of light at the aperture stop in the far field of the mask. The direction of the diffracted light depends on the diffraction angles θ_x, θ_y or on the spatial frequencies $f_x = \sin \theta_x / \lambda$, and $f_y = \sin \theta_y / \lambda$. The resulting complex field amplitude in the image plane for an illumination of the mask by a plane wave (direction $q_x = \sin \theta_x^\circ / \lambda$, $q_y = \sin \theta_y^\circ / \lambda$) is given by the second equation. The F and F^{-1} symbolize the 2D Fourier transformation and the inverse Fourier transformation, respectively. The P is the pupil function of the projection system, which depends on the numerical aperture and on the wave aberrations of the projection lens. The third equation describes the incoherent superposition of contributions resulting from different source directions θ_x°, θ_y°. Generalization of Equation (31.1) for high numerical aperture systems is straightforward [5,6]. Such generalization covers both a pupil apodization to fulfill energy conservation between entrance and exit pupil and vector effects in the interference of arbitrary polarized light waves.

Figure 31.2 symbolizes the evaluation of Equation (31.1). The smallest dimension of the chromium features on the binary mask is 250 nm. The mask is illuminated with light ($\lambda = 248$ nm) from a single source point on the optical axis ($\theta_x^\circ = 0$, $\theta_y^\circ = 0$). Fourier transformation of the layout $T(x, y)$ results in a diffraction spectrum $s(f_x, f_y)$ entering the projection lens. The numerical aperture (NA = 0.7) blocks the components of the diffraction spectrum propagating at large angles with respect to the optical axis. Inverse Fourier transformation of the resulting band limited diffraction spectrum gives the aerial image of a single source point. The images corresponding to other source points of the partial coherent source ($\sigma = 0.52$) are obtained by shifting the diffraction spectrum according to the coordinates of the source point [viz., the 2nd row of Equation (31.1)]. The final aerial image is obtained by an incoherent superposition of the contributions from all source points. A closer look at the aerial image reveals that the small contacts-like openings in the

Modeling and Simulation

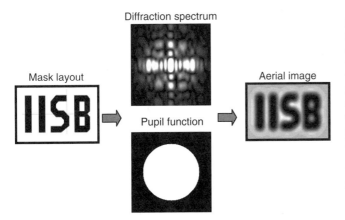

FIGURE 31.2
Schematic presentation of aerial image computation: Fourier transformation of the mask layout produces the diffraction spectrum. Inverse Fourier transformation of the product of the diffraction spectrum and the pupil function produces the aerial image resulting from a single source point. Performing this operation for all source points by appropriate shift of the object spectrum (not shown here) produces the total aerial image. Bright areas: high transmission/intensity; dark areas: low transmission/intensity.

letter B are not well resolved. The line end shortening observed at the letters "I" presents another important imaging artifact.

It is worth mentioning that other mathematical formulations of the imaging problem exist. In contrast to the so-called Abbe method described earlier, the Hopkins formulation of the imaging problem characterizes the performance of the system by a transmission cross-correlation (TCC) matrix of the illumination pattern with the projection pupil and its complex conjugate. In many cases, the Hopkins formulation gives a better computing performance than the Abbe method. For a more complete comparison of the imaging performances of the Abbe method and of the Hopkins formulation, the reader is referred to Ref. [7]. Further enhancements in the speed of the aerial image computation can be achieved by alternative representations of partially coherent light sources and by the application of pre-calculated imaging kernels [8,9].

The aerial image provides very useful information regarding the performance of a lithographic process. A simple threshold model can be applied to determine the shape of the photoresist after development and to perform line-width or critical dimension (CD) measurements. The slope of the intensity profile at the nominal edge position of the resist and the contrast of the image for dense/semi-dense features provide additional measures to evaluate the performance of a process. A general aerial image metric, which unifies different aerial image evaluation methods, was recently proposed [10].

Although the aerial image can be used as a first indicator on the process performance, the full exploration of lithographic processes requires a more detailed description of the interaction of the incident light with the photoresist/wafer-stack. The aerial image described above specifies the intensity distribution at a certain focus position. Variations of this focus position within the photoresist cannot be ignored. The interference between the incident light and light that is reflected from substrate layers below the resist result in standing wave phenomena, such as a variation of the total amount of energy coupled into the photoresist versus the resist thickness. In practice, line-width variations resulting from standing wave phenomena are reduced by the application of anti-reflecting coatings (ARC). Finally, the deposition of energy during lithographic exposures results in a modification of the chemical composition of the photoresist.

If the resist is deposited over a planar substrate and the optical properties of the photoresist are not changed during the exposure, the light propagation in the resist/wafer-stack can be described by relatively simple methods. For moderate numerical apertures NA < 0.65 focusing effects within the resist are described by the scaled defocus approach (SDFA) [11]. Higher NA requires the combination of vector aerial image computation with transfer-matrices for the analytical description of light propagation within homogeneous layer stacks [12]. The resulting expressions describing the intensity distribution inside the resist are valid for arbitrary NA [5,13]. Significant changes of the optical properties of the photoresist during the exposure require the application of more advanced methods to compute the deposited energy within the resist [13,14]. Finally, the modeling of lithographic exposures over non-planar substrates requires the application of rigorous EMF solvers, such as explained in Section 31.2.2. The application of EMF-solvers for the modeling of exposures over non-planar wafers is beyond the topic of this book. The interested reader is referred to Ref. [15], and references therein.

31.1.2 Physical/Chemical Modifications during the Processing of the Photoresist

From a simplified modeling point of view, a modern chemically amplified photoresist can be considered to consist of dissolution inhibitors and photoacid generators. The dissolution inhibitors are long chains of molecules, which do not dissolve in a standard developer. The photoacid generators are other molecules that, when hit by a photon, produce a photoacid.

Figure 31.3 demonstrates the status of the photoresist during the different processing steps. After the resist spin on and a pre-exposure bake, both the dissolution inhibitors M and the photoacid generators PAG are uniformly distributed within the resist [Figure 31.3(a)]. Light incident on the central part of the resist initializes photoacid generators PAG to produce an acid A [see Figure 31.3(b)]. The resulting first-order reaction kinetics is described by Dill's law:

$$\frac{\partial [PAG]}{\partial t} = -C_{Dill} \cdot I \cdot [PAG]$$
$$\alpha = -A_{Dill} \cdot [PAG] + B_{Dill}$$
(31.2)

Here [PAG] symbolizes the concentration of PAG normalized with respect to the uniform concentration of photoacid generator immediately before the exposure. I is the intensity of light inside the photoresist. A_{Dill}, B_{Dill}, and C_{Dill} are characteristic material properties that describe the bleachable and unbleachable absorption and the photosensitivity of the resist. The concentration of acid is determined from $[A] = 1 - [PAG]$, where the symbol [... indicates the specification of a normalized concentration.

As can be seen from Figure 31.3(c), supply of thermal energy during the PEB triggers an acid catalyzed deprotection of inhibitor molecules. The long chains of molecules that

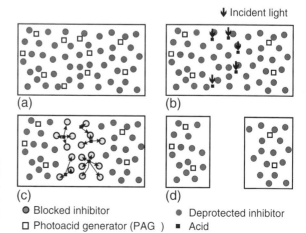

FIGURE 31.3
Schematic presentation of the functionality of a CAR: (a) status of the resist before exposure, (b) light-induced generation of photoacid in exposed areas, (c) acid-catalyzed thermal deprotection of inhibitor during PEB, and (d) chemical development of the resist.

○ Blocked inhibitor   ● Deprotected inhibitor
□ Photoacid generator (PAG)   ■ Acid

inhibit the solubility of the resist during the development process are deprotected. Note that the acid is not consumed in such reaction. One acid molecule may contribute to the deprotection of several inhibitor molecules. In its simplest form such deprotection reaction can be described by

$$\frac{\partial [M]}{\partial t} = -k_{\text{amp}} \cdot [A]^m \cdot [M] \qquad (31.3)$$

where $[M]$ symbolizes the normalized inhibitor concentration. k_{amp} and m are a reaction constant and order, respectively. The real processes in chemical amplified resist (CAR) systems can be much more complicated. Modern CARs incorporate also a quencher base into the resist design to control the amount of acid that is generated in certain parts of the resist. The supply of thermal energy and the deprotection of inhibitor increase the mobility of the acid and of other resist components resulting in complex diffusion phenomena. For a more complete discussion of the modeling of CAR, the reader is referred to Ref. [16], and references therein.

The final step in the processing sequence [Figure 31.3(d)] is the chemical development of the resist. Parts of the resist with a reduced concentration of inhibitor are dissolved. The characteristics of a development process are specified by a rate function $R([M])$ that describes the dependency of the development speed on the concentration of inhibitor. A review on the performance of different rate functions can be found in Ref. [17].

The user of lithography simulation programs is not always interested in details of the resist chemistry. Often he does not have access to the necessary simulation parameters. In such cases, various forms of approximate models for resist processing effects or lumped parameter models provide interesting alternatives to include resist effects into efficient simulation flows (see Refs. [18–21], for example).

31.1.3 General Simulation Strategies

The modeling approaches for the individual processing steps can be summarized in a general simulation scheme (see Figure 31.4). Once such a general simulation scheme is

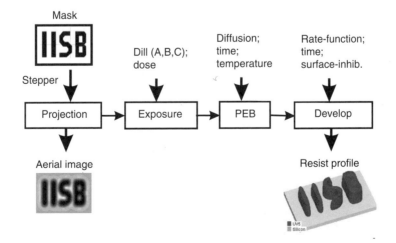

FIGURE 31.4
Schematic presentation of a lithography simulator.

established, the impact of various parameters on the lithographic process parameters can be investigated. According to the multitude of simulation parameters and model options, a virtually infinite amount of data for the characterization of lithographic processes can be generated. The most important standard measures, which will be used in the following sections, are:

- First, the size or CD of a feature has to be extracted from the simulation result. This can be done by the application of a threshold value to simulated aerial images or by the analysis of simulated resist profiles after development.
- Exposure/defocus-matrices (ED-matrices) [22] or Bossung curves characterize the sensitivity of a CD on the wafer with respect to the axial (z-) position of the wafer and the exposure dose, respectively. The dose, which is required to obtain the target feature width at a certain focus position, is specified as the dose to size. The isofocal dose is the dose with the smallest sensitivity of the CD with respect to defocus. A good process should be tuned to have a dose to size close to the isofocal dose.
- A PW specifies the dose/defocus range, which can be used to print a feature within a certain tolerance limit around the target feature size. It is not only important to achieve maximum PW for single feature types but also to guarantee maximum areas of overlapping PWs between different feature types.
- The mask error enhancement factor (MEEF) is a measure of the linearity of a process [23]. It describes how small changes of feature dimensions on the mask ΔCD_{mask} translate into changes of feature dimensions on the wafer ΔCD_{wafer}:

$$\text{MEEF} = m \cdot \frac{\Delta CD_{wafer}}{\Delta CD_{mask}} \qquad (31.4)$$

where m specifies the reduction factor of the projection system (generally 4 or 5).
- Another important parameter for the performance of a lithographic process is the placement error, that is the deviation between the real and the nominal position of a certain feature on the wafer. Usually, placement errors are attributed to alignment problems or to certain wave aberrations of the projection system. Section 31.3 will show that the mask topography may also contribute to a placement error.

31.2 Rigorous Modeling of Light Diffraction from Masks

31.2.1 Motivation

One of the assumptions concerning the standard modeling approaches for the aerial image formation, which were made in Section 31.1.1, was the description of the mask as an infinitely thin object (Kirchhoff approach). This approach suggests that sidewalls, shapes of etched areas on PSM, and other geometrical details of the mask have virtually no impact on the mask performance. This may be true for large areas on binary chromium on glass masks, but what about sub-wavelength assist features in aggressive OPC, edges on alternating PSM, or complicated multilayer structures on EUV masks? To explore the limits of the Kirchhoff approach, a more rigorous approach for the modeling of light diffraction from masks will be introduced.

Modeling and Simulation 657

Figure 31.5 compares the alternative approaches for the modeling of light diffraction from lithographic masks. Figure 31.5(a) illustrates the traditional Kirchhoff approach. The complex valued mask transmission is directly derived from the design of the mask. In contrast to that, Figure 31.5(b) shows the EMF, which results, when an incident plane wave hits the real mask. A part of the incident wave is reflected back into the glass substrate. Interference between the incident plane wave and the reflected light creates a typical standing wave pattern. The light, which is transmitted through the mask, resembles this one that is specified by the Kirchhoff approach. However, a closer look at the transmitted near fields directly below the absorber reveals some important differences. The Kirchhoff approach suggests a sharp transition between the bright and the dark regions on the mask. In contrast to that, both the amplitude and the phase of the "real electromagnetic field" show a much smoother transition from the dark to the bright area. How does the observed effect scale with feature size? Are the observed differences equally important for all mask technologies? How important are these differences in the near field for the general imaging performance of a mask? The remaining part of this section will introduce extensions of standard simulation methods, which are necessary to provide answers to these questions.

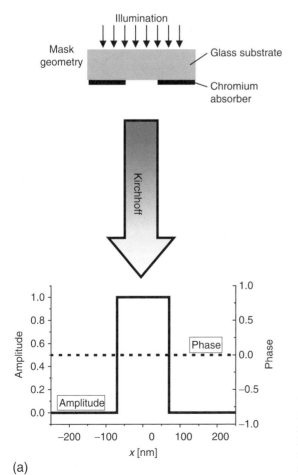

FIGURE 31.5
Alternative modeling approaches for a 4×, KrF mask with a 140-nm wide isolated space: (a) Kirchhoff approach, the amplitude and phase of the transmission function are directly derived from mask layout.

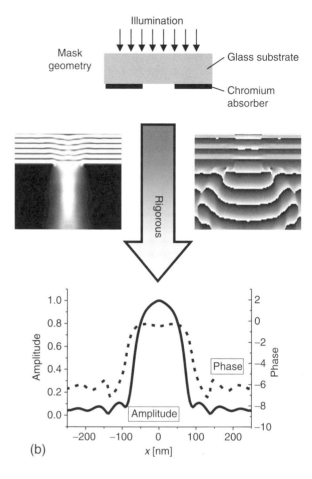

FIGURE 31.5 cont'd
(b) Rigorous approach, the amplitude and phase of the transmission function are extracted from the EMF in the vicinity of the mask. All dimensions are given on wafer scale.

31.2.2 General Remarks on Rigorous Electromagnetic Field Solvers

In general, the diffraction of light from a mask has to be considered as an electromagnetic boundary value problem. The electric and magnetic fields $\vec{E}$ and $\vec{H}$ in the vicinity of the mask are described by Maxwell's equations. Most materials used in semiconductor lithography are isotropic, non-magnetic, and source-free. In this case Maxwell's equations are given as

$$\vec{\nabla} \times \vec{E} = -\mu_0 \frac{\partial \vec{H}}{\partial t}$$
$$\vec{\nabla} \times \vec{H} = \varepsilon_0 \varepsilon \frac{\partial \vec{E}}{\partial t} + \rho \cdot \vec{E} \quad (31.5)$$
$$\vec{\nabla}\left(\varepsilon \vec{E}\right) = 0$$
$$\vec{\nabla} \vec{H} = 0$$

Here ε_0 and μ_0 specify electric permittivity and the magnetic permeability of vacuum, respectively. The electric permittivity ε and the electric conductivity ρ contain the information on the optical properties of the mask materials. In general, the system of partial differential equations specified by Equation (31.5) has to be solved by numerical methods.

The resulting time-and-space-dependent fields are obtained for a certain area. The behavior of the fields outside this area is specified by appropriate boundary conditions.

In optical lithography, the mask is illuminated by quasi-monochromatic light. The average rate of the absorption of the incident light by the photoresist occurs on a timescale that is approximately 7 or more orders of magnitude slower than the periodicity of the incident light [24]. Therefore, lithographers are mostly interested in steady-state solutions of Equation (31.5) with a periodic time dependence $\vec{E}, \vec{H} \propto \exp(i \cdot \omega t)$.

The described problem can be addressed by different numerical methods. Besides finite-difference time-domain method (FDTD) algorithms, which will be described in detail in the following section, the most important representatives of such methods are listed below.

- The rigorous coupled wave analysis (RCWA) [25] is a well-established method for the rigorous modeling of light diffraction from optical gratings. In this method, the optical material properties of the materials and the EMFs are expanded in Fourier series. The solution of Maxwell's equations is reduced to the solution of an eigenvalue problem in discrete Fourier space. Schiavone et al. [26] applied RCWA to the modeling of light diffraction from EUV masks. Also, RCWA is commonly used in scatterometry applications.

- Another representative of the frequency domain methods, which are based on Fourier expansions of the EMF and of the optical material properties, is the differential method (DIM). In contrast to RCWA, DIM transforms the partial differential equations [Equation (31.5)] into sets of coupled ordinary differential equations, which are solved by standard integration schemes. Yeung [13] employed this method to the modeling of light scattering from inhomogeneously bleached photoresists. Kirchhauer and Selberherr [27] extended DIM to the modeling of 3D geometries. The resulting algorithms were successfully applied to the modeling of EUV masks [28].

- The waveguide method (WGM) was originally developed by Nyyssonen [29] for a rigorous modeling of optical linewidth measurement. In this method, the fields within the object, which is assumed to be periodic in one direction and to have a rectangular sidewall profile, are expanded in the eigenmodes of a waveguide having the same horizontal cross-sections. Yuan [30] applied WGM to the modeling of light diffraction from 2D PSMs. Lucas et al. [31] extended WGM to the modeling of 3D geometries. Recently, the method was applied to the rigorous modeling of EUV masks [32].

- Finite-element modeling (FEM) is an another standard method of computational electromagnetics. FEM formulates electromagnetic problem as a variational principle, which is evaluated at an appropriate mesh of elements. Wojcik et al. [33] applied FEM to the modeling of light scattering from PSMs.

Due to the slow convergence of Fourier series expansions, the application of frequency domain methods, such as RCWA, DIM, and WGM, is limited to small periods. In general, both 2D and 3D codes using these methods are available; however, 3D cases require at least 512 MB of CPU. The possibility to adapt the mesh of finite elements to a specific geometry makes FEM very efficient. On the other hand, due to meshing, FEM methods are difficult to use without in-depth knowledge of theoretical electromagnetics.

The most popular method for the rigorous simulation of light diffraction from photomasks is the FDTD method. The next section will introduce the basic ideas of this method. There are thousands of publications on special physical/numerical aspects of the FDTD and on various applications. The web page [34] can be used as a starting point for an extensive recherché of the method. Taflove [35] provides an excellent guideline for the implementation of effective FDTD codes. In addition to that, several theses from the University of California at Berkeley provide very useful information regarding the practical application of FDTD in lithography modeling [36–38]. The following three sections are intended to present a review of the most important aspects of FDTD with special emphasis on applications for lithography simulations. Readers who are less interested in the mathematics and in the theoretical background of FDTD may skip Section 31.2.3 and continue with Section 31.2.4.

31.2.3 Finite-Difference Time-Domain Method

31.2.3.1 A Short History of FDTD

The original version of the FDTD was published in 1966 by Yee [39]. At the time of this publication, the performance of available computers was rather limited. Therefore, the interest in the computationally extensive scheme proposed by Yee was only limited. This situation changed in the following years. In the 1980s, FDTD was applied to all types of electromagnetic scattering problems. For example, Wojcik et al. [40] computed the scattering of light by small particles on the wafer surface. Alfred Wong was the first to apply a FDTD simulator TEMPEST, developed by students of Neureuther [41], at University of California, Berkeley, to rigorous simulation of light diffraction from photomasks [42]. He generalized TEMPEST from 2D to 3D and identified many important problems, which are associated with the topography of PSM [43]. In the late 1990s, other people developed their own FDTD programs for the rigorous EMF simulation of light scattering from photomasks [44,45]. Meanwhile, various commercial software packages, such as SOLID-CM [46], TEMPESTpr [47], and PROMAX/2D [48], have become available.

31.2.3.2 The FDTD Algorithm

The basic idea of FDTD is to numerically integrate the first two equations in Equation (31.5) in discrete time steps. Both, electric and magnetic fields are composed of three field components: $\vec{E} = (E_x, E_y, E_z)$, $\vec{H} = (H_x, H_y, H_z)$. In other words, six mutually dependent field components have to be integrated over time. This integration is performed on equidistant numerical grids. Every field component is specified on an individual grid. Appropriate positioning of the meshes of the different field components guarantees that the last two equations of Equation (31.5) are also fulfilled during the numerical integration procedure.

To demonstrate the basic ideas of FDTD, in the following, a relative simple diffraction problem will be considered: An incident plane light wave propagating along the z-axis hits an isolated phase edge on an alternating PSM (see Figure 31.6). The geometry of the mask, which is the electric permittivity $\varepsilon(x,z)$, the electric conductivity $\rho(x,z)$, and the incident light do not depend on the coordinate y. In that case, the general system of six coupled field components specified in Equation (31.5) decouple into two systems of independent field components: TE-polarization (E_y, H_x, H_z) and TM-polarization (H_y, E_x, E_z), respectively.*

For the described geometry and naming conventions, Equation (31.5) can be used to derive the equation of the TE-polarized EMF components:

---

*The specified naming convention is used in most of the lithography related literature. However, there exist also alternative naming conventions in other fields of electromagnetics.

Modeling and Simulation

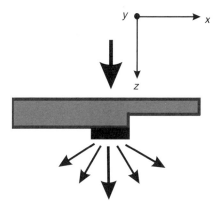

FIGURE 31.6
Schematic presentation of light diffraction from an isolated phase edge: black, chromium; gray, glass.

$$\frac{\partial H_x}{\partial t} = \frac{1}{\mu_0}\left(\frac{\partial E_y}{\partial z}\right)$$

$$\frac{\partial H_z}{\partial t} = -\frac{1}{\mu_0}\left(\frac{\partial E_y}{\partial x}\right) \quad (31.6)$$

$$\frac{\partial E_y}{\partial t} = \frac{1}{\varepsilon_0 \varepsilon}\left(\frac{\partial H_x}{\partial z} - \frac{\partial H_z}{\partial x} - \rho \cdot E_y\right)$$

Figure 31.7 shows the staggered grids that are used for the numerical integration of Equation (31.6). The structure of the staggered grid follows the symmetry of the upper differential equations. According to the direction of the spatial derivatives, the field component E_y is specified above/below the H_x field component and left/right to the H_z field components. Similar relations hold for other field components. With the upper choice of grids and a corresponding discretization in time, Equation (31.6) can be rewritten as:

$$H_x|_{i,j}^{n+\frac{1}{2}} = H_x|_{i,j}^{n-\frac{1}{2}} + \frac{\Delta t}{\mu_0 \Delta x}\left(E_y|_{i,j+1}^{n} - E_y|_{i,j}^{n}\right)$$

$$H_z|_{i,j}^{n+\frac{1}{2}} = H_z|_{i,j}^{n-\frac{1}{2}} + \frac{\Delta t}{\mu_0 \Delta x}\left(E_y|_{i,j}^{n} - E_y|_{i+1,j}^{n}\right) \quad (31.7)$$

$$E_y|_{i,j}^{n+1} = C_a|_{i,j} \cdot E_y|_{i,j}^{n} + C_b|_{i,j}\left(H_x|_{i,j}^{n+\frac{1}{2}} - H_x|_{i,j-1}^{n+\frac{1}{2}} + H_z|_{i-1,j}^{n+\frac{1}{2}} - H_z|_{i,j}^{n+\frac{1}{2}}\right)$$

Here Δx and Δt are the discrete steps in space and time, respectively. For a simplified discussion of the results, the same spatial discretization in x and z was assumed $\Delta z = \Delta x$. The coefficients $C_a|_{i,j}$ and $C_b|_{i,j}$ depend on the material properties at the position i,j in the grid of the E_y-field component:

$$C_a|_{i,j} = \left(1 - \frac{\rho_{i,j}\Delta t}{2\varepsilon_0 \varepsilon_{i,j}}\right) \cdot \left(1 + \frac{\rho_{i,j}\Delta t}{2\varepsilon_0 \varepsilon_{i,j}}\right)^{-1}$$

$$C_b|_{i,j} = \left(\frac{\Delta t}{2\varepsilon_0 \varepsilon_{i,j}}\right) \cdot \left(1 + \frac{\rho_{i,j}\Delta t}{2\varepsilon_0 \varepsilon_{i,j}}\right)^{-1} \quad (31.8)$$

According to Equation (31.7), the EMF components are not only staggered in space but also in time. The magnetic field components are sampled half a time step between the

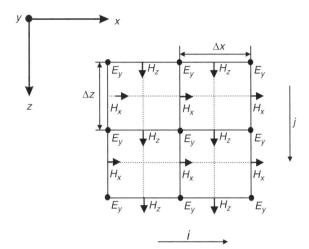

FIGURE 31.7
Position of the electric and magnetic field vector components for the numerical integration of the TE-field equations [Equations (31.6)].

electric field components and vice versa. Equations (31.7) and (31.8) provide the updating equations for the TE-polarized field components inside a y-independent geometry. Similar equations can be derived for the TM-polarized case and for the general case of six coupled field components.

Before a first example of an EMF field computation using FDTD is presented, some additional remarks on the algorithm are necessary as follows.

31.2.3.2.1 Accuracy and Stability Condition

Equations (31.7) and (31.8) are given for arbitrary discretizations in space Δx and in time Δt. The accuracy of a numerical solution obtained with FDTD is governed by the spatial discretization Δx. Often, it is convenient to specify the number of meshpoints per wavelength in an optical material with the refractive index n by $N = \lambda/(n \cdot \Delta x)$. As a rule of thumb a relative accuracy better than 2% requires $N > 15$ meshpoints per wavelength in the material with the largest refractive index n [36]. The stability conditions for the upper finite difference scheme in equations limits the magnitude of discrete time steps to $\Delta t < \Delta x/c$, where c is the vacuum velocity of light. Standard FDTD programs propose the application of a specific discretization in space Δx to achieve a certain accuracy of the numerical solution. The magnitude of the time step Δt is automatically determined.

31.2.3.2.2 Boundary Conditions

Equation (31.7) specifies the updating equations inside the staggered grid. The values of the EMF components have to be specified everywhere in the neighborhood of the meshpoint. This is not possible for field components at the boundary of the finite computing window. Meshpoints on the boundary require specific updating equations, which determine the behavior of the EMF at the boundaries of the computing window. Possible choices are as follows.

- *Reflecting boundary conditions* (RBCs): The numerically simplest choice of boundary conditions is perfectly conducting electric (or magnetic) walls. This type of boundary conditions is implemented by setting the values of all electric (or magnetic) field components outside the computing window to zero. The EMF incident on such type of boundary is simply reflected. The phase of the reflected field depends on whether perfectly conducting electric or magnetic walls are implemented. Although this type of boundary conditions is easy to implement, it does not describe the physical reality. Therefore, RBCs are used only for some software tests and for a few special geometries.

- *Periodic boundary conditions* (PBC): In this case, the electric (or magnetic) field component on the boundary is computed using the magnetic (or electric) field components from both sides of the computing window. Figure 31.8(a) shows the practical implementation of a PBC. A Gaussian-shaped electromagnetic pulse that hits the left boundary appears on the right-hand side of the computing window. Such types of boundary conditions are especially useful for the modeling of periodic structures, such as dense lines and spaces.

- *Transparent boundary conditions* (TBC): Since the first paper of Yee on the FDTD algorithm was published, several attempts were made to formulate boundary conditions in such a way that the field that strikes the boundary of the computing window leaves that window without any reflection from this boundary. In fact, such boundary conditions are also required for the mask modeling. The diffracted light leaves the computing window in direction of the entrance pupil of the projection system without any "virtual" back reflection. The basic idea of TBC is to surround the computing window by a material that absorbs the light. Unfortunately, standard materials that absorb the light do also reflect it. Therefore, the construction of TBC has always been a compromise between the absorption of the outgoing light and residual reflection. The best TBC published so far are Berenger's perfectly matched layers (PML) [49]. These are artificially constructed layers that absorb propagating waves but do not reflect them. Figure 31.8(b) shows the propagation of a Gaussian-shaped electromagnetic pulse through Berengers PML. The residual reflectivity from the boundary of the computing window is lower than 70 dB.

Standard FDTD programs for rigorous EMF simulation of light diffraction from mask employ TBC directly above and below the mask. In most cases, the boundaries perpendicular to the mask plane are specified as PBC. In special cases, such as certain mask symmetries, other specific boundary conditions, such as mirror boundary conditions, are applied.

31.2.3.2.3 Introduction of Field

The updating equations [Equations (31.7)] describe the evaluation of the EMF inside the computing window in time. To excite an EMF inside this computing window, an appropriate formulation of the incident wave has to be found. Care is to be taken, to avoid an interaction of the incident wave and of the propagating field. For example, the formulation of the incident wave as a "hard source" by simply overwriting the updating

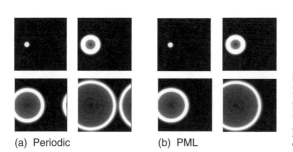

(a) Periodic   (b) PML

FIGURE 31.8
FDTD simulation of the propagation of a Gaussian pulse in vacuum: four figures starting from the upper left to demonstrate the propagation of the pulse in time: (a) PBCs; (b) absorbing boundary conditions using Berenger's PML.

equations at certain meshpoints by the field values of the incident wave will result in reflections of the propagating waves from the "hard source." To avoid this problem, Taflove [35] introduced the total/scattered field formulation. In this concept, the total electric and magnetic fields $\vec{E}_{tot}$ and $\vec{H}_{tot}$ in the computing domain are decomposed as

$$\vec{E}_{tot} = \vec{E}_{inc} + \vec{E}_{scat}$$
$$\vec{H}_{tot} = \vec{H}_{inc} + \vec{H}_{scat} \quad (31.9)$$

where $\vec{E}_{inc}$ and $\vec{H}_{inc}$ are the known field values of the incident wave. These are the field values that would exist in vacuum or in the material in the input region if there were no other materials in the computing domain. $\vec{E}_{scat}$ and $\vec{H}_{scat}$ are the initially known field values of the scattered field. These are the fields that result from the interaction of the incident wave with other materials in the computing domain. The FDTD algorithm can be applied to either the incident, scattered and total field components. In general, the form of the incident wave can be arbitrary. For the simulation of light scattering from photomasks, usually plane waves emerging from the condenser system are assumed. In the total/scattered field formulation, the computing domain is zoned into two distinct regions, which are separated by a non-physical virtual surface that serves to connect the field in each region.

31.2.3.2.4 Specification of Material Properties/Numerical Stability for Absorbing Materials
According to Equations (31.7) and (31.8) the optical properties of the materials inside the simulation window are given by the electric permittivity ε and conductivity ρ_c, respectively. Contrary to that, lithographers specify the material properties by a complex refractive index $\tilde{n} = n - j \cdot \kappa$. The relation between these magnitudes is given by

$$\varepsilon = n^2 - \kappa^2$$
$$\rho_c = 4\pi c_0 \varepsilon_0 \cdot n \cdot \frac{\kappa}{\lambda} \quad (31.10)$$

where c_0 is the vacuum velocity of light. The absorption constant α is related to the imaginary part of the refractive index κ by

$$\alpha = 4 \cdot \pi \cdot \frac{\kappa}{\lambda} \quad (31.11)$$

In general, the user of FDTD programs for mask simulations has only to specify the n and κ values. Often these values are available from databases. The web pages in Refs. [50,51] present interesting alternatives for the optical and EUV-spectral ranges. The user of rigorous EMF simulation programs should also be aware of the fact that the optical parameters may strongly depend on deposition conditions.

Chromium absorbers on optical mask consist of a thin adhesion layer, of bulk chromium and of a chromium oxide layer, which is used to reduce the reflections from the chromium surface. The concentration of oxide is continuously increased during the deposition process. Therefore, the transition from chromium to chromium oxide is more or less homogeneous. Fortunately, the details of the transition between chromium and chromium oxide are not very important for the lithographic performance. Especially for attenuated PSM, it is important to check the consistency of the material parameters with the transmission values specified by the mask shop.

A last but very important issue in the discussion of the material properties is connected to the numerical stability of the FDTD scheme presented in Equation (31.7). This scheme is only stable for positive permittivities $\varepsilon_{i,j}$. In other words, the standard FDTD formulation will fail for materials with a strong absorption ($\kappa > n$). Luebbers et al. [52] introduced a special updating scheme for absorbing materials. The Luebbers scheme is stable for virtually all materials. Unfortunately, it requires the updating of up to three additional field quantities. Therefore, strongly absorbing materials in the computing window require more memory and computing time. Alternatively, the original set of equations with complex-valued fields/constants can be used. In this case, the stability criterion demands smaller Δx, but memory demands are not increased [53].

31.2.3.2.5 Extraction of Steady-State Fields

The updating equation [Equation (31.7)] is used to compute the progress of the real part of the EMF in time. The extraction of the complex amplitude of the near field implies two additional requirements as follows.

- *Control of the convergence*: In general, at the beginning of the FDTD time stepping algorithm the fields everywhere in the computing window are initialized with zeros. The propagation of the incident wave through the whole simulation window requires a certain amount of time. The time or the necessary number of time steps depends on the optical properties of the materials inside the simulation window and on the geometrical configuration. Materials with large refractive indices (or low speed of light) and certain resonant geometries with multiple reflections inside the simulation window will require a larger number of time steps than standard materials/geometries. The convergence of FDTD algorithms is either controlled by comparison with the field values during previous time steps or by certain experience-based rules. The comparison of the field values with field values at previous time steps is objective, but it requires additional memory. A proper weighting of the differences between the field values at different time steps is by far not trivial. Moreover, evanescent modes, which never appear in the far field, can have a drastic impact on the convergence of the near field. Rule-based convergence criteria, such as: "propagate the wave five times through the simulation area," are more easily used. However, they always bear the risk of not detecting certain effects, which are strongly connected to resonances in certain mask configurations.
- *Extraction of amplitude and phase*: The extraction of the amplitude and phase of the EMF from the temporal behavior of the real part of the corresponding field component uses the *a priori* knowledge on the time dependence of the field [35], which is sinusoidal with a certain frequency $\omega = 2 \cdot \pi \cdot c_0/(n \cdot \lambda)$.

Next the described modeling approach is applied to the rigorous simulation of the light diffraction from an isolated phase edge on an alternating PSM. The geometry of the mask is shown in Figure 31.9(a). Figures 31.9(b)–(e) demonstrate the propagation of the EMF in the computing window. After a few time steps [Figure 31.9(b)], the incident field appears only in a small area directly below the position $z = -0.09$ μm, where the incident field is introduced. This position corresponds to the boundary between the areas with the total and the scattered fields. The intensity of the scattered field $\vec{E}_{\text{scat}}$ above this position is zero. The intensity of the total field $\vec{E}_{\text{tot}}$ below this position indicates the propagating plane wave. After a certain number of time steps the incident wave has reached the boundary of the chromium absorber. The chromium reflects the incident light and a typical standing wave interference pattern appears [Figure 31.9(c)]. Figure 31.9(d) demonstrates the delay of the propagating field in the unetched region of the mask. Figure 31.9(e) shows the

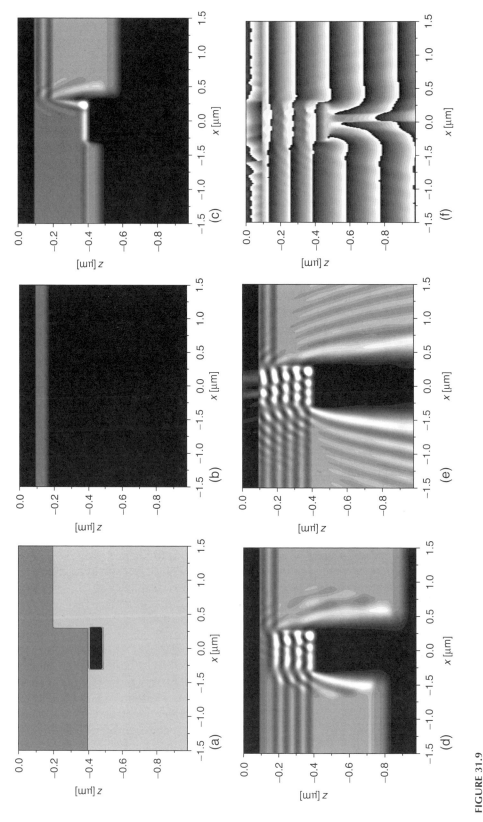

FIGURE 31.9
FDTD simulation of light scattering from an isolated phase edge on an alternating PSM: $\lambda = 193$ nm, TE-polarization, width of chromium = 4×150 nm, all dimensions are given on mask scale. (a) geometry: black — chromium, dark gray — glass, light gray — air; (b)–(e) intensity of scattered light after 30, 120, 190, and 400 time steps; (f) phase of scattered light after 400 time steps.

Modeling and Simulation 667

intensity of the light in the vicinity of the phase edge when the FDTD algorithm is converged. The corresponding phase of the E_y-component of the EMF can be seen in Figure 31.9(f). It shows the 180° phase shift between the transmitted field to the left and the right of the edge.

31.2.4 Coupling with Standard Lithography Simulations and Performance Considerations

The result of the rigorous EMF simulation described in the previous section is the intensity and the phase in the vicinity of the mask. This section describes how the result of EMF simulation is coupled to the aerial image simulation and the other parts of a photolithography simulator. The basic procedure is shown in Figure 31.10. It compares the "classical" Kirchhoff simulation with a rigorous simulation. First, the EMF directly below the mask is used to construct an equivalent mask, that is a mask with the amplitude and phase transmission values, which were obtained as the result of the rigorous EMF simulation. Figure 31.10(a) compares the transmission values of the "classical" Kirchoff-type mask with the rigorously simulated mask. The scattering of light from the mask edges results in a transmission curve, which is much more complicated than predicted by

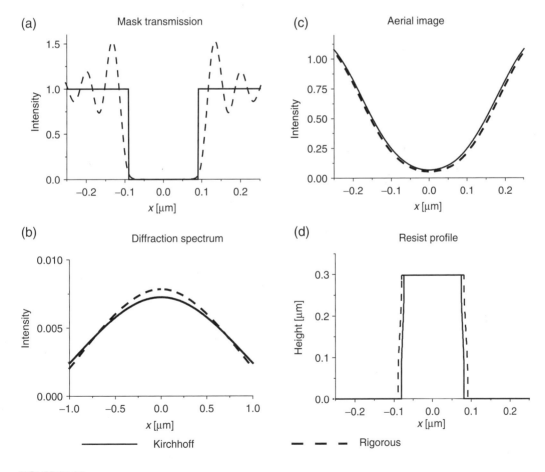

FIGURE 31.10
Comparison of Kirchhoff and rigorous simulation of a 180 nm isolated line on a binary mask, $\lambda = 248$ nm, NA = 0.6, $\sigma = 0.3$, 4× reduction; resist: 300-nm thick UV6. All dimensions are given on wafer scale.

the Kirchhoff approach. One should note, however, that the oscillations in the transmission curve have very high spatial frequencies. Therefore, they will not be transferred to the imaging projection system. The equivalent mask is used to compute the diffraction spectrum, the aerial image, and the resist profile after development. The differences between the results obtained by the different simulation approaches [Figures 31.10(b)–(d)] are rather small. The most dominating effect is an almost constant offset in the aerial images and resist profiles. The Kirchhoff approach neglects the diffraction of light from the edges on the mask and overestimates the amount of light that exposes the resist.

The simulations in Figure 31.10 were performed for a standard binary mask. The rigorous simulation predicts a slightly wider line. The same result could be obtained with the Kirchhoff approach and a slightly reduced dose. The scaling factor of the dose to match Kirchhoff and rigorous simulation results for standard binary masks is close to 1.0 and almost independent on the feature size. This is the main reason why the Kirchhoff approach is considered to be valid for binary masks. Section 31.3 of this chapter will present several examples where a dose scaling is not sufficient to match Kirchhoff and rigorous simulation results.

The simplified modeling scheme from Figure 31.10 neglects some important details that have to be taken into account in the coupling of rigorous EMF simulations with standard lithography simulation schemes. The equivalent mask can be constructed for 2D geometries with TE polarization where only one electric field component E_y is present. But what about TM-polarization and 3D mask geometries? In the general case, three electric field components are present. The vector properties of the light, which is diffracted from the mask have to be taken into account in the mathematical formulation of the imaging problem. Therefore, the scalar diffraction spectrum, which was introduced in Section 31.1.1, has to be replaced by vector scattering coefficients. Details about this can be found in the Ph.D. thesis of Pistor [37].

Another important issue is related to the simulation of the illumination system. So far, the diffraction from the mask was computed only for a vertically incident plane wave. In real systems, the mask is illuminated by a spectrum of plane waves within a certain range of incidence angles, which is defined by the spatial coherence parameter σ of the illumination system. In standard lithography simulation, the diffraction spectrum for nonvertical incidence is obtained by a shift of the diffraction spectrum obtained for vertical incidence. In many practical cases, this shift invariance of the diffraction spectrum can also be applied to rigorously simulated diffraction spectra. If this is done, one speaks of the so-called Hopkins approach. Several publications have shown that the Hopkins approach is valid for standard binary and PSMs and for typical optical settings of 5× or 4× reduction projection systems [54,55]. However, the application of rigorous EMF simulation for the modeling of mask inspection techniques and for extreme off-axis illumination techniques requires the separate simulation of diffraction spectra for at least several representative angles of incidence. Note that this recalculation of diffraction spectra for various angles of incidence can be quite time-consuming and should be applied only in important cases. Rigorous image modeling without the Hopkins approach is also necessary for accurate simulation of EUV-lithography [56].

The performance of rigorous EMF simulation depends strongly on the algorithmic implementation, on the discretization, and on the convergence of the FDTD codes. Appropriate initialization of the field components at the beginning of the FDTD time stepping can strongly improve the convergence of the algorithms. Running rigorous EMF simulation of the aerial image for a 2D geometry (lines/spaces) on a standard PC should take a few seconds and require <64 MB memory. Three-dimensional problems may take many minutes or hours of computing time. At least 512 MB of memory is advisable for 3D calculations.

Modeling and Simulation

Besides from the computing performance and the user-friendliness of the employed methods, the accuracy is another very important issue in the application of rigorous EMF simulation. Experimental data are often too noisy or even not available. The comparison with experimental results from aerial image measurement systems (AIMS) is far from trivial. Flare and aberrations of AIMS require the application of special conversion algorithms for the interpretation of AIMS data [57]. The comparison with results obtained by other EMF simulation methods is very useful and important to evaluate the accuracy of FDTD simulations. This includes the comparison with analytical solutions for special cases [58] and the comparison with special methods, which can be applied for certain geometries only [59].

31.2.5 Field Decomposition Techniques and other Enhancements

Despite of the steady improvement of computing performance over the last years and the considerable improvement of FDTD algorithms, the simulation of full 3D problems is still very time- and memory-consuming. The application of EMF simulations was restricted to small mask areas. Recently, Adam and Neureuther [60,61] proposed the application of domain/field decomposition methods (DDMs) to make large mask areas accessible to rigorous EMF simulation. The idea of these methods is to split the full 3D problem into several 2D problems.

Figure 31.11 demonstrates the application of the DDM for the simulation of light diffraction from a contact hole on a binary mask. First, the transmitted near field is computed with the assumption that the dielectric properties of the mask vary only in horizontal direction (x-axis). The resulting intensity of the transmitted near field is shown on the left-hand side of Figure 31.11. In the upper and lower parts the chromium absorber transmits almost no light. In the central part of the mask, one obtains an intensity pattern of a space. Next, 2D simulations for vertical configurations with constant dielectric properties along the x-axis are performed. The intensity of the resulting near field is shown second from the left on Figure 31.11. No intensity is transmitted through the left and right part of the mask. In the central part of the mask, the near field of a slit is obtained. Because of the changed orientation of the slit, the near field thus generated by polarized light is different from that obtained for the horizontal configurations. Finally, the transmission of the mask is computed under the assumption of an infinitely thin mask

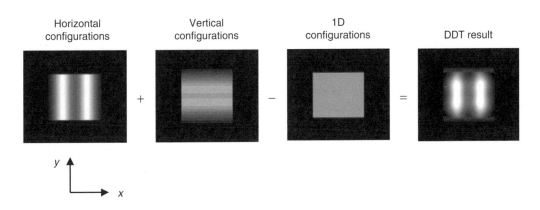

FIGURE 31.11
Schematic presentation of the DDTs for the rigorous simulation of light transmission through a 150-nm isolated contact hole on a binary mask ($\lambda = 248$ nm).

(second from the right in Figure 31.11). The transmission inside the contact hole is close to 1, whereas the transmission through the chromium-covered area is almost zero. The transmission of the 3D mask is obtained by a superposition of the complex transmission values of the horizontal and vertical configurations and the thin mask transmission. The resulting intensity of the transmitted light in the near field of the mask is shown on the right in Figure 31.11.

Although there is still some difference between the near fields obtained with full 3D simulation and with field/domain decomposition, these differences almost vanish in the aerial image. The reason for that is the spatial frequency filtering of the near field by the numerical aperture of the projection system. Domain/field decomposition techniques reduce the computing time and the memory requirements of 3D rigorous EMF simulations at least by a factor of 100. Some applications of domain/field decomposition will be presented in the following sections.

In certain cases the accuracy of FDTD calculations for a given spatial discretization can be strongly improved by so-called "average material properties techniques" [35,62]. These techniques introduce additional materials at the material boundaries to improve the numerical representation of the geometry. The accuracy of FDTD for coarse grids can be considerably improved by the application of wavelet-based algorithms [63,64].

31.3 Application of Simulation to Different Mask Types

Optical resolution enhancement techniques are employed to push optical lithography to its limits [65]. Most of these resolution enhancements introduce more complex mask patterns, such as OPC and PSM. The importance of rigorous EMF simulation of light diffraction from photomasks increases for smaller and thicker mask features. Obviously, this tendency towards smaller and thicker mask features applies to all mask-related resolution enhancement technologies. The size of subresolution assist features used in OPC is comparable to the wavelength of light λ. The depth of etched features on alternating PSM is comparable to λ. Finally, the thickness of absorber layers in EUV masks exceeds the exposure wavelength by several times. This section will review imaging effects of different lithographic techniques, which can be only understood in terms of rigorous EMF simulation.

31.3.1 Binary Masks

For standard binary masks used in 4× or 5× reduction projection systems, the features on the mask are large compared to the exposure wavelength λ. The thickness of chromium absorbers in the order of 80 nm is smaller compared to λ. Therefore, rigorous EMF simulation is only of limited importance for this type of masks. As shown in Section 31.2.4, the rigorous EMF simulation of diffraction from chromium-on-glass masks results only in a small rescaling of the aerial image. This rescaling is more or less independent from the feature size and from optical settings. In contrast to the Kirchhoff approach, which assumes the same mask transmission for TE and TM-polarized light, rigorous EMF simulation predicts certain polarization dependence. In general, TM polarization shows higher transmission and lower sidelobes. For 5× reduction projection systems, differences in mask transmission values up to 3% were observed [66]. Such polarization effects can produce eccentricities in the aerial images of small contact holes [67].

One should keep in mind, however, that such statements have to be re-evaluated when new lithography techniques are applied. The absorber thickness on masks for 157-nm technology will be almost the same one that was applied for 193- and 248-nm technologies. This relative growth of the absorber thickness compared to the exposure wavelength will result in a larger sensitivity with respect to rigorous EMF effects [68]. Recently, Pierrat and Wong [69] reported that mask topography effects could represent a significant part of the MEEF even for binary masks.

The most prominent rigorous EMF effects for binary optical masks are observed when subresolution OPC features are applied [70]. Figure 31.12 compares the results of Kirchhoff and of rigorous simulations for the imaging of 150-nm features with 50-nm subresolution assists on bright and dark field masks, respectively. The optical settings of the ArF stepper are specified in the figure caption. In both cases, the aerial images obtained by the different simulation approaches are normalized with respect to the maximum intensity. Both, for the bright and the dark field masks, Kirchhoff and the rigorous EMF simulation predict the same characteristics of the main features. However, a closer look at the images of the sidelobes reveals important differences.

Rigorous EMF simulation predicts that dark assists on bright field masks print stronger than expected. Simulations using the Kirchhoff approach underestimate the printability of the assist. On the other hand, to achieve a certain effect of the assist, the size of the assist can be smaller than suggested by the Kirchhoff approach. The smaller assists are easier to place within the limited space between the main features. In contrast to that, rigorous EMF simulation shows that bright assists on dark field masks have a smaller impact than expected by the Kirchhoff approach. To achieve the desired effect on the imaging, bright assists have to be larger than specified by the Kirchhoff approach. The described effect may have important design implications, especially if there is not enough space to place the enlarged assists.

Similar effects can also be observed for other types of subresolution features, such as serifs. The Kirchhoff approach suggests that the effects of subresolution features more or less scale with their size/area and their distance from main features. In contrast to that, rigorous EMF simulation will reveal more complicated amplitude and phase transmission effects in the near field of such subresolution features. The resulting "wave front deformations" have a certain impact on the size, placement, and through focus imaging

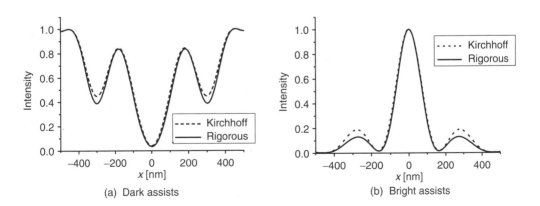

FIGURE 31.12
Simulated aerial images of (a) a 150-nm isolated line with 50-nm dark assists and (b) a 150-nm isolated space with 50-nm bright assists: $\lambda = 193$ nm; NA $= 0.75$; annular illumination, 0.5/0.75, 4× reduction. All dimensions are given on wafer scale.

characteristics. Rigorous EMF simulation will help to explore these imaging effects and their impact on the practical implementation of aggressive OPC technologies.

31.3.2 Phase Shift Masks

The depth of the etched areas in alternating PSM is comparable to the exposure wavelength. Obviously, the assumption of an infinitely thin mask is not an appropriate approach that can be used to understand the imaging details of advanced PSM. This was first noticed by Wong and Neureuther [42], who applied rigorous EMF simulation to simulate the imaging performance of PSM.

Figures 31.13 and 31.14 compare the simulation of dense lines/spaces on an alternating PSM with the Kirchhoff assumption and with rigorous EMF simulation, respectively. The geometry and the rigorously computed EMF are given in Figure 31.13. The mask transmission, aerial images, and resist profiles predicted by the Kirchhoff approach and by the rigorous approach are compared in Figure 31.14. The Kirchhoff approach characterizes the mask by a uniform transmission in different parts of the mask. Two different phase values of 0° and 180° are assigned to the bright features with a uniform transmission 1. Imaging of the mask results in a perfect lines space pattern. The peak intensities, corresponding to the unshifted ($\phi = 0°$) and the shifted ($\phi = 180°$) regions of the mask, have the same height. The resulting resist profile is symmetric with respect to the target.

In the rigorous EMF simulation, the full information on the geometrical shape or the topography of the mask is taken into account. Figure 31.13 shows the intensity and the phase of the field in the vicinity of the mask. The pronounced standing waves in the upper part of the intensity distribution result from the interference between the incident light and the light that is reflected from the backside of the chromium absorber. The shape of the transmitted light below the unetched and the etched openings is similar but not identical. A similar statement can be made concerning the phase of the transmitted light. In first-order approximation, the light transmitted through both types of openings is shifted by 180°. However, a closer look at the near fields in Figure 31.14 reveals further subtleties, especially for the field values close to the vertical edges of the mask. The comparison of the aerial images and resist profiles obtained with the different simulation approaches shows that the Kirchhoff approach overestimates the transmission of light through the mask and the projection system, respectively. The peak intensities of the

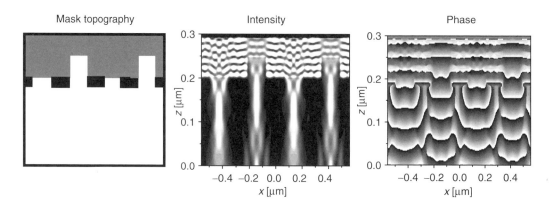

FIGURE 31.13
Rigorous simulation of light scattering from 140-nm dense lines and spaces on an alternating PSM: $\lambda = 248$ nm, TE-polarization, 4× reduction, mask topography: gray—glass substrate, black—chromium, white—air. All dimensions are given on wafer scale.

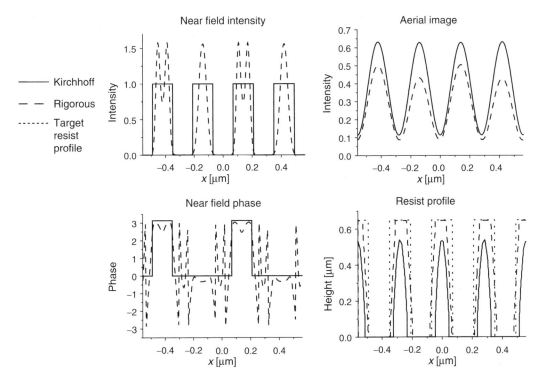

FIGURE 31.14
Near field phase and intensity, aerial images, and resist profiles of 140-nm dense lines and spaces on an alternating PSM for rigorous and Kirchhoff simulation: NA = 0.6, σ = 0.4, defocus 0.3 μm inside resist; resist: 0.65 μm UV6, all other simulation parameters are as given in Figure 31.13. All dimensions are given on wafer scale.

aerial image obtained with the Kirchhoff approach are much higher than this one obtained with the rigorous approach. After the development, this results in smaller resist lines. Furthermore, the imaging result obtained with the rigorous EMF simulation suggests a pronounced imaging artifact. The intensity peak below the unetched feature is higher than one below the etched feature. As can be seen from the rigorously simulated resist profile, the intensity imbalance causes a placement error. Soon after the theoretical prediction, the described effect also was observed in the experiment [71,72].

Numerous papers discuss the origin and the dependencies of the described intensity imbalancing effect. Ferguson et al. [73] attributed this imbalance in peak intensity to the effective transmission and phase errors associated with the glass edges. Wong and Neureuther [43] pointed out that the imbalancing results from a non-vanishing zero diffraction order of the light below the mask. Adam and Neureuther [60] and later Cheng et al. [74] explained the intensity imbalancing in terms of light, which is scattered from the bottom of the etched feature towards the open feature. The polarization of the light and moderate off-axis illumination have only a minor impact on the described imbalancing effect. However, a strong focus dependence of the imbalancing was observed. Rigorous EMF simulation predicts a pronounced focus asymmetry of phase-shifted or etched openings [66].

Several mask design strategies have been explored to reduce intensity imbalancing. The most obvious option would be a resizing of the etched opening. Although this resizing reduces the imbalancing, it does not provide the necessary through focus performance [58]. Extensive rigorous EMF simulations were used to devise new balancing strategies,

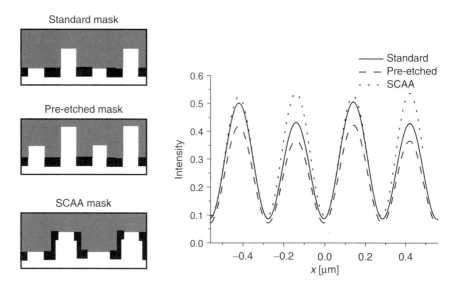

FIGURE 31.15
Alternative mask geometries for the reduction of intensity imbalancing effects in alternating PSM and simulated aerial images. All parameters are as given in Figure 31.14.

such as combinations of pre-etches and quartz underetches [58,75], or the SCAA concept [76]. Geometries and corresponding simulation results are shown in Figure 31.15. A pre-etching of all transparent areas on the PSM reduces the pronounced imbalancing, which was observed for the standard mask. The depth of the pre-etch was not optimized for the simulations in Figure 31.15. Therefore, a residual imbalancing can be seen in the corresponding aerial image. The SCAA mask presents another interesting possibility to reduce the imbalance. The deposition of the chromium over the glass etches reduces the impact of light scattering from the glass edges and gives an almost perfect balancing. It is important to mention that the optimization of geometry parameters of the mask for the reduction of intensity imbalancing effects does not always require full aerial image simulations. As demonstrated in Ref. [77], the analysis of the diffraction spectra generated by alternating PSM can help to achieve the same goal with much less numerical effort.

So far, the discussion of imaging effects in PSM in this section was restricted to dense lines/spaces on alternating PSM. In recent years, several other important mask topography-induced imaging artifacts were observed. As stated before, the importance of mask topography effects increases with a decreasing ratio between feature size and wavelength. A severe impact of the rigorous diffraction effects on the process linearity for the printing of features with a small k_1 can be expected. The described behavior can be translated into a strong sensitivity of the MEEF [69].

The light diffraction from thick features on alternating PSM results in phase effects with an impact on the imaging performance, which is very similar to that one of wave aberrations [77]. For example, the PWs of isolated lines may exhibit a pronounced tilt. Previously tilted PWs were considered to be an indicator of spherical aberration. However, mask topography-induced phase effects will produce tilted PWs even for a perfect aberration-free lens. The described phenomenon was observed both in simulated and in experimental data (see Figure 31.16).

During recent years, several proposals were made to use PSM in aberration-monitoring [78,Nakao,Robins]. Rigorous EMF simulation can be used to explore the impact of mask

Modeling and Simulation 675

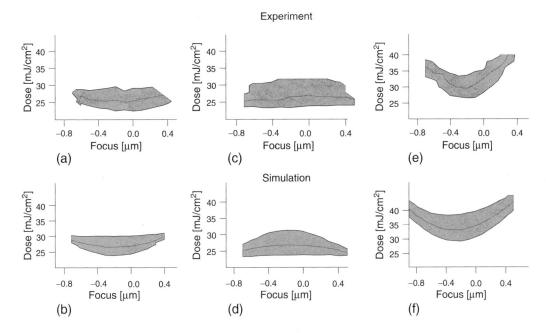

FIGURE 31.16
Comparison of measured [upper: (a), (c), and (e)] and simulated [lower: (b, (d), and (f)] PWs for 130-nm lines on alternating PSM at different pitches: (a), (b)—280 nm; (c), (d)—320 nm; (e), (f)—780 nm, unpolarized light, mask without pre-etch, scanner: $\lambda = 193$ nm, 4×, NA = 0.63; $\sigma = 0.3$; resist: 395-nm PAR 810, BARC: 89-nm AR25; substrate: Si.

topography-induced phase effects on the performance of the proposed techniques. For example, it was shown that aberration-monitoring using a phase dot [78] might result in inaccurate interpretations of the measurement results for spherical aberrations [77].

Due to their smaller thickness, attenuated PSMs are less sensitive with respect to mask topography effects than alternating or chromeless PSMs. Nevertheless, issues, such as sidelobe printability and linearity, have to be carefully evaluated by rigorous EMF simulation.

31.3.3 EUV-Masks

Masks for EUV lithography are fundamentally different from the masks that are applied in conventional optical lithography (see Figure 31.17, Table 31.1). EUV masks are used in reflection. The high reflectivity of the mask blank is achieved with a multilayer system

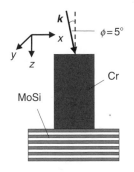

FIGURE 31.17
Basic geometry of a EUV mask: dark gray—Cr-absorber, bright gray/white—MoSi-multilayer (schematically only). The arrow indicates the direction of the incident light. Mask parameters for the simulations presented in this chapter: MoSi-multilayer—40 bilayers of silicon (thickness = 4.0 nm) and of molybdenum (thickness = 2.9 nm), Si top, $n_{Si} = 0.999931 - 0.00182109j$, $n_{Mo} = 0.922737 - 0.0062202j$, Cr-absorber—height = 80 nm, $n_{Cr} = 0.933328 - 0.0381982j$. Mask substrate (not shown in the figure): Si.

TABLE 31.1

Comparison between Optical Masks (Binary and PSMs — PSM) and EUV Masks

	Optical Masks	EUV Masks
Operating mode	Used in transmission, glass substrate	Used in reflection, multilayer substrate
Feature thickness	0.25–0.5λ (binary); 1.0–1.5λ (PSM)	6–8λ
Illumination	Centered on axis	Centered off-axis

of molybdenum and silicon (MoSi) on top of a mask substrate. The height of Cr or TaN absorbers on top of the multilayers is between 50 and 100 nm, which is larger compared to the operating wavelength $\lambda \approx 13.4$ nm. Furthermore, the EUV-masks are illuminated off axis. These peculiarities of EUV-masks have a number of consequences for their lithographic process performance. Predictive process simulation can be very helpful to investigate these consequences, to identify potential problems, and to find appropriate technological solutions.

However, the simulation of EUV masks is a challenging problem. Taking into account the special properties of EUV masks, it is obvious that the traditional assumption of an infinitely thin mask (Kirchhoff approach) cannot be used to explore the peculiarities of the process performance of EUV masks. The large ratio absorber thickness/operating wavelength and the off-axis illumination of the mask require the application of rigorous EMF solvers for the simulation of the light diffraction from the mask. The FDTD method, which is the standard method for the simulation of thick PSMs in optical lithography, was used for both, the simulation of 2D (lines/spaces) and for 3D (contacts/posts, etc.) EUV masks [79–81]. Due to the small wavelength and the large thickness of the mask, full 3D simulations typically require dozens of hours and several GB of memory.

Several authors applied other rigorous EMF simulation methods, such as RCWA [26], the differential method [28], and the WGM [32] to the modeling of EUV masks. These alternative methods show very good performance for 2D features, such as lines/spaces with small pitches. Some of the methods have been extended to the simulation of 3D mask features, such as contact holes.

31.3.3.1 Efficient Modeling of EUV-Masks

In certain cases, the efficiency of the modeling of EUV masks can strongly be improved. This includes the modeling of defect-free multilayer mask blanks, which consist of parallel, homogeneous layers and the modeling of Manhattan-type features where all mask edges are oriented along a rectangular grid. Modeling approaches that make use of these assumptions are described in the following.

From a simulation point of view, a EUV mask consists of two fundamentally different parts. The light propagation in a defect-free and homogeneous stack of parallel multilayers, including the mask substrate, can be described by analytical methods, such as transfer-matrices. The modeling of light diffraction from the thick absorber requires the application of rigorous EMF simulation. Figure 31.18 demonstrates a general simulation scheme, which combines analytical transfer-matrices and rigorous EMF simulation for the modeling of EUV-masks [82]. It resembles the Fourier boundary conditions proposed by Pistor et al. [80].

First, the diffraction of the incident light from the absorber is computed by FDTD. Directly below the absorber a perfect matched absorbing layer (PML), as proposed by Berenger [49], is applied. This PML guarantees that all diffracted light leaves the simulation area without any back-reflection from the boundaries. Figures 31.18(a) and (b) show

Modeling and Simulation 677

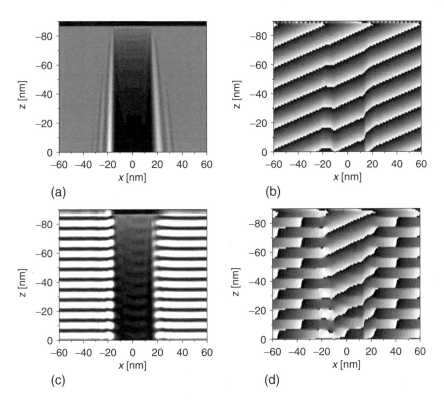

FIGURE 31.18
Intensity and phase of the EMF in the vicinity of an EUV mask: 30-nm isolated line, other mask parameters are as specified in Figure 31.17, $\lambda = 13.4$ nm, incident light from the upper left $\phi = 5°$, TE-polarization, (a) and (b) intensity and phase for incident light only; (c) and (d) intensity for incident light + back-reflection from multilayer stack. Dark — low intensity, bright — high intensity. The x-dimension is given wafer scale (4× system).

the resulting intensity and the phase of the light in the vicinity of the mask. The direction of the incident TE-polarized light is as specified by the arrow in Figure 31.17. The angle ϕ between the z-axis and the wave vector of the incident light is 5°. The sharp border between the black and gray areas on top of Figure 31.18(a) indicates the position where the incident light is introduced into the simulation area. Due to the oblique incidence, the intensity peak on the left-hand side of the absorber bottom is slightly higher than that on the right-hand side. The real part of the refractive index of chromium is smaller than that of vacuum. This results in a larger phase velocity of the light inside the chromium than in the other part of the simulation area. This effect can be seen in Figure 31.18(b).

Next, the diffracted light directly below the absorber or in the uppermost layer of the homogeneous multilayer stack is decomposed into plane waves. The reflectivity of the resulting plane waves from the multilayer stack is computed by analytical transfer-matrices. The reflected plane waves are coupled back into the simulation area for FDTD. Superposition of the diffracted downward propagating incident light and of the diffracted upward propagating reflected light results in an EMF distribution in the environment of the mask. The intensity and the phase of this field are shown in Figures 31.18(c) and (d). The intensity plot in Figure 31.18(c) shows a pronounced standing wave pattern, which results from the interference between incident and reflected lights. In contrast to Figure 31.18(a), Figure 31.18(c) also shows a non-zero intensity in the very

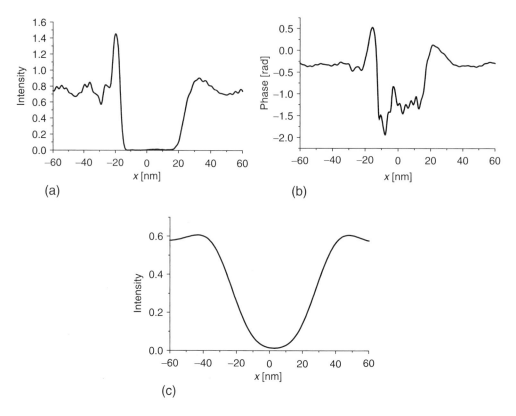

FIGURE 31.19
Intensity (a) and phase (b) of the reflected near field directly above the absorber of the EUV mask. In contrast to Figure 31.18, the phase tilt to the off-axis illumination is compensated in Figure 31.19(b). Aerial image (c) at best focus, NA = 0.25, σ = 0.5, reduction: 4×, all other parameters are as given in Figure 31.18. The x-dimension is given wafer scale (4× system).

upper part of the simulation area. This is the reflected light from the EUV-mask, which propagates towards the projection optics.

The amplitudes and phases of the EMF directly above the absorber are used to construct an equivalent thin mask, which serves as input for the following image calculation (see Section 31.2.4). Figure 31.19(a) shows the intensity of the reflected light in the near field. Due to the off-axis illumination, a pronounced peak on the left-hand side of the absorber and a reduced intensity on the right-hand side can be observed. Note also the strong phase variation of the near field in the vicinity of the absorber edges, which is shown in Figure 31.19(b). This phase variation, which can be translated into a phase deformation of the object spectrum or aberration like effects, is the main contributor to EUV-mask-related imaging artifacts. The aerial image of the simulated 30-nm line for a 4× reduction stepper with $\lambda = 13.4$ nm, NA = 0.25, $\sigma = 0.5$ can be seen in Figure 31.19(c). Note the asymmetry and the placement error of the aerial image.

Despite of the separation of the EUV-mask simulation into a rigorous EMF simulation of the absorber part and into the analytical modeling of the multilayer stack, the full simulation of 3D features, such as contact holes or post, is still very demanding. This is demonstrated by a comparison of memory, which is required by rigorous EMF simulations of optical and EUV-masks, respectively. A typical contact hole on a PSM has a height of about 1λ and transversal dimensions of $4\lambda \times 4\lambda$. This translates to memory requirements of about 5 MB. In contrast to that, the height of the EUV-mask is about 6–8λ with transversal dimensions of 30λ × 30λ, requiring about 2 GB memory. Therefore, fully rigorous EMF simulation of EUV mask with FDTD is virtually impossible on standard personal computers nowadays.

However, the application of domain/field decomposition techniques allows the computation of even larger EUV-mask areas on standard computers [82]. In addition to the general concept of these techniques, which was introduced in Section 31.2.5, the application of domain/field decomposition to EUV masks requires a consistent handling of off-axis illumination and the combination of the rigorous EMF simulations with the analytical handling of the multilayer stack. Comparison of full 3D EMF simulation with simulations using DDMs demonstrated a very good performance of the decomposition techniques. The DDM approach simulates EUV mask-related image artifacts, such as placement errors and focus shifts, with a very good accuracy (< 0.2 nm). There is still a certain offset between absolute feature sizes computed with DDM and full 3D EMF simulation. Nevertheless, the DDM approach predicts tendencies, such as the imaging bias between bright and dark features, and the dependency of the feature size on the orientation with respect to the off axis illumination. For the simulation of a 30-nm contact hole, the DDM approach reduces the computing time by a factor of 200 and the memory requirement by a factor of 250. The improvement in the computing performance depends on the size of the simulation area. Larger simulation areas will result in larger gain factors.

31.3.3.2 Mask-Induced Imaging Artifacts in EUV Lithography

Since the first rigorous EMF simulations of EUV-masks, several more or less EUV-specific imaging artifacts were observed. Bollepalli et al. [83] demonstrated that the off axis illumination results in both an imaging bias and a feature displacement due to the image asymmetry. The shadowing of certain parts of the reflective multilayer mask blank depends strongly on the orientation of the features with respect to the oblique illumination of the mask. This phenomenon transfers into an orientation-dependent placing and sizing of the features. For typical mask and optical system parameters, orientation-dependent placement errors in the order of 2–5 nm were observed [56]. These placement errors become more pronounced for larger absorber heights and for more oblique incidence. Similar tendencies can be observed regarding the variation of typical CDs or feature sizes versus the orientation of the illumination. To a larger extent the described placement and sizing effects can be considered in the design of EUV-masks. This is not the case for focus-dependent effects. For example, Otaki [56] observed focus-dependent placement error, which translates into telecentricity errors in the order of a few milli-radians.

The most prominent EUV-mask-related imaging artifacts can be observed by a comparison of PWs of different feature types. Figure 31.20 compares aerial image-based PWs of typical 30-nm features imaged with $\lambda = 13.4$ nm, NA $= 0.25$, and $\sigma = 0.5$. Note the different focus/threshold intensity scaling for the different feature types. The PWs of isolated lines and isolated spaces are considerably tilted. This effect was first observed by Krautschik et al. [84]. As discussed in Section 31.3.2, these tilted PWs are an indicator for mask-induced phase effects, which are similar to spherical aberrations. Such phase effects may result both from the angular dependence of the phase of the reflected light from EUV-multilayer stacks [85] and from phase deformations of the reflected light due to the diffraction at the edges of the mask. Yan [86] correlated the asymmetry of the Bossung curves with phase errors at the mask edge. Another important effect that can be observed in Figure 31.20 is the considerable shift of the best focus position with respect to the nominal focus position occurring for certain feature types. In general, this shift is most

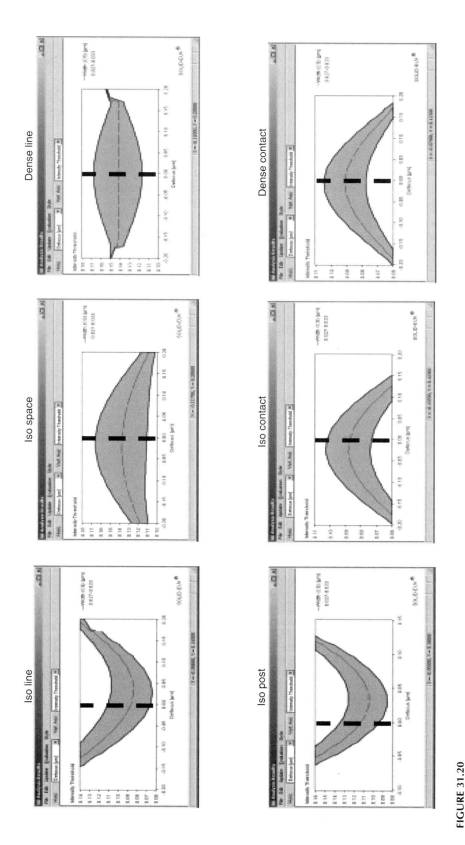

FIGURE 31.20
Simulated aerial image-based PWs for different feature types with feature sizes of 30 nm; mask parameters as specified in Figure 31.17, illumination: $\lambda = 13.4$ nm, incident light tilted to the left $\phi = 5°$, TE-polarization, projection system: NA $= 0.25$, $\sigma = 0.5$, $4\times$ reduction. The thick dashed line indicates the nominal zero defocus position.

pronounced for isolated dark features. The shift virtually vanishes for dense features. Bright features show a small focus shift in the opposite direction. The described focus shift for isolated contacts and posts can consume a considerable part of the available focus budget. More details on that phenomenon can be found in Ref. [82].

Due to the shorter wavelength, light scattering from "optically rough" surfaces presents another challenge for EUV lithography. These scattering phenomena occur in all parts of the optical system. Light scattering in the projection system produces a long-range flare around bright features. Therefore, the average amount of flare observed in aerial images depends on the density of bright features in the environment of the feature to be printed. Krautschik et al. [87] applied the power spectral density concept to the modeling of corresponding phenomena. Deng et al. [88] applied different simulation approaches for the investigation of the impact of multilayer mask roughness on EUV lithography.

Extensive simulation studies are necessary to understand the sources of and the issues resulting in complications of EUV-mask-induced imaging artifacts. The angular dependence of the complex reflectivity of the multilayer stack and diffraction of the light from the absorber edges result in phase deformations of the reflected near field. Symmetric deformations lead to asymmetric Bossung curves and focus shifts, whereas antisymmetric phase deformations result in a telecentricity error. The performance of alternative EUV-mask concepts, such as the inverse technology, in terms of mask-induced imaging artifacts is yet to be evaluated.

31.4 Simulation of Mask Defects

The improvement of the resolution capabilities of lithographic projection systems by using shorter wavelengths, advanced off-axis illumination schemes, and higher numerical apertures increases the sensitivity of lithographic processes with respect to small mask defects. Pellicles are used to reduce the deposition of particles directly on the patterned mask structures. Typical pellicle standoffs are in the order of 6–8 mm. Therefore, defective particles on the pellicle are located in the far field of the light, which is diffracted from the features on the mask. Flamholz [89] proposed the treatment of pellicle defects as a pupil filtering effect. The remaining part of this section will be focused on defects that are located directly on the mask pattern.

Mask defects may arise from various steps in the mask making process. Certain types of defects are difficult to find with available inspection tools. Mask making and inspection is very expensive. A predictive simulation of defect printability helps to evaluate the performance of produced masks and to establish specs for future mask generations.

31.4.1 Standard Defect Printability Considerations

The majority of defects on optical masks are chromium defects. Traditionally, decisions on the printability and repair of defects were taken on the basis of size criteria. For the subwavelength technology nodes, there is a gradual change towards including information on the printability properties of the defect along with its position with respect to optical design features [90]. It is not only the printability in terms of the defect appearance in the aerial/resist image that matters, but also its impact on the CD on the printed feature.

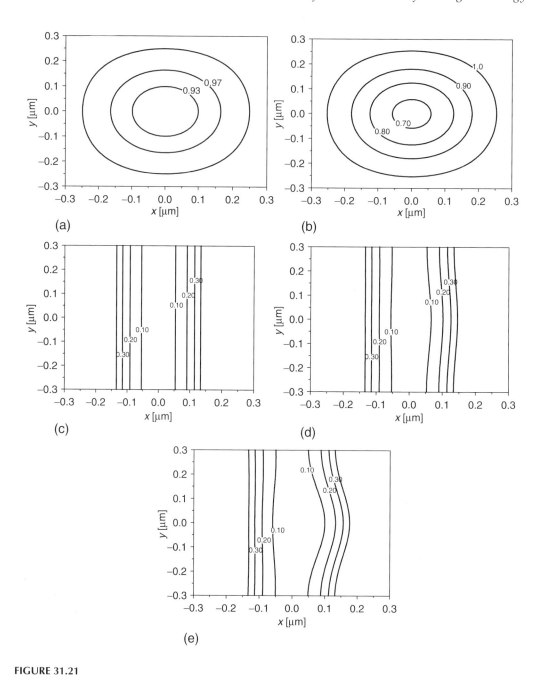

FIGURE 31.21
Simulated intensity contour plots of aerial curve images for isolated chromium defects: (a) 50 nm × 50 nm, (b) 100 nm × 100 nm, and for chromium defects centered 100 nm from the center of a 180-nm isolated line, (c) no defect, (d) 50 nm × 50 nm defect, (e) 100 nm × 100 nm defect, $\lambda = 248$ nm, 4× reduction, NA = 0.6, $\sigma = 0.4$. All simulations are performed with the Kirchhoff approach.

This statement becomes clear from Figure 31.21, which shows the impact of chromium defects at different positions on a printed isolated line. The upper part of the figure shows simulated aerial image intensity contour lines for isolated square chromium defects in a clear mask area. A 100 nm × 100 nm defect results in a drop of the intensity to about 70% of the open frame intensity. The threshold intensity of standard processes is between 20%

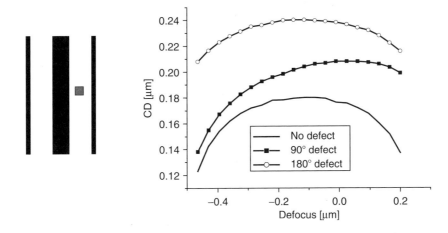

FIGURE 31.22
Simulated Bossung curve of defect-affected CD: (left mask geometry) 180-nm line with 40-nm assists, 80 nm × 80 nm phase defect (transmission: 90%); (right) simulated CD versus focus, $\lambda = 248$ nm, 4× reduction, NA = 0.6, $\sigma = 0.4$, dose = 23.9 mJ/cm^2, resist: 300 nm thick UV6. All simulations are performed with the Kirchhoff approach.

and 30%. Therefore, the defect will not print. The lower row of Figure 31.21 shows the impact of the same defect close to a 180-nm isolated line. Even the small (50 nm × 50 nm) defect produces a certain shift of the 20% and 30% contour lines of the isolated line close to the defect area. This shift becomes very pronounced for the 100 nm × 100 nm defect and will result in unacceptable CD variations of the line. Driessen et al. [90] performed extensive simulations to determine CD deviations in 193-nm lithography for varying defect distances.

Apart from chromium defects there are also phase defects. Such defects can be found not only on PSMs but also standard binary masks (polymer residues, chemical contamination, etc.). The printing of phase effects depends strongly on the focus setting [91]. This is shown in Figure 31.22 where the impact of a phase defect on an isolated line with subresolution assists at different focus positions is simulated. The CD versus defocus curve with the 180° defect is shifted with respect to CD versus defocus curve without defect. The impact of the defect on the produced line-width is almost independent of the defocus value. In contrast to that, the 90° defect has a much stronger impact for positive defocus values than it has for negative defocus values.

As was shown before, printability studies using lithography simulation have to be performed for many defect positions and defocus settings. Socha and Neureuther [92] proposed a perturbation model to reduce the computing time for these extensive simulations. The basic idea of this method is to separate the image computations of the defect-free structures and of defects and superpose them afterwards. Special care has to be taken to model the cross term between features and defect according to the theory of mutual coherence.

31.4.2 Rigorous EMF Simulation of Defects on Optical Masks

In the last subsection, the defect was considered to be an infinitely thin object. Defects are usually the smallest features on masks. Therefore, the applicability of the Kirchhoff approach for the very accurate modeling of defect printability is of special concern. In general, the statements from Section 31.3.1 regarding the modeling of subresolution assist features in OPC can be transferred to the simulation of chromium defects. The Kirchhoff approach will underestimate the printability of chromium dots. Pinholes in chromium absorbers are another concern. The Kirchhoff approach overestimates the light transmission through these pinholes. Moreover, the diffraction from the edges of the pinhole will result in a certain modification of the phase of the transmitted light. Similar to phase defects, the printablity of these defects has to be investigated through focus.

A realistic simulation of phase defects on alternating and chromeless PSM requires rigorous EMF simulation as well. The most famous example is a 360° phase defect. According to the Kirchhoff assumption, such defect would not have any impact on the process. Wong and Neureuther [67] demonstrated by rigorous EMF simulations that such defects cause a drop of the aerial image intensity in the defect area. For the special settings used in his simulations, the drop of intensity was too small to be printed. However, this statement has to be re-evaluated for other NA/σ settings.

Figure 31.23 shows another rigorous EMF simulation of defect printability for an alternating PSM [45]. Both a quartz bump in a nominally etched area of a PSM and an etched defect in a nominally not etched region of the alternating PSM produce a 180° phase defect. The Kirchhoff approach suggests the same result for both types of defects. The center row of Figure 31.23 shows the near fields of the transmitted light obtained with rigorous EMF simulation. Similar to a fiber tip, the bump defect "focuses" the light in the defect area. Contrary to that, the light is scattered away from the etch defect. The impact of the results on the final resist performance can be seen on the right-hand side of Figure 31.23. The

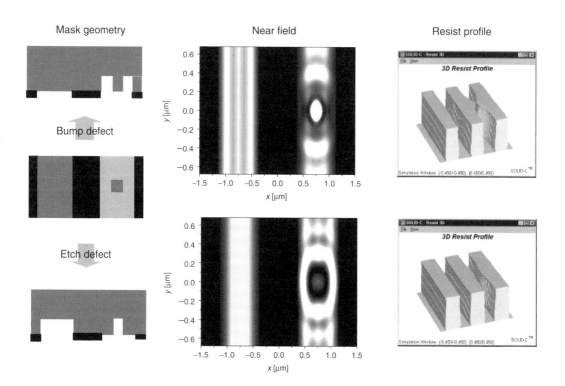

FIGURE 31.23
Simulation of the printability of bump and etch defects on an alternating PSM: (left) mask geometry 150-nm dense lines spaces, defect size 60 nm × 60 nm (wafer side specification); (center) simulated near fields directly below a 5× mask, $\lambda = 248$ nm, TE-polarization; (right) resist profiles after projection imaging, NA = 0.63, $\sigma = 0.3$, resist, UV6 (thickness = 420 nm). All simulations are performed with the rigorous approach.

bump defect produces a very small intensity spot in the near field. The Fourier transform of this small spot gives a very broad diffraction spectrum. Most of the light emerging from the defect will not be transferred through the numerical aperture of the projection system. As a result, there is a pronounced intensity loss in the defect area, which produces a bridging between neighbored resist lines. The light scattering from the etch defect results in a larger bright "object" with a smaller diffraction spectrum. More light from the defect arrives at the image plane. There is no bridging. In summary, rigorous EMF simulation of 180° phase defects predicts that bump defects are more severe than etch defects. This simulation result was also confirmed by the experimental data [45,93].

Lam et al. [94] applied DDMs for a more effective rigorous EMF simulation of PSM defect printability. Similar to the perturbation approach of Socha and Neureuther [92], the simulation is split into a simulation of the defect-free mask and a simulation of the defect. The comparison of DDM defect simulations and full 3D EMF simulations shows a good accuracy of DDM for features lager than 2λ (mask scale). The simulation of smaller features and the exact simulation of off axis illumination in $1\times$ mask inspection systems require further research.

31.4.3 EUV Multilayer Defects

Defects in the multilayer part of EUV-masks are of special concern because they are difficult to repair. The physical structure a defect produced by a particle within the multilayer coating of a EUV mask can be complex. Particularly important are defects nucleated by particles or pits on the mask substrate [95]. Although various methods for the modeling of these defects were proposed, the accuracy and performance of these approaches are still not sufficient. Therefore, the modeling of defective EUV multilayers is an important subject of current research.

The light reflection from defective multilayers cannot be described by straightforward application of the analytical transfer-matrix method as it was done in the case of defect-free multilayers (see Section 31.3.3A). Gullikson et al. [95] proposed the representation of the complex multilayer structure by a single reflecting surface with the shape of the top surface multilayer. This single surface approximation (SSA) covers phase variations in the reflectivity of the multilayer, but it neglects variations of the reflectivity due to local variation of the multilayer configuration.

Recently, Evanschitzky et al. [96] proposed a decomposition of the multilayer stack into segments with homogeneous multilayers (Fresnel method). The reflectivity of light from the segments is computed with transfer-matrices. The reflectivity of the full defective multilayer stack is composed from the results of the individual segments. Finally, the method is coupled with the FDTD simulation of light propagation in the absorber area. Figure 31.24 compares the aerial images of a dot with and without defect. The defect deforms the aerial image close to the right edge of the dot. Comparisons between the Fresnel method and rigorous simulation of the defective multilayer with RCWA demonstrate the good performance of the method. However, for certain defect geometries still some differences were observed.

Pistor et al. [62] and Pistor and Neureuther [97] modeled the light reflection from defective multilayers by application of FDTD to the full multilayer stack. The highly resonant structure of the multilayer system requires a very accurate modeling of the phase of the propagating light. To achieve the required accuracy, a very fine meshing has to be used in the FDTD algorithm. This transfers into almost impracticable memory requirements and computing times. To some extent, the application of material averaging techniques alleviates this problem, but the accuracy of the results remains a concern.

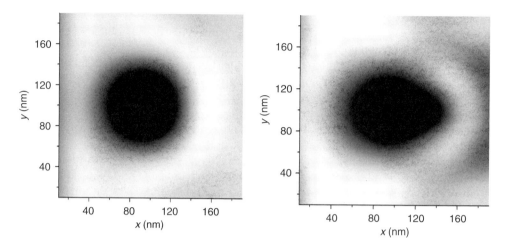

FIGURE 31.24
3D rigorous simulation of aerial images of a EUV mask with a 50 nm × 50 nm chromium dot (absorber) at $x = 100$ nm, $y = 100$ nm: (left) without defect; (right) with a defect at the position $x = 150$ nm, $y = 100$ nm. Stepper parameters 1($ρεδυχτιον$, $λ = 13.4$ nm, illumination at $θ = 5°$. From right-hand side, NA = 0.2, $σ = 0.5$, nominal best focus.

Recently, alternative methods, such as RCWA [26,98] and the WGM [32], were applied to the modeling of defective multilayer stack (see Figure 31.25, for example). Extensive comparisons between these and other methods have to be performed to identify the most accurate and most effective modeling approaches.

31.5 Summary

Lithography simulation has been proven to be very useful for the simulation of the performance of masks in lithographic processes. Advanced mask technologies, such as OPC, PSM, and EUV masks, require the application of rigorous EMF simulation for the description of light diffraction from these masks. The FDTD method is the most popular method that is used for this task. Domain decomposition techniques can be used to achieve considerable improvements of the performance of EMF simulation methods.

Rigorous EMF simulation is used for a more accurate prediction of the performance of subresolution assists and of chromium defects on binary masks. Light diffraction from alternating and chromeless PSM produces phase effects that result in imaging artifacts, such as intensity imbalancing, placement errors, and tilted PWs, which cannot be understood by the standard assumption of an infinitely thin mask. Similar statements apply to EUV masks.

In future, more efficient and faster simulation methods will increase the application of lithography simulation in mask design. A stronger connection between metrology and simulation is necessary to increase the predictivity of the simulation results. This includes the optical characterization of the mask material, the measurement, and specification of real mask geometries and the more accurate description of optical system parameters, such as real illumination geometries, flare, polarization properties of projection systems, and last but not the least, predictive resist models and model parameters.

Modeling and Simulation

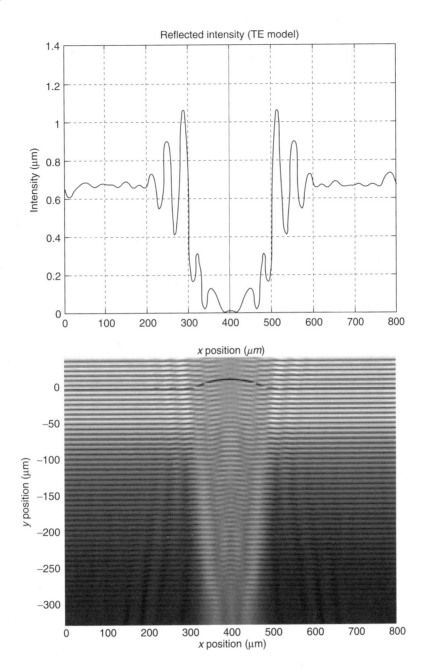

FIGURE 31.25
RCWA simulation results for a 30-nm circular defect covered by a conformal deposition of the multilayer (from Ref. [98]): (lower part) intensity inside the multilayer system; (upper part) intensity of the near field directly above the multilayer.

Acknowledgments

The author would like to thank Peter Evanschitzky from Fraunhofer Institute IISB, Ron Gordon from IBM Fishkill, Patrick Schiavone from CNRS LTM, and Alfred Wong from the University of Hong Kong for their helpful comments.

References

1. A. Erdmann and W. Henke, Simulation of optical lithography in optics and optoelectronics — theory, devices and applications, *Proc. SPIE*, 3729, 480 (1999).
2. H.J. Levinson, *Principles of Lithography*, vol. PM97, SPIE Press, 2001.
3. C.A. Mack, in: J.R. Sheats and B.W. Smith (eds.), *Optical Lithography Modeling, Microlithography, Science and Technology*, Marcel Dekker, New York, 1998.
4. A.R. Neureuther and C.A. Mack, Optical lithography modeling, in: P. Rai-Choudhury (ed.), *Handbook of Microlithography, Micromachining and Microfabrication, vol. I: Microlithography*, SPIE Press, PM39, 1997.
5. D.G. Flagello and T.D. Milster, High numerical aperture effects in photoresist, *Appl. Opt.*, 36, 8944 (1997).
6. M. Mansuripur, Distribution of light at and near the focus of high numerical aperture objectives, *J. Opt. Soc. Am.*, A3, 2086 (1986).
7. R. Schlief, A. Liebchen, and J.F. Chen, Hopkins vs. Abbe, a lithography simulation matching study, *Proc. SPIE*, 4691, 1106 (2002).
8. Y.C. Pati and T. Kailath, Phase-shifting masks for optical microlithography: automated design and mask requirements, *J. Opt. Soc. Am.*, A11, 2438 (1994).
9. N. Cobb and A. Zakhor, Fast sparse aerial image calculation for OPC, *Proc. SPIE*, 2621, 534 (1995).
10. P. Tat and A.K. Wong, Quantification of image quality, *Proc. SPIE*, 4691, 169 (2002).
11. D.A. Bernard, Simulation of focus effects in photolithography, *IEEE Trans. Semicond. Manufact.*, 1, 85 (1988).
12. M.V. Klein and T.E. Furtak, *Optics*, John Wiley & Sons, New York, 1986.
13. M.S. Yeung, Modeling high numerical aperture lithography, *Proc. SPIE*, 922, 149 (1988).
14. A. Erdmann and W. Henke, Simulation of light propagation in optical linear and nonlinear layers by finite difference beam propagation and other methods, *J. Vac. Sci. Technol.*, 14, 3743 (1996).
15. A. Erdmann, C.K. Kalus, T. Schmöller, Y. Klyonova, T. Sato, A. Endo, T. Shibata, and Y. Kobayashi, Rigorous simulation of exposure over nonplanar wafers, *Proc. SPIE*, 5040, 101 (2003).
16. A. Erdmann, W. Henke, S. Robertson, E. Richter, B. Tollkühn, and W. Hoppe, Comparison of simulation approaches for chemically amplified resists, *Proc. SPIE*, 4404, 99 (2001).
17. S. Robertson, C. Mack, and M. Maslow, Towards a universal resist dissolution model for lithography simulation, *Proc. SPIE*, 4404, 111 (2001).
18. C.A. Mack, Enhanced lumped parameter model for photolithography, *Proc. SPIE*, 2197, 501 (1994).
19. T.A. Brunner and R.A. Ferguson, Approximate models for resist processing effects, *Proc. SPIE*, 2726, 198 (1996).
20. D. Fuard, M. Besacier, and P. Schiavone, Assesment of different simplified resist models, *Proc. SPIE*, 4691, 1266 (2002).
21. D. van Steenwickel and J.H. Lammers, Enhanced processing: sub-50 nm features with 0.8 micron DOF using a binary reticle, *Proc. SPIE*, 5039 (2003) (in print).
22. B.J. Lin, The exposure-defocus forest, *Jpn. J. Appl. Phys.*, 33, 6756 (1994).
23. A.K. Wong, R.A. Ferguson, and S.M. Mansfield, The mask error factor in optical lithopgraphy, *IEEE Trans. Semicond. Manuf.*, 13, 235 (2000).
24. D.C. Cole, E. Barouch, E.W. Conrad, and M. Yeung, Using advanced simulation to aid microlithography development, *Proc. IEEE*, 89, 1194 (2001).
25. M.G. Moharam and T.K. Gaylord, Rigorous coupled-wave analysis of planar grating diffraction, *J. Opt. Soc. Am.*, 71, 811 (1981).
26. P. Schiavone, G. Granet, and J.Y. Robic, Rigorous electromagnetic simulation of EUV-mask defects: influence of the absorber properties, *Microelectronic Eng.*, 57–58, 497 (2001).
27. H. Kirchhauer and S. Selberherr, Three-dimensional photolithography simulator including rigorous non-planar exposure simulation for off axis illumination, *Proc. SPIE*, 3334, 764 (1998).

28. C. Krautschik, M. Ito, I. Nishiyama, and T. Mori, Quantifying EUV imaging tolerances for the 70, 50, and 35 nm nodes through rigorous aerial image simulations, *Proc. SPIE*, 4343, 524 (2001).
29. D. Nyyssonen, The theory of optical edge detection and imaging of thick layers, *J. Opt. Soc. Am.*, 72, 1425 (1982).
30. C.M. Yuan, Calculation of one-dimensional lithographic aerial images using the vector theory, *IEEE Trans. Electron Devices*, 40, 1604 (1993).
31. K. Lucas, H. Tanabe, and A.J. Strojwas, Efficient and rigorous three-dimensional model for optical lithography simulation, *J. Opt. Soc. Am.*, A13, 2187 (1996).
32. Z. Zhu, K. Lucas, J.L. Cobb, S.D. Hector, and A.J. Strojwas, Rigorous EUV-mask simulator using 2D and 3D waveguide methods, *Proc. SPIE*, 5037 (2003) (in print).
33. G.L. Wojcik, J. Mould, R. Ferguson, R. Martino, and K.K. Low, Some image modeling issues for i-line, 5× phase shifting masks, *Proc. SPIE*, 2197, 455 (1994).
34. www.fdtd.org, www.borg.umn.edu/toyfdtd.
35. A. Taflove, *Computational Electrodynamics: The Finite-Difference Time-Domain Method*, Artech House, Boston, 1995.
36. A.K. Wong, Rigorous Three-Dimensional Time-Domain Finite-Difference Electromagnetic Simulation, Ph.D. thesis, University of California at Berkley, 1994.
37. T.V. Pistor, Electromagnetic Simulation and Modeling with Applications in Lithography, Ph.D. thesis, University of California at Berkley, 2001.
38. K. Adam, Domain decomposition methods for electromagnetic simulation of scattering from three-dimensional structures with applications in lithography, Ph.D. thesis, University of California at Berkeley, 2001.
39. K.S. Yee, Numerical solution of initial boundary value problems involving Maxwell's equations in isotroptic media, *IEEE Trans. Antennas Propagation*, 14, 302 (1966).
40. G.L. Wojcik, D.K. Vaughan, and L. Galbraith, Calculation of light scatter from structures on silicon surfaces, *Proc. SPIE*, 774, 21 (1987).
41. www.eecs.berkeley.edu/~neureuth
42. A.K. Wong and A.R. Neureuther, Edge effects in phase shifting masks for 0.25 μm lithography, *Proc. SPIE*, 1809, 222 (1992).
43. A.K. Wong and A.R. Neureuther, Mask topography effects in projection printing of phase-shifting masks, *IEEE Trans. Electron Devices*, 41, 895 (1994).
44. R. Gordon and C.A. Mack, Lithography simulation employing rigorous solutions to Maxwell's equations, *Proc. SPIE*, 3334, 176 (1998).
45. A. Erdmann and C.H. Friedrich, Rigorous diffraction analysis for future mask technology, *Proc. SPIE*, 4000, 684 (2000).
46. www.sigma-c.com.
47. www.panoramictech.com.
48. www.kla-tencor.com/products/promax_2d/promax_2d.html.
49. J.P. Berenger, A perfectly matched layer for the absorption of electromagnetic waves, *J. Computational Phys.*, 14, 185–200 (1994).
50. www.rit.edu/~635dept5.
51. www-cxro.lbl.gov/optical_constants.
52. R. Luebbers, F. Hunsberger, K.S. Kunz, R.B. Standler, and M. Schneider, A frequency dependent finite-difference time-domain formulation for dispersive materials, *IEEE Trans. Electromagnetic Compatibility*, 32, 222 (1990).
53. R. Gordon, Personal communication.
54. T.V. Pistor, A.R. Neureuther, and R.J. Socha, Modeling oblique incidence effects in photomasks, *Proc. SPIE*, 4000, 228 (2000).
55. A. Erdmann and N. Kachwala, Enhancements in rigorous simulation of light diffraction from phase shift masks, *Proc. SPIE*, 4691, 1156 (2002).
56. K. Otaki, Asymmetric properties of the aerial image in extreme ultraviolet lithography, *Jpn. J. Appl. Phys.*, 39, 6819 (2000).
57. R. Gordon, D.G. Flagello, and M. McCallum, Deducing aerial image behavior from AIMS data, *Proc. SPIE*, 4000, 734 (2000).

58. C. Friedrich, L. Mader, A. Erdmann, S. List, R. Gordon, C. Kalus, U. Griesinger, R. Pforr, J. Mathuni, G. Ruhl, and W. Maurer, Optimising edge topography of alternating phase shift masks using rigorous mask modelling, *Proc SPIE*, 4000, 1323 (2000).
59. B.H. Kleemann, J. Bischoff, and A.K. Wong, Alternate rigorous method for photolithographic simulation based on profile sampling, *Proc. SPIE*, 2726, 334 (1996).
60. K. Adam and A.R. Neureuther, Simplified models for edge transitions in rigorous mask modeling, *Proc. SPIE*, 4346, 331 (2001).
61. K. Adam and A.R. Neureuther, Domain decomposition methods for the rapid electromagnetic simulation of photomask scattering, *J. Microlithography, Microfabrication, Microsystems*, 1, 253 (2002).
62. T.V. Pistor, Y. Deng, and A.R. Neureuther, Extreme ultraviolett mask defect simulation: low profile defects, *J. Vac. Sci. Technol. B.*, 18, 2926 (2000).
63. A. Taflove, *Advances in Computational Electrodynamics: The Finite-Difference Time-Domain Method*, Artech House, Boston, 1998.
64. M.S. Yeung, Fast rigorous three-dimensional mask diffraction simulation using Battle-Lemarie wavelet-based multiresolution time-domain method, *Proc. SPIE*, 5040 (2003) (in print).
65. A.K. Wong, *Resolution Enhancement Techniques in Optical Lithography, Tutorial Texts in Optical Engineering*, TT47, SPIE Press, 2001.
66. A.K. Wong and A.R. Neureuther, Polarization effects in mask transmission, *Proc. SPIE*, 1674, 193 (1992).
67. A.K. Wong and A.R. Neureuther, Examination of polarization and edge effects in photolithographic masks using three-dimensional rigorous simulation, *Proc. SPIE*, 2197, 521 (1994).
68. M.S. Yeung and E. Barouch, Limitation of the Kirchhoff boundary conditions for aerial image simulation in 157-nm optical lithography, *IEEE Electron Device Lett.*, 21, 385 (2000).
69. C. Pierrat and A.K. Wong, The MEF revisited: low k1 effects versus mask topography effects, *Proc. SPIE*, 5040 (2003) (in print).
70. A. Erdmann and R. Gordon, *Mask topography Effects in Reticle Enhancement Technologies and Next-Generation Lithography*, SPIE Short Course SC482.
71. R. Kostelak, C. Pierrat, J. Garofalo, and S. Vaidya, Exposure characteristics of alternate aperture phase-shifting masks fabricated using a subtractive process, *J. Vac. Sci. Technol. B.*, 10, 3055 (1992).
72. C. Pierrat, A.K. Wong, and S. Vaidya, Phase-shifting mask topography effects on lithographic image quality, *Proc. SPIE*, 1927, 28 (1993).
73. R. Ferguson, R. Martino, R. Budd, G. Hughes, J. Skinner, J. Stables, C. Ausschnitt, and J. Weed, Etched-quartz fabrication issues for a 0.25 µm phase-shifted DRAM application, *J. Vac. Sci. Technol. B.*, 11, 2645 (1993).
74. M. Cheng, B. Ho, and D. Guenther, The impact of mask topography and resist effects on optical proximity correction in advanced alternating phase shift process, *Proc. SPIE*, 5040 (2003) (in print).
75. R. Gordon, C.A. Mack, and J.S. Petersen, Design and analysis of manufacturable alternating phase-shifting masks, *Proc. SPIE*, 3546, 606 (1998).
76. M. Levenson, T. Ebihara, and M. Yamachika, SCAA mask exposures and Phase Phirst Design for 110 nm and below, *Proc. SPIE*, 4346, 817 (2001).
77. A. Erdmann, Topography effects and wave aberrations in advanced PSM-technology, *Proc. SPIE*, 4346, 345 (2001).
78. P. Dirksen, C. Juffermans, A. Engelen, P. de Bisschop, and H. Muellerke, Impact of high order aberrations on the performance of the aberration monitor, *Proc. SPIE*, 4000, 9 (2000).
79. K.B. Nguyen, A.K. Wong, A.R. Neureuther, and D.T. Attwood, Effects of absorber topography and multilayer coating defects on reflective masks for soft x-ray/EUV projection lithography, *Proc. SPIE*, 1924, 418 (1993).
80. T.V. Pistor, K. Adam, and A. Neureuther, Rigorous simulation of mask corner effects in extreme ultraviolet lithography, *J. Vac. Sci. Technol. B.*, 16, 3449 (1998).
81. R. Gordon and C.A. Mack, Mask topography simulation for EUV-lithography, *Proc. SPIE*, 3676, 283 (1999).

82. A. Erdmann, C.K. Kalus, T. Schmöller, and A. Wolter, Efficient simulation of light diffraction from 3-dimensional EUV-masks using field decomposition techniques , *Proc. SPIE*, 5037, 482 (2003).
83. B.S. Bollepalli, M. Khan, and F. Cerrina, Imaging properties of the extreme ultraviolet mask, *J. Vac. Sci. Technol. B.*, 16, 3444 (1998).
84. C. Krautschik, M. Ito, I. Nishiyama, and K. Otaki, The impact of EUV mask phase response on the asymmetry of Bossung curves as predicted by rigorous EUV mask simulations, *Proc. SPIE*, 4343, 392 (2001).
85. C. Liang, M.R. Descour, J.M. Sasian, and S.A. Lerner, Multilayer-coating-induced aberrations in extreme-ultraviolet lithography optics, *Appl. Opt.*, 40, 129 (2001).
86. P.Y. Yan, Understanding Bossung curve asymmetry and focus shift effect in EUV lithography, *Proc. SPIE*, 4562, 279 (2001).
87. C. Krautschik, M. Ito, I. Nishiyama, and S. Okaszaki, Impact of EUV light scatter on CD control as a result of mask density changes, *Proc. SPIE*, 4688, 289 (2003).
88. Y. Deng, T. Pistor, and A.R. Neureuther, Effects of multilayer mask roughness on extreme ultraviolet lithography, *J. Vac. Sci. Technol. B.*, 20, 344 (2002).
89. A. Flamholz, An analysis of pellicle parameters for step-and-repeat projection, *Proc. SPIE*, 470, 138 (1984).
90. F. Driessen, P. van Aldrichem, V. Philipsen, R. Jockheere, H.Y. Liu, and L. Karklin, Aerial image simulations of soft and phase defects in 193 nm lithography for 100 nm node, *Proc. SPIE*, 4691, 1180 (2002).
91. H. Watanabe, E. Suguira, T. Imoriya, and M. Inoue, Detection and printability of shifter defects in phase shifting masks II: defocus characteristics, *Jpn. J. Appl. Phys.*, 31B, 4155 (1992).
92. R. Socha and A. Neureuther, The role of illumination and thin film layers on the printability of defects, *Proc. SPIE*, 2440, 532 (1995).
93. J. Kim, W.P. Mo, R. Gordon, and A. Williams, Alternating PSM defect printability for 100 nm KrF lithography, *Proc. SPIE*, 3998, 308 (2000).
94. M. Lam, K. Adam, and A. Neureuther, Demain decomposition methods for simulation of printing and inspection of phase defects, *Proc. SPIE*, 5040, 1492 (2003).
95. E.M. Gullikson, C. Cerjan, D.G. Srearns, P.B. Mirkarimi, and D.W. Sweeney, Practical approach for modeling extreme ultraviolet lithography mask defects, *J. Vac. Sci. Technol.*, B20, 81 (2002).
96. P. Evanschitzky, A. Erdmann, M. Besacier, and P. Schiavone, Simulation of extreme ultraviolet masks with defective multilayers, *Proc. SPIE*, 5130, 1035 (2003).
97. D. Pistor and A. Neureuther, Extreme ultraviolet mask defect simulation, *J. Vac. Sci. Technol. B.*, 17, 3019 (1999).
98. P. Schiavone and R. Payerne, Rigorous simulation of line-defects in EUV masks, *Microprocess and Nanotechnology*, Matsue, 2001.

Index

A
Abbe method, 653
Absorber, 309
  defect inspection, 253
    repair, 255
  materials, 310, 389
  stack etch, 253
Absorption
  interband, 557
  intraband, 557
  Urbach, 559–560
Accuracy, IP
  factors, 82
Acousto-optical deflector (AOD), 55, 101
Acousto-optical modulator, 106, 111
Acousto-optics, 110–111
Adaptive metrology, 547
Address grid, 105
Advanced optical mask, 157–186
  alt.PSM, 165, 168, 171, 173, 452
  HT.PSM, 175
  OPC, 182
  RET, 186
  subresolution assist features, 179
Aerial image, 107, 126
  analysis, 612
  measurement technique, 609–612, 810
Aerial image measurement system (AIMS), 609
  applications for, 615–616
  as phase indicator, 619
    defect analysis, 615
  Carl Zeiss AIMS tools, 609
  in IC manufacturing, 625
AFM
  measurement of silicon mask–aperture
    structure, 284
  selection, 528
AIMS, 615
Algorithm(s)
  finite-difference time-domain, 660–667
Alignment
  many point, 546
  n-point, 546–547
ALTA, 55, 106, 113
  tools, 326
Anaheim pattern, 202
Analysis, aerial image, 612–615
Analyzer, linear-dimensional, 531

Angle, Azimuth, 106
Anodic bonding, 314
Antireflection plate, 90–91
Approximation, nonorthogonal, 35–36
APSM materials, 385
AR coating, ARC, 402
Arbitrarily shaped beam system, 49
Artwork, 9
Atomic force microscope, 8, 500
Atomic force microscopy (AFM), 169
Auger electrons, 343
Autoloader, 604
Automatic operation function, 586

B
Back end of line (BEOL), 7–8
  defect inspection, 8
    automated, 10
  defect repair, 8
  pellicle application, 8
Back scattered electrons (BSE), 345
Beam
  arbitrarily shaped, 49
  cell-projection, 48
  computer aided design (CAD), 60–61
  current, 54
  edge, 51
  IPL, 272
  microcolumn
    concept of, 49
  modulation, 111
  multibeam, 48–49
  multicolumn cell lithography, 50
  point beam, 67–68
  PXL, 305–306
  raster, 49–50
    beam edge, 52
    scan, 45–46
    shaped, 47–48
  scanning, 111
  shaping lens system, 72–73
  sigma, 102
  triangle, 84–85
  variable shaped, 46–47
  vector scan, 46
Binary intensity mask (BIM), 5
Binary mask, 443, 500
  simulation, 670

Blank(s)
  defect free, 240
  fabrication, 232
  low thermal expansion material (LTEM), 232
  mask, 204
    silicon-on-insulator (SOI), 234
Boolean mask operations, 22
Bosch process, 211–212
Bossung process window tilt, 263–264, 619–622
Bounding box, 415
Bragg('s)
  condition, 110, 112
  law, 238
Buffer layer
  defect inspection and repair, 257
  etch, 257
Butting error, 52, 108

C
CAD, 60, 105
Calma stream, 21
Capping layer, 241
Carbon cage experiment, 90
Carrier, mask, 259f
Cauchy model, 557
CD errors, 39
CD-SEM, 216
Cell-projection systems, 48
Character projection lithography
  masks for, 200
    structures, materials, and processes, 203
Chrome, 3–4, 568
  etching of, 369–371; *see also* Etching
    dry, 370
    wet, 370
Chrome film, 322
Chromeless PSM (CL PSM), 173
Chrome-on-glass mask (COG), 5, 367
Chromium, 5, 7, 10
Clean container, 223
Climate chamber, 106
Combined corrections, 537
Congruence, concept of, 535
Contact lithography, 137
Contact print era, 8
Contamination, 389
Continuous membrane, 206
Conventional mask
  making, 159
  mask processing, 160
  mask qualification, 160
  mask writing, 160
  starting material, 159–160
Converter, digital to analog, 78–79
Coordinate system traceability path, 535

COSMOS, 209–211, 227
Coulomb effect, 66
Critical dimension, 26, 145, 149–150, 216
  accuracy, 83
    technology for, 83–92
  measurement of, 150–152
Curve, sinusoid, 542
  phase, 542

D
Data
  augmentation, 6
  GDS, 33
  manipulation, 25
    pattern of, 25
    rotation of, 25
    scaling, 25
    sizing of, 25, 33
    tone reversal, 25
  nonrandom, 359
  path, 105
  preparation, 5, 92–93
    complementary split, 226
    for EPL masks, 224–225
  processing architecture
    storage area network (SAN), 93
  transformation, 5–6
  verification, 6
    processes of, 6
Data path, 105
Data preparation, 5, 20–26, 92, 224–227
Data slicing, 34
DEAL, 51
Defect (s), 152, 170, 177
  absorber, 255
  amplitude, 246
  detection sensitivity, 254
  e-beam repair, 223
  electron beam detection of, 219
  hard, 145
  inspection, 8
    by transmission image, 218
    electron beam, 219
    optical detection, 218–219
  opaque, 293
  particle, 220–221
  phase, 246
  printability, 217
  protrusion, 255
  repair, 8, 221
    FIB, 221–222
Definition, mask, 3
Deflection architecture, multistage, 76–78
Deflector, dual shaping, 85
De-slivering, 23

Index

Diffractive optical elements (DOE), 114
Digital light processor (DLP), 101–102
Digital signal processor (DSP), 89
Digital to analog converter, 78
Dirac pulses, 119f
Dispersion, 557
    models, 557
        Cauchy, 557–559, 562–563
        Drude, 559
        Lorentz oscillator, 559–560
Distortion control, 281–283
Distributed variable shaped beams (DiVa), 50–51
Donut problem, 195
DRAM, 48
Dry-etching, 39, 312, 370
Dynamic charging model, 345

E

EAPSM, 373
EB writing, 312
E-beam, 36–37, 39, 48–51, 55, 60–94
    CD error factors, 84
    data, 65
    data flow, 92f
    inspection tool, 292
    laser, 36
    lithography, 44, 60–94
        system, 63–64
        uses, 61
    mask writers, 60–94
        history, 62
        role of, 60
        structure, 65–70
        technology, 71–82
    optical, 23
    parameters
        accuracy, 70
        resolution, 70
        throughput, 70
    point beam system, 68–69
    proximity correction, 37
    raster scanning mode, 66–67
    resist patterning, 253
    scanning mode, 66
    stage movement, 65
        continuous stage moving method, 65
        step and repeat method, 65
    systems, 48–51, 55, 62
        arbitrarily shaped, 49
        DEAL, 51
        DiVa, 50
        MAPPER, 51
        microcolumn, 49
        multicolumn cell lithography, 50
        multicolumn multibeam, 50
        raster multibeam, 49
    temperature balance, 81–82
    vector scanning mode, 66–67
    writer, 13
EBPC, 37, 38
Eddy current, 80
Effects
    etch-loading, 39–40
    Coulomb, 66
    mask shadowing, 261–262
    proximity, 67–68, 84, 211
        correction, 85–86
    resist-heating
        strategies to mitigate, 363–364
Elastic scattering event, 343
Electron beam
    charging effects, 349
        characterization of, 349
    energy and charge disposition, 343
    lithography, 44, 195, 201, 341
    masks for, 201–202
    projection lithography, 194, 199–230
    proximity effects, 37
    resist charging, 345
    secondary electron emission, 345
Electron gun, 71–72
Electron(s)
    emitters, 51
    secondary, 345
Electronic design automation (EDA), 5
Electronic order format
    SEMI-P10, 26
Ellipsometry, spectroscopic, 552, 555–556
EPL masks
    cleaning, 214–215
    data preparation, 224–227
    inspection and repair, 217–223
    making of, 211–212
    metrology, 215
Equations
    Fresnel's, 553–554
    Kramers-Kronig, 557
    Maxwell's, 553, 657
Erlangen program, 535
Error(s)
    CD, 82, 215, 615
    effect of substrate on, 323–324
    ESE masks, 273
    image placement, 152, 215, 279–280; *see also* Metrology
    IP, 79, 82, 204, 215
    phase, 153–154
    stitching, 82, 108
    surface slope, 236

ESEM detectors, 642
Etch(ing), 7, 160
　absorber stack, 253
　buffer layer, 257
　dry, 39
　electron cyclotron resonance (ECR), 312
　gas assisted etching (GAE), 637
　of chrome, 369–371
　reactive ion (RIE), 212–213
Euclidean group, 535
EUV multilayer defects, 685
Experiment(s)
　Michelson–Morley, 532
　Young's two slit, 554
Exposure control program, 65
Extreme ultraviolet lithography (EUVL), 196, 232–267, 644; see also Lithography
　absorber defect inspection, 253
　　defect repair, 255
　　stack etch, 253
　alternating, 266
　attenuated, 264
　buffer layer etch, 257
　e-beam resist patterning, 253
　mask absorber material, 251
　mask blank fabrication, 233–252
　　defect reduction, 240
　mask cleaning, 257
　mask patterning, 252–258
　mask processing tools, 259–261; see also Tools
　mask protection, 258–259
　mask substrate fabrication, 234
　　material requirement, 234–235
　　surface defect requirement, 237
　　surface flatness requirement, 235
　　surface roughness requirement, 236
　ML blank fabrication, 237
　multilayer defects, 685
　　alternating, 266
　　attenuated, 264
　　embed, 264
　phase-shift masks, 264–266
　reflective mask performance, 261
　　Bossung process window tilt, 263–264
　　focus shift, 263–264
　　mask shadowing effect, 261–262

F

Fabrication
　of masks, 5–8, 44, 55, 211, 238–246
　　back end of line (BEOL), 7–8
　　data preparation, 5, 20
　　EPL, 211
　　front end of line (FEOL), 6
　　mask process flow, 143
　　process technology, 145
　　of stencil masks, 274
FE model, 283
Features, optical proximity correction, 6
Finite-difference time-domain method, 660
　algorithm, 660
Focus control, 105
Focused ion beam techniques, 293
　repair tools, 294, 295
Fogging, 53
　effect correction, 91–92
　electrons, 90
Formula(e)
　Rayleigh, 118
Fourier
　coefficients, 541
　transform(s), 119, 541, 653
Fracturing, 5
Frame
　bonding, 314
　generation, 20
Frequency, Nyquist, 544
Fresnel's equations, 553–554
Fringe scan, 582–583
Front end of line (FEOL), 6
　masks writing, 7
　processes of, 6–7
　　negative working, 7
　　positive working, 7
Fused silica for mask blanks, 382

G

Gallium staining, 637
Gas-assisted etching (GAE), 637
Gaussian beam, 62
GHOST, 37, 47, 52, 89
Grating light valve (GLV), 101–102
Gray scaling, 109–110
Green's function, 358
Grid snapping, 33, 35, 105
Group, Euclidean, 535

H

Hard defects, 591
Hard pellicles, 388, 409
Hard shifter masks, 631
High spatial frequency roughness (HSFR), 236, 240, 243
Hopkins approach, 668
HT.PSM, 175
　3-tone, 177–178
　basic functionality, 175–177
　high transmission, 177
　manufacture, 175
Hydrogen silsesquioxane (HSQ), 331–332

Index

I

Image
  aerial, 107
  diffraction effects on, 118–122
  etched, 100
Image placement
  analysis, 537
    Fourier transforms, 541
    regressions, 537
  coordinate system traceability path, 535
  error, 152, 279, 286
  metrology, 531–547
  sampling, 546
  x–y metrology, 532
Imaging
  1:1 lithography, 137
Inspection
  defect, 8
  die-to-die, 8
Interband absorption, 557
Interferogram, 236
Interferometer, 532, 632
  optics, 579
Intraband absorption, 557
Ion beam projection lithography, 196, 271–303
  cleaning masks, 295
  defects inspection repairs, 291
    defect repairs, 293–295
    e-beam inspection tools, 292–293
    optical inspection tools, 291
  masks, 271–303
    cleaning, 295
    defects and repairs, 291
    fabrication, 274
    metrology, 283
    pattern placement, 279
    stability, 297–298
  metrology of stencil masks, 283
    CD, 289
    image placement metrology, 286–287
    measurement method, 285–286
    repeatability, CD, 289
    tools for measurement, 284–285
    uniformity, CD, 289–290
  pattern placement, 279
  principle, 272
  process-induced distortions, 279–280
  stencil mask fabrication, 274–278
  stitcher configuration, 273f
  system, 272
  test mask, 284
IPL
  exposure station, 300
  stitcher, 273
Isometry, 535

J

Job deck, 6, 27, 37
  concept, 27–28
  example, 27
  frame generation, 20

K

K1 factor
  definition, 164
Kelvin probe experiment, 353
Kirchhoff approach, 656–657
Kramers-Kronig relation, 557

L

Laser durability, 381
Laser mask writer, 100–129
  history, 100
  laser pattern generator, 100–114
Laser pattern generators (LPGs), 55, 100
  amplitude modulating, 122
  matrix exposure, 56
  piston arrangement, 123
  pivot micro mirrors, 125
  raster scan, 55
    architecture, 106
    beam modulation, 111
    formation of aerial image, 107
    scan separation, 107–108
    stripe boundaries, 108–109
  SLM, 115, 122
Lasertec MD 2500, 599–601
  application, 601–602
  outlines and features, 600–601
  reliability, 604
  technology, 602–604
    autoloader, 604
Layer
  absorber stack, 250–251
  ARC, 250–251
  bottom antireflective, 571–572
  BOX, 204, 211
  buffer, 250
  capping, 241–242
  deviation, 23
  top antireflective, 570–571
Limitations, beam current, 54
Linear-dimensional analyzer, 531
Lithographic exposure tool
  illumination, 118
    coherent, 118
    incoherent, 118
    partially coherent, 118
    numerical aperture (NA), 116–118, 164, 608
Lithographic projection stepper, 651
Lithographic tolerance, 137–138

Lithography
  aerial and resist image formation, 651
  DEAL, 51
  electron projection (EPL), 194, 195, 200
  electronic beam, 60–61
    example of, 63–65
  extreme ultraviolet (EUVL), 194, 196
  ion beam projection (IPL), 194, 196; see also Ion projection lithography
  low energy electron beam (LEEPL), 194, 197, 208
  nano-imprint, 194, 197
  optical, 61, 136
  physical/chemical modifications, 654
  proximity x-ray, 196–197
  simulation, 650
  strategies, 655
  sub-wavelength, 141
Localized heating on resists, 361
Lorentz oscillator model, 559
Low-energy electron beam proximity lithography, 197

M
Manhattan-type pattern, 47
Manufacturing electron beam exposure system (MEBES), 13–14
MAPPER, 51
Marangoni drying, 214
Mask
  data
    creation, 25
    preparation, 20–26, 33, 40
  defect types, 591–593, 630
    hard defects, 591
    PSM defects, 593
  inspection, 589–597
    technology, 602
  pattern, 44
  repair, 629–644
    AFM nanomachining repair, 640
    electron beam repair, 641
    focused ion beam repair, 635
    laser defect repair, 633
    qualification, 632
    repair challenges, 644
    substrate, 321–323, 379
Mask error enhancement factor (MEEF), 164–165, 590
Mask making, 211
  cleaning, 214
    wet, 214
  dry cleaning, 215
  fabrication of blanks, 211
  patterning, 211
  trench etching, 212
Mask materials, 377–389
  APSM, 385
  bulk materials, 378
  fused silica, 383, 386
  pellicles, 383
  thin films on substrates, 379
Mask processing, 367–373
  cleaning, 371–373
  pattern transfer, 369
  resist stripping, 371
  resists and developers, 368
Mask shadowing effect, 261
Mask stability, 297
Mask writers, 43–56
  cell projection systems, 48
  raster scan systems, 45
  raster-shaped systems, 47
  variable-shaped beam systems, 46
  vector scan systems, 46
Mask-writing tools
  requirements of, 20
Mask(s), 3
  advance, 6
  applying OPC and PCM
  binary intensity mask (BIM), 5
  Boolean mask operations, 22
  chrome-on-glass mask (COG), 5
  cleaning, 257–258
  data preparation flow, 20
  defect types, 591–594
    hard, 591–592
    PSM, 593
  definition, 3
  emulsion, 141
  error factor, 104
  fabrication of, 5–8, 44, 143
    process of, 143–148
  flatness, 82
  hard shift, 631
  history, 8
  inspection of, 594–595
    aerial image based, 597
  layout interchange formats, 20
  maker's holiday, 13
  making of, 3–16, 211–212
  mask data creation, 25
  measurements performed on, 26
  membrane, 200
  optical, 135–186
    defects, 629–631
    for IPL, 271–303
    inspection, 589–597
    NGL, 193–315

Index

repair, 629–644
substrate, 321
phase shift, 165
photolightographic, 589
programmable, 115, 122; *see also* SLM
stencil, 273
technology history of, 8
  1x projection era, 10–13
  contact print era, 8–10
  sub-wavelength era, 14–15
  wafer stepper era, 13–14
technology, 30, 145
ultrathin membrane, 206–208
  bi-layer, 206
  tri-layer, 206
writers, 43
writing, 7, 19–130, 160
  accuracy, 51
  e-beam, 60–94
  laser, 100–129
  principles, 28–30
Material, mask, 141–143
  glass substrate, 141
  pattern film material, 141–143
Measurement
  phase-shift, 578
  transmittance, 578–579
MEBES, 32, 44–45
Mechanics, stage and chamber, 80–81
MEEF, 104, 164, 656
Membrane(s)
  continuous, 206–208
  diamond-like-carbon (DLC), 206, 221
  stencil, 203–206
  stress, 282
Method(s)
  boundary conditions
    periodic (PBCs), 663
    reflecting (RBCs), 662–663
    transparent (TBCs), 663
  cleaning
    dry laser, 295
    Huang/piranha, 295
  combined corrections, 537
  dose correction, 86–89
  finite difference time-domain method (FDTD), 660
  finite element modeling, 659
  GHOST, 89
  least squares, 537
  multi-layer resist method, 89–90
  PAT, 312
  PSE, 312
  shape modification, 89
  waveguide, 659

Metrology, 6–7, 9, 160, 215, 378, 414
  adaptive, 547
  image placement, 215
  mask feature
    costs and benefits of, 415–420
  measurement uncertainty, traceability, 420–421
  registration/position, 7
  sampling, 546
    characterization, 546
    dispositioning, 546–548
    economics, 546
    frequency, 546
    objectives, 546
    time, 546
  thin film, 551
    spectroscopic ellipsometry, 552, 555–556
    spectroscopic reflectometry, 552
  x–y, 532
Michelson–Morley experiment, 532
Micralign projection aligner, 11
Microcolumn, 49
Micro-mirrors
  piston, 123
  pivot, 124
Mid-spatial frequency roughness (MSFR), 236
ML
  capping layer, 241–242
  defect
    compensation, 249
    repair, mitigation, 246–250
  deposition processes and tools, 240–241
  EUVL blank fabrication, 238–240
  interface engineering, 241
  smoothing, 242
  stress control, 245–246
  thermal stability, 244–245
Model calibration, OPC, 38
Modeling and simulation, 649–686
Modified fused silica, 386
Molybdenum silicide film, 323
Monte-Carlo method, 343–344
Moore's law, 44, 102
MoSi film, 323
MPM series, 579
Multibeam strategies, 113–114
Multicolumn cell lithography system, 50
Multicolumn multibeam system, 50
Multiple exposure, 105
Multistage deflection architecture, 76
Mylar, 307

N

Nanoimprint lithography, 197
Next-generation lithography, 16, 186, 193–315

Next-generation lithography (*Continued*)
  electron beam projection lithography, 200–227
  extreme ultraviolet lithography, 232–267
  ion projection lithography, 271–303
Numerical aperture (NA), 163, 272, 649
Nyquist frequency, 544

O
OASIS, 21
Objective lens system, 75–76
Off-axis illumination, 165, 179
Opaque defects, 193
Open artwork system interchange standard (OASIS), 21–22
Optical critical dimension metrology, 433–454
  applications, 441
  methodology, 435
    nonpolarized reflectance, 438
    polarized reflectance, 438
    scatterometry, 435
    spectroscopic ellipsometry, 436
  rigorous coupled-wave analysis, 440
Optical inspection tools, 291
Optical lithography, 136
  contact/proximity, 137
  function of optical mask, 139
  projection, 137
    reduction projection optics, 138–139
Optical mask, 61, 135–186
  classification of, 140, 158
    advanced, 159
    conventional, 159
  fabrication, 143
  function of, 139
  lithography, 136–137
  material, 141, 159
  phase shift, 141
  quality of, 149–153
  resolution enhance techniques for, 141
Optical models, 557–560
  Cauchy model, 557
  Lorentz oscillator model, 559
Optical pattern generator, 10
Optical properties of materials, 556
Optical proximity correction, 23–26, 31, 38, 49, 61, 159
  model calibration, 38
Optical proximity effects, 23, 211
  correction, 182
    correction strategies (OPC), 182, 590, 608
    manufacturing of OPC,
    principle of OPC, 183*f*
Optical techniques, 552–556
  reflection of light, 552
  spectroscopic ellipsometry, 552, 555–556
  spectroscopic reflectance, 556
Out-plane-distortion (OPD), Mask, 235
Outgassing, resists, 336

P
Pattern fidelity, 32
Pattern-induced distortions, 280–281
  process induced, 280
Pattern stitching, 299
PBS, 328
PEL masks,
  data preparation for, 227
Pellicle(s), 8, 147, 160, 379, 383, 387–389, 396–409
  anatomy, 399
  AR coating, 402
  cleaning, 407
  film, 400
  frame, 402
  future of, 409
  handling, 406
  hard, 409
  inspection, 404
  optical requirements, 396
  removable, 409
  soft, 409
  use of, 397
Penetration depth, 557, 560
Phase defects, ML, 246
  repair, 247–248
Phase-shift masks (PSM), 141, 264–266, 500, 577, 672; *see also* EUVL
  alternating, 165–168, 593
    manufacturing of, 168–171
  effective phase, 170
  history, functionality, 165
  phase
    balancing, 170
    conflict, 171
    phirst strategy, 171
  rim, 174; *see also* CL PSM
  special cases, 173–174
Phase measurement tool, 577–586
  fringe scan, 582
  interferometer optics, 579
  MPM series, 579
  phase shift measurement, 578
  transmittance measurement, 578
Phasor, 542
Photolithographic masks, 589
Photolithography, 377
Photomask critical dimension metrology, 458–494
  architecture, 460
  CD-SEM metrology, 467

Index

instrument calibration, 485
modeling, 478
particle metrology, 475
signals, 462
Photomasks, 4, 44; *see also* Masks
  emulsion, 9
  feature metrology, 413–430
  focus control, 105
  history, 100–101
  making of, 100
  modeling of thin films, 560–574
  resist requirements, 325–327
  thin films, 860
Photorepeater, 9
Photoresist, 3–4, 137, 326, 567–568
  ESCAP, 334
  processing of
    physical/chemical modifications, 654–655
Piston micro-mirrors, 123–124
Pitch, 413
Pixel, 35, 105
Plasma reactors, 370
Point beam system, 68
Positional error vector (PEV), 537
Postexposure decay, 351
Principle(s)
  mask writing, 25
  raster scanning writing principle, 29
  variable shape beam writer, 30
  vector scanning writing principle, 30
Probeam3D/1PL simulation, 301
Process(es)
  induced distortions, 280
  iterative etching, 7
Processing, mask, 367
Profilometer, 552
Program, Erlangen, 535
Projection lithography, 137
Proximity effect, 67–68
Proximity effect correction, 52, 85
Proximity electron lithography
  masks for, 208
Proximity X-ray lithography, 196, 305–315
  fabrication, 310
  membrane requirements, 307
  system, 305
  X-ray mask structure, 306
PSM defects, 593
PXL system, 305–306

R

Raster
  reticles, 4, 40, 143; *see also* Optical masks
  scan systems, 44
  scanning mask writer, 29

Raster multibeam system, 49
Raster scan
  laser pattern generators, 106
  principle, 29
  systems, 45, 52
Raster-shaped system, 47
RCWA
  implementation, 440
  simulation, 687
Reactor(s)
  ICP, 371
  plasma, 370
Reflection
  electromagnetic theory, 552
  light, 552
  spectroscopic, 556
Regression, 537–541
  analysis, 541
  first order, 546
Repair
  AFM nanomachining mask, 640
    theory of operation, 640
  defect classification, 631–632
  defect types, 630–631
  defects and mask, 629–630
  electron beam mask, 641
    theory of operation, 641–643
  focused ion beam mask, 635
    theory of operation, 636–639
  laser defect, 633–635
    theory of operation, 633–634
  repair qualification, 632–633
Resist(s)
  –blank interactions, 335
  charging, 345
    dynamic charging model, 345–347
  chemically amplified (CARs), 326–327, 332
  COP, 331
  DNQ-novolak, 329–330
  footing, 335
  heating, 357–364
  implementation challenges, 335–337
  mask making, 325
    heating, 326
    sensitivity, 326
  materials, 325–336, 368
    chemically amplified, 332–334
    mask shop issues, 336
    nonchemically amplified resists (NCA), 327
    properties, 368
    resist–blank interactions, 335
    resist implementation, 335
    resist outgassing, 336
  negative, 368
  outgassing, 336

Resist(s) (*Continued*)
　PEB free, 334
　positive, 368
　stripping and cleaning, 371–374
　temperature evolution, 358
Resolution enhancement, 31
Resolution enhancement techniques (RETs), 14–15, 23, 104, 165, 590, 608
　strategies, 186
　sub-resolution assist features, 179
Retrograde trenches, 278
Riverbed, 638
Roadmap
　mask technology, 61
Rotation, 538
Rubylith, 9–10

S

Sampling, 546
Scale, 535, 538, 546
Scan
　field stitching, 34
　fringe, 583–584
　separation, 107–108
Scanner, E–beam, 219–220
Scanning
　beam, 111–112
　　AOD, 112
　　rotating mirror polygon, 111
Scanning electron microscope (SEM)
　architecture, 460
　calibration, 485
　CD-SEM metrology
　　high pressure/environmental, 492
　　in photomask dimension metrology, 458–492
　　instrument performance, 488
　　linewidth standard, 489
　　modeling of SEM signed, 478
　　sharpness, 488
　　signals, 462
Scanning principles, 500–504
　scanner calibration, 501
　scanning algorithms, 503
　scanning modes, 501
　sensors, 500
Scanning tools
　raster scanning mask writer, 29
　　writing principle, 29
　variable shape beam writer, 30
　vector scanning writer, 30
Scanning tunneling microscopy, 500
Scatterometry, 435
Scission, polymer chain, 327
Self-calibration, 536
Settling time, 53

Shear, 538, 546
Shot noise, 273
Sidelobes, 177–178
Sigma pattern generator
　aerial image, formation of, 126
Simulation(s)
　application to different mask types, 670
　　binary masks, 670–672
　　EUV masks, 675–676
　　phase shift masks, 672–675
　finite element method (FEM), 357
　general strategies, 655–656
　lithography, 650
　　aerial and resist image formation, 651–654
　Monte-Carlo electron trajectory, 37, 354–355
　of mask defects, 681–686
　temperature, 357–359
SLM chip, 127
Soft pellicles, 387, 409
Software, TEMPTATION, 359–360
SOI wafer flow process, 276
Spatial light modulator (SLM), 56, 100, 115–128
Spectroscopic ellipsometry, 436, 555
Spectroscopic reflectance, 556
　spin coat, 368
　spin develop, 369
SR lithography system, 306
Stage and chamber mechanics
　stage and writing-chamber materials, 80
　structure
　　chamber, 80
　　stage, 81
Stage deformation, 82
Stage interferometer of x–y metrology, 533
Standard mechanical interface (SMIF), 223–224
Stencil mask, 274, 279, 644
　fabrication, 274
　membrane, 203
　metrology of, 283–290
Stepper, 115, *see* Lithographic exposure tool
Stitching error, 297
Strategy
　Matrix OPC, 184
Stripe boundaries, 108
Sub-resolution assist features (SRAF), 165, 174, 179–181
　manufacturing of, 181
Substrate, mask, 321; *see also* mask blanks
　material
　　chrome film, 322
　　glass, 322
　　molybdenum silicide film, 323
Subwavelength era, 14
Superpose writing, 82

Surface, patterned, 3
Systematic inclusion, 547

T
TaN, 251
   absorber, 253
Technique(s)
   aerial image measurement, 609–612
   Ar aerosol cleaning, 215
   deposition
      ML smoothing, 237–238, 242
   dry cleaning
      PLASMAX, 215, 373
   field decomposition, 669
   GHOST, 37, 52
   ion-beam figuring (IBF), 233
   ion-beam sputtering, 240
   magnetron sputtering, 240
   multiple exposure, 105
   off contact/proximity mode, 137
   pattern area density, 87
      principle, 87
   per pixel deflection, 46
   resolution enhancement, 104, 141
   reticle enhancement, 608
Technology(ies)
   advanced, 135
   IBF, 236
   mask blank polishing, 236
      batch, 236
      single sample, 236
   matured, 135
   phase-shift mask, 577
Temperature balance, 81
   simulation, 357
TEMPTATION, 359–360
Terpolymer, 328
Test masks, 285
Theory, electromagnetic, 553
Thermal stability, ML, 244–245
Thin film substrates, 379
   thermocouple, 342
Tip, 504–516
   calibration, 514
   characterization, 518
   tip–sample interaction, 505
   tip-shape removable, 516
   width, 516
Tool(s)
   AFM, 284
   ALTA, 326
   Carl Zeiss AIMS, 609
   cell projection (CP), 200
   EUVL
      AIMS, 260–261

      blank inspection, 260
      e-beam repair, 260
      mask fabrication, 233
      mask processing, 259–261
      mask reflector, 260
      ML deposition, 259
   focused ion beam (FIB), 145
   for inspecting masks, 599–605
      application, 601
      reliability, 604
      safety, 605
      system outlines and features, 600
      technology, 602
   for mask image evaluation, 607–626
      aerial image analysis, 612
      aerial image measurement, 609
      application of AIMS, 615
      reticle enhancement, 608
   lithographic exposure, 101
   modeling and simulation, 649
   projection reduction exposure with variable axis
      immersion lens (PREVAIL), 200
   regression analysis, 541
   scattering with angular limitation projection
      electron lithography (SCALPEL),
      200, 206
   x–y metrology, 533
Transitivity, 536
Translation, 538
Transmission mask, 128
Transmittance measurement, 578, 583
Trench etching, 212
Triangle beam, 84
Trim mask, 171
Types of defects, 630–631

U
Uniformity, CD, 289–290
   effect of substrate, 323–324

V
Variation, CD, 363
Vector scan systems, 46
Vector scanning, 30, 46
   systems, 46
   writer principle, 30
Vectors of residues, 538

W
Wafer(s)
   IDP distortion
      sources of, 252
   SOI, 275
Wafer exposure using mask, 4
Wafer stepper era, 13

Waveguide method, 659
Wet etching of chrome, 370
Writers
   electronic beam, 7, 13
   history of evolution, 62
   wafers
      ADVANTEST, 212
Writing, superpose, 82–83

X

X-ray mask, 315f
   defect inspection and repair, 314
   fabrication of, 310–314
      dry etching, 312–313
      EB writing, 312
      frame bonding, 314
      mask processes, 311–312
   for LSI fabrication, 314–315
   structure, 306
      absorber, 309–310
      membrane, 307–308
x–y metrology, 532, 534
   analysis, 537–540
   system, 534f
   tool, 533

Y

Young's
   modulus, 308
   two-slit experiment, 554

Z

ZEP 7000, 329